Routledge Handbook of Biomechanics and Human Movement Science

The Routledge Handbook of Biomechanics and Human Movement Science is a landmark work of reference. It offers a comprehensive and in-depth survey of current theory, research and practice in sports, exercise and clinical biomechanics, in both established and emerging contexts.

Including contributions from many of the world's leading biomechanists, the book is arranged into eight thematic sections:

- Modelling and computational simulation
- Neuromuscular system and motor control
- Methodology and systems of measurement
- Engineering, technology and equipment design
- Biomechanics in sports
- Injury, orthopaedics and rehabilitation
- Health and motor performance
- Training, learning and coaching

Drawing explicit connections between the theoretical, investigative and applied components of sports science research, this book is both a definitive subject guide and an important contribution to the contemporary research agenda in biomechanics and human movement science. It is essential reading for all students, scholars and researchers working in sports biomechanics, kinesiology, ergonomics, sports engineering, orthopaedics and physical therapy.

Youlian Hong is Professor and Head of the Department of Sports Science and Physical Education at the Chinese University of Hong Kong. He is President of the International Society of Biomechanics in Sports.

Roger Bartlett is Professor of Sports Biomechanics in the School of Physical Education, University of Otago, New Zealand. He is an Invited Fellow of the International Society of Biomechanics in Sports (ISBS) and the European College of Sports Sciences.

Routledge Handbook of Biomechanics and Human Movement Science

Edited by
Youlian Hong and
Roger Bartlett

Routledge
Taylor & Francis Group
LONDON AND NEW YORK

First published 2008
by Routledge
2 Park Square, Milton Park, Abingdon, Oxon, OX14 4RN

Simultaneously published in the USA and Canada
by Routledge
270 Madison Avenue, New York, NY 10016

Routledge is an imprint of the Taylor & Francis Group, an informa business

Typeset in Goudy by Keyword Group Ltd.
Printed and bound in Great Britain by Antony Rowe Ltd., Chippenham, Wilts

British Library Cataloguing in Publication Data
A catalogue record for this book is available from the British Library

Library of Congress Cataloging in Publication Data
A catalog record for this book has been requested

ISBN 10: 0-415-40881-4 (hbk)
ISBN 10: 0-203-88968-1 (ebk)

ISBN 13: 978-0-415-40881-3 (hbk)
ISBN 13: 978-0-203-88968-8 (ebk)

Contents

Introduction

It is now widely recognized that biomechanics plays an important role in the understanding of the fundamental principles of human movement; the prevention and treatment of musculoskeletal diseases; and the production of implements and tools that are related to human movement.

To the best of our knowledge, the biomechanics books that have been published in the past decade mainly focus on a single subject, such as motor control, body structure and tissues, the musculoskeletal system, or neural control. As biomechanics has become an increasingly important subject in medicine and human movement science, and biomechanics research is published much more widely, it is now time to produce a definitive synthesis of contemporary theories and research on medicine and human movement related biomechanics. For this purpose, we edited the *Handbook of Biomechanics and Human Movement Science*, which can act as a comprehensive reference work that covers injury-related research, sports engineering, sensorimotor interaction issues, computational modelling and simulation, and other sports and human performance related studies. Moreover, this book was intended to be a handbook that will be purchased by libraries and institutions all over the world as a textbook or reference for students, teachers, and researchers in related subjects.

The book is arranged in eight sections, each containing four to seven chapters, totalling 41 chapters. The first section is titled 'Modelling and simulation of tissue load'. Under this title, David Lloyd and colleagues, Clark Dickerson, and Navid Arjmand and colleagues address issues of modelling and simulation of tissue load in the lower extremity, upper extremities, and human spine, respectively, using traditional mechanical methods. Wolfgang Schöllhorn and colleagues describe the artificial neural network (ANN) models for modelling and simulation of sports motions. Finally, Jason Tak-Man Cheung introduces the finite element modelling and simulation of the foot and ankle and its application.

The title of the second section is 'Neuromuscular system and motor control'. Albert Gollhofer introduces muscle mechanics and neural control, which is followed by the chapter discussing the amount and structure of human movement variability by Karl Newell and Eric James. Then Jamie Lukos and colleagues address the issue of planning and control of object grasping focusing on kinematics of hand pre-shaping, contact and manipulation. The fourth chapter by Dario Liebermann, describes the biomechanical aspects of motor control in human landings. Finally, Matt Dicks and colleagues discuss the ecological psychology and task representativeness, focusing on the implications for the design of perceptual-motor training programmes in sport.

The third section is devoted to methodologies and system measurement. The opening chapter, by Ewald Hennig, highlights the measurement of pressure distribution. Young-Hoo Kwon addresses the use of numerical methods in measurements for deriving kinematic parameters in sports biomechanics. The methods used in Alpine and Nordic skiing biomechanics is explained by Hermann Schwameder and colleagues. The fourth chapter looks at issues in measurement and estimation of human body segment parameters, as contributed by Jennifer Durkin. The section ends with the chapter about the use of electromyography in studying human movement contributed by Travis Beck and Terry Housh.

The fourth section is for engineering technology and equipment design and has four chapters. The section begins with the chapter in biomechanical aspects of footwear written by Ewald Hennig, which is followed by the chapters in biomechanical aspects of the tennis racket by Duane Knudson, sports equipment – energy and performance by Darren Stefanyshyn and Jay Worobets, and biomechanical aspects of artificial sports surface properties by Sharon Dixon.

The seven chapters in the fifth section introduce the application of biomechanics in sports. The biomechanics in throwing, snowboarding, in striking and kicking, in swimming, in long jump, and in sprinting running are discussed by Roger Bartlett and Matthew Robins; Greg Woolman; Bruce Elliott and colleagues; Ross Sanders and colleagues Nick Linthorne and Joseph Hunter. The section concludes with the chapter that addresses the biomechanical simulation models of sports activities as contributed by Maurice Yeadon and Mark King.

The sixth section focuses on the biomechanical aspects of injury, orthopaedics, and rehabilitation. The opening chapters highlight the biomechanical aspects of injuries in the lower extremity (William Whiting and Ronald Zernicke), upper extremity (Ronald Zernicke and colleagues), and spine (Brian Stemper and Narayan Yoganandan). The fourth chapter introduces in vivo biomechanical studies for injury prevention contributed by Mario Lamontagne and colleagues. Finally, Rosanne Naunheim shows how impact attenuation and injury are affected by artificial turf.

The contribution of biomechanical study to health promotion is demonstrated in the seventh section. Youlian Hong and colleagues look at the influence of backpack weight on biomechanical and physiological responses of children during treadmill walking. Jing Xian Li and Youlian Hong introduce ankle proprioception in young ice hockey players, runners, and sedentary people. De Wei Mao explains the plantar pressure characteristics during Tai Chi exercise. The last two chapters are related to falls and posture control. Daina Sturnieks and Stephan Lord introduce the biomechanical study of falls in older adults, while Stephan Turbanski discusses postural control in Parkinson's disease.

The eighth and last section is devoted to biomechanics in training, learning, and coaching. For this purpose, W.S. Erdmann addresses the application of biomechanics in soccer training; Chris Button explores the perceptual-motor workspace using new approaches to skill acquisition and training; and Manfred Vieten looks at the application of biomechanics in martial art training. The book ends with two chapters on motor learning. Jin Yan and colleagues describe developmental and biomechanical characteristics of motor skill learning and Bruce Abernethy and colleagues consider the use of biomechanical feedback to enhance skill learning and performance.

Contributions to this book are original articles or critical review papers written by leading researchers in their topics of expertise. Many recognized scholars participated in this book project, to the extent that some eminent biomechanists have been omitted. We offer our apologies. Finally we acknowledge our deep appreciation of the authors of this book who devoted their precious time to this endeavour.

Youlian Hong and Roger Bartlett

Section I

Modelling and simulation of tissue load

1

Neuromusculoskeletal modelling and simulation of tissue load in the lower extremities

David G. Lloyd[1], Thor F. Besier[2], Christopher R. Winby[1] and Thomas S. Buchanan[3]
[1]University of Western Australia, Perth; [2]Standford University, Stanford; [3]University of Delaware, Newark

Introduction

Musculoskeletal tissue injury and disease in the lower extremities are commonly experienced by many people around the world. In many sports anterior cruciate ligament (ACL) rupture is a frequent and debilitating injury (Cochrane *et al.*, 2006). Patellofemoral pain (PFP) is one of the most often reported knee disorders treated in sports medicine clinics (Devereaux and Lachmann, 1984), and is a common outcome following knee replacement surgery for osteoarthritis (Smith *et al.*, 2004). Osteoarthritis (OA) is one of the most common musculoskeletal diseases in the world (Brooks, 2006), with the knee and hip the most often affected joints (Brooks, 2006). All of these musculoskeletal conditions are associated with large personal and financial cost (Brooks, 2006).

Clearly, orthopaedic interventions and pre- and rehabilitation programs are needed to reduce the incidence and severity of these injuries and disorders. However, an understanding of the relationship between the applied forces and resultant tissue health is required if appropriate programmes are to be designed and implemented (Whiting and Zernicke, 1998). Estimating tissue loads during activities of daily living and sport is integral to our understanding of lower extremity injuries and disorders. Many acute injuries such as ACL ruptures occur during sporting movements that involve running and sudden changes of direction (Cochrane *et al.*, 2006). The progression of some joint disorders is also influenced by tissue loading during walking or running, such as PFP (Smith *et al.*, 2004) and tibiofemoral OA (Miyazaki *et al.*, 2002). It is also important to appreciate that for similar, or even identical tasks, people use different muscle activation and movement patterns depending on the type of control (Buchanan and Lloyd, 1995), experience (Lloyd and Buchanan, 2001), gender (Hewett *et al.*, 2004), and/or underlying pathologies (Hortobagyi *et al.*, 2005). Therefore, to examine tissue loading for some injury or disorder, people from different cohorts must be assessed performing specific tasks.

Understanding the action of muscles is important in regard to the loading of ligaments, bone and cartilage. For example, muscles can either load or unload the knee ligaments depending on the knee's external load and posture (O'Connor, 1993), and muscle co-contraction can potentially generate large joint articular loading (Schipplein and Andriacchi, 1991).

Therefore, models that estimate tissue loading need to estimate the forces and moments that muscles produce.

So how do we assess the action of muscles and other tissue loading in the lower extremities, accounting for subject-specific movement and muscle activation patterns? This is possible with new measurement methods coupled with neuromusculoskeletal modelling techniques. Neuromusculoskeletal modelling simply means modelling the actions of muscle on the skeletal system as controlled by the nervous system. This chapter will explore the development and application of our neuromusculoskeletal modelling methods to assess the loads, stresses and strain of tissues in the lower extremities.

Models to assess tissue loading during motion

Our neuromusculoskeletal models are driven by experimental data measured from three-dimensional (3D) motion analysis of people performing various sport-specific tasks or activities of daily living. In this, stereophotogrammetry techniques and kinematic models (Besier *et al.*, 2003) are used to measure the lower limb motion, electromyography (EMG) to assess muscle activation patterns, and force plates and load cells to measure the externally applied loads. These experimental data are used with the neuromusculoskeletal model to estimate joint loading and tissue loading. Joint loading in this case refers to the net joint moments and forces. Tissue loading is the force or moment applied to a group of tissues (e.g. all knee extensors), or the force borne by an individual tissue (e.g. the ACL, cartilage, muscle). With the addition of finite element models, the distribution of load (i.e. stress) or deformation (i.e. strain) within an individual tissue (Besier *et al.*, 2005a) is also assessed. However, let's first review the estimation of joint and muscle loading.

Joint loading and, in theory, muscle forces, can be estimated by inverse (Figure 1.1) or forward dynamics (Figure 1.2). Inverse dynamics is most commonly used to determine the joint

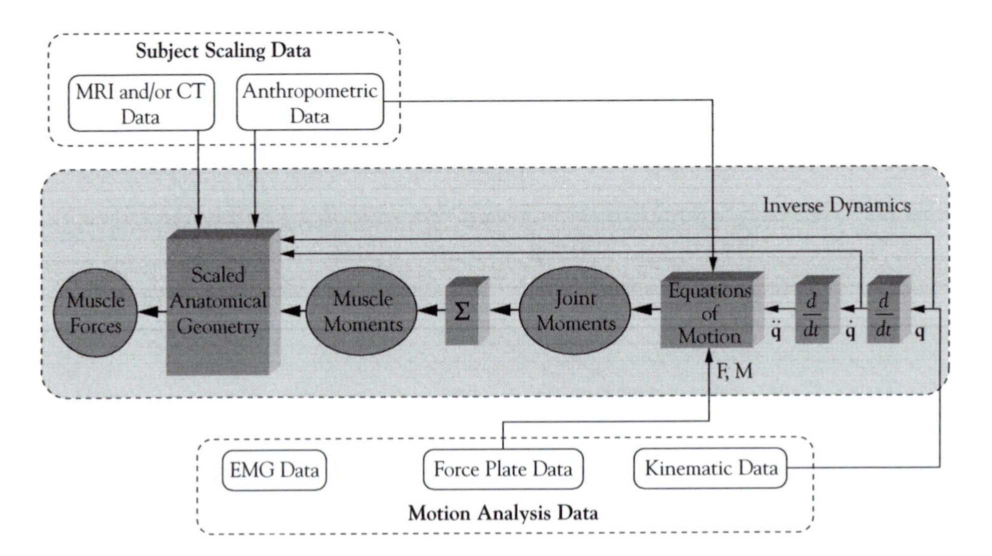

Figure 1.1 Schematic of an inverse dynamic model.

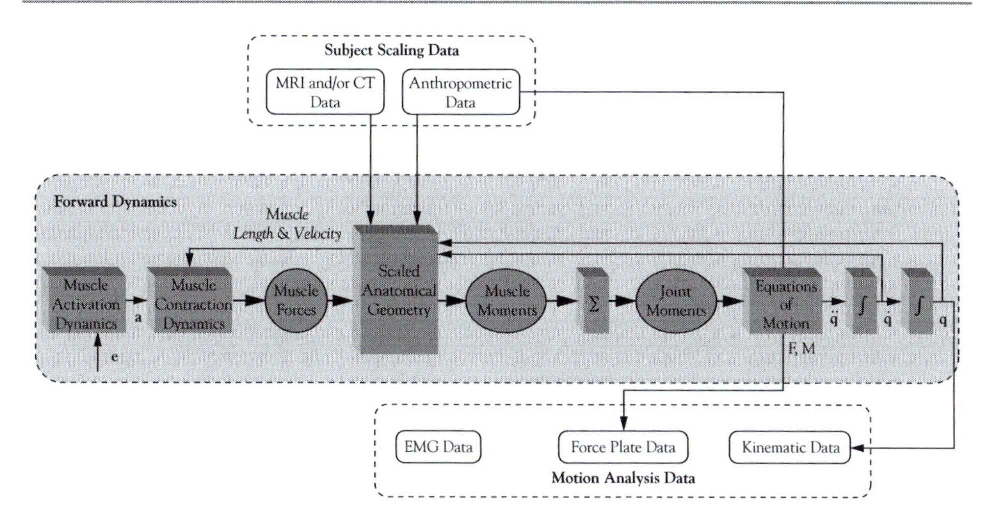

Figure 1.2 Schematic of a forward dynamic model, including full neuromusculoskeletal modelling components.

moments ($\mathbf{M}^J$) and forces based on data collected in a motion analysis laboratory. With the segments' mass and inertial characteristics ($\mathbf{m}(q)$), the joint motions (the joint angles (q), angular velocities ($\dot{q}$), and angular accelerations ($\ddot{q}$)), and the external loading (moments ($\mathbf{M}_{Ext}$), forces ($\mathbf{F}_{Ext}$), gravitational loading ($\mathbf{G}(q)$), centrifugal and coriolis loading ($\mathbf{C}(q,\dot{q})$)), equation 1 is solved for the joint moments, that is:

$$-\mathbf{M}^J = \mathbf{m}(q)\ddot{q} + \mathbf{C}(q,\dot{q}) + \mathbf{G}(q) + \mathbf{F}_{Ext} + \mathbf{M}_{Ext} \quad (1)$$

However, the joint moment is the summation of the individual muscles moments (Figure 1.1). In addition, the joint reaction forces, which are calculated as by-product of solving equation 1, is the summation of the muscle, ligament and articular forces and provides no information on their individual contributions. What is required is a means of decomposing the net joint moments and forces into the individual muscle moments, and subsequently the forces sustained by the muscles, ligaments and articular surfaces. Forward dynamics can provide the solution.

Forward dynamics solves the equations of motion for the joint angular accelerations ($\ddot{q}$) by specifying the joint moments, that is:

$$\ddot{q} = \mathbf{m}^{-1}(q)[\mathbf{C}(q,\dot{q}) + \mathbf{G}(q) + \mathbf{F}_{Ext} + \mathbf{M}_{Ext} + \mathbf{M}^J] \quad (2)$$

which can be numerically integrated to produce joint angular velocities and angles. In the primary movement degrees of freedom the joint moments are the sum of the all muscle moments ($\Sigma\mathbf{M}^{MT}$), i.e. $\Sigma\mathbf{M}^{MT} = \mathbf{M}^J$. Therefore, a complete forward dynamics approach models incorporates the action of each muscle driven by neural commands (Figure 1.2). In this, the muscle activation dynamics determines the matrix of muscles' activation profiles (**a**) from the neural commands (**e**); the muscle contraction dynamics determines the muscle forces ($\mathbf{F}^{MT}$) from muscle activation and muscle kinematics. Finally, the matrix of individual muscle moments ($\mathbf{M}^{MT}$) are calculated using the muscle moment arms ($\mathbf{r}(q)$) estimated from a model of musculoskeletal geometry. Summing the individual muscle moments provides the net muscle moments at each joint (ΣM_j^{MT}), i.e.

$$\Sigma M_j^{MT} = \sum_1^m M_m^{MT} = \sum_1^m r_m(q) \times F_m^{MT} \tag{3}$$

where m is number muscles at a joint, j. The matrix of net muscle moments across all joints ($\Sigma\mathbf{M}^{MT}$) are used in equation 2 to solve for the joint angular accelerations. However, there is yet another solution to determine the muscle contributions to the net muscle moments, which is a hybrid of forward and inverse dynamics.

Hybrid forward and inverse dynamics

The hybrid scheme can be used to calibrate and validate a neuromusculoskeletal model based on the estimation of joint moments determined by both inverse and forward dynamics (Figure 1.3) (Buchanan *et al.*, 2004; Buchanan *et al.*, 2005; Lloyd and Besier, 2003; Lloyd *et al.*, 2005). For forward dynamics to estimate individual muscle forces and thereby calculate net muscle moments, the neural command to each muscle has to be estimated. This can be achieved using a numerical optimization with an appropriately selected cost function (e.g. minimize muscle stress) (Erdemir *et al.*, 2007). Even though there is much debate over the best choice of cost function, in some instances the resulting neural commands from optimization reflect the EMG signals. However, this begs the question; why not use EMG to specify the neural command and drive the neuromusculoskeletal model?

EMG-driven models (Buchanan *et al.*, 2004; Lloyd and Besier, 2003) or EMG-assisted models (McGill, 1992) of varying complexity have been used to estimate moments about

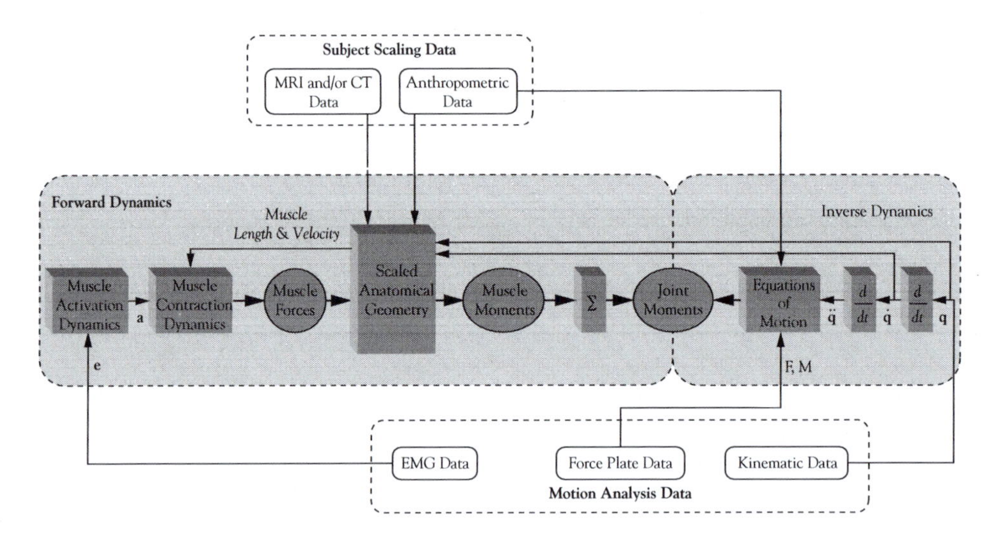

Figure 1.3 Schematic of a hybrid inverse dynamic and forward dynamic model, including full neuromusculoskeletal modelling components.

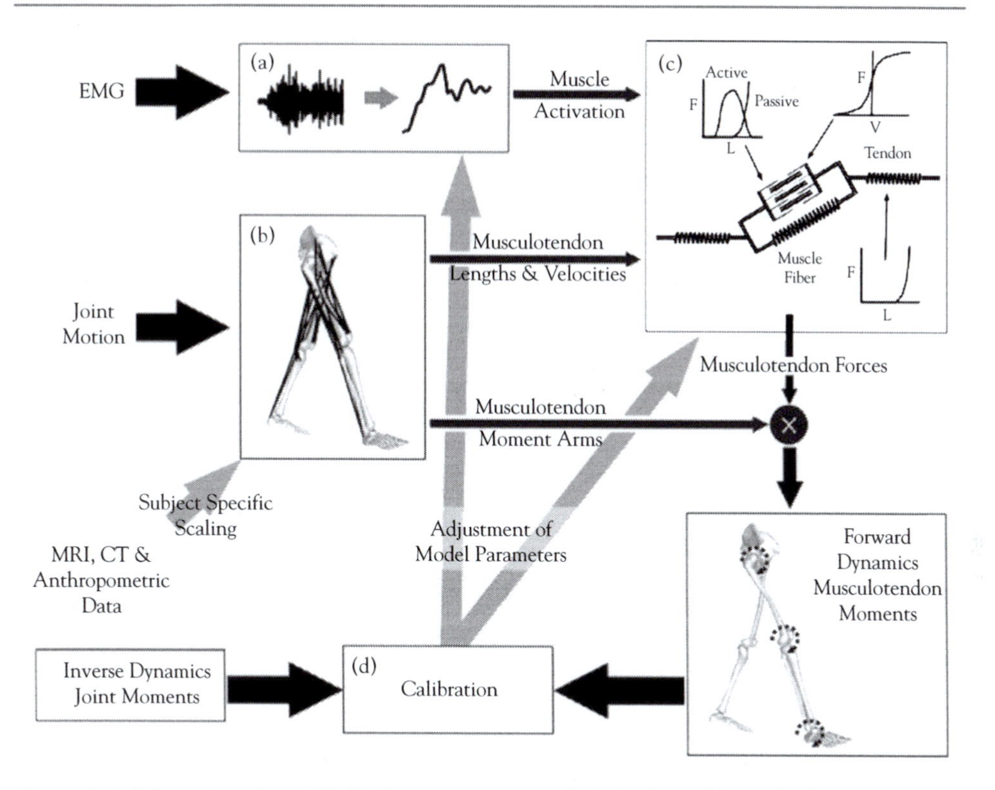

Figure 1.4 Schematic of our EMG-driven neuromusculoskeletal model, with four main parts: (a) EMG-to-muscle activation dynamics model; (b) model of musculoskeletal geometry; (c) muscle contraction dynamics model; and (d) model calibration to a subject.

the knee (Lloyd and Besier, 2003; Olney and Winter, 1985), ankle (Buchanan *et al.*, 2005), lower back (McGill, 1992), wrist (Buchanan *et al.*, 1993), and elbow (Manal *et al.*, 2002). Our EMG-driven neuromusculoskeletal model (Figure 1.4), has four main components:

(a) the EMG-to-muscle activation dynamics model;
(b) the model of musculoskeletal geometry;
(c) muscle contraction dynamics model; and
(d) model calibration to a subject.

EMG-to-muscle activation dynamics

In EMG-driven models, the neural commands (**e** in Figure 1.3) are replaced by the EMG linear envelope. These neural signals are transformed into muscle activation (Zajac, 1989) by the activation dynamics represented by either a first (Zajac, 1989) or second (Lloyd and Besier, 2003) order differential equation. These assume a linear mapping from neural activation to muscle activation, and even though linear transformations provide reasonable results (Lloyd and Buchanan, 1996; McGill, 1992), it is more physiologically correct to use non-linear relationships (Fuglevand *et al.*, 1999). For this we have additionally used either power or exponential functions, which results in a damped second-order non-linear dynamic process, controlled by three parameters per muscle (Buchanan *et al.*, 2004; Lloyd and Besier, 2003). This produces a muscle activation time series between zero and one.

Musculoskeletal geometry model

In addition to muscle activation, muscle forces depend on the muscle kinematics. Muscle kinematics, typically muscle-tendon moment arms and lengths, are estimated using a model of the musculoskeletal geometry. The implementation of these musculoskeletal models are made easier with the availability of modelling software such as SIMM (Software for Interactive Musculoskeletal Modeling – MusculoGraphics Inc. Chicago, USA) or AnyBody (AnyBody Technology, Denmark), which use graphical interfaces to help users create musculoskeletal systems that represents bones, muscles, ligaments, and other tissues. These models have been developed using cadaver data (Delp *et al.*, 1990) and represent the anatomy of a person with average height and body proportions.

These average 'generic' musculoskeletal models should be scaled to fit an individual's size and body proportions. This is not a trivial task (Murray *et al.*, 2002) and the techniques used depend on the technology available. At the simplest level, allometric scaling of bones can be performed based upon anatomical markers placed on a subject during a motion capture session. The attachment points of each musculotendon unit are scaled with the bone dimensions, thereby altering the operating length and moment arm of that muscle. Regions of bone can also be deformed based on anthropometric data to generate more accurate muscle kinematics (Arnold *et al.*, 2001). Alternatively, musculoskeletal models can be created from medical imaging data.

Medical imaging such as computed tomography (CT) or magnetic resonance imaging (MRI) can be used to obtain geometry of the joints and soft tissue. Although CT provides high resolution images of bone with excellent contrast, MRI is a popular choice as it is capable of differentiating soft tissue structures and bone boundaries, without ionizing radiation. Creating 3D models from MRI images involves segmentation to identify separate structures such as bone, cartilage, or muscle (Figure 1.5a). Depending on image quality, this is achieved using a combination of manual and semi-automated methods that produce structures represented as 3D point clouds (Figure 1.5b). These data are processed with commercial software packages (e.g. Raindrop Geomagic, Research Triangle Park, NC) to create triangulated surfaces of the anatomical structures (Figure 1.5c). We have used this technique to generate subject-specific finite element models of the patellofemoral (PF) joint to estimate contact and stress distributions (Besier *et al.*, 2005a).

By imaging muscles and joints in different anatomical positions, it is possible to determine how muscle paths change with varying joint postures. In these approaches muscles are typically represented as line segments passing through the centre of the segmented muscle, which provides muscle lines of actions throughout the joint range of motion. This has been shown to accurately determine muscle-tendon length and moment arms of the lower limb (Arnold *et al.*, 2000) and upper limb (Holzbaur *et al.*, 2005). Although this provides a good estimate of the total musculotendon length and can answer many research questions, it does not reflect the muscle fibre mechanics. To address this, several researchers have generated 3D muscle models using the finite element method (Blemker and Delp, 2006; Fernandez and Hunter, 2005). Unfortunately, these are computationally expensive and difficult to create, limiting their use in large scale musculoskeletal models, i.e. modelling many degrees of freedom with many muscles.

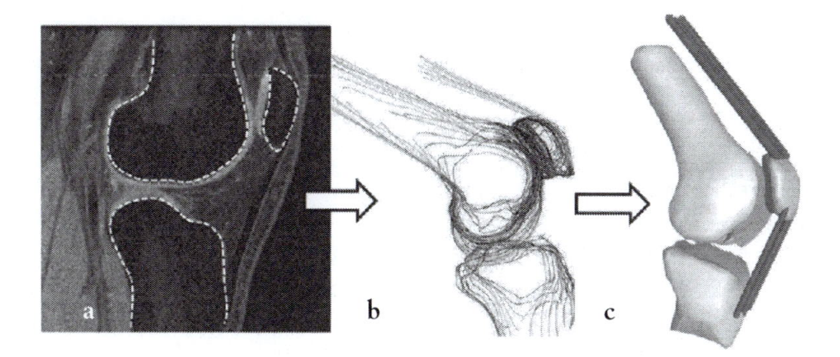

Figure 1.5 Creating a subject-specific musculoskeletal model of the knee involves segmentation of the medical images to define the boundaries of the various structures: (a) the segmented structures, representing 3D point clouds; (b) are then converted to triangulated surfaces and line segments; (c) that are used to determine musculotendon lengths and moment arms or converted to a finite element mesh for further analysis.

Another approach to create subject-specific models is to deform and scale an existing model to match a new data set. Free form deformation morphing techniques are capable of introducing different curvature to bones and muscles and have been used to individualise musculoskeletal models based on a CT or MRI data set (Fernandez *et al.*, 2004). These techniques make it possible to quickly generate accurate, subject-specific models of the musculoskeletal system. However, one still needs to model how muscles generate force.

Muscle contraction dynamics model

Muscle contraction dynamics govern the transformation of muscle activation and musculotendon kinematics to musculotendon force (F^{MT}). Musculotendon models include models of muscle, the contractile element, in series with the tendon. Most large scale neuromusculosketelal models employ 'Hill-type muscle models' (Zajac, 1989; Hill, 1938) because these are computationally fast (Erdemir *et al.*, 2007). Typically, a Hill-type muscle model has a generic force-length ($f(l_m)$), force-velocity ($f(v_m)$), parallel passive elastic force-length ($f_p(l_m)$) curves (Figure 1.4C), and pennation angle ($\phi(l_m)$) (Buchanan *et al.*, 2004, 2005). These functions are normalized to the muscle properties of maximum isometric muscle force (F^{max}), optimal fibre length (L_m^o), and pennation angle at optimal fibre length (ϕ^0). The non-linear tendon function ($f^t(l_t)$) relates tendon force ($F^t(t)$) to tendon length (l_t) and depends on F^{max} and the tendon property of tendon slack length (L_T^S) (Zajac, 1989). The general equations for the force produced by the musculotendon unit ($F^{mt}(t)$) are given by:

$$\begin{aligned} F^{mt}(t) &= F^t(t) = F^{max} f^t(l_t) \\ &= F^{max}\left[f(l_m)f(v_m)a(t) + f_p(l_m)\right]\cos(\phi(l_m)) \end{aligned} \tag{4}$$

The muscle and tendon properties (F^{max}, L_m^o, ϕ^0, and L_T^S) are muscle-specific and person dependent. Values for these can be set to those reported in literature (Yamaguchi *et al.*, 1990). However, when the anatomical model is scaled to an individual this alters the operating length of each musculotendon unit, which necessitates scaling of the muscle and tendon properties. For simplicity, muscle pennation angles are assumed to be constant across subjects. Since people have different strengths, scaling factors for the physiological cross-sectional area (PCSA), and, therefore, F^{max}, of the flexors (δ_{flex}) and extensors (δ_{ext}) are used to adjust the model to the subject's strength (Lloyd and Besier, 2003). Alternatively regression equations can be used to scale PCSA according to body mass or MRI images of the muscles (Ward *et al.*, 2005). However, L_m^o and L_T^S do not scale proportional to body dimensions, such as bone length (Ward *et al.*, 2005), yet strongly influence the force outputs of the musculotendon model (Heine *et al.*, 2003). Therefore, it is important to scale L_m^o and L_T^S to an individual.

In scaling it is important to appreciate that is a constant property of a musculotendon unit irrespective of musculotendon length (Manal and Buchanan, 2004), that is:

$$L_T^S = \frac{l_{MT} - L_m^o \cdot l_m^{norm} \cos\alpha}{1+\varepsilon_T} = \text{constant}$$

$$\varepsilon_T = \frac{\tilde{F}^m \cos\alpha + 0.2375}{37.5} \qquad \text{for } \varepsilon_T \geq 0.0127$$

$$\text{and} \qquad \varepsilon_T = \frac{\ln\left(\dfrac{\tilde{F}^m \cos\alpha}{0.06142}+1\right)}{124.929} \qquad \text{for } \varepsilon_T \leq 0.0127 \qquad (5^1)$$

In this equation l_{MT} is musculotendon length, l_m^{norm} normalized muscle fibre length, ε_T tendon strain, and $\tilde{F}^m$ the normalized muscle force ($F^m(t)/F^{max}$) that would be generated by the maximally activated muscle at l_m^{norm}. For a given L_m^o multiple solutions for L_T^S exist that satisfy equation 5 (Winby *et al.*, 2007). However, a unique solution for L_m^o and L_T^S can be found if one assumes that the relationship between joint angle and the isometric musculotendon force, and therefore l_m^{norm}, is the same for all subjects regardless of size (Garner and Pandy, 2003; Winby *et al.*, 2007). If the l_m^{norm} for a range of joint postures in the unscaled anatomical model is known, with corresponding L_{MT} obtained from the scaled model at the same postures, a non-linear least squares fit provides solutions for unique values of both L_m^o and L_T^S (Winby *et al.*, 2007). These values will create a scaled l_m^{norm} – joint angle relationship that is the closest match to the unscaled model across the entire range of joint motion. These values can either be treated as final muscle and tendon properties for a particular subject or used as a starting point for a calibration process.

Calibration and validation

Calibration of a neuromusculoskeletal model is performed to minimize the error between the joint moments estimated from inverse dynamics (equation 1) and the net musculotendon moments estimated from the neuromusculoskeletal model (Buchanan *et al.*, 2005; Lloyd and Besier, 2003), that is:

$$\min \sum_1^n \left(\Sigma M^{MT} - M^J\right)^2 \qquad (6)$$

where n is the number trials that we include in the calibration process. If, for example, we are implementing a knee model (Lloyd and Besier, 2003; Lloyd and Buchanan, 1996), we would use the knee flexion-extension moments since these are primarily determined by the muscles. These moments are determined from walking, running and/or sidestepping trials, but we may also use trials from an isokinetic dynamometer, such as passively moving knee through a range of motion, or maximal or submaximal eccentric and concentric isokinetic strength tests (Lloyd and Besier, 2003).

An optimization scheme, such as simulated annealing (Goffe *et al.*, 1994) is used to adjust the activation dynamics parameters and musculotendon properties (δ_{flex}, δ_{ext}, L_T^S) (Lloyd and Besier, 2003). Activation parameters are chosen as these have day-to-day variation due to placement and positioning of the EMG electrodes. The musculotendon properties chosen are those that are not well defined or easily measured (Lloyd and Besier, 2003). The initial values are based on those in literature or scaled to the individual, and adjusted to lie within a biologically acceptable range (Lloyd and Besier, 2003; Lloyd and Buchanan, 1996). In this way, the model can be calibrated to each subject, and then validation of the calibrated model performed.

Validation is one of the strengths of this neuromusculoskeletal modelling method. This allows one to check how well the calibrated model can predict joint moments of other trails not used in the calibration. For example, the calibrated neuromusculoskeletal knee model produced exceptionally good predictions of the inverse dynamics knee flexion-extension moments from over 200 other trials and predicted trials two weeks apart without loss in predictive ability (average $R^2 = 0.91 \pm 0.04$) (Lloyd and Besier, 2003). Once a calibrated model is shown to well predict joint moments, there is increased confidence in the estimated musculotendon forces, which can then be used to assess the loads, stresses and strains experienced by other tissues.

Assessing tissue load, stress and strain during motion

Two applications will be presented to illustrate the use of EMG-driven models to assess tissue loads; one for the tibiofemoral joint and the other for the PF joint. Refer to our previous papers for other applications in the lower limb (Buchanan *et al.*, 2005; Lloyd and Buchanan, 1996; Lloyd and Buchanan, 2001; Lloyd *et al.*, 2005).

Tibiofemoral joint contact forces while walking

This research is assessing how muscle activation patterns affect loading of the medial and lateral condyles of the tibiofemoral joint during walking. Large knee adduction moments in gait predict fast progression of medial compartment knee osteoarthritis (Miyazaki *et al.*, 2002) and rapid reoccurrence of varus deformity of the knee after high tibial osteotomy (Prodromos *et al.*, 1985). It is believed that large articular loading in the medial compartment of the tibiofemoral joint, produced by the adduction moments, causes the rapid osteoarthritic changes (Prodromos *et al.*, 1985). Using a simple knee model, articular loading in the medial relative to the lateral compartment also predicts the bone density distribution in the proximal tibia (Hurwitz *et al.*, 1998). However, muscle contraction may change the loading of the medial and lateral condyles of the tibiofemoral joint (Schipplein and Andriacchi, 1991) and must be taken into account. Indeed, altered muscle activations patterns, which include high levels of co-contraction of the

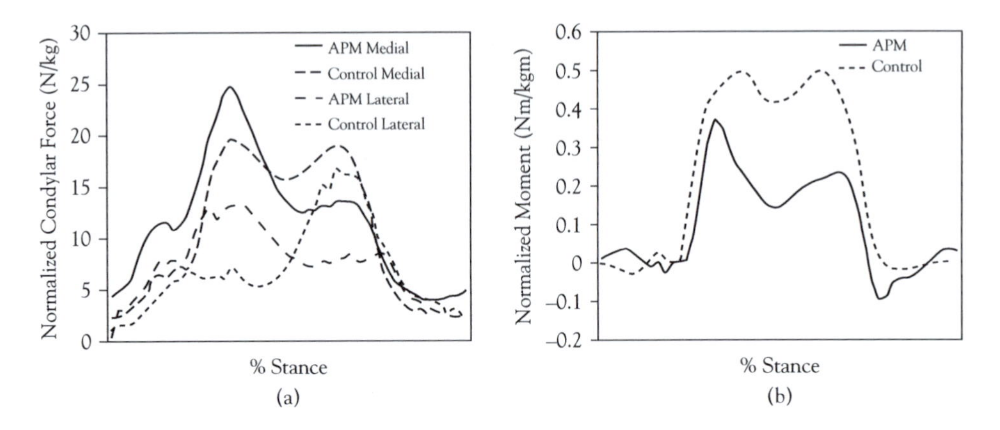

Figure 1.6 Condylar contact forces: (a) and external adduction moments; (b) for an APM subject and age-matched control. Condylar forces normalized to body mass and a dimensionless scaling factor related to knee joint size. External adduction moments normalized to body mass and distance between condylar contact points.

hamstrings and quadriceps, have been observed in people with knee osteoarthritis (Hortobagyi *et al.*, 2005) and in those who have undergone arthroscopic partial meniscectomy (Sturnieks *et al.*, 2003). High levels of co-contraction may increase articular loading and hasten tibiofemoral joint degeneration (Lloyd and Buchanan, 2001; Schipplein and Andriacchi, 1991).

Subsequently we have examined the gait patterns and loading of the tibiofemoral joint in patients who have had arthroscopic partial meniscectomy. Ground reaction forces, 3D kinematics and EMG during gait were collected from these patients and control subjects with no history of knee joint injury or disease. A scaled EMG-driven neuromuscular skeletal model of the lower limbs and knee was used, calibrated using the knee flexion moments of several walking trials of different pace. Another set of walking trials were processed using the calibrated model in order to obtain estimates of the musculotendon forces and to estimate the adduction/abduction moments generated by the muscles about both the medial and lateral condyles. Using these muscle moments, with the external adduction moments estimated from inverse dynamics, dynamic equilibrium about each condyle permitted the condylar contact forces to be calculated.

Preliminary results of this research suggest that contact forces on both the medial and lateral condyles during early stance for the APM patients may be larger than those in the control subject (Figure 1.6). This is despite the larger external adduction moments exhibited by the control subject during the same phase, indicating that muscle forces may confound estimates of internal joint loads based solely on external loading information. More importantly, models that rely on lumped muscle assumptions (Hurwitz *et al.*, 1998) or dynamic optimization (Shelburne *et al.*, 2006) to estimate muscle activity suggest that unloading of the lateral condyle occurs during the stance phase of gait, necessitating ligament load to prevent condylar lift off. This is contrary to our findings, and those from instrumented knee implant research (Zhao *et al.*, 2007), that indicate no unloading of the lateral condyle occurs throughout stance. This example illustrates the importance of incorporating measured muscle activity in the estimation of tissue loading.

Patellofemoral cartilage stresses in knee extension exercises

Muscle forces obtained from an EMG-driven model can also provide input for finite element (FE) models to determine the stress distribution in various tissues in and around the knee. This framework is being used to estimate the cartilage stress distributions at the PF joint (Besier *et al.*, 2005b).

It is believed that elevated stress in the cartilage at the PF joint can excite nociceptors in the underlying subchondral bone and is one mechanism by which patients may experience pain. The difficulty in testing this hypothesis is that many factors can influence the stresses in the cartilage, including: the geometry of the articulating surfaces; the orientation of the patella with respect to the femur during loading; the magnitude and direction of the quadriceps muscle forces; and the thickness and material properties of the cartilage. These complexities are also likely reasons as to why previous research studies on PFP have failed to define a clear causal mechanism. The FE method is well-suited to examine if pain is related to stress, as it can account for all factors that influence the mechanical state of the tissue. An EMG-driven musculoskeletal model is also essential to this problem, as a subject with PFP may have very different muscle activation patterns and subsequent stress distributions than a subject who experiences no pain.

To create subject-specific FE models of the PF joint, the geometry of the bones and other important soft tissues (cartilage, quadriceps tendon and the patellar tendon) are defined using high resolution MRI scans, as outlined previously. Using meshing software, these structures are discretized into 'finite elements', which are assigned material properties (e.g. Young's modulus and Poisson's ratio) to reflect the stress–strain behaviour of that tissue. Since we are interested primarily in the cartilage, and we know that bone is many times stiffer than cartilage, we assume the bones to be rigid bodies to reduce the computational complexity. Various constitutive models exist to represent the behaviour of cartilage, however, for transient loading scenarios (greater than 0.1 Hz), such as walking and running, the mechanical behaviour of cartilage can be adequately modelled as a linear elastic solid (Higginson and Snaith, 1979).

The MR images also allow the quadriceps muscle and patellar tendon attachments to be defined. These structures can be modelled as membrane elements over bone (Figure 1.7),

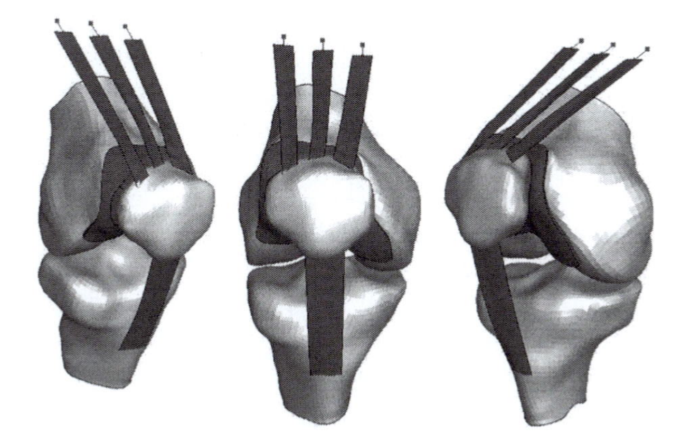

Figure 1.7 Finite element model of the PF joint showing the connector element used to apply the quadriceps forces determined from the EMG-driven neuromusculosketelal model.

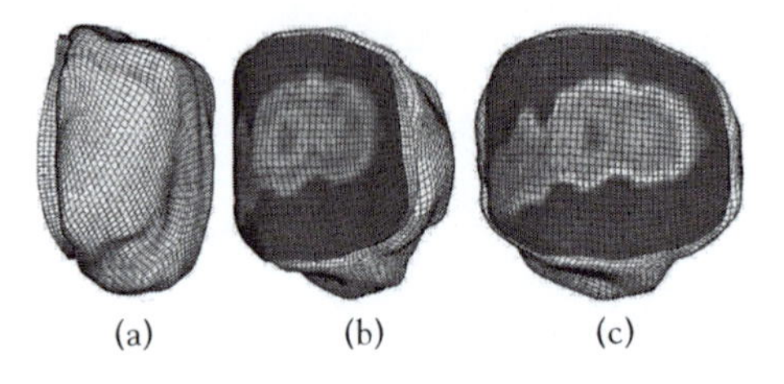

Figure 1.8 Finite element mesh of a patella showing solid, hexahedral elements that make up the patellar cartilage: (a) These elements are rigidly fixed to the underlying bone mesh. Exemplar contact pressures; (b) and; (c) developed during a static squat held at 60° knee flexion.

actuated by a connector element that represents the line of action of the musculotendon unit. For quasi-static loading situations, the tibiofemoral joint is fixed to a specific joint configuration (e.g. stance phase of gait) and muscle forces from the scaled EMG-driven neuromusculoskeletal model for that subject are applied to the PF model. In these simulations, the patella, with it six degrees of freedom, will be a positioned in the femoral trochlear to achieve static equilibrium, following which the stresses and strains throughout the cartilage elements are determined (Figure 1.8).

To validate these EMG-driven FE simulations, subjects are imaged in an open-configuration MRI scanner, which allows volumetric scans to be taken of the subject's knee in an upright, weight-bearing posture. These images can be used to provide accurate joint orientations for each simulation as well as PF joint contact areas (Gold *et al.*, 2004). The same upright, weight-bearing squat is performed in a motion capture laboratory, where EMG, joint kinematics and joint kinetics and subsequent muscle forces are estimated using an EMG-driven model. The simulation of these weight-bearing postures permits a validation of the FE model in two ways. Firstly, we compare the contact areas of the simulation with those measured from the MR images, which should closely match. Secondly, we compare the final orientation of the patella from the simulation to the orientation within the MR images, and these should also match. Future development can include an optimisation-feedback loop in which information from the FE model is then given back to the EMG-driven model in order to improve the estimation of contact area and patella orientation.

This approach provides a framework for the estimation of muscle forces that accounts for muscle activations, joint contact mechanics and tissue-level stresses and strains. The validated EMG-driven FE model can then be used to simulate other activities by placing the tibiofemoral joint into other configurations and using muscle forces from the EMG-driven model.

In summary, the development of medical imaging modalities combined with EMG-driven musculoskeletal models are providing researchers with tools to estimate the loads, stresses, and strains throughout various biological tissues. These advancements have created avenues for biomedical researchers to gain valuable insight regarding the form and function of the musculoskeletal system, and to prevent or treat musculoskeletal tissue injury and disease in the lower extremities.

Acknowledgements

Financial support for this work has been provided in part to DGL from the National Health and Medical Research Council, Western Australian MHRIF, and University of Western

Australia Small Grants Scheme, to TFB from the Department of Veterans Affairs, Rehabilitation R&D Service (A2592R), the National Institute of Health (1R01-EB005790), and Stanford Regenerative Medicine (1R-90DK071508), and to TSB from the National Institutes of Health (NIH R01-HD38582).

Note

1 This equation is modified from Manal and Buchanan (2004) to include the toe region of the tendon stress/strain curve.

References

1. Arnold, A.S., Blemker, S.S. and Delp, S.L. (2001) 'Evaluation of a deformable musculoskeletal model for estimating muscle-tendon lengths during crouch gait'. *Ann Biomed Eng*, 29: 263–74.
2. Arnold, A.S., Salinas, S., Asakawa, D.J., and Delp, S.L. (2000) 'Accuracy of muscle moment arms estimated from MRI-based musculoskeletal models of the lower extremity'. *Computer Aided Surgery*, 5: 108–19.
3. Besier, T.F., Draper, C.E., Gold, G.E., Beaupre, G.S. and Delp, S.L. (2005a) 'Patellofemoral joint contact area increases with knee flexion and weight-bearing'. *Journal of Orthopaedic Research*, 23: 345–50.
4. Besier, T.F., Gold, G.E., Beaupre, G.S. and Delp, S.L. (2005b) 'A modeling framework to estimate patellofemoral joint cartilage stress in-vivo'. *Medicine & Science in Sports and Exercise*, 37: 1924–30.
5. Besier, T.F., Sturnieks, D.L., Alderson, J.A. and Lloyd, D.G. (2003) 'Repeatability of gait data using a functional hip joint centre and a mean helical knee axis'. *J Biomech*, 36: 1159–68.
6. Blemker, S.S. and Delp, S.L. (2006) 'Rectus femoris and vastus intermedius fiber excursions predicted by three-dimensional muscle models'. *Journal of Biomechanics*, 39: 1383–91.
7. Brooks, P.M. (2006) 'The burden of musculoskeletal disease – a global perspective'. *Clin Rheumatol*, 25: 778–81.
8. Buchanan, T.S. and Lloyd, D.G. (1995) 'Muscle activity is different for humans performing static tasks which require force control and position control'. *Neuroscience Letters*, 194: 61–4.
9. Buchanan, T.S., Lloyd, D.G., Manal, K. and Besier, T.F. (2004) 'Neuromusculoskeletal Modeling: Estimation of Muscle Forces and Joint Moments and Movements From Measurements of Neural Command'. *J Appl Biomech*, 20: 367–95.
10. Buchanan, T.S., Lloyd, D.G., Manal, K. and Besier, T.F. (2005) 'Estimation of muscle forces and joint moments using a forward-inverse dynamics model'. *Med Sci Sports Exerc*, 37: 1911–6.
11. Buchanan, T.S., Moniz, M.J., Dewald, J.P. and Zev Rymer, W. (1993) 'Estimation of muscle forces about the wrist joint during isometric tasks using an EMG coefficient method'. *J Biomech*, 26: 547–60.
12. Cochrane, J.L., Lloyd, D.G., Buttfield, A., Seward, H. and McGivern, J. (2006) 'Characteristics of anterior cruciate ligament injuries in Australian football'. *J Sci Med Sport*.
13. Delp, S.L., Loan, J.P., Hoy, M.G., Zajac, F.E., Topp, E.L. and Rosen, J.M. (1990) 'An interactive graphics-based model of the lower extremity to study orthopaedic surgical procedures'. *IEEE Trans Biomed Eng*, 37: 757–67.
14. Devereaux, M.D. and Lachmann, S.M. (1984) 'Patello-femoral arthralgia in athletes attending a Sports Injury Clinic'. *Br J Sports Med*, 18: 18–21.
15. Erdemir, A., Mclean, S., Herzog, W. and Van Den Bogert, A.J. (2007) 'Model-based estimation of muscle forces exerted during movements'. *Clin Biomech (Bristol, Avon)*, 22: 131–54.
16. Fernandez, J.W. and Hunter, P.J. (2005) 'An anatomically based patient-specific finite element model of patella articulation: towards a diagnostic tool'. *Biomech Model Mechanobiol*, 4: 28–38.

17. Fernandez, J.W., Mithraratne, P., Thrupp, S.F., Tawhai, M.T. and Hunter, P.J. (2004) 'Anatomically based geometric modelling of the musculo-skeletal system and other organs'. *Biomech Model Mechanobiol*, 2: 139–155.
18. Fuglevand, A.J., Macefield, V.G. and Bigland-Ritchie, B. (1999) 'Force-frequency and fatigue properties of motor units in muscles that control digits of the human hand'. *J Neurophysiol*, 81: 1718–29.
19. Garner, B.A. and Pandy, M.G. (2003) 'Estimation of musculotendon properties in the human upper limb'. *Ann Biomed Eng*, 31: 207–20.
20. Goffe, W.L., Ferrieir, G.D. and Rogers, J. (1994) 'Global optimization of statistical functions with simulated annealing'. *Journal of Econometrics*, 60: 65–99.
21. Gold, G.E., Besier, T.F., Draper, C.E., Asakawa, D.S., Delp, S.L. and Beaupre, G.S. (2004) 'Weight-bearing MRI of patellofemoral joint cartilage contact area'. *J Magn Reson Imaging*, 20: 526–30.
22. Heine, R., Manal, K. and Buchanan, T.S. (2003) 'Using Hill-Type muscle models and emg data in a forward dynamic analysis of joint moment:evaluation of critical parameters'. *Journal of Mechanics in Medicine and Biology*, 3: 169–186.
23. Hewett, T.E., Myer, G.D. and Ford, K.R. (2004) 'Decrease in neuromuscular control about the knee with maturation in female athletes'. *J Bone Joint Surg Am*, 86–A: 1601–8.
24. Higginson, G.R. and Snaith, J.E. (1979) 'The mechanical stiffness of articular cartilage in confined oscillating compression'. *Eng Medicine*, 8: 11–14.
25. Hill, A.V. (1938) 'The heat of shortening and the dynamic constants of muscle'. *Proceedings of the Royal Society of London Series B*, 126: 136–195.
26. Holzbaur, K.R., Murray, W.M. and Delp, S.L. (2005) 'A model of the upper extremity for simulating musculoskeletal surgery and analyzing neuromuscular control'. *Ann Biomed Eng*, 33: 829–40.
27. Hortobagyi, T., Westerkamp, L., Beam, S., Moody, J., Garry, J., Holbert, D. and Devita, P. (2005) 'Altered hamstring-quadriceps muscle balance in patients with knee osteoarthritis'. *Clin Biomech (Bristol, Avon)*, 20: 97–104.
28. Hurwitz, D.E., Sumner, D.R., Andriacchi, T.P. and Sugar, D.A. (1998) 'Dynamic knee loads during gait predict proximal tibial bone distribution'. *Journal of Biomechanics*, 31: 423–30.
29. Lloyd, D.G. and Besier, T.F. (2003) 'An EMG-driven musculoskeletal model to estimate muscle forces and knee joint moments in vivo'. *J Biomech*, 36: 765–76.
30. Lloyd, D.G. and Buchanan, T.S. (1996) 'A model of load sharing between muscles and soft tissues at the human knee during static tasks'. *Journal of Biomechanical Engineering*, 118: 367–76.
31. Lloyd, D.G. and Buchanan, T.S. (2001) 'Strategies of muscular support of varus and valgus isometric loads at the human knee'. *J Biomech*, 34: 1257–67.
32. Lloyd, D.G., Buchanan, T.S. and Besier, T.F. (2005) 'Neuromuscular biomechanical modeling to understand knee ligament loading'. *Med Sci Sports Exerc*, 37: 1939–47.
33. Manal, K. and Buchanan, T.S. (2004) 'Subject-specific estimates of tendon slack length: a numerical method'. *Journal of Applied Biomechanics*, 20: 195–203.
34. Manal, K., Gonzalez, R.V., Lloyd, D.G. and Buchanan, T.S. (2002) 'A real-time EMG-driven virtual arm'. *Comput Biol Med*, 32: 25–36.
35. McGill, S.M. (1992) 'A myoelectrically based dynamic three-dimensional model to predict loads on lumbar spine tissues during lateral bending'. *J Biomech*, 25: 395–414.
36. Miyazaki, T., Wada, M., Kawahara, H., Sato, M., Baba, H. and Shimada, S. (2002) 'Dynamic load at baseline can predict radiographic disease progression in medial compartment knee osteoarthritis'. *Ann Rheum Dis*, 61: 617–22.
37. Murray, W.M., Buchanan, T.S. and Delp, S.L. (2002) 'Scaling of peak moment arms of elbow muscles with upper extremity bone dimensions'. *Journal of Biomechanics*, 35: 19–26.
38. O'Connor, J.J. (1993) 'Can muscle co-contraction protect knee ligaments after injury or repair?' *Journal of Bone & Joint Surgery – British Volume*, 75: 41–8.
39. Olney, S.J. and Winter, D.A. (1985) 'Predictions of knee and ankle moments of force in walking from EMG and kinematic data'. *J Biomech*, 18: 9–20.

40. Prodromos, C.C., Andriacchi, T.P. and Galante, J.O. (1985) 'A relationship between gait and clinical changes following high tibial osteotomy'. *J Bone Joint Surg (Am)*, 67: 1188–94.
41. Schipplein, O.D. and Andriacchi, T.P. (1991) 'Interaction between active and passive knee stabilizers during level walking'. *Journal of Orthopaedic Research*, 9: 113–9.
42. Shelburne, K.B., Torry, M.R. and Pandy, M.G. (2006) 'Contributions of muscles, ligaments, and the ground-reaction force to tibiofemoral joint loading during normal gait'. *J Orthop Res*, 24: 1983–90.
43. Smith, A.J., Lloyd, D.G. and Wood, D.J. (2004) 'Pre-surgery knee joint loading patterns during walking predict the presence and severity of anterior knee pain after total knee arthroplasty'. *J Orthop Res*, 22: 260–6.
44. Sturnieks, D.L., Besier, T.F., Maguire, K.F. and Lloyd, D.G. (2003) 'Muscular contributions to stiff knee gait following arthroscopic knee surgery'. Sports Medicine Australia Conference. Canberra, Australia, Sports Medicine Australia.
45. Ward, S.R., Smallwood, L.H. and Lieber, R.L. (2005) 'Scaling of human lower extremity muscle architecture to skeletal dimensions'. ISB XXth Congress. Cleveland, Ohio.
46. Whiting, W.C. and Zernicke, R.F. (1998) 'Biomechanics of musculoskeletal injury,' Champaign, Illinois, Human Kinetics, p. 177.
47. Winby, C.R., Lloyd, D.G. and Kirk, T.B.K. (2007) 'Muscle operating range preserved by simple linear scaling'. ISB XXIst Congress. Taipei, Taiwan.
48. Yamaguchi, G.T., Sawa, A.G.U., Moran, D.W., Fessler, M.J., Winters, J.M. and Stark, L. (1990) 'A survey of human musculotendon actuator parameters'. In Winters, J.M. and Woo, S.L.Y. (Eds.) Multiple Muscle Systems: Biomechanics and Movement Organization. New York, Springer-Verlag.
49. Zajac, F.E. (1989) 'Muscle and tendon: properties, models, scaling, and application to biomechanics and motor control'. *Critical Reviews in Biomedical Engineering*, 17: 359–411.
50. Zhao, D., Banks, S.A., D'lima, D.D., Colwell Jr., C.W. and Fregly, B.J. (2007) 'In vivo medial and lateral tibial loads during dynamic and high flexion activities'. *Journal of Orthopaedic Research*, 25: 593–602.

2

Modelling and simulation of tissue load in the upper extremities

Clark R. Dickerson
University of Waterloo, Waterloo

Introduction

Biomechanical analyses of the upper extremities are at a comparatively early stage compared to other regional investigations, including gait and low back biomechanics. However, upper extremity use pervades life in the modern world as a prerequisite for nearly all commonly performed activities, including reaching, grasping, tool use, machine operation, typing and athletics. Such a crucial role demands systematic study of arm use. This chapter highlights methods used to analyze tissue loads in the shoulder, elbow and wrist.

Increased recognition of upper extremity disorders amongst both working (Silverstein *et al.*, 2006) and elderly (Juul-Kristensen *et al.*, 2006) populations has resulted in numerous targeted exposure modelling efforts. Many approaches focus on easily observable quantities, such as gross body postures and approximations of manual force or load (McAtamney and Corlett, 1993; Latko *et al.*, 1997). While popular due to ease of use and low computational requirements, they cannot estimate tissue-specific loads. However, computerized biomechanical analysis models exist to describe both overall joint loading and for loading of specific tissues. Efforts to include biomechanical simulation using these tools in surgical planning and rehabilitation are also gaining traction.

This chapter focuses primarily on the shoulder and elbow joints, which have received minimal treatment in recent summaries of upper extremity biomechanics and injuries (Keyserling, 2000; Frievalds, 2004). It reviews classic and current methods for analyzing tissue loads, and highlights upcoming challenges and opportunities that accompany the development of new approaches to better analyze, describe, and communicate the complexity of these loads.

The shoulder

Three core concepts define shoulder biomechanical research: (1) fundamental shoulder mechanics; (2) shoulder modelling approaches; and (3) unanswered shoulder function questions.

Fundamental shoulder mechanics: The shoulder allows placement of the hand in a vast range of orientations necessary to perform a spectrum of physical activities. The joint's intricate morphology, including the gliding scapulothoracic interface, enables this versatility. Table 2.1 contains descriptions of the roles of the principal shoulder components. However, high postural flexibility comes at the cost of intrinsic joint stability, as is widely reported (an overview is provided in Veeger and Van der Helm, 2007). The shallowness of the glenoid fossa requires additional glenohumeral stability generation from one or more mechanisms: active muscle coordination, elastic ligament tension, labrum deformation, joint suction, adhesion/cohesion, articular version, proprioception, or negative internal joint pressure (Schiffern *et al.*, 2002; Cole *et al.*, 2007). Systematic consideration of the impact of these many mechanisms is arguably required in order to replicate physiological shoulder muscle activity. Beyond this concern, the mechanical indeterminacy in the shoulder must be addressed, as there are more actuators (muscles) than there are degrees of freedom (DOF), by most definitions.

Shoulder biomechanical modelling: Although a historic focus on defining shoulder kinematics has existed (Inman *et al.*, 1944; others), large-scale musculoskeletal models of the shoulder did not emerge until the 1980s. Early methods for estimates of shoulder loading focused on calculation of external joint moments using an inverse, linked-segment modelling kinematics approach (as described in Veeger and van der Helm, 2004). These models require kinematic or postural data along with external force data as inputs and output joint external moments. Specific muscle force and stress levels, or internal joint force levels, can provide expanded insight into the potential mechanisms of specific shoulder disorders. Unfortunately, this capability is currently lacking in applied tools, though software does exist for the calculation of externally generated static joint moments and loads (Chaffin *et al.*, 1997; Norman *et al.*, 1994; Badler *et al.*, 1989). Several attempts have been made to model the musculoskeletal components of the shoulder. In general, determining tissue loads requires four stages: (1) musculoskeletal geometry reconstruction; (2) calculation of external forces and moments; (3) solving the load-sharing problem while considering joint stability; and (4) communicating results. Recent refinements mark each of these areas. Figure 2.1 demonstrates this progression, with intermediate outputs.

Geometric reconstruction: Several descriptions of shoulder musculoskeletal geometric data exist (Veeger *et al.*, 1991; van der Helm *et al.*, 1992; Klein Breteler *et al.*, 1999; Johnson *et al.*, 1996; Hogfors *et al.*, 1987; Garner and Pandy, 1999). These typically consist of locally (bone-centric) defined muscle attachment sites. The list is limited due to the considerable difficulty in both obtaining and measuring the many elements. Although recent attempts to characterize shoulder geometry using less invasive measurements (Juul-Kristensen *et al.*, 2000 a&b) have had success, they have been limited in scope to specific muscles, and thus the cadaveric data sources remain the most completely defined for holistic modelling purposes. Additionally, several authors (van der Helm and Veenbaas, 1991; Johnson *et al.*, 1996; and Johnson and Pandyan, 2005) have approached conversion of measured cadaveric muscle attachment data to mechanically relevant elements. Presently, no consensus for their definition exists across model formulations. The attachment sites and mechanical properties of many shoulder girdle ligaments have also been described (Pronk *et al.*, 1993; Debski *et al.*, 1999; Boardman, 1996; Bigliani, 1992; Novotny *et al.*, 2000). However, their limited contributions for the majority of midrange shoulder postures have minimized their implementation as elastic elements in most approaches.

Beyond definition of muscle attachment sites, many models recognize the need to represent physiological muscle line-of-action paths through 'muscle wrapping'. Amongst the most commonly used techniques to generate wrapped muscles are the centroid method

Table 2.1 Components of the shoulder and associated mechanical functions

Component	*Tissue*	*DOF*	*Location in shoulder*	*Mechanical function*
Sternum	Bone	N/A	Base of SCJ	Base of upper arm kinematic chain
Clavicle	Bone	N/A	Bridge between SCJ and ACJ	Enables positioning of scapula relative to torso, attachment site for some shoulder muscles
Scapula	Bone	N/A	Connects torso, clavicle and humerus	Site of proximal glenohumeral joint, enables multiple DOF for shoulder girdle, attachment site for many postural and articulating muscles
Humerus	Bone	N/A	Bony component of upper arm	Attachment site for articulating muscles, beginning of torso-independent arm linkage
Sternoclavicular Joint (SCJ)	Joint	6	Base of clavicular segment	Allows movement of clavicle
Acromioclavicular Joint (ACJ)	Joint	6	Base of scapular segment	Allows modest movement of scapula relative to clavicle
Scapulothoracic Joint (STJ)	Joint	5	Floating movement base of the scapula	Allows changes in scapular position to enable glenoid placement
Glenohumeral Joint (GHJ)	Joint	6	Interface between scapula and humerus	Greatest range of motion of all shoulder joints, also highest instability
Postural Muscles	Muscle	1	Throughout shoulder complex; primarily between the scapula, clavicle and torso	Enable scapular positioning and provide postural stability. Examples include trapezius, rhomboids, serratus anterior, and levator scapulae
Humeral Articulating Muscles	Muscle	1	Muscles connecting humerus to proximal sites	Enable humeral movement relative to the torso. Examples include the deltoid, latissimus dorsi, pectoralis major, and the rotator cuff.
Rotator Cuff	Muscle	1	Set of four muscles that attach humerus to scapula	Enable humeral movement while contributing to stabilizing glenohumeral joint forces. Includes the supraspinatus, infraspinatus, subscapularis, and teres minor
Multi-joint Muscles	Muscle	1	Throughout shoulder	Multi-function muscles, an example of which is biceps which flexes the elbow while also providing some glenohumeral joint stability
Ligaments	Soft tissue	1	SJC, ACJ, GHJ	Contribute to joint stability when under tension, can also produce countervailing forces
Labrum	Soft Tissue	N/A	Over the articular surface of the glenoid	Increases surface area of the glenoid, provides stability through suction and adhesion mechanisms

(Garner and Pandy, 2000), geometric geodesic wrapping (van der Helm, 1994a; Hogfors *et al.*, 1987; Charlton and Johnson, 2001), and the via point method (Delp and Loan, 1995). The validity of using extrapolations of cadaveric musculoskeletal data for model applications, including muscle-wrapping, has seen limited testing, although recent evidence supports its use in principle (Dickerson *et al.*, 2006b; Gatti *et al.*, 2007).

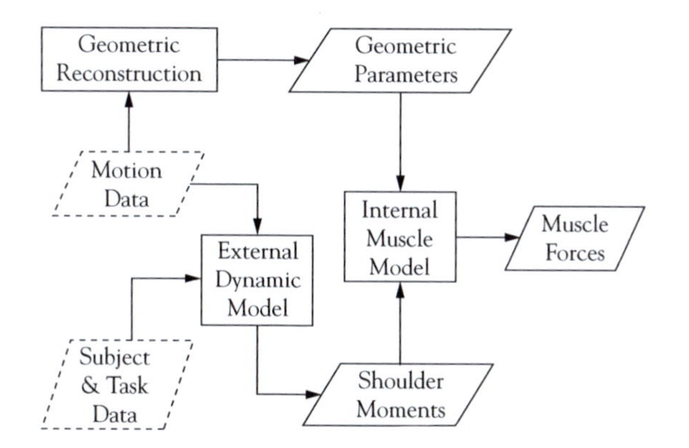

Figure 2.1 Typical components of a musculoskeletal shoulder model. Inputs include body motion, task characteristics and personal characteristics. While the first two models are independent, the muscle force prediction model depends on the outputs of these models.

Arm and shoulder kinematics must also be represented in biomechanical models. Due to substantial bony motion beneath and independent of the skin, palpation and tracking of scapular and clavicular kinematics in vivo is difficult without invasive techniques. The shoulder rhythm (often attributed to Inman *et al.*, 1944) refers to the phenomenon that in a given humerothoracic position, scapular and clavicular movement contribute to achieving that posture in a reproducible fashion. Generalized rhythm descriptions have been observed (Karduna *et al.*, 2001; McClure *et al.*, 2001; Barnett *et al.*, 1999) and described mathematically (Inman *et al.*, 1944; de Groot *et al.*, 1998, 2001; Ludewig *et al.*, 2004; Borstad and Ludewig, 2002; Hogfors *et al.*, 1991; Pascoal *et al.*, 2000). However, when using derived shoulder rhythms, it is important to consider that multiple scapular positions can achieve identical humerothoracic postures (Matsen *et al.*, 1994), and that biological variation in shoulder rhythms exists.

Geometric models produce descriptions of the internal geometry of the shoulder, which are primary inputs required for the load-sharing indeterminacy problem in an internal muscle model.

Calculation of external forces and moments: An external arm model generally includes three defined segments: hand, forearm and upper arm. Segmental inertial parameters can be estimated using a variety of methods (Durkin and Dowling, 2003; Chapter 14 of this handbook). The segmental forms for translational and rotational equilibrium, respectively, for the inverse dynamics approach are:

$$\sum_{1-m} F_{EXT} + F_J = m_s * a_s \tag{1}$$

And:

$$\sum_{1-n} M_{EXT} + M_J = H_s \tag{2}$$

Where F_{EXT} values represent the m external forces acting on the segment, F_J the reactive force at the proximal joint, m_s the mass of the segment, and a_s the acceleration of the segment in Eq. 1. In Eq. 2, M_{EXT} represents the n external moments acting on the segment, M_J the reactive moment at the proximal joint, and H_s the first derivative of the angular momentum.

These equations yield joint reactive forces and moments. Inverse models used to calculate external dynamic shoulder loading are frequently discussed (Hogfors *et al.*, 1987; Dickerson *et al.*, 2006a), and analogues exist for gait analysis (Vaughan, 1991).

Calculation of individual muscle forces: The contributions of all shoulder musculoskeletal components to generating joint moments must equal the external moment to achieve equilibrium. Most models consider only active muscle contributions, due to the incompletely documented contributions of shoulder ligaments across postures (Veeger and van der Helm, 2007; Matsen *et al.*, 1994). The musculoskeletal moment equilibrium equation, for each segment, is:

$$\sum_{1-k} l_i * F_i = M_J \quad (3)$$

Where l_i is the magnitude of the moment arm for muscle i, F_i the force produced in muscle i, M_J the net reactive external moment, and k the number of muscles acting on the segment.

Models often use one of two popular methods to estimate muscle forces based on this equilibrium: optimization-based and electromyography (EMG)-based. Optimization techniques are frequently used to estimate muscular function in other body regions (McGill, 1992; Hughes *et al.*, 1994, 1995; Crowninshield and Brand,1981 a&b). Essentially, an optimization approach assumes that the musculoskeletal system, through the central nervous system (CNS), allocates responsibility for generating the required net muscle moment to the muscles in a structured, mathematically describable manner. This usually involves minimizing an objective function based on specific or overall muscle activation, such as overall muscle stress, fatigue, specific muscle stress, or other quantities (Dul, 1984 a&b), with the generic form of the solution as follows:

Minimize Θ (4)

s.t. $\mathbf{A}\,\mathbf{x} = \mathbf{B}$ (linear equality constraints) (5)

$0 \le f_i \le u_i \; for\; i = 1,\ldots,38$ (muscle force bounds) (6)

Where Θ is the objective function, **A** and **B** coefficients in the segmental equilibrium equations, **x** a matrix of unknown quantities including muscle forces, f_i the individual muscle forces, and u_i the maximum forces generated by each muscle.

One popular objective function in musculoskeletal optimization models is the sum of the muscles stresses raised to a power:

$$\Theta = \sum_{i=1}^{n} \left(\frac{f_i}{PCSA_i} \right)^m \quad \text{for } m = 1, 2, 3 \ldots \quad (7)$$

Where $PCSA_i$ is the physiological cross-sectional area of muscle i, n the number of muscles considered, and m the exponent on the stress calculation. Most models use second or third order formulations.

Optimization approaches have dominated model development due to their mathematical tractability and extensive theoretical foundations. A criticism of these models is their chronic underestimation of muscle activity as measured by EMG (Laursen *et al.*, 1998; Hughes *et al.*, 1994). Predicting shoulder muscle activation patterns is complicated by the importance of glenohumeral stability to muscular recruitment and its appropriate mathematical representation. Current approaches include requiring the net humeral joint reaction force vector to be directed into the glenoid defined as an ellipse (van der Helm, 1994 a&b; Hogfors *et al.*, 1987) and applying directional empirically-derived stability ratios to limit the amount of joint shear force relative to compressive force at the glenohumeral joint (Dickerson, 2005, 2007 a&b).

Conversely, EMG-based approaches (such as Laursen *et al.*, 1998, 2003) use physiologically measured electrical signals measured at defined anatomical sites to set force levels in individual muscles. As it is very difficult to measure all muscles in the shoulder simultaneously, due to limited accessibility and their number, assumptions are needed to assign activity to unmonitored muscles. One drawback of the EMG approach is its impotency for proactive analyses of future tasks. By relying on individual measurements, however, this approach assures the physiological nature of the outputs. Caution is required, as the EMG signal is highly sensitive to changes in muscle length, posture, velocity and other factors (Chaffin *et al.*, 2006; Basmajian and De Luca, 1985) that can hinder interpretation and affect the consistency of model predictions, particularly in dynamic analyses. The limited utility of EMG-based models for simulation studies has led to their more frequent use in clinical and rehabilitation applications.

The outputs of internal muscle models are muscle forces, net joint reaction forces, ligament tension, and muscle stresses. These may be reported as either time series or instantaneous values.

Communication of results: Historically, the output of many biomechanical shoulder models has been made available only through archival literature. Unfortunately, this has also led to limited use outside academia. This has changed with recent inclusions of shoulder models in commercial software (Dickerson, 2006b) and independent, accessible implementations (see Figure 2.2).

Forward dynamics approaches: Forward biomechanical models have also been developed for several applications (Stelzer and von Stryk, 2006; van der Helm *et al.*, 2001; Winters and van der Helm, 1994), including in the shoulder. In a forward solution, a seeding pattern or scheme of muscular activations is determined a priori or based upon the constraints of a given exertion. This pattern produces a specific end effector force or generates a specific movement. A final activation pattern results from the matching of these system outputs with the desired function. The final pattern is useful for applications such as estimating joint loading, powering prostheses (Kuiken, 2006) or specifying functional electrical stimulation (FES) sites (Kirsch, 2006).

Unanswered shoulder modelling questions: Although many shoulder models exist, each has advantages and limitations (Table 2.2). For simulation, few interface with motion analysis or ergonomics software, especially those dependent on experimental measurements. Beyond this functional requirement, several other aspects of shoulder biomechanics are underdeveloped, specifically: population scalability, pathologies that impact shoulder function, and muscle use strategy modifiers, including arm stiffness and precision demands. Systematic validation of many shoulder models would expand their implementation, as many evaluations have been limited to constrained, static, planar exertions. However, many arm-dependent activities are complex and include movement and inertial components that

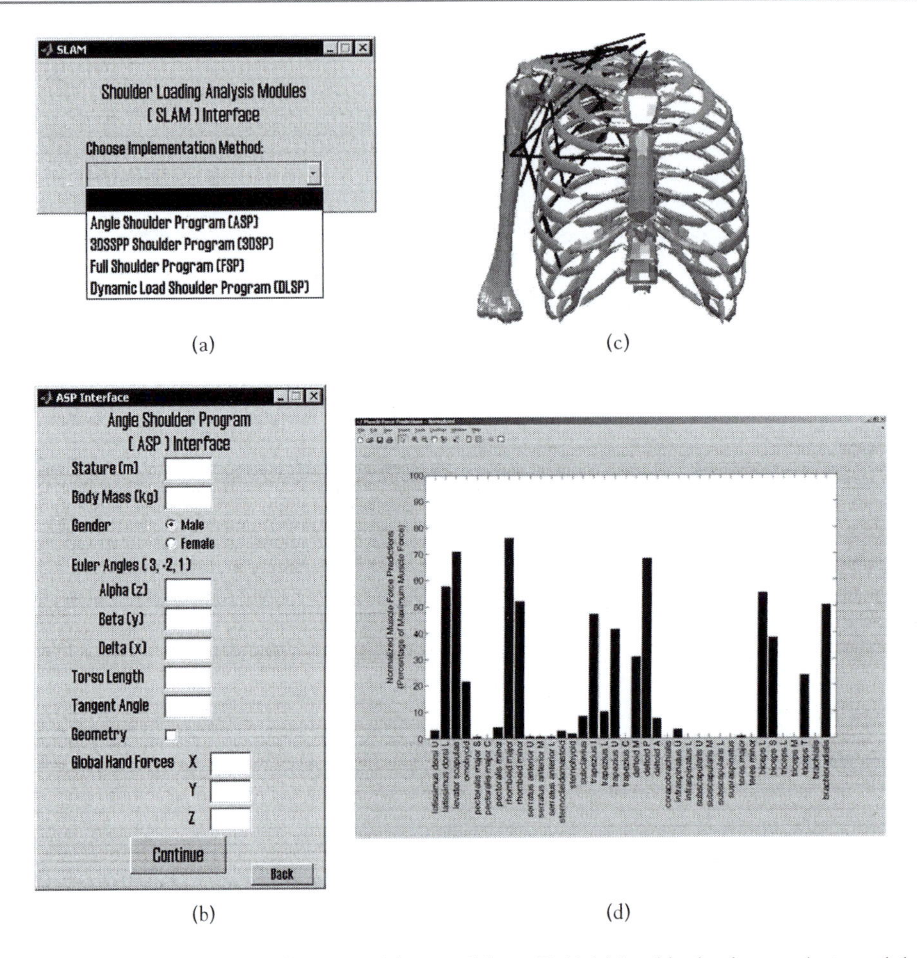

Figure 2.2 Screenshots of several aspects of the standalone SLAM (shoulder loading analysis modules) software (Dickerson *et al.*, 2006; 2007 a&b). (a) Beginning screen, which allows a choice of four input methods; (b) angle-based input screen, in which anthropometrics, posture, and hand force levels are specified; (c) geometric shoulder representation; and (d) muscle force predictions for a given posture-load combination.

alter exposures. Further, of the multiple contributors to glenohumeral stability, few are incorporated in existing models, with the exception of active muscle contraction. This may partially explain predictions that do not match experimentally measured data, particularly for antagonists (Karlsson *et al.*, 1992; Garner and Pandy, 2001). Finally, although attempts have been made to understand reflexive control in the shoulder (de Vlugt *et al.*, 2002, 2006; van der Helm *et al.*, 2002), the field is still in its infancy. Ultimately, despite many fundamental findings, much work remains to accurately describe shoulder function and mechanics.

The elbow

The human elbow, while often modeled as a single DOF hinge joint, is a multi-faceted intersection of the humerus, radius and ulna. The elbow combines ligamentous, bony and capsular contributions to achieve a high level of stability.

Table 2.2 A partial list of available shoulder biomechanical models and some of their characteristics

Model	*Dynamics included/ population scalable?*	*2-D or 3-D*	*Dependent on experimental data*	*Number of muscle elements*	*Available in popular public software*	*Online Realistic visualization of musculoskeletal geometry*	*Graphical user interface developed*	*Interfaces directly with ergonomic analysis software*
Dutch Shoulder and Elbow Model (DSEM) [Van der Helm 1991, 1994a&b]	Yes/No	3-D	No	95+	No	Yes (SIMM*)	Unknown	No
Gothenburg Model [Hogfors 1987, 1991, 1993; Karlsson, 1992]	No/Yes	3-D	No	38	Yes	No	Yes	No
Mayo Model [Hughes, 1997]	No/No	3-D	No	10	No	No	No	No
Utah Model [Wood, 1989a&b]	No/No	3-D	No	30	No	Unknown	Unknown	No
Texas Model [Garner & Pandy 1999, 2001, 2002, 2003]	Yes/No	3-D	No	30	No	Yes	Unknown	No
Copenhagen Model [Laursen 1998, 2003]	No/No	3-D	Yes	13	No	No	Unknown	No
Dul [1988]	Yes/No	2-D	No	2	No	No	No	No
Minnesota Model [Soechting and Flanders, 1997]	Yes/No	3-D	Yes	4	No	No	Unknown	No
Stanford Model [Holzbaur, 2005, 2007a&b]	Yes/ Unknown	3-D	No	50	Yes	Yes (SIMM*)	Yes	No
Shoulder Loading Analysis Modules (SLAM) [Dickerson 2005, 2006a&b, 2007a&b]	Yes/Yes	3-D	No	38	Yes	Yes	Yes	Yes

*SIMM is a commercial product, Software for Interactive Musculoskeletal Modeling, and is distributed by Musculographics, Inc.
•In cases where there are more than two authors, only the first author's name is used.

Elbow models: Two-dimensional (2D) models of elbow flexion are common examples in biomechanics textbooks (i.e. Hughes and An, 1999; Chaffin, 2006). Much elbow biomechanical research, however, has focused on producing three-dimensional (3D) elbow models. An and Morrey (1985) presented 3D potential moment contributions of 12 muscles for multiple arm configurations. They concurrently required equilibrium about both the flexion/extension and varus/valgus axes. Their model revealed differential biceps contributions to varus or valgus moments with changing forearm position, while the flexor digitorum superficial is showed similar postural oscillations between flexion and extension. EMG studies of elbow flexion have complemented optimization approaches in concluding the following: (1) biceps activity decreases in full pronation; (2) the brachialis is nearly always active in flexion; and (3) the heads of the biceps and triceps activate in the same manner during most motions (An and Morrey, 1985).

Computerized models of the elbow: Enhanced computerized models of the elbow have built on these concepts, and allow simulation of all possible arm placements and provide rapid feedback regarding tissue loading. They contain full descriptions of muscle elements crossing the elbow (Murray, 1995, 2000, 2002) and consider geometric scaling (Murray, 2002), and postural variation (Murray, 1995). Another area offering novel solutions to elbow biomechanical tissue load quantification is stochastic modelling approaches, in which distributed inputs yield multiple solutions and generate ranges of outputs (Langenderfer *et al.*, 2005). This approach is sensitive to population variability, whereas deterministic models typically use only average data as input, limiting application to a diverse population.

The wrist and hand

The wrist is an intricate network of carpal bones and ligaments that form a bridge between the distal ulna and radius and the metacarpals. It enables high flexibility and an extraordinary range of manual activities.

Pulley models: Significant attention has focused on the estimation of tissue loads on the flexor tendons, synovium and median nerve, as these are the sites of common cumulative wrist pathologies. Tendon pulley models (Landsmeer, 1962; Armstrong and Chaffin, 1978) were early approaches to modelling the influence of hand-supplied forces on flexor tendon loads. In these models, flexor tendons are supported by the flexor retinaculum in tension and the carpal bones in extension. They proposed that the radial reaction force on the support surfaces is describable by:

$$F_R = 2F_T e^{\mu\theta} \sin\left(\frac{\theta}{2}\right) \tag{8}$$

where F_R is the radial reaction force, F_T the force or tension in the tendon, μ the coefficient of friction between the tendon and support tissue, and θ the wrist deviation angle.

The normal forces acting on the tendon are characterized by:

$$F_N = \frac{F_T e^{\mu\theta}}{R} \tag{9}$$

where R is the radius of curvature of the tendon around the supporting tissues.

In healthy synovial joints, the value of μ is small (about 0.003–0.004, according to Fung, [1981]). Small values cause Eq. 4 and 5 to simplify to:

$$F_R = 2F_T \sin\left(\frac{\theta}{2}\right) \tag{10}$$

and:

$$F_N = \frac{F_T}{R} \tag{11}$$

Finally, the shear force in the synovial tunnels through which the tendons slide is calculated as:

$$F_s = F_N \mu \tag{12}$$

This value approaches zero in a healthy wrist, but increases markedly in the case of an inflamed synovium. Wrist angular accelerations have been included to extend these static models into dynamic models (Schoenmarklin and Marras, 1993). However, some of these models remain 2D representations and neglect muscular coactivation (Frievalds, 2004).

Advanced wrist models: The pulley models have been followed by intricate wrist representations that recognize the wrist's mechanical indeterminacy. Methods to resolve the indeterminacy include reduction, optimization and combination (Freivalds, 2004; Kong *et al.*, 2006). Reduction solutions exist for both 2D (Smith *et al.*, 1964) and 3D idealized solutions. Optimization techniques are typically used for more complex representations (Penrod *et al.*, 1974; Chao and An, 1978a; An *et al.*, 1985). Combination approaches have also been used (Chao and An, 1978b; Weightman and Amis, 1982). Current advanced models include enhanced geometric musculoskeletal representations. For example, in the 2D model of Kong (2004), the fingers are four distinct phalanges. The model identifies the optimal handle size for gripping to minimize tendon forces (Kong *et al.*, 2004) and models handgrip contact surfaces for multiple segments. Magnetic resonance imaging (MRI) techniques have been leveraged to create physiological representations of flexor tendon geometry and changes that occur with postural and applied load variation (Keir *et al.*, 1997, 1998; Keir and Wells, 1999; Keir, 2001). These efforts have revealed the combination effect of tendon tension on influencing the radius of curvature, which results in a marked increase in the tendon normal force (Figure 2.3). EMG coefficient models also exist to estimate wrist muscle forces (Buchanan *et al.*, 1993).

Experimental complements to wrist models: Experimental studies have assessed the function of wrist flexors and dynamics for computer activities (An *et al.*, 1990; Dennerlein, 1999, 2005, 2006; Keir and Wells, 2002; Sommerich *et al.*, 1998). Some results suggest that certain tendon model formulations may underestimate tissue tensions (Dennerlein, 1998). Goldstein *et al.* (1987) demonstrated that residual strain occurs with repeated exposure and may lead to flexor tendon damage and inflammation. Fatigue has also been implicated as influencing muscular activation patterns (Mogk and Keir, 2003).

Composite arm models

Multi-joint upper extremity models also exist. Several shoulder models include the elbow and multi-articular muscles, while others additionally include the wrist and hand. Examples of

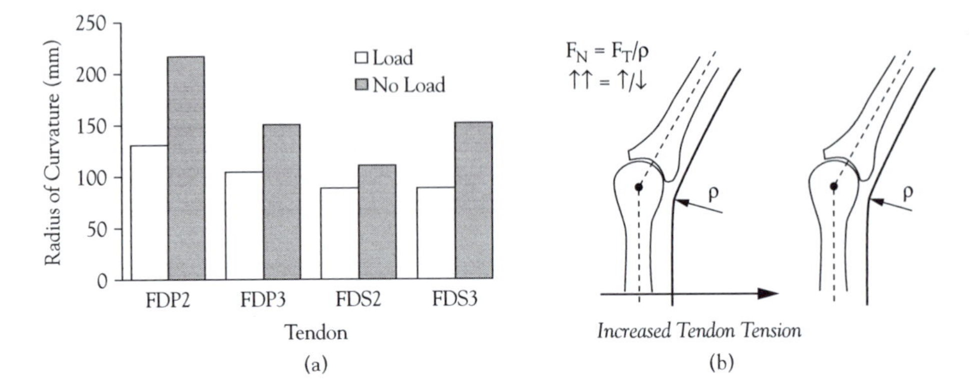

Figure 2.3 (a) The effect of tendon load on the radius of curvature for the load bearing flexor tendons [pinch grip, flexed 45°]. Deep (FDP) and superficial (FDS) tendons are shown for digits 2 and 3. (b) Implications of the force increase on normal force applied to tendon as defined by the equations of Armstrong and Chaffin (1978). (Figure courtesy of P. Kier.)

composite models include the Dutch Shoulder and Elbow Model (van der Helm, 1991, 1994 a&b), the Stanford Model (Holzbaur *et al.*, 2005, 2007 a&b), and the Texas Model (Garner and Pandy, 1999, 2001, 2002, 2003).

Clinical biomechanical simulations

Though the importance of shoulder biomechanics for effective orthopedics has long been recognized (Fu *et al.*, 1991), researchers have recently focused on refining or developing surgical techniques with biomechanics. An investigation of massive rotator cuff repair techniques showed tendon transfer using teres major to be more effective than using latissimus dorsi (Magermans *et al.*, 2004). Quantitative geometric shoulder models exist for the surgical planning of shoulder arthroplasty, reducing the likelihood of post-surgical implant impingement (Krekel *et al.*, 2006). Studies of rotator cuff pathologies have focused on simulation of muscle deficiency to predict post-injury activation patterns and capacity (Steenbrink, 2006a; 2006b), with moderate success. Similar approaches have evaluated specific muscle transfers in the elbow by adjusting model-defined muscle attachment parameters (Murray *et al.*, 2006). Despite the novelty of biomechanically assisted surgery, rapid advances make this a promising field.

Conclusion

Upper extremity biomechanical models provide valuable information regarding specific tissue exposures. The models are useful for answering questions in the areas of ergonomics, athletics, injury pathogenesis, rehabilitation, surgical techniques, and biomimetic applications. Model selection depends on the fidelity of output dictated by the scientific question. Although many primary questions have answers, gaps remain. Two priorities are assessing cumulative tissue loading during longer activities and seamless integration of biomechanical models into workplace and product design software. Despite numerous future challenges, biomechanical modelling promises to produce many insights into arm function and dysfunction.

References

1. An, K.N. and Morrey, B.F. (1985) 'Biomechanics of the Elbow, The Elbow and its Disorders' (Morrey, B.F., ed.) W.B. Saunders: Toronto.
2. An, K.N., Chao, E.Y.S., Cooney, W.P., Linscheid, R.L. (1985) 'Forces in the normal and abnormal hand', *J Orth Res*, 3:202–11.
3. An, K.N., Bergland, L., Cooney, W.P., Chao, E.Y.S., Kovacevic, N. (1990) 'Direct in vivo tendon force measurement system', *J Biomech*, 23:1269–71.
4. Armstrong, T.J., Chaffin, D.B. (1978) 'An Investigation of the Relationship between Displacements of the Finger and Wrist Joints and the Extrinsic Finger Flexor Tendons', *J Biomech*, 11:119–28.
5. Badler, N., Lee, P., Phillips, C. and Otani, E. (1989) 'The Jack interactive human model', *Concurrent Engineering of Mechanical Systems*, Vol. 1, First Annual Symposium on Mechanical Design in a Concurrent Engineering Environment, Univ. of Iowa, Iowa City, IA, October 1989: 179–98.
6. Barnett, N.D., Duncan, R.D., Johnson, G.R. (1999) 'The measurement of three dimensional scapulohumeral kinematics – a study of reliability', *Clin Biomech*, 14(4):287–90.
7. Basmajian, J.V., De Luca, C.J. (1985) *Muscles Alive* (5th Edition). Williams and Wilkins: Baltimore, MD.
8. Bigliani, L.U., Pollock, R.G., Soslowsky, L.J., Flatow, E.L., Pawluk, R.J., Mow, V.C. (1992) 'Tensile properties of the inferior glenohumeral ligament', *J Orth Res*, 10:187–97.
9. Boardman, N.D., Debski, R.E., Warner, J.J.P., Taskiran, E., Maddox, L., Imhoff, A.B., Fu, F.H., Woo, S.L. (1996) 'Tensile properties of the superior glenohumeral and coracohumeral ligaments', *Journal of Shoulder and Elbow Surgery*, 5:249–54.
10. Borstad, J.D., Ludewig, P.M. (2002) 'Comparison of scapular kinematics between elevation and lowering of the arm in the scapular plane', *Clin Biomech*, 17:650–9.
11. Buchanan, T.S., Moniz, M.J., Dewald, J.P., Zev Rymer, W. (1993) 'Estimation of muscle forces about the wrist joint during isometric tasks using an EMG coefficient method', *J Biomech*, 26:547–60.
12. Chaffin, D.B. (1997) 'Development of computerized human static strength simulation model for job design', *Human Factors and Ergonomics in Manufacturing*, 7:305–22.
13. Chaffin, D.B., Andersson, G.B.J., Martin, B.J. (2006) *Occupational Biomechanics* (4th Edition). J. Wiley & Sons: New York, NY.
14. Chao, E.Y.S., An, K.N. (1978a) 'Determination of internal forces in human hand', *Journal of Engineering Mechanics Division*, ASCE, 104:255–72.
15. Chao, E.Y.S., An, K.N. (1978b) 'Graphical interpretation of the solution to the redundant problem in biomechanics', *J Biomech*, 100:159–67.
16. Charlton, I.W. and Johnson, G.R. (2001) 'Application of spherical and cylindrical wrapping algorithms in a musculoskeletal model of the upper limb', *J Biomech*, 34:1209–16.
17. Cole, B.J., Rios, C.G., Mazzocca, A.D., and Warner, J.J.P. (2007) 'Anatomy, biomechanics, and pathophysiology of glenohumeral instability, in Disorders of the Shoulder': *Diagnosis and Management* (Iannotti, J.P. and Williams, G.R., ed.) Lippincott Williams and Wilkins: New York.
18. Crowninshield, R.D., Brand, R.A. (1981a) 'The prediction of forces in joint structures: Distribution of intersegmental resultants', *Exercise Sports Science Reviews*, 9:159.
19. Crowninshield, R.D. and Brand, R.A. (1981b) 'A Physiologically Based Criterion of Muscle Force Prediction in Locomotion', *J Biomech*, 14:793–801.
20. Debski, R.E., Wong, E.K., Woo, S.L., Fu, F.H., Warner, J.J. (1999). 'An analytical approach to determine the in situ forces in the glenohumeral ligaments', *J of Biomechanical Engineering*, 121:311–15.

21. de Groot, J.H., Valstar, E.R., Arwert, H.J. (1998) 'Velocity effects on the scapulo-humeral rhythm', *Clin Biomech*, 13:593–602.
22. de Groot, J.H., Brand, R.,(2001) 'A three-dimensional regression model of the shoulder rhythm', *Clin Biomech*, 16:735–43.
23. Delp, S.L., Loan, J.P. (1995) 'A graphics-based software system to develop and analyze models of musculoskeletal structures', *Computers in Biology and Medicine*, 25:21–24.
24. Dennerlein, J.T., Diao, E., Mote, C.D., Rempel, D.M. (1998) 'Tensions of the flexor digitorum superficialis are higher than a current model predicts', *J Biomech*, 31:296–301.
25. Dennerlein, J.T., Diao, E., Mote, C.D., Rempel, D.M. (1999) 'In vivo finger flexor tendon force while tapping on a keyswitch', *J Orthop Res*, 17:178–84.
26. Dennerlein, J.T. (2005) 'Finger flexor tendon forces are a complex function of finger joint motions and fingertip forces', *J Hand Ther*, 18:120–7.
27. Dennerlein, J.T., and Johnson, P.W. (2006) 'Different computer tasks affect the exposure of the upper extremity to biomechanical risk factors', *Ergonomics*, 49(1):45–61.
28. de Vlugt, E., Schouten, A.C., van der Helm, F.C.T. (2002) 'Adaptation of reflexive feedback during arm posture to different environments', *Biol Cybern*, 87:10–26.
29. de Vlugt, E., Schouten, A.C., van der Helm, F.C.T. (2006) 'Quantification of intrinsic and reflexive properties during multijoint arm posture', *J Neurosci Methods*, 155:328–49.
30. Dickerson, C.R. (2005) A biomechanical analysis of shoulder loading and effort during load transfer tasks, Doctoral Dissertation, University of Michigan, Ann Arbor, MI, USA.
31. Dickerson, C.R., Martin, B.J., and Chaffin, D.B. (2006a) 'The relationship between shoulder torques and the perception of muscular effort in load transfer tasks', *Ergonomics*, 49(11):1036–51.
32. Dickerson, C.R. (2006b) Issues facing integration of high-fidelity models in functional job analysis, in: The state and future of upper extremity biomechanics in the workplace. A symposium, Canadian Society for Biomechanics Meeting, Waterloo, ON.
33. Dickerson, C.R., Martin, B.J., Chaffin, D.B. (2007a) 'Predictors of perceived effort in the shoulder during load transfer tasks', *Ergonomics*, 50(7):1004–16.
34. Dickerson, C.R., Hughes, R.E., and Chaffin, D.B. (2007b) 'A computational shoulder muscle force model designed for ergonomic analysis', *Computer Methods in Biomechanics and Biomedical Engineering*, 10(6):389–400.
35. Dul, J., Townsend, M.A., Shiavi, R., Johnson, G.E. (1984) 'Muscular synergism – I. On criteria for load sharing between synergistic muscles', *J Biomech*, 17:663–73.
36. Dul, J., Johnson, G.E., Shiavi, R., Townsend, M.A. (1984) 'Muscular synergism – II. A minimum-fatigue criterion for load sharing between synergistic muscles', *J Biomech*, 17:675–84.
37. Durkin, J.L. and Dowling, J.J. (2003) 'Analysis of body segment parameter differences between four human populations and the estimation errors of four popular mathematical models', *J Biomech Eng*, 125:515–22.
38. Ebaugh, D.D., McClure, P.W., Karduna, A.R. (2005) 'Three-dimensional scapulothoracic motion during active and passive arm elevation', *Clin Biomech*, 20:700–9.
39. Freivalds, A. (2004) 'Biomechanics of the Upper Limbs, Mechanics, Modelling and Musculoskeletal Disorders', CRC Press: New York.
40. Fu, F.H., Seel, M.J., Berger, R.A. (1991) 'Relevant shoulder biomechanics', *Operative Techniques in Orthopaedics*, 1:134–46.
41. Fung, Y. C. (1981) '*Biomechanics*'. Springer-Verlag: New York.
42. Garner, B.A., Pandy, M.G. (1999) 'A kinematic model of the upper limb based on the Visible Human Project (VHP) image dataset', *Computer Methods in Biomechanics and Biomedical Engineering*, 2:107–24.
43. Garner, B.A., Pandy, M.G. (2000) 'The obstacle-set method for representing muscle paths in musculoskeletal models', *Computer Methods in Biomechanics and Biomedical Engineering*, 3:1–30.
44. Garner, B.A. and Pandy, M.G. (2001) 'Musculoskeletal model of the upper limb based on the visible human male dataset', *Computer Methods in Biomechanics and Biomedical Engineering*, 4:93–26.

45. Garner, B.A., Pandy, M.G. (2003) 'Estimation of musculotendon properties in the human upper limb', *Annals of Biomedical Engineering*, 31:207–20.
46. Gatti, C., Dickerson, C.R., Chadwick, E.K., Mell, A.G., Hughes, R.E. (2007) 'Comparison of model-predicted and measured moment arms for the rotator cuff muscles', *Clin Biomech*, 22:639–644.
47. Goldstein, S.A., Armstrong, T.J., Chaffin, D.B., and Matthews, L.S. (1987) 'Analysis of cumulative strain in tendons and tendon sheaths', *J Biomech*, 20:1–6.
48. Hogfors, C., Sigholm, G., Herberts, P. (1987) 'Biomechanical model of the human shoulder – I. Elements', *J Biomech*, 20:157–66.
49. Hogfors, C., Peterson, B., Sigholm, G., Herberts, P. (1991) 'Biomechanical model of the human shoulder – II. The shoulder rhythm', *J Biomechanics*, 24:699–709.
50. Hogfors, C., Karlsson, D., and Peterson, B. (1995) 'Structure and internal consistency of a shoulder model', *J Biomech*, 28: 767–77.
51. Holzbaur, K.R.S., Murray, W.M., Delp, S.L. (2005) 'A model of the upper extremity for simulating musculoskeletal surgery and analyzing neuromuscular control', *Annals of Biomedical Engineering*, 33:829–40.
52. Holzbaur, K.R., Delp, S.L., Gold, G.E., Murray, W.M. (2007) 'Moment-generating capacity of upper limb muscles in healthy adults', *J Biomech*, in press.
53. Holzbaur, K.R., Murray, W.M., Gold, G.E., Delp, S.L. (2007) 'Upper limb muscle volumes in adult subjects', *J Biomech*, 40(11):2442–9.
54. Hughes, R.E., Chaffin, D.B., Lavender, S.A., and Andersson, G.B.J. (1994) 'Evaluating muscle force prediction models of the lumbar trunk using surface electromyography', *J Orthopaedic Research*, 12:698–98.
55. Hughes, R.E., Bean, J.C., Chaffin, D.B. (1995) 'Evaluating the effect of co-contraction in optimization models', *J Biomech*, 28:875–8.
56. Hughes, R.E. and An, K.N. (1999) 'Biomechanical models of the hand, wrist and elbow in ergonomics', in Kumar, S. (ed.) *Biomechanics in Ergonomics:* Taylor and Francis. London, UK
57. Inman, V.T., Saunders, J.B.C.M., Abbott, L.C. (1994) 'Observations of the function of the shoulder', *J Bone J Surg*, 26:1–30.
58. Johnson, G.R., Spalding, D., Nowitzke, A., Bogduk, N. (1996) 'Modelling the muscles of the scapula morphometric and coordinate data and functional implications', *J Biomech*, 29:1039–51.
59. Johnson, G.R., Pandyan, A.D. (2005) 'The activity in the three regions of the trapezius under controlled loading conditions – an experimental and modelling study', *Clin Biomech*, 20:155–61.
60. Juul-Kristensen, B., Bojsen-Møllerb, B., Holsta, E. and Ekdahl, C. (2000) 'Comparison of muscle sizes and moment arms of two rotator cuff muscles measured by Ultrasonography and Magnetic Resonance Imaging', *European J of Ultrasound*, 11(3):161–73.
61. Juul-Kristensen, B., Bojsen-Møller, F., Finsen, L., Eriksson, J., Johansson, G., Ståhlberg, F., Ekdahl, C. (2000) 'Muscle sizes and moment arms of rotator cuff muscles determined by Magnetic Resonance Imaging', *Cells Tissues Organs*, 167:214–22.
62. Juul-Kristensen, B., Kadefors, R., Hansen, K., Bystrom, P., Sandsjo, L., Sjogaard, G. (2006) 'Clinical signs and physical function in neck and upper extremities among elderly female computer users': the NEW study, *Eur J Appl Physiol*, 96:136–45.
63. Karlsson, D. and Peterson, B. (1992) 'Towards a model for force predictions in the human shoulder', *J Biomech*, 25:189–99.
64. Keir, P.J., Wells, R.P., Ranney, D.A., Lavery, W. (1997) 'The effects of tendon load and posture on carpal tunnel pressure', *J Hand Surg [Am]*, 22:628–34.
65. Keir, P.J., Bach, J.M., Rempel, D.M. (1998) 'Effects of finger posture on carpal tunnel pressure during wrist motion', *J Hand Surg [Am]*, 23:1004–9.
66. Keir, P.J., Wells, R.P. (1999) 'Changes in geometry of the finger flexor tendons in the carpal tunnel with wrist posture and tendon load: an MRI study on normal wrists', *Clin Biomech*, 14:635–45.
67. Keir, P.J. (2001) *Seminars in Musculoskeletal Radiology*, 5:241–49.
68. Keir, P.J., and Wells, R.P. (2002) 'The effect of typing posture on wrist extensor muscle loading', *Hum Factors*, 44:392–403.

69. Keyserling, W.M. (2000) 'Workplace risk factors and occupational musculoskeletal disorders, Part 2: A review of biomechanical and psychophysical research on risk factors associated with upper extremity disorders', *AIHAJ*, 61(2): 231–43.
70. Kirsch, R.F. (2006) 'Shoulder biomechanics and movement resoration', Keynote Address. 2006 Meeting of the International Shoulder Group, Chicago, IL.
71. Klein Breteler, M.D., Spoor, C.W., Van der Helm, F.C.T. (1999) 'Measuring muscle and joint geometry parameters of a shoulder for modelling purposes', *J Biomech*, 32:1191–7.
72. Kong, Y.K., Freivalds, A., Kim, S.E. (2004) 'Evaluation of handles in a maximum gripping task', *Ergonomics*, 47:1350–64.
73. Kong, Y.K., Jang, H., Freivalds, A. (2006) 'Wrist and tendon dynamics as contributory risk factors in work-related musculoskeletal disorders', *Human Factors in Ergonomics and Manufacturing*, 16:83–105.
74. Krekel, P.R., Botha, C,P., Post, F.H., Valstar, E.R., Rozing, P.M. (2006) 'Pre-operative impingement detection for shoulder arthroplasty', 2006 Meeting of the International Shoulder Group, Chicago, IL.
75. Kuiken, T. (2006) 'Controlling artificial shoulders in shoulder disarticulation amputees: possibilities and challenges'. 2006 Meeting of the International Shoulder Group, Chicago, IL.
76. Landsmeer, J.M.F. (1962) 'Power Grip and Precision Handling', *Ann. Rheum. Dis.*, 21:164–70.
77. Langenderfer, J.E., Hughes, R.E., Carpenter, J.E. (2005) 'A stochastic model of elbow flexion strength for subjects with and without long head biceps tear', *Computer Methods Biomech Biomed Engin*, 8:315–22.
78. Latko, W.A., Armstrong, T.J., Foulke, J.A., Herrin, G.D. (1997) 'Development and evaluation of an observational method for assessing repetition in hand tasks', *AIHA*, 58:278–85.
79. Laursen, B., Jensen, B.R., Nemeth, G., Sjogaard, G. (1998) 'A model predicting individual shoulder muscle forces based on relationship between electromyographic and 3D external forces in static position', *J Biomech*, 31:731–39.
80. Laursen, B., Sogaard, B., Sjogaard, G. (2003) 'Biomechanical model predicting electromyographic activity in three shoulder muscles from 3D kinematics and external forces during cleaning work', *Clin Biomech*, 18:287–95.
81. Lippitt, S. and Matsen, F. (1993) 'Mechanisms of glenohumeral joint stability', *Clinical Orthopaedics and Related Research*, 291:20–8.
82. Ludewig, P.M., Behrens, S.A., Meyer, S.M., Spoden, S.M., Wilson, L.A. (2004) 'Three-dimensional clavicular motion during arm elevation: reliability and descriptive data', *J Orthop Sports Phys Ther*, 34:140–9.
83. Magermans, D.J., Chadwick, E.K., Veeger, H.E.J., Rozing, P.M., van der Helm, F.C.T. (2004) 'Effectiveness of tendon transfers for massive rotator cuff tears: a simulation study', *Clin Biomech*, 19:116–22.
84. Makhsous, M. (1999) 'Improvements, Validation and Adaptation of a Shoulder Model', Doctoral Dissertation, Chalmers University of Technology, Gothenburg, Sweden.
85. Matsen, F.A., Lippitt, S.B., Sidles, J.A., and Harryman, D.T. (1994). Practical Evaluation and Management of the Shoulder, W.B. Saunders: Philadelphia.
86. McAtamney, L., Corlett, E.N., RULA (1993) 'A survey method for the investigation of work-related upper limb disorders', *Applied Ergonomics*, 24:91–99.
87. McClure, P.W., Michener, L.A., Sennett, B.J., Karduna, A.R. (2001) 'Direct 3-dimensional measurement of scapular kinematics during dynamic movements in vivo', *J Shoulder Elbow Surg*, 10:269–77.
88. McGill, S.M. (1992) 'A myoelectrically based dynamic three-dimensional model to predict loads on lumbar spine tissues during lateral bending', *J Biomech*, 25:395–414.
89. Mogk, J.P.M., Keir, P.J. (2003) 'The effects of posture on forearm muscle loading during gripping', *Ergonomics*, 46:956–75.
90. Murray, W.M., Delp, S.L., Buchanan, T.S. (1995) 'Variation of muscle moment arms with elbow and forearm position', *J Biomech*, 28:513–25.

91. Murray, W.M., Buchanan, T.S., Delp, S.L. (2000) 'The isometric functional capacity of muscles that cross the elbow', *J Biomech*, 33:943–52.
92. Murray, W.M., Buchanan, T.S., Delp, S.L. (2000) 'Scaling of peak moment arms of elbow muscles with upper extremity bone dimensions', *J Biomech*, 35:19–26.
93. Murray, W.M., Hentz, V.R., Friden, J. and Lieber, R.L. (2006) 'Variability in surgical technique for brachioradialis tendon transfer: evidence and implications', *J Bone Joint Surg Am*, 88:2009–16.
94. Norman, R.W., McGill, S.M., Lu, W., Frazer, M. (1994) 'Improvements in biological realism in an industrial low back model', 3D WATBAK, in Proceedings of the 12th Triennial Congress of the International Ergonomics Association, published by the Association of Canadian Ergonomists, Toronto, 2:299–301.
95. Novotny, J.E., Beynnon, B.D., Nichols, C.E. (2000) 'Modelling the stability of the human glenohumeral joint during external rotation', *J Biomech*, 33:345–54.
96. Pascoal, A.G., van der Helm, F.F., Pezarat Correia, P., Carita, I. (2000) 'Effects of different arm external loads on the scapulo-humeral rhythm', *Clin Biomech*, 15:S21–4.
97. Penrod, D.D., Davy, D.T., Singh, D.P. (1974) 'An optimization approach to tendon force analysis', *J Biomech*, 7:123–29.
98. Pronk, G.M., van der Helm, F.C.T., Rozendaal, L.A. (1993) 'Interaction between the joints in the shoulder mechanism: the function of the costoclavicular, conoid and trapezoid ligaments', *Proc Institution of Mechanical Engineers*, 207:219–29.
99. Schiffern, S.C., Rozencwaig, R., Antoniou, J., Richardson, M.L., Matsen, F.A. (2002) 'Anteroposterior centering of the humeral head on the glenoid in vivo', *Am J Sports Med*, 30:382–7.
100. Schoenmarklin, R.W., Marras, W.S. (1993) 'Dynamic capabilities of the wrist joint in industrial workers', *International Journal of Industrial Ergonomics*, 11:207–24.
101. Sommerich, C.M., Marras, W.S., Parnianpour, M. (1998) 'A method for developing biomechanical profiles of hand-intensive tasks', *Clin Biomech*, 13:261–71.
102. Silverstein, B.A., Viikari-Juntura, E., Fan, Z.J., Bonauto, D.K., Bao, S., Smith, C. (2006) 'Natural course of nontraumatic rotator cuff tendinitis and shoulder symptoms in a working population', *Scand J Work Environ Health*, 32:99–108.
103. Smith, E.M., Juvinall, R.C., Bender, L.F., Pearson, J.R. (1964) 'Role of the finger flexors in rheumatoid deformities of the metacarpophalangeal joints', *Arthritis and Rheumatism*, 7:467–80.
104. Steenbrink, F., de Groot, J.H., Veeger, H.E.J., Meskers, C.G.M., van de Sande, M.A., Rozing PM (2006) 'Pathological muscle activation patterns in patients with massive rotator cuff tears, with and without subacromial anasthetics', *Man Ther*, 11:231–37.
105. Steenbrink, F., de Groot, J.H., Meskers, C.G.M., Veeger, H.E.J., Rozing, P.M. (2006) 'Model simulation of pathological muscle activation in cuff tears during abduction force exertion'. 2006 Meeting of the International Shoulder Group, Chicago, IL.
106. Stelzer, M., von Stryk, O. (2006) 'Efficient forward dynamics simulation and optimization of human body dynamics', ZAMM: *Journal of Applied Mathematics and Mechanics / Zeitschrift für Angewandte Mathematik und MechanikVolume* 86, Issue 10:828–40.
107. van der Helm, F.C.T., Veenbaas, R. (1991) 'Modelling the mechanical effect of muscles with large attachment sites: application to the shoulder mechanism', *J Biomech*, 24:1151–63.
108. van der Helm, F.C.T., Veeger, H.E.J., Pronk, G.M., Van der Woude, L.H., Rozendal, H. (1992) 'Geometry parameters for musculoskeletal modelling of the shoulder system', *J Biomech*, 25:129–44.
109. van der Helm, F.C.T. (1994a) 'A finite-element musculoskeletal model of the shoulder mechanism', *J Biomech*, 27:551–69.
110. van der Helm, F.C.T. (1994b) 'Analysis of the kinematic and dynamic behavior of the shoulder mechanism', *J Biomech*, 27:527–50.
111. van der Helm, F.C.T., Chadwick, E.K.J., Veeger, H.E.J. (2001) 'A large-scale musculoskeletal model of the shoulder and elbow: The Delft Shoulder (and elbow) Model'. Proceedings of the XVIIIth Conference of the International Society of Biomechanics, Zurich, 8–13 July.

112. van der Helm, F.C., Schouten, A.C., de Vlugt, E., Brouwn, G.G. (2002) 'Identification of intrinsic and reflexive components of human arm dynamics during postural control', *J Neurosci Methods*, 119:1–14.
113. Vaughan, C.L., Davis, B.L., O'Connor, J.C. (1992) Dynamics of Human Gait, Human Kinetics.
114. Veeger, H.E.J., Van der Helm, F.C.T., Van der Woude, L.H.V., Pronk, G.M., Rozendal, R.H. (1991) 'Inertia and musculoskeletal modelling of the shoulder mechanism', *J Biomech*, 24:615–29.
115. Veeger, H.E.J., van der Helm, F.C.T. (2004) Shoulder Girdle, in Working Postures and Movements: Tools for Evaluation and Engineering. (Delleman, N.J., Haslegrave, C.M. and Chaffin, D.B., eds) CRC Press.
116. Veeger, H.E.J., van der Helm, F.C.T. (2007) 'Shoulder function: The perfect compromise between mobility and stability', *J Biomech*, 40(10):2119–29.
117. Weightman, B., Amis, A.A. (1982) 'Finger joint force predictions related to design of joint replacements', *Journal of Biomedical Engineering*, 4:197–205.
118. Winters, J.M., Van der Helm, F.C.T. (1994) 'A field-based musculoskeletal framework for studying human posture and manipulation in 3D', Proc. of the Symp. on Modelling and Control of Biomed. Sys., Intern. Fed. on Autom. Control, pp. 410–15.

3

Modeling and simulation of tissue load in the human spine

N. Arjmand, B. Bazrgari and A. Shirazi-Adl
École Polytechnique, Montréal

Introduction

Low back disorders (LBDs) are highly prevalent worldwide, affecting up to 85 per cent of adults at some time in their lives, causing suffering, disability and loss of productivity (Frymoyer, 1996). The one-year prevalence rate in the US, Germany, Norway, and Sweden, for instance, has been reported to be as high as 56 per cent, 59 per cent, 61 per cent and 70 per cent, respectively (Ihlebaek *et al.*, 2006; Manchikanti, 2000; Schneider *et al.*, 2007). They remain as a major economic burden on individuals, industries and societies as a whole. In the US during 2005 alone, the total cost associated with LBDs has been suggested to vary from $100 to $200 billion (Katz, 2006) that is comparable with an estimated $81.2 billion in damage associated with the *Hurricane Katrina*, recognized as the costliest natural disaster in the US history, that took place in the same year (Wikipedia Encyclopedia, 2006).

Although low back pain (LBP) could originate from different musculoskeletal structures such as vertebrae, ligaments, facet joints, musculature, and disc annulus fibrosis, in most cases the exact cause of the symptoms remains, however, unknown (Diamond and Borenstein, 2006). In a large survey, lifting or bending episodes accounted for 33 per cent of all work-related causes of LBP (Damkot *et al.*, 1984). Combination of lifting with lateral bending or twisting that occurs in asymmetric lifts has been identified as a frequent cause of back injury in the workplace (Andersson, 1981; Hoogendoorn *et al.*, 2000; Kelsey *et al.*, 1984; Marras *et al.*, 1995; Troup *et al.*, 1981; Varma and Porter, 1995). Among various work-related activities, lifting, awkward posture, and heavy physical work have strong relationship with lumbar musculoskeletal disorders (NIOSH, 1997). Lifting, therefore, is one of the major documented risk factors for LBDs (Burdorf and Sorock, 1997; Ferguson and Marras, 1997; Frank *et al.*, 1996).

The foregoing studies confirm an association between manual material handling tasks and LBDs and suggest that excessive and repetitive mechanical loads acting on the spine could play major causative roles in LBDs. Proper knowledge of ligamentous loads, muscle forces, and trunk stability in the normal and pathologic human spine under various recreational and occupational activities, hence, becomes crucial towards appropriate and effective management of LBDs. Prevention, rehabilitation, treatment, and performance enhancement programs stand to substantially benefit from an improved understanding on the load partitioning in the human spine. Infeasibility of direct measurement of muscle forces and spinal loads in human beings and the limitations in extrapolation of such data collected from animal studies have led to indirect quantification of loads on spine by measuring representative biomedical indicators (e.g. intra-discal pressure, muscle electromyographic (EMG) activity). However, apart from invasiveness, cost concerns, limitations and difficulties, the validity of such indicators to adequately represent spinal loads has also been questioned (van Dieen *et al.*, 1999). Biomechanical models have, thus, been recognized as indispensable tools for estimation of muscle forces, spinal loads, and trunk stability during various occupational and recreational activities.

Single-level biomechanical models

Biomechanical models, both static and dynamic, use basic principles of mechanics to estimate muscle forces and spinal loads under different loading conditions. Forces in various active (i.e., trunk muscles) and passive (e.g. posterior ligaments and discs) structures are calculated by consideration of equilibrium equations. A free body diagram of the trunk (typically cut by an imaginary plane through the L4-L5 disc) is employed to maintain equilibrium between known external moments (due to gravity/inertia/external loads usually estimated by a link segment model) and unknown internal moments (due to spinal active and passive structures at the plane of cut). Unfortunately, such equilibrium equations cannot be resolved deterministically, as the number of unknowns significantly exceeds that of available equations (*kinetic redundancy problem*). A number of biomechanical models have been introduced to tackle the foregoing kinetic redundancy in equilibrium equations and to estimate spinal and muscle loads. Three approaches that have often been used in the analysis of different joint systems are: single-equivalent muscle, optimization-based, and EMG-assisted approach (Gagnon *et al.*, 2001; Shirazi-Adl and Parnianpour, 2001; van Dieen and Kingma, 2005).

In the reduction or equivalent muscle approach, the role of muscles is simplified by neglecting some and grouping others into synergistic ones, assuming a priori known activation levels; thus reducing the number of unknown muscle forces to the available equilibrium equations. In the optimization approach it is assumed that there is one cost (objective) function (or many cost functions) that may be minimized or maximized by the central nervous system (CNS) while attempting to satisfying the equilibrium conditions. Constraint equations on muscle forces are introduced in parallel enforcing that muscle forces remain greater than zero and smaller than some maximum values corresponding to the maximum allowable stress in muscles. Various linear (e.g., related to axial compression) and nonlinear (e.g., related to muscle fatigue) cost functions have been employed. The nonlinear cost function of the sum of cubed or squared muscle stresses has been suggested to adequately match collected EMG data (Arjmand and Shirazi-Adl, 2006a). In the EMG-assisted approach, electromyography signals of limited and often only superficial trunk muscles are first measured. A relationship between normalized EMG activity of a trunk muscle and its force is subsequently presumed, allowing for the estimation of individual trunk muscle forces while satisfying the existing equilibrium equations (Gagnon *et al.*, 2001; Marras and Granata, 1997; McGill and Norman, 1986). Each of these three approaches has its own advantages and drawbacks (van Dieen and Kingma, 2005; Reeves and Cholewicki, 2003).

Multi-level biomechanical models

Single-level models have been and remain to be very popular in biomechanical model investigations of different multi-joint studies (e.g., Cholewicki *et al.*, 1995; Granata *et al.*, 2005; Marras *et al.*, 2006; McGill and Norman, 1986; Parnianpour *et al.*, 1997; Schultz *et al.*, 1982; van Dieen *et al.*, 2003; van Dieen and Kingma, 2005). These models have widely been employed in ergonomic applications and in injury prevention and treatment programs. A major shortcoming with these models, however, lies in the consideration of the balance of net external moments only at a single joint or cross-section (typically at lowermost lumbar discs) rather than along the entire length of the spine. This drawback naturally exists in dynamic and quasi-static model studies alike while simulating either sagittaly symmetric two-dimensional (2D) or asymmetric three-dimensional (3D) movements. It has been demonstrated that the muscle forces evaluated based on such single-level equilibrium models, once applied on the system along with external loads, will not satisfy equilibrium at remaining levels along the spine (Arjmand *et al.*, 2007). They will neither yield the same deformed configuration based on which they were initially evaluated.

To overcome the foregoing major shortcoming, multi-level stiffness model studies, along with optimization (Gardner-Morse *et al.*, 1995) or EMG-assisted (Cholewicki and McGill, 1996) approaches, have been developed and used to evaluate muscle recruitment, internal loads, and stability margin. The former model neglects nonlinearities in spinal behavior whereas the latter overlooks translational DoF at various joint levels and, hence, associated shear/axial equilibrium equations. These omissions have been found to adversely influence predictions on muscle forces and spinal loads (Arjmand, 2006).

Kinematics-driven approach

For more than a decade, our group has been developing a novel iterative Kinematics-driven approach in which a-priori measured vertebral/pelvis rotations of the spine (i.e., movement trajectory as much as available) are prescribed a priori into a nonlinear finite element model to evaluate muscle forces, internal loads, and spinal stability in static and dynamic analyses of lifting tasks with and without loads in hands (Arjmand and Shirazi-Adl, 2006b; Bazrgari *et al.*, 2007, El-Rich *et al.*, 2004; Kiefer *et al.*, 1997; Shirazi-Adl *et al.*, 2002). This iterative approach (see Figure 3.1) not only satisfies the equilibrium equations in all directions along the entire length of the spine but yields spinal postures in full accordance with external/inertia/gravity loads, muscle forces, and passive ligamentous spine with nonlinear properties. Using this approach, the role of the intra-abdominal pressure (IAP) (Arjmand and Shirazi-Adl, 2006c), lumbar posture (Arjmand and Shirazi-Adl, 2005) and lifting technique (Bazrgari *et al.*, 2007; Bazrgari and Shirazi-Adl, 2007) on loading and stabilization of the spine during lifting activities has also been examined. Moreover, wrapping of thoracic extensor muscles around vertebrae while taking into account the contact forces between muscles and remaining spinal tissues in between has been simulated for the first time by using the Kinematics-driven approach (Arjmand *et al.*, 2006).

In this approach, a sagittally-symmetric nonlinear finite element model of the entire thoraco-lumbar spine has been considered. This is a beam-rigid body model comprising of six deformable beams to represent T12-S1 discs and seven rigid elements to represent T1-T12 (as a single body) and lumbosacral vertebrae (L1 to S1) (Figure 3.2). The beams model the overall nonlinear stiffness of T12-S1 motion segments (i.e., vertebrae, disc, facets and ligaments)

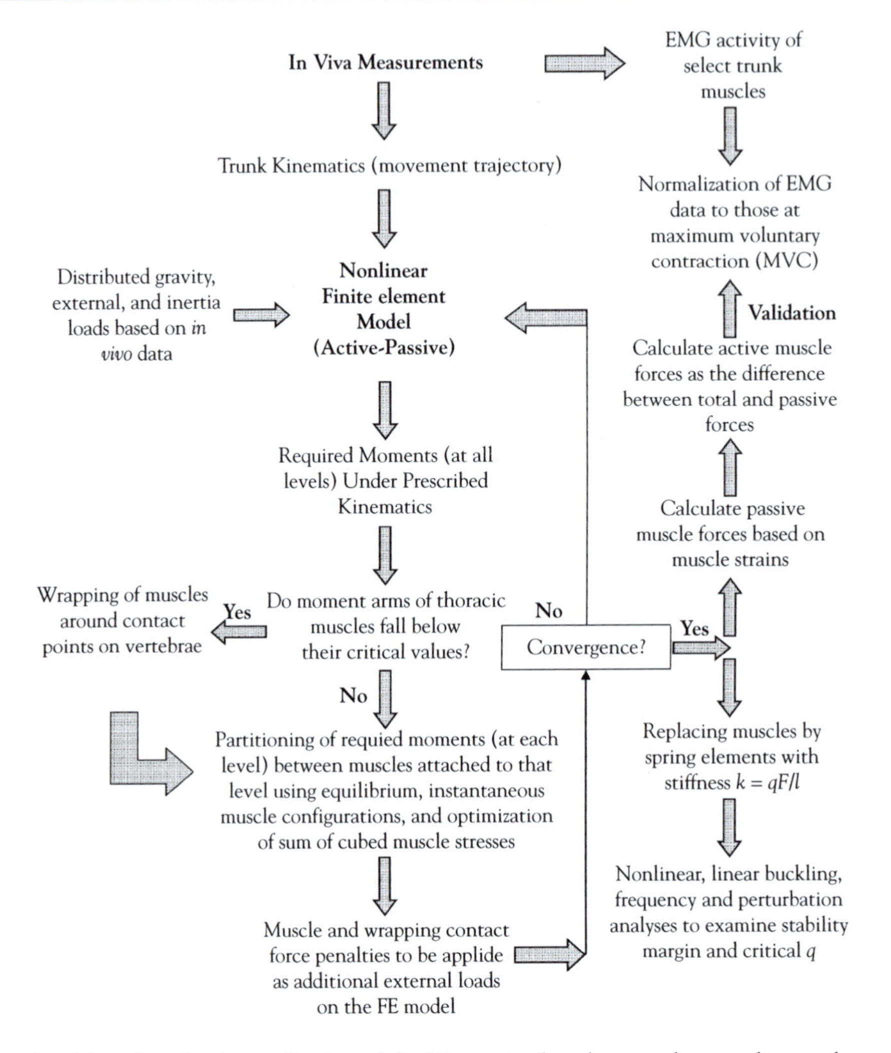

Figure 3.1 Flow-chart for the application of the Kinematics-based approach to predict trunk muscle forces, spinal loads, and stability (critical lever arms for thoracic muscles can be defined as a function of their initial values in upright posture; anatomy of muscles used in the model is depicted in Figure 3.2, convergence is attained if calculated muscle forces in two successive iterations remain almost the same).

at different directions and levels. The nonlinear load-displacement response under single and combined axial/shear forces and sagittal/lateral/axial moments, along with the flexion versus extension differences, are represented in this model based on numerical and measured results of previous single- and multi-motion segment studies (Arjmand and Shirazi-Adl, 2005, 2006b). The trunk mass and mass moments of inertia are assigned at gravity centers at different levels along the spine based on published data for trunk segments and head/arms. Connector elements parallel to deformable beams are added to account for the intersegmental damping using measured values; translational damping = 1,200 N s/m and angular damping = 1.2 N m s/rad (Bazrgari *et al.*, 2007).

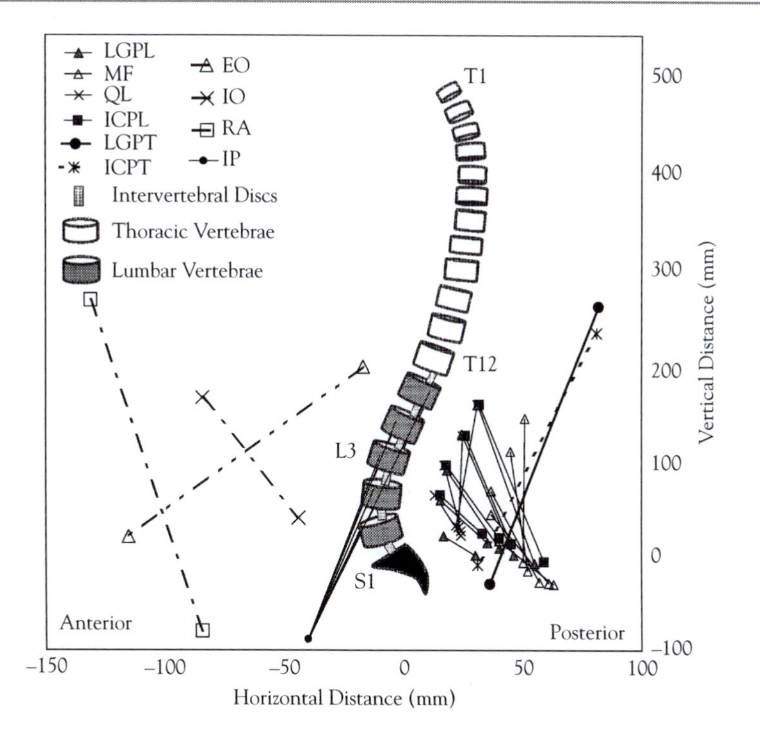

Figure 3.2 The FE model as well as global and local musculatures in the sagittal plane (only fascicles on one side are shown) in upright standing posture at initial undeformed configuration. ICPL: iliocostalis lumborum pars lumborum; ICPT: iliocostalis lumborum pars thoracic; LGPL: longissimus thoracis pars lumborum; LGPT: longissimus thoracis pars thoracic; MF: multifidus; QL: quadratus lumborum; IP: iliopsoas; IO: internal oblique; EO: external oblique; and RA: rectus abdominus.

In the present study, the Kinematics-driven approach is applied to analyze the steady state flexion relaxation phenomenon that remains as a controversial issue in biomechanics of the spine in forward flexion postures.

Application: flexion relaxation phenomenon (FRP)

Upon progressive forward flexion of the trunk from the upright standing posture towards the peak flexion, a partial or complete silence in EMG activity of superficial extensor muscles has been recorded. This phenomenon has been well documented in healthy asymptomatic subjects and is called the flexion–relaxation phenomenon (FRP) (Floyd and Silver, 1951) that may persist even in the presence of weights carried in hands. The FRP has been recorded to occur at about 84–86 per cent of peak voluntary flexion in slow movements, irrespective of the magnitude of load carried in hands (Sarti *et al.*, 2001). The presence and absence of the FRP could be used as a signature to discriminate LBP patients from healthy controls, as in the former group the FRP is frequently absent (Kaigle *et al.*, 1998; Kippers and Parker, 1984; Watson *et al.*, 1997). The FRP assessment has, thus, been suggested as a valuable clinical tool to aid in the diagnosis and treatment of LBP patients (Colloca and Hinrichs, 2005).

In order to explain the partial or full relaxation in back muscles in large trunk flexion postures, several hypotheses have been put forward in which the load is transferred from extensor muscles to passive tissues (Floyd and Silver, 1951; McGill and Kipper, 1994), from superficial muscles to deeper ones (Andersson *et al.*, 1996), or from lumbar extensors to thoracic ones (Toussaint *et al.*, 1995). Since the FRP is likely related to the relatively large axial strain (or elongation) in extensor muscles during forward flexion, it is expected to also depend on the lumbar rotation and pelvic-lumbar rhythm. The relative activity of various back muscles in deep flexion movements remain controversial as some suggest relaxation in global extensor muscles (Mathieu and Fortin, 2000; Sarti *et al.*, 2001) while others report relaxation only in lumbar extensor muscles (McGill and Kippers, 1994; Toussaint *et al.*, 1995). Using deep wire electrodes, Andersson *et al.* (1996) reported silence only in superficial lumbar erector spinae muscles with activity remaining in deeper ones.

Method

The trunk and pelvic rotations required in the Kinematics-driven approach to analyze the trunk movement in deep flexion was obtained from our own parallel ongoing in vivo measurements performed on 14 healthy male subjects with no recent back complications. For this purpose, infrared light emitting markers, along with a three-camera Optotrak system (NDI International, Waterloo/Canada), were used to track movement trajectory at different joint levels. From upright standing postures, subjects were instructed to slowly flex the trunk forward as much as possible. Total trial duration lasted about 12 seconds, including three seconds of rest at upright and peak flexion positions and six seconds of slow forward movement in between. Measured trunk and pelvic rotation trajectories for a typical subject were prescribed into the transient finite element model at the T12 and S1 levels, respectively. As for the individual lumbar vertebrae, the total lumbar rotation evaluated as the difference between foregoing two rotations was partitioned between different levels in accordance with proportions reported in earlier investigations (Arjmand and Shirazi-Adl, 2006b). Kinematics-driven approach was subsequently employed to calculate muscle forces, spinal loads and system stability margin throughout forward flexion movement (Figure 3.1).

In the current study, the cost function of the minimum sum of the cubed muscle stresses was considered in the optimization with inequality equations of muscle forces remaining positive and greater than their passive force components (calculated based on muscle strain and a tension-length relationship (Davis *et al.*, 2003)) but smaller than the sum of maximum physiological active forces (taken as 0.6 MPa × PCSA where PCSA is the physiological cross-sectional area) and the passive force components. The finite element program ABAQUS (Hibbit, Karlsson and Sorensen, Inc., Pawtucker, RI, version 6.5) was used to carry out nonlinear transient structural analyses while the optimization procedure was analytically solved using an inhouse program based on Lagrange multipliers method (LMM). The total computed muscle force in each muscle was partitioned into active and passive components with the latter force evaluated based on a length-tension relationship (Davis *et al.*, 2003).

Parametric studies on passive properties

The choice of a passive muscle length-force relationship used in the model for all muscles would influence the magnitude of muscle activity and, hence, the appearance or absence of partial or full flexion relaxation. Due to the existence of a number of rather distinct curves in the literature on the muscle passive length-tension relationship, it was decided to alter the reference curve used in the model (Davis *et al.*, 2003) and compute its effect on results. In this work, the passive curve was shifted either by +5 per cent (softened curve) or by –5 per cent (stiffened curve) in muscle axial strain and the effects of such decrease or increase in passive muscle resistance on results were analyzed. Furthermore, to investigate the relative importance of ligamentous spine passive properties on predictions, the bending stiffness of motion segments was also altered by ±20 per cent and the analyses were repeated. A decrease in bending stiffness (–20 per cent) could approximately simulate segmental degeneration, injury or tissue viscoelasticity, whereas an increase (+20 per cent) could simulate stiffer segments or ones with constructs or bone fusion.

Results

The trunk, pelvis, and lumbar rotations used in the model reached their peak values, though not at the same time, of 127.4°, 76.4°, and 51.9°, respectively (see Figure 3.3 for temporal variations of these prescribed rotations). Starting from the upright posture, the initial trunk rotation is due primarily to the lumbar rotation that reaches its maximum early in flexion phase of the movement. Subsequently, further increase in the trunk rotation at the mid- and final phases of flexion movement is found to be due solely to the pelvic rotation.

Starting from the upright standing posture towards full trunk flexion, temporal variation of active force in abdominal and global extensor muscles are also shown in Figure 3.3. The global extensor muscles experience an initial increase in activity as the forward flexion

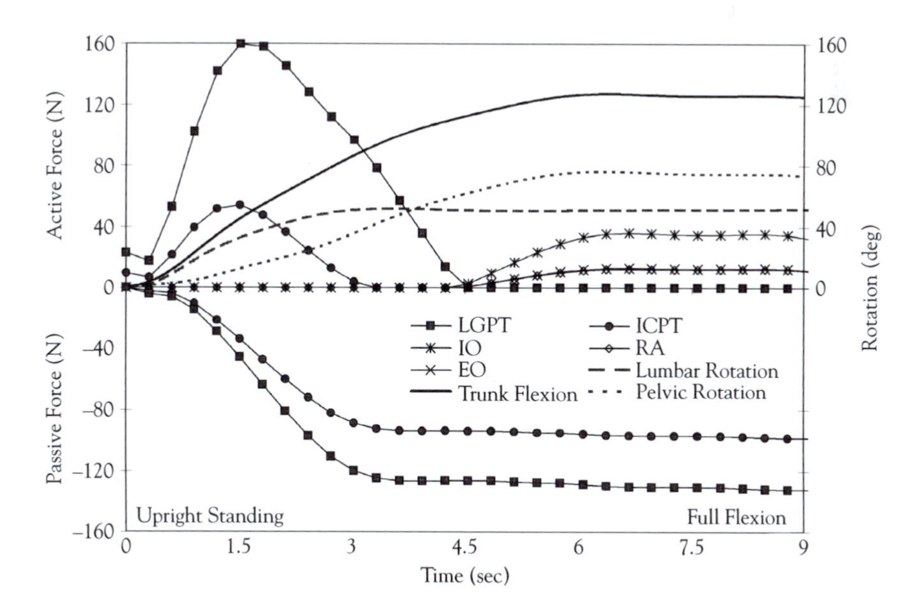

Figure 3.3 Predicted temporal variation of active and passive forces in global trunk muscles (on each side) as well as measured trunk and pelvic rotations with time advancing from upright standing posture (left) toward full flexion (right). Total lumbar rotation is the difference between trunk and pelvic rotations. The initial drop in global extensor muscle forces is due to the inertia of the upper body at the onset of the task.

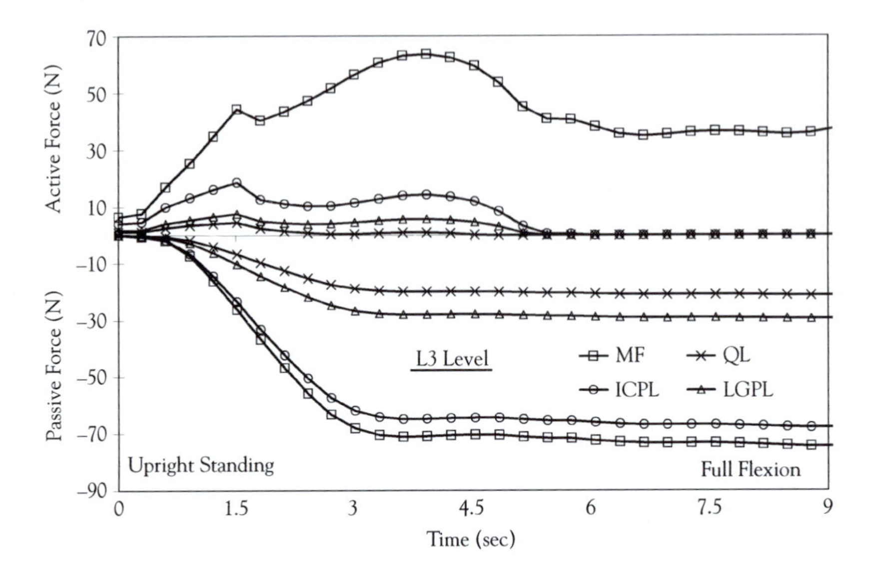

Figure 3.4 Predicted temporal variation of active force in local trunk extensor muscles (on each side) attached to the L3 level with time during forward flexion movement. Muscle forces at other lumbar levels (not shown) follow similar trends.

initiates followed by a decrease and finally a complete silence at larger flexion angles. In contrast, the abdominal muscles are active only at larger flexion angles when the thoracic extensor muscles are silent. The passive contribution of thoracic extensor muscles monotonically increases with flexion reaching their maximum at full lumbar flexion (Figure 3.3). Local lumbar muscles, with the exception of the multifidus, demonstrate a pattern similar to that of global thoracic muscles but with the full relaxation at larger trunk angles. The multifidus, on the other hand, undergoes only a partial relaxation (see Figure 3.4 for the L3 level).

The ligamentous spinal loads in the local segmental directions (i.e., axial compression, anterior-posterior shear force, and bending moment), as well as net external moment at the L5-S1 disc level, are also shown in Figure 3.5 for the entire duration of movement. The spinal loads increase downward reaching their maximum values at the distal L5-S1 level. The net external moment increases with the trunk flexion except at larger values where it drops slightly. The local axial compression and shear forces also follow the same trend, reaching peak values of 2629 N and 689 N, respectively.

As the passive contribution of muscles increases (i.e., case with –5 per cent shift in passive force-length relationship shown in Figure 3.6), the activity in thoracic extensor muscles substantially diminishes, demonstrating the FRP at smaller trunk flexion angles while activity in abdominal muscles initiates earlier at smaller trunk flexion and increases further at larger flexion angles. In contrast, activity in thoracic muscles increases as the passive contribution decreases, resulting in a delay in the FRP and a residual activity in the longissimus muscle at peak flexion angles. Similar trends, but to a lesser extent, in activities of thoracic extensor muscles and abdominal muscles are predicted when the passive properties of the ligamentous spine are increased or decreased by 20 per cent, respectively (Figure 3.7).

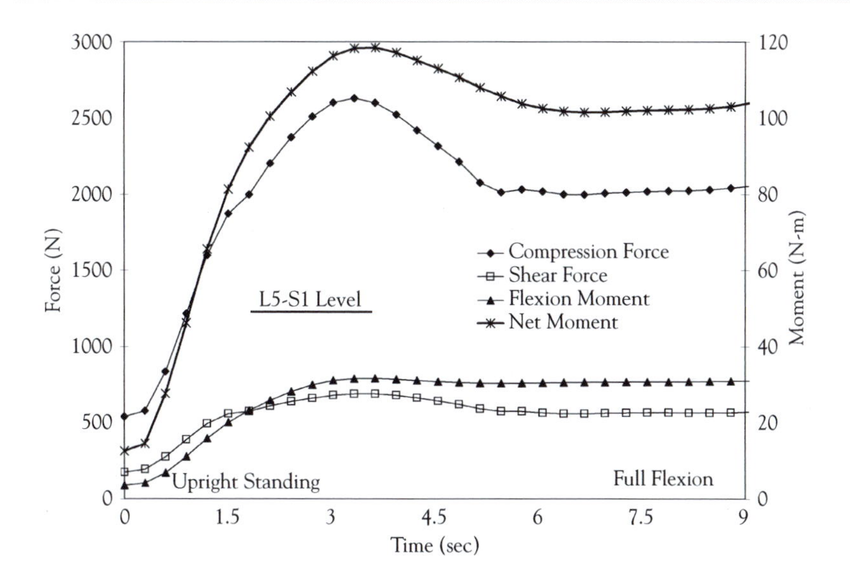

Figure 3.5 Predicted temporal variation of passive ligamentous loads as well as net external moment at the L5-S1 disc level with time during forward flexion movement.

Discussion

Excessive mechanical loading of the spine is recognized as a major cause of LBDs. To reduce the risk of back injuries, existing ergonomic guidelines aim to limit compressive force on the spine (Waters *et al.*, 1993). In the absence of any direct method to measure spinal loads and muscle forces, biomechanical modeling of the spine has become an indispensable tool in evaluation, prevention, and proper management of back disorders. A number of mathematical biomechanical models of the spine with diverse simplifying assumptions have been

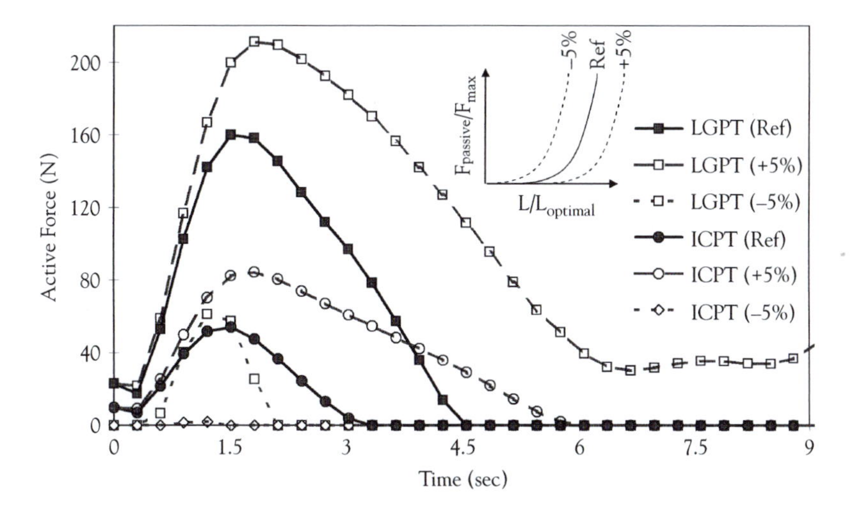

Figure 3.6 Predicted temporal variation of active force (on each side) in extensor global muscles with time during forward flexion movement as passive force-length curve is altered ±5% for all muscles.

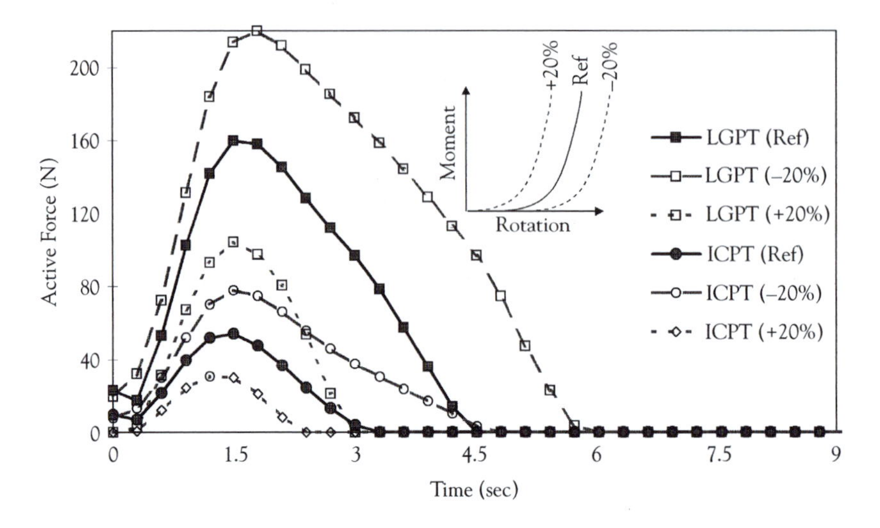

Figure 3.7 Predicted temporal variation of active force (on each side) in extensor global muscles with time from upright standing posture toward full flexion as the bending stiffness of motion segments is altered by ±20% at all levels.

introduced to address related biomechanical and clinical issues. Our earlier studies have demonstrated the crucial role of consideration of equilibrium at all levels and not just one, both translational and rotational degrees of freedom at each joint, wrapping of global extensor muscles, and nonlinearity of passive ligamentous spine on predictions.

The novel Kinematics-driven model takes into account the nonlinear behavior of the ligamentous spine while satisfying equilibrium conditions at all thoracolumbar spinal levels and directions. This approach has previously been applied to address a number of biomechanical issues during static and dynamics lifting activities in both upright and forward flexed standing postures. In the current study, though we used a full transient analysis, due to the rather slow pace of the forward flexion movement, a quasi-static approach would have yielded almost the same results. Consideration of inertia is particularly essential when the task is performed at movement velocities much faster than that considered in this study.

Physical activities involving trunk flexion are very common in regular daily, occupational, and athletic activities. An improved understanding of spinal functional biomechanics during full trunk flexion (i.e., partitioning of loads between passive spine and active trunk muscles and between active-passive components of muscles) are essential in proper analysis of risks involved. The kinematics required as input data into the finite element model have been taken from an ongoing in vivo study on the effect of movement velocity on trunk biomechanics. Throughout a forward flexion movement from upright standing posture to full flexion, muscle forces (active and passive), ligamentous spinal loads (axial compression force, anterior-posterior shear force, and sagittal moment) at all levels have been estimated. The effect of alterations in muscle passive force-length relationship and in bending properties of the ligamentous spine on results is also investigated.

During forward flexion movement simulated in this study, a sequential lumbar-pelvic rotation is observed in which greater lumbar rotation is apparent at the beginning of the task followed by pelvic rotation at the final phase (Figure 3.3). Similarly, as compared to the lumbar rotation, the pelvic rotation has been reported to become predominant at the end of flexion and beginning of extension phase during flexion-extension movements

(Paquet *et al.*, 1994). Others, however, suggest that lumbar and pelvic rotations act simultaneously during flexion and/or extension phases (Nelson *et al.*, 1995; Sarti *et al.*, 2001). The lumbar-pelvic rhythm affects results by influencing the wrapping of global extensor muscles and the relative passive contribution of muscles and ligamentous spine.

As the trunk flexes forward from upright posture, initially both active and passive components of forces in global extensor muscles increase with the formers reaching their peak values at about 45° (Figure 3.3). Thereafter, up to the trunk flexion of about 95°, active forces in thoracic extensor muscles diminish despite the continuous increase in net external moment reaching its maximum of 118 Nm. On the contrary, passive muscle forces as well as passive ligamentous moment increase throughout the movement to peak lumbar flexion (Figures 3.3 and 3.5). The progressive relief in activity of global back muscles is due, therefore, to higher passive contribution of muscles and ligamentous spine as the lumbar rotation increases. As the trunk flexion exceeds about 95° (at about 3.3 sec), lumbar rotation (Figure 3.3) and consequently both passive muscle force and moment resistance of the ligamentous spine, remain nearly unchanged, while the activity of back muscles continues to drop. In this case, the reduction in net external moment due to the decrease in the effective lever arm of the trunk centre of mass (COM) is the primary cause in progressive decrease in back muscle activities. Global longissimus [LGPT] and iliocostalis [ICPT] become completely silent at trunk flexion angles of about 114° and 95°, respectively.

With the exception of the multifidus that only partially relaxed, local lumbar muscles also demonstrated full relaxation in activity, but at larger flexion angles as compared with global extensor muscles (Figure 3.4). Measurements reported in the literature indicate silence at superficial lumbar muscles (Andersson *et al.*, 1996; Dickey *et al.*, 2003; Mathieu and Fortin, 2000; McGill and Kipper, 1994; Sarti *et al.*, 2001; Solomonov *et al.*, 2003; Toussaint *et al.*, 1995). As for superficial thoracic extensor muscles at larger trunk flexion angles, some report continuation of activity (McGill and Kipper, 1994; Toussaint *et al.*, 1995) while others suggest relaxation (Dickey *et al.*, 2003; Mathieu and Fortin, 2000; Andersson *et al.* 1996) reported activity in deep lumbar muscles as well.

Abdominal muscles remain silent up to trunk flexion angles of about 115° at which angles global extensor muscles become inactive. Subsequently, abdominal muscles (especially internal oblique [IO], Figure 3.3) initiate activation up to the peak rotation, generating flexor moments that offset the moments produced by the passive component of back muscle forces. In other words, abdominals are activated to increase and maintain the large flexion angles. Activities in abdominal muscles have also been reported in earlier studies during full flexion as extensor muscles become silent (Mathieu and Fortin, 2000; McGill and Kipper, 1994; Olson *et al.*, 2006).

The effect of changes in passive properties of muscles and ligamentous spine on the results in general and the FRP in particular is found to be substantial. A decrease in passive contribution of extensor muscles (case with +5 per cent) (Figure 3.6) markedly increased activity in global extensor muscles and diminished that in abdominal muscles at larger flexion angles. A reverse trend was computed when the passive contribution was increased resulting in an earlier and greater activity in abdominal muscles but flexion relaxation in extensor muscles. Similar effects were also predicted as the bending rigidity of the ligamentous spine was altered (Figure 3.7). A decrease in passive stiffness due to an injury or joint relaxation could delay flexion relaxation in extensor muscles. The abdominal muscles are also affected by such changes.

In conclusion, existing biomechanical models of the spine rarely take into account the equilibrium equations simultaneously at all directions and levels (Arjmand *et al.*, 2006, 2007) and, hence, risk violating essential equilibrium conditions at levels and directions different from the ones considered. Flexion relaxation in global thoracic and local lumbar extensor muscles at larger trunk flexion angles is a direct consequence of passive resistance of extensor muscles and ligamentous spine that both increase with the lumbar rotation. Alterations in these passive properties and in the relative lumbar-pelvic rhythm could, hence, influence the load redistribution and flexion relaxation phenomenon. Future Kinematics-driven model studies should, amongst others, account for asymmetry in movements and for system stability in the optimization procedure.

Acknowledgment

The work is supported by the Natural Sciences and Engineering Research Council of Canada (NSERC-Canada) and Aga Khan Foundation.

References

1. Andersson, G.B. (1981) 'Epidemiologic aspects on low-back pain in industry'. *Spine*, 6: 53–60.
2. Andersson, E.A., Oddsson, L.I., Grundstrom, H., Nilsson, J. and Thorstensson A. (1996) 'EMG activities of the quadratus lumborum and erector spinae muscles during flexion-relaxation and other motor tasks'. *Clin Biomech*, 11: 392–400.
3. Arjmand, N. (2006) 'Computational biomechanics of the human spine in static lifting tasks'. PhD thesis, Ecole Polytechnique de Montreal.
4. Arjmand, N. and Shirazi-Adl, A. (2005) 'Biomechanics of changes in lumbar posture in static lifting'. *Spine*, 30: 2637–48.
5. Arjmand, N. and Shirazi-Adl, A. (2006a) 'Sensitivity of kinematics-based model predictions to optimization criteria in static lifting tasks'. *Med Eng Phys*, 28: 504–14.
6. Arjmand, N. and Shirazi-Adl, A. (2006b) 'Model and in vivo studies on human trunk load partitioning and stability in isometric forward flexions'. *J Biomech*, 39: 510–21.
7. Arjmand, N. and Shirazi-Adl, A. (2006c) 'Role of intra-abdominal pressure in the unloading and stabilization of the human spine during static lifting tasks'. *Eur Spine J*, 15: 1265–75.
8. Arjmand, N., Shirazi-Adl, A. and Bazrgari, B. (2006) 'Wrapping of trunk thoracic extensor muscles influences muscle forces and spinal loads in lifting tasks'. *Clin Biomech*, 21: 668–75.
9. Arjmand, N., Shirazi-Adl, A. and Parnianpour, M. (2007) 'Trunk biomechanical models based on equilibrium at a single-level violate equilibrium at other levels'. *Eur Spine J*, 16: 701–9.
10. Bazrgari, B. and Shirazi-Adl, A. (2007) 'Spinal stability and role of passive stiffness in dynamic squat and stoop lifts'. *Computer Methods in Biomechanics and Biomedical Engineering*, 10: 351–60.
11. Bazrgari, B., Shirazi-Adl, A. and Arjmand, N. (2007) 'Analysis of squat and stoop dynamic liftings: muscle forces and internal spinal loads'. *Eur Spine J*, 16: 687–99.
12. Burdorf, A. and Sorock, G. (1997) Positive and negative evidence of risk factors for back disorders, *Scand J Work Environ Health*, 23: 243–56.
13. Cholewicki, J. and McGill, S.M. (1996) 'Mechanical stability of the in vivo lumbar spine: Implications for injury and chronic low back pain'. *Clin Biomech*, 11: 1–15.
14. Cholewicki, J., McGill, S.M. and Norman, R.W. (1995) 'Comparison of muscle forces and joint load from an optimization and EMG assisted lumbar spine model: towards development of a hybrid approach'. *J Biomech*, 28: 321–31.
15. Colloca, C.J. and Hinrichs, R.N. (2005) 'The biomechanical and clinical significance of the lumbar erector spinae flexion-relaxation phenomenon: a review of literature'. *J Manipulative Physiol Ther*, 28: 623–31.

16. Damkot, D.K., Pope, M.H., Lord, J. and Frymoyer, J,W. (1984) 'The relationship between work history, work environment and low-back pain in men'. *Spine*, 9: 395–9.
17. Davis, J., Kaufman, K.R. and Lieber, R.L. (2003) 'Correlation between active and passive isometric force and intramuscular pressure in the isolated rabbit tibialis anterior muscle'. *J Biomech*, 36: 505–12.
18. Diamond, S. and Borenstein, D. (2006) 'Chronic low back pain in a working-age adult'. *Best Pract Res Clin Rheumatol*, 20: 707–20.
19. Dickey, J.P., McNorton, S. and Potvin, J.R. (2003), 'Repeated spinal flexion modulates the flexion-relaxation phenomenon'. *Clin Biomech*, 18: 783–9.
20. El-Rich, M., Shirazi-Adl, A. and Arjmand, N. (2004) 'Muscle activity, internal loads and stability of the human spine in standing postures: combined model in vivo studies'. *Spine*, 29: 2633–42.
21. Ferguson, S.A. and Marras, W.S. (1997) 'A literature review of low back disorder surveillance measures and risk factors'. *Clin Biomech*, 12: 211–26.
22. Floyd, W.F. and Silver, P.H. (1951) 'Function of erectores spinae in flexion of the trunk'. *Lancet*, 1: 133–4.
23. Frank, J.W., Kerr, M.S., Brooker, A.S., DeMaio, S.E., Maetzel, A., Shannon, H.S., Sullivan, T.J., Norman, R.W. and Wells, R.P. (1996) 'Disability resulting from occupational low back pain. Part I: What do we know about primary prevention? A review of the scientific evidence on prevention before disability begins'. *Spine*, 21: 2908–17.
24. Frymoyer, J. (1996) *The magnitude of the problem in the lumbar spine*, Philadelphia: WB Saunders, PP. 8–15
25. Gagnon, D., Lariviere, C. and Loisel, P. (2001) 'Comparative ability of EMG, optimization, and hybrid modelling approaches to predict trunk muscle forces and lumbar spine loading during dynamic sagittal plane lifting'. *Clin Biomech*, 16: 359–72.
26. Gardner-Morse, M., Stokes, I.A.F. and Laible, J.P. (1995) 'Role of muscles in lumbar spine stability in maximum extension efforts'. *Journal of Orthopaedic Research*, 13: 802–8.
27. Granata, K.P., Lee, P.E. and Franklin, T.C. (2005) 'Co-contraction recruitment and spinal load during isometric trunk flexion and extension'. *Clin Biomech*, 20: 1029–37.
28. Hoogendoorn, W.E., Bongers, P.M., de Vet, H.C., Douwes, M., Koes, B.W., Miedema, M.C., Ariens, G.A. and Bouter, L.M. (2000) 'Flexion and rotation of the trunk and lifting at work are risk factors for low back pain: results of a prospective cohort study'. *Spine*, 25: 3087–92.
29. Ihlebaek, C., Hansson, T.H., Laerum, E., Brage, S., Eriksen, H.R., Holm, S.H., Svendsrod, R. and Indahl, A. (2006) 'Prevalence of low back pain and sickness absence: a "borderline" study in Norway and Sweden'. *Scand J Public Health*, 34: 555–8.
30. Kaigle, A.M., Wessberg, P. and Hansson, T.H. (1998) 'Muscular and kinematic behavior of the lumbar spine during flexion-extension'. *J Spinal Disord*, 11: 163–74.
31. Katz, J.N. (2006) 'Lumbar disc disorders and low-back pain: socioeconomic factors and consequences'. *J Bone Joint Surg Am*, 88, Suppl 2: 21–4.
32. Kelsey, J.L., Githens, P.B., White, A.A. 3rd, Holford, T.R., Walter, S.D., O'Connor, T., Ostfeld, A.M., Weil, U., Southwick, W.O. and Calogero, J.A. (1984) 'An epidemiologic study of lifting and twisting on the job and risk for acute prolapsed lumbar intervertebral disc'. *J Orthop Res.*, 2: 61–6.
33. Kiefer, A., Shirazi-Adl, A. and Parnianpour, M. (1997), 'Stability of the human spine in neutral postures'. *Eur Spine J*, 6: 45–53.
34. Kippers, V. and Parker, A.W. (1984) 'Posture related to myoelectric silence of erector spinae during trunk flexion'. *Spine*, 9: 740–5.
35. Manchikanti, L. (2000) 'Epidemiology of low back pain'. *Pain Physician*, 3: 167–92.
36. Marras, W.S. and Granata, K.P. (1997) 'The development of an EMG-assisted model to assess spine loading during whole-body free-dynamic lifting'. *Journal of Electromyography and Kinesiology*, 7: 259–68.
37. Marras, W.S., Lavender, S.A., Leurgans, S.E., Fathallah, F.A., Ferguson, S.A., Allread, W.G. and Rajulu, S.L. (1995) 'Biomechanical risk factors for occupationally related low back disorders'. *Ergonomics*, 38: 377–410.

38. Marras, W.S., Parakkat, J., Chany, A.M., Yang, G., Burr, D. and Lavender, S.A. (2006) 'Spine loading as a function of lift frequency, exposure duration, and work experience'. *Clin Biomech*, 21: 345–52.
39. Mathieu, P.A. and Fortin, M. (2000) 'EMG and kinematics of normal subjects performing trunk flexion/extensions freely in space'. *J Electromyogr Kinesiol*, 10: 197–209.
40. McGill, S.M. and Kippers, V. (1994) 'Transfer of loads between lumbar tissues during the flexion-relaxation phenomenon'. *Spine*, 19: 2190–6.
41. McGill, S.M. and Norman, R.W. (1986) 'Partitioning of the L4-L5 dynamic moment into disc, ligamentous, and muscular components during lifting'. *Spine*, 11: 666–78.
42. National Institute for Occupational Safety and Health (NIOSH) (1997) *Musculoskeletal disorders and workplace factors*, US Dept. of Health and Human Services.
43. Nelson, J.M., Walmsley, R.P. and Stevenson, J.M. (1995) 'Relative lumbar and pelvic motion during loaded spinal flexion/extension'. *Spine*, 20: 199–204.
44. Olson, M., Solomonow, M. and Li, L. (2006) 'Flexion-relaxation response to gravity'. *J Biomech*, 39: 2545–54.
45. Paquet, N., Malouin, F. and Richards, C.L. (1994) 'Hip-spine movement interaction and muscle activation patterns during sagittal trunk movements in low back pain patients'. *Spine*, 19: 596–603.
46. Parnianpour, M., Wang, J.L., Shirazi-Adl, A., Sparto, P. and Wilke, H.J. (1997) 'The effect of variations in trunk models in predicting muscle strength and spinal loads'. *Journal of Musculoskeletal Research*, 1: 55–69.
47. Reeves, N.P. and Cholewicki, J. (2003) 'Modeling the human lumbar spine for assessing spinal loads, stability, and risk of injury'. *Crit Rev Biomed Eng*, 31: 73–139.
48. Sarti, M.A., Lison, J.F., Monfort, M. and Fuster, M.A. (2001) 'Response of the flexion-relaxation phenomenon relative to the lumbar motion to load and speed'. *Spine*, 26: E421-6.
49. Schneider, S., Mohnen, S.M., Schiltenwolf, M. and Rau, C. (2007) 'Comorbidity of low back pain: Representative outcomes of a national health study in the Federal Republic of Germany'. *Eur J Pain*, 11: 387–97.
50. Schultz, A., Andersson, G., Ortengren, R., Haderspeck, K., Ortengren, R., Nordin, M. and Bjork, R. (1982) 'Analysis and measurement of lumbar trunk loads in tasks involving bends and twists'. *J Biomech*, 15: 669–75.
51. Shirazi-Adl, A. and Parnianpour, M. (2001) 'Finite element model studies in lumbar spine biomechanics'. *Computer Techniques and Computational Methods in Biomechanics*. New York: CRC Press, PP. 1–36.
52. Shirazi-Adl, A., Sadouk, S., Parnianpour, M., Pop, D. and El-Rich, M. (2002) 'Muscle force evaluation and the role of posture in human lumbar spine under compression'. *European Spine Journal*, 11: 519–26.
53. Solomonow, M., Baratta, R.V., Banks, A., Freudenberger, C., Zhou, B.H. (2003) 'Flexion-relaxation response to static lumbar flexion in males and females'. *Clin Biomech*, 18: 273–9.
54. Toussaint, H.M., de Winter, A.F., de Haas, Y., de Looze, M.P., Van Dieen, J.H., Kingma, I. (1995) 'Flexion relaxation during lifting: implications for torque production by muscle activity and tissue strain at the lumbo-sacral joint'. *J Biomech*, 28: 199–210.
55. Troup, J.D., Martin, J.W. and Lloyd, D.C. (1981) 'Back pain in industry. A prospective survey'. *Spine*, 6: 61–9.
56. van Dieen, J.H., Hoozemans, M.J., Toussaint, H.M. (1999) 'Stoop or squat: a review of biomechanical studies on lifting technique'. *Clin Biomech*, 14: 685–96.
57. van Dieen, J.H. and Kingma, I. (2005) 'Effects of antagonistic co-contraction on differences between electromyography based and optimization based estimates of spinal forces'. *Ergonomics*, 48: 411–26.
58. van Dieen, J.H., Kingma, I. and van Der Bug, P. (2003) 'Evidence for a role of antagonistic cocontraction in controlling trunk stiffness during lifting'. *J Biomech*, 36: 1829–36.
59. Varma, K.M. and Porter, R.W. (1995) 'Sudden onset of back pain'. *Eur Spine J*, 4: 145–7.

60. Waters, T.R., Putz-Anderson, V., Garg, A. and Fine, L.J. (1993) 'Revised NIOSH equation for the design and evaluation of manual lifting tasks'. *Ergonomics*, 36: 749–76.
61. Watson, P.J., Booker, C.K., Main, C.J. and Chen, A.C. (1997) 'Surface electromyography in the identification of chronic low back pain patients: the development of the flexion relaxation ratio'. *Clin Biomech*, 12: 165–71.
62. Wikipedia Encyclopedia (2006) *Publishing on the Internet*. Available HTTP: <http://en.wikipedia.org/wiki/Hurricane_Katrina> (accessed 12 Nov 2006).

4

Artificial neural network models of sports motions

W.I. Schöllhorn, J.M. Jäger and D. Janssen
Johannes Gutenberg University of Mainz, Mainz

Introduction

Describing and analysing phenomena in sports is necessarily connected with the process of modelling. Modelling starts with the description of the phenomena and it is accompanied by the active and subjective decision of researchers for a limited number of its criteria. Thus, a reduced or simplified mapping of the real object is achieved. Accordingly, the description of a phenomenon means mapping the model original (real object) to a (simplified) model and is part of the process of scientific modelling. In this process of scientific modelling the determination of the model purpose is followed by the selection of adequate model variables that serve as in- and output variables for different forms of models. The criteria for the selection of these variables are manifold and reach from 'just measurable' up to highly sophisticated mathematical extraction procedures.

In the context of this book two classes of models are of special interest, one class of more theory driven models and another class of more data driven models. In stronger theory driven or deterministic models knowledge about the subsystems and their interaction is a prerequisite and data are mainly used to verify what we know about the system. But very often in biological or social systems not all necessary information about the details of the system is given. In this case an alternative approach is pursued. If no or little is known about the subsystems characteristics and/or internal structures typically a more data-driven approach is chosen to develop a model for the relation between input and output of a system. In contrast to traditional models, which are more *theory-rich and data-poor*, the data driven artificial neural network (ANN) approaches are *data-rich and theory-poor* in a way that little or no a priori knowledge of the system exists.

Independent of the type of model, once a model is constructed it serves for simulating reality. The difference between the simulated result and reality is often taken as a measure for the quality of the model. Typically mainly linear approaches of statistics have been suggested for data rich models. More recently nonlinear alternatives like ANNs are applied in great number in neighboured research areas, e.g. in healthcare (Begg, Kamruzzaman and Sarker, 2006) or clinical biomechanics (Schöllhorn, 2004). Originally ANNs have been developed during the late 80th and early 90th with enormous success in different fields

(Kohonen, 2001) and are applied by a growing number of researchers in the domain of sports science. In the literature ANNs are also referred to as connectionist networks, neurocomputers, parallel distributed processors, etc. The purpose of this chapter is to explain possibilities of applying ANNs for modelling movements in sports.

Modelling with ANN

ANNs are adaptive models that are able to learn from the data and generalize things learned. They extract only the essential characteristics from numerical data instead of memorizing all of it. This offers a convenient way to reduce the amount of stored data and to form an implicit model without having to construct a traditional, physical model of the underlying phenomenon.

An ANN consists of simple processing units, called 'neurons' or 'nodes', which are linked to other units by connections of different strength or weight (Figure 4.1). According to biological structures: (a) an artificial neuron consists of dendrites, a cell body and an axon; (b) synapses (the connection entity between the axon and the dendrites of other neurons)

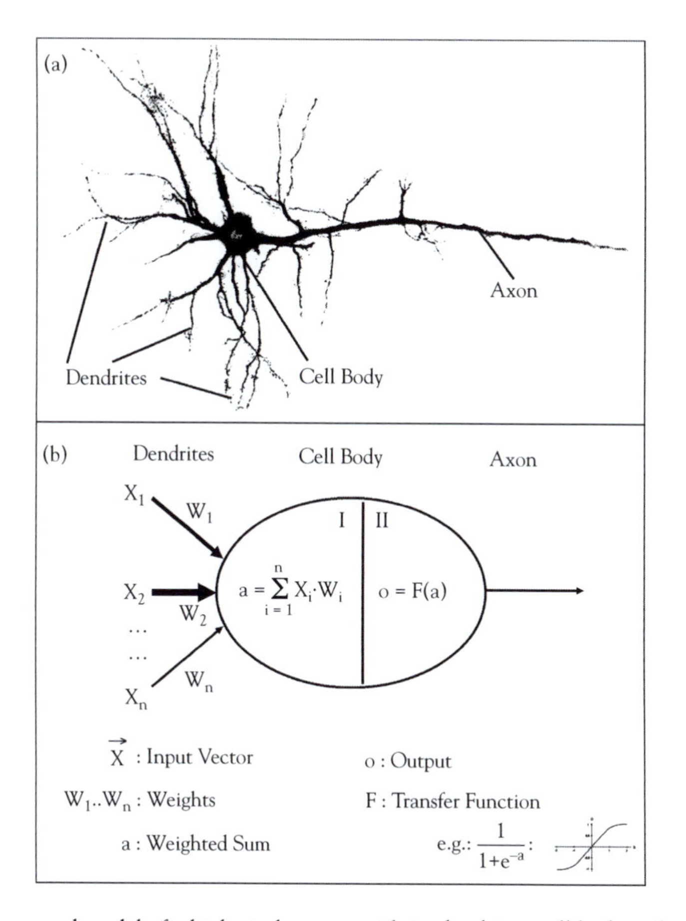

Figure 4.1 Picture and model of a biological neuron with its dendrites, cell body and axon.

are modelled by means of numerical values, the so-called weights (w_1, ..., w_n) illustrated by arrows of different strengths, representing different values in the figure. An artificial neuron uses the input values (x_1, ..., x_n) and computes the weighted sum 'a' (Figure 4.1). This is done by summing up all products of the inputs and their weights (I). This value 'a' is then passed to the transfer function that computes the final output value of the neuron (II). The neurons are typically arranged in a series of layers. After early intentions applying linear transfer functions to the production of output (McCulloch and Pitts, 1943) more recent ANNs use linear and nonlinear transfer functions, and either binary or continuous activations (Hertz, Krogh and Palmer, 1994)

The term 'neural' is thus associated in two ways with biological neurons. First, knowledge is acquired by the network through a learning process and second, synaptic weights, i.e, the strengths of connections between neurons are used to store knowledge (Haykin, 1994).

Selection of model variables

In many applications of ANNs, the modelling is a multiple stage process, and some of the stages are closely coupled. Modelling with ANNs is preceded by a feature selection/extraction procedure, wherein two stages may be distinguished (Figure 4.2): the identification of description variables (variables are identified that describe the original data the best from the investigator's point of view) and the dimensionality reduction (the number of variables is reduced according to certain statistical criteria). Within the first stage certain criteria are selected to describe the model original (Table 4.1). Both stages serve to retain only relevant information and are subject to the investigator's experience, intention, philosophy and point of view. Alternative approaches are suggested by Fukunaga (1990) and Englehart *et al.* (1999).

At least in the first, second, and one of the sub-domains 3a, 3b and 3c we have to decide what input variables should be selected. On the one hand – in case of classification problems where no distinct biomechanical model serves for a coarse orientation of the relation between input and output variables – very often the number of variables is still high in order to ensure that the measurements carry all of the information contained in the original data. On the other hand a classifier with fewer inputs has less adaptive parameters to be determined and often better generalization properties. Common techniques to reduce the number of input variables are principal component analysis (PCA) or factor analysis (FA) and Karhunen-Loeve transformation (KLT). All techniques aim to produce uncorrelated

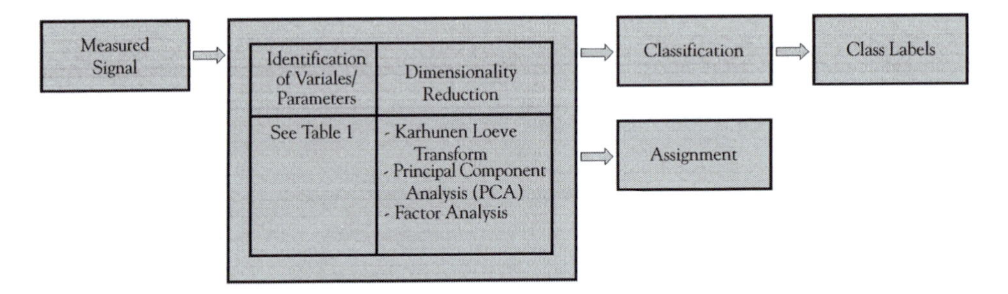

Figure 4.2 Flow chart for signal processing and classification.

Table 4.1 Criteria for variable selection within the feature extraction procedure

1. General domain
   - Biomechanical data – test scores / questionnaires, etc.
   - Individual (data from unique events) – Group (averaged over several samples)
2. Space domain
   - Kinematic, dynamic, EMG or image variables

3a. Time domain
   - Raw data – filtered data
   - Product-oriented (intensities of variables at certain instants in time)
   - Process-oriented (whole time course of a variable)

3b. Frequency domain
   - Fourier coefficients

3c. Frequency – Time domain
   - Wavelet coefficients
   - Windowed Fourier coefficients

feature sets by mapping original data onto the eigenvectors of the covariance or correlation matrix (Watanabe, 1985). However, this feature extraction is the most elaborate and time-consuming part of most studies (Barton and Lees, 1995).

Two types of ANN models

Dependent on the learning procedure, ANNs can be grouped into two main classes: *supervised* and *unsupervised learning*. Hybrid systems like counter propagation networks (Hecht-Nielsen, 1991) using both strategies have also been developed and are described in more detail in Hertz *et al.* (1994). Statistically, these two approaches can be assigned to hypothesis verifying (supervised) and hypotheses generating (unsupervised) approaches. Closely related to the statistical analogy is the epistemological basis for both approaches. Accordingly each approach corresponds to different grades of influence of the investigator on the result. The application of supervised ANNs presumes the classes of outcome and generates a generalized relationship between the input data and the assumed output classes. For instance, subjects may be divided into normal and pathological walkers. However, new classes are typically not identified applying this approach. In case of unsupervised ANNs only the criteria for classification are given and no output classifications are expected.

From a statistical point of view data processing by means of ANNs is considered as non-parametric. Accordingly, input data in most cases here stem from biomechanical measurements on the basis of metric scales as well as from test batteries for motor control or questionnaires based on ordinal or nominal scales. Sometimes neural nets are considered as nonlinear factor analysis (Haykin, 1994).

ANNs are rapidly finding new fields of application within sports biomechanics, from classification procedures over time-series analysis up to performance prediction. Some applications will be discussed in the following two subsections: The first includes applications of supervised ANNs while the second subsection can be assigned to applications of unsupervised ANNs.

Supervised learning

Within supervised ANNs the multilayer feed-forward neural network (Rumelhart and McClelland, 1986) is of specific interest and has been used in a wide range of applications.

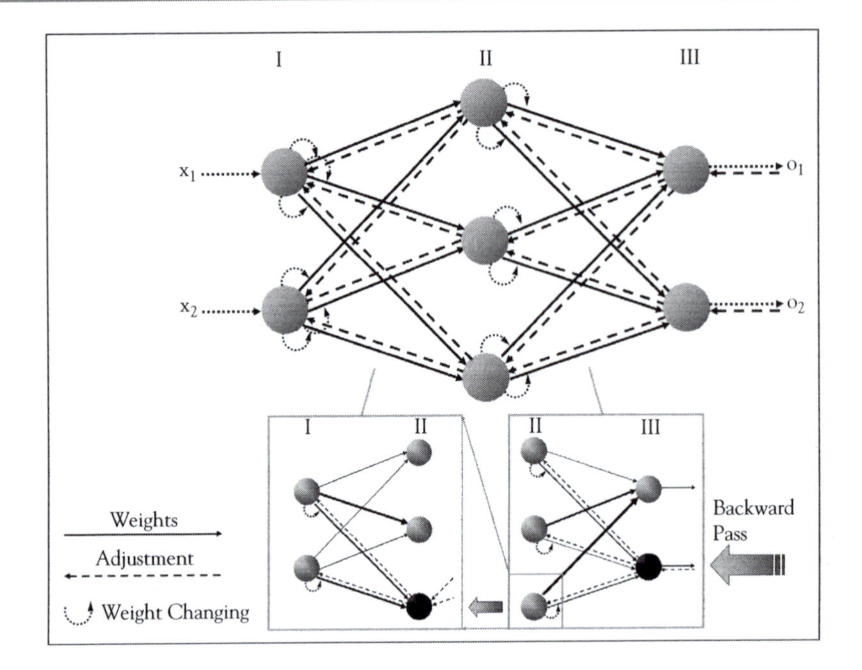

Figure 4.3 Schematic illustration of a 2-3-2 multilayer perceptron.

Equivalent names are multilayer perceptron (MLP) or back propagation network. For a general introduction see Haykin (1994). Originally, the elementary processor employed in MLPs is the linear threshold unit first proposed by McCulloch and Pitts in 1943 (McCulloch *et al.*, 1943) and for which Rosenblatt (Rosenblatt, 1957) designed his iterative perceptron training algorithm.

MLPs usually have input and output layers, with some processing or hidden layers in between (Figure 4.3). In the upper part of Figure 4.3, a two-dimensional (2D) input vector (x_1, x_2) is propagated through the network computing the 2D output (o_1, o_2). The lower part of the figure shows the backward pass with two neurons in more detail: The dark neurons pass adjustment values to the connected neurons in the neighboured layer one after another. Depending on the chosen algorithms the weights between the two considered layers (III, II and II, I) are modified in a specific way. The procedure continues until the input neurons are reached. However, during the training phase pairs of inputs and corresponding desired outputs are simultaneously presented to the MLP, which iteratively self-adjusts by back propagation. The error difference between the generated and desired output is used for modifying the ANN connections until learning is accomplished. Training is completed when certain criteria are met (for instance, the prediction error remains under a preset threshold) (Chau, 2001). Other supervised types include the probabilistic neural networks (Specht, 1990), the radial basis function (c.f. Bishop, 1995) and higher order neural networks like recurrent ANNs (Pandya and Macy, 1995). Specific recurrent ANNs are Elman- (Elman, 1990), Jordan- (Jordan, 1986) and Hopfield-Networks (Hopfield, 1982). Elman- and Jordan-Networks are three-layered back propagation networks, with a further feedback connection from the output of the hidden layer to its input (Elman) or hidden (Jordan) layer. This feedback path allows the networks to learn to recognize and generate temporal patterns, as well as spatial patterns. The Hopfield network is used to store one or more stable target vectors. These stable vectors can be viewed as memories that the network recalls when provided with similar vectors that act as a cue to the network memory.

Applications of supervised ANN in sports

First applications of supervised nets in sports can be found in Herren, Sparti, Aminian and Schutz (1999). Triaxial body accelerations during running were recorded at the low back and at the heel. Selected parameters of the body accelarations served as input variables in order to train two MLPs to recognize each outdoor running pattern and determine speed and incline. Ten parameters were selected as input variables according to their correlations with speed or incline. The mean square error for speed was 0.12 m/s and 0.014 (1.4 per cent in absolute value) radiant for incline. In comparison the multiple regression analysis allowed similar accurate prediction of speed but worse for incline (2.6 per cent in absolute value).

A similar ANN approach was chosen by Maier (Maier, Wank, Bartonietz and Blickhan, 2000; Maier, 2001; Maier, Meier, Wagner and Blickhan, 2000; Maier, Wank and Bartonietz, 1998) for the prediction of distances in shot put and javelin throwing on the basis of release parameters. By neglecting wind speed and aerodynamics of the shot in combination with a constant release height of 2 m an MLP was trained with two input and one output neuron as well as two hidden layers with five and three neurons. Input variables were the angle and the velocity of release. Output variable was the shot put distance. Data patterns with resulting distance were generated for release angles of $30° < \alpha < 55°$and release velocities of 9 m/s < v < 15 m/s. Angles were varied in steps of 1° and velocity in steps of 1 m/s. From all calculated combinations 100 shot put trials were selected at random to create an ANN model. The simulated flight distance of the ANN included errors of only 2.5 per cent on average. This ANN model serves as an example for implicit physical modelling without describing the explicit physical principle.

The ANN model for predicting the flight of javelins used three release angles and the velocity at release as input variables, while the distance reached was the output variable. Inputs (velocity, angle of release, angle of attack, side angle of attack) and outputs were recorded from 98 throws. Several ANNs with two hidden layers and two to eight neurons in each layer were tested for effectiveness. The ANN model was able to predict actual flights with mean differences of 2.5 per cent between the model and real throws.

Yan and Li (2000) and Yan and Wu (2000) investigated the relationship between the athletes' movement technique and the release parameter of the shot put. Twenty global and 33 local technique parameters were chosen as input variables. Output variables of the MLP with one hidden layer were angle and speed. The errors between network outputs and the measured release parameters were compared to those obtained by regression analysis. Results show that ANN errors were typically 25–35 per cent less than those from regression analysis and even smaller than the uncertainties in release parameter values that occur during manual digitizing (Bartlett, 2006).

The influence of training interventions on competition performance in swimming was analyzed by Edelmann-Nusser, Hohmann and Henneberg (2002) by means of three MLPs each with ten input- two hidden- and one output-neuron. The input variables were training contents during the two to four weeks prior competition. Training contents were quantified in swam kilometres in a certain intensity as well as hours of specific strength and conditioning training. Output variable was the competitive performance collected from

19 competitions over approximately three years. The analysis resulted in a prediction error of 0.05 s in a total swim time of 2:12:64 sec.

In sports medicine and sport sociology ANNs have also been established. Ringwood (1999) modelled the anaerobic threshold on the basis of work and heart rate data acquired during Conconi test. On the basis of demographic and heart rate variables an aerobic fitness approximation was modelled by Vainamo and co workers by means of two serially connected MLPs (Vainamo, Makikallio, Tulppo and Röning, 1998; Vainamo, Nissila, Makikallio, Tulppo and Röning, 1996). The success of nations at the Summer Olympics was predicted by Condon, Golden and Wasil (1999) on the basis of a nations sociological parameters like life expectancy, electric capacity or infant mortality rate.

Unfortunately, back propagation and other supervised learning algorithms may be limited by their poor scaling behaviour. With increasing differentiation of the output that leads to the use of more neurons, the time required to train the network grows exponentially and the learning process becomes unacceptably slow (Haykin, 1994). One possible solution to the scaling problem is to use an unsupervised learning procedure.

Unsupervised learning

In the unsupervised learning procedure the training data only consist of input patterns, no expected outputs exist, and no information about correct or false classifications is given by the net. Here, the learning algorithm on its own attempts to identify clusters of similar input-vectors and maps them on groups of similar or neighboured neurons. One of the best known classes of unsupervised learning procedures is called self-organizing map (SOM), the most famous one is the Kohonen feature map (KFM) (Kohonen, 1982; Kohonen, 2001). A SOM can be thought of a pair of layers: first, an input layer that receives the data, and second, a competitive layer with neurons that compete with each other to respond to features contained in the input data (Figure 4.4). A 5 × 5 network is shown with a three-dimensional (3D) input and its full connections in (a). Parts (b)–(g) of the figure show the training process, when a pattern (X) is applied. The lower parts thereby show the changes of a neuron's weights as a point (with the three weights as x, y and z coordinates) in a 3D space whereas the upper parts show which neurons are affected at a certain point in time. The weight vectors are modified in the way that they become more similar to the input vector of the pattern X. Therefore, the most similar weight vector (of the winner neuron 2 in part [b]) is modified strongest (see part [e]). The weight vectors of neuron 3 and all other dark-shaded neurons nearest to the so-called winner neuron 2 are modified less strong (parts [c, f]). The procedure continues so that weight vectors of neurons farther away (e.g. neuron 1 in parts [d, g]) are only pushed slightly to the weight vector of neuron 2.

Once the network has become tuned to the statistical regularities of the input data, it develops the ability to form internal representations for encoding features of the input and thereby generates classes automatically (Becker, 1991). Learning vector quantization (LVQ) is a supervised learning extension of the Kohonen network methods (Kohonen, 2001). Another form of unsupervised learning network is based on the adaptive resonance theory (ART) developed by Carpenter and Grossberg (1987). ART-1 Network, as introduced by Grossberg (Grossberg, 1976b; Grossberg, 1976a) is a self-organizing and self-stabilizing vector classifier where the environment modulates the learning process and thereby performs an unsupervised teaching role. Input vectors are added to existing feature clusters and are, therefore, able to learn new examples after the initial training has been completed. Finally, the approach of dynamically controlled networks (DyCoN) (Perl, 2000) allows

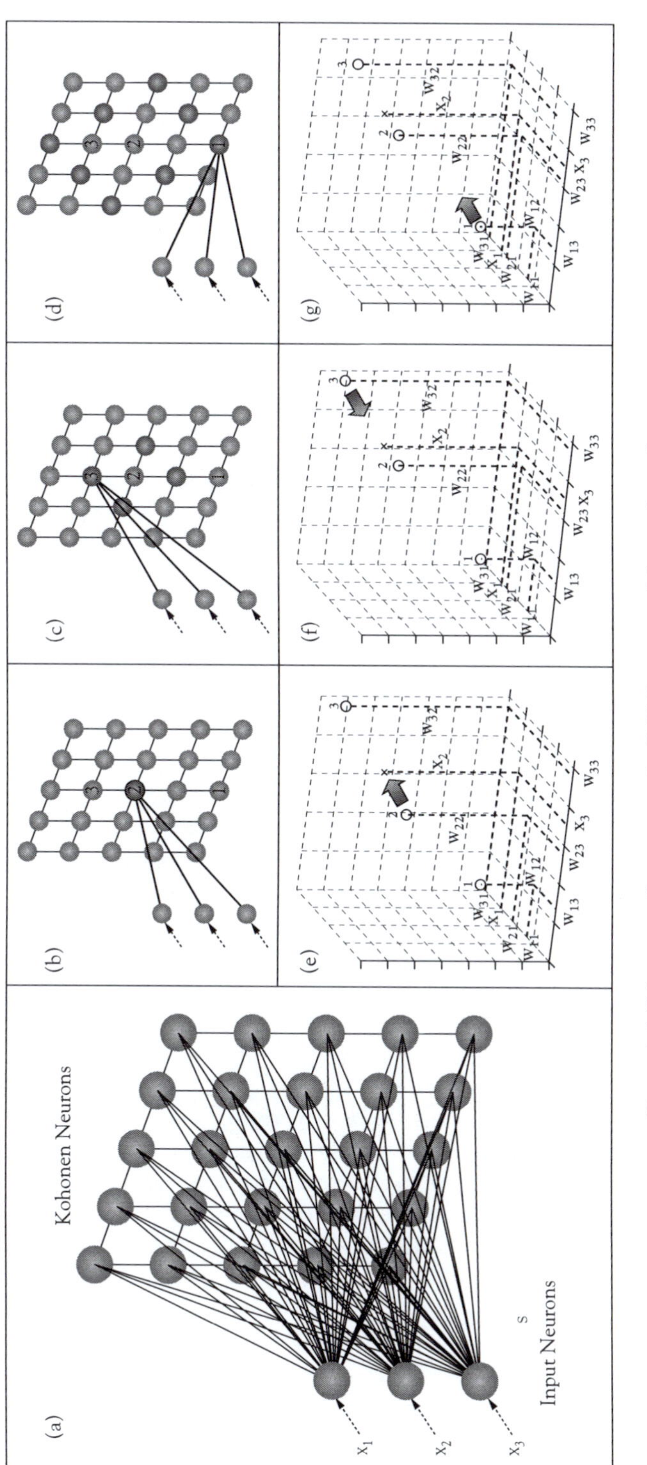

Figure 4.4 Schematic illustration of a Self-Organizing Map and its adjustments.

for dynamically self-adapting the training process to continuously changing situations. One advantage of DyCoNs is seen in their trainability. Once a SOM is trained it can be applied quite successfully for interpolation. If data occur that are outside the trained area the SOM has to be trained again from the beginning. In contrast to this a trained DyCoN can be trained again with new data sets without starting from the beginning. This provides an additional advantage, since it allows to simulate qualitatively learning and adaptation processes as well (Perl and Baca, 2003). Most recently, Dynamically Controlled Networks (Perl, 2002) are combined with Neural Gases (Martinetz and Schulten, 1991; Martinetz and Schulten, 1994; Fritzke, 1995) to DyConG-models (Memmert and Perl, 2005).

Applications of unsupervised ANN in sports

Self organizing ANNs are mainly used in order to reduce the dimensionality of high dimensional data sets, classifying these data sets, or comparing them with each other.

The majority of self-organizing ANN applications in sports are found in the area of classification of time course patterns. These applications seem to be appropriate for ANN because of their alternative point of view on complex phenomena. Usually several time series of intensities (for instance, movement patterns) are nonlinearly classified by their relative similarity.

By modelling time series dependent qualities two approaches can be distinguished (Schöllhorn and Bauer, 1995) (see schematic explanation in Figure 4.5).

In one approach (e.g, Bauer and Schöllhorn, 1997) the input vectors of the SOM are formed by the intensities of all variables at the same instant. These high dimensional input vectors are mapped by means of a SOM to two trajectories in a low dimensional feature map space. In Figure 4.5 the SOMs consist of 11 × 11 nodes, each node representing one state of the athlete at a certain instant. The trajectories represent the time series of an athlete's movement. In this low dimensional map the similarity of the two modelled movements are typically compared by measuring the distances between two trajectories. In a further step the structure of distances is either analyzed by means of cluster analysis techniques (Bauer *et al.*, 1997) or the trajectories are taken as input vectors for a second SOM (Barton, 1999). In both cases the movements or sports skills are clustered by their similarity and, therefore, illustrate models of movement techniques.

In the other approach the input vectors are constructed by the time courses of single variables. For the comparison of two movements the time series of variables are mapped on trajectories that are compared on similarity by their relative distances. The subsequent possibilities of data analyses are similar to the first.

Most intriguingly, both approaches provide the possibility to analyze quantitatively movement qualities that have been mostly neglected. Especially, for learning and developmental processes, where qualitative changes in the movements occur frequently, this approach seems to be promising. Furthermore this approach of movement modelling allows classifying: (a) movement classes like walking, jumping, running throwing; (b) modes of movement classes like springy or creepy walking; and (c) individual styles of movement classes and their modes. A transfer to game analysis seems to offer the possibility of analyzing tactical behaviour patterns quantitatively as well (Jäger, Peal and Schöllhorn 2007).

Analysis of movement techniques by means of SOMs

The application of self-organizing ANNs on sports movement analysis may consider the development of two high performance athletes during one year of training and competition (Bauer *et al.*, 1997). Kohonen SOMs were used to analyze 53 discus throws of two athletes,

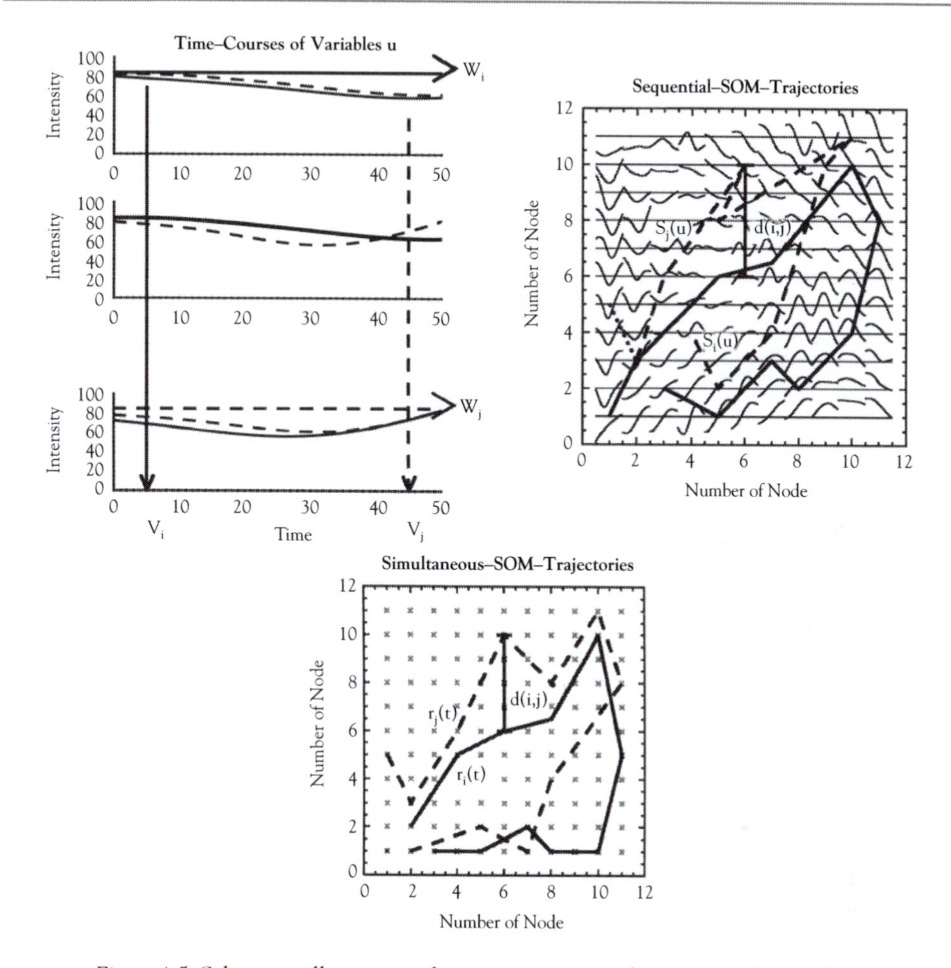

Figure 4.5 Schematic illustration of two process oriented structures of input data.

each based on 34 kinematic time series. The results revealed a qualitative learning progress for one athlete that was characterized by a disjoint separation of pre- and post-intervention trials. The analysis of the other athlete's training process showed daily dependant throwing techniques.

In a similar approach 51 javelin throws of 37 national and international world class athletes were analysed (Schöllhorn and Bauer, 1998). Smaller variations in the cluster of the national athletes in comparison to international athletes were held to contradict the existence of an 'ideal movement pattern'. According to the stable fingerprint-like throwing patterns of two athletes in this study, a similar stability of individual gait patterns with high heels can be found up to certain heights (Schöllhorn, Nigg, Stefanyshyn, and Liu, 2002).

Further evidence for the individuality of movement patterns in general and running patterns in particular was found by (Schöllhorn *et al.*, 1995). By analysing three to five double steps of 20 runners during a 2000m run it was possible to separate automatically contact phases of the right/left foot and the flight phase as well as to identify the athletes by means of their kinematical data during a single ground contact phase. Identification was based on a 91 per cent recognition rate for the left foot and 96 per cent for the right foot but only 75 per cent for the flight phase.

Lees, Barton, and Kershaw (2003) reported the results of a study that applied SOMs with 12 x 8 neurons to analyze instep kicks by two soccer players for maximum speed and accuracy. The output matrix showed dependencies on the task and players. In a similar approach, Lees and Barton (2005) modelled several kicks by six soccer players. Similar to Schöllhorn *et al.* (1995) the output maps distinguished well between right and left footed groups.

DyCoNs were applied to compare rowing techniques measured in a rowing boat and in a rowing simulator (Perl *et al.*, 2003). Results led to an identification of different movement patterns.

Partly similar modelling approaches in a closely related field of research (gait analysis) provide evidence for the dependence of movement patterns: (a) on the grade of muscular fatigue (Jäger, Alichmann and Schöllhorn, 2003); (b) on emotional states; and (c) even on music which the subjects are listening to (Janssen, Fölling and Schöllhorn, 2006; Janssen, Schöllhorn, Lubinefzki, Fölling, Kokenge and Davids 2008). In consequence these results offer a more differentiated type of movement diagnostics and build the basis for a more individualized movement therapy and training.

Eimert (1997) modelled the evaluation of shot put movements with a SOM on the basis of two different sets of variables. One set contained grades ranging from 1 (best) to 6 (worst), and the other set included variables of selected body angles that were assumed to be crucial for successful shot put movements. In both cases the data were acquired from the same videos of athletes. Physical education teachers evaluated the quality of the recorded shot put movements qualitatively on the basis of previously given criteria. The same criteria were quantified by means of intensities of kinematic variables at selected instants. Both data sets were provided to an 11 × 11 SOM. The clustering or neuron activation levels of both sets were compared with each other and showed fairly good coincidence. However, this study rather provides evidence for the quality of the ANN model than the superiority of any shot put technique.

Analysis of tactical behaviour in game sports by means of SOMs

An application of Kohonen feature maps in game sports is introduced by Perl and Lames (2000) in volleyball. They model the process of complete rallies. A rally is described as a sequence of coded key states of the game (= activities during a game), like serve – reception – set – attack – point. The classification separated game processes into classes that represent the characteristic structure of the game.

By analyzing the positions of squash players, i.e. from where they hit the ball, and coding them as a sequence of zones on the court, Perl (2002) transferred the modelling by means of ANNs from volleyball to racket sports. The results provided evidence for opponent specific game behaviour that was relatively stable against the same opponent but differed substantially against changing opponents.

In order to avoid the time-consuming coding of the human actions or their corresponding zones, Schöllhorn (2003) suggested several quantitative approaches for the analysis of game sports that are able to transfer the above mentioned movement analysis approach to game analysis. Therefore, the movements of the players should be video-recorded and described by their *x* and *y* coordinates on the court. For instance, in volleyball the players' movements are mapped to six (one team) or twelve (both teams) times two time-series with their *x*- and *y*-coordinates. These are taken as input variables, either for the training of a self-organizing map in order to automatically identify characteristic team moves, or for the training of an MLP in order to identify individual teams by their relative moves (Jäger, 2006; Jäger, et al 2007).

Model validation by estimating its accuracy

It is a key issue to validate the selected ANN model and to estimate its accuracy in classification tasks, since the classification error is the ultimate measure of a classifying ANN (Jain *et al.*, 2000). Accuracy in this case means the probability of correctly classifying a randomly selected instance or sample (Kohavi, 1995) and is closely connected to the performance of the classifier.

The most common methods for accuracy estimation are *cross-validation* and *bootstrap*. The *hold out* and *leave-one-out* methods may be considered as versions of the cross-validation (Jain *et al.*, 2000).

If a sample size of *n* is given the *hold out method* (test sample estimation) partitions the data into two mutually exclusive subsets called training set and test set (hold out set). Usually 2/3 of the data are assigned to the training set while the remaining 1/3 is used as the test set (Kohavi, 1995). The accuracy of the classified test set data is used as an estimator of the ANN's performance and – since the elements of the two subsets are chosen randomly – it is depending on the data in the training set. Sometimes a *random subsampling* is used and the hold out method is repeated a few times. The accuracy of the ANN in each repetition can then be used to calculate the mean accuracy and its standard deviation. However, one assumption is violated in the random subsampling method: The data of the training and the test set are not independent from each other and this may influence the estimation of the accuracy (Kohavi, 1995).

The *leave-one-out method* designs the classifying ANN using (n-1) samples of the data and evaluates the performance on the basis of the remaining sample (Jain *et al.*, 2000). This procedure, which is repeated *n*-times with different training sets of size (n-1), is a special case of the *k*-fold cross-validation that is sometimes called rotation estimation.

The *k-fold cross-validation* splits the whole data set into *k* mutually disjoint subsets, the so-called folds, which are usually of equal size. Then one-fold is excluded and the ANN is trained on the other folds. This procedure is repeated *k* times while every single fold is excluded once. The cross-validation estimate of accuracy is then the number of correct classifications, divided by the number of examples in the data set (Kohavi, 1995).

In some cases the folds are also *stratified*, i.e., they contain approximately the same proportions of patterns as the original data set. One disadvantage is connected to these cross-validation approaches, because not all the available *n* samples are used in the training phase of the classifier. Particularly when sample sizes are small, as in many studies, this can be a decisive problem. Therefore, bootstrapping has been proposed to overcome this limitation (Jain *et al.*, 2000). The bootstrap method re-samples the available patterns / samples in the data with replacement to generate a large number of the so-called bootstrap samples which are of the same size as the training set. These bootstrap samples are then used as new training sets in the learning phase of the classifying ANN.

Both approaches – the cross-validation and its versions as well as the bootstrapping – have their specific advantages and disadvantages in certain cases. However, some indications exist that a stratified ten-fold cross-validation is sufficiently appropriate and should be used for model selection (Kohavi, 1995).

Conclusion

Due to their characteristics artificial neural nets show a broad and versatile field of application in sport science. Nevertheless, careful handling and knowledge about the specificity of this kind of modelling tool is necessary for a responsible relation and interpretation of the results. According to Bartlett (2006) Kohonen mapping will become commonplace in sports biomechanics and human movement sciences, particularly if the technique elements captured by the mapping can be identified. Dynamically controlled networks will become more widely used in studying learning of movement patterns. Multi-layer ANNs will have an important role in technique analysis. According to applications in healthcare (Begg *et al.*, 2006), gait analysis (Schöllhorn *et al.*, 2002), clinical biomechanics (Schöllhorn, 2004) and other fields, it can be assumed that ANN applications have an enormous potential in sports sciences and biomechanics.

References

1. Bartlett, R. (2006) 'Artificial intelligence in sports biomechanics: new dawn or false hope?' *Journal of Sports Science and Medicine*, *5*: 474–79.
2. Barton, G. (1999) 'Interpretation of gait data using kohonen neural networks.' *Gait and Posture*, *10[1]*: 85–6.
3. Barton, J. G. and Lees, A. (1995) 'Development of a connectionist expert system to identify foot problems based on under-foot pressure patterns.' *Clinical Biomechanics*, *10*: 385–91.
4. Bauer, H.U. and Schöllhorn, W. I. (1997) 'Self-Organizing Maps for the Analysis of Complex Movement Patterns.' *Neural Processing Letters*, *5*: 193–99.
5. Becker, S. (1991) 'Unsupervised learning procedures for neural networks.' *International Journal of Neural Systems*, *2*: 17–33.
6. Begg, R. K., Kamruzzaman, J. and Sarker, R. (2006) '*Neural Networks in Healthcare*.' Hershey: Idea Group Publishing.
7. Bishop, C. M. (1995) '*Neural Networks for Pattern Recognition*.' Oxford: Oxford University Press.
8. Carpenter, G. and Grossberg, S. (1987) 'A massively parallel architecture for a self organising neural pattern recognition machine.' *Computer Vision, Graphics and Image Processing*, *37*: 54–115.
9. Chau, T. (2001) 'A review of analytical techniques for gait data. Part 2: neural network and wavelet methods.' *Gait and Posture*, *13*: 102–20.
10. Condon, E. M., Golden, B. L., and Wasil, E. A. (1999) 'Predicting the success of nations at the Summer Olympics using neural networks.' *Computers and Operations Research*, *26*: 1243–65.
11. Edelmann-Nusser, J., Hohmann, A., and Henneberg, B. (2002) 'Modeling and prediction of competitive performance in swimming upon neural networks.' *European Journal of Sport Science*, *2*.
12. Eimert, E. (1997) '*Beobachten und Klassifizieren von sportlichen Bewegungen mit neuronalen Netzen*. University of Tübingen.
13. Elman, J. L. (1990) 'Finding Structures in Time.' *Cognitive Science*, *14*: 179–211.
14. Fritzke, B. (1995) 'Growing Grid: A self organizing network with constant neighborhood range and adaptation strength.' *Neural Processing Letters*, *2*: 9–13.
15. Grossberg, S. (1976a) 'Adaptive pattern classification and universal recoding. II: feedback, expectation, olfaction, and illusions.' *Biological Cybernetics*, *23*: 187–202.
16. Grossberg, S. (1976b) 'Adaptive pattern classification and universal recoding: I. parallel development and coding of neural feature detectors.' *Biological Cybernetics*, *23*: 121–34.
17. Haykin, S. (1994) '*Neural Networks*.' New York: Macmillan College Publishing Company.
18. Hecht-Nielsen, R. (1991) '*Neurocomputing*.' Reading, Mass.: Addison-Wesley Publishing Company.

19. Herren, R., Sparti, A., Aminian, K., and Schutz, Y. (1999) 'The prediction of speed and incline in outdoor running in humans using accelerometry.' *Medicine and Science in Sports and Exercise, 31*: 1053–59.
20. Hertz, J., Krogh, A., and Palmer, R. G. (1994) '*Introduction to the theory of neural computation*' (9 ed.) (vol. 1). Reading: Addison-Wesley Publishing Company.
21. Hopfield, J. J. (1982) 'Neural Networks and Physical Systems with Emergent Collective Computational Abilities.' *Proceedings of the National Academy of Sciences USA, 79*: 2554–58.
22. Jäger, J. M. (2006) '*Mustererkennung im Sportspiel: eine topologisch-geometrische Analyse spieltaktischer Muster im Volleyball*. Marburg: Tectum.
23. Jäger, J. M., Alichmann, M., and Schöllhorn, W. (2003). 'Erkennung von Ermüdungszuständen anhand von Bodenreaktionskräften mittels neuronaler Netze.' In G. P. Brüggemann and G. Morey-Klapsing (Eds.) (pp. 179–183). Hamburg: Czwalina.
24. Jäger, J.M., Perl, J. and Schöllhorn, W. I. (2007). Analysis of players' configurations by means of artificial neural networks. *International Journal of Performance Analysis in Sport*, 7(3): 90–103.
25. Jain, A. K., Duin, R. P. W. and Mao, J. (2000). Statistical pattern recognition: a review. *IEEE Transactions on Pattern Analysis and Machine Intelligence*, 22(1): 4–37.
26. Janssen, D., Fölling, K., and Schöllhorn, W. I. (2006). 'Recognising emotions in biomechanical gait patterns with neural nets.' *Journal of Biomechanics, 39*: S118.
27. Janssen, D., Schöllhorn, W. I., Lubienefzki, J. Fölling, K., Kokenge, H., and Davids, K. (2008). Recognition of Emotions in Gait Patterns by Means of Artificial Neural Nets. *Journal of Nonverbal Behaviour*, (Online).
28. Jordan, M. I. (1986). 'Attractor dynamics and parallelism in a connectionist sequential machine.' In (pp. 531–546). NJ: Erlbaum.
29. Kohavi, R. (1995). A study of cross-validation and bootstrap for accuracy estimation and model selection. In C. S. Mellish (Ed.), *Proceedings of the Fourteenth International Joint Conference on Artificial Intelligence* (pp. 1137–1143). San Mateo, CA: Morgan Kaufmann.
30. Kohonen, T. (1982). 'Self-Organized Formation of Topological Correct Feature Maps.' *Biological Cybernetics, 43*: 59–69.
31. Kohonen, T. (2001). '*Self-Organizing Maps*.' (3 ed.) (vol. 30). Berlin: Springer.
32. Lees, A., Barton, B., and Kershaw, L. (2003) 'The use of Kohonen neural network analysis to establish characteristics of techniques in soccer.' *Journal of Sports Sciences, 21*: 243–44.
33. Lees, A. and Barton, G. (2005) 'A characterisation of technique in the soccer kick using a Kohonene neurla network analysis.' In T.Reilly, J. Cabri and D. Araujo (Eds.), *Science and Football V* (pp. 83–88). London UK: Routledge.
34. Maier, K. D. (2001) 'Neural Network Modelling studied on the example of shot-put flight.' In J. Mester, G. King, H. Strüder and E. Tsolakidis (Eds.) (pp. 332). Cologne: Sport und Buch Strauß.
35. Maier, K. D., Meier, P., Wagner, H. and Blickhan, R. (2000) 'Neural network modelling in sports biomechanics based on the example of shot-put flight.' In Y. Hong and D. P. Johns (Eds.) (pp. 550–553). Chinese University of Hong Kong: ISBS.
36. Maier, K. D., Wank, V. and Bartonietz, K. (1998) 'Simulation of the Flight Distances of Javelins based on a Neural Network Approach.' In H. J. Riehle and M. Vieten (Eds.) (pp. 363–366) Konstanz: Universitätsverlag Konstanz.
37. Maier, K. D., Wank, V., Bartonietz, K. and Blickhan, R. (2000) 'Neural network based models of javelin flight: prediction of flight distances and optimal release parameters.' *Sports Engineering, 3*: 57–63.
38. Martinetz, T. M. and Schulten, K. J. (1994) 'Topology representing networks.' *Neural Networks, 7*: 507–22.
39. Martinetz, T. M. and Schulten, K. J. (1991) 'A neural gas network learns topologies.' In T. Kohonen, K. Mäkisara, O. Simula and J. Kangas (Eds.), *Artificial Neural Networks* (pp. 397–402). Amsterdam: North Holland.
40. McCulloch, W. S. and Pitts, W. (1943) 'A logical calculus of the ideas immanent in nervous activity.' *Bulletin of Mathematical Biophysics, 5*: 113–15.

41. Memmert, D. and Perl, J. (2005) 'Game Intelligence analysis by means of a combination of variance-analysis and neural network.' *International Journal of Computer Science in Sport, 4*: 29–39.
42. Pandya, A. S. and Macy, R. P. (1995) *'Pattern Recognition with Neural Networks in C++.'* Boca Raton, Florida: CRC Press.
43. Perl, J. (2000) 'Artificial neural networks in Sports: New Concepts and Approaches.' *International Journal of Performance Analysis in Sport, 1*: 106–21.
44. Perl, J. (2002) 'Game analysis and control by means of continuously learning networks.' *International Journal of Performance Analysis of Sport, 2*: 21–35.
45. Perl, J. and Baca, A. (2003) 'Application of neural networks to analyze performance in sport.' In E. Müller, H. Schwameder, G. Zallinger and F. Fastenbauer (Eds.) (pp. 342). Salzburg: ECSS.
46. Perl, J. and Lames, M. (2000) 'Identifikation von Ballwechselverlaufstypen mit neuronalen Netzen am Beispiel Volleyball.' In W. Schmidt and A. Knollenberg (Eds.) (pp. 211–216) "Sport – Spiel – Forschung: Gestern. Heute. Morgen". Hamburg: Czwalina.
47. Ringwood, J. V. (1999) 'Anaerobic threshold measurement using dynamic neural network models.' *Computers in Biology and Medicine, 29*: 259–71.
48. Rosenblatt, F. (1957) *'The perceptron – a perceiving and recognizing automaton.'* (Rep. No. Technical Report 85–460–1). Ithaca, New York, Cornell Aeronautical Laboratory.
49. Rumelhart, D. E. and McClelland, J. L. (1986) *'Parallel Distributed Processing: Explorations in the Microstructure of Cognition.'* (vols. 1 and 2). Cambridge, Mass.: MIT Press.
50. Schöllhorn, W. I. (2004) 'Applications of artificial neural nets in clinical biomechanics.' *Clinical Biomechanics, 19*: 876–98.
51. Schöllhorn, W. I. (2003) 'Coordination Dynamics and its consequences on Sports.' *International Journal of Computer Science in Sport, 2*: 40–6.
52. Schöllhorn, W. I. and Bauer, H.U. (1995) 'Linear–nonlinear classification of complex time course patterns.' In J. Bangsbo, B. Saltin, H. Bonde, Y. Hellsten, B. Ibsen, M. Kjaer and G. Sjogaard (Eds.) (pp. 308–309). Copenhagen: University of Copenhagen.
53. Schöllhorn, W. I. and Bauer, H.U. (1998) 'Identifying individual movement styles in high performance sports by means of self-organizing Kohonen maps.' In H. J. Riehle and M. Vieten (Eds.), *Proceedings of the XVI ISBS 1998* (pp. 574–577). Konstanz: University Press.
54. Schöllhorn, W. I., Nigg, B. M., Stefanyshyn, D., and Liu, W. (2002) 'Identification of individual walking patterns using time discrete and time continuous data sets.' *Gait and Posture, 15*: 180–186.
55. Schöllhorn, W. I. and Perl, J. (2002) 'Prozessanalysen im Sport.' *Spectrum der Sportwissenschaften, 14*: 30–52.
56. Specht, D. F. (1990) 'Probabilistic neural networks.' *Neural Networks, 3*: 109–18.
57. Vainamo, K., Makikallio, T., Tulppo, M., and Röning, J. (1998) 'A Neuro-Fuzzy approach to aerobic fitness classification: a multistructure solution to the context sensitive feature selction problem.' In (pp. 797–802). Anchorage, Alaska, USA, May 4–9, 1998.
58. Vainamo, K., Nissila, S., Makikallio, T., Tulppo, M., and Röning, J. (1996) 'Artificial neural networks for aerobic fitness approximation.' In (pp. 1939–1949). Washington DC, June 3-6, 1996.
59. Watanabe, S. (1985) *'Pattern Recognition: human and mechanical'* (1st edn.). New York: John Wiley and Sons.
60. Yan, B. and Li, M. (2000) 'Shot Put Technique Analysis using ANNAMT Model.' In Y. Hong and D. Johns (Eds.) (pp. 580–584). Hong Kong SAR: The Chinese University of Hong Kong.
61. Yan, B. and Wu, X. (2000) 'The ANN-based analysis model of the sport Techniques.' In Y. Hong and D. Johns (Eds.) (pp. 585-589). Hong Kong SAR: The Chinese University of Hong Kong.
62. Zell, A. (1996)' *Simulation Neuronaler Netze.*' (1 ed.). Bonn; Reading, Mass.: Addison Wesley.

5

Biomechanical modelling and simulation of foot and ankle

Jason Tak-Man Cheung
University of Calgary, Calgary

Introduction

Many experimental techniques were developed to study ankle-foot biomechanics. Due to the lack of technology and invasive nature of experimental measurements, experimental studies were often restricted to study the plantar pressure and gross motion of the ankle-foot complex and the evaluation of internal bone and soft tissue movements and load distributions are rare. Many researchers have turned to the computational approach, such as the finite element (FE) method, in search of more biomechanical information. Continuous advancement in numerical techniques as well as computer technology has made the FE method a versatile and successful tool for biomechanical research due to its capability of modelling irregular geometrical structures, complex material properties, and complicated loading and boundary conditions in both static and dynamic analyses. Regarding the human foot and ankle, the FE approach allows the prediction of joint movement and load distribution between the foot and different supports, which offer additional information such as the internal stress and strain of the modeled structures. Although the FE method has been widely used in studying the intervertebral, shoulder, knee, and hip joints, the development of detailed FE foot model has just been sparked off in the late 1990s. In this chapter, the current establishments, limitations, and future directions of the FE modelling technique for the biomechanical research of the foot and ankle are discussed.

Finite element analysis of the foot and ankle

A number of two-dimensional (2D) and three-dimensional (3D) FE models have been built to explore the biomechanics of the ankle-foot structures. These FE analyses provided further insights into different clinical[1,3,7,10–13,20–24,25,27,28,38,41] and footwear[5,8,9,14–17,26,32,34,36,39] conditions. The model characteristics and applications of existing FE foot models are tabulated in Table 5.1.

Table 5.1 Configurations and applications of existing FE foot models in the literature

Years	*Authors*	*Geometrical properties*	*Material and loading conditions*	*Parameters of interest*
1981	Nakamura *et al.*	2D, engineering sketch (unified foot bones, plantar soft tissue, shoe sole)	Bones (linearly elastic), plantar tissue (nonlinearly elastic) Shoe sole (linearly/nonlinearly elastic) Ankle joint & Achilles tendon forces to simulate midstance	Shoe sole stiffness on stress in plantar soft tissue
1995 1995	Chu *et al.* Chu and Reddy	3D, engineering sketch (unified ankle-foot bones, encapsulated soft tissue, ankle-foot orthosis)	Bones, ligaments, soft tissue, orthosis (linearly elastic) Ground reaction, Achilles, flexor, extensor tendons forces to simulate heel strike & toe-off	Drop foot, stiffness of orthosis & soft tissue on stress distribution in ankle-foot orthosis
1997 2005	Lemmon *et al.* Erdemir *et al.*	2D, video image of specimen (metatarsal bone, encapsulated soft tissue, insole, midsole)	Bone (linearly elastic), encapsulated tissue, insole, midsole (hyperelastic) Contact simulation of foot-support interface Vertical load on metatarsal bone to simulate push-off	6 insole thicknesses, 2 tissue thicknesses, 36 plug designs of midsole (3 materials, 6 geometries, 2 locations of placement) on peak plantar pressure
1999 1999 2004	Jacob and Patil Jacob and Patil Thomas *et al.*	3D, X-rays of subject (foot bones, plantar soft tissue)	Bones, cartilages, ligaments, plantar tissue (linearly elastic) Ankle joint & muscular forces to simulate heel strike, midstance & push-off	Stress distribution in foot bones & soft tissues with dysfunction of muscles, reduction in cartilage thickness, plantar soft tissue stiffening, varying plantar tissue thickness & hardness
2000	Giddings *et al.*	2D, CT image of subject (foot bones)	Bones, cartilages, ligaments, tendon (linearly elastic) Contact simulation of major joints Distributed ground reaction forces at foot bones to simulate 6 stance phases	Load distribution of foot joints & ligaments during stance phases of gait
2000	Kitagawa *et al.*	3D, CT images of cadaveric foot (ankle-foot bones)	Bones, ligaments, plantar tissue (linearly elastic) Contact simulation of major joints & plantar support Static & impact compression on foot with & without Achilles tendon load	Force & deformation of foot under static & dynamic compression load
2000 2001 2002	Gefen *et al.* Gefen Gefen	3D, MR images of subject (foot bones, plantar soft tissue)	Bones, cartilages (linearly elastic) Ligaments, plantar tissue (nonlinearly elastic) Ankle joint & major muscular forces to simulate 6 stance phases of gait	Stress distribution in foot bones & soft tissues during stances with varying locations of ground reaction forces & magnitudes of muscles forces

2002 2003	Gefen Gefen	2D, MR images of subject (foot bones, plantar soft tissue)	Bones, cartilage (linearly elastic) Ligaments, plantar tissue (nonlinearly elastic) Ankle joint & Achilles tendon forces to simulate balanced standing	Partial and complete plantar fascia release Plantar soft tissue stiffening
2001	Bandak *et al.*	3D, CT images of cadaveric foot (ankle-foot bones, plantar soft tissue)	Bones, cartilages, plantar tissue (viscoelastic), ligaments (linearly elastic) Contact simulation of major joints & plantar support Vertical impact load at various initial velocities	Stress distribution in rearfoot bones & ankle ligaments under impact loading
2001 2003	Chen *et al.* Chen *et al.*	3D, CT images of subject (ankle-foot bones, encapsulated soft tissue)	Bones, cartilages, ligaments, encapsulated tissue (linearly elastic) Insole, midsole (hyperelastic) Contact simulation of plantar support Displacement control of plantar foot & support to simulate midstance to push-off	Plantar foot pressure & bone stress during stances Flat & total-contact insoles with different material combinations on plantar pressure distribution
2002	Camacho *et al.*	3D, CT images of cadaveric foot (ankle-foot bones, plantar soft tissue)		Relative positions of foot bones
2004	Verdejo & Mills	2D, engineering sketch (heel bone, heel pad, midsole)	Bone (linearly elastic), heel pad, midsole (hyperelastic) Contact simulation of heel-support interface Vertical deformation on plantar heel to simulate heel strike	Stress distribution in heel pad with & without midsole support
2004 2005 2005 2006 2006 2006 2007	Cheung *et al.* Cheung *et al.* Cheung and Zhang Cheung *et al.* Cheung *et al.* Cheung and Zhang Cheung and Zhang	3D, MR images of subject (ankle-foot bones, encapsulated soft tissue)	Bones, cartilages, ligaments (linearly elastic) Encapsulated tissue, foot orthosis (linearly elastic / hyperelastic) Contact simulation of most foot joints (except the toes) & plantar support Ground reaction & Achilles tendon or other extrinsic muscles tendon forces to simulate balanced standing / midstance	Stress distribution in foot bones & soft tissues with varying stiffness of plantar fascia, partial & complete fasciotomy, pathologically stiffened soft tissue & varying magnitudes of Achilles tendon load, posterior tibial tendon dysfunction, flat & custom-moulded foot orthosis with different combination of material stiffness, arch height & thickness on plantar pressure & bone stress distribution
2005	Spears *et al.*	3D, CT images of cadaveric foot (heel bone, heel pad tissue)	Heel bones (rigid), plantar heel pad (hyperelastic & viscoelastic) Contact simulation of heel support Vertical ground reaction force on heel support to simulate heel strike	Heel tissue stress with varying force, loading rates & foot-ground inclination

Continued

Table 5.1 Configurations and applications of existing FE foot models in the literature—cont'd

Years	*Authors*	*Geometrical properties*	*Material and loading conditions*	*Parameters of interest*
2006	Goske *et al.*	2D, MR image of subject (heel bone, heel pad tissue, heel counter, insole, midsole)	Bone (rigid), heel counter (linearly elastic) Heel pad, insole, midsole (hyperelastic) Contact simulation of foot-shoe interface Vertical load on heel bone to simulate heel strike	3 insole conformity levels, 3 different materials, 3 insole thicknesses on heel pressure distribution
2006	Erdemir *et al.*	2D, MR image of subject (heel bone, heel pad tissue)	Bone (rigid), heel pad (hyperelastic) Contact simulation of foot-support interface Vertical load on heel bone to simulate loading on heel	5 heel pad thickness & 40 subject-specific nonlinear tissue properties on peak heel pressure
2006	Actis *et al.*	2D, CT images from 6 subjects (foot bones, encapsulated soft tissue, insole, shoe sole)	Bones, cartilages, plantar fascia, flexor tendon (linearly elastic), encapsulated tissue, insole, shoe sole (nonlinearly elastic) Contact simulation of foot-support interface Vertical forces & moment on ground support to simulate push-off	Modulus of elasticity of bone, cartilage, fascia, flexor tendon & the use of total contact insert on peak metatarsal pressure in diabetic feet
2007	Wu	2D, MR & CT images (foot bones, encapsulated soft tissue, major plantar ligaments & muscles)	Bones, cartilages, plantar ligaments, fascia & tendons (linearly elastic), Muscles & cartilages within toes (hyperelastic) Contact simulation of foot-support interface Bodyweight & Achilles tendon forces to simulate balanced standing	Varying stiffness of passive intrinsic muscle tissue, plantar fasciotomy & plantar ligament injuries on tissue stress & planar pressure distributions.
2007	Spears *et al.*	2D, MR image of subject (heel bone, heel pad tissue, heel counter, insole, midsole)	Bone (rigid), heel counter (rigid / linearly elastic) Skin, fat pad, insole, midsole (hyperelastic) Contact simulation of foot-shoe interface Vertical load on heel bone to simulate standing	Heel counter on tissue stress distribution in skin & fat pad of heel

Two-dimensional finite element model

The first known FE analysis of the foot was reported by Nakamura *et al.*[34] in 1981, who developed a 2D FE foot model of a unified bony structure of the foot, plantar soft tissue and a shoe sole. A sensitivity analysis on the shoe sole material with varied Young's modulus (0.08 to 1000 MPa) suggested an optimum range of 0.1 to 1 MPa for stress reduction in the plantar soft tissue. A further simulation using a nonlinearly elastic foamed shoe sole showed similar responses in stress reduction as compared to the predictions with the optimized linearly elastic shoe material.

In 1997, Lemmon *et al.*[32] developed a 2D model of the second metatarsal bone and encapsulated soft tissue to investigate the metatarsal head pressure as a function of six insole thicknesses and two tissue thicknesses. The plantar soft tissue, polyurethane insole, and cloud crepe foamed midsole were defined as hyperelastic. Frictional contact between the foot and support was considered and a vertical load was applied at the metatarsal bone to simulate push-off. Orthosis with relatively soft material was found to reduce peak plantar pressure, which also decreased with an increase in insole thickness. The pressure reduction for a given increase of insole thickness was greater when plantar tissue layer was thinner. Using the same model, Erdemir *et al.*[17] investigated 36 plug designs of a Microcell Puff midsole including a combination of three materials (Microcell Puff Lite, Plastazote medium, Poron), six geometries (straight or tapered with different sizes), and two locations of placement. Plugs that were placed according to the most pressurized area were more effective in plantar pressure reduction than those positioned based on the bony prominences. Large plugs (40 mm width) made of Microcell Puff Lite or Plastazote Medium, placed at peak pressure sites, provided the largest peak pressure reductions of up to 28 per cent.

Several 2D full-length FE foot models in the sagittal section were reported. Giddings *et al.*[25] developed a linearly elastic model from CT image, which included the talus, calcaneus, fused midfoot and forefoot bony structures, major plantar ligaments and Achilles tendon. The interactions of the calcaneotalar and calcaneocuboid joints were defined as contacting surfaces. The measured ground reaction forces and major ankle moments were applied to simulate walking and running at speeds of 1.6 and 3.7 m/s, respectively. Maximum rearfoot joint forces were predicted at 70 per cent and 60 per cent of the stance phase during walking and running, respectively. The tensions on the Achilles tendon and plantar fascia increased with gait velocities in similar scales.

Gefen[21] built five 2D nonlinearly elastic models from MR images to represent the five rays of foot to study the effect of partial and complete plantar fascia release on arch deformations and soft tissue tensions during balanced standing. Simulation of partial and total release of the plantar fascia was done by gradually decreasing the fascia's thickness in intervals of 25 per cent. The vertical arch deformation during weightbearing increased from 0.3 mm to about 3 mm with complete fascia release. Removal of the plantar fascia increased about two to three times the tensions of long plantar ligament and up to 65 per cent of metatarsal stress. Gefen[23] further investigated the effects of soft tissue stiffening in the diabetic feet and predicted about 50 per cent increase in forefoot contact pressure of the standing foot with five times the stiffness of normal tissue. An increase in soft tissue stiffness increased the peak plantar pressure but with minimal effects on the bones.

Actis *et al.*[1] developed six subject-specific 2D FE models of the second and third metatarsal rays from CT images to study the plantar pressure distribution in diabetic feet during push-off. The FE model considered a unified rearfoot and midfoot structures, one metatarsal and three phalangeal bones connected by cartilaginous structures, flexor tendon, fascia, and encapsulated soft tissue. A total-contact plastizote insert with a rubber shoe sole was incorporated to study its effect on forefoot pressure. Subject-specific bulk soft tissue material properties were obtained and vertical load and moment were applied at the ground support. Frictionless contact interface between the foot-sole and shoe-ground surfaces was assumed. The sensitivity analysis on bone Young's modulus showed a minimal effect on pressure distribution while the corresponding effect of cartilage has a stronger influence. An increase in fascia and flexor tendon Young's modulus produced minimal effects on pressure distribution. Peak metatarsal pressure was reduced by about 46 per cent with the use of a total-contact insert. The authors suggested that incorporating the bony segments of the rearfoot, metatarsal, and toes, tendon and fascia with linear material properties, and surrounding bulk soft tissue with nonlinear material properties, are plausible simplified configurations for determining metatarsal head pressure distribution during push-off.

A relatively detailed 2D foot model in terms of soft tissue modelling was developed by Wu[41] using CT and MR images. The second and fifth rays were chosen to model the medial and lateral longitudinal foot arches, which included the cortical and trabecular bones, cartilages, fat pad, major plantar ligamentous structures and 11 associated muscles and tendons. Except the intrinsic muscle tissue and the cartilaginous structures between the toes were defined as hyperelastic, the rest of the modelled structures were assumed to be linearly elastic. Three different stiffness of intrinsic muscle tissue, plantar fasciotomy, and major plantar ligament injuries were analyzed in balanced standing. Fasciotomy increased the peak stresses in the bones, and shifted the maximum stresses to the second metatarsal and the long plantar ligament attachment area for more than 100 per cent. About 65 per cent increases in maximum strain of the long plantar ligament was predicted. With simulated plantar ligament injuries, the plantar fascia and flexor tendon sustained increased peak stresses of about 40 per cent and 30 per cent, respectively while the bony structures sustained similar stress increases as with fasciotomy. The predicted intrinsic muscle stresses in the intact foot was minimal. Increasing the passive tensions of intrinsic muscles in the injured foot decreased the stress levels to close to the intact foot but resulted in about 20 times increase in peak muscle stresses of the flexor digitorum brevis and abductor digiti minimi. It was speculated that fasciotomy may lead to metatarsal stress fractures, whereas, strengthening intrinsic muscle passive tensions may reduce the risk of developing plantar fasciitis and related forefoot pain and metatarsal fracture.

Several studies focused on the loading response of the plantar heel pad. Verdejo and Mills[39] developed a 2D hyperelastic FE model of the heel to study the stress distribution in the heel pad during barefoot running and with EVA foamed midsole. The heel pad had a higher order of nonlinearity but a lower initial stiffness than the foam material. The predicted peak bareheel plantar pressure was about two times the pressure during shod.

Using a hyperelastic heel model, Erdemir *et al.*[18] studied the effect of five heel pad thicknesses and the nonlinear stress-strain properties of 40 normal and diabetic subjects on the predicted peak heel pressure under vertical compression. The subject-specific heel pad tissue properties were calculated using a combined FE and ultrasound indentation technique to predict the stress-strain behaviour. The heel pad thickness and stiffness of diabetic subjects were not significantly different from normal subjects. Root mean square errors of up to 7 per cent were predicted by comparing the predicted peak plantar pressure obtained from the average and subject-specific hyperelastic material models during simulated heel weightbearing. Goske *et al.*[26] incorporated a shoe counter and sole into this heel model to investigate the effect of three insole conformity levels (flat, half conforming, full conforming), three insole thickness values (6.3, 9.5, 12.7 mm), and three insole materials (Poron Cushioning, Microcel Puff Lite and Microcel Puff) on pressure distribution during heelstrike. Conformity of insole was a more important design factor than insole material in terms of peak pressure reduction. The model predicted a 24 per cent reduction in peak plantar pressure compared to the barefoot condition using flat insoles while the pressure reduced up to 44 per cent for full conforming insoles. Increasing the insole thickness provided further pressure reduction.

A similar 2D heel model was developed by Spears *et al.*[37] to study the influence of heel counter on the stress distribution during standing. Considering a distinction between the material properties of fat pad and skin rather than a unified bulk soft tissue provided a better match with the measured barefoot plantar pressure. The predicted stresses in the skin were higher and predominantly tensile in nature, whereas the stress state in the fat pad was hydrostatic. Inclusion of a heel counter to the shod model resulted in an increase in compressive stress of up to 50 per cent and a reduction in skin tension and shear of up to 34 per cent and 28 per cent, respectively. The compressive and shear stresses in the fat pad reduced up to 40 per cent and 80 per cent, respectively, while minimal changes were found in tension. A properly fitted heel counter was suggested to be beneficial in terms of heel pad stress relief.

Three-dimensional finite element model

A number of 3D FE models, considering partial, simplified or geometrically detailed foot structures were reported. Chu *et al.*[14,15] developed a linearly elastic model with simplified geometrical features of the foot and ankle to study the loading response of ankle-foot orthosis. In heel strike and toe-off, peak compressive and tensile stresses concentrated at the heel and neck regions of the ankle-foot orthosis, respectively. The highly stressed neck region reflected the common site of orthoses breakdown. The peak compressive stress in the orthosis increased with increasing Achilles tendon force, whereas the peak tensile stress decreased with increasing stiffness of the ankle ligaments. The stress distribution in the orthosis was more sensitive to the stiffness of orthosis than that of the soft tissue.

Jacob and Patil[28] developed a linearly elastic foot model from X-rays, taking into consideration bones, cartilages, ligaments, and plantar soft tissue. Simulations of heelstrike, midstance and push-off were achieved by aligning the metatarsophalangeal joint angle and the associated plantar soft tissue and applying the ankle joint forces and predominant muscle forces at the points of insertion. The response of Hansen's disease with muscle paralysis was studied by reducing the thickness of the cartilages between the talus, navicular and the cuneiforms and by neglecting the action of peroneal and dorsiflexor muscles. Highest stress was predicted in the dorsal shaft of the lateral and medial metatarsals, the dorsal junction of the calcaneus and cuboid, and the plantar shaft of the lateral metatarsals during push-off. Stresses in the tarsal bones during push-off increased with simulated muscle paralysis. A further study on the effect of soft tissue stiffening on stress distribution in the diabetic foot[27] was done by representing the elastic moduli of normal and stiffened soft tissue as 1 MPa and 4 MPa, respectively. The maximum normal stresses of the plantar foot were higher with tissue stiffening and the forefoot increased with a larger extent compared to the heel. Soft tissue stiffening had a negligible effect on bone stress distribution. The FE model was further utilized to study the effect of plantar tissue thickness and hardness on the stress distribution of the plantar foot during push-off[38]. Comparing to the normal forefoot sole of thickness 13 mm, the predicted normal and shear stress of the foot sole increased more than 50 per cent in the diabetic forefoot sole of thickness 7.8 mm.

Gefen *et al.*[24] developed a nonlinearly elastic foot model, including cartilage and ligament connection for 17 bony elements and plantar soft tissue. Each toe was unified as a single unit and 38 major ligaments and the plantar fascia were incorporated. The ankle joint and major muscular forces were applied to simulate six stance phases. The model structures were adapted for each stance phase by alternating the inclination of the foot and alignment of the phalanges and plantar soft tissue. The model predicted the highest stress in the dorsal mid-metatarsals from midstance to toe-off. Other high stress regions were found at the posterior calcaneus. Gefen[20] further investigated the change in location of centre of pressure under the heel during foot placement with force reductions in tibialis anterior and extensor digitorum longus. Reductions of greater than 50 per cent in the tibialis anterior force caused a medialization of centre of pressure, which was thought to compromise stability and a possible cause of falling in the elderly. Gefen[22] also examined the changes in tissue deformations and stresses with muscle fatigue and suggested a possible cause of stress fractures in military recruits. It was found that a 40 per cent reduction in pretibial muscle force resulted in a 50 per cent increase in peak calcaneal stress, and that a similar force reduction in triceps surae force during push-off increased metatarsal stresses by 36 per cent.

A 3D linearly elastic ankle-foot model, consisting of the bony, encapsulated soft tissue and major plantar ligamentous structures, was developed by Chen *et al.*[6] using CT images to estimate the plantar foot pressure and bone stresses. The joint spaces of the metatarsophalangeal joints and ankle joint were connected with cartilaginous structures while the rest of the bony structures were merged. Frictional contact between the plantar foot and a rigid support was considered. The peak stress region was found to shift from the second metatarsal to the adjacent metatarsals from midstance to push-off. Chen *et al.*[5] further studied the efficiency for stress and plantar pressure reduction and redistribution using flat and total-contact insoles with different material combinations. Nonlinear elasticity for the insoles and the frictional contact interaction between the foot and support were considered. The predicted peak and average normal stresses were reduced in most plantar regions except the midfoot and hallux regions with total-contact insole. The percentage of pressure reduction by total-contact insole with different combination of material varied with plantar foot regions and the difference was minimal.

A 3D dynamic ankle-foot model was developed by Bandak *et al.*[3] to investigate the impact response. The model was constructed from CT images of cadaver specimen and the same specimens were subjected to impulsive axial impact loading for model validation. Distinction between the material stiffness of cortical and cancellous bone was taken into account for the tibia, fibula, calcaneus and talus, and these bones, together with the major ligaments, plantar soft tissue and cartilages were defined as linearly viscoelastic material. Relative bone movements were allowed in the rearfoot while the rest of the midfoot and forefoot bone were fused and modelled as rigid bodies. Forces and accelerations measured at the impactor were generally comparable to the FE predictions, especially at lower impact velocities. The predicted stress distributions showed localization of stress at the attachment sites of the anterior talofibular and deltoid ligaments, which were consistent with clinical reports of injuries sites and calcaneal fractures. A similar FE model was developed by Kitagawa *et al.*[29] to study the static and impact response of the foot under compression. They concluded from their experimental validation on cadavers that the ligamentous and tendon structures were important in achieving a realistic simulation of the foot response in both static and dynamic loading conditions.

Spears *et al.*[37] developed a viscoelastic FE model of the heel from CT images to quantify the stress distribution in heel tissue with different forces, loading rates, and angles of foot inclination. The material model was based on force-displacement data derived from *in vitro* experiments. The highest internal compressive stress was predicted in the heel pad inferior to the calcaneal tuberosity. During heelstrike, the internal stresses were generally higher than the plantar pressures. Increasing the loading rate caused a decrease in contact area and strains, and an increase in tissue stress with the internal stress increased with a greater extent than plantar pressure. The general stress levels were higher when the heel was loaded in an inclined position and a posterior shift of induced stress was predicted.

Recent efforts have been directed to the development of geometrically detailed ligamentous joint contact model of the foot and ankle. Camacho *et al.*[4] described a method for creating an anatomically detailed, 3D FE ankle-foot model, which included descriptions of bones, plantar soft tissue, ligaments and cartilages. The employed contact modelling technique permitted calculation of relative bone orientations based on the principal axes for individual bone, which were established from the inertia matrix. The developed model allowed predictions of stress/strain distribution and joint motions of the foot under different loading and boundary conditions.

Cheung *et al.*[11] developed a linearly elastic FE ankle-foot model from MR images, which took into consideration large deformations and interfacial slip/friction conditions, consisted of 28 bony structures, 72 ligaments and the plantar fascia embedded in a volume of encapsulated soft tissue. A sensitivity study was conducted on the Young's modulus (0–700 MPa) of plantar fascia, which was found to be an important stabilizing structure of the longitudinal arch during weight-bearing. Decreasing the stiffness of plantar fascia reduced the arch height and increased the strains of the long and short plantar and spring ligaments. Fasciotomy increased the strains of the plantar ligaments and intensified stress in the midfoot and metatarsal bones. Fasciotomy did not lead to the total collapse of foot arch even with additional dissection of the long plantar ligaments. The effect of partial and complete plantar fascia release was also investigated[7] and it was implicated that plantar fascia release may provide relief of focal stresses and associated heel pain. However, these surgical procedures, especially complete fascia release, were compromised by increased strains of the plantar ligaments and intensified stress in the midfoot and metatarsal bones and may lead to arch instability and associated midfoot syndromes.

Hyperelastic properties of the encapsulated soft tissue were incorporated into the same model to study the effect of pathologically stiffened tissue with increasing stages of diabetic neuropathy[13]. An increase in bulk soft tissue stiffness from 2 to 5 times the normal values led to a decrease in total contact area between the plantar foot and its support with pronounced increases in peak plantar pressure at the forefoot and rearfoot regions. The effect of bulk soft tissue stiffening on bone stress was minimal. The effect of Achilles tendon loading was also investigated using the same model,[12] which showed a positive correlation between Achilles tendon loading and plantar fascia tension. With the total ground reaction forces maintained at 350 N to represent half body weight, an increase in Achilles tendon load from (0–700 N) resulted in a general increase in total force and peak plantar pressure at the forefoot. There was a lateral and anterior shift of the centre of pressure and a reduction in arch height with an increasing Achilles tendon load. Achilles tendon forces of 75 per cent of the total weight on the foot provided the closest match of the measured centre of pressure of the subject during balanced standing. Both the weight on the foot and Achilles tendon loading resulted in an increase in plantar fascia tension, with the latter showing a two-times larger straining effect. Overstretching of the Achilles tendon, and tight Achilles tendon were suggested to be possible mechanical factors for overstraining of the plantar fascia and associated plantar fasciitis or heel pain.

Another study by Cheung and Zhang[10] utilized the same model to investigate the load distribution and spatial motion of the ankle-foot structure with posterior tibial tendon dysfunction on intact and flat-arched foot structures during midstance. The musculotendon forces for the achilles, tibialis posterior, flexor hallucis longus, flexor digitorum longus, peroneus brevis and peroneus longus were applied at their corresponding points of insertion while the ground reaction force was applied underneath the ground support. Cadaveric foot measurements were generally comparable to the FE predicted strains of the plantar fascia and major joint movements. Unloading the posterior tibial tendon increased the arch deformations and strains of the plantar ligaments, especially the spring ligament, while plantar fasciotomy had a larger straining effect on the short and long plantar ligaments. The arch-flattening effect of posterior tibial tendon dysfunction was smaller than that of fasciotomy. It was speculated that fasciotomy and posterior tibial tendon dysfunction may lead to attenuation of surrounding soft tissue structures and elongation of foot arch, resulting in a progressive acquired flatfoot deformity.

Cheung and Zhang[9] employed their linearly elastic model to study the interactions between the foot and orthosis. Custom-moulded shape was found to be a more important design factor in reducing peak plantar pressure than the stiffness of orthotic material. A further study considering the hyperelastic material properties of the encapsulated soft tissue and shoe sole was conducted to identify the sensitivity of five design factors (arch type, insole and midsole thickness, insole and midsole stiffness) of foot orthosis assigned with four different levels on peak plantar pressure relief[8]. The Taguchi experimental design method, which utilizes a fractional factorial design approach to assess the sensitivity of each design factor of a system and to determine its optimal quality level, was used. The custom-moulded shape was the most important design factor in reducing peak plantar pressure while the insole stiffness ranked the second most important factor. Other design factors, such as insole thickness, midsole stiffness and midsole thickness, contributed to less important roles in peak pressure reduction in the given order. The FE predictions suggested that custom pressure-relieving foot orthosis providing total-contact fit of the plantar foot is important in the prevention of diabetic foot ulceration. The cushioning insole layer of an orthosis should contribute to the majority of the thickness of the foot orthosis to maximize peak pressure reduction.

Dai *et al.*[16] used a simplified version of the FE model developed by Cheung *et al.*[11] to study the biomechanical effects of socks wearing during normal walking. The simplified model pertained the geometries of the foot bones and encapsulated soft tissue but the bony structures were merged by cartilaginous structures. A foot-sock-insole contact model was developed to investigate the effect of wearing socks with different combinations of frictional properties on the plantar foot contact. A nylon sock was defined as a 5×10^{-4} thick shell conforming to the foot soft tissue surface and assigned with orthotropic and linearly elastic material according to the wale and course directions. The dynamic responses from foot-flat to push-off during barefoot walking and with sock were simulated. Three different sock conditions were simulated: a barefoot with a high frictional coefficient against the insole (0.54) and two socks, one with a high frictional coefficient against the skin (0.54) and a low frictional coefficient against the insole (0.04) and another with an opposite frictional properties assignment. The FE model predicted that wearing sock with low friction against the foot skin was found to be more effective in reducing plantar shear force on the skin than the sock with low friction against the insole, which might reduce the risk of developing plantar shear related blisters and ulcers.

Limitations of existing finite element models

Many FE foot models were developed under certain simplifications and assumptions, including a simplified or partial foot shape, assumptions of linear material properties, and linear boundary conditions without considering friction and slip. Because of the high demanding accuracy for simulation of geometry, material behaviour, loading and boundary conditions for realistic biomechanical simulation of the human foot and ankle, improvements on certain categories are needed for the existing FE models in order to serve as an objective tool for clinical applications and footwear designs. In this section, the major drawbacks of existing FE models and potential advancements of the modelling techniques are discussed.

Geometrical properties

A number of FE models were built on a simplified or partial foot shape. For instance, some models are 2D in nature and incorporated only parts, symmetric structure of the foot in which the out of plane loading and joint movement cannot be accounted. Several 3D models considered only simplified foot bone structures representing the medial and lateral arch without differentiation of individual metatarsal bones in the medial-lateral directions. Meanwhile, fused tarsal and toe bones were often considered without allowing relative bone movements. Many foot models considered only the major ligaments and the plantar soft tissue, resulting in an inaccurate representation of the structural integrity and stiffness of the ankle-foot complex. As of 2006, the foot and ankle model developed by Cheung and colleagues (Figure 5.1) was the most geometrically detailed among the 3D FE models in the literature. However, realistic structural modelling of the hyaline cartilages, joint capsules, muscles, tendons, and ligaments, as well as differentiation of the skin and fat tissue layer has not been implemented. In addition, upper shoe structure has not been incorporated into the 3D ankle-foot model (Figure 5.2a) and the effects of motion control and pressure distribution by the shoe, insole and foot interactions cannot be addressed. Above all, knee and shank structures have not been included to study the coupling mechanism of the lower leg and foot (Figure 5.2b).

Material properties

The material properties were assumed to be homogeneous, isotropic and linearly elastic in some foot models. This is surely an approximated situation of the biological tissues, which exhibit non-homogeneous, anisotropic, nonlinear, and viscoelastic behaviour. In addition, the justification of some of the parameter selection was not provided and the values assumed were deviated from experimental observations. Several foot models have considered the use of nonlinear elasticity, hyperelasticity, and viscoelasticity in modelling the material behaviour of ligaments[3,20–24], plantar soft tissue[8,10,12,13,17,18,20–24,32,34], bony structures[3] and various orthotic material[5,8,17,32]. However, some of the assigned material parameters were not extracted from the ankle-foot structures and were originally from other parts of the body instead. Comprehensive site, direction and tissue dependent material models of the human ankle-foot structures should be obtained via experimental measurements on cadavers and live subjects. The mechanical properties of cartilage,[2] ankle ligaments,[19,35] plantar fascia,[30,40] plantar soft tissue[18,31–33] have been reported.

Loading and boundary conditions

A certain number of FE models considered only the ground reaction forces or vertical compression forces, with the stabilizing muscular forces ignored or lumped as a resulting ankle moment. Only a limited number of FE analyses of the foot considered the physiological loadings involving musculotendon forces[8,10,22,24,27,28,38]. In these models, muscles forces were approximated by normalized electromyographic data by assuming a constant muscle gain

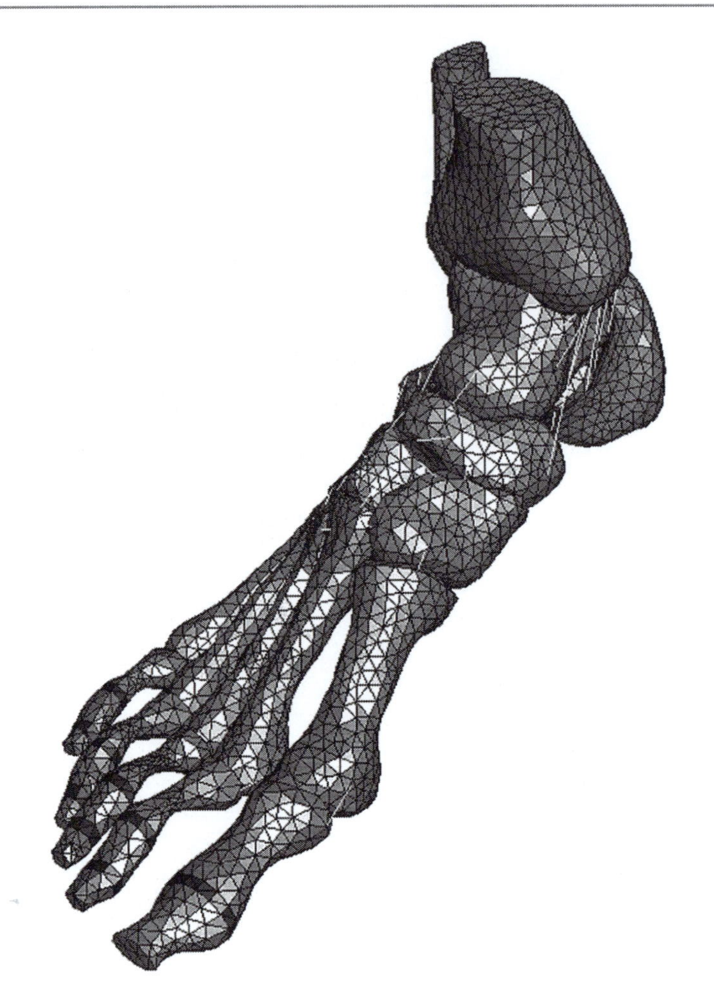

Figure 5.1 Finite element meshes of encapsulated soft tisuue, bony and ligamentous structures.

and cross-sectional area relationship and by force equilibrium consideration. Except for some vertical impact analyses, most models ignored the dynamic or inertia effects of ankle-foot motion. The quasi-static simulations may hinder the dynamic biomechanical behaviour of the foot whenever acceleration of the lower limb, such as during heel strike and push-off, is significant.

Finite joint and bone movements have not been accounted for in many FE models. The adjacent articulating surfaces were usually fused by connective tissue without considering the relative joint movement. Although relative articulating surfaces movements were allowed in some models,[3,25] these models consisted of only reduced number of distinct bony segments. Consequently, many of the existing models were limited in predictions of relatively small ankle-foot motions and should have resulted in a representation of an overly-stiffened ankle-foot structure. The FE model developed by Cheung and colleagues[7–13] considered the relative foot-support interface and joint motions in most foot joints except the phalangeal joints.

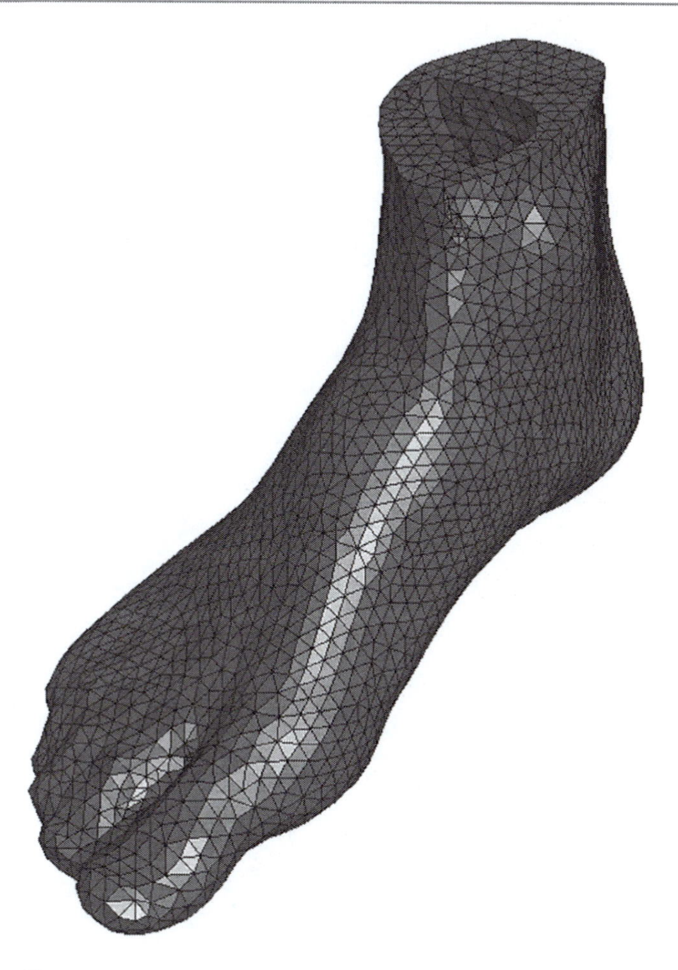

Figure 5.1 – Cont'd

To enable a more realistic simulation of the foot and its supporting interfaces, a contact modelling approach should be considered among all the bony segments, foot-ground and foot-shoe interfaces. In addition, ligament-bone and tendon-bone interfaces can be implemented to simulate more accurately the gliding and stabilizing mechanism of these soft tissues and their interactions with the bony structures.

Summary

Foot-related problems or diseases, such as plantar foot pain, diabetic foot, arthritic foot, pathological flatfoot, ankle sprain, bone fractures, or other sports-related injuries, have been costing a significant amount of medical expenditure and hospitalization. A clear picture of the internal stress/strain distribution and force transfer mechanism of the foot can provide fruitful information on ankle-foot pathomechanics and the biomechanical rationale and consequences behind different functional footwear designs and treatment plans. Existing FE

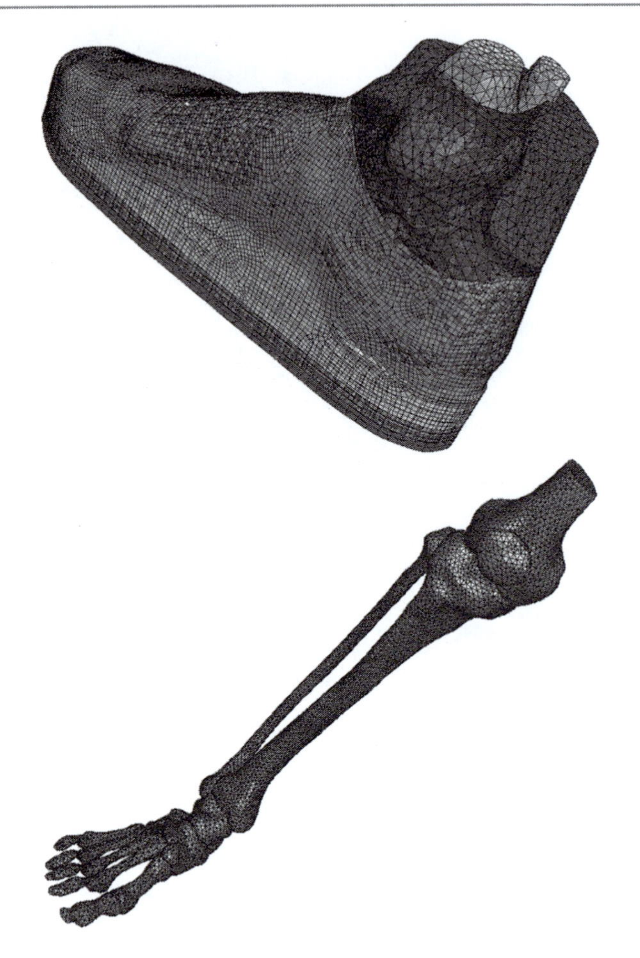

Figure 5.2 Finite element models of: (a) foot-shoe interface; and (b) knee-ankle-foot structures.

foot models have shown their contributions to the understanding of foot biomechanics and performance of different foot supports. The ability of the FE model to identify vulnerable skeletal and soft tissue components of the foot can serve as a tool for development of novel clinical decision-making and foot treatment approaches. In terms of orthotics or footwear design, it can allow efficient parametric evaluations for the outcomes of the shape modifications and other design parameters of the orthosis without the prerequisite of fabricated orthosis and replicating subject trials. Despite tremendous potentials for the existing FE models, more realistic geometry and material properties of both bony, soft tissue and footwear components, in addition to realistic physiological loadings, are required to provide a better representation of the foot and its supporting conditions and more accurate predictions of ankle-foot biomechanics.

Acknowledgement

The Croucher Foundation Fellowship of Hong Kong.

References

1. Actis, R.L., Ventura, L.B., Smith, K.E., Commean, P.K., Lott, D.J., Pilgram, T.K., Mueller, M.J. (2006) 'Numerical simulation of the plantar pressure distribution in the diabetic foot during the push-off stance'. *Med Biol Eng Comput*, 44:653–63.
2. Athanasiou, K.A., Liu, G.T., Lavery, L.A., Lanctot, D.R., Schenck, R.C. (1998) 'Biomechanical topography of human articular cartilage in the first metatarsophalangeal joint'. *Clin Orthop*, 348:269–81.
3. Bandak, F.A., Tannous, R.E., Toridis, T. (2001) 'On the development of an osseo-ligamentous finite element model of the human ankle joint'. *Int J Solids Struct*, 38:1681–97.
4. Camacho, D.L.A., Ledoux, W.R., Rohr, E.S., Sangeorzan, B.J., Ching, R.P. (2002) 'A three-dimensional, anatomically detailed foot model: A foundation for a finite element simulation and means of quantifying foot-bone position'. *J Rehabil Res Dev*, 39:401–10.
5. Chen, W.P., Ju, C.W., Tang, F.T. (2003) 'Effects of total contact insoles on the plantar stress redistribution: a finite element analysis'. *Clin Biomech*, 18:S17–24.
6. Chen, W.P., Tang, F.T., Ju, C.W. (2001) 'Stress distribution of the foot during mid-stance to push-off in barefoot gait: a 3-D finite element analysis'. *Clin Biomech*, 16:614–20.
7. Cheung, J.T., An, K.N., Zhang, M. (2006) 'Consequences of partial and total plantar fascia release: a finite element study'. *Foot Ankle Int.*, 27:125–32.
8. Cheung, J.T., Luximon, A., Zhang, M. (2007) 'Parametrical design of pressure-relieving foot orthoses using statistical–based finite element method', *Med Eng Phys*; in press.
9. Cheung, J.T, Zhang, M. (2005) 'A 3-dimensional finite element model of the human foot and ankle for insole design'. *Arch Phys Med Rehabil*, 86:353–58.
10. Cheung, J.T., Zhang, M. (2006) 'Finite element and cadaveric simulations of the muscular dysfunction of weightbearing foot' *HKIE Transactions*, 13:8–15.
11. Cheung, J.T., Zhang, M., An, K.N. (2004) 'Effects of plantar fascia stiffness on the biomechanical responses of the ankle-foot complex'. *Clin Biomech*, 19:839–46.
12. Cheung, J.T., Zhang, M., An, K.N. (2006) 'Effect of Achilles tendon loading on plantar fascia tension in the standing foot'. *Clin Biomech*, 21:194–203.
13. Cheung, J.T., Zhang, M., Leung, A.K., Fan, Y.B. (2005) 'Three-dimensional finite element analysis of the foot during standing a material sensitivity study'. *J Biomech*, 38:1045–54.
14. Chu, T.M., Reddy, N.P. (1995) 'Stress distribution in the ankle-foot orthosis used to correct pathological gait'. *J Rehabil Res Dev*, 32:349–60.
15. Chu, T.M., Reddy, N.P., Padovan, J. (1995) 'Three-dimensional finite element stress analysis of the polypropylene, ankle-foot orthosis: static analysis'. *Med Eng Phys*, 17:372–79.
16. Dai, X.Q., Li, Y., Zhang, M., Cheung, J.T. (2006) 'Effect of sock on biomechanical responses of foot during walking'. *Clin Biomech*, 21:314–21.
17. Erdemir, A., Saucerman, J.J., Lemmon, D., Loppnow, B., Turso, B., Ulbrecht, J.S., Cavanagh P.R. (2005) Local plantar pressure relief in therapeutic footwear: design guidelines from finite element models. *J Biomech*, 38:1798–1806.
18. Erdemir, A., Viveiros, M.L., Ulbrecht, J.S., Cavanagh, P.R. (2006) 'An inverse finite-element model of heel-pad indentation'. *J Biomech*, 39:1279–86.
19. Funk, J.R., Hall, G.W., Crandall, J.R., Pilkey, W.D. (2000) 'Linear and quasi-linear viscoelastic characterization of ankle ligaments'. *J Biomech Eng*, 122:15–22.
20. Gefen, A. (2001) 'Simulations of foot stability during gait characteristic of ankle dorsiflexor weakness in the elderly'. *IEEE Trans Neural Syst Rehabil Eng*, 9:333–37.
21. Gefen, A. (2002) 'Stress analysis of the standing foot following surgical plantar fascia release'. *J Biomech*, 35:629–37.
22. Gefen, A. (2002) 'Biomechanical analysis of fatigue-related foot injury mechanisms in athletes and recruits during intensive marching'. *Med Biol Eng Comput*, 40:302–10.
23. Gefen, A. (2003) 'Plantar soft tissue loading under the medial metatarsals in the standing diabetic foot'. *Med Eng Phys*, 25:491–99.

24. Gefen, A., Megido-Ravid, M., Itzchak, Y., Arcan, M. (2000) 'Biomechanical analysis of the three-dimensional foot structure during gait: a basic tool for clinical applications'. *J Biomech Eng*, 122:630–39.
25. Giddings, V.L., Beaupre, G.S., Whalen, R.T., Carter, D.R. (2000) 'Calcaneal loading during walking and running'. *Med Sci Sports Exerc*, 32:627–34.
26. Goske, S., Erdemir, A., Petre, M., Budhabhatti, S., Cavanagh, P.R. (2006) 'Reduction of plantar heel pressures: Insole design using finite element analysis'. *J Biomech*, 39:2363–70.
27. Jacob, S., Patil, M.K. (1999) 'Stress analysis in three-dimensional foot models of normal and diabetic neuropathy'. *Front Med Biol Eng*, 9:211–227.
28. Jacob, S., Patil, M.K. (1999) 'Three-dimensional foot modelling and analysis of stresses in normal and early stage Hansen's disease with muscle paralysis'. *J Rehabil Res Dev*, 36:252–63.
29. Kitagawa, Y., Ichikawa, H., King, A.I., Begeman, P.C. (2000) 'Development of a human ankle/foot model'. In: Kajzer J. Tanaka E. Yamada H. (Eds.), *Human Biomechanics and Injury Prevention*. Springer: Tokyo, 117–22.
30. Kitaoka, H.B., Luo, Z.P., Growney, E.S., Berglund, L.J., An, K.N. (1994) 'Material properties of the plantar aponeurosis'. *Foot Ankle Int*, 15:557–60.
31. Klaesner, J.W., Hastings, M.K., Zou, D., Lewis, C., Mueller, M.J. (2002) 'Plantar tissue stiffness in patients with diabetes mellitus and peripheral neuropathy'. *Arch Phys Med Rehabil*, 83:1796–1801.
32. Lemmon, D., Shiang, T.Y., Hashmi, A., Ulbrecht, J.S., Cavanagh, P.R. (1997) 'The effect of insoles in therapeutic footwear a finite element approach'. *J Biomech*, 30:615–20.
33. Miller-Young, J.E., Duncan, N.A., Baroud, G. (2002) 'Material properties of the human calcaneal fat pad in compression: experiment and theory'. *J Biomech*, 35:1523–31.
34. Nakamura, S., Crowninshield, R.D., Cooper, R.R. (1981) 'An analysis of soft tissue loading in the foot—a preliminary report'. *Bull Prosthet Res*, 18: 27–34.
35. Siegler, S., Block, J., Schneck, C.D. (1988) 'The mechanical characteristics of the collateral ligaments of the human ankle joint'. *Foot Ankle*, 8:234–42.
36. Spears, I.R., Miller-Young, J.E., Sharma, J., Ker, R.F., Smith, F.W. (2007) 'The potential influence of the heel counter on internal stress during static standing: A combined finite element and positional MRI investigation'. *J Biomech*; in press.
37. Spears, I.R., Miller-Young, J.E., Waters, M., Rome, K. (2005) 'The effect of loading conditions on stress in the barefooted heel pad'. *Med Sci Sports Exerc*, 37:1030–36.
38. Thomas, V.J., Patil, K.M., Radhakrishnan, S. (2004) 'Three-dimensional stress analysis for the mechanics of plantar ulcers in diabetic neuropathy'. *Med Biol Eng Comput*, 42:230–235.
39. Verdejo, R., Mills, N.J. (2004) 'Heel-shoe interactions and the durability of EVA foam running-shoe midsoles'. *J Biomech*, 37:1379–86.
40. Wright, D., Rennels, D. (1964) 'A study of the elastic properties of plantar fascia'. *J Bone Joint Surg Am*, 46:482–92.
41. Wu, L. (2007) 'Nonlinear finite element analysis for musculoskeletal biomechanics of medial and lateral plantar longitudinal arch of Virtual Chinese Human after plantar ligamentous structure failures'. *Clin Biomech*, 22:221–9.

Section II

Neuromuscular system and motor control

6

Muscle mechanics and neural control

Albert Gollhofer
Albert-Ludwigs-University of Freiburg, Freiburg

Introduction

The human skeletal muscle represents in a distinct manner an example of a direct relationship between a biological structure and its function. Muscular function is basically converting chemical energy into force. Force is necessary in order to compensate external loads acting on the human body and moreover, allows us to move our limbs and body. Structure elements in the muscle are responsible for the synthesis of protein, the storage of energy substrates and for the contraction of the muscle itself.

The first section focuses on the human skeletal muscle as the final executive organ for generating movement. The functional properties of muscles cannot be understood without deeper knowledge of the architectural features of the different muscles. Only basic mechanisms are described in order to give an insight into the human skeletal muscle and to understand how force is generated by this unique biological structure. The chapter contains three points which are subjected to the anatomy of skeletal muscle, its force-length and force-velocity relationship, respectively.

In the second section the neuromuscular aspects are addressed. For understanding human movement and for studying motor function, a thorough understanding of the mechanisms controlling the force output is essential. Two basic motor tasks, locomotion and balance, are described. Specific emphasis is given to the function of sensory-motor interaction in these types of movements. New technologies give insight into the modulation mechanisms of centrally and spinally organized influences on the activation of the motor neurons.

Muscle mechanics

Muscle anatomy

The largest organ in our body is the skeletal muscle system. Each of our muscles is specifically designed or architectured to meet the functional requirement for the adequate contribution in posture and human movement behaviour. The entire muscle is a composite of tightly packed substructures: muscle fibres, myofibrils and finally sarcomeres infolded in structural connective tissue have integral function in force production, force transmission and in energy substrate supply.

The diameters of muscle fibres (muscle cells) range from 10–100 micrometres up to a length of a few centimetres. The fluid part inside the muscle cell, the sarcoplasm substance, houses nutrients as primarily glycogen, fats, proteins and minerals for the metabolism of the muscle contraction. Extensions of the sarcolemma which serve as plasma membrane are known as transverse tubules (T tubules). The notation results from their alignment transverse to the long axis of the fibre with a radial inward orientation. T tubules build an extensive network going from the surface to the centre of the muscle cell. These structures transmit nerve impulses and nutritive substances through the muscle fibre. The second network of tubules is longitudinally arranged within the muscle fibre and is called sarcoplasmatic reticulum. Within this system a huge amount of calcium ions is stored. Once an action potential from the motor nerve depolarizes the postsynaptic membrane of the muscle these ions are released in order to initiate muscle contraction. Apart from the knowledge that these two parts of the muscle fibre are necessary for the generation of force, one special biological structure has been shown to be closely related to the fundamental function: the mechanical process of muscle action. This process takes place in the myofibrils which are the contractile elements of the skeletal muscle. Based on their function, the orientation of the hundreds to thousands of myofibrils in one muscle fibre follows straight the way towards the production of force. They mainly extend along the entire length of the muscle fibre. In a closer view, this biological structure appears as a series of subunits – the sarcomeres. The sarcomeres consist of thin and thick contractile elements – myosin and actin. Under the light microscope a characteristic sequence is easily observable, the striated pattern of dark and bright bands. This sequence contains five structures: an A-band, an H-band, an I-band, an M-band and the two Z-bands, which display the borders of each sarcomeres. Due to the differential refraction lights under microscopy the A (anisotropic) band shows the dark zone and represents the zone of overlapping thin and thick filaments. The H-band is located in the middle of the A-band. It compromises only thin filaments and is thus less dark than the A-band. The I-band was labelled in respect to its isotropic property. The thin filaments are attached to the Z-band and the thick filaments are directly connected to the M-band. Thereby, each thin filament stands in relationship with three thick filaments whereas one thick filament interplays with six thin filaments. A third fine filament composed of titin is a further component of the filament network that was suspected to stabilize the thick filaments in a longitudinal direction during both sequences of muscle function, in relaxation and during contraction. Moreover, a direct contribution to the elasticity of the entire myofibrils is apparently given due to its structural role. The highlighted special organization of the filament network displays a banded pattern which produces the characteristic appearance of muscle striation.

Muscle contraction

Within the sarcomeres generation of force is performed by the interaction between the thin and thick filaments. This process is an energy demanding procedure and classically described by the sliding filament theory: Both filaments slide against each other which causes a reduction of the sarcomere length. It has been shown that two proteins are substantially responsible for this fundamental process. Whereas actin is the major protein for the thin filaments, myosin is the one for the thick filaments. The actin filament is fixed at the

Z-band while the opposite end lies in the centre of the sarcomeres without inserting into any contractile structure. The structure of the actin filament is characterized by its helical pattern formed by strands of actin molecules. In this – pearl necklet – alignment each molecule of actin has a contact area for possible binding of the myosin head. Attached to the actin strand is tropomyosin in a twisted manner and troponin that is located in regular distance to the actin strands. The structural arrangement of the myosin filament contains two twisted heavy chains that terminate in two large globular heads. Each of them has two different sites of function in order to bind adenosine triphosphate (ATP) on the one site and actin on the other site. The contraction itself is initiated by the release of calcium ions in the sarcoplamatic reticulum. Ca^{2+} travels into the sarcoplasm and accumulates to the troponin. This binding procedure removes the tropomyosin from active sites of the actin filament. Now, the myosin heads are able to attach to the actin filament. This step represents the beginning of the cross-bridge cycle. The cross-bridge cycle can be further subdivided in four temporary sections. In the first section, the myosin head is weakly attached to the actin filament and carries an ATP bound at the ATPase enzyme site. In the second step, the myosin head is straightened by the distal part of the myosin head, while the ATP is cleaved in ADP and phosphate without releasing these energy substances, since they remain at the ATPase site. It is important to note that these two first stages are reversible. In the third section the myosin head is going to bind strongly to the actin filament. Concurrently, the lever arm of the myosin head is twisted to the original position. For this process the release of the phosphate is necessary. Due to the strong connection with the myosin head this process realises the shift. The actin filament is pulled towards the middle of the sarcomere. This tilting of the head which generates force is labelled as a powerstroke. Immediately after tilting, the myosin head detaches from the actin filament and rotates back to its original position. For this detaching process, as well as for the transfer of the calcium ions back to the sacroplasmic reticulum, ATP is required. The removal from calcium deactivates the troponin and tropomyosin and thus blocks the cross-bridging between the actin and myosin filaments.

Force-length relationship (for the muscle fibre)

The force capacity of one actin-myosin complex is fairly low: One million cross-bridges are able to produce a force of just one milliNewton! Therefore, the overall amount of force is obviously connected with the ability of the muscle fibre to provide a tremendous number of cross-bridges. In this context Gordon and co-workers stated that: '*tension varies with the amount of overlap between thin and thick filaments with the sarcomere*' (Gordon, Huxley and Julian, 1966). This relationship is illustrated in Figure 6.1A, which shows four different lengths and the corresponding forces. From a sarcomere length of approx. 3.6 μm onwards, no force can be generated because there is almost no overlap between actin and myosin filaments. At a sarcomere length of approx. 2.0–2.2 μm a maximum overlapping of the thin and thick filaments results in a maximal force generation. Sarcomeres shortening as shown on the ascending part of the curve declines force generation due to interference of the binding sites (< 2.0 μm) which is fortified by the thick filaments striking against the Z-bands (<1.7 μm). In conclusion, the typical force-length curve reflects the strength of the acto-mysin cross-bridges that are bound in the various lengths: In small muscle lengths the spacing of myosin and actin filaments is too narrow to allow the cross-bridge for much tension. However, in large lengths there is too little overlap between both filaments to allow for many attachments.

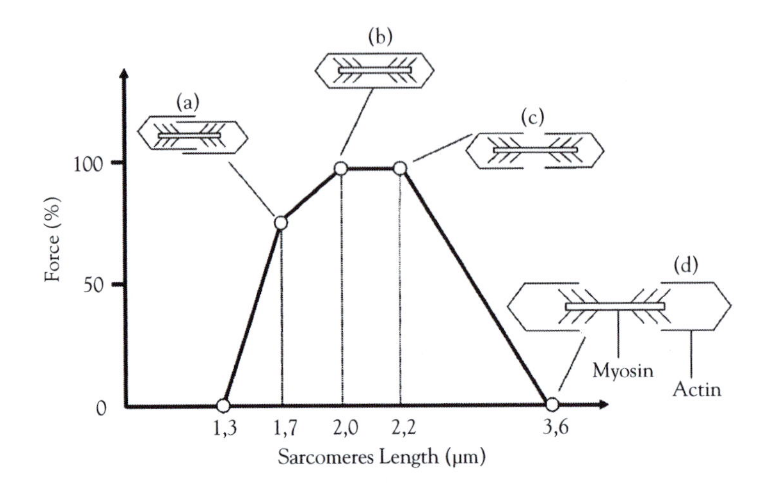

Figure 6.1.A The four different lengths and the corresponding forces.

Force-length relationship (for the entire muscle)

For the entire muscle tendon complex, the amount of force capacity is dependent on the number of sarcomeres in series within the muscle fibres, their pinnation angle and on the relative length of the entire configuration with respect to their resting length. Figure 6.1B schematically shows the active and the passive contributions to the entire muscle force as a function of length. In small muscle length the active component contributes exclusively to the entire force capacity. Here, the force of the muscle-tendon complex is determined by

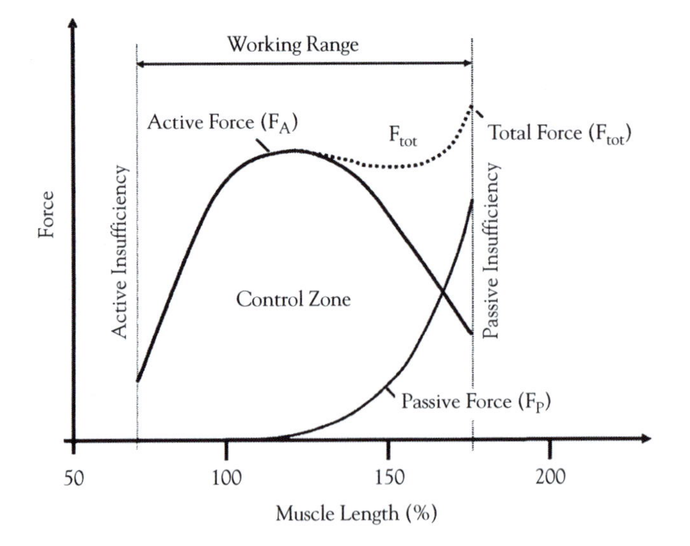

Figure 6.1.B Schematic illustration of the active and the passive contributions to the entire muscle force as a function of length.

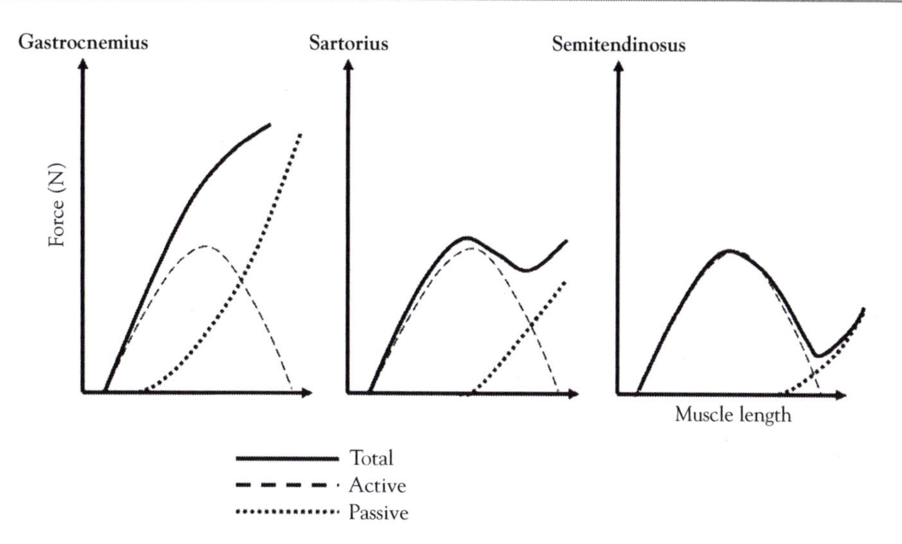

*Figure 6.1.*C Due to differences in muscle architecture the shape and the relative contribution of both components vary largely among different muscles.

the state of muscular activation. In large muscle length, however, the contribution of the passive components increases exponentially. Due to the viscoelastic properties of the connective tissue of the parallel elastic system (e.g. sarcolemn, endo- and perimysium) and of the tissue in series (e.g. tendon, aponeurosis), the resistant forces increase as the muscle-tendon complex is stretched extensively.

In conclusion, active force production by motor neuron recruitment and firing frequencies determine the amount of the contribution of the active component on each initial length. Active insufficiency can be observed in small lengths when the sarcomeres are contracted and spacing between the contractile proteins is very narrow, passive insufficiency is characteristic for large lengths when the sarcomeres are stretched and filament overlap is decreased (Elftman, 1966).

The sum of active and passive components describes the functional force-length relationship of a given muscle. However, due to the differences in muscle architecture, the shape and the relative contribution of both components varies largely among different muscles (Figure 6.1C). For functional interpretations of force-length behaviour in human movement it is important to note that the area underneath the sum of active and passive components describes the 'control zone' in which the neuronal system can 'determine' force output of the muscle tendon complex.

In multi-muscle systems like in M. quadriceps or in the hamstring muscles the muscle-tendon complex encompasses two or even more joints. Thus, under natural movements the force-length relation of each single muscle system may be compromised by the other muscles that may have completely different architectural properties. Moreover, in natural movements (e.g. locomotion) the distal joint complex (hip) may be flexed whereas the proximal joint (knee) extends. In consequence, the large variations in force capacity derived from the force-length relation of a single sarcomere are smoothed down by the various compensatory mechanisms under natural conditions. Therefore, for complex systems, like M. quadriceps, the force-length relation is rather constant within natural movement.

Force-velocity relationship

Apart from the force-length interaction the force-velocity relationship affects the amount of force. Figure 6.2 shows that the generation of force highly depends on the velocity of muscle contraction. In detail, the curve clearly demonstrates that the greater velocity of contraction, the lesser the generation of force. Within this curve three key points should be mentioned. The maximum generation of force is given under isometric condition. However, this state changes with increasing velocity of muscle contraction and results in a decrease of force (point 1). In contrast, when the velocity of contraction is near its maximum, the generation of force declines to a minimum (point 2). In the case of negative velocity, which happens during lengthening contractions (equivalent of 'eccentric' actions), the force can be increased up to 40–60 per cent above the maximum isometric force capacity (point 3). How can the phenomenon of the force-velocity relationship by the use of the cross-bridging cycle be explained? As aforementioned, the amount of force generation is based on the number of cross-bridges between actin and myosin filaments. Thus, at high shortening velocities the number is active cross-bridges limited and therefore the force output is decreased. The maximum contraction velocity is determined by the maximum cycling rate of the cross-bridges. As the rate of ATPase activity is closely related to the cycling rate, muscles with a high content of fast muscle fibres, with enormous enzymatic capacities of ATPase, have faster contraction velocities.

At low shortening velocities the ratio between changes in muscle length and changes in shortening velocity is more sensitive. Only small changes in shortening velocity produce large alterations in force capacity. Under isometric, static conditions ($v = 0$) the maximum voluntary contraction (MVC) capacity is achieved. Under eccentric actions some of the actually bound cross-bridges are stretched, leading to an additional force per cross-bridge. Furthermore, Lombardi and Piazzesi 1990 reported that cross-bridges will reattach more rapidly after they have been stretched.

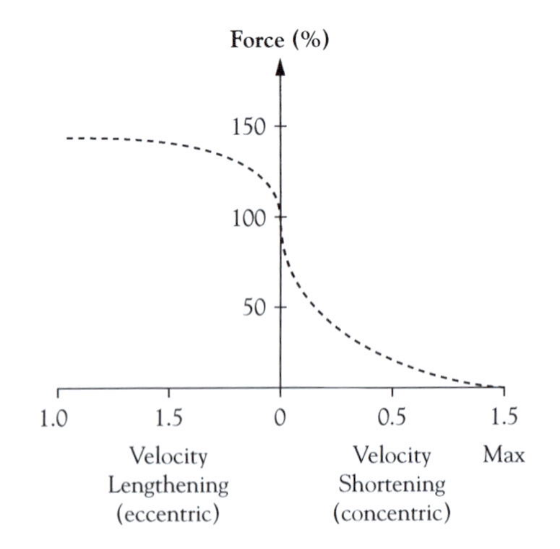

Figure 6.2 Force v velocity.

Neural control of skeletal muscle contraction

The motor unit

The relative distribution of slow and fast twitch motor units within a given muscle determines its force, power and endurance characteristics. In sprinters, for example, the gastrocnemius muscle may consist of 25 per cent slow motor units and 75 per cent fast motor units, whereas in marathon runners this relation may even be reversed. However, large differences in the relative distribution exist in the various muscles, due to their functional tasks. Intraindividually, the relative distribution is basically preset by genetic parameters.

All muscle fibres activated by one motor neuron have the same properties. They are called the motor unit. Thus, muscles that need power- and forceful action consist predominantly of fast motor units, whereas muscles with endurance and postural tasks are composed of a major portion of slow motor units. Another important parameter, the size of the motor unit, can differ largely with the functional properties of the muscles: In the eye muscles only few muscle fibres form a motor unit, whereas in the lower limb muscles a motor neuron may activate up to 2000 muscle fibres.

From an electrophysiological point of view, both types of motor units have different thresholds when recruited. According to the size-principle (Hennemann, 1965) slow motor units with their low thresholds are recruited first, whereas fast units with higher thresholds are involved later as force development is progressed. This principle ensures that, depending on the demands of a muscular action, a specific set of motor neurons is involved. Only in situations when full muscular actions are required (MVC) all the motor units are activated. In literature it has been shown that all muscle activations are related to this principle. However, due to the differences in fatigue capacities, a selective population of motor units may be observed when large forces must be sustained over long periods: Here the fast muscle units may quite soon be exhausted due to their specific enzyme profiles, whereas slow units with their high oxidative capacities may act over longer periods.

Once a motor unit is recruited the firing frequency provided by the nervous system determines its force output. In contrast to slow motor units, the fast units can be driven by a great increase in firing frequency leading to the fused tetanus. The enhancement of force output per motor unit can be graded by modulation of the excitation frequency. The range is much higher in the fast motor units. Recent research revealed that in fast, ballistic type of muscular actions the activation of some motor units occurs in doublets. These doublets have high efficacy for a high rate of force development (Van Cutsem *et al.*, 1998).

The sensory-motor interaction

The force-generating cross-bridge cycle is initiated by nerve impulses arriving from motoneurons located at the spinal cord. The arriving impulse at the terminal axon causes a release of acetylcholine (ACh) as a neurotransmitter substance into the synaptic cleft which is bound by the receptors of the muscle fibre sarcolemma. If the amount of bound ACh is above a certain threshold, the excitation depolarizes the muscle fibre membrane over the entire length and also enters into the tubules network inside the muscle cell. Calcium ions are poured out and the cross-bridge cycle is started. Because the motoneuron, together with the muscle fibre, builds up a functional unit it is considered as the motor unit (MU). However, coordinated human movement is never based on a single muscle that produces force. A complex sensory-motor interaction is needed to balance and optimize motion tasks around single joint systems. Therefore, every human movement has to be controlled by the central nervous system (CNS).

To understand how the nervous system controls movement, it is not sufficient to focus solely on the 'motor system' which activates the muscles but also on the 'sensory system'. Sensory input provides an internal representation of our body and the external surrounding and thus is essential to guide movement. Spindle afferents, for example, behave like passive stretch receptors, increasing their firing frequency during stretch and decreasing it during shortening. Therefore, stretch applied to one muscle is detected by muscle spindles which propagate the information via la-afferents to the CNS. When the excitation is sufficiently high to exceed a certain threshold at the alpha-motoneuron, an efferent volley is released which in turn leads to muscular contraction of the previously stretched muscle. This example shows two things: First, motor action directly depends on sensory information and second, movement initiation can emanate from the spinal level without making demands on higher centres. In this way, a variety of different motor actions can be processed by motor circuits in the spinal cord. Besides the possibility of achieving control of muscle length, for example to stabilize posture, the nervous system has many possibilities to use afferent input. Control mechanisms that adequately modulate the gain of postural reflexes should include peripheral information to describe the influence of gravity as well as inputs from muscle spindles and the vestibular system. Moreover, mechanoreceptors located in the in joint capsules, joint ligaments and skin provide sensory feedback. All these inputs form the characteristic and highly adaptable reflex pattern.

However, if the complexity of the movement increases, control of motor activity is shifted to higher centres like the brain stem or cortical regions. Locomotion and posture are two excellent examples to clarify the importance of sensory information and the contribution of spinal and supraspinal sources in central motor control.

Locomotion

Motor circuits in the spinal cord are not only relevant to execute stretch and withdrawal reflexes but also to enable natural locomotion. Animal studies provide strong evidence that the spinal cord contains the basic circuitry to produce locomotion. As early as 1911, Graham-Brown observed in the cat that coordinated flexor-extensor alternating movements could be generated in the absence of descending or afferent input to the spinal cord (Graham-Brown, 1911). The neural network in the spinal cord which has the capacity to produce this basic locomotor rhythm is called central pattern generator (CPG). The original half-centre model proposed by Graham-Brown consists of a flexor and an extensor half centre that individually possess no rhythmogenic ability, but which produce rhythmic output when reciprocally coupled. However, based on this model it is difficult to explain the diverse patterns which can be generated by spinal CPGs (Stein *et al.*, 1998; Burke *et al.*, 2001). To overcome this problem, it was proposed that multiple oscillators are flexibly coupled to create different patterns (Grillner, 1981). According to this model, spinal CPGs are able to realize many different motor behaviours like walking, swimming, hopping, flying and scratching. The basic pattern produced by a CPG is influenced by signals from other parts of the CNS and sensory information arising from peripheral receptors. This sensory feedback can help to increase the drive to the active motoneurons and is also needed for corrective responses which may be reflectively or voluntarily performed.

With respect to animal locomotion it can be concluded that descending signals from the brain stem are sufficient to activate the rhythmic generating network, which in turn activates the muscles in the correct way to obtain a coordinated gait pattern. Motor cortical lesions or damages of the corticospinal tract provoke only minor impairments of the locomotor system in the cat or the rat whereas humans suffer great functional deficits (Porter and Lemon, 1993). Thus, locomotion of humans obviously relies much more on supraspinal and especially cortical influences than animal locomotion does. For this reason it was difficult to obtain clear evidence for CPGs in humans. In subjects with complete spinal cord injury it was shown that electrical stimulation of the spinal cord elicited alternating activity in leg muscles (Dimitrijevic *et al.*, 1998). Furthermore, experiments in healthy humans revealed strong coupling of arm and leg movements during gait but not during standing or sitting (Dietz *et al.*, 2001). These and other observations suggest that cyclic activity in humans is also generated by a spinal CPG (Dietz and Harkema, 2004).

Obviously there are many similarities in the organization of coordinated locomotion patterns in humans and animals. However, there are also great differences. As mentioned above, humans more strongly rely on supraspinal influences than animals do. Recent studies provided accumulative evidence for cortical contribution during gait (Schubert *et al.*, 1997; Petersen *et al.*, 2001). Thereby, direct corticospinal projections descending from the primary motor cortex to the spinal cord seem to play an important role (Petersen *et al.*, 1998). The existence of such monosynaptic projections make up one of the major differences in the organization of the motor system in primates compared to other species. As these direct corticospinal pathways are most pronounced to distal arm and finger muscles they are believed to be responsible for the high level of skilled motor control in primates (Porter and Lemon, 1993). The assignment of direct corticospinal projections is, however, not restricted to the performance of fine motor skills of the upper extremity. Activity in lower limb muscles during locomotion and postural regulation is also influenced by their input.

Postural regulation

Similarly to the organization of locomotion, postural regulation relies on the integration of afferent information to allow for appropriate muscular activation. Sensory input in order to maintain balance is mainly provided by the proprioceptive, the vestibular and the visual system. Depending on the type of balance task, the relative importance of these sources is differently weighted. For example, the compensatory responses following slow rotational disturbances of the feet are primarily triggered by vestibular information (Allum and Pfaltz, 1985), whereas fast horizontal displacements are predominantly processed via spinal proprioceptive pathways (Dietz *et al.*, 1988). Regarding the organization of unperturbed stance, it was shown that despite excluding vestibular, visual and cutaneous information, equilibrium could still be maintained (Fitzpatrick *et al.*, 1994). Afferents from the lower limb therefore seem to provide sufficient information to control quiet stance. After the Ia afferents were considered to be most relevant to enable upright stance, nowadays there is strong evidence that type II afferents are even more important (Nardone and Schieppati, 2004).

The sensory information obtained by the proprioceptive, vestibular and visual system is processed at spinal and supraspinal levels to ensure appropriate muscular activation in order to maintain balance. Commonly, it is assumed that postural control is restricted to subcortical centres and reflex action. However, based on results obtained in recent studies using transcranial magnetic stimulation, it is supposed that the primary motor cortex is involved at least in some postural tasks requiring fast compensatory responses (Taube *et al.*, 2006) and might even be of relevance for the control of quiet stance (Soto *et al.*, 2006). These results indicate that, in humans, not only muscular activation during skilful movements is controlled by the cortex but also locomotion and postural regulation.

References

1. Allum, J. H., Pfaltz, C.R. (1985) 'Visual and vestibular contributions to pitch sway stabilization in the ankle muscles of normals and patients with bilateral peripheral vestibular deficits'. *Exp Brain Res*, 58: 82–94.
2. Burke, R. E., Degtyarenko, A.M., Simon, E.S. (2001) 'Patterns of locomotor drive to motoneurons and last-order interneurons: clues to the structure of the CPG'. *J Neurophysiol*, 86: 447–462.
3. Dietz, V., Fouad, K., Bastiaanse, C.M. (2001) 'Neuronal coordination of arm and leg movements during human locomotion'. *Eur J Neurosci*, 14: 1906–1914.
4. Dietz, V., Harkema, S.J. (2004) 'Locomotor activity in spinal cord-injured persons'. *J Appl Physiol*, 96: 1954–1960.
5. Dietz, V., Horstmann, G., Berger, W. (1988) 'Involvement of different receptors in the regulation of human posture'. *Neurosci Lett*, 94: 82–87.
6. Dimitrijevic, M.R., Gerasimenko, Y., Pinter, M.M. (1998) 'Evidence for a spinal central pattern generator in humans'. *Ann N Y Acad Sci*, 860: 360–376.
7. Elftmann, H. (1966). 'Biomechanics of muscle'. *Journal of Bone and Joint Surgery*, 48A(2): 363–377.
8. Fitzpatrick, R., Rogers, D.K., McCloskey, D.I. (1994) 'Stable human standing with lower-limb muscle afferents providing the only sensory input'. *J Physiol*, 480(2): 395–403.
9. Gordon, A.M., Homsher, E., Regnier, E. (2000) 'Regulation of contraction in striated muscle'. *Physiological Reviews*, 80: 853–924.
10. Graham-Brown, T. (1911) 'The intrinisic factors in the act of progression in the mammal'. *Proc R Soc Lond B Biol Sci*, 84: 308–319.
11. Grillner, S. (1981) 'Control of locomotion in bipeds, tetrapods and fish'. In: Brooks, V.B. (ed) *Handbook of Physiology*: Section 1, The Nervous System II. Motor control. American Physiological Society, Waverly Press, Bethesda, pp. 1179–1236.
12. Lombardi and Piazzesi (1990) in Enoka-Buch.
13. Nardone, A., Schieppati, M. (2004) 'Group II spindle fibres and afferent control of stance. Clues from diabetic neuropathy'. *Clin Neurophysiol*, 115: 779–789.
14. Petersen, N., Christensen, L.O.D., Nielsen, J.B. (1998) 'The effect of transcranial magnetic stimulation on the soleus H reflex during human walking'. *Journal of Physiology*, 513: 599–610.
15. Petersen, N.T., Butler, J.E., Marchand-Pauvert, V., Fisher, R., Ledebt, A., Pyndt, H.S., Hansen, N.L., Nielsen, J.B. (2001) 'Suppression of EMG activity by transcranial magnetic stimulation in human subjects during walking'. *J Physiol*, 537: 651–656.
16. Porter, R. and Lemon, R. N. (1993) 'Corticospinal function and voluntary movement'. Monographs of the Physiological Society, No. 45, New York, Oxford University Press.
17. Schubert, M., Curt, A., Jensen, L., Dietz, V. (1997) 'Corticospinal input in human gait: Modulation of magnetically evoked motor responses'. *Experimental Brain Research*, 115: 234–246.
18. Soto, O., Valls-Sole, J., Shanahan, P., Rothwell, J. (2006) 'Reduction of intracortical inhibition in soleus muscle during postural activity'. *J Neurophysiol*, 96: 1711–1717
19. Stein, P. S., McCullough, M.L., Currie, S.N. (1998) 'Spinal motor patterns in the turtle'. *Ann N Y Acad Sci*, 860: 142–154.
20. Taube, W., Schubert, M., Gruber, M., Beck, S., Faist, M., Gollhofer, A. (2006) 'Direct corticospinal pathways contribute to neuromuscular control of perturbed stance'. *J Appl Physiol*, 101: 420–429.
21. Van Cutsem, M., Duchateau, J., Heinault, K. (1998) 'Changes in single motor unit behaviour contribute to the increase in contraction speed after dynamic training in humans'. *Journal of Physiology*, 315: 295–305.

7

The amount and structure of human movement variability

Karl M. Newell and Eric G. James
The Pennsylvania State University, Pennsylvania

Introduction

Variability is a hallmark property of biological systems within and between species. In many disciplines variability has been a fundamental source of information for the development of theory but the importance of variability has been less central in human biomechanics and motor control research. Generally, the field of human movement has focused on the invariance of movement properties as opposed to its variance for the primary source of information about the organization of motor control.

The fields of biomechanics and motor control have begun to systematically investigate the variability of movement with the view that it is much more than just the reporting of a standard deviation of a movement-related variable (cf. Davids, Bennett, and Newell, 2006; Newell and Corcos, 1993). Biomechanics with its basis in Newtonian mechanics has traditionally given emphasis to the deterministic aspects of movement with little formal involvement or even mention of stochastic processes (e.g., Özkaya and Nordin, 1991). Motor control by contrast has long recognized movement variability and stochastic processes but traditionally assumed that the variability was no more than uncorrelated random noise (Fitts, 1954; Schmidt *et al.*, 1979), and a nuisance to be minimized or eliminated in system control.

The introduction of the construct of self-organization to biological processes (Glass and Mackey, 1988; Yates, 1987), together with the accompanying tools and methods of dynamical systems and chaos theory (Kaplan and Glass, 1995; Thompson and Stewart, 1986), have provided a new framework to consider a variety of biological phenomena, including human movement, the cooperative and competitive influences of deterministic and stochastic processes. The theoretical and experimental influence of this approach has been growing in the study of complex biological systems, sometimes under the umbrella label of fractal physiology (Bassingthwaighte, Liebovitch and West, 1994; West, 2006). The human in movement and action is clearly a complex system of systems the control of which also needs to include the contributing roles of the environment and task in adaptively harnessing deterministic and stochastic processes.

The complexity of human movement is manifest in the consideration of Bernstein's (1967) degrees of freedom problem in motor control: namely, that skill acquisition is the mastery of the redundant DoF. This mastery resides in exploiting the deterministic and stochastic forces present in system control and the execution of a movement sequence that exhibits movement variability of a variety of kinds while still realizing the goal of the action. Figure 7.1 shows Bernstein's classic example of the variability present in the repetitive hammering of skilled Italian stone workers. The figure captures the essence of Bernstein's insightful notion of practice, even in the highly skilled performer, as repetition without repetition.

This chapter focuses on one central issue in the within-subject variability of the repetition of movement and maintenance of posture: namely, the relation between the amount of variability and the time- and frequency-dependent structure of variability. We consider this issue across the task categories of posture, locomotion and manipulation together with the influence of lifespan development, learning and disease states on variability. The amount and structure of movement variability can be considered at the level of the individual effectors, the coupling between effectors and the outcome dimension of the task goal.

Amount and structure of movement variability

Figure 7.1 shows that there is a movement form or an invariance to the action that is present from repetition to repetition or trial to trial even though each cycle of hammering has some variability in details of the kinematic trajectories. This invariance can be used to categorize the stability of the action and the perceptual labelling of a given movement activity class, such as perceiving an action to be that of walking, running, hopping or skating. In contrast, the movement variance when expressed in terms of kinematics is referring to variability in the either the spatial or temporal dimensions of some aspect of the trajectory(ies) of movement over repetitions of the movement or maintenance over time of posture.

The variability is usually expressed as an amount and measured as the standard deviation of the distribution of like measures from the repetition of performance. For example, in Figure 7.1 one could measure the variability of the peak spatial location of the trajectory of the hammer across the cycles of hammering. In addition, one could also measure the variability of the time to peak spatial location in the limb trajectory. In either case, the point measure of space or time in the movement cycle is taken with respect to a time or place, respectively. These measures could be made in the framework of either an absolute extrinsic frame of reference or an intrinsic body relative reference.

The standard deviation of a distribution may be sufficient to reflect the dispersion or variability of the scores when the distribution is normal but the distribution of movement properties often departs from normality when scaled, in particular, to extreme spatial and temporal demands in an action. In these cases, skewness and kurtosis measures of the distribution can be useful to capture a fuller appreciation of the variability (Kim, Carlton, Liu and Newell, 1999), though these higher order moments of the distribution tend to be less stable. One can also measure directly the entropy (an index of disorder) of a distribution through the probability assumptions of information theory (Shannon and Weaver, 1949).

The measures of the amount of variability of some discrete spatial-temporal point of a movement property, trajectory or its outcome can be contrasted with measures of the time- or frequency-dependent properties of the movement time series that capture the structure of the variability. Thus, the repetitions of the spatial-temporal trajectories of hammering in Figure 7.1 could also be analyzed as a position time series or through frequency analysis procedures. These approaches provide clues as to the structure of variability in that they reveal indices of the time or frequency domain properties of the signal. The application of this approach to movement analysis has clearly shown that the variability of the motor output across a variety of tasks is typically not that of white Gaussian noise (Newell and Slifkin, 1998; Newell *et al.*, 2006; Riley and Turvey, 2002).

A central finding to emerge from recent research on the variability of movement is that the variability reflects considerable time- and frequency-dependent structure. Of special significance to this chapter is the emerging view of an apparent inverse relation between the amount (standard deviation) and the degree of structure in movement variability. Specifically, that the reduction in the amount of outcome variability in task space or the coupling of effectors *tends* to be related to a higher dimension or complexity of the motor output while in contrast an enhanced amount of variability is related to a reduction in the dimension or complexity of the output. This inverse relation between the amount and structure of movement variability has been shown in a wide range of motor tasks, but there are also important examples of exceptions to this relation that provide insights for a more broad based coherent account of the organization of movement variability.

Posture

The study of postural variability has been dominated by the experimental protocol where the subject stands on a force platform as still as possible and looks ahead to a marker at eye height.

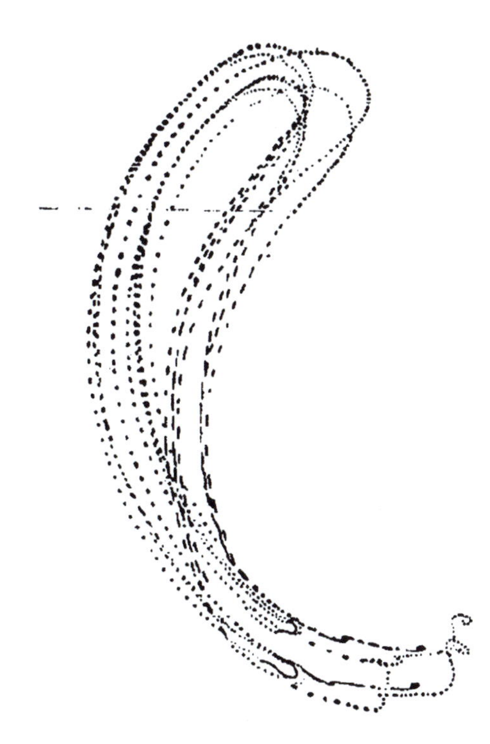

Figure 7.1 Cyclogram of a series of hammering trajectories (adapted from Bernstein, 1967, Figure 21).

The platform records the force and moments in the *x*, *y*, and *z* planes at the surface of support from which the centre of pressure is derived. The centre of pressure time series provides the basis to estimate the amount of variability in the anterior and posterior directions of motion, together with the derivation of a number of other kinematic properties. The prototypical measure of postural variability has been the standard deviation of the location of the centre of pressure in the medial-lateral or anterior-posterior dimension. Using this measure a host of studies of the centre of pressure profile of upright stance has shown that the variability of centre of pressure changes as a function of many variables.

Analysis of the time- or frequency-dependent properties of the time series of the motion of the centre of pressure reveals additional features to the distributional properties on a given dimension of centre of pressure motion (Bensel and Dzendolet, 1968; Myklebust and Myklebust, 1989). Some of these approaches provide information that can be used to address directly the relation between the amount and structure of the variability of the centre of pressure. Newell, van Emmerik, Lee and Sprague (1993) examined the relation between the variability and dimension of the centre of pressure profile in healthy young adults and those diagnosed with tardive dyskinesia (a movement disorder that arises from the prolonged regimen of neuroleptic medication). Dimension (D) is a nonlinear measure of the type of attractor dynamic (point attractor, limit cycle, and so on) formed from the centre of pressure trajectories in state space and provides an index of the number of dynamical degrees of freedom required to produce the output of the system.

Figure 7.2 shows, from Newell *et al.* (1993), some sample centres of pressure trajectories from individuals standing still: (a) normal healthy adult; (b) tardive dyskinetic profile that appears as unstructured as the profile of the healthy subject; and (c) a rhythmical tardive dyskinetic profile. The analysis showed that there is more structure to the location of the centre of pressure over time in healthy subjects than has typically been inferred, in that it was of a much lower dimension (D < 2.50) than that of a random white noise profile (D tending to infinite – uncorrelated random process). The mean dimension of the subjects with tardive dyskinesia (rhythmic, D = 1.30; unstructured to the eye, D = 1.75) was systematically lower than that of the healthy control group (D = 2.20). The centre of pressure profiles in Figure 7.2 are consistent with the finding that the standard deviation of the centre of pressure was higher in the tardive dyskinesia groups (rhythmic, SD = 1.12 cm; unstructured to the eye, SD = 0.21 cm) than the healthy control group (SD = 0.11 cm). Thus, there was an inverse relation between dimension and the variability of the motion of the centre of pressure, in that higher levels of variability of motion were associated with a lower dimension of the centre of pressure time series, and vice versa.

The findings of this study show that the amount of variability of the centre of pressure (standard deviation) is not a sufficient index of postural stability without a determination of the dimension of the attractor. This is because different attractors can have different degrees of variability and stability. Furthermore, centre of pressure profiles could have the same amount of variability (standard deviation) but different geometries to the attractor dynamic leading to different levels of stability.

The inverse relation between the amount and structure of the centre of pressure is strongly influenced by the age of the subjects (Newell, 1998) that can be indexed respectively, by the area of the centre of pressure (amount of motion) and ApEn (a measure of the regularity of the centre of pressure time series – Pincus, 1991) as a function of age (3 years, 5 years, young (18–25 years) and elderly (60–80 years)). The amount of variability decreases with advancing age in childhood through young adulthood but then increases again in the elderly. The opposite trend is apparent in the relation of age to regularity of the time series.

These two examples of centre of pressure profiles (Newell *et al.*, 1993; Newell, 1998) are consistent with the position that a reduction in degrees of freedom in control space is associated with an increase in the amount of variability exhibited in task space. The degrees of freedom of the collective organization of the output of the centre of pressure was enhanced in development but reduced in ageing and tardive dyskinesia (and other disease states – e.g., Parkinson's disease). The ageing effect is consistent with the loss of complexity and ageing hypothesis of Lipsitz and Goldberger (1992) and the tardive dyskinesia effect is consistent with the dynamical disease concept of Glass and Mackey (1988). The general position is that variability of the centre of pressure is driven by deterministic *and* stochastic processes (Newell, Slobounov, Slobounova and Molenaar, 1997) over multiple time scales (Blaszczyk and Klonowski, 2001; Collins and De Luca, 1994; Duarte and Zatsiorski, 2000) that are dependent on the constraints to action.

Locomotion

The traditional emphasis in the study of locomotion was on the amount of variation (standard deviation) in the step and stride intervals and the associated footfall locations (Gabell and Nayak, 1984; Owings and Grabiner, 2004). The general finding is that the amount of variation in the stride interval can change with a variety of individual and task properties. For example, stride interval variability decreases with advancing developmental age up to maturity (Hausdorff, Zemany, Peng and Goldberger, 1999) but that there is an increase in the variability of the stride interval during ageing (Hausdorff *et al.*, 1997). When the

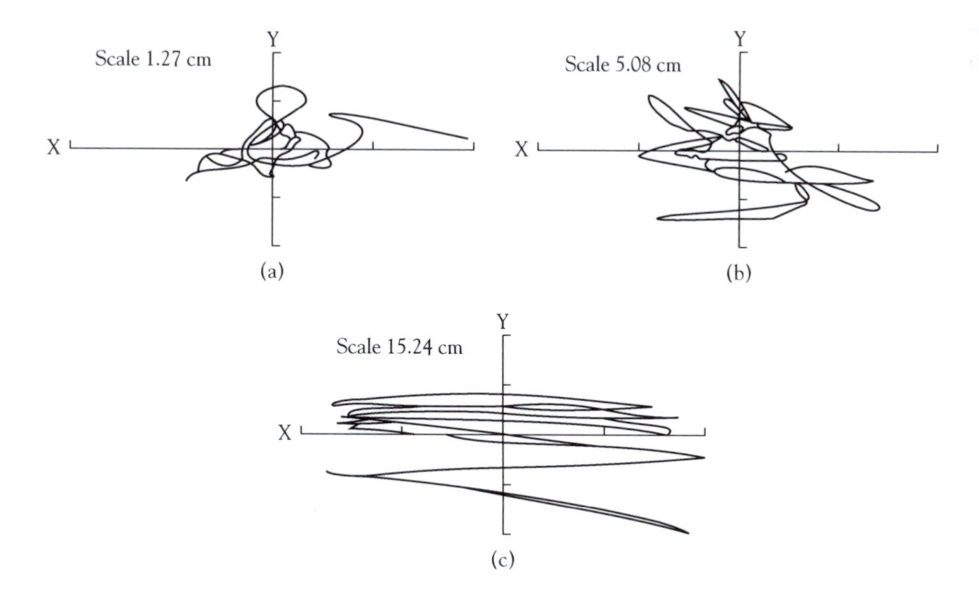

Figure 7.2 Example centres of pressure profiles from individual subjects of three groups: (a) healthy subject; (b) tardive dyskinetic profile that appears as unstructured as the healthy subject; (c) a rhythmical tardive dyskinetic profile. The scale refers to the interval between tick marks on the axes of each respective figure. Adapted from Newell *et al.* (1993) with permission.

coefficient of variation of the temporal and spatial footfall properties is considered across motor tasks it is the case that human walking and running have the lowest level of relative variability, in contrast to discrete movements, particularly in locomotion under the constraints of a treadmill. This low level of variability has supported the established view that locomotion has repeatable movement patterns with small levels of white noise driven variability.

The structure of the variability of step and stride intervals in walking has been extensively studied by Hausdorff and colleagues (1995, 1996, 1997) and West and Griffin (1998, 1999). They have used spectral analysis and detrended fluctuation analysis to reveal the structure in the gait (walking) cycle as a function of age and disease state. The central finding is that the fluctuations of the gait cycle exhibit long range correlations whereby the stride properties of any given gait stride are dependent on a stride that occurred previously, perhaps even hundreds of cycles earlier. The long range correlations in the stride time series have the form of a power law, indicating that the variability of the gait cycle is a self-similar fractal process (Hausdorff *et al.*, 1995, 1997).

Hausdorff *et al.* (1997) compared the detrended fluctuation score of a young adult group with that of an elderly group. The long range correlations were less strong in the elderly ($\alpha = 0.68$) than the young adults ($\alpha = 0.87$) and weaker still in elderly adults with Huntington's disease ($\alpha = 0.60$). Expressed another way the fractal slope is closer to 0.5 (white noise) in the Huntington's disease group followed by the elderly group whereas the young adult group has a fractal slope closer to pink noise (1.0). Thus, the structure of gait variability is progressively influenced by both ageing and disease, a finding consistent with the earlier studies reported on posture.

Jordan, Challis, and Newell (2006, 2007) have examined the variability of the running gait in young adults. They used experimental techniques similar to Hausdorf and colleagues but the timing properties, spatial locations, and contact forces of the footfalls were recorded from a force platform embedded in a treadmill. The detrended fluctuation analysis revealed the fractal nature of running variability in that the DFA was U-shaped over running speed with the minimum DFA anchored at the preferred speed. Thus, the preferred speed produced the most flexible structure (lower DFA) over the stride intervals. This pattern of findings was evident in a range of kinematic and kinetic variables of both the step and stride intervals.

The experiments on walking and running clearly show that there is a relation between the amount and structure of the variability of stride intervals. An important point arising, however, is that the directional relation is opposite to that shown previously for posture. In locomotion, ageing, for example, creates an increase in the complexity of the stride interval dynamics (and a lowering of the influence of long range correlations) along with a larger amount of stride interval variability, whereas in posture, ageing was associated with a greater amount of variability and a loss of complexity, as evidenced by the dimension or ApEn value generally being lower than that of young adults.

Hausdorff and colleagues showed that the self-similar pattern is not present across the stride intervals of walking when subjects are trying to time their footfalls to the regular beat of a metronome. In this situation the time-dependent variability approximates an uncorrelated random process in that the time dependent structure of natural walking is 'lost'. Given that the auditory cue of the metronome is being picked up by higher centres of the brain the suggestion is that the fractal nature of walking is mediated at the level of the brain (West, 2006). This also demonstrates that humans can adapt the output of movement to mimic a range of coloured noise and model type qualities to the movement variability.

Walking and running are made up of a sequence of steps that move the individual through the dynamic cycle of postural stability, instability and stability. Even the act of moving the body from a standing posture via one step to another standing posture captures this dynamic cycle. Johnson, Mahalko, and Newell (2003) examined the time it took young (20–29 years) and older (60–89 years) adults to regain postural stability in upright stance after performing just a single step forward. Stability was here defined as regaining the baseline level of variability of the centre of pressure motion by having the velocity of the centre of pressure below the criterion of 4 cm/s for 4 s. The length of the step forward was varied as a function of the percentage of the subject's own preferred step length.

Figure 7.3 shows that there was a systematic increase in the mean time and its variability required to regain stability as a function of advancing age. Thus, even in the taking of a single step there is an age-related increment in the variability of motion. Haibach, Slobounov, Slobounova and Newell (in press) have shown a similar age-related finding in the regaining of stability from a perturbation to posture that was induced by movement of a virtual reality room. West (2006) has proposed that the studies of the structure of variability in posture, regaining postural control from a single step and the repetitive processes of locomotion all point to a common control process for activities that have traditionally been interpreted as distinct activities requiring unique theoretical accounts.

Prehension and manipulation

The final class of motor tasks that we examine here with respect to the relation between the amount and structure of movement variability is that of prehension and manipulation tasks. These tasks usually involve the arm and hand neuromuscular systems and have smaller muscle groups than the major effectors supporting posture and locomotion. Nevertheless, there is also evidence of a task-driven pattern to the relation between the amount and structure of finger, hand and arm movement variability when considered in terms of attractor dynamics.

Arutyunyan, Gurfinkel and Mirsky (1968, 1969) provided an often quoted example of an inverse relation in the variability of the outcome of the action and that of the movement itself. They showed in the task of pistol shooting that the skilled marksman has less postural variability in the upright stance than unskilled marksman. However, the skilled marksman

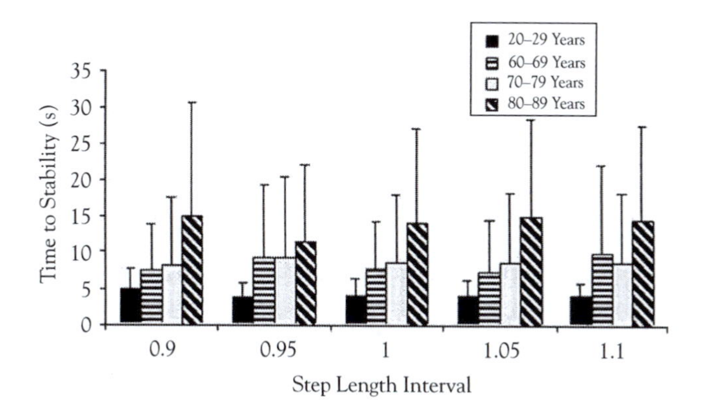

Figure 7.3 Mean time to stability (with between subject group standard deviation) as a function of age group and step length (expressed as a ratio of preferred step length). (Adapted from Johnson *et al.*, 2003, with permission.)

shows more motion than the unskilled in the joints of the shooting arm, although this is of a compensatory nature so as to reduce the motion variability of the pistol and ultimately the shots hitting the target. In the skilled marksman, the joints of the arm act in a compensatory cooperative way, to reduce variability at the periphery. In contrast, the unskilled shooter freezes the peripheral joints of the arm and in effect tries to control the variation in pistol from the proximal shoulder joint, without any compensatory movements of the joints of the arm and hand neuromuscular complex. Stuart and Atha (1990) have shown similar results in an investigation of skilled and unskilled archers. These examples point up the need to be careful in understanding 'which' category of variability one is measuring, task space, coupling between effectors, or that of the motion of a single effector.

The variability of motor output in an effector is strongly influenced by the level of force to be produced in both discrete (Schmidt *et al.*, 1979; Carlton and Newell, 1993) and continuous tasks (Slifkin and Newell, 1999). In the production of discrete force impulses, Carlton and Newell (1993) have shown that the variability of peak force, time to peak force and impulse all increase at a negatively accelerating rate over force level and map to the equation, $C \text{ of } V = 1/(PF^{1/2} \times T_{pf}^{1/4})$, where C of V is the coefficient of variation of the force, PF is peak force and T_{pf} is the time to peak force. There has, however, been no study of the sequential trial-to-trial structure of force variability, though in discrete movement tasks the sequential structure to the trial outcome has been shown to be modest to nonexistent (Newell, Liu and Mayer-Kress, 1996). That is, about 50 per cent of the individual trial sequences showed no systematic time dependent relation while in the other 50 per cent of the trials the relation was limited in most cases to the next trial only (n to $n + 1$).

In the continuous isometric force task of flexing the index finger to exert a constant level of force output to match target level on a computer screen the standard deviation of force increased at an increasing rate over increments of force level from 5–90 per cent maximal voluntary contraction, whereas the irregularity of the force output (ApEn) showed an inverted U-shaped function over the force range (Slifkin and Newell, 1999). The regularity function is consistent with the interpretation that the highest number of dynamical degrees of freedom in force output is in the 30–40 per cent range of the individual's force maximal output with least flexibility at the extremes of the force range. Thus, force level mediates the relation between the amount and structure of force variability.

The study of age on the variability of a constant level of isometric force output has clearly demonstrated the inverse pattern of the amount to structure relation to variability in the same isometric force task. Deutsch and Newell (2001) showed that as children age (10–18 years) they decrease the amount of force variability (standard deviation) but increase the dimension of the force output (dimension, ApEn). In contrast, ageing elderly adults increase the amount of force variability but decrease the dimension of the force output (Vaillancourt and Newell, 2003). Furthermore, the pattern of amount to structure of variability effects is magnified in elderly adults that have a low level of Parkinson's disease (Vaillancourt, Slifkin, and Newell, 2001).

Vaillancourt and Newell (2003) provided further evidence that ageing mediates the relation between the amount and structure of motor output variability in a finger force flexion task. They conducted a direct test of the hypothesis that the direction of the age-related change in the variability of the degrees of freedom is task dependent. Young adults (20–24 years), old (60–69 years) and older-old (75–90 years) adults produced either a constant force level target or that of a sine wave. As anticipated from previous work the force variability on each task increased with advancing age. But significantly the time and frequency analyses showed that the force output of the old and older-old subjects was progressively less complex in the constant force level task but progressively more complex in the sine wave force task (see Figure 7.4).

These findings show that the age-driven amount to complexity relation in movement variability is mediated by the short-term demands of the task in addition to the long timescale of changes from ageing. Expressed another way the general slow loss of complexity that comes with healthy ageing can be masked or magnified by the directional changes in behaviour that are driven by the immediate task demands. The more general hypothesis is that ageing leads to a loss of adaptation (Newell and Vaillancourt, 2001; Vaillancourt and Newell, 2002) rather than necessarily a loss of complexity (Lipsitz and Goldberger, 1992).

Concluding comments

The synthesis of experimental findings across posture, locomotion and manipulation tasks shows consistent trends in the relation between the amount of variability and the time- and frequency-dependent structure of variability. In most tasks the amount of variability in the task outcome measure is inversely related to the dimensionality of the organization of the effector output. Namely, a lower level of task outcome variability arises from a more adaptive

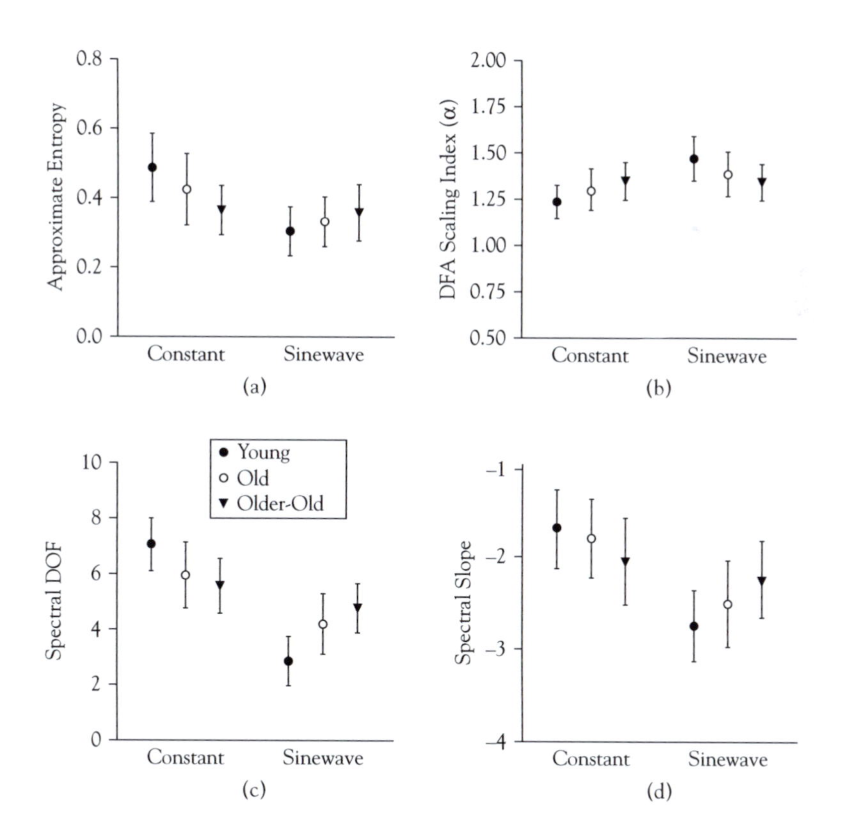

Figure 7.4 Structure of force output variability. A–D are the time and frequency analysis using ApEn, detrended fluctuation analysis, spectral DoF, and the spectral slope analysis, respectively. The closed circle is the young, open circle the old, and closed triangle the older-old group. Each symbol represents the average of all ten subjects collapsed across all four force levels, and the error bars represent the standard deviation. (Adapted from Vaillancourt and Newell, 2003, with permission.)

organization over the individual degrees of freedom, while, in contrast, the higher amounts of outcome variability are related to lower dimensional output of the collective from individual effectors.

The inverse relation of low task variability being associated with high complexity of movement dynamics (and vice versa) holds for many tasks, but it breaks down when the dimension of the to-be-achieved task dynamics is lower than that of the intrinsic dynamics (Vaillancourt and Newell, 2003). The collective pattern of findings invites the proposition that the amount of variability in task space is driven by the confluence of organismic, environmental and task constraints, and the direction of change in the dimension of the to-be-produced attractor dynamics. This adaptive relation places the regulation of the degrees of freedom (Bernstein, 1967) as a central problem in driving the variability at the different levels: task space, coupling between degrees of freedom, collective or order parameter, and at each individual degree of freedom.

Recent experimental work has provided considerable evidence for the position that the variability of motor output rarely approximates uncorrelated random processes or white Gaussian noise (Newell and Slifkin, 1998; Newell *et al.*, 2006; Riley and Turvey, 2002). The only task category that shows some evidence of an uncorrelated random process is the trial to trial outcome behaviour of discrete movements or short bursts of serial movements when the task is well learned, but this is probably only attainable over short durations of practice before nonstationarities set in. Clearly, the task constraints strongly influence the structure of variability and even the potential for uncorrelated random processes to be a significant part of the movement variability.

That the variability of motor output is not white Gaussian noise does not mean that the numerous subsystems of the individual do not display uncorrelated processes, although the strong interconnectivity of the subsystems of the organism suggests that even these levels of variability will exhibit time- and frequency-dependent structure. Furthermore, this finding does not mean that the intrinsic noise of a subsystem(s) is a negative influence in the regulation of movement. Rather, the evidence suggests that uncorrelated noise of the output is a very small component of the outcome variability that includes a small contribution of equipment and environmental noise. This small uncorrelated white noise component is embedded in a signal output with multiple time scales, the structure of which changes with the constraints on movement and action (Newell, Liu and Mayer-Kress, 2001).

Finally, it should be noted that the patterns of the variability of movement shown here are consistent with those found at other levels of biological analysis, including heart rate, temperature regulation, brain wave activity and muscle activity (Bassingthwaighte *et al.*, 1994; West, 2006). Furthermore, the findings are consonant with those shown in a range of disciplinary analyses of non-living systems. However, the intentional and adaptive capacities of the human system allow individuals to mimic a range of model types of variability and relations to outcome variability, according to the constraints imposed on action. A theory of movement variability needs to be able to generalize across the changing constraints to action and the multiple patterns of system organization arising as revealed in the amount and structure of movement variance.

This paper was supported in part by grants NIH RO1 HD046918, RO3 AG023259 and NSF #0114568.

References

1. Arutyunyan, G. H., Gurfinkel, V. S. and Mirskii, M. L. (1968) 'Investigation of aiming at a target'. *Biophysics*, 13: 536–538.
2. Arutyunyan, G. H., Gurfinkel, V. S. and Mirskii, M. L. (1969). 'Organization of movements on execution by man of an exact postural task'. *Biophysics*, 14: 1162–1167.
3. Bassingthwaighte, J. B., Liebovitch, L. S. and West, B. J. (1994). *Fractal physiology*. Oxford: Oxford University Press.
4. Bensel, C. K. and Dzendolet, E. (1968). 'Power spectral density analysis of the standing sway of males'. *Perception and Psychophysics*, 4: 285–288.
5. Bernstein, N. (1967). *The co-ordination and regulation of movements*. London: Pergamon.
6. Blaszczyk, J. W. and Klonowski, W. (2001). 'Postural stability and postural dynamics'. *Acta Neurobiology Exp*, 61: 105–112.
7. Carlton, L. G. and Newell, K. M. (1993). 'Force variability and characteristics of force production'. In K. M. Newell and D. M. Corcos (Eds.), *Variability and motor control* (pp. 15–36). Champaign: Human Kinetics.
8. Collins, J. J. and De Luca, C. J. (1994). 'Random walking during quiet standing'. *Physical Review Letters*, 73: 764–767.
9. Davids, K., Bennett, S. and Newell, K. (eds.) (2006) *Variability in the movement system: A multidisciplinary perspective*. Champaign, Ill: Human Kinetics.
10. Deutsch, K. M. and Newell, K. M. (2001). 'Age differences in noise and variability of isometric force production'. *Journal of Experimental Child Psychology*, 80: 392–408.
11. Duarte, M. and Zatsiorsky, V. M. (2000). 'On the fractal properties of natural human standing'. *Neuroscience Letters*, 283: 173–176.
12. Fitts, P. M. (1954). 'The information capacity of the human motor system in controlling the amplitude of movement'. *Journal of Experimental Psychology*, 47: 381–391.
13. Gabell, A. and Nayak, U. S. (1984). 'The effect of age on variability in gait'. *Journal of Gerontology*, 39: 662–666.
14. Glass, L. and Mackey, M. C. (1988). *From clocks to chaos*, New Jersey: Princeton University Press.
15. Haibach, P. S., Slobounov, S. M., Slobounova, E. S. and Newell, K. M. (in press). 'Aging and time-to-postural stability following a visual perturbation'. *Aging Clinical and Experimental Research*.
16. Hausdorff, J. M., Zemany, L., Peng, C.K. and Goldberger, A. L. (1995). 'Maturation of gait dynamics: stride-to-stride variability and its temporal organization in children'. *Journal of Applied Physiology*, 86: 1040–1047.
17. Hausdorff, J.M., Peng, C.K., Lading, Z., Ladin, J. Y., Wei, J. Y. and Goldberger, A. L. (1995). 'Is walking a random walk? Evidence for long-range correlations in stride interval of human gait'. *Journal of Applied Physiology*, 78: 349–358.
18. Hausdorff, J. M., Mitchell, S. L., Firtion, R., Peng, C. K., Cudkowicz, M. E., Wei, J. Y. and Goldberger, A. L. (1997). 'Altered fractal dynamics of gait: reduced stride interval correlations with aging and Huntington's disease'. *Journal of Applied Physiology*, 82: 262–269.
19. Johnston, C. B., Mihalko, S. L. and Newell, K. M. (2003). 'Aging and the time needed to reacquire postural stability'. *Journal of Aging and Physical Activity*, 11: 419–429.
20. Jordan, K. J., Challis, J. H. and Newell, K. M. (2006). 'Long range correlations in the stride interval of running'. *Gait and Posture*, 24: 120–125.
21. Jordan, K. J., Challis, J. H. and Newell, K. M. (2007). 'Speed influences on the scaling behavior of gait cycle fluctuations during treadmill running'. *Human Movement Science*, 26: 87–102.
22. Kaplan, D. and Glass, L. (1995). *Understanding nonlinear dynamics*. New York: Springer-Verlag.
23. Kim, S., Carlton, L. G., Liu, Y.T. and Newell, K. M. (1999). 'Impulse and movement space-time variability'. *Journal of Motor Behavior*, 31: 341–357.
24. Lipsitz, L. A. and Goldberger, A. L. (1992). 'Loss of "complexity" and aging'. *Journal of the American Medical Association*, 267: 1806–1809.
25. Myklebust, J. B. and Myklebust, B. M. (1989). 'Fractals in kinesiology'. *Society for Neuroscience Meeting abstract*. No 243.2.

26. Newell, K. M. (1998). 'Degrees of freedom and the development of postural centre of pressure profiles'. In K. M. Newell and P. C. M. Molenaar (Eds.), *Applications of nonlinear dynamics to developmental process modeling* (pp. 63–84). Hillsdale, NJ: Erlbaum.
27. Newell, K. M. and Corcos, D. M. (Eds.). (1993). *Variability and motor control* Champaign, IL: Human Kinetics.
28. Newell, K. M. and Slifkin, A. B. (1998). 'The nature of movement variability', In J. P. Piek (Ed.), *Motor behavior and human skill: A multidisciplinary approach* (pp. 143–160). Champaign, IL: Human Kinetics.
29. Newell, K. M. and Vaillancourt, D. E. (2001). 'Dimensional change in motor learning'. *Human Movement Science*, 4–5: 695–716.
30. Newell, K. M., Liu, Y.T. and G. Mayer-Kress. (1996). 'The sequential structure of movement outcome in learning a discrete timing task'. *Journal of Motor Behavior*, 29: 366–382.
31. Newell, K. M., Mayer-Kress, G. and Liu, Y-T. (2001). 'Time scales in motor learning and development'. *Psychological Review*, 108: 57–82.
32. Newell, K. M., Deutsch, K. M., Sosnoff, J. J. and Mayer-Kress, G. (2006). 'Motor output variability as noise: A default and erroneous proposition?' In K. Davids, S. Bennett and K. Newell (Eds.), *Variability in the movement system: A multidisciplinary perspective* (pp. 3–23). Champaign, Ill: Human Kinetics.
33. Newell, K. M., van Emmerik, R. E. A., Lee, D. and Sprague, R. L. (1993). 'On postural stability and variability'. *Gait and Posture*, 4: 225–230.
34. Newell, K. M., Slobounov, S. M., Slobounova, E. and Molenaar, P. C. M. (1997). 'Deterministic and stochastic processes in centre of pressure profiles'. *Experimental Brain Research*, 113: 158–164.
35. Owings, T. M. and Grabiner, M, D. (2004). 'Variability of step kinematics in young and older adults'. *Gait and Posture*, 20: 26–29.
36. Özkaya, N. and Nordin, M. (1991). *Fundamentals of biomechanics: Equilibrium, notion and deformation*. New York: Van Nostrand Reinhold.
37. Pincus, S. M. (1991). 'Approximate entropy as a measure of system complexity'. *Proceedings of the National Academy of Sciences*, 88: 2297–2301.
38. Riley, M. A. and Turvey, M. T. (2002). 'Variability and determinism in motor behavior'. *Journal of Motor Behavior*, 34: 99–125.
39. Schmidt, R. A., Zelaznik, H., Hawkins, B., Frank, J. S. and Quinn, J. T. (1979). 'Motor-output variability: A theory for the accuracy of rapid motor acts'. *Psychological Review*, 86: 415–451.
40. Shannon, C. E. and Weaver, W. (1949). *The mathematical theory of communication*. Urbana, Ill.: University of Illinois Press.
41. Slifkin, A. B. and Newell, K. M. (1999). 'Noise, information transmission, and force variability'. *Journal of Experimental Psychology: Human Perception and Performance*, 25: 837–851.
42. Stuart, J. and Atha, J. (1990). 'Postural consistency in skilled archers'. *Journal of Sports Sciences*, 8: 223–234.
43. Thompson, J. M. T. and Stewart, H. B. (1986). *Nonlinear dynamics and chaos*. Chichester: Wiley.
44. Vaillancourt, D. E. and Newell, K. M. (2002). 'Changing complexity in behavior and physiology through aging and disease'. *Neurobiology of Aging: Experimental and Clinical Research*, 23: 1–11.
45. Vaillancourt, D. E. and Newell, K. M. (2003). 'Aging and the time and frequency structure of force output variability'. *Journal of Applied Physiology*, 94: 903–912.
46. Vaillancourt, D.E., Slifkin, A.B. and Newell, K.M. (2001). 'Regularity of force tremor in Parkinson's disease'. *Clinical Neurophysiology*, 112: 1594–1603.
47. West, B. J. (2006). *Where medicine went wrong: Rediscovering the path to complexity*. New Jersey: World Scientific.
48. West, B. J. and Griffin, L. (1998). 'Allometric control of human gait'. *Fractals*, 6: 101–108.
49. West, B. J. and Griffin, L. (1999). 'Allometric control, inverse power laws and human gait'. *Chaos, Solitons and Fractals*, 10: 1519–1527.
50. Yates, F. E. (1987). *Self-organizing systems: The emergence of order*. New York: Plenum Press.

8

Planning and control of object grasping: kinematics of hand pre-shaping, contact and manipulation

Jamie R. Lukos[1], *Caterina Ansuini*[2], *Umberto Castiello*[2] *and Marco Santello*[1]
[1]*Arizona State University, Tempe;* [2]*Università di Padova, Padova*

Introduction

The hand is a uniquely complex sensory and motor structure of fundamental importance to our motor behaviour, whether it is used for artistic expression and communication, tool making and use, or perception. To address the seemingly simple question of: 'How does the central nervous system (CNS) control the hand?' requires understanding of its biomechanical structure and neural mechanisms. The complex nature of such organization has prompted scientists over the past three decades to use a wide range of multi-disciplinary approaches – on human and non-human primates – to unravel the intricate mechanisms underlying hand function, ranging from recording from motor and sensory cortical neurons to recording the activity of motor units of hand muscles, from imaging neural activity of the brain during object grasping to measuring the movement and force coordination patterns of the digits.

Due to the invasive nature of some of the above research approaches, most inferences about neural mechanisms underlying the control of the hand in humans have been obtained indirectly through detailed analysis of the hand's behaviour such as coordination of digit motions and forces, i.e., kinematics and kinetics, respectively. These research efforts have probed the hand sensory and motor capabilities by challenging it with tasks that require individual or multiple digit actions, sub-maximal to maximal forces, constraining the type of sensory modalities normally involved in sensorimotor transformations, or perturbing an ongoing behaviour (e.g., reaching or object lift) by unpredictably changing task conditions (e.g., eliminating vision or changing object weight or shape) to examine the interaction between anticipatory and feedback-based control mechanisms.

Research on hand kinematics has proved to be particularly insightful in revealing important aspects of sensorimotor transformations responsible for modulating the shape of the hand to object geometry and intended use of the object as the hand approaches the object, culminating with a hand configuration and distribution of contact points appropriate for object manipulation. Although object manipulation is by definition characterized by exerting forces and torques on an object, *how* the CNS prepares the hand to make contact at specific locations on the object also plays a crucial role for successful interactions with the object. The objective of this chapter is to give an overview of research on human hand kinematics with emphasis on studies of reach-to-grasp tasks, the state of our current understanding in this field and questions that remain to be addressed. The first section of the chapter is devoted to a brief literature review of the main findings obtained by kinematic studies of whole hand grasping. The second section focuses on our recent work on hand pre-shaping in response to object shape perturbation and planned manipulation as well as on contact point selection as a function of object properties and their predictability. We conclude the chapter by discussing research questions that remain to be addressed, in particular with regard to the relation between hand kinematics (hand posture, contact points) and kinetics (between-digit force coordination). Readers interested in topics that are complementary to the content of this chapter are referred to the following review articles: Jeannerod *et al.* (1995) for review on kinematic synergies and cortical control; Latash *et al.* (2002, 2004), Johansson and Cole (1994) and Johansson (1998) for review on force synergies underlying force production and grasping tasks; Schieber and Santello (2004) for review on peripheral and central constraints to hand motor control; Castiello (2005) for review on the cognitive aspects of grasping.

Control of hand shape during reach-to-grasp

Effect of object geometry on hand shaping

Research on two-digit grasping has shown that during hand transport to an object, the distance between thumb and index finger (grip aperture) gradually increases and reaches a peak at about 70 per cent of the reach, maximum grip aperture being linearly related to object size (for review see Jeannerod, 1995; Castiello, 2005). The relation of whole hand shape during the reach to object geometry was first examined by Santello and Soechting (1998). Subjects were asked to reach, grasp and lift objects whose contours were varied to elicit different hand configurations at contact. The shape of the hand was defined as the pattern of angular excursions at the joints of the thumb and fingers while discriminant analysis was used to quantify the extent to which hand shape resembled object shape at different epochs of the reach. The general features of whole hand shaping were similar to those described for two-digit grasping, i.e., all digits extended throughout most of the reach, then closed as the digits were about to close on the object. However, it was also found that the extent to which object geometry could predict hand shape increased in a monotonic fashion throughout the reach. This finding was later confirmed by subsequent studies addressing the effect(s) of sensory cues on hand shaping (Santello *et al.*, 2002) where it was found that gradual pre-shaping of the hand occurs also when subjects perform grasping to remembered objects as well as virtual objects, i.e., objects that they can perceive through vision but not touch. A subsequent study (Winges and Santello, 2003) found that removing vision of both the hand and the object did not affect the gradual pre-shaping of the hand to objects with different contours. Schettino *et al.* (2003) defined two dimensions of pre-shaping (flexion-extension and adduction-abduction) that are differentially affected by the availability of vision of the hand and/or object. Specifically, pre-shaping would be characterized by two epochs: an early predictive phase during which grip selection is attained regardless of availability of visual feedback and a subsequent phase during which subjects may use visual feedback to finely modulate hand posture. The analysis of whole hand pre-shaping has also been applied to quantify the effect of neurological disorders on processes of sensorimotor integration associated with modulation of hand shape to object geometry (Parkinson's disease: Schettino *et al.*, 2004, 2006; stroke: Raghavan *et al.*, 2006).

Analysis of digit kinematics, both during the reach and at contact with the object, also revealed that hand postures elicited by objects of different sizes and shapes can be described by a few linear combinations of kinematic coordination patterns as revealed by principal components analysis (PCA) (Santello *et al.*, 1998, 2002). These studies found that motion at many joints of the digits covaried in a similar way regardless of object geometry. For example, digit extension and flexion were generally accompanied by digit abduction and adduction, respectively. Similarly, motion at the metacarpal-phalangeal joints, as well as at the proximal interphalangeal joints, were characterized by strong linear covariation across all digits despite wide differences in the size and shape of virtual and real objects being grasped (Santello *et al.*, 1998, 2002). Subsequent work on humans (Mason *et al.*, 2001) and non-human primates (Mason *et al.*, 2004; Theverapperuma *et al.*, 2006) confirmed and extended this finding by showing that multiple degrees of freedom of the hand are controlled as a unit during reach-to-grasp movements. Although it is recognized that these kinematic coordination patterns are the net result of biomechanical as well as neural constraints (for review see Schieber and Santello, 2004), the neural bases of such dimensionality reduction in how the hand is controlled is still not well understood and is a subject of ongoing investigation.

New insights into hand shaping during reach-to-grasp tasks

Shape perturbation effects

As previously reported in this chapter, a characteristic finding within the multi-digits grasping literature is the relationship between the configuration assumed by the hand during reaching and the shape of the to-be-grasped object (Santello and Soechting, 1998). A feature of such phenomenon is that the strength of this relationship increases gradually and monotonically during reaching (Santello and Soechting, 1998). Until recently a question which has remained unsolved is whether the need for a fast hand shaping reorganization would elicit a breakdown in the relationship. This issue has been addressed by Ansuini and colleagues (2007). They assessed how hand shaping responds to a perturbation of object shape. In *blocked* trials (80 per cent of total), participants were instructed to reach, to grasp and lift a concave or a convex object. In *perturbed* trials (20 per cent of total), a rotating device allowed for the rapid change from the concave to the convex object or vice versa as soon as the reach movement started (see Figure 8.1a). In this situation participants grasped the last presented object. For both directions of perturbation (i.e., 'convex → concave' and 'concave → convex'), linear regression analysis revealed that the presence of the perturbation reduced the strength of the relationship between hand shape during reaching and hand configuration at the end of the movement (see Figure 8.1b). Nevertheless, such reduction in correlation did not fully compromise gradual hand shaping (if compared to that found for the same object during the unperturbed conditions; see Figure 8.1b).

These results suggest that the degree of correlation between hand posture during reaching and at contact is modulated by the time the to-be-grasped object is presented; and that the system reacts to the object's shape perturbation by applying a control strategy that involves all fingers. In these terms, the hand is adaptively pre-shaped as to fulfil the grasp requirements dictated by the newly presented object.

Contextual effects on hand-shaping

Contextual factors seem to have an effect on motor behaviour as ongoing movements are influenced by the manipulation of forthcoming task demands (Cohen and Rosenbaum, 2004).

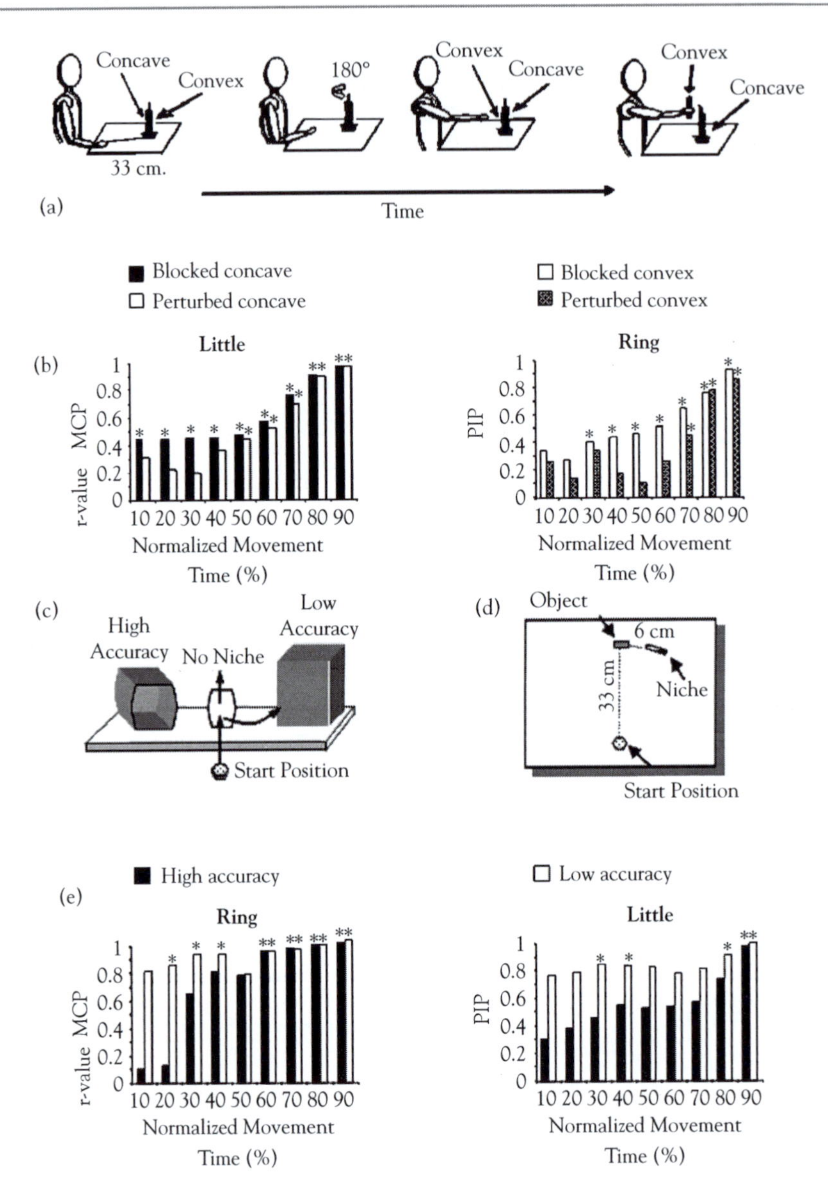

Figure 8.1 Panel a. Schematic representation of the subject's posture and an example of the time course for a perturbed trial (i.e. from concave to convex object; Panel b. Correlation coefficients between joint angles during the reach vs. joint angles at the end of the movement for 'blocked' concave and 'perturbed' concave trials, and for 'blocked' convex and 'perturbed' convex, respectively. The *mcp* and *pip* labels denote metacarpal-phalangeal and proximal interphalangeal joints, respectively. Asterisks indicate significant correlation values ($P < 0.05$). Panels c and d represent the experimental set-up (front and top view, respectively) and the three experimental conditions for the 'context' experiment. Panel e represents the correlation coefficients between joint angles during the reach vs. joint angles at the end of the movement for high and low accuracy trials. The *MCP* and *PIP* labels denote metacarpal-phalangeal and proximal interphalangeal joints, respectively. Asterisks indicate significant correlation values ($P < 0.01$).

Surprisingly, there has been little research on how the hand approaches an object depending on where and for what purpose the object will be used.

This has recently been examined in a study in which participants were asked to reach towards and grasp a convex object between the thumb and the four fingers and to perform one of the following actions: (1) lift the object; (2) insert the object into a niche of a similar shape and size as the object; or (3) insert the object into a rectangular niche much larger than the object (Ansuini *et al.*, 2006; see Figure 8.1 c-d). Although in all experimental conditions subjects were asked to grasp the same object, a gradual preshaping of the hand occurred only when planning object lift or when the end-goal required a great level of accuracy, i.e., object placement into the tight niche. In contrast, when the end-goal did not require an accurate manipulation, i.e., object placement into the large niche, the hand posture used to grasp the object was attained early in the reach and did not change significantly during the reaching (see Figure 8.3 e). These results were interpreted as evidence that in order to execute the task, an internal model of the whole motor sequence (i.e., reach to grasp and the action following object contact) could be used. Such a model may include both object features and the forces needed for the manipulatory experience following grasping. Along these lines, hand shaping would be functional not only for the successful grasp of a particular shaped object but also for the accomplishment of the action goal following grasping.

Choice of contact points for multi-digit grasping

Choice of fingertip contact points as a function of object centre of mass (COM) and its predictability

Although both kinematic and kinetic studies have provided significant insight about neural control of grasping, several important questions remain. One of these questions concerns how digit placement is chosen when planning and executing grasping and manipulation. How the hand makes contact with the object, in particular how the fingertips are spatially distributed relative to the object as well as to each other, is a crucial control variable as it affects how forces can be transmitted to the object. This becomes a particularly important factor when considering grasp tasks that are characterized by precision constraints. Within this context, it should be emphasized that contact point distribution on an object does not depend only on its geometry, but also on its intended use, i.e., the *same* object could be grasped differently depending on how it is going to be used (Friedman and Flash, 2007).

The problem of how forces are distributed among the digits has been extensively studied (for review see Zatsiorsky and Latash, 2004). A well-studied feature of force coordination during grasping is the phenomenon of anticipatory force control. This refers to the ability to predict the forces required to successfully manipulate an object based on the sensory information acquired during previous manipulations of the same object. Subjects have been shown to implement such anticipatory force control mechanisms by studies examining two-digit grasping of objects with different weights (Johansson, Backlin, and Burstedt, 1999), sizes (Gordon, *et al.*, 1991; Reilmann, Gordon, Henningsen, 2001), textures (Quaney and Cole, 2004), and COMs (Salimi, *et al.*, 2000). It is important to note that anticipatory force control mechanisms allow subjects to plan the temporal development of forces *before* properties such as object weight can be perceived through manipulation (e.g., Forssberg *et al.*, 1991; Forssberg *et al.*, 1992), as opposed to having to rely primarily on proprioceptive and tactile input that would become available only later in the task. This has been assessed in the literature showing that the rate of tangential force development before the object is lifted is modulated as a function of the expected object properties (for review see Johansson and Cole, 1992). Therefore, sensorimotor memories built through previous interactions with an object allow the formation of an internal representation for effective digit force planning.

The large number of studies on grasp kinetics has significantly improved our understanding of how the CNS coordinates the many degrees of freedom of the hand. However, a major limitation of these studies is that force distribution among the digits has been assessed through force sensors that constrain digit placement on the object. As noted above, however, choice of contact point distribution appears to be a very important component of grasp control. Therefore, examining how subjects choose digit placement on an object as a function of its intended use could provide significant information about grasp control and, most importantly, allow for a better comparison with more 'natural' scenarios encountered during everyday manipulative tasks. Surprisingly, however, very little research has been devoted to this problem. Even though hand kinematics during reach-to-grasp tasks has been extensively studied (as described in the second section of this chapter), the focus has been on quantifying hand shape in terms of joint angles and hand postures while neglecting *where* individual digits are placed on an object or the extent to which contact points are modulated as a function of object properties.

Therefore, we examined how subjects choose contact point distributions on an object as a function of its COM and its predictability (Lukos *et al.*, 2007a). For this study, we focused our analysis on where the fingertips were placed on the object (vertical dimension), how this interacted with the ability to anticipate object properties and how both of these factors affected grasp performance (object roll; see below). We addressed this question by asking subjects to grasp, lift and replace a centrally located cylinder atop a long horizontal base (an inverted 'T'-shaped object) while attempting to minimize object roll (Figure 8.2). The object COM was altered by placing a weight in slots located to the right, centre, or left of the graspable cylinder. The object was positioned parallel to the frontal plane of the subject so that clockwise and counter-clockwise torques were generated from the added mass. Reflective markers were placed on the subject and the object to record hand and arm kinematics as well as object movement. Subjects were not told or shown where the COM was located. The task was performed in blocks where either the COM was placed in a random order (unpredictable condition) or the same COM location was repeated over five trials (predictable condition). In the predictable condition, subjects were told that COM would be the same across the block of trials associated with each COM location.

When object COM location could be anticipated (i.e., predictable condition), subjects altered their contact points with changes in COM location and were able to minimize object roll (~3° from vertical in predictable versus ~8° in unpredictable when object COM was asymmetrical, $P < 0.001$; 'no cue', Figure 8.3). However, when subjects could not predict the COM location on a trial-to-trial basis (i.e., unpredictable condition), they used the same fingertip contact points regardless of COM location. Interestingly, the fingertip contact point distributions elicited by all COM locations for the unpredictable condition were statistically indistinguishable from that elicited by the centre COM for the predictable condition. This indicates that a default distribution of contact points was used when the COM location could not be anticipated. Adopting a spatial distribution similar to that used for the predictable centre COM location allowed subjects to minimize roll most effectively, regardless of COM location. If subjects had distributed digit contact points in preparation for manipulating the object with a left or right COM location, a much greater roll would have occurred if their anticipation was incorrect, i.e., by wrongly anticipating a right or left

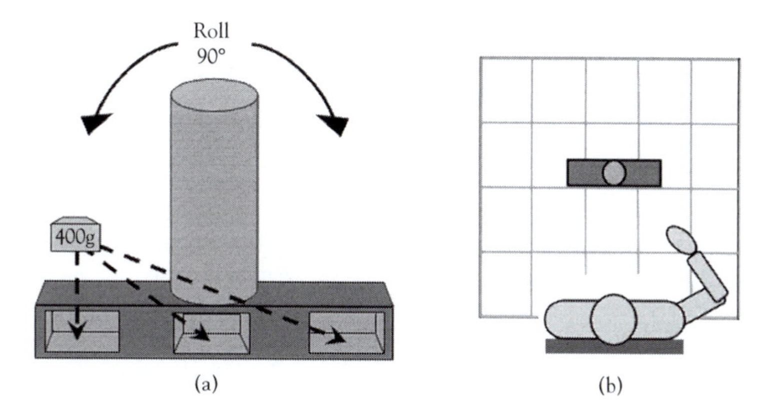

Figure 8.2 Experimental object and setup. Panel a. Schematic representation of the object used in the experimental task. A 400 g weight was added to the left, center, or right slot in the base of the object to alter the location of its COM (total object mass = 810 g). Behavioral performance was measured by object roll that was defined as the initial roll of the object in the frontal plane succeeding lift (>90° to the right and < 90° to the left of the subject). Panel b. Depiction of the subject relative to the object during the experiment. Subjects sat ~30 cm away from the object with the hand at a 35° angle from the object. For more information about the methodology see Lukos, *et al.* (2007a).

COM, respectively, subjects would have generated a torque in the same direction as that caused by the added mass. By placing their digits as if the COM location was in the centre, subjects could prevent the risk of excessively large object rolls when COM location was in the left or right, while effectively minimizing object roll when COM was in fact in the centre (see Figure 8.3, 'no cue').

The greatest amount of modulation across COM locations was observed for the thumb and index finger. During the predictable condition, the thumb contact point was significantly higher when the COM was located in the left (121.4 mm) than in the centre or right (104.8 mm and 98.0 mm, respectively; $P < 0.0001$). The index finger exhibited an opposite trend, the contact point being higher for right than left COM locations (145.7 mm and 132.4 mm, respectively; $P < 0.0001$). The other digits did not reveal any significant differences, though they exhibited similar trends, i.e., when the COM was located on the right, the digits were placed higher than when in the left or centre COM locations. Therefore, different digit placements were chosen but only when object properties could be anticipated.

To determine the extent to which whole hand placement could be predicted by COM location, discriminant analysis was performed and a SensoriMotor efficiency (SME) index was computed (Sakitt, 1980; Santello and Soechting, 1998). It was found that when subjects could not anticipate COM location, object COM could not reliably predict the spatial distribution of all contact points (SME Index ~ 15 per cent), indicating that similar digit placements were used regardless of COM location. In contrast, when COM location could be predicted, COM location could be discriminated to a much greater extent (SME Index ~ 55 per cent) resulting in two distinct spatial distributions of contact points associated with left and right COM locations. In this latter condition, the ability to discriminate COM location was not perfect (i.e., SME Index ≠ 100 per cent) due to a higher degree of overlap between the distribution of contact points associated with the centre COM location and those on the left and right of the object.

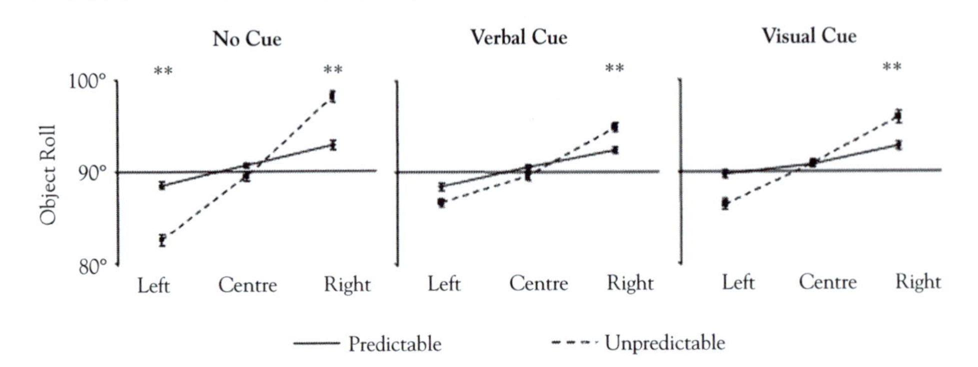

Figure 8.3 Object roll. Grasp performance quantified by peak object roll. When CM location was changed from trial-to-trial, subjects were able to minimize object roll to a greater extent when given a verbal or visual cue about CM location than in the no-cue condition. Nevertheless, neither cue allowed subjects to fully compensate for CM randomization as revealed by significant difference between the blocked and random conditions for the right CM location. Asterisks indicate significant correlation values ($P < 0.01$).

It is important to note that the modulation of fingertip contact points as a function of object properties and their predictability is a novel finding and extends previous literature on anticipatory control of digit forces. Specifically, in agreement with the above results, two-digit grasp studies have shown that when object COM location can be predicted, subjects can effectively minimize object roll (Salimi *et al.*, 2000). However, this study constrained contact points on the sensors mounted on the object. Similarly, many studies of whole-hand grasping of objects with different COMs (for review see Latash *et al.*, 2004) have demonstrated that subjects can share forces across the digits to maintain an object aligned with the vertical. Taken together, these findings indicate that subjects could have performed our task *without having to alter contact points* on the object. Therefore, the fact that they chose different spatial distributions of contact points demonstrates that grasp planning includes both force control and kinematic anticipatory mechanism. We speculate that contact point modulation in response to changes in object COM was performed to facilitate object manipulation (Lukos *et al.*, 2007a).

The effect of cues about object properties on the selection of fingertip contact points

We had found that previous experience with grasping and lifting an object with invariant COM locations (i.e., predictable condition; see above) allowed subjects to predict the necessary forces to minimize roll, this further leading to choosing distinct contact points. However, when object COM was changed from trial to trial, subjects could not use sensory feedback from previous trials, i.e., somatosensory feedback primarily from proprioceptive and tactile input and visual feedback about direction of object roll, to plan subsequent manipulations. To further understand the relationship between a priori knowledge of object properties and grasp planning, we tested whether providing cues about object COM location could reduce the previously observed decrement in grasp performance (large object rolls) when object COM was changed randomly from trial-to-trial (random condition) compared to fixing the COM location for five trials consecutively (blocked condition). This question was addressed by testing the effect(s) of providing either verbal or visual cues about COM location (Lukos *et al.*, 2007b).

For the verbal cue condition, the experimenter told the subjects where the added mass was, but no vision of this mass was allowed. For the visual cue, subjects were allowed to see where the added mass was but no additional verbal cues were given. The effect(s) of verbal and visual cues was tested by comparing object roll using two experimental conditions: (1) 'blocked', where subjects grasped and lifted the object with the same COM for five consecutive trials; and (2) 'random' where object COM location was randomly changed from trial to trial (note that, since cues about object properties were provided in this study, 'blocked' and 'random' better describe the experimental conditions than the previously used definitions of 'predictable' and 'unpredictable', respectively). If peak object roll in the blocked and random conditions was the same when providing either verbal or visual cues, this would indicate that these cues allow subjects to plan contact point locations and forces even when sensory information about the previous trial could not be used to plan a subsequent one. Conversely, different magnitudes of object roll across predictability conditions for both cues would indicate that the cues we used were not sufficient to remove the uncertainty associated with trial-to-trial changes in object COM location.

We found that both visual and verbal cues allowed subjects to perform object grasp and lift similarly across blocked and random presentations of COM locations. With regard to contact points, during the unpredictable no cue condition subjects chose a 'default' positioning of the fingertips regardless of COM location, as opposed to distinct contact point distributions found for the predictable condition. In contrast, for the 'random' condition subjects adopted a spatial distribution of the fingertips (primarily for the thumb and index finger) that was more similar to that adopted for the 'blocked' condition when they either saw or were told where the COM was located. With regard to object roll, the effect of either verbal or visual cue was not uniform across all COM locations. Specifically, although object roll was significantly affected by COM location and predictability condition for both cue experimental conditions ($P < 0.001$), we also found significant interaction between these two factors ($P < 0.05$). This interaction resulted from larger object rolls for right COM locations during 'random' but not 'blocked' trials. In contrast, subjects minimized object roll similarly in both predictability conditions when object COM was on the left or centre. Therefore, the efficacy of sensorimotor transformations – from either verbal or visual cues to contact points and forces – depended on whether the external torques caused by the added mass were on the thumb or finger side of the object. Such an asymmetry might be due to a different degree to which the thumb contact point can be changed compared to those of the fingers, as well as a higher computational load associated with distributing forces across four fingers vs. the thumb to counteract external torques.

Salimi and colleagues (2003) applied a paradigm where they altered the COM of an object and presented no cue or a visual cue as to where the COM was located, and measured digit forces during two-digit grasping. They found that when the COM location was unpredictable, subjects could not anticipate the proper force distribution even when a visual cue (a dot on the object or view of the added mass) was presented, and, therefore, exhibited a significant object roll similar to that of the no cue condition. However, we have found that given either a verbal or a visual cue as to where the COM was located, subjects were able to minimize object roll to a greater extent than in the no cue condition.

Methodological differences prevent us from making a direct comparison between our study and that by Salimi *et al.* (2003). First, the tasks were performed using five vs. two digits, respectively. Second, the manipulanda used differed in their mass, torques and geometry. These two factors could have affected the degree of difficulty in minimizing object roll. However, there is another factor that might have contributed more significantly to the different results between the two studies: in our study, subject could choose where to place their digits on the object. Therefore, we speculate that the better grasp performance reported by our study could have been due to the subjects' ability to combine anticipatory force *and* kinematic mechanisms to effectively manipulate the object.

Discussion

The studies discussed in this chapter provide converging evidence that mechanisms underlying planning and execution of grasping can be revealed by examining important kinematic variables. In particular, studies of hand shaping during reach-to-grasp indicate that multiple degrees of freedom of the hand are spatially and temporally coordinated in a continuous fashion as the hand approaches the object. Importantly, kinematic coordination patterns reduce the large dimensionality of hand control. It has also been found that the relationship between hand shaping and the shape of the to-be-grasped object is preserved even when the CNS is forced to suddenly change the initially planned hand shaping and/or when the same object is grasped for different purposes. Furthermore, hand shaping when reaching to grasp a given object appears to be sensitive to the manipulation being planned. Taken together, these considerations highlight a fundamental issue that any model of grasping should consider: hand shaping cannot be justified or explained solely in terms of object geometry as the scope of the action and task context also need to be considered. Analysis of contact point selection further demonstrates that the same object is grasped in a different manner, depending on the extent to which its object properties can be predicted and the sensory modality through which such anticipation can occur. Therefore, aspects concerned with intentionality, end-goal accuracy and anticipation of object properties appear to play a fundamental role in grasp planning and execution.

Despite the advances made by studies of hand kinematics, more work is needed to address fundamental questions concerning the control of grasping. In particular, many studies have been devoted to understanding the coordination of multi-digit forces even though this has been done by constraining contact point locations. The recent studies on contact point selection discussed above have addressed new questions on this important aspect of grasp control, but lack information about digit forces due to the nature of the experimental design (i.e., no force sensors can be used on the object). Therefore, a major gap in our knowledge remains: how does the selection of a given contact point distribution correlate with choosing a given force distribution among the digits? Another key question is the extent to which hand shaping is crucial for the placement of the digits at specific contact points. As a given contact point distribution can be attained using different hand configurations (i.e., by varying the degree of flexion/extension or adduction/abduction of the digits), we do not know whether the spatial distribution of the digits can be predicted by hand shaping or whether the two phenomena are independent. These questions limit our understanding of grasp control and warrant further investigation. The results of this research have tremendous potential for enhancement of several fields such as neural prosthetics, hand rehabilitation, robotics, and object design.

Conclusion

Our knowledge of the sensorimotor control of the hand for grasping and manipulation has improved significantly in the past three decades thanks to the application of many complementary and interdisciplinary research approaches. Studies of hand kinematics have revealed important aspects of grasp planning as quantified by the relationship(s) between hand posture and contact points on the object as a function of object properties, their predictability and planned manipulation. However, several important questions remain to be addressed. Although hand kinematics has proved to be a very useful avenue for studying hand control and its underlying neural mechanisms, further progress will be made by combining complementary research techniques, in particular digit force measurements, to fully understand the relation(s) between hand configuration and forces generated for skilled object manipulation.

References

1. Ansuini, C., Santello, M., Tubaldi, F., Massaccesi, S. and Castiello, U. (2007) 'Control of hand shaping to object shape perturbation'. *Exp Brain Res*, 180: 85–96.
2. Ansuini, C., Santello, M., Massaccesi, S. and Castiello, U. (2006) 'Effects of end-goal on hand shaping'. *J Neurophysiol*, 95: 2456–65.
3. Castiello, U. (2005) 'The neuroscience of grasping'. *Nat Rev Neurosci*, 6: 726–36.
4. Cohen, R. G. and Rosenbaum, D. A. (2004) 'Where grasps are made reveals how grasps are planned: generation and recall of motor plans'. *Exp Brain Res*, 157: 486–95.
5. Forssberg, H., Eliasson, A. C., Kinoshita, H., Johansson, R. S. and Westling, G. (1991) 'Development of human precision grip. I: Basic coordination of force'. *Exp Brain Res*, 85: 451–7.
6. Forssberg, H., Kinoshita, H., Eliasson, A. C., Johansson, R. S., Westling, G. and Gordon, A. M. (1992) 'Development of human precision grip. II. Anticipatory control of isometric forces targeted for object's weight'. *Exp Brain Res*, 90: 393–8.
7. Friedman, J. and Flash, T. (2007) 'Task-dependent selection of grasp kinematics and stiffness in human object manipulation'. *Cortex*, 43: 444–60.
8. Gordon, A. M., Forssberg, H., Johansson, R. S. and Westling, G. (1991) 'Visual Size Cues in the Programming of Manipulative Forces during Precision Grip'. *Experimental Brain Research*, 83: 477–482.
9. Jeannerod, M., Arbib, M. A., Rizzolatti, G. and Sakata, H. (1995) 'Grasping objects: the cortical mechanisms of visuomotor transformation'. *Trends Neurosci*, 18: 314–20.
10. Johansson, R. S. and Cole, K. J. (1994) 'Grasp stability during manipulative actions'. *Can J Physiol Pharmacol*, 72: 511–24.
11. Johansson, R. S. (1998) 'Sensory input and control of grip'. *Novartis Found Symp*, 218: 45–59; discussion 59–63.
12. Johansson, R. S., Backlin, J. L. and Burstedt, M. K. (1999) 'Control of grasp stability during pronation and supination movements'. *Exp Brain Res*, 128: 20–30.
13. Johansson, R. S. and Cole, K. J. (1992) 'Sensory-motor coordination during grasping and manipulative actions'. *Curr Opin Neurobiol*, 2: 815–23.
14. Latash, M. L., Scholz, J. P. and Schoner, G. (2002) 'Motor control strategies revealed in the structure of motor variability'. *Exerc Sport Sci Rev*, 30: 26–31.
15. Latash, M. L., Shim, J. K. and Zatsiorsky, V. M. (2004) 'Is there a timing synergy during multi-finger production of quick force pulses?' *Exp Brain Res*, 159: 65–71.
16. Lukos, J. R., Ansuini, C., and Santello, M. (2007a) 'Choice of contact points during multidigit grasping: Effect of predictability of object center of mass location'. *J Neurosci*, 27(14): 3894–903.
17. Lukos, J.R., Ansuini, C., and Santello, M. (2007b) 'Effect of cue about object centre of mass location on fingertip contact point selection for multidigit grasping'. *Soc Neurosci Abst* 818.23.
18. Mason, C. R., Gomez, J. E. and Ebner, T. J. (2001) 'Hand synergies during reach-to-grasp'. *J Neurophysiol*, 86: 2896–910.

19. Mason, C. R., Theverapperuma, L. S., Hendrix, C. M. and Ebner, T. J. (2004) 'Monkey hand postural synergies during reach-to-grasp in the absence of vision of the hand and object'. *J Neurophysiol*, 91: 2826–37.
20. Quaney, B. M. and Cole, K. J. (2004) 'Distributing vertical forces between the digits during gripping and lifting: the effects of rotating the hand versus rotating the object'. *Exp Brain Res*, 155: 145–55.
21. Raghavan, P., Santello, M., Krakauer, J. W. and Gordon, A. M. (2006) 'Shaping the hand to object contours after stroke, the control of fingertip position during whole hand grasping'. *Soc Neurosci Abst* 655:14.
22. Reilmann, R., Gordon, A. M. and Henningsen, H. (2001) 'Initiation and development of fingertip forces during whole-hand grasping'. *Exp Brain Res*, 140: 443–52.
23. Sakitt, B. (1980) 'Visual-motor efficiency (VME) and the information transmitted in visual-motor tasks'. *Bul. Psych Soc* 16: 329–32.
24. Salimi, I., Frazier, W., Reilmann, R. and Gordon, A. M. (2003) 'Selective use of visual information signaling objects' centre of mass for anticipatory control of manipulative fingertip forces'. *Exp Brain Res*, 150: 9–18.
25. Salimi, I., Hollender, I., Frazier, W. and Gordon, A. M. (2000) 'Specificity of internal representations underlying grasping'. *J Neurophysiol*, 84: 2390–7.
26. Santello, M. (2002) 'Kinematic synergies for the control of hand shape'. *Arch Ital Biol*, 140: 221–8.
27. Santello, M., Flanders, M., Soechting, J. F. (1998) 'Postural hand synergies for tool use'. *J Neurosci* 18: 10105–15.
28. Santello, M., Flanders, M. and Soechting, J. F. (2002) 'Patterns of hand motion during grasping and the influence of sensory guidance'. *J Neurosci*, 22: 1426–35.
29. Santello, M. and Soechting, J. F. (1998) 'Gradual molding of the hand to object contours'. *J Neurophysiol*, 79: 1307–20.
30. Schettino, L. F., Adamovich, S. V., Hening, W., Tunik, E., Sage, J., Poizner, H. (2006) 'Hand preshaping in Parkinson's disease: effects of visual feedback and medication state'. *Exp Brain Res* 168: 186–202
31. Schettino, L. F., Adamovich, S.V. and Poizner, H. (2003) 'Effects of object shape and visual feedback on hand configuration during grasping'. *Exp Brain Res*, 151: 158–66.
32. Schettino, L. F., Rajaraman, V., Jack, D., Adamovich, S. V., Sage, J., Poizner, H. (2004) 'Deficits in the evolution of hand preshaping in Parkinson's disease'. *Neuropsychologia*, 42: 82–94.
33. Schieber, M. H. and Santello, M. (2004) 'Hand function: peripheral and central constraints on performance'. *J Appl Physiol*, 96: 2293–300.
34. Theverapperuma, L. S., Hendrix, C. M., Mason, C. R. and Ebner, T. J. (2006) 'Finger movements during reach-to-grasp in the monkey: amplitude scaling of a temporal synergy'. *Exp Brain Res*, 169: 433–48.
35. Winges, S. A., Weber, D. J. and Santello, M. (2003) 'The role of vision on hand preshaping during reach to grasp'. *Exp Brain Res*, 152: 489–98.
36. Zatsiorsky, V. M. and Latash, M. L. (2004) 'Prehension synergies'. *Exerc Sport Sci Rev*, 32: 75–80.

9

Biomechanical aspects of motor control in human landing

Dario G. Liebermann
Tel Aviv University, Tel Aviv

Introduction

Free falls trigger protective landing actions that concern diverse areas of research because of the variety of responses elicited and the mechanisms involved. From a functional vantage point, the ability to land softly is required in many daily locomotor tasks of animals and humans, e.g., after sudden unexpected falls (Greenwood and Hopkins, 1976a, b), while stepping downstairs (Greenwood and Hopkins, 1980), and when humans (Liebermann and Hoffman, 2006), cats (McKinley, Smith and Gregor, 1983) or monkeys (Dyhre-Poulsen and Laursen, 1984) jump down. Research interest has also focused on specific applications such as parachute landings (Hoffman, Liebermann and Gusis, 1997), landings after releasing a horizontal bar (Liebermann and Goodman, 2007), landings onto a gymnastics mat (Arampatzis, Morey-Klapsing and Brüggemann, 2003) as well as in physical activities such as dancing (Cluss *et al.*, 2006), whereas the inherent practical goal is to improve the quality of the landing performance in order to prevent injuries.

Preventive landing actions may be built upon basic reflexes (e.g., Moro's reflex) that disappear within the first months after birth (usually four months). Babies respond to a free fall with extensions of the arms (Schaltenbrand, 1925) or with extensions of arms and legs together when they are raised in supine position and dropped in the air (Irwin, 1932). Such actions evoke startle as well as vestibulo-spinal reflexes (labyrinthine responses). Through development, learning and experience these early reflexive reactions may eventually combine with more complex voluntary responses. For example, infants learn from continuously falling when they start walking. They accumulate landing experiences (Joh and Adolph, 2006) and, at some stage, the basic responses are suppressed while a well-structured set of protective reactions appear during a fall and after landing. Such responses in healthy adults become pre-programmed and voluntarily triggered (Craik, Cozzens and Freedman, 1982).

The current chapter focuses primarily on these motor responses during the pre- and post-landing periods. It is argued that during the pre-landing stage, subjects follow a plan of action based on a temporal framework defined by the moment of release and the moment of touchdown. In the post-landing stage, when the fall involves small magnitude impacts, the motor responses are assumed to follow the laws of biomechanics. If impact forces are high, the responses may obey anticipatory motor strategies with the goal of preventing injuries. For the purposes of this chapter, the landing event is subdivided into four different stages:

1 reflexive post-release stage;
2 voluntary pre-landing stage;
3 passive post-landing stage; and
4 active post-landing stage.

The first two stages take place during the flight, while the last two take place after touchdown.

Reflexive post-release stage: Unexpected free-falls from low heights

Matthews (1956) first introduced the free-fall paradigm to investigate neurophysiologic reactions to sudden changes from 1 g to zero gravity. In continuation with this initial attempt, Matthews and Whiteside (1960) showed that tendon reflexes and voluntary responses were suppressed during free falling if subjects were unexpectedly released while sitting on an elevated chair. Matthews and Whiteside (1960) observed also a relatively long silent-period until an initial 'startle reflex' was elicited.

Melvill Jones and Watt (1971a, b) argued that such reflexive responses after unexpected falls, at a latency of ≈75 msec, were of vestibulo-otolith origin and seemed to be independent of the height of fall. Melvill Jones and Watt also noticed that sudden falls from low heights (i.e., heights <5 cm) would not allow time for the reflex response and its subsequent build up of muscle tension to reduce landing impact. Considering an electro-mechanical delay of around 40 msec after the muscle response begins, the vestibulo-otolith reflex could only play a protective role in unexpected falls that last longer than 105 msec.

Greenwood and Hopkins (1976a, b) reported that under the constant acceleration of gravity, changes in height of fall did not affect the amplitude of the first EMG burst. They argued that the response is rather a startle reflex, specifically observed in the unexpected falls when acceleration was larger than 0.2 g. A lower acceleration level did not elicit the reflex because it did pass the sensitivity threshold of the otolith organs but more recent results showed that this excitation threshold for changes in linear acceleration is ≈0.1 g, depending on the direction of the change (Benson, Kass and Vogel, 1986).

In order to test whether the observed first burst is indeed a startle response in landing, an experiment was carried out (Liebermann, D.G., unpublished data) where six subjects (mean age = 26.16 years; range = 22–32 years) performed vertical free falls (self-initiated trials) in vision and no vision conditions. EMG from the Frontalis muscle and lower-limb extensor muscles were recorded during the flight. Obviously, the Frontalis is functionally irrelevant to the landing responses, and, therefore, its activity shortly after the start of the fall was hypothesized to reflect a general startle response.

After signing the consent forms, subjects performed six blocked free-fall trials from each of four flight-time ranges (0.10–0.15 sec, 0.20–0.22 sec, 0.32–0.37 sec and 0.40–0.44 sec) in vision and no vision conditions (48 landings in counterbalanced order). A horizontal bar that could be set at a variety of heights from the floor was used as a free-fall device. Upon release of the bar the fall started and with it the recording process. Surface EMG was collected, amplified (gain x 1000) and sampled via A/D at 1 KHz. The data were stored in a PC and analyzed offline. The frequency of cases where the Frontalis muscle was activated within 100 msec after the start of the fall was computed. Analysis of variance (with 'visual condition' and 'flight-time range' as the within-subjects factors) was carried out using the percentage of trials that presented Frontalis activity. The results showed that the Frontalis muscle was activated in more than 30% of the cases for flight times greater than 0.20 sec in visual landings, while in blindfold landings the chance of activation increased for short as well as long fight times up to 100% of the trials for highest falling heights (Table 9.1).

A two-way ANOVA with repeated measures (2 'conditions' x 4 flight-time range 'categories') using percentages of cases showing Frontalis activation resulted in significant effects of both factors and a significant interaction (**p** $\leq$ 0.05), whereas most pair-wise comparisons showed significant differences with the exception of the lowest landing heights pair in vision conditions. That is, when the falls lasted longer than 0.32 sec (>0.50 m) the probability was high that a startle response was evoked regardless of the visual condition. Such responses were unlikely in falls shorter than 0.22 sec (>0.24 m).

The Frontalis onset occured simultaneously with activation of the lower limb extensors shortly after release. Thus, the first burst of activity may be considered a startle response particularly in conditions where visual input is occluded when the falling height is high enough. It was also observed that self-released landings elicited the same response even when vision was available. That is, the first burst of EMG activity is not exclusively related to sudden unexpected falls, as suggested by Greenwood and Hopkins (1976a). This, and similar landing-related responses, serve a wide range of tasks.

Stretch reflexes may co-exist with pre-programmed landing actions voluntary triggered. Muscle spindles may be set to a higher sensitivity before touchdown and tuned to the actual conditions during the fall (McDonagh and Duncan, 2002). When unexpected falls last longer than 85 msec descending gating mechanisms facilitate the reflex, and thus, its amplitude may double (McDonagh and Duncan, 2002).

In summary, the first EMG after release is related to the initiation of the fall and it has a vestibulo-otolith origin. It may be a default startle response, which is facilitated when vision is occluded. As with other reflex mechanisms, free fall reflexes may be coupled with the voluntary responses that follow.

For significant heights of falls (>200 msec or around 0.20 m), Greenwood and Hopkins (1976a, b) showed that at the end of the reflexive responses a silent period followed until a voluntary preparatory activity was observed. In the next section, we shall review the main findings regarding the influence of visual perception, cognition and passive biomechanical properties on the voluntary preparatory responses during this second period.

Table 9.1 Percentage of trials in which the Frontalis muscle was activated within a 100 msec time-window after the start of self-initiated falls in two visual conditions and four flight-time ranges*

	Flight time category ranges			
Condition	*0.10–0.15*	*0.20–0.22*	*0.32–0.37*	*0.40–0.44*
Vision	27.78	30.55	86.11	94.44
No-vision	41.66	86.11	100	100

*(Liebermann, D.G., unpublished data)

Voluntary pre-landing stage

Is online visual perception important for the preparation to land?

In the voluntary pre-landing stage, individuals implement different strategies to reduce the forthcoming impact with the ground. Greenwood and Hopkins (1976 a, b) observed that voluntary EMG bursts started in the lower-limb muscles at an invariant time of ≈40–140 msec before touchdown. During these moments before landing, availability of vision was generally assumed to be essential for the landing preparation. However, this is not clear yet, even though the role of vision in guiding interceptive actions (Lee, 1976; Tresilian, 1990, 1991, 1993; Regan 2002), and in particular during landing events (Lee and Reddish, 1981; Goodman and Liebermann, 1992), has been a focus of interest for several years.

The initiation of the preparatory actions has traditionally been related to the moment of touchdown. The Tau (τ) visual strategy based on time-to-contact (Lee, 1976, 1980a, b) could indeed be a reasonable heuristic for explaining the preparatory landing actions in humans as it has been hypothesized in birds (Lee and Reddish, 1981; Lee, Reddish and Rand, 1991; Lee *et al*., 1993) and flies (Wagner, 1982). In honeybees, velocity of approach obtained visually, rather than time to contact information, has been suggested as the relevant information used in guiding landings (Srinivasan *et al*., 2000). Humans may also control landing actions via continuous visual input, however, it has been observed that when falls are self-initiated, the voluntary EMG pattern in vision and blindfold trials is not very different (Greenwood and Hopkins, 1976b, p. 380) even when standing and falling forward on the hands (Dietz and Noth, 1978, p. 579). Such observations are consistent with the preparatory landing behaviour reported also in blindfolded cats (McKinley and Smith, 1983) and primates (Dyhre-Poulsen and Laursen, 1984), which suggests that the timing of EMG onset before landing is not affected by elimination of vision during the flight. A preliminary study in human subjects (Liebermann, Goodman and Chapman, 1987) showed also that the timing of the preparatory landing actions was not affected by visual occlusion during self-release vertical falls. To further investigate the control of preparatory landing actions, a more detailed analysis was carried out, first to test if such actions were based on the τ visual strategy based on time-to-contact as suggested by Lee (1980a,b), and second to assess if online visual inputs were used while landing. If a τ strategy were implemented, the relation between time-to-collision and the flight time should have followed a negatively accelerated exponential curve reaching saturation at long free-fall durations. An alternative strategy was that subjects would start acting at a constant time from the beginning of the fall or at a constant distance. In the latter case, the relationship would also yield an exponential curve following the increase in falling speed under constant gravity. The results did not present any exponential increase in time-to-contact with increases in the duration of the fall. For the height of fall ranges in the above-mentioned experiment (Liebermann, Goodman and Chapman, 1987), it seemed that the relationship between time-to-contact and flight time followed a linear trend. This is illustrated in the following plots (Figure 9.1) obtained in vision and no vision trials for one subject that participated in those trials (Liebermann, D.G., unpublished data).

These findings argue against comparable research presented by Sidaway, McNitt-Gray and Davis (1989), which suggests that the use of a τ heuristic is implemented by humans when vision is available. In fact, the preliminary results presented by Liebermann and colleagues suggested that online visual input may not even be necessary because vision and blindfolded trials generated similar outcomes before and after landing (Liebermann and Goodman, 1991).

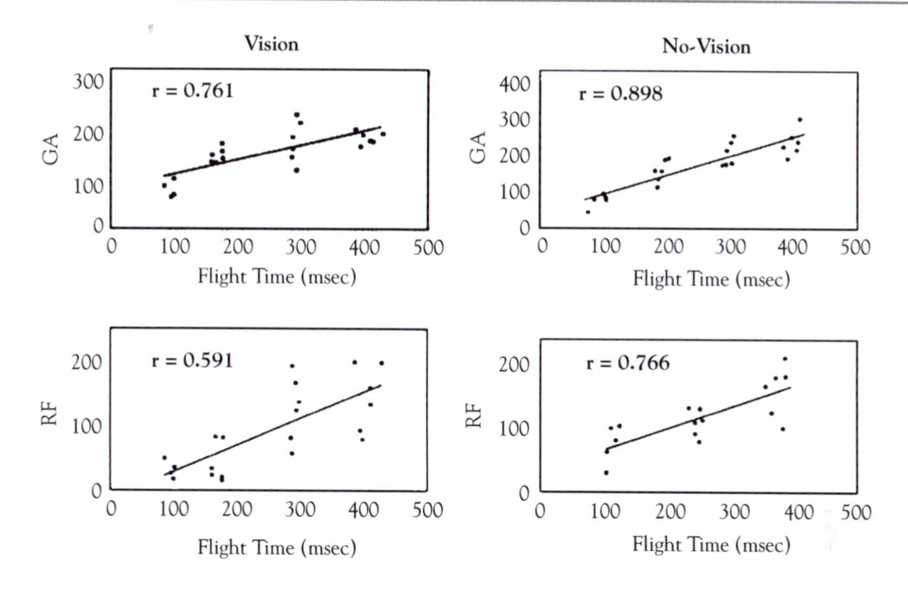

Figure 9.1 The plots shows the linear fit and the coefficients of correlation between fight-time (x-axis) and the time-to-contact (y-axis) in milliseconds for Subject#1 during self-released landings trials in vision (left column) and no-vision conditions (right column). Time to contact is defined as the period from the moment of onset of the second burst of EMG in the gastrocnemius (GA) and the rectus femoris muscles (RF) until the moment of contact with the ground.

There could be several explanations for the discrepancies between the above reported findings. Mammals and insects land head-down and depend on an efficient use of continuous visual information for landing. However, humans land first on their feet, while the landing surface and the legs are not always seen. In addition, humans cannot focus visual attention on several events simultaneously (Gazzaniga *et al.*, 2002). Visual perception and attention may be based on the sampling of events, particularly during interceptive actions such as landing on surfaces or catching a flying baseball (McLeod, Reed, and Dienes, 2003). Intermittent visual inputs, sampled at critical times, may be enough to control movement (Elliott, 1990), and consequently, humans may not need continuous visual information for landing, as some authors argue (Lee, 1981; Sidaway, McNitt-Gray and Davis, 1989).

Is cognitive input relevant for the control of landing actions?

It has been shown that appropriate timing of voluntary pre-landing responses may be triggered during the flight whether vision is available or completely occluded (Goodman and Liebermann, 1992). When people are aware about the forthcoming event, they may organize motor actions based on prior knowledge about the environment and the height of fall. Cognitive rather than perceptual visual input may thus be used to trigger a set of pre-programmed responses during the flight relative to the moment of the initiation of the fall. The time of onset relative to release is not constant, and increases linearly (proportionally) with increases in the flight time. Such a linear increase in latency of onset has been observed within a wide range of falling heights (from 10 cm up to a limit of 130 cm free-falls), when vision was occluded or available (Liebermann and Goodman, 2007), and when gaze was oriented in different directions (Liebermann and Hoffman, 2006). In the latter two studies, Liebermann and his colleagues allowed subjects to have prior knowledge of the landing height, approximately 10 sec before wearing opaque goggles. As found also by other authors, this prior knowledge may significantly influence the landing strategy (Thompson and McKinley, 1995). However, knowledge of the height of the fall was not allowed in similar research (Santello and McDonagh, 1998); Santello, McDonagh, and Challis, 2001), and therefore, findings were different.

From the above studies, it transpires that cognitive information may play a fundamental role in the formation of a strategy for preventing impacts at touchdown. In order to reduce the impact forces, humans may use their ability to organize prospective actions based upon their previous experiences. This does not imply that visual information is completely irrelevant but it cannot be consolidated with direct visual perception (Gibson, 1979; Lee, 1980b). Visual representation of the heights of fall and information about the surrounding environment may be very important. However, this visual information is not necessarily perceived online and may not rely on optic flow. In brief, it appears that pre-landing actions are pre-programmed and may be triggered relative to the moment of release. Amplitude and duration of the muscular activity may be tuned, at least partially, based on some estimates of the height of fall or the time of flight, and not on online visual information about the moment of touchdown. The implementation of such plan of action implies a feedforward mode of control.

Passive post-landing stage: Free-fall mechanics and reliance on passive tissue characteristics

During the first post-landing period subjects are rather passive (Nigg and Herzog, 1999). However, it was argued that responses may already be structured before landing, i.e., they are anticipatory. Anticipatory strategies may play a role during the passive landing stage by allowing to an increase in joint stiffness and the passive absorption of kinetic energy.

The rise time of the vertical force to a first peak (the peak of the first peak PFP) is in the range of 10–20 msec (the time to first peak TFP), which is not enough time for implementing a voluntarily response and hardly for initiating a reflex. The total duration of this passive stage is often shorter than 50 msec (Nigg and Herzog, 1999) depending on the actual landing momentum p (i.e., the longer the flight time and the larger the body mass, the larger the momentum and the shorter the duration of the first stage). Some biological adaptation and accommodation may be observed during this brief landing stage (James, Bates and Dufek, 2003), although the most likely alternative is still that subjects prevent potential injuries by adopting anticipatory strategies.

Anticipatory landing actions in the first passive stage

Increasing joint stiffness is a strategy that has been observed in monkeys while landing on their arms (Dyhre-Poulsen and Laursen, 1984) as well as in humans when landing on their feet (Horita *et al.*, 1996). In the case of humans, such an increase in joint stiffness is particularly observed for the knee joint in landings from low falling heights (<30 cm); the implication is that the heels absorb the impacts (Self and Payne, 2001). Such an increase in lower-limb joint stiffness starts before touchdown and continues after contacting the ground (Horita *et al.*, 2002). Passive tissue properties of the calf muscles and the Achilles tendon (e.g., viscoelastic and elastic elements, in parallel and in series) may be sufficient to dissipate the kinetic energy in low landings when the knee joint stiffness is increased (Self and Payne, 2001).

The unilateral tensile forces acting in vivo on the Achilles tendon–calf muscle system have been estimated at approximately 1895 N during maximal squat jumps of 0.33 m (Fukashiro *et al.*, 1995). Assuming that as far as lower-limb muscle forces are concerned, jumping and landings are similar (Hoffman, Liebermann and Gusis, 1997), it is possible to extrapolate from the stress on the Achilles tendon during jumping to the stress exerted on the tendon during landings.

A body mass (m) of 76.5 kg falling vertically from a height (H) of 0.33 m will generate forces that will not surpass the injury threshold of the tendon even if the landing is on one leg (landings are often on both legs). The impact forces at the instant of touchdown can be estimated from simple Newtonian mechanics. The final velocity V_1 of the falling mass can be calculated using the formula $V_I = A \cdot Tf$, where the acceleration is $\underline{A = (V_1 - V_0)}$ / dt, and the time of flight *Tf* can be calculated from the formula $Tf = \sqrt{2H/g}$, where gravity $g = 9.81$ m.sec^{-2}. For a falling height $H = 0.33$m, the flight time is $Tf = 0.259$ sec and the final velocity of the mass *is* $V_1 = 9.81$ m.sec^{-2} x 0.259 sec; i.e., $V_1 = 2.545$ m/s.

The force *F* is the difference between quantity of movement (i.e., the mass momentum *p*) before and after the application of the force within some interval *dt* (e.g., 0.1 sec). Using Newton's second law, $F = (m{\cdot}V_1 - m{\cdot}V_0)/dt$, where the initial velocity in our falling mass is $V_0 = 0$ and mass is constant. Thus, the vertical ground reaction force (VGRF) may simply be F_z = (76.5 kg x 2.545 m/s)/0.1 sec = 1947 N (N unit is equivalent to kg.m.sec^{-2}), which is distributed between the two legs.

The jolts experienced during the post-landing stage depend on the evolution of the momentum over time. The momentum *p* is conserved, and thus, a collision with the stiff ground returns the force to the body (action-reaction principle), and may be differently experienced as a function of the post-landing durations, i.e., as a function of the product of F_z and the time interval *dt*, called the impulse $I = \int_{t0}^{t1} Fz\, dt$ (the integral of the VGRF). The impulse is a physical quantity that may be used to express the effect of the landing collision. If *dt* were too short, a non-impulsive force would be generated and, if it were long enough the effects of the landing force F_z would be dissipated. In the first instance after touchdown, passive components absorb the impact forces. The length of this period depends on the velocity of landing, and thus, if the velocity is high, the time to the first peak of force is short.

Based on a pilot field experiment in parachute performances with a standard size canopy from 400 m dives, the final velocity at the moment of touchdown has been estimated between 5.0–6.0 m/s, reaching occasionally speeds of ≈8.0 m/s (Liebermann, D.G., unpublished data). In laboratory conditions, such final velocities were simulated during free-fall landings after drop jumps from a platform of adjustable heights up to 2.00 m. In the experiment, a group of seven experienced male subjects carried out five consecutive trials from each of seven landing heights. The TFP values for this group are presented in Table 9.2.

There is little that can be done in order to extend the time to the first increase of force. As previously suggested, the adoption of anticipatory strategies is a most likely option. Increasing muscle power is another option. Hoffman, Liebermann and Gusis (1997) reported that experienced and well-trained subjects may have an advantage in the reduction of the impacts during the first phase as compared with inexperienced subjects. Such ability is related to the higher levels of lower-limb muscular strength and power. Experience and training may be significant for the reduction of landing injuries during the first moments after touchdown, which are particularly observed at the ankle joint of inexperienced individuals (Amamilo *et al.*, 1987).

In brief, the suggestion that motor control strategies may take advantage of the biomechanical properties of the muscle-tendon system is not new (Bach, Chapman and Calvert,

Table 9.2 Mean and SD for the time to first peak values (TFP) for a group of experienced subjects performing drop jumps from different height of fall ranges

HEIGHTS	*TFP*
Range (n trials)	*Mean ± SD*
20–25 (n = 27)	0.0190 ± 0.0070
45–50 (n = 28)	0.0217 ± 0.0068
70–75 (n = 27)	0.0170 ± 0.0036
95–100 (n = 29)	0.0164 ± 0.0032
120–125 (n = 26)	0.0144 ± 0.0030
145–150 (n = 32)	0.0142 ± 0.0021
170–175 (n = 30)	0.0137 ± 0.0033

1983; Chapman, Caldwell and Selbie, 1985; Alexander, 1988). Given enough practice, performers may take advantage of passive properties of the system such as visco-elasticity and elasticity (piston-like behaviour or spring-like behaviour, respectively). For example, by adopting a slight flexion of the knees during the flight (shortly before touchdown) and dorsal flexion at the ankle joint, landing impacts can be reduced by maintaining joint stiffness and letting the system respond like a pre-loaded spring. It is argued that implementation of anticipatory actions based on the use of mechanical properties is important mainly during the passive post-landing stage.

Active post-landing stage: More than just mechanics

The second phase starts at periods larger than 50 msec from the moment of touchdown (Nigg and Herzog, 1999) and it ends when zero changes in VGRF are achieved. The major peak of the active post-landing stage (the peak of the second peak [PSP]) can reach high magnitudes, but such forces can be reduced by voluntary motor control. During this active part of the post-landing stage, the performer may avoid sudden impacts by using smooth multi-joint interactions and by increasing the duration of the post-landing period (recall that if the landing force F_z were dissipated within a long period, the effects of the impulsive forces would be reduced). A discontinuous rise in force F_z within brief time-intervals (i.e., the occurrence of sudden peaks) implies potentially high impulses (e.g., high PSP) and a potential risk of injury. The time of dissipation of F_z and the evolution of the forces are therefore critical for the reduction of collision impacts during this second post-landing phase.

The time between the first peak and the second peak may be an indicator of how long subjects spend actively reducing the impacts (Liebermann and Goodman, 1991). It was hypothesized that the longer the interval between the first and the second peak, the smaller the magnitude of the second peak. From data obtained from the previous sample of experienced subjects, the time difference between the peaks ($\Delta_{TSP\text{-}TFP}$) was calculated and correlated with the impacts at touchdown. Table 9.3 presents the values obtained for different heights of fall categories. The results suggest that when the time interval $\Delta_{TSP\text{–}TFP}$ increases, the peak of the second peak of force decreases (Pearson coefficient of correlation $\rho = -0.684$; $p<0.05$ for $N=134$). Nevertheless, such a trend is not observed for all landing heights. For low heights of fall, impact forces may have been dissipated by passive biomechanical properties, and thus, an expansion of the active landing stage was of no use. For very high landing heights, the results suggest that other strategies are needed; e.g., a better distribution of

Table 9.3 Mean and SD for time difference between the peaks ($\Delta_{TSP\text{-}TFP}$), the amplitudes of the second peak (PSP in times body weight TBW units), and the corresponding Pearson correlations (ρ) between the time interval and the peak force during the active post-landing period

HEIGHTS	$\Delta_{TSP\text{-}TFP}$	*PSP*	*Correlation*
Range (n trials)	*Mean ± SD*	*Mean ± SD*	*ρ values*
20–25 (n = 27)	0.0016 ± 0.0558	3.8470 ± 0.5432	–0.467
45–50 (n = 28)	0.0171 ± 0.0425	4.3552 ± 0.8308	–0.663*
70–75 (n = 27)	0.0295 ± 0.0072	4.8882 ± 0.7189	–0.689*
95–100 (n = 29)	0.0305 ± 0.0060	4.9735 ± 0.7759	–0.743**
120–125 (n = 26)	0.0257 ± 0.0051	5.1548 ± 0.8092	–0.618*
145–150 (n = 32)	0.0246 ± 0.0046	5.6460 ± 0.7443	–0.689*
170–175 (n = 30)	0.0239 ± 0.062	5.8553 ± 0.8015	–0.367

* $p \leq 0.05$
** $p \leq 0.01$

the forces over a larger body area by rolling sideways, as commonly recommenced in military parachute landings.

It is suggested that good landing performances involve the use of multiple joints in order to decompose the VGRF into several vector subcomponents. Since the area under the feet is rather small, the impulse of force at landing may create relatively high pressures. Increasing body contact-area with the landing surface is one way to decrease the effect of the landing impacts. For example, judo falls are characterized by using multiple joints and lateral rolling actions that decrease pressures and prolongs post-landing time (Groen, Weerdesteyn and Duysens, 2007).

Finally, in the active post-landing stage, cognitive information may play an important role in determining the strategy of action required for meeting specific task demands. Contextual information (e.g., landings in sport situations or landing from a stair), knowledge about the type of landing surface (stiff versus soft; Fukuda, Miyashita and Fukuoka, 1987; Gollhofer, 1987) and the instructions given to a performer (McNair, Prapavessis and Callender, 2000) may have significant effects on the way people reduce impacts at landing.

Conclusions

The present chapter dedicated special attention to the different stages of falling and landing. In the first stages, reflexive actions seem to serve for the purpose of landing by providing a build up of muscle tension that will later allow taking advantage of passive mechanics. In the second stage, the reduction of impacts with the ground is based on a voluntary and active implementation of biomechanical principles, perhaps internalized upon practice and experience (Mussa-Ivaldi and Bizzi, 1997).

The use of continuous visual information during the flight was also of concern. It was argued that continuous vision is not used during landings because online optic flow information may not convey the inputs required to organize action within the limits of normal landing performances. Cognitive mechanisms were instead assumed to provide the basis for the formation of landing plans that could be triggered at any time relative to the moment of the start of a fall in anticipation of the impact.

Future research should focus on single-legged landings. The existing literature concentrates mostly in landings performed on both legs and in near vertical conditions. It may be presumed that a certain percentage of injuries (ankle sprains and ACL rupture in the knee joint) may be related to landing asymmetrically on one leg. In addition, most landing situations encountered in natural daily conditions are not strictly vertical, i.e., forces applied in the three directions are exerted while landing (e.g., the wind velocity in parachute landings or in basketball performance after jumping). Therefore, future research should focus on gathering information in field conditions under normal daily living constraints.

Finally, the role of cognition (e.g., the role of visual attention, the use of visual representations and the sense of duration and timing) is not resolved yet. These issues should be explored because they may have a relevance in the way people (e.g., elderly) could be trained to prevent falls and to avoid landing-related injuries.

Acknowledgment

The author wishes to thank Prof. L. Katz from the University of Calgary for his comments on an early version of this manuscript.

References

1. Alexander McNeil, R. (1988) 'Elastic Mechanisms in Animal Movement'. Cambridge: Cambridge University Press.
2. Amamilo, A. C., Samuel, A. W., Hesketh, A. T. and Moynihan, F. J. (1987) 'A prospective study of parachute injuries in civilians', *J. Bone Joint Surg. (Britain)*, 69–B: 17–19.
3. Arampatzis, A., Morey-Klapsing, G. and Brüggemann, G.P. (2003) 'The effect of falling height on muscle activity and foot motion during landings', *J. Electromyo Kines*, 13: 533–544.
4. Bach, T. M., Chapman, A. E. and Calvert, T. W. (1983) 'Mechanical resonance of the human body during voluntary oscillations about the ankle joint', *J. Biomech*, 16: 85–90.
5. Benson, A. J., Kass, J. R. and Vogel, H. (1986) 'European vestibular experiments on the Spacelab-1 mission: 4. Thresholds of perception of whole-body linear oscillation', *Ex. Brain Res.*, 64: 264–271.
6. Chapman, A. E., Caldwell, G. E. and Selbie, W. S. (1985) 'Mechanical output following muscle stretch in forearm supination against inertial loads', *J. Appl. Physiol.*, 59: 78–86.
7. Cluss, M., Laws, K., Martin, N. and Nowicki, T. S. (2006) 'The indirect measurement of biomechanical forces in the moving human body', *Am. J. Phys.*, 74: 102–108.
8. Craik, R. L., Cozzens, B. A. and Freedman, W. (1982) 'The role of sensory conflict on stair descent performance in humans', *Ex. Brain Res.*, 45(3): 399–409.
9. Dietz, V. and Noth, J. (1978) 'Pre-innervation and stretch responses of the triceps brachii in man falling with and without visual control', *Brain Res.*, 142: 576–579.
10. Dyhre-Poulsen, P. and Laursen, A. M. (1984) 'Programmed electromyographic activity and negative incremental stiffness in monkeys jumping downward', *J. Physiol.*, 350: 121–136.
11. Elliott, D. (1990) 'Intermittent visual pickup and goal directed movement: A review', *Hum. Mov. Sci.*, 9: 531–548.
12. Fukashiro, S., Komi, P. V., Järvinen, M. and Miyashita, M. (1995) 'In vivo Achilles tendon loading during jumping in humans', *Eur. J. Appl. Physiol.*, 71: 453–458.
13. Fukuda, H., Miyashita, M. and Fukuoka, M. (1987) 'Unconscious control of impact force during landing', In: B. Jonsson (Ed.), *Biomechanics* X-A. Champaign, Ill: Human Kinetic Publishers, pp. 301–305.

14. Gazzaniga, M., Ivry, R. and Mangun, G. (2002) 'Cognitive Neuroscience: The Biology of the Mind' (2nd Ed.). New York: W.W. Norton and Company, Inc., pp. 247–252.
15. Gibson, J. J. (1979) 'The ecological approach to visual perception', Boston: Houghton Mifflin.
16. Gollhofer, A. (1987) 'Innervation characteristics of the m. gastrocnemius during landing on different surfaces', In B. Jonsson (Ed.), *Biomechanics X-B*. Champaign, Ill: Human Kinetic Publishers, pp. 701–706.
17. Goodman, D. and Liebermann, D. G. (1992) 'Time to contact as a determiner of action: vision and motor control'. In: D. Elliott and L. Proteau (Eds.), *Vision and Motor Control*. Amsterdam: North Holland, pp. 335–349
18. Greenwood, R. and Hopkins, A. (1976 a) 'Landing from an unexpected fall and a voluntary step', *Brain*, 99: 375–386.
19. Greenwood, R. and Hopkins, A. (1976 b) 'Muscle responses during sudden falls in man', *J. Physiol. (London)*, 254: 507–518.
20. Greenwood, R. and Hopkins, A. (1980) 'Motor control during stepping and falling in man'. In: J. E. Desmedt (Ed.) Spinal and supraspinal mechanisms of voluntary motor control and locomotion. *Progress in Clinical Neurophysiology*, 8. Basel: Karger, pp. 294–309.
21. Groen, B.E., Weerdesteyn, V. and Duysens, J. (2007) 'Martial arts fall techniques decrease the impact forces at the hip during sideways falling', *J. Biomechanics*, 40: 458–462.
22. Hoffman, J.R., Liebermann, D. G. and Gusis, A. (1997) 'Relationship of leg strength and power to ground reaction forces in both experienced and novice parachute jump trained personnel', *Av. Space Environ. Med.*, 68: 710–714.
23. Horita, T., Komi, P. V., Nicol, C. and Kyröläinen, H. (1996) 'Stretch shortening cycle fatigue: interactions among joint stiffness, reflex and muscle mechanical performance in the drop jump', *Eur. J. Appl. Physiol.*, 73: 393–403.
24. Horita, T., Komi, P. V., Nicol, C. and Kyröläinen, H. (2002) 'Interaction between pre-landing activities and stiffness regulation of the knee joint musculoskeletal system in the drop jump: implications to performance', *Eur. J. Appl. Physiol.*, 88: 76–84.
25. Irwin, O. C. (1932) 'Infant Responses to Vertical Movements', *Child Development*, 3(2): 167–169.
26. James, R. C., Bates, B. T. and Dufek, J. S. (2003) 'Classification and comparison of biomechanical response strategies for accommodating landing impact', *J. Appl. Biomechanics*, 19: 106–118.
27. Joh, A. S. and Adolph, K. E. (2006) 'Learning from falling', *Child Development*, 77: 89–102.
28. Lee, D. N. (1976) 'A theory of visual control of braking based on information about time-to-collision', *Perception*, 5: 437–459.
29. Lee, D. N. (1980a) 'Visuo-motor coordination in space-time', In: G. E. Stelmach and J. Requin (Eds.), *Tutorials in Motor Behaviour*. Amsterdam: North-Holland.
30. Lee, D. N. (1980b) 'The optic flow field: the foundation of vision', *Phil. Transac. Royal Soc of London*, 290: 169–179.
31. Lee, D. N. and Reddish, P. E. (1981) 'Plummeting gannets: a paradigm of ecological optics', *Nature*, 293: 293–294.
32. Lee, D. N., Davies, M. N. O., Green, P. R. and van der Weel, F. R. (1993) 'Visual control of velocity of approach by pigeons when landing', *J. Exp. Biol.*, 180: 85–104
33. Lee, D. N., Reddish, P.E. and Rand, D.T. (1991) 'Aerial docking by hummingbirds', *Naturwissenschaften*, 78: 526–527.
34. Liebermann, D. G. and Goodman, D. (1991) 'Effects of visual guidance on the reduction of impacts during landings', *Ergonomics*, 34 (11): 1329–1406.
35. Liebermann, D. G. and Goodman, D. (2007) 'Pre- and Post-landing effects of falling with and without vision', *J Electromyogr. Kines.*, 17: 212–227.
36. Liebermann, D. G. and Hoffman, J. R. (2005) 'Changes in the timing of the preparatory muscle responses to free falls as a function of visually mediated information' *J. Electromyogr. Kines.*, 15: 120–130.

37. Liebermann, D. G., Goodman, D. and Chapman, A. E. (1988) 'Implications of the time-to-contact variable on timing the preparatory landing response', In: *Proceedings of the 19th Annual Meeting of the Canadian Society for Psychomotor Learning and Sport Psychology*, Collingwood-Ontario, p. 29.
38. Lishman, J. R. and Lee, D. N. (1973). 'The autonomy of visual kinaesthesis.' *Perception*, 2: 287–294.
39. Matthews, B. (1956) 'Tendon reflexes in free falls', *J. Physiol.*, 133: 31–32P.
40. Matthews, B. and Whiteside, T. C. D. (1960) 'Tendon Reflexes in Free Fall'. *Proc. Royal Soc. London, Series B, Biological Sciences*, 153 (951): 195–204.
41. McDonagh, M. J. N. and Duncan, A. (2002) 'Interaction of pre-programmed control and natural stretch reflexes in human landing movements', *J. Physiol.*, 544.3: 985–944.
42. McKinley, P. A. and Smith, J. L. (1983) 'Visual and vestibular contributions to prelanding EMG during jump downs in cats', *Ex. Brain Res.*, 52: 439–448
43. McKinley, P. A., Smith, J. L. and Gregor, R. J. (1983) 'Responses of elbow extensors to landing forces during jump downs in cats', *Ex. Brain Res.*, 49: 218–228.
44. McLeod, P., Reed, N. and Dienes, Z. (2003) 'How fielders arrive in time to catch the ball', *Nature*, 426: 244–245.
45. McNair, P.J., Prapavessis, H. and Callender, K. (2000) 'Decreasing landing forces: effect of instruction', *British J. Sport Med.*, 34: 293–296.
46. Melvill Jones, G. and Watt, D. G. D. (1971) 'Muscle control of landing from unexpected falls in man', *J. Physiol. (London)*, 219: 729–737.
47. Mussa-Ivaldi, F. A. and Bizzi, E. (1997) 'Learning Newtonian mechanics'. In: P. Morasso, and V. Sanguineti (Eds.), *Self-organization, Computational Maps, and Motor Control*, Amsterdam: Elsevier, pp. 491–501.
48. Nigg, B. M. and Herzog, W. (1999), 'Biomechanics of the Musculo-skeletal system', Wiley, New York, pp: 218–221.
49. Regan, D. (2002) 'Binocular information about time to collision and time to passage', *Vision Res.*, 42: 2479–2484.
50. Santello, M. and McDonagh, M. J. N. (1998) 'The control of timing and amplitude of EMG activity in landing movements in humans', *Exp Physiol.*, 83: 857–874.
51. Santello, M., McDonagh, M. J. N. and Challis, J. H. (2001) 'Visual and non-visual control of landing movements in humans', *J. Physiol.*, 2001; 537: 313–327.
52. Schaltenbrand (1925) 'Normale Bewegungsund Lagereaktion beim Kinde', *Dtsch. Zeitschr. f. Nervenheilk.*, 87: 23.
53. Self, B. P. and Paine, D. (2001) 'Ankle biomechanics during four landing techniques', *Med. Sci. Sports Exerc.*, 33: 1338–1344.
54. Sidaway, B., McNitt-Gray, J. and Davis, G. (1989) 'Visual timing of muscle preactivation in preparation for landing', *Ecol. Psych.*, 1: 253–264.
55. Srinivasan, M. V., Zhang, S. W., Chahl, J. S., Barth, E. and Venkatesh S. (2000) 'How honeybees make grazing landings on flat surfaces', *Biol. Cybern.*, 83: 171–183.
56. Thompson, H. W. and McKinley, P. A. (1995) 'Landing from a jump: the role of vision when landing from known and unknown heights', *Neuroreport*, 6:581–584.
57. Tresilian, J. R. (1990) 'Perceptual information for the timing of interceptive action', *Perception*, 19: 223–239.
58. Tresilian, J. R. (1991) 'Empirical and theoretical issues in the perception of time to contact', *J. Exp. Psychol.: Hum. Percep. Perform.*, 17: 865–876.
59. Tresilian, J. R. (1993) 'Four questions of time to contact: a critical examination of research on interceptive timing', *Perception*, 22: 653–680.
60. Wagner, H. (1982) 'Flow-field variables trigger landing in flies', *Nature*, 297: 14–15.

10

Ecological psychology and task representativeness: implications for the design of perceptual-motor training programmes in sport

Matt Dicks[1], Keith Davids[2] and Duarte Araújo[3]
[1]University of Otago, Dunedin; [2]Queensland University of Technology, Queensland; [3]Technical University of Lisbon

Introduction

In this chapter we present theoretical posits from ecological psychology that inform understanding of the acquisition of perceptual skill in sport, and consider how these ideas can underpin the design of training protocols for acquiring perceptual-motor skills. For example, we examine the information presented by one athlete to another in team sports and discuss how these informational variables might guide perceptual-motor behaviour and inform programmes for perceptual skill acquisition. Central to ecological insights on programme structuring is Brunswik's (1956) concept of *representative task design*, which provides a framework to critique traditional laboratory-based analyses of perceptual behaviour in sport performance. We start by overviewing existing research on perceptual skill acquisition in sport.

Investigations of perceptual skill and its acquisition in sport

Two of the most popular experimental paradigms in this research area include spatial and temporal occlusion techniques (Williams, Davids and Williams, 1999). In spatial occlusion, body parts of a projected opponent are concealed to prevent the pickup of different informational variables (e.g. Müller, Abernethy and Farrow, 2006). In temporal occlusion paradigms, video footage of an action is edited to terminate the display at critical moments (e.g., foot-ball contact in the football penalty kick) requiring participants to anticipate performance with varying degrees of information (e.g., McMorris and Colenso, 1996). Typically, these studies have reported that experts are significantly better at anticipating performance outcomes of projected footage than novices, a finding which is exacerbated under impoverished information conditions such as temporal occlusion prior to foot-ball contact (Abernethy, 1991; Starkes *et al.*, 2001; Williams *et al.*, 1999).

Technological developments have led to a third method to determine perceptual information usage in sports: analysis of visual search strategies with eye-movement registration systems. Measuring visual search behaviours of participants directed towards a projected display implies the recording of information picked up in foveal vision (e.g., Savelsbergh *et al.*, 2005). Visual search behaviours measured typically include the location and duration of visual fixations on features of the display. As with studies using occlusion paradigms, research has demonstrated significantly better anticipatory performance by experts coupled with distinct differences in visual search behaviour compared to novices (e.g., Savelsbergh *et al.*, 2002; Williams and Davids, 1998). The body of evidence compiled within these three experimental paradigms demonstrates the significant role that perceptual skill plays in sport. Consequently, there has been a steady increase in empirical investigations aimed at developing training programmes to promote perceptual skill acquisition.

One of the most popular perceptual skill training methods implemented as a result of extant research is video simulation of sporting scenarios. These programmes typically use video projection to simulate performance that participants repeatedly view, coupled with instructional constraints to focus attention on informational variables presented by movements of an individual, such as an opponent in squash (e.g., Abernethy, Wood and Parks, 1999). Despite the potential of this form of intervention, the contribution of much research in this area is undermined by limitations in experimental design, questioning the efficacy of this type of perceptual training programme (Williams *et al.*, 2002). Some methodological limitations include the failure to use placebo or control groups, record the transfer of performance from the laboratory to sport environments, and conflict over the instructions given to participants (Williams and Ward, 2003). Research progress has addressed some of these experimental limitations (e.g. Farrow and Abernethy, 2002; Jackson and Farrow, 2005; Smeeton *et al.*, 2005) although doubts still remain over the experimental paradigms used to measure perceptual skill in sports.

Task representativeness in the study of perceptual skill in sport

An important feature of the environment that shapes perceptual skill acquisition is practice task constraints (Araújo *et al.*, 2004, 2006). Central to concerns over experimental designs has been the impoverished nature of informational and instructional constraints on the actions of participants, resulting in passive processes of perception without action (Marsh, Richardson, Baron and Schmidt, 2006; Neri, Luu and Levi, 2006; Shim, Carlton, Chow and Chae, 2005; Thornton and Knoblich, 2006). These concerns may be best exemplified by research in which film and video presentations have been used to simulate sport performance. Discrepancies between these contrived task constraints and behavioural contexts include: (i) divergences in size and the dimensionality of the film display with performance contexts; (ii) differences between the view in a presented film clip and an athlete's perspective during actual performance; and (iii), the prevention of opportunities for exploring the performance context to search for more relevant information (see Williams *et al.*, 1999). Montagne (2005) identified the importance of an 'information-based' *prospective* control in many sports predicated on a circular link between information and movement. During prospective control of movement, performers need to pick up information on differences between current and required behaviour to support ongoing movement adaptations in the environment. Clearly, the implications of prospective control are of particular importance to research designs examining perceptual skill (i.e., occlusion paradigms and visual search studies) and the perceptual training programme methodologies currently used.

Questions exist over the representative nature of these task designs relative to performance contexts (Brunswik, 1956). The *representativeness* of particular task constraints implies that performers can achieve their goals by acting to find information to guide action (Gibson, 1979). From the perspective of ecological psychology, experimental tasks should be designed so that picking up an informational variable that specifies a property of interest in the environment should allow a participant to make reliable judgements about this property in achieving performance outcomes. Brunswik (1956) argued that neurobiological systems detect environmental information in the form of multiple imperfect indicators of some unobservable state of the environment. These indicators, or perceptual variables, are those features of objects or events that can be used to infer those aspects of the objects or events that are not directly available. There are many of them, and they rarely are perfectly dependable in their capacity to indicate those performance aspects that one is trying to infer (Hammond, 2000). Representativeness refers to the generalization of task constraints in specific experimental paradigms to performance constraints away from the respective research designs (Araújo, *et al.*, 2007; Davids, *et al.*, 2006). The perceptual variables presented in an experimental laboratory need to reflect those available for pick up and use in a specific performance context, towards which one is attempting to generalize.

This and other insights in ecological psychology are pertinent to our understanding of the emergent, and adaptive perceptual-anticipatory behaviours exemplified in sports environments (Araújo *et al.*, 2006; Marsh *et al.*, 2006; Montagne, 2005; Warren, 2006). In the remainder of this chapter we discuss how these theoretical ideas can provide a conceptual framework for designing perceptual training programmes in sport. To achieve this aim, we review research on interpersonal dynamics in sports like rugby union where other performers form an integral part of the perceptual environment.

Perception and action in sports environments

Ecological psychology emphasizes the importance of a synergetic relationship between a performer and the environment. It predicates that performers can gain information for action from the surrounding distributions of energy to specify action-relevant properties of the world (Fajen, Riley and Turvey, in press; Montagne, 2005). An athlete with a diminished capacity to act on the surrounding environment would not only have fewer possibilities to change the structure of the surrounding environment, he/she would also have fewer opportunities to know about the environment within which they are performing. For example, on an attacking break, rugby union players constantly search the environment for information to guide actions: Should they pass to a team-mate? If so, when? Should they attempt to go past an opponent? Which opponent? Whatever decision the player takes, the energy flows available in the surrounding environment (e.g., optic or acoustic) will alter, presenting new information and thus new opportunities for action. Actions can result in the presentation of otherwise unavailable perceptual information.

In his theory of direct perception, Gibson (1979) proposed that what animals perceive, and act in respect to, are the substances, surfaces, places, objects, and events of the environment. From a Gibsonian perspective, the specific informational variables perceived within the ambient energy arrays hold a tightly coupled relationship with the surrounding environment (Fajen, 2005). Such environmental properties are opportunities or possibilities for behaviour: that is, affordances for action. Analyzing and interpreting affordances with respect to what humans perceive and how they act is a fundamental research goal in ecological psychology and has implications for understanding the acquisition of perceptual motor skills.

A programme of perceptual motor training in sport, therefore, should be designed with affordances for learners as a relevant constraint on their actions (Fajen *et al.*, in press).

The regulation of action depends on the continual coupling of movements to perceptual informational variables (Savelsbergh, van der Kamp, Oudejans and Scott, 2004; Montagne, 2005). Planned actions can be adapted ongoingly under the guidance of perceptual information to account for environmental changes (Thornton and Knoblich, 2006). Changes in action by a performer, such as a re-orientation in space or an increase in movement velocity, can yield new affordances for actions (Turvey and Shaw, 1999). Such subtle changes of action can lead to multiple and marked variations in opportunities for subsequent actions. Some opportunities for action persist, some newly arise, and some dissolve even though the surrounding environmental properties and their relations remain the same (Turvey and Shaw, 1999). A central theme in ecological psychology is that action not only has a perceptually guided function but also an epistemic role in gaining knowledge about the environment. Acquiring knowledge in sport involves understanding affordances for action.

From this perspective, the design of training programmes should be based on manipulating task and informational constraints that force learners to seek and exploit different affordances for action. In this way, each individual learns to act in order to pick up environmental properties. With perceptual skill acquisition, the pick up of environmental properties will become more specific to an intended outcome (Shaw and Turvey, 1999). It is important that training programmes should be predicated on the notion that an individual uses perceptual skill to regulate behaviour by detecting informational constraints specific to intended goal paths (Araújo *et al.*, 2004). Goal constraints, in comparison to physical constraints take the form of rules that prescribe how one should act if a particular performance outcome is intended and, therefore, are considered extraordinary (Kugler and Turvey, 1987; Shaw and Turvey, 1999). Given the epistemic role of action in human perception, it is important to design pedagogical programmes that permit participants to act upon the environment and obtain further information to enhance performance (Farrow and Abernethy, 2003; Warren, 2006).

Perceptual skill acquisition in ecological psychology

Clearly, it is imperative to understand the informational variables that learners use to guide action in different performance contexts. The process of learning to attend to relevant environmental properties in specific performance contexts has been referred to as the *education of attention* or *perceptual attunement* (Fajen *et al.*, in press). If the process of perception is specific to informational variables, which are in turn specific to properties of the environment, then perception needs to be specific to those environmental properties (Beek, Jacobs, Daffertshofer and Huys, 2003). Informational variables picked up by a performer are uniquely and invariantly tied to their sources in the environment. This interpretation implies that informational variables are not ambiguous, indeed they relate one-to-one with environmental properties (Withagen, 2004). Gibson (1979) termed the tight coupling between informational variables in the environment and movement as *specification*. Specifying variables in the environment provide information on key environmental properties to support the actions of performers. To exemplify, a well-known proposed specifying environmental property is the optic variable tau (Lee, 1976). This informational variable constitutes the inverse of the relative dilation of an approaching object that specifies time-to-contact (under certain constraints) with an intended target (Savelsbergh, Whiting and

Bootsma, 1991; Withagen, 2004). By learning to pick up this optic variable, performers in ball games can time a whole range of interceptive actions such as catching a ball and tackling a moving player in rugby (Williams *et al.*, 1999).

Empirical evidence suggests that certain task constraints can direct performers to rely on nonspecifying (as well as specifying) variables in their perception of surrounding environmental properties. Nonspecifying variables are less regulatory of actions and do not provide the same informational constraint as specifying variables. To exemplify, Oudejans *et al.* (2000) investigated errors of assistant referees in association football when judging offside and documented a nonspecifying variable as the relative optical projection of players onto the assistant referee's retina. In some circumstances, Oudejans *et al.* (2000) showed how the assistant referees' viewing angle of attacking and defending players may lead to perceptual errors. An attacking player, in line with a defender, when viewed from the opposite side of the pitch at an angle, can appear offside when the image of the two players is projected onto the assistant referee's retina. Regardless of the abilities of the assistant referee, players in this position when viewed on the pitch will cause an incorrect decision due to a limitation of the perceptual system. These findings imply that the process of specification of informational variables as proposed by Gibson (1979) may actually explain only part of the perceptual process that guides action. Oudejans *et al.*'s (2000) work suggests it is possible that assistant referees could be trained to find a more reliable informational variable to help them judge offside decisions, an implication that can be generalized to perceptual training programmes in sport.

Withagen (2004) argued that not every environmental property may be specified by an informational variable and that learners, in particular, may have no option other than to perceive and act upon nonspecifying variables in certain situations, especially early in learning. This interpretation implies that nonspecifying variables are not necessarily irrelevant variables for learners. However, they do not support actions as effectively as specifying informational variables. To summarize, the acquisition of perceptual skill involves learning to pick up and use *specifying* variables rather than *nonspecifying* variables. Practitioners need to design practice task constraints which allow learners to become attuned to specifying variables and to acquire skill in picking up these information sources to support their decisions and actions.

Research has demonstrated that, following an initial reliance on nonspecifying variables, participants can become attuned to specifying information through practice coupled with feedback (e.g. Jacobs and Michaels, 2001; Runeson, Juslin, and Olsson, 2000). Jacobs and Michaels (2000) investigated novices attempting to determine the mass of colliding balls, under three different practice conditions which all included feedback. It was reasoned that all individual environment property manipulations would direct participants to seek out specifying information sources. Practice conditions consisted of: (1) no-variation practice with repeated exposure to the same colliding ball mass examples; (2) zero-correlation practice in which nonspecifying variables were uncorrelated with the mass of the colliding balls rendering the information less effective; and (3) a second type of zero-correlation practice in which only one of the nonspecifying variables did not correlate with the colliding ball mass. Their findings indicated that the third condition was the most successful in training participants to attend to specifying informational variables. In the first condition some participants altered the variable used during practice, although they reverted back to the original nonspecifying variable in a post-test, depicting the intervention as unsuccessful. The second condition directed participants to seek out other sources of information, although some were unable to perceive the specifying variable, decreasing post-test performance. The third condition was successful in changing the informational variable used by all participants, although not all altered their perception of the specifying variable, rather using alternative sources of nonspecifying information. The authors suggested that adaptation of perceptual variables used by learners should be monitored during practice to change the environmental context accordingly (Beek *et al.*, 2003).

These findings imply that it is important to vary the information presented to learners to influence the success of such training intervention protocols. Typically, many perceptual training studies have relied on presenting learners with the same sources of perceptual information (e.g., the same video footage), but have varied the length of exposure to the same clips (e.g. McMorris and Hauxwell, 1997) or given participants differing instructions (e.g. Farrow and Abernethy, 2002). Through the manipulation of task constraints, coaches can alter the environment so that athletes are required to attend to varying information in a specific coaching context. In this way, athletes may learn to differentiate between specifying and nonspecifying sources of information and thus improve their ability to correctly identify and satisfy task constraints that offer successful information-movement couplings.

Ecological psychology and perceptual skill in sport: specification of information at the interpersonal level

As noted, a key point in ecological psychology to guide perceptual training is the representativeness of task design (Brunswik, 1956). Current design concerns over empirical investigations of perceptual skill in sport include the inability of video projections to accurately replicate the level of ambient information available in performance contexts and the failure to afford the tight coupling of perception and action in these environments. A better understanding of how perceptual skill is acquired in performance contexts, leading to more relevant training programmes, could be provided by research which replicates the constraints of behavioural settings in its design.

The first step of any research design on perceptual training could be to identify the informational variables present within a specific sports environment, after which perceptual training programmes can be developed. For example, a key informational constraint guiding perception and action in team sports is the presence of opponents or team-mates (Araújo *et al.*, 2004; Schmidt *et al.*, 1999). Despite theoretical and empirical advances, very little is known about how the perceptual-motor dynamics of one individual affect another (Marsh *et al.*, 2006). This form of visual perception is abundant in dynamic team sports where players are required to interact, perceive and act upon the behaviours of teammates and opponents during skill execution.

In ecological psychology, Marsh *et al.* (2006) argued that informational constraints incorporate the mutuality of one person's perception-action system with another person's perception-action system. For this reason it is important to understand the nature of interpersonal emergent adaptive behaviours in sports, exemplified in 1 v 1 dyadic systems (e.g., attacker-defender systems or goalkeeper-penalty taker systems). These theoretical insights are harmonious with research from a dynamical systems perspective analyzing emerging interactive processes between opposing competitive players during dyadic exchanges in sports such as tennis (Palut and Zanone, 2005), basketball (Araújo *et al.*, 2004), rugby (Passos *et al.*, 2006), and boxing (Hristovski *et al.*, 2006).

Traditionally, investigations aimed at identifying the information presented by one player to another in mutually constrained sports environments have largely relied upon verbally reported data (McMorris *et al.*, 1993; Williams and Burwitz, 1993). Analysis of visual search behaviours (Savelsbergh *et al.*, 2005; Williams and Davids, 1998) has enabled experimenters to further understand potentially specifying variables used by experts, although limitations of these designs include the use of two-dimensional (2D) video projection images that do not allow perception for action. The need to measure the 'live' presentation of informational variables between opponents was highlighted by Shim *et al.* (2005) in an analysis of information used by expert tennis players. They examined the anticipatory performance of skilled and novice tennis players who acted upon observed tennis ground strokes whilst viewing a point-light display, a full-sized 2D video display, or a three-dimensional (3D) live action. Each respective presentation condition observed by the participants comprised the opponents' hitting actions which were occluded from the participant's view after ball-racquet contact. Participants were required to anticipate shot direction by initiating a movement to either hit a forehand or backhand shot whilst vision was occluded. The experts' anticipatory behaviours became more accurate with the presentation of more information across conditions, whereas the novices' accuracy decreased. Significant differences were noted between groups in both the live and 2D conditions, but not for the point-light display condition, suggesting that it is essential to measure the fully interactive behavioural exchanges between opponents in sport. The presentation of kinematic information from one player to another was deemed most important for organizing a motor response.

Sports such as tennis, baseball, volleyball, handball, cricket, hockey, and badminton would benefit from identification of information conveyed by opponents. Identification of a specifying informational variable implies that the environmental property is always available for perception by an individual (Fajen, 2005; Gibson, 1979; Withagen, 2004). For example, empirical evidence on anticipatory information provided by penalty-takers in association football (Franks and Harvey, 1997; McMorris *et al.*, 1993; Savelsbergh *et al.*, 2005; Savelsbergh *et al.*, 2002; Williams and Burwitz, 1993) suggests that specifying variables might include the placement of the non-kicking foot, or the angle of the kicking foot and hip prior to ball contact. Possible nonspecifying variables could comprise the penalty-taker's point of gaze, position of run-up initiation, the arc of leg on approach to the ball, and the angle of approach or the angle of the player's trunk at foot-ball contact. Future research is required to identify the properties of these potential sources of information across a number of participants with kicks aimed at varying locations. Technological advances now support the measurement of kinematic variables to record the usefulness of information conveyed by each variable.

In line with the concept of task representativeness (Brunswik, 1956), in perceptual training programmes informational variables should be explored by learners in simplified task conditions (Araújo *et al.*, 2004). In perceptual training programmes, 'task simplification' is preferred to 'task decomposition' which tends to dominate traditional pedagogical approaches (Araújo *et al.*, 2004). Task decomposition occurs when skills such as the volleyball serve, tennis forehand drive and long jump, are broken up into more manageable smaller units for the purposes of practice (e.g., the toss and strike components of the serve or the run-up and jump components of the long jump). Unfortunately, this pedagogical method also tends to decompose the link between information and movement, breaking up potential information-movement couplings (Araújo *et al.*, 2004). Task simplification, on the other hand, induces manageability by providing more time for learners to pick up and use key perceptual variables to support actions. This practice strategy ensures that perception-action cycles remain intact during practice, with key informational variables remaining available to support actions in simplified task settings. Examples of task simplification include badminton or tennis shots fed by a coach as opposed to a ball machine, hitting a baseball thrown gently to a learner compared to hitting a static ball on a tee, cricket bowlers practising the run-up by walking towards the popping crease before delivering a ball to a batter, and young catchers learning to intercept a rolling ball before moving on to a bouncing ball. Task simplification allows the use of prospective control mechanisms to control actions in sport. Failure to faithfully reflect ecological constraints in practice design (e.g., use of ball machines) may provide learners with different informational variables that are not specific to the performance environment.

It is also important to understand how expert performers learn to converge on specifying information sources through experience. Every learner will have experienced different perceptual constraints when learning how to regulate movements, and hence, a wide variety of information-movement couplings is likely to exist between individuals (Savelsbergh *et al.*, 2004). Indeed, one approach that performers may utilize in gaining information for movement is through perceptually (re-)tuning to alternative informational variables as task constraints are altered (Fajen *et al.*, in press). As Jacobs and Michaels (2001) suggested, as well as the education of attention, perceptual motor skill acquisition also requires the calibration of the link between an informational variable and movement (see also Montagne, 2005). This calibration process depends on practitioners implementing representative task designs so that athletes learn to develop information and movement couplings in training that are reflective of those experienced during competition. Withagen (2004) proposed that identification of specifying and nonspecifying variables within the optic array could be conceptualized as a continuum that accounts for the level of epistemic contact (expertise of) that each performer has with the environment. Interpreting Gibson's (1966) proposal of perceptual learning, Withagen (2004) suggested that the degree to which performers learn to converge upon specifying variables for a given situation depends upon the level of direct epistemic contact with the environment. A novice with a low level of epistemic contact with the environment in a given performance context may initially rely on nonspecifying variables. With practice they can learn to rely upon specifying informational variables, thereby increasing their level of epistemic contact with the environment (e.g. Fajen and Devaney, 2006; Jacobs, Runeson, and Michaels, 2001). A low level of epistemic contact with the environment may be one consequence of implementing practice task designs which are not representative of natural performance environments. Learners may only be offered a low level of epistemic contact with the environment as a consequence of practice task designs which decouple the complementary processes of perception and action. Epistemic action represents a relevant strategy for overcoming perceptually impoverished environments.

In many studies using the traditional experimental paradigms highlighted earlier, an ongoing adaptation of a motor response through the coupled perception and action systems was not possible, unlike in sport performance environments (Savelsbergh *et al.*, 2006). Consequently, these experimental designs have been criticized for focusing upon 'telling rather than acting' processes in anticipatory behaviour (Shim *et al.*, 2005). Typically, required action from participants has consisted of a written or verbal response, pressing a button (Abernethy *et al.*, 1999), moving a joystick (Savelsbergh *et al.*, 2002) or a simulated movement response (Farrow *et al.*, 1998). Savelsbergh *et al.* (2006) attempted to overcome some of these design limitations when studying spatio-temporal displacements and visual search behaviours of amateur association footballers during a ball passing task. Following task completion, participants were assigned to either a low- or high-score group which reflected their ability to anticipate ball location. No between-group differences were observed for visual search behaviours, which may be explained by the impoverished nature of the 2D video projection protocol used. However, results indicated that only the high-score group was able to exploit advanced sources of information to predict final ball location. Additionally, they continued to detect and interpret information whilst acting on perceived footage, implying that under those specific task constraints, successful action was continuously coupled to perceptual information. Future research is needed to verify these results in different sports.

Conclusion

In this chapter we presented work from ecological psychology as a potential framework for the design of perceptual motor skill acquisition programmes in sport, which would provide learners with opportunities to couple their actions to information picked up in the environment. Ensuring that tasks are representative in design would enable a tight fit between training and performance environments, enabling learners to pick up relevant environmental properties to support perception and action. Ecological psychology predicates that information in the environment can be specifying or nonspecifying, determining their relative utility in guiding behaviour. Evidence suggests that initially novices may sub-optimally guide action using nonspecifying variables and coaches should design task constraints to enable learners to search the environment for specifying perceptual variables. Experimental designs in studies of perceptual skill acquisition should offer participants opportunities for emergent, adaptive behaviour by ensuring that perceptual and movement processes are enabled to function without being biased by restrictive, arbitrary and artificial task constraints lacking in representative design (Brunswik, 1956).

References

1. Abernethy, B. (1991) 'Visual search strategies and decision making in sport'. *International Journal of Sport Psychology, 22*: 189–210.
2. Abernethy, B., Wood, J. M. and Parks, S. L. (1999) 'Can the anticipatory skills of experts be learned by novices?' *Research Quarterly for Exercise and Sport, 70*(3): 313–318.
3. Araújo, D., Davids, K., Bennett, S. J., Button, C. and Chapman, G. (2004) 'Emergence of sport skills under constraints'. In A. M. Williams and N. J. Hodges (Eds.), *Skill acquisition in sport: research, theory and practice* (pp. 409–433). London: Routledge, Taylor and Francis.
4. Araújo, D., Davids, K. and Hristovski, R. (2006) 'The ecological dynamics of decision making in sport'. *Psychology of Sport and Exercise, 7*: 653–676.
5. Araújo, D., Davids, K. and Passos, P. (2007) 'Ecological Validity, Representative Design and Correspondence between Experimental Task Constraints and Behavioral Settings'. *Ecological Psychology, 19*: 69–78.
6. Beek, P. J., Jacobs, D. M., Daffertshofer, A. and Huys, R. (2003) 'Expert performance in sport: views from joint perspectives of ecological psychology and dynamical systems theory'. In J. L. Starkes and K. A. Ericsson (Eds.), *Expert performance in sport: advances in research on sport expertise* (pp. 321–344). Champaign, IL: Human Kinetics.
7. Brunswik, E. (1956) '*Perception and the representative design of psychological experiments*' (2nd ed.) Berkeley: University of California Press.

8. Davids, K., Button, C., Araújo, D., Renshaw, I. and Hristovski, R. (2006). 'Movement models from sports provide representative task constraints for studying adaptive behavior in human movement systems'. *Adaptive Behavior, 14*: 73–95.
9. Fajen, B. R. (2005) 'Perceiving possibilities for action: on the necessity of calibration and perceptual learning for the visual guidance of action'. *Perception, 34*: 717–740.
10. Fajen, B. R. and Devaney, M. C. (2006) 'Learning to control collisions: the role of perceptual attunement and action boundaries'. *Journal of Experimental Psychology: Human Perception and Performance, 32*: 300–313.
11. Fajen, B. R., Riley, M. A. and Turvey, M. T. (in press) 'Information, affordances and the control of action in sport'. *International Journal of Sport Psychology*.
12. Farrow, D. and Abernethy, B. (2002) 'Can anticipatory skills be learned through implicit video-based perceptual training?' *Journal of Sports Sciences, 20*: 471-485.
13. Farrow, D. and Abernethy, B. (2003) 'Do expertise and the degree of perception-action coupling affect natural anticipatory performance?' *Perception, 32*: 1127–1139.
14. Farrow, D., Chivers, P., Hardingham, C. and Sachse, S. (1998) 'The effect of video-based perceptual training on the tennis return of serve'. *International Journal of Sport Psychology, 29*: 231–242.
15. Franks, I. M. and Harvey, T. (1997) 'Cues for goalkeepers: high-tech methods used to measure penalty shot response'. *Soccer Journal, 42*: 30–38.
16. Gibson, J. J. (1979) '*The ecological approach to visual perception*'. Boston, MA: Houghton Mifflin.
17. Hristovski, R., Davids, K., Araújo, D. and Button, C. (2006) 'How boxers decide to punch a target: Emergent behaviour in nonlinear dynamical movement systems'. *Journal of Science and Medicine in Sport, 5*: 60–73.
18. Hammond, K. (2000) '*Judgement under stress*'. New York: Oxford University Press.
19. Jackson, R. C. and Farrow, D. (2005) 'Implicit perceptual training: How, when, and why?' *Human Movement Science, 24*: 308–325.
20. Jacobs, D. M. and Michaels, C. F. (2001) 'Individual differences and the use of nonspecifying variables in learning to perceive distance and size: Comments on McConnell, Muchisky, and Bingham'. *Perception & Psychophysics, 63*: 563–571.
21. Jacobs, D. M., Runeson, S. and Michaels, C. F. (2001) 'Learning to visually perceive the relative mass of colliding balls in globally and locally constrained task ecologies'. *Journal of Experimental Psychology: Human Perception and Performance, 27*: 1019–1038.
22. Lee, D. N. (1976) 'A theory of visual control of braking based on information about time-to-collision'. *Perception, 5*: 437–459.
23. Marsh, K. L., Richardson, M. J., Baron, R. M. and Schmidt, R. C. (2006) 'Contrasting approaches to perceiving and acting with others'. *Ecological Psychology, 18*: 1–38.
24. McMorris, T. and Colenso, S. (1996) 'Anticipation of professional soccer goalkeepers when facing right- and left-footed penalty kicks'. *Perceptual and Motor Skills, 82*: 931–934.
25. McMorris, T. and Hauxwell, B. (1997) 'Improving anticipation of soccer goalkeepers using video observation'. In T. Reilly, J. Bangsbo and M. Hughes (Eds.), *Science and Football III* (pp. 290–294). London: E. and F. N. Spon.
26. Montagne, G. (2005) 'Prospective control in sport'. *International Journal of Sport Psychology, 36*: 127–150.
27. Müller, S., Abernethy, B. and Farrow, D. (2006) 'How do world-class cricket batsmen anticipate a bowler's intention?' *The Quarterly Journal of Experimental Psychology, 59*(12): 2162–2186.
28. Neri, P., Luu, J. Y. and Levi, D. M. (2006) 'Meaningful interactions can enhance visual discrimination of human agents'. *Nature Neuroscience, 9*: 1186–1192.
29. Oudejans, R. R. D., Verheijen, R., Bakker, F. C., Gerrits, J. C., Steinbruckner, M. and Beek, P. J. (2000) 'Errors in judging 'offside' in football'. *Nature, 404*: 33.
30. Palut, Y. and Zanone, P. G. (2005) 'A dynamical analysis of tennis: Concepts and data'. *Journal of Sports Sciences, 23*: 1021–1032.
31. Runeson, S., Juslin, P. and Olsson, H. (2000) 'Visual perception of dynamic properties: cue heuristics versus direct-perceptual competence'. *Psychological Review, 107*: 525–555.

32. Savelsbergh, G. J. P., Onrust, M., Rouwenhorst, A. and van der Kamp, J. (2006) 'Visual search and locomotion behaviour in a four-to-four football tactical position game'. *International Journal of Sport Psychology*, *37*: 248–264.
33. Savelsbergh, G. J. P., van der Kamp, J., Oudejans, R. R. D. and Scott, M. A. (2004) 'Perceptual learning is mastering perceptual degrees of freedom'. In A. M. Williams and N. J. Hodges (Eds.), *Skill acquisition in sport: research, theory and practice* (pp. 374–389). London: Routledge, Taylor and Francis.
34. Savelsbergh, G. J. P., van der Kamp, J., Williams, A. M. and Ward, P. (2005) 'Anticipation and visual search behaviour in expert soccer goalkeepers'. *Ergonomics*, *48*: 1686–1697.
35. Savelsbergh, G. J. P., Whiting, H. T. A. and Bootsma, R. J. (1991) 'Grasping tau'. *Journal of Experimental Psychology: Human Perception and Performance*, *17*: 315–322.
36. Savelsbergh, G. J. P., Williams, A. M., van der Kamp, J. and Ward, P. (2002) 'Visual search, anticipation and expertise in soccer goalkeepers'. *Journal of Sports Sciences*, *20*: 279–287.
37. Schmidt, R. C., O'Brien, B. and Sysko, R. (1999) 'Self-organization of between-persons cooperative tasks and possible applications to sport'. *International Journal of Sport Psychology*, *30*: 558–579.
38. Shaw, R. E. and Turvey, M. T. (1999) 'Ecological foundations of cognition II. Degrees of freedom and conserved quantities in animal-environment systems'. *Journal of Consciousness Studies*, 6: 111–123.
39. Shim, J., Carlton, L. G., Chow, J. W. and Chae, W. K. (2005) 'The use of anticipatory visual cues by highly skilled tennis players'. *Journal of Motor Behavior*, *37*: 164–175.
40. Smeeton, N. J., Williams, A. M., Hodges, N. J. and Ward, P. (2005) 'The relative effectiveness of various instructional approaches in developing anticipation skill'. *Journal of Experimental Psychology: Applied*, *11*: 98–110.
41. Starkes, J., Helsen, W. and Jack, R. (2001) 'Expert performance in sport and dance'. In R. Singer, H. Hausenblas and C. Janelle (Eds.), *Handbook of Sport Psychology* (pp. 174–201). Chichester: Wiley.
42. Thornton, I. M. and Knoblich, G. (2006) 'Action perception: seeing the world through a moving body'. *Current Biology*, *16*: R27–R29.
43. Turvey, M. T. and Carello, C. (1995) 'Some dynamical themes in perception and action'. In R. F. Port and T. van Gelder (Eds.), *Mind as motion: explorations in the dynamics of cognition* (pp. 373–402). London: The MIT Press.
44. Turvey, M. T. and Shaw, R. E. (1999) 'Ecological foundations of cognition I. Symmetry and specificity of animal-environment systems'. *Journal of Consciousness Studies*, 6: 95–110.
45. Warren, W. H. (2006) 'The dynamics of perception and action'. *Psychological Review*, *113*: 358–389.
46. Williams, A. M. and Davids, K. (1998) 'Visual search strategy, selective attention, and expertise in soccer'. *Research Quarterly for Exercise and Sport*, 69: 111–128.
47. Williams, A. M., Davids, K. and Williams, J. G. (1999) 'Visual perception and action in sport'. London: E. and F. N. Spon.
48. Williams, A. M. and Ward, P. (2003) 'Perceptual expertise: development in sport'. In J. L. Starkes and K. A. Ericsson (Eds.), *Expert performance in sports: advances in research on sport expertise* (pp. 219–250). Champaign, IL: Human Kinetics.
49. Williams, A. M., Ward, P., Knowles, J. M. and Smeeton, N. J. (2002) 'Anticipation skill in a real-world task: measurement, training, and transfer in tennis'. *Journal of Experimental Psychology: Applied*, 8: 259–270.
50. Withagen, R. (2004) 'The pickup of nonspecifying variables does not entail indirect perception'. *Ecological Psychology*, *16*(3): 237–253.

Section III

Methodologies and systems of measurement

11

Measurement of pressure distribution

Ewald M. Hennig
University of Duisburg-Essen, Essen

Introduction

Movements originate from forces acting on the human body. For events of short duration, cinematographic techniques can not be used to estimate the forces and accelerations experienced by the body's centre of mass (COM) or any one of its parts. Therefore, transducers are necessary to register forces, accelerations and pressure distributions that occur during human locomotion. Ground reaction force and centre of pressure information can be determined with force platforms. These data are important to estimate internal and external loads on the body during locomotion and sport activities. Ground reaction forces represent the accelerations experienced by the COM of a moving body. These forces provide little information about the actual load under defined anatomical structures of the foot. To understand the etiology of stress fractures, for example, a more detailed analysis of foot loading is necessary. This is only possible by many separate force measuring sensors that cover the area of contact between the foot and the ground. Researchers in anatomy and human movement science have been interested in the load distribution under the human foot during various activities for more than 100 years. Early methods estimated plantar pressures from impressions of the foot in plaster-of-Paris and clay. Later techniques included optical methods with cinematographic recording. Only in recent years, the availability of inexpensive force transducers and modern data acquisition systems has made the construction of various pressure distribution measuring systems possible. Force transducers rely on the registration of the strain induced in the measuring element by the force to be measured. Pressures are calculated from recorded forces across a known area. Because all force measurements are based on the registration of the same phenomenon – strain – *similar* technical characteristics apply to all types of transducers. Desirable transducer characteristics for biomechanical applications may differ from the characteristics which are advantageous for engineering usage. Measurement of pressure during sitting or lying on a bed requires a soft and pliable transducer mat that will adapt to the shape of the human body. However, such a transducer will normally not have good technical specifications concerning linearity, hysteresis and frequency response. Based on different technologies, several pressure distribution systems are sold today. Most of them are used for the analysis of the foot to ground interaction as pressure platforms or as insoles for in-shoe measurements.

Sensor technologies

Pressure is calculated as force divided by the contact area on which this force acts. It is measured in units of kPa (100 kPa = 10 N/cm^2). Force measuring elements are needed to determine pressures and pressure distributions. There are only a few measuring principles that are commonly used to measure force. In the following section these principles will be described and their advantages and disadvantages will be discussed. Electro-mechanical transducers produce a change in electrical properties when subjected to mechanical loads. Depending on the type of transducer, forces can create electrical charges, cause a change in capacitance, modify the electrical resistance or influence inductance. Because inductive transducers are based on relatively large displacements, they are rarely used in measuring instruments for biomechanical applications.

Resistive transducer

A variety of different sensor types belongs to the category of resistive transducers. The electrical resistance of the sensor material changes under tension or compression. Volume conduction has been used as a method for measuring forces. Silicone rubber sensors, filled with silver or other electrical conducting particles, were produced in the past. With increasing pressure the conducting particles are pressed closer together, increasing the surface contact between the conducting particles and thus lowering the electrical resistance. These transducers show large hystereses, especially for higher frequency impact events. Resistive contact sensors work on a similar principle to the volume conduction transducers (Tekscan Inc., Boston, MA). Typically, two thin and flexible polyester sheets with electrically conductive electrodes are separated by a semi-conductive ink layer between the electrical contacts (rows and columns). Exerting pressure on two intersecting strips causes an increased and more intense contact between the conductive surfaces, thus causing a reduction in electrical resistance.

Piezoelectric transducers

Most high precision force transducers use quartz as the sensor material. The electrical charge that is generated on the quartz surfaces is very low (2.30 pC/N for the longitudinal piezoelectric effect) and charge amplifiers have to be used for electronic processing. Piezoceramic materials show very small and highly elastic deformations. As compared to quartz, piezoceramic materials generate approximately 100 times higher charges on their surfaces when identical forces are applied. The high charge generation allows the use of inexpensive charge amplifiers. For rapid loads, these transducers combine good linearity and a very low hysteresis. Whereas temperature has only a minor influence on the piezoelectric properties of quartz, piezoceramics also exhibit pyroelectric properties. Therefore, for these transducers, thermal insulation or a temperature equilibrium, as is normally present inside shoes, is necessary. High precision and inexpensive piezoceramic materials have been used for pressure distribution measurements (Hennig *et al.*, 1982). Piezoelectric polymeric films (polyvinylidene fluoride (PVDF)) and piezoelectric rubbers have also been developed and have been proposed for the measurements of forces. However, due to large hysteresis effects and unreliable reproducibility, these methods have been less successful.

Capacitive transducers

An electrical capacitor typically has two metal plates in parallel to each other with a dielectric material in between. Capacitance varies when the distance between the two plates and the properties of the dielectric material change. Applied forces can thus be determined by detecting the change in capacitance through compression of the elastic material (dielectric) which will result in a reduction of the distance between the two plates. The simple construction and low material costs allow the manufacturing of inexpensive pressure distribution mats with up to several thousand discrete capacitive transducers (Nicol and Hennig, 1976; Hennig and Nicol, 1978). As compared to the rigid nature of piezoelectric transducers with very low deformations of the transducer ($<10^{-7}\%$), large relative displacements ($>10\%$) are necessary for capacitive transducers. Large displacements and the viscoelastic nature of the dielectric limit measurement accuracy. Depending on the elastic properties of the dielectric, recoil of the material will require time and thus introduces hysteresis effects. Silicone rubber mats with good elastic characteristics have been used for the production of commercially available pressure distribution measuring platforms and insoles (Novel Inc., München, Germany). With this technology, capacitive pressure mats show a good accuracy for dynamic loads under the foot in walking and running. Capacitive transducers can be built flexible to accommodate body curvatures and contours. Due to their soft and pliable nature, capacitive transducers can be employed for the measurements of seating pressures and the registration of body pressures on mattresses (Nicol and Rusteberg, 1993).

Data acquisition and visualization of pressure distribution data

Uniform pressures across a surface are rare for technical applications and uncommon for contacts of the human with the ground. Many pressure sensors in a defined area are desirable for most biomechanical applications. However, sensor costs as well as the amount of data will limit the desired resolution of pressure mats. Rosenbaum and Lorei (2003) compared the pressure distribution patterns from 27 participants between 2 and 33 years during gait across two capacitive pressure distribution platforms (Novel GmbH, Munich) with resolutions of 4 and 9 sensors/cm^2. Although they found differences in peak pressures and contact area values between the two platforms, these differences were for most foot regions below 10%. The authors concluded that these differences were clinically not relevant, and that even for the smaller feet of the children a resolution of 4 sensors/cm^2 appeared to be sufficient. Even a resolution of 4 sensors/cm^2 on a 40 cm by 60 cm pressure mat would result in 9,600 single transducers. At a data collection rate of 100 Hz close to one million measuring values will have to be recorded in only one second. Therefore, data reduction techniques and simplified visualization of the data are essential in the use of pressure distribution information. Graphical representation of pressure distribution is commonly achieved through colour-coded matrix graphs, wire frame diagrams or isobarographs (Figure 11.1). These pressure maps can be obtained for each sampling interval or at specific instants during the foot-ground contact.

Recently, a program was developed that visually combines three-dimensional (3D) surface contour data (Figure 11.2) with pressure distribution information under the foot (Hömme *et al.*, 2007). Data from a 3D Laser foot scanning device (INFOOT, IWare-Laboratories Inc., Minoh City, Osaka, Japan) were combined with pressure distribution information from a capacitive pressure mat with a spatial resolution of 4 sensors/cm^2 (Emed ST, Novel Inc.,

Figure 11.1 Three graphical peak pressure representations for a subject walking across a pressure distribution platform with a resolution of four sensors/cm^2.

Munich, Germany). Both independent sets of data were recorded during standing of a participant and through transformations the data were matched to origin, orientation and in size for the visual representation in a single graph. With the development of dynamic foot scanners this kind of representation may become a standard for gait analysis, because it offers the clinician a simple tool to recognize shape and function of the foot in dynamic loading.

Pressure variables

As mentioned above, more than 100.000 measuring values are common for the recording of plantar pressures of a single walking trial, even if the spatial resolution and the measuring frequency are not very high. Therefore, data reduction becomes important for a meaningful presentation of results. A peak pressure graphical representation can be used to illustrate individual foot contact behaviour with the ground. This image is created by presenting the highest pressures under the foot, as they have occurred at any time during the ground

Figure 11.2 Visual combination of foot morphology and pressure distribution data from a 3D laser foot scanning device (LFSD) and a pressure platform.

contact. To include time information, a pressure-time integral graph can also be produced by time integration of the forces under each single transducer and displaying these impulse values in a single graph. However, this kind of presentation is less commonly used. For numerical and statistical analyses a division of the plantar contact of the foot into meaningful anatomical areas (masks) is usually performed.

Within selected anatomical areas different pressure variables can be defined: regional peak pressures (kPa), maximum pressure rates (kPa/s), regional impulses (Ns), and relative loads (%). The regional peak pressure during foot contact reveals information about the highest pressures in this anatomical region. Regional impulses are calculated by determining the local force-time integral (force (F) = pressure * area) under the specific anatomical region. These impulse values can then be taken to further calculate a relative impulse distribution under the foot. Dividing the foot into n different anatomical regions, the following equation is used to determine the relative load RL_i of the foot region i.

$$RL_i(\%) = \frac{\int F_i(t)dt}{\sum_{j=l\text{-}n} \int F_j(t)dt} * 100$$

This procedure allows a comparison of the load distribution pattern between individuals that is less dependent on the participants' weight and anthropometric dimensions. Therefore, a better understanding of the load-bearing role of individual anatomical structures can be obtained using a relative load analysis technique.

Reliability of pressure distribution measurements

Using a Novel Pressure platform Hughes *et al.* (1991) investigated the reliability of pressure distribution measurements during gait. These authors concluded from their study that a determination of peak pressures can be achieved with a reliability coefficient of $R = 0.94$, if the average of 5 trial repetitions is used. More recently, McPoil *et al.* (1999) also investigated the number of repetitive trials needed to obtain a reliable representation of plantar pressure patterns. Using plantar pressure data from 20 trials of each of the 10 participating volunteers, the authors concluded that three to five walking trials are sufficient to obtain reliable peak pressure and pressure-time integral values, when a two-step data collection protocol is used.

With 25 participants Kernozek *et al.* (1996) studied the reliability of in-shoe foot pressure measurements (PEDAR insole, NOVEL Inc.) during walking at three speeds on a treadmill. The study revealed not only an effect of walking speed on the magnitude of the plantar pressures under the various anatomical regions of the foot but also a change in load distribution was found. Therefore, no linear interpolation is possible to correct for differences in walking speed. The authors emphasize the need to control speed for in-shoe pressure measurements. Depending on the pressure variable, a maximum of eight steps was needed to achieve an excellent reliability (> 0.90). Whereas peak force, peak pressures and pressure time integrals tended to need fewer trials for a good representation of foot loading, timing variables tended to be least reliable. Using the same instrumentation (Pedar Insole, NOVEL GMBH, Munich), Kernozek and Zimmer (2000) found good to excellent test retest reliabilities for slow treadmill running (2.24 m/s to 3.13 m/s). Depending on foot region and the measurement variable the two day test-Intraclass correlation coefficients (ICCs) ranged from 0.84–0.99. Furthermore, with running speed all pressure variables (peak pressure, peak force, pressure time integrals) increased. Similar to their previous findings for walking, the authors emphasize that the control of running speed is essential in obtaining reproducible plantar pressure results.

Results of plantar pressure distribution studies

Pressure distribution measurement technology was first used by researchers, primarily interested in the basic function of the foot structure during weight bearing and during various locomotion activities. In the area of sport biomechanics pressures have been recorded to analyze various parts of the body during bicycling, horse riding, skiing, skating, basketball, soccer, and golfing. Pressure distribution measurements are also widely used in the design and individual adaptation of seats and mattresses for better sitting and lying comfort. However, the major use of plantar pressure measurements within the last 20 years has been in medical applications. The diabetic and rheumatoid feet have been the focus of extensive research using pressure distribution technology. More recently, the diagnosis of balance control in the elderly, as well as in patients with neurological disorders, have received considerable attention.

Pressure patterns under adult feet

Using a capacitance pressure distribution mat (1 sensor/cm^2), Cavanagh, P. R. *et al.* (1987) investigated the plantar pressures under the symptom-free feet of 107 participants during bipedal standing. They found approximately 2.6 times higher heel (139 kPa) against forefoot pressures (53 kPa). The highest forefoot pressures were located under the second and third metatarsal heads. There was almost no load sharing contribution of the toes during standing. Furthermore, peak plantar pressures did not show a significant relationship to the body weight of participants. Using a capacitive pressure distribution platform EMED F01

(NOVEL Inc.) with a higher resolution (2 sensors/cm^2), Hennig *et al.* (1993) presented the pressures during bipedal standing and walking (approx. 1 m/s) under the symptom-free feet of 49 women and 62 men (Figure 11.3). During both standing and walking the highest pressures under the forefoot were found under the third metatarsal head. The authors concluded that during weight bearing no transverse arch across the metatarsal heads is apparent from the pressure data.

Therefore, a tripod theory of foot loading, as it had been advocated in many anatomy text books, cannot be confirmed by plantar pressure measurements. No differences in plantar pressures were found between the women and men except for the pressures under the longitudinal arch. Under the midfoot the women showed highly significantly ($p < 0.01$) reduced peak pressures during standing.

Morag and Cavanagh (1999) identified structural predictors of peak pressures under the foot during walking. This research aimed towards recognizing potential etiological factors associated with elevated plantar pressures, to identify at-risk patients (e.g. diabetic patients) early enough to provide preventive measures against excessive plantar pressure development. Fifty-five asymptomatic participants between 20 and 70 years participated in the study. Physical characteristics, anthropometric data, joint range of motion, radiographs, plantar soft tissue properties, kinematic variables, and electromyography (EMG) data were compared to the plantar pressure variables during walking. As to be expected midfoot pressure prediction was mainly a function of longitudinal arch structure. Heel peak pressures were also dependent on arch structure, thickness of the plantar soft tissue, and age. Pressures under the first metatarsal head could be predicted by radiographic measurement results, talo-crural joint motion, and EMG activity of the gastrocnemius. Overall, foot structure and function could only predict approximately 50% of the variance in peak pressures.

Pressures under the feet of children

On radiographs the infant foot skeleton appears as a loose assembly of ossified diaphyses of the phalangeal and metatarsal bones and the nuclei of the talus and calcaneus. The connection of

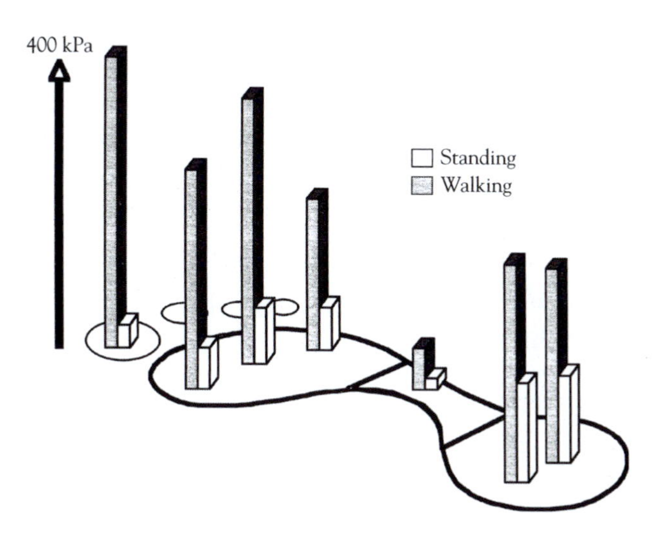

Figure 11.3 Peak pressures (kPa) under the foot of 111 adults during standing and walking.

the bones to a skeletal structure takes place with the proceeding transformation of cartilage to bone. At birth the mechanical structure of the foot is predominantly soft tissue. Only by age six the major structural changes of the foot have been completed, and the appearance is similar to an adult foot. A plantar pressure study with 15 infants (7 male, 8 female) between the ages of 14 and 32 months was conducted by Hennig and Rosenbaum in 1991. The pressures underneath the foot were recorded with an EMED F01 (NOVEL Inc.) pressure platform during barefoot walking and slow running at self-selected speeds. The children were too young to adjust to a prescribed speed. The results were compared with data on plantar pressures from adults. Considerably reduced peak pressures in the infant group could be attributed to a softer foot structure and a lower body-weight to foot-contact area ratio. An almost three times higher relative load under the midfoot of the infant foot shows that the longitudinal foot arch is still a weak structure in this age group. Across the age range of the children it was observed that with increasing age the relative load under the midfoot decreases, while it becomes more pronounced under the third and fifth metatarsals. Within only a few months a rapid development of the growing foot towards an adult loading pattern was observed. Bertsch *et al.* (2004) studied in a longitudinal study the plantar pressures of 20 male and 22 female infants every three months over a period of one year. The mean age of the infants at the beginning of this study was 16.1 months. After a few weeks of independent walking, substantial changes in the foot structure and loading behaviour were observed. The development of the longitudinal arch correlated well with reduced midfoot pressures during ground contact in walking. Even at this early age Unger and Rosenbaum (2004) found differences between boys and girls with regard to foot shape as well as foot loading behaviour. The boys had a broader midfoot due to a lower longitudinal arch, and the girls showed a more pronounced loading under the heel and forefoot regions during walking. The authors concluded that the differences seen in their study should be taken into account by the shoe industry to provide adequate footwear for young boys and girls. For older children pressure data were collected and compared to those from adults (Hennig *et al.*, 1994). Peak pressures and relative loads under the feet of 125 children (64 boys, 61 girls) between 6 and 10 years of age were determined. The school children showed considerably lower peak pressures under all anatomical structures as compared to the adults. These lower pressures could mainly be attributed to the larger foot dimensions per kg body mass for the children. With increasing age a medial load shift in the forefoot was observed for the older children. Reduced loading of the first metatarsal head in the younger children was attributed to a valgus knee condition with hyperpronation of the foot and a reduced stability of the first ray. No reduction in pressures under the longitudinal arch with an increase of age suggests that foot arch development is almost complete before the age of six. Between boys and girls no differences in the peak pressure or relative load patterns were present. Dowling *et al.* (2004) investigated whether inked footprints would be good predictors of plantar pressures in children. If this would be the case large scale screenings of children's feet would be possible without expensive equipment. However, in the 51 primary school age children the authors only found weak relationships between foot data from the ink footprints towards the plantar pressure information from the pedobarograph. Therefore, the authors suggest that in spite of the costs pedobarographs have to be used to identify excessive pressures under children's feet.

Plantar pressures in obese persons

Obesity is a major health problem in many parts of the world. Among numerous other medical conditions, a high incidence of osteoarthrits, painful feet, and symptomatic complaints in the joints of the lower extremities are frequently reported for overweight persons. Plantar pressure analysis may provide additional insight into the etiology of pain and lower extremity complaints. Hills *et al.* (2001) investigated the plantar pressure differences between obese and non-obese adults during standing and walking. Thirty-five males (67–179 kg) and 35 females (46–150 kg) were divided into an obese (body mass index (BMI) 38.8 kg/m^2) and a non-obese (BMI 24.3 kg/m^2) group. A BMI of greater than or equal to 30kg/m^2 was used to classify individuals as obese. Using this criterion, 17 men and 19 women were categorized as obese. All participants were otherwise healthy with no locomotor limitation such as symptomatic osteoarthritis. During standing and walking data were collected with a capacitive pressure distribution platform (Novel Inc., München, Germany). Pressures were evaluated for eight anatomical sites under the feet. The obese participants showed an increase in the forefoot width to foot length ratio, suggesting a broadening of the forefoot under increased weight loading conditions. In spite of the increased load bearing contact area of the foot with the ground, the obese men and women had substantially higher pressures under the heel, midfoot, and forefoot during standing. For both obese groups the dynamic peak pressures during walking were also significantly increased under most foot areas, especially for the women under the midfoot and across the metatarsal heads (Figure 11.4).

In static as well as dynamic loading, the highest pressure increases for the non-obese group were found under the longitudinal arch of the foot. Increases in pressure under the mid-foot were higher during standing for the obese women as compared to the obese men. A strong relationship between BMI and the peak pressures ($r^2 = 0.66$) under the longitudinal foot arch is apparent across the 35 women. This relationship is much weaker for the 35 men ($r^2 = 0.26$). The clear gender related influence of body weight on the flattening of the arch may be the consequence of a reduced strength of the ligaments in women's feet. The authors concluded that their findings may have implications for lower extremity pain and discomfort in the obese, the choice of footwear and predisposition to participation in activities of daily living such as walking.

Dowling *et al.*, 2001 compared the effects of obesity on plantar pressure distributions in prepubescent children. Thirteen obese children with a body mass index of 25.5 were compared to 13 age and height matched non-obese children (BMI 16.9). Footprints structural differences were observed between the groups, identifying a lower longitudinal arch, a flatter cavity, and a broader midfoot area for the obese children. For walking the forefoot pressures as well as the forefoot contact area were significantly increased for the obese group. The authors hypothesized that increased forefoot plantar pressures may lead to foot discomfort and may hinder obese children from participating in physical activity.

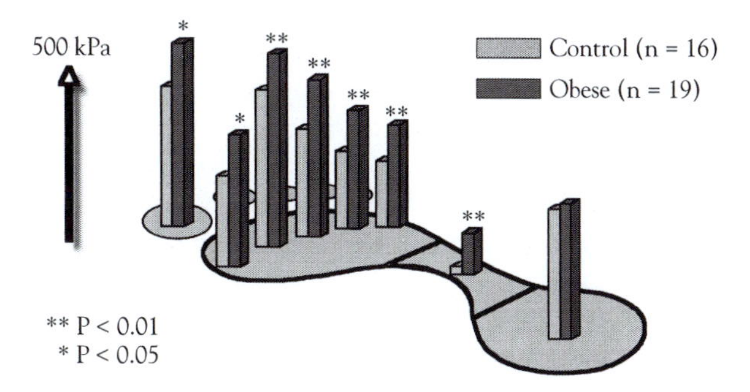

Figure 11.4 Peak pressures (kPa) under the foot of obese and non-obese women during walking.

Plantar pressures in sports

Pressure distribution measurements have been used to understand the foot loading behaviour and to estimate injury risks in many sports disciplines. One of the earliest applications of pressure distribution measurements in sports was the recording of in-shoe pressures during downhill skiing (Schaff and Hauser, 1987). Numerous other sports have been investigated since with pressure distribution technology such as running, bicycling, track and field events, indoor sports, ice and inline skating and even in horse riding. The two following studies are noteworthy since they demonstrate the wide spectrum of applications for pressure distribution measurements to prevent overuse injuries in sports. Nagel *et al.* (2007) measured barefoot gait of 200 marathon runners before and after a race. They found increased pressures under the metatarsal heads and lower toe pressures after the race. The authors speculated that this higher forefoot loading may explain the high incidence of metatarsal stress fractures in long distance runners. With 17 football players, Ford *et al.* (2006) investigated the effect of playing football on natural grass and synthetic turf on foot loading. They found increased peak pressures in the central forefoot region on the synthetic turf surface. The authors suggested more extensive research on the load distribution under the foot to understand specific overuse injury mechanisms on different playing surfaces.

Pressure studies for clinical applications

Depending on the severity of the disease and at a later stage diabetes mellitus may lead to neuropathic and vascular changes in patients. Sensory neuropathy will result in a progressive, distal to proximal, loss of sensation in the lower extremities, often resulting in ulceration at locations of high plantar pressures. Ulceration can result in partial or total amputation of the foot. Numerous studies have proven the usefulness of pressure distribution measurements for the prescription of therapeutic footwear. As soon as modern pressure distribution instrumentation became available, research groups realized the potential of this technology for the diagnosis and treatment of the diabetic feet. Good reviews about the possibilities and limitations of plantar pressure measurements for the diagnosis and therapy of the diabetic foot have been published by Cavanagh *et al.* in 1993 and in 2000. Recently, Lott *et al.* (2007) combined pressure distribution measurements with computer tomography (CT) scans on the feet of 20 diabetic patients with a history of plantar ulcers. They investigated the effect of therapeutic footwear and custom-made orthotic inserts on pressure and tissue strain along 16 different points along the second ray of the plantar foot. Correlation coefficients showed strong relationships between pressure and strain in the barefoot and shoe conditions. They found that footwear and orthotic devices decreased pressure and soft tissue strain, and that these two variables were strongly related. The authors concluded that a better understanding of tissue strain effects in distributing plantar pressures may lead to better orthotic device designs.

Plantar pressure studies on the rheumatoid foot have also received attention by research groups for adults (Woodburn and Helliwell, 1996; Otter *et al*, 2004) and children (Orlin *et al.*, 1997). Hallux valgus, a fairly common forefoot deformity at the metatarsophalangeal joint, is more frequently seen in women. Despite congenital factors, poorly fitting footwear and fashionable high-heeled shoes, worn by many women, are mentioned by many authors to be the etiologic factor in many cases of hallux valgus. A lateral forefoot load shift in hallux valgus feet has been described by Blomgren *et al.* (1991), the effect of high-heeled shoes by Nyska *et al.* in 1996. Different orthotic materials were investigated by Brown *et al.* (1996) to demonstrate the beneficial effect of orthotics in relieving high pressures in shoes. However, as the authors point out, stress relief can only occur at the cost of increasing pressure in other areas of the plantar surface. Using the centre of pressure path and other pressure variables, plantar pressure research has also been used in the analysis of postural control (Nurse *et al.*, 2001; Tanaka *et al.*, 1999) and the analysis of neurological disorders such as Parkinsonian gait (Kimmeskamp and Hennig, 2001; Nieuwboer *et al.*, 1999).

Conclusion

The foot loading behaviour of the human foot has been the subject of interest for more than 100 years. Different sensor technologies are available today to measure pressure distributions in biomechanics. The availability of modern transducers, the possibilities of rapid data acquisition, and an efficient processing of large data volumes have resulted in easy to use, commercially available pressure distribution instrumentation. Starting from basic research to understand the mechanical function of the human foot during gait, running and in different sport events, more and more applied studies were performed in recent years. In sport biomechanics, it is used to analyze the loading characteristics of the human foot in athletics and understand the etiology of overuse injuries. It is also extensively used in athletic footwear research for injury prevention and performance enhancement. During recent years, more and more clinical studies have been performed in the diagnosis and treatment of foot and lower extremity problems and postural control. Plantar pressure distribution instrumentation has become a standard clinical tool for diagnostics and therapeutic interventions.

References

1. Bertsch, C., Unger, H., Winkelmann, W. and Rosenbaum, D. (2004) 'Evaluation of early walking patterns from plantar pressure distribution measurements. First year results of 42 children', *Gait Posture*, 19: 235–42.
2. Blomgren, M., Turan, I. and Agadir, M. (1991) 'Gait analysis in hallux valgus', *Journal of Foot Surgery*, 30: 70–1.
3. Brown, M., Rudicel, S. and Esquenazi, A. (1996) 'Measurement of dynamic pressures at the shoe-foot interface during normal walking with various foot orthoses using the FSCAN system', *Foot & Ankle*, 17: 152–6.
4. Cavanagh, P. R., Rodgers, M. M. and Iiboshi, A. (1987) 'Pressure distribution under symptom-free feet during barefoot standing', *Foot & Ankle*, 7: 262–76.
5. Cavanagh, P. R., Simoneau, G. G. and Ulbrecht, J. S. (1993) 'Ulceration, unsteadiness, and uncertainty: the biomechanical consequences of diabetes mellitus', *Journal of Biomechanics*, 26: 23–40.
6. Cavanagh, P. R., Ulbrecht, J. S. and Caputo, G. M. (2000) 'New developments in the biomechanics of the diabetic foot', *Diabetes/Metabolism Research and Reviews*, 16 Suppl 1: S6-S10.
7. Dowling, A. M., Steele, J. R. and Baur, L. A. (2001) 'Does obesity influence foot structure and plantar pressure patterns in prepubescent children?', *International Journal of Obesity and Related Metabolc Disorders*, 25: 845–52.
8. Dowling, A. M., Steele, J. R. and Baur, L. A. (2004) 'Can static plantar pressures of prepubertal children be predicted by inked footprints?', *Journal of the American Podiatric Medical Association*, 94: 429–33.
9. Ford, K. R., Manson, N. A., Evans, B. J., Myer, G. D., Gwin, R. C., Heidt, R. S., Jr. and Hewett, T. E. (2006) 'Comparison of in-shoe foot loading patterns on natural grass and synthetic turf', *Journal of Sports Science & Medicine*, 9: 433–40.

10. Hennig, E. M. and Nicol, K. (1978) 'Registration methods for time-dependent pressure distribution measurements with mats working as capacitors', in E. Asmussen and K. Joergensen (eds.) *Biomechanics VI-A* (pp. 361–7). Baltimore: University Park Press.
11. Hennig, E. M., Cavanagh, P. R., Albert, H. and Macmillan, N. H. (1982) 'A piezoelectric method of measuring the vertical contact stress beneath the human foot', *Jounal of Biomedical Engineering*, 4: 213–22.
12. Hennig, E. M. and Rosenbaum, D. (1991) 'Pressure distribution patterns under the feet of children in comparison with adults', *Foot & Ankle*, 11: 306–11.
13. Hennig, E. M. and Milani, T. L. (1993) 'The tripod support of the foot. An analysis of pressure distribution under static and dynamic loading', *Zeitschrift für Orthopädie und Ihre Grenzgebiete*, 131: 279–84.
14. Hennig, E. M., Staats, A. and Rosenbaum, D. (1994) 'Plantar pressure distribution patterns of young school children in comparison to adults', *Foot & Ankle*, 15: 35–40.
15. Hills, A. P., Hennig, E. M., Mcdonald, M. and Bar–or, O. (2001) 'Plantar pressure differences between obese and non-obese adults: a biomechanical analysis', *Journal of Obesity*, 25: 1674–9.
16. Hömme, A.K., Hennig, E. M. and Hartmann, U. (2007) '3-dimensional foot geometry and pressure distribution analysis of the human foot. Visualization and analysis of two independent foot quantities for clinical applications', in T.M. Buzug, D. Holz, S. Weber, J. Bongartz, M. Kohl-Bareis and U. Hartmann (eds.) *Advances in Medical Engineering – Springer Proceedings in Physics*, 114 (pp. 308–13), Berlin. Heidelberg, New York: Springer.
17. Hughes, J., Pratt, L., Linge, K., Clark, P., and Klenerman, L. (1991) 'Reliability of pressure measurements: the EMED F system', *Clinical Biomechanics*, 6: 14–18.
18. Kernozek, T. W., Lamott, E. E. and Dancisak, M. J. (1996) 'Reliability of an in–shoe pressure measurement system during treadmill walking', *Foot & Ankle*, 17: 204–9.
19. Kernozek, T. W. and Zimmer, K. A. (2000) 'Reliability and running speed effects of in-shoe loading measurements during slow treadmill running', *Foot & Ankle*, 21: 749–52.
20. Kimmeskamp, S. and Hennig, E. M. (2001) 'Heel to toe motion characteristics in Parkinson patients during free walking', *Clinical Biomechanics*, 16: 806–12.
21. Lott, D. J., Hastings, M. K., Commean, P. K., Smith, K. E. and Mueller, M. J. (2007) 'Effect of footwear and orthotic devices on stress reduction and soft tissue strain of the neuropathic foot', *Clinical Biomechanics*, 22: 352–9.
22. Mcpoil, T. G., Cornwall, M. W., Dupuis, L. and Cornwell, M. (1999) 'Variability of plantar pressure data. A comparison of the two-step and midgait methods', *Journal of the American Podiatric Medical Association*, 89: 495–501.
23. Morag, E. and Cavanagh, P. R. (1999) 'Structural and functional predictors of regional peak pressures under the foot during walking', *Journal of Biomechanics*, 32: 359–70.
24. Nagel, A., Fernholz, F., Kibele, C. and Rosenbaum, D. (2007) 'Long distance running increases plantar pressures beneath the metatarsal heads A barefoot walking investigation of 200 marathon runners', *Gait Posture*.
25. Nicol, K. and Hennig, E. M. (1976) 'Time-dependent method for measuring force distribution using a flexible mat as a capacitor', in P. V. Komi (ed.) *Biomechanics V-B* (pp. 433–40). Baltimore: University Park Press.
26. Nicol, K. and Rusteberg, D. (1993) 'Pressure distribution on mattresses', in S. Bouisset, S. Métral and H. Monod (eds.) *Biomechanics XIV* (pp. 942–3). Paris: International Society of Biomechanics.
27. Nieuwboer, A., De Weerdt, W., Dom, R., Peeraer, L., Lesaffre, E., Hilde, F. and Baunach, B. (1999) 'Plantar force distribution in Parkinsonian gait: a comparison between patients and age-matched control subjects', *Scandinavian Journal of Rehabilitation Medicine*, 31: 185–92.
28. Nurse, M. A. and Nigg, B. M. (2001) 'The effect of changes in foot sensation on plantar pressure and muscle activity', *Clinical Biomechanics*, 16: 719–27.
29. Nyska, M., McCabe, C., Linge, K. and Klenerman, L. (1996) 'Plantar foot pressures during treadmill walking with high-heel and low-heel shoes', *Foot & Ankle*, 17: 662–6.

30. Orlin, M., Stetson, K., Skowronski, J. and Pierrynowski, M. (1997) 'Foot pressure distribution: methodology and clinical application for children with ankle rheumatoid arthritis', *Clinical Biomechanics*, 12: S17.
31. Otter, S. J., Bowen, C. J. and Young, A. K. (2004) 'Forefoot plantar pressures in rheumatoid arthritis', *Journal of the American Podiatric Medical Association*, 94: 255–60.
32. Rosenbaum, D. and Lorei, T. (2003) 'Influence of density of pressure sensors on parameters of plantar foot load – a pedographic comparison of 2 pressure distribution recording platforms', *Biomed Tech (Berl)*, 48: 166–9.
33. Schaff, P. and Hauser, W. (1987) 'Dynamic measurement of pressure distribution with flexible measuring mats – an innovative measuring procedure in sports orthopedics and traumatology', *Sportverletzung Sportschaden*, 1: 185–222.
34. Tanaka, T., Takeda, H., Izumi, T., Ino, S. and Ifukube, T. (1999) 'Effects on the location of the centre of gravity and the foot pressure contribution to standing balance associated with ageing', *Ergonomics*, 42: 997–1010.
35. Unger, H. and Rosenbaum, D. (2004) 'Gender-specific differences of the foot during the first year of walking', *Foot & Ankle*, 25: 582–7.
36. Woodburn, J. and Helliwell, P. S. (1996) 'Relation between heel position and the distribution of forefoot plantar pressures and skin callosities in rheumatoid arthritis', *Ann Rheum Dis*, 55: 806–10.

12

Measurement for deriving kinematic parameters: numerical methods

Young-Hoo Kwon
Texas Woman's University, Texas

Introduction

Kinematics is defined as 'study of motion' dealing with a body or system in motion without reference to the cause (force). The phenomena (motions) rather than the causes (forces) are the subject of analysis (measurement) in a kinematic study. Kinematic quantities of interest include position and orientation (attitudes) of the segments, positions of the joints, and their time-derivatives (linear and angular velocities and accelerations).

The human body is often considered as a linked segment system, a system of rigid bodies linked at joints. Two observations may be drawn from this consideration. Firstly, the rigid body assumption simplifies analysis of the segment motions. It reduces a complex segment motion to two mutually independent motion components: translation of the centre of mass of the segment and rotation of the segment about its centre of mass (COM). A segment rarely shows pure translation or pure rotation, but a combination of both. Secondly, since the joint motions are the building blocks of the whole body motion, the whole body motion can be decomposed to individual joint motions. The essence of the joint motion is the relative motion of the distal segment of the joint to its proximal counterpart. (Mechanically speaking, the term 'joint motion' is misleading since joints are not the actors but segments are.) Thus, the joint motion can be derived from the rigid body motions of the segments, which are of primary importance in deriving kinematic parameters. The type of segment motion allowed at a joint is determined by the type of joint: ball-and-socket, hinge, etc.

Analysis methods commonly used to derive kinematic parameters include motion analysis, goniometry, and accelerometry. The immediate outcome varies from one method to another. Motion analysis provides coordinates of markers placed on the body. Active marker systems may also provide the orientation of the motion sensors. The immediate outcome of goniometry and accelerometry is electric signals representing joint angles or body accelerations. These immediate outcomes must be converted to meaningful primary parameters such as position and orientation of the segment (motion analysis), joint angle (goniometry), and acceleration (accelerometry). Additional parameters may be derived from the primary parameters based on the mechanical relationships among the kinematic quantities.

Among the kinematic analysis methods, motion analysis provides the most comprehensive set of kinematic parameters, involving multiple steps such as camera calibration, camera synchronization and reconstruction of the object space coordinates of the markers, digital filtering, calculation of the kinematic parameters, etc. In goniometry and accelerometry, once the primary data are obtained, the data are treated no differently from those obtained from motion analysis. Therefore, this chapter will mainly focus on motion analysis and, more specifically, key numerical methods and procedures of deriving kinematic parameters.

Camera calibration and reconstruction

Imaging an object point located in the object space with a camera is the process of mapping the point to a corresponding image point on the image plane (Figure 12.1). The object point (O), the image point (*I*), and the projection centre (*N*) are collinear. There exists a direct relationship between the object space coordinates, $[x,y,z]$, and the image plane coordinates, $[u,v]$ (Abdel-Aziz and Karara, 1971; Marzan and Karara, 1975; Walton, 1981):

$$u_i - u_o - \Delta u_i = -\lambda_u w_o \cdot \frac{t_{21}(x_i - x_o) + t_{22}(y_i - y_o) + t_{23}(z_i - z_o)}{t_{11}(x_i - x_o) + t_{12}(y_i - y_o) + t_{13}(z_i - z_o)} \tag{1}$$

$$v_i - v_o - \Delta v_i = -\lambda_v w_o \cdot \frac{t_{31}(x_i - x_o) + t_{32}(y_i - y_o) + t_{33}(z_i - z_o)}{t_{11}(x_i - x_o) + t_{12}(y_i - y_o) + t_{13}(z_i - z_o)} \tag{2}$$

where i is the control point number, $[0,u_i,v_i]$ and $[w_o,u_o,v_o]$ are the image plane coordinates of the image point (*I*) and the projection centre (*N*), respectively, $[x_i, y_i, z_i]$ and $[x_o, y_o, z_o]$ are the object space coordinates of the object point (O) and the projection centre, respectively, $[\Delta u_i, \Delta v_i]$ are the optical errors involved in the image coordinates, and $[\lambda_u, \lambda_v]$ are the scaling factors for the unit conversion from the real-life unit to the digitizer unit (DU).

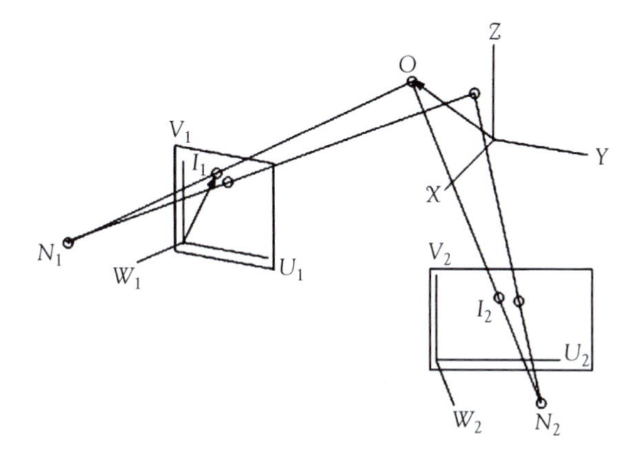

Figure 12.1 Object-to-image mapping in imaging and reconstruction.

$t_{11} - t_{33}$ in equations 1–2 are the elements of a 3 x 3 matrix constructed by three attitude angles (ϕ, θ, ψ) of the camera with respect to the laboratory reference frame:

$$\mathbf{T}(\phi,\theta,\psi) = \begin{bmatrix} t_{11} & t_{12} & t_{13} \\ t_{21} & t_{22} & t_{23} \\ t_{31} & t_{32} & t_{33} \end{bmatrix} = \begin{bmatrix} c(\theta)c(\psi) & s(\phi)s(\theta)c(\psi)+c(\phi)s(\psi) & -c(\phi)s(\theta)c(\psi)+s(\phi)s(\psi) \\ -c(\theta)s(\psi) & -s(\phi)s(\theta)s(\psi)+c(\phi)c(\psi) & c(\phi)s(\theta)s(\psi)+s(\phi)c(\psi) \\ s(\theta) & -s(\phi)c(\theta) & c(\phi)c(\theta) \end{bmatrix} \quad (3)$$

where c( ) and s( ) are the cosine and sine functions, respectively.

Optical errors (optical distortion and de-centring distortion) have traditionally been modelled

$$\Delta u_i = \xi_i \left(K_1 r_i^2 + K_2 r_i^4 + K_3 r_i^6\right) + K_4 \left(r_i^2 + 2\xi_i^2\right) + K_5 \left(2\xi_i \eta_i\right) \quad (4)$$

$$\Delta v_i = \eta_i \left(K_1 r_i^2 + K_2 r_i^4 + K_3 r_i^6\right) + K_4 \left(2\eta_i \xi_i\right) + K_5 \left(r_i^2 + 2\eta_i^2\right) \quad (5)$$

(Marzan and Karara, 1975; Walton, 1981), where:

$$[\xi_i, \eta_i] = [u_i - u_o, v_i - v_o] \quad (6)$$

$$r_i^2 = \xi_i^2 + \eta_i^2 \quad (7)$$

$K_1 - K_3$ and $K_4 - K_5$ are the optical distortion and de-centring distortion parameters, respectively. Equations 1 and 2 are essentially functions of 15 independent camera and optical error parameters, $[x_o, y_o, z_o, u_o, v_o, d_u, d_v, \phi, \theta, \psi, K_1, K_2, K_3, K_4, K_5]$, where $[d_u, d_v] = [\lambda_u w_o, \lambda_v w_o]$. Camera calibration is the process of determining the camera and optical error parameters based on a set of control points, the object spaces coordinates of which are known.

The direct linear transformation method

The direct linear transformation (DLT) method (Abdel-Aziz and Karara, 1971; Marzan and Karara, 1975; Walton, 1981) has been one of the most widely used camera calibration and reconstruction algorithms. The relationship between the object space coordinates and the image plane coordinates are described by a set of 11 DLT parameters in this method. The DLT parameters are mutually dependent since 11 parameters are derived from 10 camera parameters (Hatze, 1988; Kwon, 2005).

One definite strength of the DLT method is that it is also applicable to two-dimensional (2D) studies: 2D DLT (Brewin and Kerwin, 2003; Kwak *et al.*, 1996; Kwon, 1999; Kwon and Lindley, 2000; Walton, 1981). A total of eight DLT parameters are used in the 2D DLT method and, for this reason, the ten camera parameters cannot be determined individually. The 2D DLT method does not require the plane of motion and the image plane to be parallel to each other. Multiple cameras can be used with a common calibration frame (Kwon and Casebolt, 2006; Kwon and Fiaud, 2002).

Modified versions of the three-dimensional (3D) and 2D DLT methods specialized for analysis of underwater motions are also available: the localized DLT (Kwon *et al.*, 2002; Kwon and Casebolt, 2006; Kwon and Lindley, 2000). The control volume (the volume occupied by the control points) is sectioned into multiple distinct or overlapping sub-volumes to minimize the effect of the light refraction in these methods.

The direct solution method

Due to the inter-dependencies among the DLT parameters, the DLT method does not guarantee an orthogonal reconstruction of the object space. This shortcoming can be overcome by directly solving Equations 1–2 with respect to the camera parameters (Kwon, 2005). Equations 1–2 give a set of nonlinear equations:

$$f_i(\mathbf{X}) = (\xi_i - \Delta u_i)W_i + d_u U_i = 0 \tag{8}$$

$$g_i(\mathbf{X}) = (\eta_i - \Delta v_i)W_i + d_v V_i = 0 \tag{9}$$

where:

$$\mathbf{X} = \left[x_o, y_o, z_o, u_o, v_o, d_u, d_v, \phi, \theta, \psi, K_1, K_2, K_3, K_4, K_5\right]^t \tag{10}$$

$$W_i = t_{11}(x_i - x_o) + t_{12}(y_i - y_o) + t_{13}(z_i - z_o) \tag{11}$$

$$U_i = t_{21}(x_i - x_o) + t_{22}(y_i - y_o) + t_{23}(z_i - z_o) \tag{12}$$

$$V_i = t_{31}(x_i - x_o) + t_{32}(y_i - y_o) + t_{33}(z_i - z_o) \tag{13}$$

Expansion of Equations 8–9 for n (≥ 8) control points yields a sufficiently determined nonlinear system of equations:

$$\mathbf{F}(\mathbf{X}) = [.\ f_i(\mathbf{X})\ g_i(\mathbf{X}).]^t$$

Solution of this nonlinear system requires the use of Newton's method (Press *et al.*, 2002), for which the Jacobian matrix must be derived:

$$\mathbf{J}(\mathbf{X}) = \begin{bmatrix} . & . & . & . & . & . & . & . & . & . \\ \frac{\partial f_i}{\partial x_o} & \frac{\partial f_i}{\partial y_o} & \frac{\partial f_i}{\partial z_o} & \frac{\partial f_i}{\partial u_o} & . & . & \frac{\partial f_i}{\partial K_2} & \frac{\partial f_i}{\partial K_3} & \frac{\partial f_i}{\partial K_4} & \frac{\partial f_i}{\partial K_5} \\ \frac{\partial g_i}{\partial x_o} & \frac{\partial g_i}{\partial y_o} & \frac{\partial g_i}{\partial z_o} & \frac{\partial g_i}{\partial u_o} & . & . & \frac{\partial g_i}{\partial K_2} & \frac{\partial g_i}{\partial K_3} & \frac{\partial g_i}{\partial K_4} & \frac{\partial g_i}{\partial K_5} \\ . & . & . & . & . & . & . & . & . & . \end{bmatrix} \tag{14}$$

See the Appendix for a complete set of partial derivatives with respect to the camera and optical error parameters.

Newton's method uses an iterative approach:

1. An initial approximation of the unknowns, $\mathbf{X}^{(0)}$, is required. The output of the DLT method may be used as the initial approximation.
2. In each iteration, update matrix **X**:

$$\mathbf{X}^{(k)} = \mathbf{X}^{(k-1)} - \boldsymbol{\Delta}^{(k)}$$

where:

$$\boldsymbol{\Delta}^{(k)} = [\mathbf{J}(\mathbf{X}^{(k-1)})^t.\mathbf{J}(\mathbf{X}^{(k-1)})]^{-1}\,[\mathbf{J}(\mathbf{X}^{(k-1)})^t.\mathbf{F}(\mathbf{X}^{(k-1)})]$$

k is the current iteration (k = 1,2,3 ...), and $X^{(k-1)}$ is the parameter matrix obtained in the previous iteration.

3. In each iteration, check the convergence of the camera and optical error parameters. Stop when all parameters converge sufficiently.
4. Assess the MSE:

$$\mathrm{MSE} = \frac{\mathbf{F}'(\hat{\mathrm{X}})^t \cdot \mathbf{F}'(\hat{\mathrm{X}})}{2N-16} \quad \text{where } \mathbf{F}'(\hat{\mathrm{X}}) = \begin{bmatrix} \cdot & \dfrac{f_i(\hat{\mathrm{X}})}{Q_i} & \dfrac{g_i(\hat{\mathrm{X}})}{Q_i} & \cdot \end{bmatrix}^t$$

and $\hat{\mathrm{X}}$ is the converged set of **X**. The unit of MSE is DU^2. Compute the DLT parameters from the camera parameters and use them in the reconstruction process.

Matrix $\mathbf{T}(\phi, \theta, \psi)$ (Equation 3) is always orthogonal in the direct solution (DS) method (Kwon, 2005). The DS method is not applicable to 2D studies since the camera parameters cannot be determined individually.

Reconstruction of the object space coordinates is the process of mapping the image points back to the object point (Figure 12.1). Two or more cameras are required to reconstruct the 3D object space coordinates. Rearrangement of the terms of Equations 1–2 for n (≥ 2) cameras yields a sufficiently determined linear system of equations of the object space coordinates of a marker placed on the body. The object space coordinates can be computed by using the least square method.

Segmental reference frame

A reference frame is a particular perspective from which the motion of a body or system is described or observed. A reference frame fixed to the laboratory is inertial while those fixed to the body segments are non-inertial. The laboratory reference frame is also global (common) since this frame can be used in describing the motion of any body segments. A reference frame fixed to a segment is local and is meaningful only in that particular segment. The object space coordinates of the markers obtained through reconstruction are 'global' since they are described in the laboratory reference frame. The acceleration measured by a tri-axial accelerometer attached to tibia is 'local' since it is based on the accelerometer reference frame. Local data may be globalized only when the relative orientation and position of the local frame to the global frame is known.

Definition of the segmental reference frame

A segmental reference frame (SRF), a local frame fixed to the segment, simplifies the analysis of the segment motion. It is advantageous to align the frame axes with the anatomical axes. For example, if the axes of the pelvis and thigh reference frames are aligned with their respective anatomical axes (mediolateral, anteroposterior, and longitudinal), the relative motion of the thigh frame to the pelvis can be translated to anatomical hip joint motions (flexion/extension, adduction/abduction, and rotation).

To define a SRF, a set of three non-collinear points on the same segment are required. For example, the pelvis reference frame (PRF) is typically defined by three markers placed on the right anterior superior iliac spine (ASIS), left ASIS and the mid-point of the posterior superior iliac spines (PSIS):

$$\mathbf{i}_L = \frac{\mathbf{r}_{RA} - \mathbf{r}_{LA}}{|\mathbf{r}_{RA} - \mathbf{r}_{LA}|}$$

$$\mathbf{k}_L = \frac{(\mathbf{r}_{MP} - \mathbf{r}_{LA}) \times \mathbf{i}_L}{|(\mathbf{r}_{MP} - \mathbf{r}_{LA}) \times \mathbf{i}_L|}$$

$$\mathbf{j}_L = \mathbf{k}_L \times \mathbf{i}_L$$

where $\mathbf{i}_L$, $\mathbf{j}_L$, $\mathbf{k}_L$ are the unit vectors of the mediolateral (rightward), anteroposterior (forward), and longitudinal (upward) axes of the pelvis, respectively, and $\mathbf{r}_{RA}$, $\mathbf{r}_{LA}$, $\mathbf{r}_{MP}$ are the position vectors. The origin is at the mid-point of the two ASIS. In general, if the markers are close to each other or one marker is close to the line connecting the other two markers, the orientation of the reference frame becomes sensitive to the errors involved in the marker position. For this reason, wands are often used in the thigh and shank to move the lateral thigh/shank markers away from the surface of the body.

Orientation (attitude) matrix

A rigid segment in the object space has a total of six DoF, three from the position of the origin of the segmental frame and three from the orientation of the frame. The vector drawn from the origin of the global frame to that of the segmental frame shows the position of the segmental frame. The orientation of the segmental frame, on the other hand, is described by a 3 x 3 orientation (attitude) matrix constructed by the axis unit vectors:

$$\mathbf{T}_{L/G} = \begin{bmatrix} t_{11} & t_{12} & t_{13} \\ t_{21} & t_{22} & t_{23} \\ t_{31} & t_{32} & t_{33} \end{bmatrix} = \begin{bmatrix} \mathbf{i}_L^{\,t} \\ \mathbf{j}_L^{\,t} \\ \mathbf{k}_L^{\,t} \end{bmatrix} = \begin{bmatrix} \mathbf{i}_G \bullet \mathbf{i}_L & \mathbf{j}_G \bullet \mathbf{i}_L & \mathbf{k}_G \bullet \mathbf{i}_L \\ \mathbf{i}_G \bullet \mathbf{j}_L & \mathbf{j}_G \bullet \mathbf{j}_L & \mathbf{k}_G \bullet \mathbf{j}_L \\ \mathbf{i}_G \bullet \mathbf{k}_L & \mathbf{j}_G \bullet \mathbf{k}_L & \mathbf{k}_G \bullet \mathbf{k}_L \end{bmatrix} \tag{15}$$

where L is the segment, $\mathbf{i}_L$, $\mathbf{j}_L$, $\mathbf{k}_L$ are the axis unit vectors of the segmental frame, $\mathbf{i}_G$, $\mathbf{j}_G$, $\mathbf{k}_G$ are the axis unit vectors of the global frame, and $t_{11} - t_{33}$ are the global coordinates of the axis unit vectors. The orientation matrix conveniently combines three axis unit vectors into one 3 x 3 matrix.

Vector transformation

The global position of the hip joint centre is the vector sum of the relative position of the pelvis to the global frame ($\mathbf{r}_o$) and the position of the hip observed in the pelvis frame ($\mathbf{x}$):

$$\mathbf{r} = \mathbf{r}_o + \mathbf{x} \tag{16}$$

The local position of the hip joint centre is often expressed as a function of the inter-ASIS distance (Andriacchi *et al.*, 1980; Bell *et al.*, 1990; Tylkowski *et al.*, 1982), for example:

$$\mathbf{x} = [x_L, y_L, z_L] = [\pm 0.36, -0.19, -0.30]d$$

where d is the inter-ASIS distance. Although $\mathbf{r}_o$ in Equation 16 can be computed from the global positions of the ASIS markers, the global position of the hip joint is not readily available because $\mathbf{x}$ and r_o are described in different reference frames. To compute $\mathbf{r}$, both vectors must be described in the same reference frame and a vector transformation is necessary. A vector transformation does not alter the vector itself, but alters the vector coordinates.

Multiplication of the transpose of matrix $\mathbf{T}_{L/G}$ (Equation 15) to the local coordinates of $\mathbf{x}$ yields:

$$\mathbf{T}_{L/G}{}^{t} \cdot \begin{bmatrix} x_L \\ y_L \\ z_L \end{bmatrix} = \begin{bmatrix} (x_L\mathbf{i}_L + y_L\mathbf{j}_L + z_L\mathbf{k}_L)\bullet\mathbf{i}_G \\ (x_L\mathbf{i}_L + y_L\mathbf{j}_L + z_L\mathbf{k}_L)\bullet\mathbf{j}_G \\ (x_L\mathbf{i}_L + y_L\mathbf{j}_L + z_L\mathbf{k}_L)\bullet\mathbf{k}_G \end{bmatrix} = \begin{bmatrix} \mathbf{x}\bullet\mathbf{i}_G \\ \mathbf{x}\bullet\mathbf{j}_G \\ \mathbf{x}\bullet\mathbf{k}_G \end{bmatrix} = \begin{bmatrix} x_G \\ y_G \\ z_G \end{bmatrix}$$

or:

$$\mathbf{x}^{(G)} = \mathbf{T}_{L/G}{}^{t} . \mathbf{x}^{(L)} \tag{17}$$

where $[x_G, y_G, z_G]$ are the global coordinates of $\mathbf{x}$, and $\mathbf{x}^{(L)}$ and $\mathbf{x}^{(G)}$ are vector $\mathbf{x}$ described in the local and global frame, respectively. The inverse transformation from the global to the local frame gives:

$$\mathbf{x}^{(L)} = \mathbf{T}_{L/G} . \mathbf{x}^{(G)} \tag{18}$$

Equation 16 can be translated to more workable forms:

$$\mathbf{r}^{(G)} = \mathbf{r}_o{}^{(G)} + \mathbf{T}_{L/G}{}^{t} \cdot \mathbf{x}^{(L)}$$

$$\mathbf{x}^{(L)} = \mathbf{T}_{L/G} \cdot \left(\mathbf{r}^{(G)} - \mathbf{r}_o{}^{(G)}\right)$$

The orientation matrix of frame L with respect to the global ($\mathbf{T}_{L/G}$) is also called as the transformation matrix from the global frame to frame L.

Equations 15, 17–18 essentially show:

$$\mathbf{T}_{G/L} = \mathbf{T}_{L/G}^{-1} = \mathbf{T}_{L/G}^{t} \tag{19}$$

$$\mathbf{T}_{L/G} \cdot \mathbf{T}_{L/G}^{t} = \mathbf{T}_{L/G}^{t} \cdot \mathbf{T}_{L/G} = \mathbf{I} \tag{20}$$

where **I** is the identity matrix. A transformation matrix is orthogonal (Equation 20) since its inverse is equal to its transpose (Equation 19). Multiple sequential transformations of a vector are equivalent to a single direct transformation from the first frame to the last frame:

$$\mathbf{v}^{(C)} = \mathbf{T}_{C/B} \cdot \mathbf{v}^{(B)} = \mathbf{T}_{C/B} \cdot \left(\mathbf{T}_{B/A} \cdot \mathbf{v}^{(A)}\right) = \left(\mathbf{T}_{C/B} \cdot \mathbf{T}_{B/A}\right) \cdot \mathbf{v}^{(A)} = \mathbf{T}_{C/A} \cdot \mathbf{v}^{(A)}$$

$$\mathbf{T}_{C/A} = \mathbf{T}_{C/B} \cdot \mathbf{T}_{B/A}$$

Thus, the transformation matrix from one local frame (*A*) to another (*B*) can be derived as

$$\mathbf{T}_{B/A} = \mathbf{T}_{B/G} \cdot \mathbf{T}_{G/A} = \mathbf{T}_{B/G} \cdot \mathbf{T}_{A/G}^{t} \tag{21}$$

$\mathbf{T}_{B/A}$ shows the relative orientation of segment *B* to segment A.

Orientation angles

The relative orientation matrix of the thigh frame to the pelvis frame can be decomposed into three successive rotation matrices about selected axes of the pelvis and thigh frames and intermediate frames. Six Cardan rotation sequences are possible in this process (XYZ, YZX, ZXY, XZY, YXZ, and ZYX); Figure 12.2 uses the XYZ (mediolateral-anteroposterior-longitudinal) sequence. The pelvis frame (A) first rotates about its X axis (mediolateral axis) (X_A) by ϕ, then about the Y axis (anteroposterior axis) of the first intermediate frame (Y') by θ, and finally about the Z axis (longitudinal axis) of the second intermediate frame

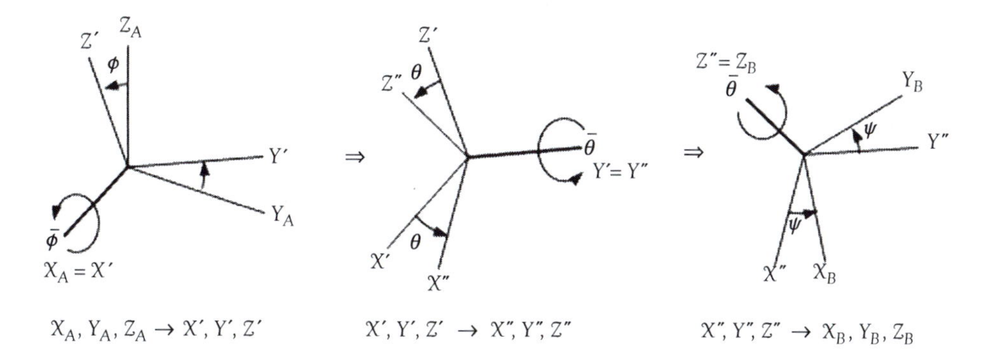

Figure 12.2 Orientation angles (XYZ rotation sequence).

(Z'') by Ψ to reach the thigh frame (B). These three rotation angles (ϕ,θ,ψ) show the relative orientation of the thigh frame to the pelvis frame, and are called orientation angles. The three successive rotations shown in Figure 12.2 are essentially equivalent to

$$\mathbf{T}_{B/A} = \begin{bmatrix} c(\psi) & s(\psi) & 0 \\ -s(\psi) & c(\psi) & 0 \\ 0 & 0 & 1 \end{bmatrix} \begin{bmatrix} c(\theta) & 0 & -s(\theta) \\ 0 & 1 & 0 \\ s(\theta) & 0 & c(\theta) \end{bmatrix} \begin{bmatrix} 1 & 0 & 0 \\ 0 & c(\phi) & s(\phi) \\ 0 & -s(\phi) & c(\phi) \end{bmatrix}$$
$$= \begin{bmatrix} c(\theta)c(\psi) & s(\phi)s(\theta)c(\psi)+c(\phi)s(\psi) & -c(\phi)s(\theta)c(\psi)+s(\phi)s(\psi) \\ -c(\theta)s(\psi) & -s(\phi)s(\theta)s(\psi)+c(\phi)c(\psi) & c(\phi)s(\theta)s(\psi)+s(\phi)c(\psi) \\ s(\theta) & -s(\phi)c(\theta) & c(\phi)c(\theta) \end{bmatrix} \tag{22}$$

where $\mathbf{T}_{B/A}$ is the known relative orientation matrix of the thigh to the pelvis (Equation 21).

If $|\theta| \neq \pi/2$, Equations 15 and 22 yield:

$$s(\theta) = t_{31} \tag{23}$$

$$c(\psi) = \frac{t_{11}}{c(\theta)} \tag{24}$$

$$s(\psi) = -\frac{t_{21}}{c(\theta)} \tag{25}$$

$$c(\phi) = \frac{t_{33}}{c(\theta)} \tag{26}$$

$$s(\phi) = -\frac{t_{32}}{c(\theta)} \tag{27}$$

Equations 23–27 provide two solutions: (ϕ_o,θ_o,ψ_o) and $(\phi_o \pm \pi, \pm\pi - \theta_o, \psi_o \pm \pi)$, where $|\theta_o| < \pi/2$. Hinge joints typically have small $|\theta|$ values, thus resulting in one solution. In ball-and-socket joints, particularly the shoulder, however, it is possible for $|\theta|$ to exceed $\pi/2$ and the correct angle combination must be selected between the two candidates. Since the two solutions differ in ϕ and ψ by π, respectively, the angle combination giving smaller changes in these angles from the previous values must be selected.

If $\theta = \pm\pi/2$, Equation 22 reduces to:

$$\mathbf{T}_{B/A} = \begin{bmatrix} 0 & \pm s(\phi\pm\psi) & \mp c(\phi\pm\psi) \\ 0 & c(\phi\pm\psi) & s(\phi\pm\psi) \\ \pm 1 & 0 & 0 \end{bmatrix} = \begin{bmatrix} t_{11} & t_{12} & t_{13} \\ t_{21} & t_{22} & t_{23} \\ t_{31} & t_{32} & t_{33} \end{bmatrix}$$

$$c(\phi \pm \Psi) = t_{22}$$

$$s(\phi \pm \Psi) = t_{23}$$

This is the so-called gimbal-lock problem in which angles ϕ and Ψ must be computed collectively as $\phi \pm \psi$. When the distal segment approaches to the gimbal-lock position ($c(\theta) \approx 0$), angles ϕ and ψ become sensitive to the experimental errors as well (Equations 24–27). One potential solution for the problems at/near the gimbal-lock position is to treat the orientation angles as missing and generate them through interpolation.

The main strength of the orientation angles is their intuitiveness. In the XYZ rotation sequence shown in Figure 12.2, for example, the first rotation shows the rotation of the distal segment (thigh) in the sagittal plane of the proximal segment (pelvis), while the second rotation quantifies the deviation of the thigh from the sagittal plane of the pelvis. The third rotation is about the longitudinal axis of the thigh. Although these angles are slightly different from the conventionally defined joint motions based on the anatomical reference position (such as flexion/extension, adduction/abduction, and rotation), orientation angles do provide intuitive orientation information of the segment. Alternatives to the orientation angle approach are also available (Cheng *et al.*, 2000; Coburn and Crisco, 2005; Ying and Kim, 2002).

Orientation angles are sequence-specific, meaning different rotation sequences result in different sets of orientation angles. For this reason, comparing two data sets based on different rotation sequences must be avoided. The rotation sequence chosen must be physically and intuitively meaningful. There is no set rule concerning the selection of the rotation sequence, but the following will serve as guidelines:

- In general, it is not recommended to place the longitudinal (Z) axis between the other two axes. Angular motion of a segment can be decomposed into two component rotations: the rotation of the longitudinal axis (transverse rotation) and the rotation of the segment about the longitudinal axis (longitudinal rotation). The transverse rotation comes from the rotations about the mediolateral (X) and anteroposterior (Y) axes. Placing the longitudinal axis between the other two axes (YZX and XZY) makes the two rotation planes not perpendicular and, thus, can exaggerate the rotations substantially.
- Between the mediolateral and anteroposterior axes, placing the mediolateral axis first is more intuitive and meaningful. This is because the sagittal plane is the symmetry plane of the central segments and sagittal plane motions are the main joint motions in the limbs. The hinge axes are generally aligned with the mediolateral axes of the segments. This excludes the YXZ and ZYX sequences as well.
- There are only two remaining sequences out of the six possible sequences: XYZ and ZXY. If the longitudinal rotation of the segment is the centrefold of the analysis, placing the Z axis after the X and Y axes (XYZ) is more intuitive and meaningful. This is because the focus of the analysis is the longitudinal orientation with a given transverse orientation. If the focus of the study is the transverse orientation with a given longitudinal orientation, the ZXY sequence is more meaningful. The central segments such as the trunk (thorax, abdomen, and pelvis) and the head are the potential clients for the ZXY sequence, but the nature of the investigation must be considered carefully in the selection process.

The orientation angle approach is not applicable to the net motion of a segment from one time point to another although it is still possible to compute the relative orientation matrix of the segment of one time point to the previous and, subsequently, the relative orientation angles:

$$\mathbf{T}_{i+1/i} = \mathbf{T}_{i+1} \cdot \mathbf{T}_i^t \;\Rightarrow\; \left(\phi_{i+1/i}, \theta_{i+1/i}, \psi_{i+1/i}\right)$$

where $\mathbf{T}_i$ is the orientation matrix of the segment at time point i, $\mathbf{T}_{i+1/i}$ is the relative orientation matrix of time point $i + 1$ to time point i, and $\phi_{i+1/i}, \theta_{i+1/i}, \psi_{i+1/i}$ are the relative orientation angles computed from $\mathbf{T}_{i+1/i}$. The relative orientation angles of one time point to another must not be added to the previous orientation angles of the segment to update the orientation:

$$\begin{bmatrix} \phi_{i+1} \\ \theta_{i+1} \\ \psi_{i+1} \end{bmatrix} \neq \begin{bmatrix} \phi_i \\ \theta_i \\ \psi_i \end{bmatrix} + \begin{bmatrix} \phi_{i+1/i} \\ \theta_{i+1/i} \\ \psi_{i+1/i} \end{bmatrix}$$

where ϕ_i, θ_i, ψ_i are the orientation angles of the segment at time point i. The rotation of the segment must be analyzed with the time history of the orientation angles $(\phi_i, \theta_i, \psi_i)$, not by the relative orientation angles between time points $(\phi_{i+1i}, \theta_{i+1/i}, \psi_{i+1/i})$.

Angular velocity

The relative angular velocity of a rigid segment to its proximal counterpart is the vector sum of the angular velocities of the three successive rotations (Figure 12.2) (Kane and Scher, 1970; Ramey and Yang, 1981):

$$\boldsymbol{\omega}_{B/A} = \dot{\phi}\mathbf{i}_A + \dot{\theta}\mathbf{j}' + \dot{\psi}\mathbf{k}'' = \dot{\phi}\mathbf{i}' + \dot{\theta}\mathbf{j}'' + \dot{\psi}\mathbf{k}_B \tag{28}$$

where $\left[\dot{\phi}, \dot{\theta}, \dot{\psi}\right]$ are the first time-derivatives of the orientation angles. The relative angular velocity must be described either in frame A or B:

$$\boldsymbol{\omega}_{B/A}^{(A)} = \begin{bmatrix} \dot{\phi} \\ 0 \\ 0 \end{bmatrix} + \begin{bmatrix} 1 & 0 & 0 \\ 0 & c(\phi) & -s(\phi) \\ 0 & s(\phi) & c(\phi) \end{bmatrix} \begin{bmatrix} 0 \\ \dot{\theta} \\ 0 \end{bmatrix} + \begin{bmatrix} 1 & 0 & 0 \\ 0 & c(\phi) & -s(\phi) \\ 0 & s(\phi) & c(\phi) \end{bmatrix} \begin{bmatrix} c(\theta) & 0 & s(\theta) \\ 0 & 1 & 0 \\ -s(\theta) & 0 & c(\theta) \end{bmatrix} \begin{bmatrix} 0 \\ 0 \\ \dot{\psi} \end{bmatrix}$$
$$= \begin{bmatrix} 1 & 0 & s(\theta) \\ 0 & c(\phi) & -s(\phi)c(\theta) \\ 0 & s(\phi) & c(\phi)c(\theta) \end{bmatrix} \begin{bmatrix} \dot{\phi} \\ \dot{\theta} \\ \dot{\psi} \end{bmatrix}$$

$$\boldsymbol{\omega}_{B/A}^{(B)} = \begin{bmatrix} c(\psi) & s(\psi) & 0 \\ -s(\psi) & c(\psi) & 0 \\ 0 & 0 & 1 \end{bmatrix} \begin{bmatrix} c(\theta) & 0 & -s(\theta) \\ 0 & 1 & 0 \\ s(\theta) & 0 & c(\theta) \end{bmatrix} \begin{bmatrix} \dot{\phi} \\ 0 \\ 0 \end{bmatrix} + \begin{bmatrix} c(\psi) & s(\psi) & 0 \\ -s(\psi) & c(\psi) & 0 \\ 0 & 0 & 1 \end{bmatrix} \begin{bmatrix} 0 \\ \dot{\theta} \\ 0 \end{bmatrix} + \begin{bmatrix} 0 \\ 0 \\ \dot{\psi} \end{bmatrix}$$
$$= \begin{bmatrix} c(\theta)c(\psi) & s(\psi) & 0 \\ -c(\theta)s(\psi) & c(\psi) & 0 \\ s(\theta) & 0 & 1 \end{bmatrix} \begin{bmatrix} \dot{\phi} \\ \dot{\theta} \\ \dot{\psi} \end{bmatrix} = \begin{bmatrix} c(\theta)c(\psi)\dot{\phi} + s(\psi)\dot{\theta} \\ -c(\theta)s(\psi)\dot{\phi} + c(\psi)\dot{\theta} \\ s(\theta)\dot{\phi} + \dot{\psi} \end{bmatrix} \tag{29}$$

Equation 29 shows that the transverse angular velocity is not affected by $\dot{\psi}$, but $\dot{\phi}$ contributes to the longitudinal angular velocity. This is because the first and third rotation planes are not perpendicular to each other. The angular velocity of segment B is the sum of its relative angular velocity to segment A and the angular velocity of segment A:

$$\boldsymbol{\omega}_B^{(G)} = \boldsymbol{\omega}_A^{(G)} + \mathbf{T}_{A/G}{}^t \cdot \boldsymbol{\omega}_{B/A}{}^{(A)} = \boldsymbol{\omega}_A^{(G)} + \mathbf{T}_{B/G}{}^t \cdot \boldsymbol{\omega}_{B/A}{}^{(B)}$$

where $\boldsymbol{\omega}_A$ and $\boldsymbol{\omega}_B$ are the angular velocities of segments A and B, respectively.

An alternative method of computation of the angular velocity can be derived from the linear velocity of a point on a rotating body:

$$\mathbf{v} = \frac{d\mathbf{r}}{dt} = \boldsymbol{\omega} \times \mathbf{r}$$

$$\tilde{\boldsymbol{\omega}}^{(G)} \cdot \left(\mathbf{T}_{L/G}{}^t \cdot \mathbf{r}^{(L)}\right) = \frac{d}{dt}\left(\mathbf{T}_{L/G}{}^t \cdot \mathbf{r}^{(L)}\right) \tag{30}$$

where $\tilde{\boldsymbol{\omega}}$ is a 3 x 3 angular velocity matrix for the cross product operation:

$$\tilde{\boldsymbol{\omega}} = \begin{bmatrix} 0 & -\omega_z & \omega_y \\ \omega_z & 0 & -\omega_x \\ -\omega_y & \omega_x & 0 \end{bmatrix}$$

Since the local position ($\mathbf{r}^{(L)}$) of the point is constant, Equation 30 gives:

$$\tilde{\boldsymbol{\omega}}^{(G)} = \frac{d}{dt}\left(\mathbf{T}_{L/G}{}^t\right) \cdot \mathbf{T}_{L/G}$$

Thus, angular velocity can be computed directly from the orientation matrix of the segment.

Rigid body method

Motion of a segment can also be described by the collective motions of the markers (≥3) attached to the segment. The rigid body method is useful where the relative positions of the markers with respect to each other must be known. Common applications of the rigid body method include generation of the coordinates of hidden markers, computation of the joint centre locations in the motion trials based on a static trial, and computation of the instantaneous axis of rotation.

Motion of a rigid body from one time point to another

The motion of a marker attached to a rigid segment from one time point to the next can be expressed in terms of a rotation about a helical (or screw) axis and a translation along the axis (Figure 12.3) (Challis, 1995; Spoor and Veldpaus, 1980; Woltring *et al.*, 1985):

$$\mathbf{r}_{i+1,j} - \mathbf{r}_o = \mathbf{R}_{i+1/i} \cdot \left(\mathbf{r}_{i,j} - \mathbf{r}_o\right) + \mathbf{L} \tag{31}$$

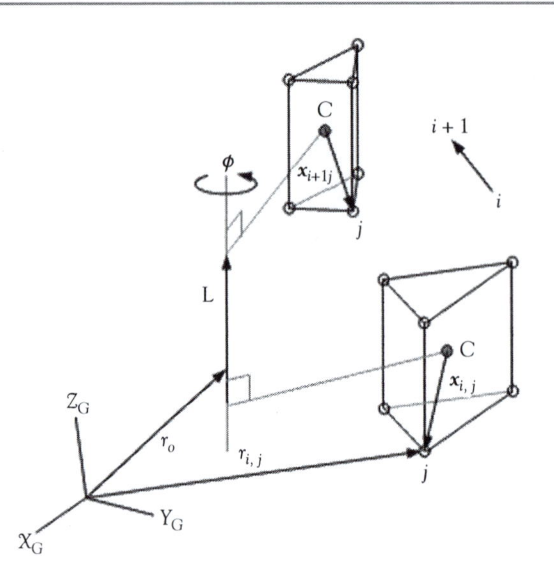

Figure 12.3 Rigid body motion about and along a screw axis.

where $\mathbf{r}_{i,j}$ is the global position of marker j at time point i, $\mathbf{r}_o$ is the position of a point on the axis of rotation, $\mathbf{R}_{i+1/i}$ is the rotation matrix from time point i to $i + 1$, and $\mathbf{L}$ is the translation vector. Averaging Equation 31 for all markers gives:

$$\overline{\mathbf{r}}_{i+1} - \mathbf{r}_o = \mathbf{R}_{i+1/i} \cdot \left(\overline{\mathbf{r}}_i - \mathbf{r}_o\right) + \mathbf{L} \tag{32}$$

where $\overline{\mathbf{r}}_i$ and $\overline{\mathbf{r}}_{i+1}$ are the positions of the centroid of the markers (C in Figure 12.3) measured at time points i and i+ 1, respectively:

$$\overline{\mathbf{r}}_i = \frac{1}{N}\sum_{j=1}^{N} \mathbf{r}_{i,j}$$

Subtracting Equation 32 from 31 yields:

$$\mathbf{x}_{i+1,j} = \mathbf{R}_{i+1/i} \cdot \mathbf{x}_{i,j} \tag{33}$$

where:

$$\mathbf{x}_{i,j} = \mathbf{r}_{i,j} - \overline{\mathbf{r}}_i$$

$\mathbf{x}_{i,j}$ is the relative position of marker j to the centroid.

Complex motions of markers attached to a rigid segment from one time point to another can be decomposed into the translation of the centroid (C in Figure 12.3) and the rotation of the marker about the centroid:

$$\mathbf{r}_{i+1,j} = \bar{\mathbf{r}}_i + \left(\bar{\mathbf{r}}_{i+1} - \bar{\mathbf{r}}_i\right) + \mathbf{R}_{i+1/i} \cdot \mathbf{x}_{i,j} \tag{34}$$

If rotation matrix $\mathbf{R}_{i+1/i}$ is known, Equation 34 can be used to compute the position of a point observed at time point i but hidden at time point $i + 1$. Equation 34 is also applicable to the static-dynamic trial setting where time point i represents the static trial and time point $i + 1$ represents a given time point in the motion trial (Schmidt *et al.*, 1999).

Computation of the rotation matrix

The rotation matrix from one time point to another can be derived from Equation 33 through a simple least square procedure (Challis, 1995):

$$\frac{1}{n}\sum_{j=1}^{n}\left(\mathbf{x}_{i+1,j} - \mathbf{R}_{i+1/i} \cdot \mathbf{x}_{i,j}\right)^t \left(\mathbf{x}_{i+1,j} - \mathbf{R}_{i+1/i} \cdot \mathbf{x}_{i,j}\right) \Rightarrow \min \tag{35}$$

where n is the marker count. Equation 35 essentially reduces to:

$$\frac{1}{N}\sum_{j=1}^{N}\left(\mathbf{x}_{i+1,j}{}^t \cdot \mathbf{R}_{i+1/i} \cdot \mathbf{x}_{i,j}\right) \Rightarrow \max$$

or:

$$Tr\left(\mathbf{R}_{i+1/i}{}^t \cdot \mathbf{C}\right) \Rightarrow \max \tag{36}$$

where:

$$\mathbf{C} = \frac{1}{N}\sum_{j=1}^{N} \mathbf{x}_{i+1,j} \cdot \mathbf{x}_{i,j}{}^t$$

Tr( ) is the trace function. Singular value decomposition (SVD) of the correlation matrix **C** yields

$$\mathbf{C} = \mathbf{U}.\mathbf{W}.\mathbf{V}^t \tag{37}$$

where **U** and **V** are 3 x 3 orthogonal matrices, and **W** is a 3 x 3 diagonal matrix containing the singular values of **C**. The diagonal elements of **W** are either positive or zero. Equations 36–37 give:

$$Tr(\mathbf{C}'.\mathbf{W}) \Rightarrow \max \tag{38}$$

where:

$$\mathbf{C}' = \mathbf{V}^t \cdot \mathbf{R}_{i+1/i}{}^t \cdot \mathbf{U} = \mathbf{I} \tag{39}$$

$\mathbf{C}'$ is orthogonal since all three matrices involved are orthogonal (Equation 39). Moreover, $\mathbf{C}'$ must have the largest positive diagonal elements to suffice Equation 38. This forces $\mathbf{C}'$ to be an identity matrix. Equation 39 further yields

$$\mathbf{R}_{i+1/i} = \mathbf{U} \cdot \mathbf{V}^t \tag{40}$$

To obtain an accurate estimation of the rotation matrix, it is important to distribute the markers throughout the segment with no apparent symmetry in shape. With existence of symmetry, Equation 40 may not provide an accurate rotation matrix and reflection may occur instead of rotation. In the case of reflection, the determinant of $\mathbf{R}_{i+1/i}$ is equal to –1.

Helical axis

As shown in Figure 12.3, motion of a segment from one time point to another can be expressed in terms of a single rotation about, and a translation along, the helical axis. To describe the complete screw motion of the segment, the unit vector of the rotation axis, rotation angle, and the location of the axis must be computed. A rotation about an arbitrary axis gives a 3 x 3 rotation matrix of the following form (Cheng *et al.*, 2000; Spoor and Veldpaus, 1980):

$$\mathbf{R}(\varphi, \mathbf{n}) = (1 - \cos\varphi)\mathbf{n} \cdot \mathbf{n}^t + \cos\varphi \mathbf{I} + \sin\varphi \tilde{\mathbf{n}} \tag{41}$$

where $\mathbf{n}$ is the unit vector of the rotation axis, ϕ is the rotation angle ($0<\phi<\pi$), and

$$\tilde{\mathbf{n}} = \begin{bmatrix} 0 & -n_z & n_y \\ n_z & 0 & -n_x \\ -n_y & n_x & 0 \end{bmatrix} \tag{42}$$

The following additional properties of $\mathbf{R}$ can be derived from Equation 41–42:

$$\mathbf{R}^t = (1 - \cos\varphi)\mathbf{n} \cdot \mathbf{n}^t + \cos\varphi \mathbf{I} - \sin\varphi \tilde{\mathbf{n}} \tag{43}$$

$$\mathbf{R} - \mathbf{R}^t = 2\sin\varphi \tilde{\mathbf{n}} \tag{44}$$

$$\frac{1}{2}(\mathbf{R} + \mathbf{R}^t) - \cos\varphi \mathbf{I} = (1 - \cos\varphi)\mathbf{n} \cdot \mathbf{n}^t \tag{45}$$

$$\mathbf{I} - \mathbf{R} = (1 - \cos\varphi)(\mathbf{I} - \mathbf{n} \cdot \mathbf{n}^t) - \sin\varphi \tilde{\mathbf{n}} = (1 - \cos\varphi)\tilde{\mathbf{n}}^t \cdot \tilde{\mathbf{n}} - \sin\varphi \tilde{\mathbf{n}} \tag{46}$$

If **R** is known (Equation 40), from Equations 41 and 44, the rotation angle can be computed:

$$\cos\varphi = \frac{Tr(\mathbf{R})-1}{2}$$

$$\sin\varphi = \frac{1}{2}\sqrt{(r_{12}-r_{21})^2+(r_{23}-r_{32})^2+(r_{31}-r_{13})^2}$$

where $r_{11} - r_{33}$ are elements of the rotation matrix.

All columns of the right-side matrix of Equation 45 are multiples of **n**, so the unit vector of the rotation axis can be obtained from any column of the left matrix, but preferably from the one with the largest magnitude. In addition, Equation 32 gives:

$$(\mathbf{I}-\mathbf{R})\cdot\mathbf{r}_o + L\mathbf{n} = \bar{\mathbf{r}}_{i+1} - \mathbf{R}\cdot\bar{\mathbf{r}}_i \tag{47}$$

where L is the translation along the helical axis. Multiplying $\mathbf{n}^t$ to both sides gives:

$$L = \mathbf{n}^t\cdot(\bar{\mathbf{r}}_{i+1} - \mathbf{R}\cdot\bar{\mathbf{r}}_i)$$

Rearranging of the terms in Equation 47 gives:

$$(\mathbf{I}-\mathbf{R})\cdot\mathbf{r}_o = (\bar{\mathbf{r}}_{i+1} - \mathbf{R}\cdot\bar{\mathbf{r}}_i) - L\mathbf{n} \tag{48}$$

If we assume $\mathbf{r}_o$ to be perpendicular to **n** for a unique solution, Equation 48 reduces to

$$\left[(1-\cos\varphi)\mathbf{I} - \sin\varphi\tilde{\mathbf{n}}\right]\cdot\mathbf{r}_o = (\bar{\mathbf{r}}_{i+1} - \mathbf{R}\cdot\bar{\mathbf{r}}_i) - L\mathbf{n}$$

which provides a unique solution for $\mathbf{r}_o$. The helical axis method is used to find the instantaneous joint axis of rotation.

Joint centre and axis

A joint is where two adjacent segments are linked to each other. Mislocation of the joint centre directly impacts kinematic and kinetic parameters significantly such as joint angles and resultant joint moments (Stagni *et al.*, 2000). In addition, misalignment of the medio-lateral and anteroposterior axes of the limb segments can substantially distort kinematic parameters such as knee valgus angle (Schache *et al.*, 2005).

Centre of rotation: ball-and-socket joint

Determining the hip joint centre is an essential but error-prone process. Although estimation methods have been the main stream (Andriacchi *et al.*, 1980; Davis III *et al.*, 1991; Seidel *et al.*, 1995; Tylkowski *et al.*, 1982), the estimated location can vary significantly from its true location, and from one method to another (Bell *et al.*, 1990; Kirkwood *et al.*, 1999). An alternative to the estimation methods is the functional method, a set of numerical procedures based on marker coordinates (Bell *et al.*, 1990; Gamage and Lasenby, 2002; Halvorsen *et al.*, 1999; Leardini *et al.*, 1999).

Figure 12.4a shows a set of markers placed on the thigh. The hip joint serves as the common centre of rotation for the markers and the distance from the joint centre to a given marker remains constant throughout the course of motion. Application of this assumption to the hip and thigh markers gives:

$$f_{ij}(\mathbf{r}_C, R_1, \ldots R_n) = \sqrt{(\mathbf{r}_C - \mathbf{r}_{ij})^t (\mathbf{r}_C - \mathbf{r}_{ij})} - R_j = 0 \tag{49}$$

where i is the time point, j is the marker, $\mathbf{r}_C$ is the pelvis-frame position of the hip joint centre, R_j is the constant distance from the joint centre to a marker, n is the total marker count, and $\mathbf{r}_{ij}$ is the pelvis-frame position of each marker at each time point. The partial derivatives of f_{ij} (Equation 49) are:

$$\begin{bmatrix} \frac{\partial f_{ij}}{\partial x_C} & \frac{\partial f_{ij}}{\partial y_C} & \frac{\partial f_{ij}}{\partial z_C} \end{bmatrix} = \frac{1}{d_{ij}} \begin{bmatrix} x_C - x_{ij} & y_C - y_{ij} & z_C - z_{ij} \end{bmatrix} \tag{50}$$

$$\frac{\partial f_{ij}}{\partial R_k} = \begin{cases} -1 & (j = k) \\ 0 & (j \neq k) \end{cases} \tag{51}$$

where:

$$d_{ij} = \sqrt{(\mathbf{r}_C - \mathbf{r}_{ij})^t \cdot (\mathbf{r}_C - \mathbf{r}_{ij})}$$

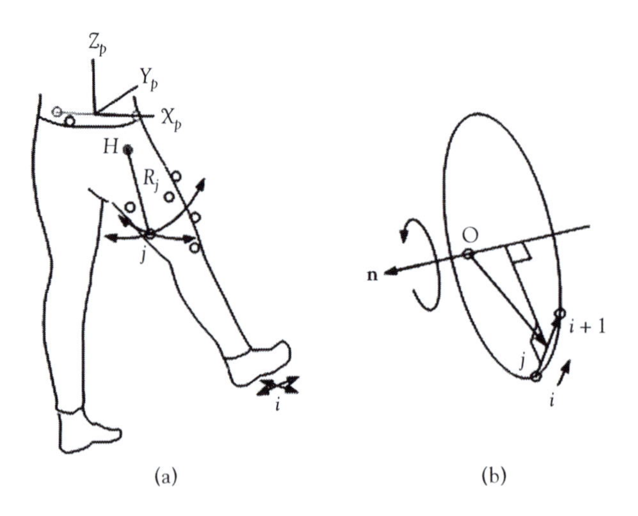

Figure 12.4 Joint centre and axis of rotation: the relative motions of the thigh markers to the hip joint centre (a); and motion of a shank marker about the knee axis (b).

(Gander *et al.*, 1994). Equations 49–50 provide a nonlinear system of equations:

$$\mathbf{X} = \left[x_C, y_C, z_C, R_1, \ldots, R_n\right]^t \tag{52}$$

$$\mathbf{F}(\mathbf{X}) = \left[\cdots \; f_{ij} \; \cdots\right]^t \tag{53}$$

$$\mathbf{J}(\mathbf{X}) = \begin{bmatrix} \frac{x_C - x_{11}}{d_{11}} & \frac{y_C - y_{11}}{d_{11}} & \frac{z_C - z_{11}}{d_{11}} & -1 & \cdot & 0 & \cdot & 0 \\ \cdot & \cdot & \cdot & \cdot & \cdot & \cdot & \cdot & \cdot \\ \frac{x_C - x_{ij}}{d_{ij}} & \frac{y_C - y_{ij}}{d_{ij}} & \frac{z_C - z_{ij}}{d_{ij}} & 0 & \cdot & -1 & \cdot & 0 \\ \cdot & \cdot & \cdot & \cdot & \cdot & \cdot & \cdot & \cdot \\ \frac{x_C - x_{mn}}{d_{mn}} & \frac{y_C - y_{mn}}{d_{mn}} & \frac{z_C - z_{mn}}{d_{mn}} & 0 & \cdot & 0 & \cdot & -1 \end{bmatrix} \tag{54}$$

where m is the total time point count (video frames). Newton's method can be used to obtain a set of converged hip joint coordinates and joint-marker distances. An initial approximation must be provided, for example:

$$\mathbf{r}_C^{(0)} = \mathbf{0}$$

$$R_j^{(0)} = \left[\frac{1}{m}\sum_{i=1}^{m} \mathbf{r}_{ij}^{\,t} \cdot \mathbf{r}_{ij}\right]^{1/2}$$

where $\mathbf{0}$ is the zero vector. The overall accuracy can be assessed by the MSE:

$$\mathrm{MSE} = \frac{\mathbf{F}(\hat{\mathbf{X}})^t \cdot \mathbf{F}(\hat{\mathbf{X}})}{mn - (3 + n)}$$

Other functional method algorithms are also available (Gamage and Lasenby, 2002; Halvorsen *et al.*, 1999; Piazza *et al.*, 2004).

Note that this method is not suitable for the shoulder joint since it is difficult to isolate the relative motion of the upper arm to the scapula. The rigid body method with a set of upper arm markers and the shoulder joint centre location obtained in a static trial is more practical (Schmidt *et al.*, 1999).

Axis of rotation: hinge joint

Shank markers rotating about the knee axis form circles perpendicular to the knee axis (Figure 12.4b). Thus, the displacement from one time point to another is perpendicular to the rotation axis (Halvorsen *et al.*, 1999):

$$\left(\mathbf{r}_{i+1,j}-\mathbf{r}_{i,j}\right)^{t}\cdot\mathbf{n}=0 \tag{55}$$

where $\mathbf{r}_{i,j}$ and $\mathbf{r}_{i+1,j}$ are the thigh-frame position of marker j at time points i and $i+1$, respectively, and $\mathbf{n}$ is the unit vector of the axis. Application of the least square method to Equation 55 yields:

$$E=\mathbf{n}^{t}\cdot\mathbf{C}\cdot\mathbf{n}\quad\Rightarrow\min \tag{56}$$

where:

$$\mathbf{C}=\frac{1}{(m-1)n}\sum_{j=1}^{n}\sum_{i=1}^{m-1}\left(\mathbf{r}_{i+1,j}-\mathbf{r}_{i,j}\right)\cdot\left(\mathbf{r}_{i+1,j}-\mathbf{r}_{i,j}\right)^{t}$$

m is the time interval count, and n is the maker count. Multiplication of $\mathbf{n}$ to Equation 56 and rearrangement of the terms gives:

$$\left(\mathbf{C}-E\mathbf{I}\right)\cdot\mathbf{n}=0 \tag{57}$$

where E is essentially the smallest Eigen value of matrix $\mathbf{C}$ with $\mathbf{n}$ being the corresponding eigenvector.

This method differs from the helical axis approach in that the knee has a common axis of rotation throughout the entire motion. Unit vector $\mathbf{n}$ obtained in Equation 57 can be used to re-orientate the mediolateral (X) axis of the thigh reference frame. This allows investigators to use a functional reference frame instead of the conventional anatomical frame. Failure to correct the mediolateral and anteroposterior axes may cause devastating results such as amplified variability in the thigh rotation and knee valgus angle and exaggeration of the knee valgus angle (Schache *et al.*, 2005). With the anatomical frame, the valgus angle can increase as a function of knee flexion angle, resulting in large non-anatomically possible values. The method outlined in this section is not suitable for the elbow joint since the forearm motion is a combination of both flexion/extension and pronation/supination.

The location of the axis can also be found. The vector drawn from a point on the rotation axis to the mid-point of the two consecutive marker positions suffices

$$\left(\mathbf{r}_{i+1,j}-\mathbf{r}_{i,j}\right)^{t}\cdot\left(\frac{\mathbf{r}_{i+1,j}+\mathbf{r}_{i,j}}{2}-\mathbf{r}_{O}\right)=0 \tag{58}$$

where $\mathbf{r}_o$ is the thigh-frame coordinates of point O (Figure 12.4b) on the axis. Rearrangement and expansion of Equation 58 for all markers and intervals yields:

$$\mathbf{A}\cdot\mathbf{r}_o=\mathbf{B}$$

$$\mathbf{r}_O = \left(\mathbf{A}^t \cdot \mathbf{A}\right)^{-1} \cdot \left(\mathbf{A}^t \cdot \mathbf{B}\right)$$

where:

$$\mathbf{A} = \begin{bmatrix} \cdot & \cdot & \cdot \\ x_{i+1,j} - x_{i,j} & y_{i+1,j} - y_{i,j} & z_{i+1,j} - z_{i,j} \\ \cdot & \cdot & \cdot \end{bmatrix}$$

$$\mathbf{B} = \begin{bmatrix} \cdot & \left(\mathbf{r}_{i+1,j} - \mathbf{r}_{i,j}\right)^t \cdot \dfrac{\mathbf{r}_{i+1,j} + \mathbf{r}_{i,j}}{2} & \cdot \end{bmatrix}^t$$

Digital filtering and differentiation

The raw position and orientation data calculated from the marker coordinates inevitably contain random experimental errors. Failure to treat random errors properly results in amplified and noisy velocity and acceleration data. The Butterworth low-pass filter is one of the most commonly used digital filters; a second-order filter gives the following filter function:

$$x_i' = a_0 x_i + a_1 x_{i-1} + a_2 x_{i-2} - b_1 x_{i-1}' - b_2 x_{i-2}' \tag{59}$$

where x_i is the raw data at time point i, x_i' is the corresponding filtered data, and $\left[a_0, a_1, a_2, b_1, b_2\right]$ are the filter coefficients which suffice:

$$a_0 + a_1 + a_2 - b_1 - b_2 = 1$$

An even-order Butterworth low-pass filter can be broken down to a cascade of second-order filters. For example, passing the data through a fourth-order Butterworth low-pass filter is equivalent to passing through two second-order filters consecutively. The filter coefficients of the k-th second-order filter are functions of the cutoff-to-sampling frequency ratio and the filter order (Oppenheim and Schafer, 1989):

$$a_{k0} = a_{k2} = \frac{\Omega_c^{\,2}}{C}$$

$$a_{k1} = 2a_{k0}$$

$$b_{k1} = \frac{2\left(\Omega_c^{\,2} - 1\right)}{C}$$

$$b_{k2} = \frac{1 - 2\cos\left(\dfrac{2k-1}{2n}\pi\right) \cdot \Omega_c + \Omega_c^{\,2}}{C}$$

where:

$$\Omega_c = \tan\left(\pi \frac{f_c}{f_s}\right) \tag{60}$$

$$C = 1 + 2\cos\left(\frac{2k-1}{2n}\pi\right) \cdot \Omega_c + \Omega_c^{\ 2}$$

n is the order of the filter (n = 2,4 …), f_s and f_c are the sampling and cutoff frequencies, respectively, and $k = 1,\ldots,n - 1$. As the order increases, the transition from the pass band to the stop band becomes steeper, more closely approximating the ideal filter.

There are two issues to consider in using a Butterworth filter. Firstly, without knowing $x'[1]$ and $x'[2]$, it is impossible to use Equation 59, because the prior filtered data are unknown. This problem is usually solved by providing an initial guess for the unknown values:

$$x'[1] = x[1]$$

$$x'[2] = x[2]$$

Due to the recursive nature (IIR filter) of the Butterworth filter, this initial guess affects all subsequent data. The effects, however, diminish rapidly as i in Equation 59 increases. The strategy often used is to pad sufficient additional data to each end of the data set and make the initial guess in the padded section so that the effects on the actual data become negligible (Giakas *et al.*, 1998; Smith, 1989).

Secondly, a Butterworth filter introduces phase delay to the data. One popular method to treat this problem is to pass the data twice, forward first and then backward, through each second-order filter (Winter, 1990). Using a second-order filter twice essentially gives a fourth-order filter (zero phase-lag filter) but its frequency response curve is slightly different from a normal fourth-order filter. With the double pass strategy, Ω_c in Equation 60 must be corrected to:

$$\Omega_c = \frac{1}{0.802}\tan\left(\pi \frac{f_c}{f_s}\right)$$

Numerical differentiation of any kinematic data can be performed by using the average difference method (Winter, 1990):

$$\frac{dx}{dt} = \frac{x_{i+1} - x_{i-1}}{2 \cdot \Delta t}$$

$$\frac{d^2x}{dt^2} = \frac{x_{i+1} - 2x_i + x_{i-1}}{\Delta t^2}$$

where Δt is the time interval between two consecutive time points.

Conclusion

This chapter focused primarily on the numerical methods and procedures for deriving kinematic parameters in motion analysis. Segments and joints are the focal points in kinematics of human motion since rigid segments are linked at the joints forming a linked segment system. The types of motions allowed at a joint are determined by the joint type. For this reason, the segment motions must be assessed in reference to the joint type and such properties as the joint centre and axis. Kinematic parameters can be derived based on the segmental reference frame properties or the collective motions of the markers placed on the segments.

Appendix: The DSM

Partial differentiation of Equation 3 with respect to the attitude angles yields:

$$\begin{bmatrix} \frac{\partial t_{11}}{\partial \phi} & \frac{\partial t_{11}}{\partial \theta} & \frac{\partial t_{11}}{\partial \psi} \\ \frac{\partial t_{12}}{\partial \phi} & \frac{\partial t_{12}}{\partial \theta} & \frac{\partial t_{12}}{\partial \psi} \\ \frac{\partial t_{13}}{\partial \phi} & \frac{\partial t_{13}}{\partial \theta} & \frac{\partial t_{13}}{\partial \psi} \end{bmatrix} = \begin{bmatrix} 0 & -s(\theta)c(\psi) & t_{21} \\ -t_{13} & s(\phi)c(\theta)c(\psi) & t_{22} \\ t_{12} & -c(\phi)c(\theta)c(\psi) & t_{23} \end{bmatrix}$$

$$\begin{bmatrix} \frac{\partial t_{21}}{\partial \phi} & \frac{\partial t_{21}}{\partial \theta} & \frac{\partial t_{21}}{\partial \psi} \\ \frac{\partial t_{22}}{\partial \phi} & \frac{\partial t_{22}}{\partial \theta} & \frac{\partial t_{22}}{\partial \psi} \\ \frac{\partial t_{23}}{\partial \phi} & \frac{\partial t_{23}}{\partial \theta} & \frac{\partial t_{23}}{\partial \psi} \end{bmatrix} = \begin{bmatrix} 0 & s(\theta)s(\psi) & -t_{11} \\ -t_{23} & -s(\phi)c(\theta)s(\psi) & -t_{12} \\ t_{22} & c(\phi)c(\theta)s(\psi) & -t_{13} \end{bmatrix}$$

$$\begin{bmatrix} \frac{\partial t_{31}}{\partial \phi} & \frac{\partial t_{31}}{\partial \theta} & \frac{\partial t_{31}}{\partial \psi} \\ \frac{\partial t_{32}}{\partial \phi} & \frac{\partial t_{32}}{\partial \theta} & \frac{\partial t_{32}}{\partial \psi} \\ \frac{\partial t_{33}}{\partial \phi} & \frac{\partial t_{33}}{\partial \theta} & \frac{\partial t_{33}}{\partial \psi} \end{bmatrix} = \begin{bmatrix} 0 & c(\theta) & 0 \\ -t_{33} & s(\phi)s(\theta) & 0 \\ t_{32} & -c(\phi)s(\theta) & 0 \end{bmatrix}$$

The partial derivatives of functions f_i and g_i with respect to the camera and optical error parameters can be obtained from Equations 3–13:

$$\begin{bmatrix} \frac{\partial f_i}{\partial x_o} & \frac{\partial g_i}{\partial x_o} \\ \frac{\partial f_i}{\partial y_o} & \frac{\partial g_i}{\partial y_o} \\ \frac{\partial f_i}{\partial z_o} & \frac{\partial g_i}{\partial z_o} \end{bmatrix} = \begin{bmatrix} -(\xi_i - \Delta u_i)t_{11} - d_u t_{21} & -(\eta_i - \Delta v_i)t_{11} - d_v t_{31} \\ -(\xi_i - \Delta u_i)t_{12} - d_u t_{22} & -(\eta_i - \Delta v_i)t_{12} - d_v t_{32} \\ -(\xi_i - \Delta u_i)t_{13} - d_u t_{23} & -(\eta_i - \Delta v_i)t_{13} - d_v t_{33} \end{bmatrix}$$

$$\begin{bmatrix} \frac{\partial f_i}{\partial u_o} & \frac{\partial g_i}{\partial u_o} \\ \frac{\partial f_i}{\partial v_o} & \frac{\partial g_i}{\partial v_o} \end{bmatrix} = \begin{bmatrix} (-1 + I_i + \xi_i^2 J_i + 6\xi_i K_4 + 2\eta_i K_5)W_i & (\xi_i \eta_i J_i + 2\eta_i K_4 + 2\xi_i K_5)W_i \\ (\xi_i \eta_i J_i + 2\eta_i K_4 + 2\xi_i K_5)W_i & (-1 + I_i + \eta_i^2 J_i + 2\xi_i K_4 + 6\eta_i K_5)W_i \end{bmatrix}$$

$$\begin{bmatrix} \frac{\partial f_i}{\partial d_u} & \frac{\partial g_i}{\partial d_u} \\ \frac{\partial f_i}{\partial d_v} & \frac{\partial g_i}{\partial d_v} \end{bmatrix} = \begin{bmatrix} U_i & 0 \\ 0 & V_i \end{bmatrix}$$

$$\begin{bmatrix} \frac{\partial f_i}{\partial \phi} & \frac{\partial g_i}{\partial \phi} \\ \frac{\partial f_i}{\partial \theta} & \frac{\partial g_i}{\partial \theta} \\ \frac{\partial f_i}{\partial \psi} & \frac{\partial g_i}{\partial \psi} \end{bmatrix} = \begin{bmatrix} (\xi_i - \Delta u_i)\frac{\partial W_i}{\partial \phi} + d_u \frac{\partial U_i}{\partial \phi} & (\eta_i - \Delta v_i)\frac{\partial W_i}{\partial \phi} + d_v \frac{\partial V_i}{\partial \phi} \\ (\xi_i - \Delta u_i)\frac{\partial W_i}{\partial \theta} + d_u \frac{\partial U_i}{\partial \theta} & (\eta_i - \Delta v_i)\frac{\partial W_i}{\partial \theta} + d_v \frac{\partial V_i}{\partial \theta} \\ (\xi_i - \Delta u_i)\frac{\partial W_i}{\partial \psi} + d_u \frac{\partial U_i}{\partial \psi} & (\eta_i - \Delta v_i)\frac{\partial W_i}{\partial \psi} + d_v \frac{\partial V_i}{\partial \psi} \end{bmatrix}$$

$$\begin{bmatrix} \frac{\partial f_i}{\partial K_1} & \frac{\partial g_i}{\partial K_1} \\ \frac{\partial f_i}{\partial K_2} & \frac{\partial g_i}{\partial K_2} \\ \frac{\partial f_i}{\partial K_3} & \frac{\partial g_i}{\partial K_3} \\ \frac{\partial f_i}{\partial K_4} & \frac{\partial g_i}{\partial K_4} \\ \frac{\partial f_i}{\partial K_5} & \frac{\partial g_i}{\partial K_5} \end{bmatrix} = \begin{bmatrix} -\xi_i r_i^2 W_i & -\eta_i r_i^2 W_i \\ -\xi_i r_i^4 W_i & -\eta_i r_i^4 W_i \\ -\xi_i r_i^6 W_i & -\eta_i r_i^6 W_i \\ -(r_i^2 + 2\xi_i^2)W_i & -2\xi_i \eta_i W_i \\ -2\xi_i \eta_i W_i & -(r_i^2 + 2\eta_i^2)W_i \end{bmatrix}$$

where:

$$\begin{bmatrix} I_i \\ J_i \end{bmatrix} = \begin{bmatrix} r_i^2 K_1 + r_i^4 K_2 + r_i^6 K_3 \\ 2K_1 + 4r_i^2 K_2 + 6r_i^4 K_3 \end{bmatrix}$$

and:

$$\begin{bmatrix} \frac{\partial W_i}{\partial \phi} \\ \frac{\partial U_i}{\partial \phi} \\ \frac{\partial V_i}{\partial \phi} \end{bmatrix} = \begin{bmatrix} \frac{\partial t_{11}}{\partial \phi}(x_i - x_o) + \frac{\partial t_{12}}{\partial \phi}(y_i - y_o) + \frac{\partial t_{13}}{\partial \phi}(z_i - z_o) \\ \frac{\partial t_{21}}{\partial \phi}(x_i - x_o) + \frac{\partial t_{22}}{\partial \phi}(y_i - y_o) + \frac{\partial t_{23}}{\partial \phi}(z_i - z_o) \\ \frac{\partial t_{31}}{\partial \phi}(x_i - x_o) + \frac{\partial t_{32}}{\partial \phi}(y_i - y_o) + \frac{\partial t_{33}}{\partial \phi}(z_i - z_o) \end{bmatrix}$$

$$\begin{bmatrix} \frac{\partial W_i}{\partial \theta} \\ \frac{\partial U_i}{\partial \theta} \\ \frac{\partial V_i}{\partial \theta} \end{bmatrix} = \begin{bmatrix} \frac{\partial t_{11}}{\partial \theta}(x_i - x_o) + \frac{\partial t_{12}}{\partial \theta}(y_i - y_o) + \frac{\partial t_{13}}{\partial \theta}(z_i - z_o) \\ \frac{\partial t_{21}}{\partial \theta}(x_i - x_o) + \frac{\partial t_{22}}{\partial \theta}(y_i - y_o) + \frac{\partial t_{23}}{\partial \theta}(z_i - z_o) \\ \frac{\partial t_{31}}{\partial \theta}(x_i - x_o) + \frac{\partial t_{32}}{\partial \theta}(y_i - y_o) + \frac{\partial t_{33}}{\partial \theta}(z_i - z_o) \end{bmatrix}$$

$$\begin{bmatrix} \frac{\partial W_i}{\partial \psi} \\ \frac{\partial U_i}{\partial \psi} \\ \frac{\partial V_i}{\partial \psi} \end{bmatrix} = \begin{bmatrix} \frac{\partial t_{11}}{\partial \psi}(x_i - x_o) + \frac{\partial t_{12}}{\partial \psi}(y_i - y_o) + \frac{\partial t_{13}}{\partial \psi}(z_i - z_o) \\ \frac{\partial t_{21}}{\partial \psi}(x_i - x_o) + \frac{\partial t_{22}}{\partial \psi}(y_i - y_o) + \frac{\partial t_{23}}{\partial \psi}(z_i - z_o) \\ \frac{\partial t_{31}}{\partial \psi}(x_i - x_o) + \frac{\partial t_{32}}{\partial \psi}(y_i - y_o) + \frac{\partial t_{33}}{\partial \psi}(z_i - z_o) \end{bmatrix}$$

References

1. Abdel-Aziz, Y.I. and Karara, H.M. (1971) 'Direct linear transformation from comparator coordinates into object space coordinates in close-range photogrammetry', in *Proceedings of Symposium on Close-Range Photogrammetry* (pp. 1–18), Falls Church, VA: American Society of Photogrammetry.
2. Andriacchi, T.P., Andersson, G.B.J., Fermier, R.W., Stern, D. and Galante, J.O. (1980) 'A study of lower-limb mechanics during stair climbing', *Journal of Bone and Joint Surgery*, 62A: 749–757.
3. Bell, A.L., Pedersen, D.R. and Brand, R.A. (1990) 'A comparison of the accuracy of several hip centre location prediction methods', *Journal of Biomechanics*, 23: 617–621.
4. Brewin, M.A. and Kerwin, D.G. (2003) 'Accuracy of scaling and DLT reconstruction techniques for planar motion analyses', *Journal of Applied Biomechanics*, 19: 79–88.
5. Challis, J.H. (1995) 'A procedure for determining rigid body transformation parameters', *Journal of Biomechanics*, 28: 733–737.
6. Cheng, P.L., Nicol, A.C. and Paul, J.P. (2000) 'Determination of axial rotation angles of limb', *Journal of Biomechanics*, 33: 837–843.
7. Coburn, J. and Crisco, J.J. (2005) 'Interpolating three-dimensional kinematic data using quaternion splines and hermite curves', *Journal of Biomechanical Engineering*, 127: 311–317.
8. Davis III, R.B., Õunpuu, S., Tyburski, D. and Gage, J.R. (1991) 'A gait analysis data collection and reduction technique', *Human Movement Science*, 10: 575–587.
9. Gamage, S.S. and Lasenby, J. (2002) 'New least squares solutions for estimating the average centre of rotation and the axis of rotation', *Journal of Biomechanics*, 35: 87–93.

10. Gander, W., Golub, G.H. and Strebel, R. (1994) 'Least-squares fitting of circles and ellipses ', *BIT Numerical Mathematics*, 34: 558–578.
11. Giakas, G., Baltzopoulos, V. and Bartlett, R.M. (1998) 'Improved extrapolation techniques in recursive digital filtering: a comparison of least squares and prediction', *Journal of Biomechanics*, 31: 87–91.
12. Halvorsen, K., Lesser, M. and Lundberg, A. (1999) 'A new method for estimating the axis of rotation and the centre of rotation', *Journal of Biomechanics*, 32: 1221–1227.
13. Hatze, H. (1988) 'High-precision three-dimensional photogrammetric calibration and object space reconstruction using a modified DLT-approach', *Journal of Biomechanics*, 21: 533–538.
14. Kane, T.R. and Scher, M.P. (1970) 'Human self-rotation by means of limb movements', *Journal of Biomechanics*, 3: 39–49.
15. Kirkwood, R.N., Culham, E.G. and Costigan, P. (1999) 'Radiographic and non-invasive determination of the hip joint centre location: effect on hip joint moments', *Clinical Biomechanics*, 14: 227–235.
16. Kwak, C.S., Kwon, Y.H., Kim, E.H., Lee, D.W. and Sung, R.J. (1996) 'A biomechanical analysis of the run-up paths of selected Korean national high jumpers', *Korean Journal of Sport Science*, 8: 39–51.
17. Kwon, Y.H. (1999) 'Object plane deformation due to refraction in two-dimensional underwater motion analysis', *Journal of Applied Biomechanics*, 15: 396–403.
18. Kwon, Y.H. (2005) 'A non-linear camera calibration algorithm: Direct Solution Method', in Q. Wang (ed.) *Scientific Proceedings of the XXIIIrd International Symposium on Biomechanics in Sports* (pp. 142), Beijing, China: The China Institute of Sport Science.
19. Kwon, Y.H., Ables, A.M. and Pope, P.G. (2002) 'Examination of different double-plane camera calibration strategies for underwater motion analysis', in K.E. Gianikellis (ed.) *Proceedings of XXth International Symposium on Biomechanics in Sports* (pp. 329–332), Caceres, Spain: Universidad de Extremadura.
20. Kwon, Y.H. and Casebolt, J.B. (2006) 'Effects of light refraction on the accuracy of camera calibration and reconstruction in underwater motion analysis', *Sports Biomechanics*, 5: 315–340.
21. Kwon, Y.H. and Fiaud, V. (2002) 'Experimental issues in data acquisition in sport biomechanics: camera calibration', in K.E. Gianikellis (ed.) *Proceedings of XXth International Symposium on Biomechanics in Sports, Applied Session in Data Acquisition and Processing* (pp. 3–15), Caceres, Spain: Universidad de Extremadura.
22. Kwon, Y.H. and Lindley, S.L. (2000) 'Applicability of four localized-calibration methods in underwater motion analysis', in R. Sanders and Y. Hong (eds) *Proceedings of XVIII International Symposium on Biomechanics in Sports. Applied program: Application of Biomechanical Study in Swimming* (pp. 48–55), Hong Kong: The Chinese University of Hong Kong.
23. Leardini, A., Cappozzo, A., Catani, F., Toksvig-Larsen, S., Petitto, A., Sforza, V., Cassanelli, G. and Giannini, S. (1999) 'Validation of a functional method for the estimation of hip joint centre location', *Journal of Biomechanics*, 32: 99–103.
24. Marzan, G.T. and Karara, H.M. (1975) 'A computer program for direct linear transformation solution of the collinearity condition, and some applications of it', In *Proceedings of Symposium on Close-Range Photogrammetric Systems* (pp. 420–476), Falls Church, VA: American Society of Photogrammetry.
25. Oppenheim, A.V. and Schafer, R.W. (1989) *Discrete-Time Signal Processing*, edn, Englewood Cliffs, NJ: Prentice Hall.
26. Piazza, S.J., Okita, N., Erdemir, A. and Cavanagh, P.R. (2004) 'Assessment of the functional method of hip joint centre location subject to reduced range of hip motion', *Journal of Biomechanics*, 37: 349–356.
27. Press, W.H., Teukolsky, S.A., Vetterling, W.T. and Flannery, B.P. (2002) *Numerical Recipes in C++: The Art of Scientific Computing*, 2nd edn, Cambridge, England: Cambridge University Press.

28. Ramey, M.R. and Yang, A.T. (1981) 'A simulation procedure for human motion studies', *Journal of Biomechanics*, 14: 203–213.
29. Schache, A.G., Baker, R. and Lamoreux, L.W. (2005) 'Defining the knee joint flexion-extension axis for purpose of quantitative gait analysis: An evaluation of methods', *Gait and Posture*, 24: 100–109.
30. Schmidt, R., Disselhorst-Klug, C., Silny, J. and Rau, G. (1999) 'A marker-based measurement procedure for unconstrained wrist and elbow motions', *Journal of Biomechanics*, 32: 615–621.
31. Seidel, G.K., Marchinda, D.M., Dijkers, M. and Soutas-Little, R.W. (1995) 'Hip joint centre location from palpable bony landmarks – a cadaver study', *Journal of Biomechanics*, 28: 995–998.
32. Smith, G. (1989) 'Padding point extrapolation techniques for the Butterworth digital filter', *Journal of Biomechanics*, 22: 967–971.
33. Spoor, C.W. and Veldpaus, F.E. (1980) 'Rigid body motion calculated from spatial co-ordinates of markers', *Journal of Biomechanics*, 13: 391–393.
34. Stagni, R., Leardini, A., Cappozzo, A., Grazia Benetti, M. and Cappello, A. (2000) 'Effects of hip joint centre mislocation on gait analysis results', *Journal of Biomechanics*, 33: 1479–1487.
35. Tylkowski, C.M., Simon, S.R. and Mansour, J.M. (1982) 'Internal rotation gait in spastic cerebral palsy', in *Proceedings of the Tenth Open Scientific Meeting of the Hip Society* (pp. 89–125).
36. Walton, J.S. (1981) 'Close-range cine-photogrammetry: a generalized technique for quantifying gross human motion', unpublished doctoral dissertation, The Pennsylvania State University.
37. Winter, D.A. (1990) *Biomechanics and Motor Control of Human Movement*, 2nd edn, New York, NY: John Wiley & Sons.
38. Woltring, H.J., Huiskes, R., De Lange, A. and Veldpaus, F.E. (1985) 'Finite centroid and helical axis estimation from noisy landmark measurements in the study of human joint kinematics', *Journal of Biomechanics*, 18: 379–389.
39. Ying, N. and Kim, W. (2002) 'Use of dual Euler angles to quantify the three-dimensional joint motion and its application to the ankle joint complex', *Journal of Biomechanics*, 35: 1647–1657.

13

Methodology in Alpine and Nordic skiing biomechanics

Hermann Schwameder[1], Erich Müller[2], Thomas Stöggl[2] and Stefan Lindinger[2]
[1]University of Karlsruhe, Karlsruhe; [2]University of Salzburg, Salzburg

Introduction

Alpine skiing, ski jumping and cross country skiing are outdoor sports typically performed on natural or artificial snow. The literature review shows that in all three disciplines many studies on biomechanical issues have been performed and published. However, due to the challenge of data collection in the field and the time-consuming data processing, several biomechanically based aspects have not yet been answered. Additionally, biomechanical field studies in Alpine and Nordic skiing have to withstand the large spatial range, exposed field conditions, low temperature, existence of appropriate measuring equipment (lightweight, mobile, temperature and moisture resistant), etc. Several research groups have developed specific and appropriate measuring devices as well as sophisticated methods for data analysis. Further developments and improvements are necessary to answer the research questions in Alpine and Nordic Skiing that are yet to be addressed.

The following sections give a short overview on the methods used in the published studies, present applications of these methods and provide perspectives and suggestions for further developments.

Alpine skiing

The literature that addresses the biomechanics of alpine skiing can be divided into three phases. Papers of the first phase primarily concentrate on the physical aspects of the gliding movements on snow and the forces involved in skiing. The monographs of Brandenberger (1974), Howe (1983) and Lind and Sanders (1997) are especially significant. The second phase has been characterized by specific measurements of alpine skiing movements using biomechanical methods. The first study of alpine skiing techniques under field conditions was that of Möser (1957). Since the 1970s, several quantitative results of alpine skiing techniques have been reported by Fukuoka (1971), Nigg *et al.* (1977), Müller (1994), Raschner *et al.* (2001) and Müller and Schwameder (2003).

Key variables of alpine skiing techniques have been detected in the third phase that started in the beginning of the 1980s. Nachbauer and Kaps analyzed aerodynamic aspects of the standing position of the skier during ski races (1991). Especially significant are also studies which deal with the relationship between ski geometry and run line (Casolo *et al.* 1997, Mössner *et al.* 1997, Niessen and Müller 1999). Müller *et al.* (1998) analyzed kinematic key variables of different ski turn techniques of experienced and inexperienced skiers.

The main reason why only very few biomechanical analyses of alpine skiing have been published so far lies in the already mentioned difficulties experienced undertaking biomechanical investigations in the alpine environment. These movements are three-dimensional (3D) in nature and take place over fairly large areas. To develop a fundamental understanding of alpine skiing techniques complex study designs using kinematic, kinetic and electromyographic methods are desirable. So far only very few such complex studies are available (Müller *et al.* 2005).

Biomechanics methodology and applications in Alpine skiing

Kinematics

In recent years some 3D studies have been published by the working group from the University of Salzburg, Austria (Raschner *et al.* 1997, Müller *et al.* 1998, Müller *et al.* 2005, Schiefermüller *et al.* 2005, Klous 2007). In all of these studies very similar methods were used to get as accurate 3D data as possible. Depending on the main research questions specific skiing slopes were prepared. Close to the run up to five camera platforms were mounted. The prepared run was marked out with 30–50 rigid calibration poles, situated about 1.5 m apart. The poles were protruded at different heights from the run. The point digitized on each pole for calibration purposes was a dark green tennis ball at the top of the pole. In addition a calibration frame was placed approximately in the centre of the calibrated area. The 3D coordinates of each of the calibration spheres on the top of the calibration poles and the coordinates of the calibration points on the calibration frame were obtained using a theodolite and standard surveying methods. The calibration frame was used to aid accurate computation of the internal camera orientation and DLT parameters. These parameters were then used to perform 3D object space coordinate reconstructions from the various sets of image space coordinates (Drenk, 1994). The pan and tilt angles were calculated from the digitized image space coordinates of at least three calibration spheres for each camera. The skier wore a tightly-fitting black and white ski suit that had been specially adapted with black and white markers which helped to aid the identification of the axes of rotation of the joints studied. The skier's movements were then recorded by up to five synchronized video cameras. While the skier was performing the run the camera operators panned and tilted the cameras to ensure the best possible positioning of the skier in the centre of the image (Figure 13.1). The interaction angle between the optical axes of the two cameras varied from about 90° to 130° during each turn.

Digitizing of the up to five synchronized video sequences was performed, using either the peak performance (PP) or the SIMI system. The 3D object space coordinates of the 20 (or 27) points, defining a 14 (or 21)-segment performer model, plus the tip and rear of each ski and the end of each pole, were reconstructed from the various sets of image coordinates using the algorithm of Drenk (1994). Using this method very high inter-operator reliability of equal or better than 0.9 for the variables analyzed could be achieved. Klous *et al.* (2006) analyzed the distance of measured and calculated values like the length of the shank of the skier and found that the error margins are in the range of 1–2 cm on a measurement range of 15 m.

Figure 13.1 Skier on the prepared ski slope with calibration poles.

Dynamics

The determination of ground reaction forces in alpine skiing is extremely important, to be able to understand the movement techniques but it is also a very challenging task. Within the last 25 years various ground reaction measurement systems have been developed. In the 1980s Müller (1994) used a system which was able to measure normal load on the skis by means of eight strain-gauge sensors. The sensors could be attached between the ski and the boot in a manner that would not impede the skier's movements. Four sensors were fixed on each ski in such a way that the forces normal to the ski could be measured independently at the heel, at the ball of the foot, and both at the inner and outer sides of the foot. The sensors were calibrated within a range of 0 to 1200 N. The strain-gauges were temperature-compensated.

The quality of measuring ground reaction forces in alpine skiing was further developed in the 1990s. Synchronous measurement of ground reaction forces and pressure distributions on the plantar surface was achieved by implementing bilateral insoles from Novel, Munich, each with

99 capacitive sensors. These insoles were specially adapted and were inserted into the skier's ski boots. The recording frequency was 50 Hz. The data were stored during the test runs in the Pedar mobile system on an exchangeable memory card, worn in a specially adapted belt around the skier's waist. Measurement errors occurred during the experiments as a result of forces which are deflected over the leg of the ski boot and are, therefore, not registered (Raschner *et al.*, 1997; Müller and Schwameder, 2003). Additional disadvantages of this system are, however, that forces are only measured in one direction and the often limited sampling rate.

It turned out that for a 3D performance analysis in alpine skiing, knowledge of force time and torque time courses in each spatial direction is necessary. Therefore, the research group at the University of Salzburg developed and validated, together with the Swiss company Kistler Instruments, a new system (Stricker *et al.*, 2007). The measurement system consists of four, six-component dynamometers that are capable of measuring forces and calculating torques in all three spatial directions (Figure 13.2). Each dynamometer weighs 0.9 kg and has a height of 36 mm. It comprises a top and bottom plate, which are connected via three 3D force sensors consisting of piezoelectric elements. The amplifiers, their power supply and the data loggers are located in a separate supply box, which is carried in a backpack by the skier. One dynamometer is mounted under the toe and one under the heel part of the binding on both skis. The system was validated under laboratory and field conditions. An accuracy of +/–3 per cent was determined for measuring the forces. The torques can be determined with an accuracy of +/–8 per cent.

Perspectives in Alpine skiing biomechanics methodology research

In future Alpine skiing biomechanics should focus on the following developments and improvements with respect to basic and applied methods:

- development of automatic tracking 3D kinematic systems;
- improvement of portable force measurement systems;

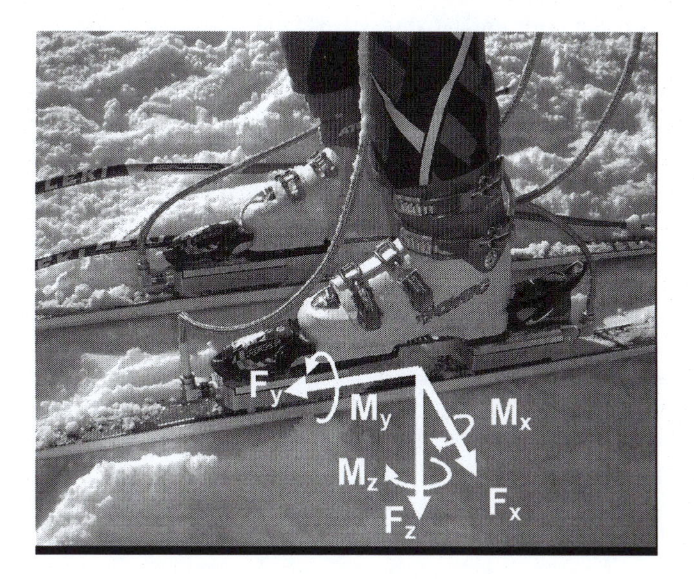

Figure 13.2 Six-component ground reaction force measurement system.

- developments and improvements of methods to assess the effect of equipment modifications on performance and safety; and
- improvement of methods to determine joint loading.

Ski jumping

Ski jumping is primarily a competitive sport and requires a sophisticated infrastructure. The main competitive season is the winter world cup from November to March; however, competitions are also performed in summer (Summer Grand Prix). Due to the competitive nature of ski jumping, it can be assumed that only a limited number of the biomechanical studies which are conducted are in fact published. The ski jumping movement sequence can be divided into the following six phases: in-run; take-off; early flight; stable flight; landing preparation and landing phase. Each has a specific function which contributes to enhancing performance by maximizing both the length and the technical quality of the jump.

Biomechanical ski jumping studies cover both simulation jumps (lab tests) and hill jumps (field studies). The large spatial dimension and the main request on non-reactive methods create many challenges to biomechanical studies. The main issues covered in biomechanical research papers are the aspects of performance enhancement (limiting factors of the take-off, specific training and conditioning, aerodynamics), injury prevention and safety.

Biomechanics methodology in ski jumping

Kinematics

Despite the methodological challenges, kinematics has been the most frequently used method in published ski jumping studies (Schwameder *et al.*, 2005). Kinematic methods have been developed for both simulated (dry land) jumps and hill jumps. Since kinematics is a non-reactive method, it can be used both during training and competition.

The methodological approaches are manifold and diverse. In simulated jumps, standard two-dimensional (2D) kinematics with fixed cameras is used to analyze the take-off movement in the sagittal plane. Usually these studies combine kinematics with methods measuring ground reaction forces and/or muscle activity (Schwameder *et al.*, 1997; Virmavirta and Komi, 2001).

Hill jump methods must be approached more carefully due to the large space covered. Even the restriction to the take-off needs a spatial view of at least 10 m, which hardly can be covered by one single camera without sacrificing accuracy. In this situation a standard PAL video format would lead to a spatial resolution of less than 13 mm. The simplest approach to study sagittal plane motion in ski jumping over a large spatial range, is recording with several synchronized cameras located perpendicular to the flight path. At first the single recordings are analyzed independently and are then transferred into the global coordinate system using specific mathematical solutions. This approach has successfully been used for many years and proven reliable (Komi *et al.* 1974, Schmölzer and Müller 2005).

Advanced and more sophisticated approaches are the kinematic methods using panned, tilted and zoomed cameras to determine 3D coordinates of body landmarks. The advantage of these methods is the combination of covering a large spatial range (up to 50 m) along with a relatively high spatial resolution (4–10 mm/pixel) of 3D coordinates. The methodological challenge of this approach is the reconstruction of the body landmark coordinates by using panned, tilted and zoomed cameras. This procedure must provide the panning and tilting angle and the zoom factor of the camera for each recorded image. Two different methods have been presented in the literature. The first approach refers to camera tripods equipped with electronic goniometers for measuring the panning and tilting angle of the camera (Virmavirta *et al.*, 2005). During filming this data is stored on the two audio tracks of the video tape. The global coordinates of selected body landmarks are then reconstructed based on this information. In using this method, zooming of the camera is not permitted. The second approach is based on computer software that relies on reference markers for calibration during data analysis (Schwameder and Müller, 1995; Schwameder *et al.*, 2005). The coordinates of the reference markers are determined in the global coordinate system using a theodolite. Along with the referred body landmarks, at least three calibration markers are digitized during the data analysis procedure. The software calculates the panning and tilting angle as well as the zooming factor of each camera. The software finally reconstructs the global coordinates of the body landmarks (Drenk, 1994). Both methods require the data acquisition from at least two synchronized cameras; however, they provide the calculation in three dimensions. In the research to date, the 3D kinematics approaches that are described above have been used to calculate the position of joint centres in space over time. This allows the calculation of specific body and segment angles, but misses other aspects of kinematics, such as torsion of the segments. So far, these aspects have not been tested and measured in research, which would provide the information needed to model the ski jumper and his equipment for 3D inverse dynamics calculations.

Dynamics and aerodynamics

Methodology measuring the dynamics during ski jumping is widely used both in simulated and hill jumps. Ground reaction force data during simulated jumps can easily be collected using force plates in a laboratory. These methods are primarily used in performance diagnostics to study basic and applied aspects of technique, power and conditioning.

The most frequently used method to measure ground reaction forces in hill jumps utilizes force plates installed in the take-off table. This method, with its first attempts in the late 1970s, is still in use at several jumping hills in middle Europe. Some of these plates only measure resultant vertical forces, while others measure each of the tracks independently in two or three directions (Virmavirta and Komi, 1993). This offers the possibility to determine the magnitudes of the force components and the direction of the resultant force for the left and right skis individually. This method provides researchers with the advantage of collecting dynamic data without interfering with the jumper's performance in any way, opening up the possibility of data collection during competition. However, only the last part of the in-run and the take-off (~12 m) can be measured. Studies using these systems have yet to be published.

Another option for measuring ground reaction forces during hill jumps is by imbedding transducers between skis and boots (Tveit and Pedersen, 1981). Pressure insoles are a more differentiating and precise system to measure force, force distribution and the force application point over time. Systems with 85 (Pedar, Novel) and 16 sensors (Paromed) have successfully been used in several studies (Schwameder and Müller, 1995; Virmavirta and Komi, 2000). With these mobile systems the entire sequence from in-run to landing can be monitored; however, they cannot be used during competitions.

The aerodynamic forces acting on the ski jumper cannot be directly measured during hill jumps. Wind tunnel measurements are used to study the effect of individual flight positions on aerodynamic forces. The measured aerodynamic and kinematic data serve as input for appropriate computer simulations for optimizing the lift/drag ratio with the purpose of increasing flight length (Hubbard *et al.*, 1989; Schmölzer and Müller, 2005). Wind tunnel measurements are also used to investigate the effect of lift and drag on the in-run position and the take-off movement (Virmavirta *et al.*, 2001).

Electromyography

Electromyography is not used very intensively in ski jumping biomechanics research. With hill jumps, subjects are required to carry the storage device during the entire event. This may present difficulties during the in-run and flight phases, restricting the acquisition of data to training jumps. Hence, muscle activation patterns are studied more thoroughly during simulated jumps in the laboratory. The contribution of electromyography (EMG), however, is limited to basic descriptions of muscle activity and coordination patterns during the entire jump sequence and to comparisons between hill and simulated jumps (Virmavirta and Komi, 2001).

Additional methods

Other available methods only play a minor role in ski jumping biomechanics research. Inverse dynamics approaches, based on kinematic and anthropometric data, are used to study the proportion of torque, power and energy produced by the structures around the hip and knee joints during hill take-offs (Sasaki *et al.*, 1997). Computer simulation is primarily used to study the ballistic and body position variables on the flight path, as well as the lower extremity joint torques needed to maintain the in-run position. Ski jumping related anthropometry, balance ability, strength and power (Hahn *et al.*, 2005) and variability are studied using specific methods.

Application of methodology in ski jumping biomechanics

One of the key issues in ski jumping biomechanics is the ability to produce a detailed biomechanical description of motion regarding kinematics, dynamics and muscle activation patterns. This is covered both in simulated and hill jumps using 2D and 3D kinematics (Arndt *et al.*, 1995; Schwameder and Müller, 1995; Schwameder *et al.*, 2005; Virmavirta *et al.*, 2005), measuring ground reaction forces (GRFs) with force platforms integrated into the take-off table (Virmavirta and Komi, 1993), pressure distribution insoles (Schwameder and Müller, 1995; Virmavirta and Komi, 2000), and muscle activation pattern (Virmavirta and Komi, 2001).

Based on the biomechanical description, researchers consequently focus on the investigation of performance-related factors in ski jumping (Arndt *et al.*, 1995; Schwameder and Müller, 1995; Virmavirta *et al.*, 2005; Schwameder *et al.*, 2005).

Further biomechanical considerations are:

- The effect of wind conditions on the take-off dynamics in simulated jumps (Virmavirta *et al.*, 2001).
- The effect of the in-run position on take-off parameters (Schwameder *et al.*, 1997).
- The effect of hill size on plantar pressure patterns and muscle activity (Virmavirta *et al.*, 2001).
- Comparison between hills and simulated jumps regarding plantar pressure and muscle activity (Virmavirta and Komi, 2001).
- Ski jumping specific performance diagnostics (e.g. Schwameder and Müller, 1997; Bruhn *et al.*, 2002; Hahn *et al.*, 2005).

Perspectives in ski jumping biomechanics methodology research

Building upon the biomechanics research methods already presented and considering key issues in the sport, it is suggested that the following methods should be improved or developed:

- Development of automatic tracking (2D and 3D kinematics) – short-term feedback device for coaches and athletes.
- Development of portable GRF feedback systems for hill jumps.
- Methods to assess the effect of equipment modifications on ski jump flight to improve safety.
- Feedback system on explosiveness and effectiveness of take-off.

Cross country skiing

Cross country skiing is one of the most demanding sports, demonstrating a large variety and a multiplicity of determinants of performance. There are two basic skiing techniques, the classical style and the skating style, including up to five sub-techniques, each applied at different terrains during a race. The mechanics involved in the differing techniques of cross country skiing locomotion include a complex interaction between the following structures: Firstly, the kinetic relationships of both the upper and lower body driving the motion, secondly, the characteristics of muscle activity generating propulsive forces, and thirdly, the kinematic characteristics of the specific movement patterns. The understanding of these interactions and the complex function mechanisms of the numerous cross country skiing techniques requires the use of kinetic, EMG and kinematic movement analysis. In addition to the diversity in skiing techniques, there is a large spectrum in competition forms, with different race distances (1 k–50 k) and modes (e.g. single start, mass start, and sprint). Another crucial factor exists in the equipment that is used. The varying nature of cross country skiing constitutes a challenge for athletes, coaches and scientists, and provides a wide scope for research. However, only 11 per cent of the 350 international research papers on cross country skiing focus on biomechanical aspects and only 4 per cent follow a mixed physiological-biomechanical study design. The enormous effort for biomechanical data collection and processing, as well as restrictions of the measurement equipment, might count for fewer publications compared to the physiological studies that are actually conducted.

Biomechanics methodology in cross country skiing

Kinematics

Kinematic analysis of how a cross country skier moves through space is of interest for comparing individuals or techniques in relation to whole body characteristics (cycle velocity, cycle rate, cycle length), or more detailed aspects, such as angle or centre of gravity parameters. Kinematic analyses in cross country skiing were developed in the late 1970s, starting with 2D single camera high speed film-analyses, investigating the relatively planar classical techniques (Marino *et al.*, 1980; Ekström, 1981; Komi *et al.*, 1982; Gervais Wronko, 1988; Norman *et al.*, 1989). Disadvantages of 2D video analyses in cross country skiing research are: (1) the restriction to classic techniques (quasi 2D movements); (2) restricted measurement space with fixed camera setting; (3) increase of perspective error at larger camera distances towards the object; and (4) decrease of digitizing accuracy at larger focal width but smaller object size on the screen. Nevertheless, depending on the aim of a study and the choice of kinematic parameters, a 2D video analysis can be well used when performing the investigation roller skiing on a treadmill in the lab. Video analysis from the back or side view can sufficiently be used for determination of cycle characteristics (poling phase, recovery phase, cycle time, poling frequency, cycle length). Additionally, the usage of electro-goniometers easily allows for determination of single joint movements and angular displacement (Holmberg *et al.*, 2005; Stöggl *et al.*, 2006a).

Three-dimensional analysis has been essential for the skating techniques introduced in the 1980s. The corresponding research method is much more sophisticated and requires a multiple-camera setup even for 'unreal' (without analyzing segmental torsion, etc.) 3D analyses (Smith *et al.*, 1989; Leppävuori *et al.*, 1993; Gregory *et al.*, 1994; Smith and Heagy, 1994; Bilodeau *et al.*, 1996; Lindinger, 2006). For expanding the measurement space, without reducing the spatial resolution of the image, systems with tilting and zooming cameras have been successfully introduced in cross country skiing research (Lindinger, 2006). The use of at least two cameras under often difficult environmental conditions, 3D calibration of the measuring space and the time-consuming digitizing of two or more camera views dramatically increases time and effort for such field analyses and prevents a quicker processing of data and a direct transfer of results to practice.

Modern and highly accurate motion capture systems, working with passive markers and infrared cameras for automatic marker recognition, have been developed primarily for lab applications (e.g. Vicon, UK; Qualisys, Sweden). These systems enable a quick and direct recording of 3D coordinates with high temporal resolution; however, they are not yet reliably usable under field conditions in winter due to different hardware and technology problems (temperature resistance, etc.). With advancing developments in equipment, it is likely that the reduction, or even elimination, of these existing problems is possible. As a result, data collection and processing could be accelerated in the area of cross country skiing field analyses. This would consequently achieve a substantial progression in kinematic data acquisition in the field. There will still remain, however, several restrictions by using these systems, such as high financial costs, data collection during the night using artificial light, inability of analyses during competition, night conditions.

Kinetics

The measurement of forces applied to skis and poles is an important element to understand cross country skiing locomotion but, compared to kinematics, this aspect is reported in less than half of cross country skiing publications.

Leg force systems

Several 2D force measurement systems (vertical and longitudinal leg forces) have been introduced by Ekström (1981) in the early 1980s using a portable load-cell system mounted on the skis. This was followed by several other portable 2D force-systems for skis or roller skis, mostly based on strain gauge or beam load cell technology (Komi, 1987; Pierce *et al.*, 1987; Bellizzi *et al.*, 1998). These systems allow force measurements during numerous consecutive strides. Unfortunately, they have not been further developed and adapted to modern shoe and ski material.

Babiel *et al.* (1997) presents a portable 3D force measurement binding consisting of one unit to measure vertical forces, and a second one for longitudinal and medio-lateral forces. The description, however, lacks details regarding accuracy, weight, practical usability, etc.

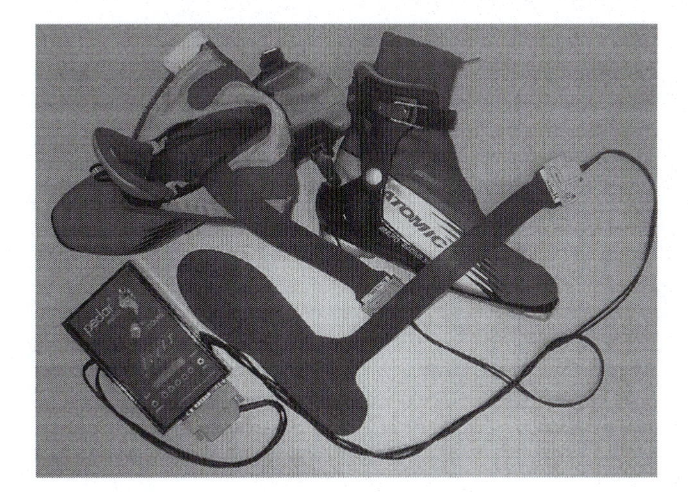

Figure 13.3 Pedar insoles and data storage unit.

In some studies pressure distribution insole-systems (Pedar, Paromed) are used to analyze forces, force distribution and force application points at the foot sole (Figure 13.3). The main limitation of these systems is the restriction to only measure one direction (vertical). The benefit of the system is its low weight and the fact that the subjects can use their own ski and boot material (Lindinger, 2006).

Another approach for 2D force measurements has been described by Komi (1987). A force plate array buried under the snow surface measures vertical and longitudinal pole and leg forces in classical techniques. The main limitations of this system are calibration difficulties for the horizontal force component and the short length of the force plates (6 m).

Leppävuori *et al.* (1993) introduces a 3D force measurement platform (2.2 m) composed of 20 separated beams buried under the ski track. Vertical, cross and longitudinal horizontal force components during skating can be measured with strain gauge bridges separately on each beam at 90 Hz; however, the short length of the system does not allow measuring several consecutive strides.

Pole force systems

Two different concepts are presented in the literature to measure pole forces. The first one is based on strain gauges mounted on the shaft of the poles (bending moment compensated) to measure axial compression. This system has the advantage of being lightweight. (Ekström, 1981; Komi, 1987; Pierce *et al.*, 1987; Bellizzi *et al.*, 1998). The second concept is based on force transducers (piezoelectric or strain gauge load cells) mounted directly under the pole grip (Millet, G.P. *et al.*, 1998; Millet, G.Y. *et al.*, 1998; Holmberg *et al.*, 2005; Lindinger, 2006; Stöggl *et al.*, 2006b). The difficulties of strain gauge systems lie in the calibration process and the measurement of incorrect force data, both of which result from the inhomogeneous bending of the poles during the poling phase. Lindinger (2006), Holmberg *et al.* (2005) and Stöggl *et al.* (2006b) use a telescope pole force system based on strain gauge load cells mounted directly below the pole grip and bed into an aluminium body for measuring axial pole forces (Fig. 13.4). The system showed mean differences of 3.8 per cent over the entire pole ground contact and –8 per cent (–20 N) regarding maximal forces compared to a commercial force plate (AMTI, USA).

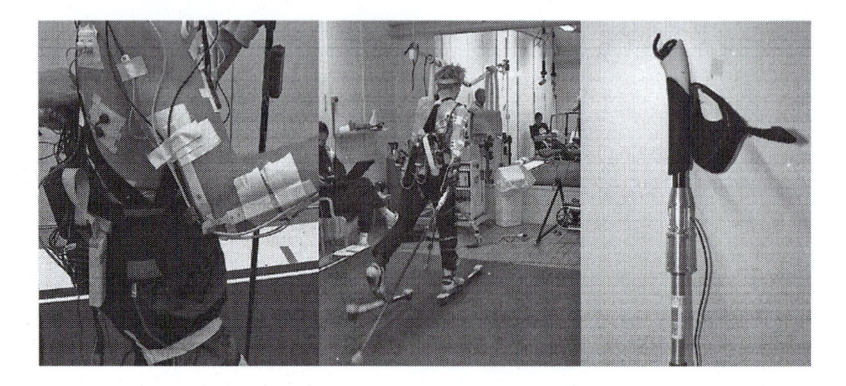

Figure 13.4 EMG, goniometers, pole force measurement device.

Electromyography (EMG)

Although knowledge on EMG activation patterns could enable a microscopic view on specific muscle activation modes (e.g. stretch-reflex behaviour during stretch-shortening cycles – (SSC)), surface EMG is the most infrequently used method in biomechanical cross country skiing research.

The lack of information concerning muscle activation patterns leaves a significant deficit in the overall knowledge and understanding about cross country skiing techniques. Few studies have discussed the occurrence of SSC, mainly focusing on specific muscle groups during classical (Komi and Norman, 1987; Holmberg *et al.*, 2005) or skating techniques. Although the switch-on and switch-off patterns, including activation intensity of several upper and lower body muscles, have been recently analyzed for the double poling technique and a double poling imitation device (Figure 13.4), muscle interaction patterns (muscle sequencing) during cross country skiing remain largely unexplored.

Application of methodology in cross country skiing biomechanics

In the internationally published biomechanical cross country skiing literature, biomechanical technique descriptions and functional analyses of both skating and classic techniques have been reported to a similar extent. To date, more complex biomechanical studies exist only for double poling (Holmberg *et al.*, 2005; Stöggl *et al.*, 2006a, b). The introduction of the sprint discipline led to technique innovations such as a jumped V2-skate named 'double-push', with two push-offs performed by each leg in one cycle (Stöggl and Lindinger, 2006). Because complex kinematic-kinetic-electromyographic study designs have rarely been implemented in cross country skiing science, there are many shortcomings in sub areas of all techniques. An essential key issue is the performance-related variables in cross country skiing. Numerous studies had a special focus on relationships between biomechanical data and performance (lab, field, and race). For example, Holmberg *et al.* (2005) suggested that the fastest athletes used a more sprinter-like double poling strategy, characterized by higher peak pole forces, and higher impulse of forces. This more explosive pattern led to shorter relative poling times and longer relative recovery phases. Several studies have shown that cycle length is a critical determinant of diagonal stride, skating and sprint performance (Smith *et al.*, 1989;

Bilodeau *et al.*, 1996; Rundell and McCarthy, 1996; Stöggl *et al.*, 2006a). The faster skiers created longer cycle lengths compared with slower ones, whereas cycle rate was similar.

Further biomechanical issues are:

- development and investigations on efficiency of cross country skiing equipment;
- joint loading and orthopaedic questions, with a special focus on sprint techniques; and
- mechanical optimization and evaluation of specific imitation drills for training and testing.

Perspectives in cross country skiing biomechanics methodology research

Based on the presented biomechanical research methods and applications, the following methods should be improved and open questions answered:

- Determination of performance-related variables and muscular demands in snow conditions to get exactly defined technique models for technique training.
- Optimization of both pole force and leg force measurement systems regarding accuracy, weight and flexible application in elite cross country skiing. The ultimate goal is a portable 3D force measurement system.
- Development of kinematic based energy analyses (inverse dynamics) methods to study biomechanical and physiological efficiency.
- Comparison of cross country skiing and roller-skiing with respect to technique aspects.
- Integrative approach to study the biomechanical-physiological interaction in cross country skiing techniques (economy and efficiency).

Conclusion

In all three ski disciplines a fairly high standard regarding biomechanical methods, both in laboratory and field study applications, has been reached. This is quite impressive considering the challenging conditions specifically in snow sport field studies. The perspectives in each of the disciplines have been outlined before. Substantial improvements can be expected based on the developments of sensor systems technology, specifically applicable in field studies, as well as data storage and signal processing.

References

1. Arndt, A., Brüggemann, G., Virmavirta, M. and Komi, P. (1995) 'Techniques used by Olympic ski jumpers in the transition from takeoff to early flight', *Journal of Applied Biomechanics*, 11: 224–237.
2. Babiel, S., Hartmann, U., Spitzenpfeil, P. and Mester, J. (1997) 'Ground-reaction forces in alpine skiing, cross-country skiing and ski jumping', in E. Müller *et al.* (eds) *Proceedings of the Science and Skiing I* (pp. 200–207), St Christoph am Arlberg: Chapman & Hall, Cambridge University Press.
3. Bellizzi, M., King, K., Cushman, S. and Weyand, P. (1998) 'Does the application of ground force set the energetic cost of cross-country skiing?', *Journal of Applied Physiology*, 85: 1736–1743.

4. Bilodeau, B., Rundell, K., Roy, B. and Boulay, R. (1996) 'Kinematics of cross-country ski racing', *Medicine and Science in Sports and Exercise*, 28: 128–138.
5. Brandenberger, H. (1974) *Ski Mechanik Methodik*, Derendingen-Solothurn: Habegger.
6. Bruhn, S., Schwirtz, A. and Gollhofer, A. (2002) 'Diagnose von Kraft- und Sprungkraftparametern zur Trainingssteuerung', *Leistungssport*, 5: 34–37.
7. Casolo, V., Lorenzi, V., Vallatta, A. and Zappa, B. (1997) 'Simulation techniques applied to skiing mechanics', in E. Müller *et al.* (eds) *Science and Skiing* (pp. 116–130), London: E&FN Spon.
8. Drenk, V. (1994) 'Bildmeßverfahren für schwenk- und neigbare sowie in der Brennweite variierbare Kameras', *Schriftenreihe zur Angewandten Trainingswissenschaft*, 1: 130–142.
9. Ekström, H. (1981) 'Force interplay in cross-country skiing', *Scandinavian Journal of Sports Science*, 3: 69–76.
10. Fukuoka, T. (1971) *Zur Biomechanik und Kybernetik des alpinen Skilaufs*, Frankfurt am Main: Limpert.
11. Gervais, P., and Wronko, C. (1988) 'The marathon skate in nordic skiing performed on roller skates, roller skis, and snow skis', *International Journal of Sport Biomechanics*, 4: 38–58.
12. Gregory, R., Humphreys, S. and Street, G. (1994) 'Kinematic analysis of skating technique of Olympic skiers in the women's 30-km race', *Journal of Sport Biomechanics*, 10: 382–392.
13. Hahn, D., Schwirtz, A., Huber, A. and Bösl, P. (2005) 'Discipline-specific biomechanical diagnosis concept in ski jumping', in E. Müller *et al.* (eds) *Science and Skiing III* (pp. 349–359), Oxford: Meyer & Meyer.
14. Holmberg, H.C., Lindinger, S., Stöggl, T., Eitzlmair, E. and Müller, E. (2005) 'Biomechanical analysis of double poling in elite cross-country skiers', *Medicine and Science in Sports and Exercise*, 37: 807–818.
15. Howe, J. (1983) *Skiing Mechanics*, Laporte, CO: Puodre Press.
16. Hubbard, M., Hibbard, R., Yeadon, M. and Komor, A. (1989) 'A multisegmental dynamic model of ski jumping', *International Journal of Sport Biomechanics*, 5: 258–274.
17. Klous, M., Schwameder, H. and Müller, E. (2006) The accuracy of 3D kinetic and kinematic data used for joint loading analysis in skiing and snowboarding, in H. Schwameder *et al.* (eds) *Proceedings of XXIV International Symposium on Biomechanics in Sports* (pp. 553), Salzburg: University of Salzburg.
18. Komi, P., Norman, R. and Caldwell, G. (eds) (1982) *Horizontal velocity changes of world class skiers using the diagonal technique*, Champaign: Human Kinetics.
19. Komi, P. (1987) 'Force measurements during cross-country skiing', *Journal of Sport Biomechanics*, 3: 370–381.
20. Komi, P. and Norman, R. (1987) 'Preloading of the thrust phase in cross-country skiing', *International Journal of Sports Medicine*, Supplement 8: 48–54.
21. Komi, P.V., Nelson, R. and Pulli, M. (1974) *Biomechanics of Skijumping*, Jyväskylä.
22. Leppävuori, A., Karras, M., Rusko, H. and Viitasalo, T. (1993) 'A new method of measuring three-dimensional reaction forces under the ski during skiing on snow', *Journal of Applied Biomechanics*, 9: 315–328.
23. Lind, D. and Sanders, S.P. (1997) *The Physics of Skiing*, Woodbury NY: AIP Press.
24. Lindinger, S. (2006) *Biomechanische Analysen von Skatingtechniken im Skilanglauf*, Aachen: Meyer & Meyer.
25. Marino, G.W., Titley, B.T. and Gervais, P. (eds) (1980) *A technique profile of the diagonal stride pattern of highly skilled female cross-country skiers*, Champaign: Human Kinetics.
26. Millet, G.P., Hoffmann, M.D., Candau, R.B. and Clifford, P.S. (1998) 'Poling forces during roller skiing: effects of grade', *Medicine and Science in Sports and Exercise*, 30: 1637–1644.
27. Millet, G.Y., Perrey, S., Candau, R., Belli, A., Borrani, F. and Rouillon, J.D. (1998) 'External loading does not change energy cost and mechanics of rollerski skating', *European Journal of Applied Physiology and Occupational Physiology*, 78: 276–282.
28. Möser, G. (1957) 'Untersuchung der Belastungsverhältnisse bei der alpinen Skitechnik', unpublished doctoral dissertation, University of Halle-Wittenberg.

29. Mössner, M., Nachbauer, W. and Schindelwig, K. (1997) 'Einfluß der Skitaillierung auf Schwungradius und Belastung', *Sportverletzung – Sportschaden*, 11: 140–145.
30. Müller, E. (1994) 'Analysis of the biomechanical characteristics of different swinging techniques in alpine skiing', *Journal of Sports Sciences*, 12: 261–278.
31. Müller, E., Bartlett, R., Raschner, C., Schwameder, H., Benko-Bernwick, U. and Lindinger, S. (1998) 'Comparisons of the ski turn techniques of experienced and intermediate skiers', *Journal of Sports Sciences*, 16: 545–559.
32. Müller, E., Schiefermüller, C., Kröll, J. and Schwameder, H. (2005) 'Skiing with carving skis – what is new?', in E. Müller *et al.* (eds) *Science and Skiing III* (pp. 15–22), Oxford: Meyer & Meyer.
33. Müller, E. and Schwameder, H. (2003) 'Biomechanical aspects of new techniques in alpine skiing and ski jumping', *Journal of Sports Sciences*, 21: 679–692.
34. Nachbauer, W. and Kaps, P. (1991) 'Fahrzeitbestimmende Faktoren beim Schußfahren', in F. Fetz and E. Müller (eds) *Biomechanik des alpinen Skilaufs* (pp. 101–111), Stuttgart: Enke Verlag.
35. Niessen, W. and Müller, E. (1999) 'Carving – biomechanische Aspekte zur Verwendung stark taillierter Skier und erhöhter Standflächen im alpinen Skisport', *Leistungssport*, 1: 39–44.
36. Nigg, B., Neukomm, P.A. and Lüthy, S. (1977) 'Die Belastung des menschlichen Bewegungsapparates beim Skifahren', in F. Fetz (ed.) *Zur Biomechanik des Skilaufs* (pp. 80–89), Innsbruck: Inn-Verlag.
37. Norman, R., Ounpuu, S., Fraser, M. and Mitchell, R. (1989) 'Mechanical power output and estimated metabolic rates of Nordic skiers during Olympic competition', *International Journal of Sport Biomechanics*, 5: 169–184.
38. Pierce, J.C., Pope, M.H., Renström, P., Johnson, R.J., Dufek, J. and Dillman, C. (1987) 'Force measurements in cross-country skiing', *International Journal of Sports Medicine*, 3: 382–391.
39. Raschner, C., Müller, E. and Schwameder, H. (1997) 'Kinematic and kinetic analysis of slalom turns as a basis for the development of specific training methods to improve strength and endurance', in E. Müller *et al.* (eds) *Science and Skiing* (pp. 251–261), London: E&FN Spon.
40. Raschner, C., Schiefermüller, C., Zallinger, G., Hofer, E., Brunner, F. and Müller, E. (2001) 'Carving turns versus traditional parallel turns – a comparative biomechanical analysis', in E. Müller *et al.* (eds) *Science and Skiing II* (pp. 203–217), Hamburg: Kovac.
41. Rundell, K.W. and McCarthy, J.R. (1996) 'Effect of kinematic variables on performance in women during a cross-country ski race', *Medicine and Science in Sports and Exercise*, 28: 1413–1417.
42. Sasaki, T., Tsunoda, K., Uchida, E., Hoshino, H. and Ono, M. (1997) 'Joint power production in take-off action during ski jumping', in E. Müller *et al.* (eds) *Science and Skiing* (pp. 309–319), London: E&FN Spon.
43. Schiefermüller, C., Lindinger, S. and Müller, E. (2005) 'The skier's center of gravity as a reference point in movement analyses for different designated systems', in E. Müller *et al.* (eds) *Science and Skiing III* (pp. 172–185), Oxford: Meyer & Meyer.
44. Schmölzer, B. and Müller, W. (2005) 'Individual flight styles in ski jumping: results obtained during Olympic Games competitions', *Journal of Biomechanics*, 38: 1055–1065.
45. Schwameder, H. and Müller, E. (1995) 'Biomechanische Beschreibung und Analyse der V-Technik im Skispringen', *Spectrum der Sportwissenschaften*, 7: 5–36.
46. Schwameder, H., Müller, E., Lindenhofer, E., DeMonte, G., Potthast, W., Brüggemann, G., Virmavirta, M., Isolehto, H. and Komi, P. (2005) 'Kinematic characteristics of the early flight phase in ski-jumping', in E. Müller *et al.* (eds) *Science and Skiing III* (pp. 381–391), Oxford: Meyer & Meyer.
47. Schwameder, H., Müller, E., Raschner, C. and Brunner, F. (1997) 'Aspects of technique–specific strength training in ski jumping', in E. Müller *et al.* (eds) *Science and Skiing* (pp. 309–319), London: E&FN Spon.
48. Smith, G.A. and Heagy, B.S. (1994) 'Kinematic analysis of skating technique of Olympic skiers in the men's 50 km race', *Journal of Applied Biomechanics*, 10: 79–88.
49. Smith, G., Nelson, R., Feldman, A. and Rankinen, F. (1989) 'Analysis of V1 skating technique of Olympic cross-country skiers', *International Journal of Sport Biomechanics*, 5: 185–207.

50. Stöggl, T. and Lindinger, S. (2006) Double-Push Skating and Klap-Skate in cross country skiing, technical developments for the future?, in H. Schwameder *et al.* (eds) *Proceedings of the 24th International Symposium on Biomechanics in Sports* (pp. 393–396), Salzburg: University of Salzburg, Austria.
51. Stöggl, T., Lindinger, S. and Müller, E. (2006a) 'Analysis of a simulated sprint competition in classical cross country skiing', *Scandinavian Journal of Medicine and Science in Sports* (epub, ahead of print).
52. Stöggl, T., Lindinger, S. and Müller, E. (2006b) 'Biomechanical validation of a specific upper body training and testing drill in cross-country skiing', *Sports Biomechanics*, 5: 23–46.
53. Stricker, G., Scheiber, P., Lindenhofer, E. and Müller, E. (2007) 'Determination of ground reaction forces in alpine skiing and snowboarding. Development and validation of a mobile data acquisition system', *Journal of Applied Biomechanics*, submitted.
54. Tveit, P. and Pedersen, P. (1981) 'Forces in the take-off in ski jumping', in A. Morecki *et al.* (eds) *Biomechanics VII-B* (pp. 472–477), Baltimore: University Park Press.
55. Virmavirta, M. and Komi, P. (1993) 'Measurement of take-off forces in ski jumping, Part I', *Scandinavian Journal of Medicine and Science in Sports*, 3: 229–236.
56. Virmavirta, M. and Komi, P. (2000) 'Plantar pressures during ski jumping take-off', *Journal of Applied Biomechanics*, 16: 320–326.
57. Virmavirta, M. and Komi, P. (2001) 'Plantar pressure and EMG activity of simulated and actual ski jumping take-off', *Scandinavian Journal of Medicine and Science in Sport*, 11: 310–314.
58. Virmavirta, M., Kivekäs, J. and Komi, P. (2001) 'Take-off aerodynamics in ski jumping', *Journal of Biomechanics*, 34: 465–470.
59. Virmavirta, M., Isolehto, J., Komi, P., Brüggemann, G., Müler, E. and Schwameder, H. (2005) 'Characteristics of the early flight phase in the Olympic ski jumping competition', *Journal of Biomechanics*, 38: 2157–2163.

14

Measurement and estimation of human body segment parameters

Jennifer L. Durkin
University of Waterloo, Waterloo

Introduction

The accurate determination of human body segment parameters has been a long-standing challenge in biomechanics. Body segment parameters are needed for inverse and forward dynamic modelling of human motion, but are also used in the development of crash-test dummies (Kim *et al.*, 2003) and in specialized applications such as cockpit design (Hanavan, 1964). Several methods for measuring or estimating body segment parameters have been developed; however, limitations in these procedures have led to continued efforts for obtaining reliable measures (Cheng *et al.*, 2000; Durkin *et al.*, 2002; Martin *et al.*, 1989).

Until the early 1970s, cadavers were most often used to obtain reasonable estimates of body segment parameters. Volumes, masses, centre of mass (COM) locations, and moments of inertia (MOI) were measured directly from the segmented limbs of specimens (Braune and Fischer, 1889; Clauser *et al.*, 1969; Dempster, 1955). Predictive equations were then developed to allow the data to be applied to a living population.

More recently, researchers have measured body segment parameters directly from living humans. Techniques such as water immersion (Drillis and Contini, 1966), stereophotogrammetry (Young *et al.*, 1983), and quick release (Drillis and Contini, 1966) have been used to estimate segment volumes, masses, COM locations and MOI. Further, an increased availability of medical imaging technology has led to direct measurements of body segment parameters being made on living human participants using computed tomography (CT) (Pearsall *et al.*, 1996), magnetic resonance imaging (MRI) (Martin *et al.*, 1989), gamma-mass scanning (Zatsiorsky and Seluyanov, 1983) and dual energy x-ray absorptiometry (DEXA) (Durkin and Dowling, 2003). The medical imaging techniques available provide the most accurate and reliable means of obtaining individual-specific body segment parameters to date.

Many of the measurement methods for body segment parameters discussed above have involved the development of predictive equations so that estimates can be made on research populations. Mathematical models have also been developed independent of these measurement techniques for the same purpose (Hanavan, 1964; Hatze, 1980). All of the methods discussed present challenges in determining accurate information about body segment parameters that can be obtained efficiently and cost-effectively. Research in this area continues with some interesting directions that combine the use of medical imaging technology and modelling techniques.

Methods of body segment parameter measurement

Cadavers

Harless (1860) and Braune and Fischer (1889) performed three of the earliest in-depth studies involving the measurement of body segment parameters. Segment volumes, masses and COM locations were measured on two (Harless, 1860) and three (Braune and Fischer, 1889) male cadavers. Harless (1860) also measured limb MOI from these specimens and in a following study defined segment volumes, absolute and relative lengths on five male and three female cadavers (Drillis and Contini, 1966). Regression equations were then developed from these measurements for the prediction of body segment parameters (Drillis and Contini, 1966).

Research increased in the 1950s when the US Air Force began investigating human segmental and whole body inertial properties. In 1955, Dempster conducted one of the most extensive cadaveric analyses of human body segment parameters to date (Drillis and Contini, 1966). Dempster (1955) examined eight male cadavers who were deceased war veterans. Segment masses were obtained through weighing and COM locations were measured using a knife-edge balance technique. Segmental MOI were measured using a pendulum method combined with parallel axis theorem. From the data, regression equations were developed for the prediction of these body segment parameters.

In 1969, Clauser *et al.* measured segment volumes, masses and COM locations on 13 male cadavers. Following this, Chandler *et al.* (1975) measured segment masses, COM locations and MOI on six male cadavers. Both studies developed regression equations for the prediction of these body segment parameters.

While cadavers provide a great opportunity to directly measure the body segment parameters of humans, difficulty in obtaining specimens of varying age and sex, as well as the cost and intricacy of the methods, preclude the analysis of large numbers of participants. The age and mass ranges of Dempster's (1955) eight specimens were 52 to 83-years-old and 51.4 and 72.5 kg, respectively. Further, Clauser *et al.* (1969) examined 13, and Chandler *et al.* (1975) dissected only six male specimens. These small sample sizes, most of which were elderly Caucasian males, make extrapolation of the results to other populations difficult. Further, the data from the various studies cannot be pooled as the segmentation methods varied between most studies, sometimes drastically, limiting the ability to compare data and predictive equations.

Using cadavers to measure human body segment parameters offers additional limitations such as fluid and tissue losses during segmentation and differences between the properties of living and deceased tissue. Harless (1860) used decapitated specimens, likely causing large losses of fluid and thus altering mass and COM measurements (Reid and Jensen, 1990). Braune and Fischer (1889) froze their specimens to prevent fluid losses; however, the resulting changes in volume may have altered segment inertial properties (Reid and Jensen, 1990). Dempster (1955) reported average segment density values in his results, but there has been some question with regards to the differences between cadaveric tissue properties and those of living tissue (Pearsall and Reid, 1994). Further, the pendulum technique that Dempster (1955) used to determine segmental MOI has been found to have a large amount of uncertainty when the objects are oscillated about an axis that is greater than the radius of gyration of the segment (Dowling *et al.*, 2006).

Living humans

Direct measurement techniques

The limitations of using cadavers for the measurement and prediction of body segment parameters prompted researchers to measure body segment parameters directly from living humans. Methods such as water immersion, reaction change, quick release, compound pendulum and stereophotogrammetry were used to approximate human body segment parameters in a non-invasive manner.

Drillis and Contini (1966) measured body segment parameters on 20 living males aged 20- to 40-years-old. Segment volume and density were measured using water immersion and reaction change methods and segmental MOI were measured using a compound pendulum method and quick release. Water immersion methods require the application of constant density values to estimate segment masses and COM locations. These density terms are obtained from cadaveric data and there has been some debate as to whether significant differences exist from living tissue properties, particularly for trunk density values (Mungiole and Martin, 1990). Further, the compound pendulum method assumes that the MOI about the longitudinal axis is negligible compared to the transverse axis and the quick release method assumes that muscle contraction is absent, that the point of release is clean and noise-free, and that all joints are frictionless (Pearsall and Reid, 1994).

Plagenhoef (1983) conducted a study that compared the values from Dempster (1955) to 135 living participants (35 males, 100 females). Segment masses and COMs were calculated using water immersion and plaster models of participants' limbs were developed to measure radii of gyration. Plagenhoef (1983) also used one male cadaver to determine the inertial parameters of the trunk segment. By using a cadaver for trunk inertial calculations, it was assumed that the properties of the tissue sampled approximated those of living tissue.

In 1983, Young *et al.* measured 46 living females for segment volumes, masses, COMs, and MOI using anthropometric and stereophotometric methods. From photographs, surface areas were reconstructed and anthropometric values were applied to arrive at these segment parameters. This process required that constant density be assumed for each segment.

The number of studies conducted on living humans is large and the limitations of the methods are well known. A detailed analysis of both cadaveric and living human studies, along with the methods used for measurement of body segment parameters and the limitations associated with each technique, is well documented by Pearsall and Reid (1994).

Medical imaging techniques

The limitations inherent in the cadaveric and direct measurement techniques mentioned previously, along with increased availability of medical imaging technology, have led to a greater use of these techniques for directly measuring body segment parameters on living humans. Technologies such as gamma mass, CT, MRI and DEXA are all known to produce accurate results and, with the exception of gamma-mass scanning, are widely available.

Zatsiorsky and Seluyanov (1983) used gamma-mass scanning to measure segment inertial properties on 100 young Caucasian males. The technology is based on the attenuation of an incident gamma-radiation beam as it passes through a sample of tissue. The attenuation of the beam provides information regarding the surface density of the mass in its path. Knowing the location of the mass element and the calculated surface density, mass distribution information can be obtained. Zatsiorsky and Seluyanov (1983) reported segment masses, COM locations and MOI about 3 axes for 16 body segments. Validation of COM and MOI calculations were possible in the frontal plane only as the other two axes were estimated from surface density calculations using a constant density parallelepiped model for each scanned element. In 1990, Zatsiorsky *et al.* used the gamma-scanner technique to develop geometrical models and regression equations from 100 male and 15 female participants. Evaluation of the two prediction methods revealed that error from the geometric models was 1.5 times larger than those from the regression equations.

Erdmann (1997), Huang (1983) and Pearsall *et al.* (1996) used CT to measure body segment parameters directly on humans. Axial scanning of biological tissue produces CT data proportional to tissue density on a pixel-by-pixel basis. These tissue densities can then be applied to the digitized pixel volumes to determine segment mass distribution information. Huang (1983) used CT imaging to measure the body segment parameters of a porcine specimen and a young female child cadaver, while Erdmann (1997) used CT to determine the mass and COM locations of human trunk sections on 15 male patients. Furthermore, Pearsall *et al.* (1996) used CT imaging to measure trunk segment mass, COM locations and MOI on two male and two female participants. Pearsall *et al.* (1996) also provided detailed information about body segment parameters at the vertebral level, something that has not otherwise been investigated. Measurement of body segment parameters using CT imaging is considered very accurate and reliable; however, the method is limited by its tediousness and its cost. Furthermore, both gamma-mass scanning and CT imaging require exposing participants to radiation, albeit in small doses.

Magnetic resonance imaging has recently been explored as a method for measuring body segment parameters directly from humans. With MRI, 3D mass distribution information can be obtained from axial scans without exposing participants to radiation. Martin *et al.* (1989) used MRI to measure the inertial properties of baboon cadaver segments by obtaining axial scans of the segments, digitizing images to obtain tissue areas and applying estimated tissue densities to determine slice mass distribution properties. They were able to measure mass, COM location and MOI with errors of 6.7, –2.4 and 4.4 per cent, respectively. Mungiole and Martin (1990) later measured the body segment parameters of 12 adult male athletes and Pearsall *et al.* (1994) used MRI to measure trunk body segment parameters on 26 males. More recently, Cheng *et al.* (2000) used MRI to determine the body segment parameters of eight Chinese men. Similar to CT scanning, the availability of MRI systems is generally limited for most researchers, data acquisition is costly, and data processing methods are time-consuming. This makes direct measurement of individuals using MRI impractical.

Recently, dual energy x-ray absorptiometry (DEXA) has been used by Durkin *et al.* (2002) and Ganley and Powers (2004a; 2004b) to determine the body segment parameters of human segments. This technology operates much in the same way as gamma-mass scanning. Two collimated beams of alternating intensity (70 keV and 140 keV) are emitted and passed through a sample of tissue. The attenuation of the high energy beam (140 keV) is directly proportional to mass (Figure 14.1), therefore, the inertial properties of the scanned material can be determined from the measured mass information and the known area dimensions of the scanned elements in the chosen scan plane. Durkin *et al.* (2002) were able to measure mass, COM location and MOI about the COM with less than 3.2 per cent error. Furthermore, the method was rapid and safe, and the technology is widely available in hospitals. The ease with which the method can be applied, accompanied by low operational costs, make the method attractive for direct body segment parameter measurement on individuals and for developing predictive equations from large databases of participants.

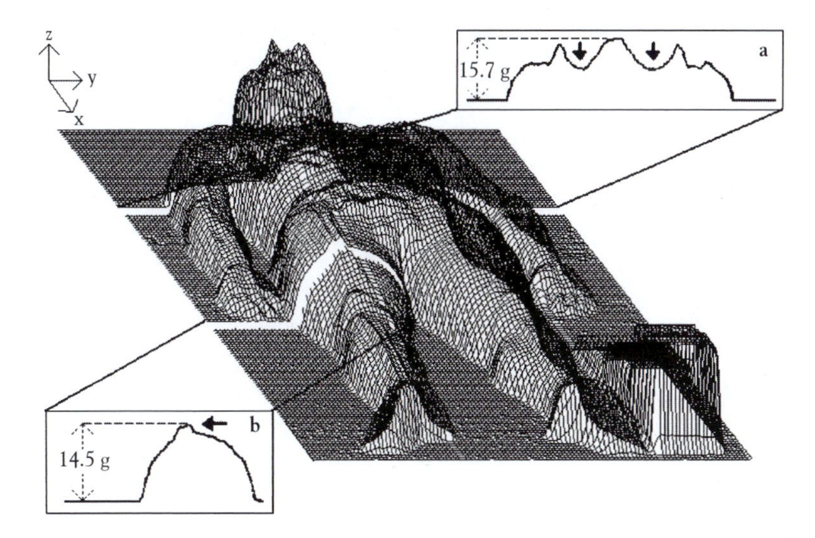

Figure 14.1 Example of a whole body scan of a human using the Hologic QDR-100/W dual energy x-ray absorptiometer. Attenuation coefficients based on x-ray absorption values are proportional to mass. Reprinted with permission from Durkin et al. (2002).

DEXA is similar to gamma-mass scanning in that it is a 2D imaging tool, therefore, mass distribution properties can be determined in one plane only.

Overall, medical imaging techniques provide a reliable and accurate means for obtaining body segment parameters directly on humans. The limitations associated with each method necessitate further exploration of a feasible source of anthropometric information. Predictive models are attractive for many researchers as they provide an expedient means for estimating body segment parameters on a research participant pool, as long as these predictive equations are based on methods that can provide accurate and reliable measures.

Mathematical models

The development of predictive models for estimating body segment parameters on living humans has been the focus of biomechanical studies for decades. These models generally fall into two categories: regression equations and geometric models.

Regression equations

Regression equations for body segment parameter estimation have been generated from a variety of data sources, including cadavers (Chandler *et al.*, 1975; Dempster, 1955), medical imaging techniques (Durkin and Dowling, 2003; Zatsiorsky *et al.*, 1990) water immersion (Drillis and Contini, 1966; Plagenhoef, 1983) and photogrammetry (Young *et al.*, 1983). The regression equations available are limited by the deficiencies of the methods used to obtain the data and are specific to the population from which they were generated. For instance, the equations generated from cadaveric data are typically based on male specimens, most of which were over 50 years of age (Pearsall and Reid, 1994) and some of which were in fairly emaciated states when they passed. Chandler *et al.* (1975) was the only study to provide equations estimating COM locations and MOI about all three segment axes (Cappozzo and Berme, 1990).

Other studies limited their experiments to segment volume, mass and COM locations (Clauser *et al.*, 1969) and many assumed symmetry between frontal and sagittal planes. Further, many researchers developed regression equations that were linear in nature (Dempster, 1955; Durkin and Dowling, 2003). Yeadon and Morlock (1989) developed non-linear regression equations based on data from Chandler *et al.* (1975) to show that linear relationships do not produce accurate predictors when relating anthropometric measurements and MOI.

Regression equations generated from reliable data measurement methods such as gamma-mass scanning and DEXA allow greater confidence in estimating human body segment parameters. Zatsiorsky *et al.* (1990) used gamma mass scanning to develop regression equations for segment mass, COM location and MOI prediction on all the segments of the body. The body was partitioned into ten segments and equations for segment MOI were provided about three axes. The multiple regression equations presented allow estimation of body segment parameters using whole body mass and whole body height squared. Further, Zatsiorsky *et al.* (1990) provide equations for both male and female populations, although the equations are based on young Russian Caucasian physical education students.

Durkin and Dowling (2003) investigated the body segment parameters of selected human body segments on male and female populations of two age categories (19–30 years old and 55+ years old). Using DEXA, segment masses, COM locations and radii of gyration were determined. Separate equations for each age-sex group were provided that were based on non-homogeneous groups varying by race and morphology. These equations were linear in nature, however, and only frontal plane analyses of a few segments were presented. Body segment representation was limited due to the nature of the DEXA technology, namely that scans are planar, making it impossible to accurately separate overlying structures such as the lateral trunk with the upper arm. The importance of this study lies in the comparison of several predictive models for estimating body segment parameters on different populations. The study showed that there are significant body segment parameter differences between age and sex groups as well as large individual differences within groups (Figure 14.2). These findings suggest that regression equations themselves may not be able to adequately account

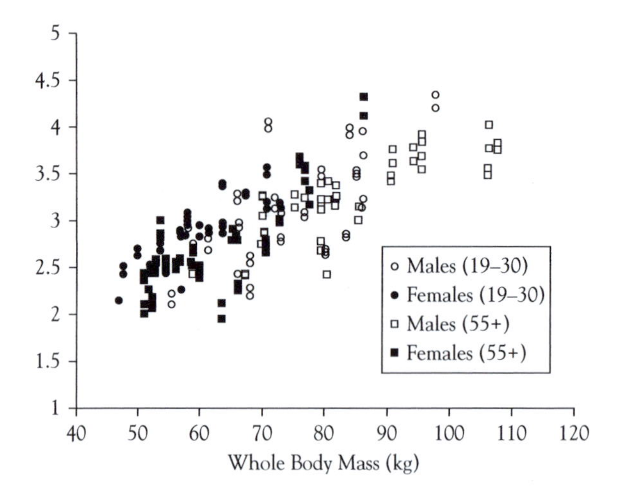

Figure 14.2 Lower leg mass differences between four human populations (YM=Young Males, OM=Older Males, YF=Young Females, OF=Older Females). Mass distributions are presented as a percent whole body mass. Reprinted with permission from Durkin (1998).

for body segment parameter differences between individuals even if the individuals are from the same age or sex group.

Geometric models

Geometric models have been developed with varying complexity to more accurately account for the differences in morphology between individuals. Hanavan (1964) developed a geometric model of the human body using a series of spheres, ellipsoids, circular cylinders and frusta. Centres of mass were predicted within 0.7 inches and MOI were predicted within 10 per cent. Segment masses were estimated using the regression equations of Barter (1957); however, Barter (1957) developed his equations using cadaveric data from two different sources, each of which used different segmentation methods. Furthermore, Hanavan's (1964) model assumed constant density throughout each segment and the estimates were validated for whole body mass and inertia estimates only.

Hatze (1980) developed a geometric model of the human body that could be applied to individuals regardless of age, sex or morphology and stated a maximum error of 5 per cent. Hatze (1980) also reports that the model can be used on pregnant women and obese individuals. The mathematical model used detailed anthropometric information and made no assumptions of constant density, yet this procedure has been criticized due to the 242 anthropometric measurements needed to obtain body segment parameter estimates.

Jensen (1978) used stereophotogrammetry to develop a geometric model of the human body using elliptical zones 2 cm thick. This method has the ability to finely mimic the volume distribution of the human body and can predict with reasonable accuracy the segmental volumes of individuals of all ages and morphologies. Segment mass is determined by applying constant density values and the COM and MOI are calculated from the geometry of the segment formed by stacking the appropriate elliptical plates. The model was validated for whole body mass only with an error of 1.8 per cent. Jensen used this method extensively to investigate the morphological differences between individuals of varying body types, sexes and ages. In fact, his research is the only resource that provides information regarding the anthropometric changes through infancy and childhood (Jensen, 1989). His contributions demonstrate that there are significant differences in the volume distribution characteristics of body segments between populations varying by age, sex and body type. Further, he demonstrates that children are not simply small adults and that the changes throughout the developmental years must be modelled carefully using the appropriate test sample. Unfortunately, validation of segmental mass, COM and MOI estimates was not possible. Further, the model requires that individual participants be photographed and the images digitized and his software is not available to the public. Applying this method to obtain individual-specific anthropometric values is, therefore, not a time-efficient process.

Zatsiorsky *et al.* (1990) developed geometric models of human body segments and validated these models against gamma-mass data. Body segments were modelled as cylinders and a constant 'pseudodensity' factor was applied to account for differences between segment and model shape and density. The models require the input of limb circumferences and lengths and separate coefficients are provided for male and female populations. The models are identified as being 1.5 times less accurate than those obtained using multiple regression equations developed from the same data set, but are advocated as appropriate for use on children. In comparison, the authors stress that the regression equations should only be used on physically fit young adults.

In an effort to more accurately mimic the volume distribution properties of human body segments, Durkin (1998) developed geometric models of selected human body segments based on composite geometric shapes. The models were applied to humans from four populations differing by age and sex using limb circumferences and lengths. The models performed poorly and it was concluded that assuming constant density while modelling according to segment volume results in large errors in body segment parameter estimation.

The use of predictive equations remains a preferred means for obtaining human body segment parameter data in biomechanical analyses of motion; however, Durkin and Dowling (2003) found that population-specific regression equations are not able to account for differences between individuals within groups. Further, the assumption of constant density in geometric modelling and its effect on inertial estimates is troublesome. Theoretically, models based on geometric approximations should provide more accurate representations of segmental body segment parameters, regardless of age, race, sex and morphology, by using anthropometric measures such as limb lengths, circumferences and breadths (Durkin, 1998; Zatsiorsky *et al.*, 1990). An alternative method for designing geometric models that accounts for the negative effects of a constant density assumption is needed. The method of Hatze (1980) addresses this issue; however, the large amount of input data required makes the model impractical. Wei and Jensen (1995) attempted to develop axial density profiles using regression equations, but they were unable to determine whether inertial estimates were improved.

Recently, Durkin *et al.* (2005) and Durkin and Dowling (2006) investigated a means for developing geometric models that mimic the mass distribution properties of body segments rather than segment volume. DEXA was used to determine the axial mass-distribution properties of the thigh and lower leg segments from four populations differing by age and sex. Segmental axial mass-distribution profiles were created by summing mass elements at intervals of 1 per cent segment length, normalizing the summed intervals to segment mass, and using the ensemble averages from a given population to investigate geometric similarity within and between populations. From these mass profiles, a geometric model was developed to mimic the segment's mass distribution properties. Geometric models were developed and validated for the thigh in the frontal plane (Durkin *et al.*, 2005) and for the leg in the frontal and sagittal planes (Durkin and Dowling, 2006). Anthropometric measurements such as segment lengths, girths and circumferences were used as input variables and model predictions were compared to DEXA benchmark values. Both studies revealed a high geometric similarity in segment mass distribution properties within the populations investigated (Figure 14.3). Additionally, high geometric similarity between groups was found (Figure 14.3) suggesting that one geometric model could be generated to fit all four groups. The models did not greatly improve on body segment parameter estimates over the other models tested, but comparisons between the mass distribution characteristics of the geometric models and those of the DEXA data revealed where the models failed and provided insight into how they might be improved (Figure 14.4). This modelling approach also allowed insight into the differences between the axial mass distribution properties of human body segments and the shortcomings of using geometric models that mimic segment volume distribution.

Reported errors in mathematical models

Many researchers find the use of mathematical models for estimating human body segment parameters preferable to the direct methods available. The reasons for this include ease of use, low cost and expediency. The limitations of the methods used to develop the models (i.e. cadavers) introduce error into the estimations. Overall, mathematical models, direct

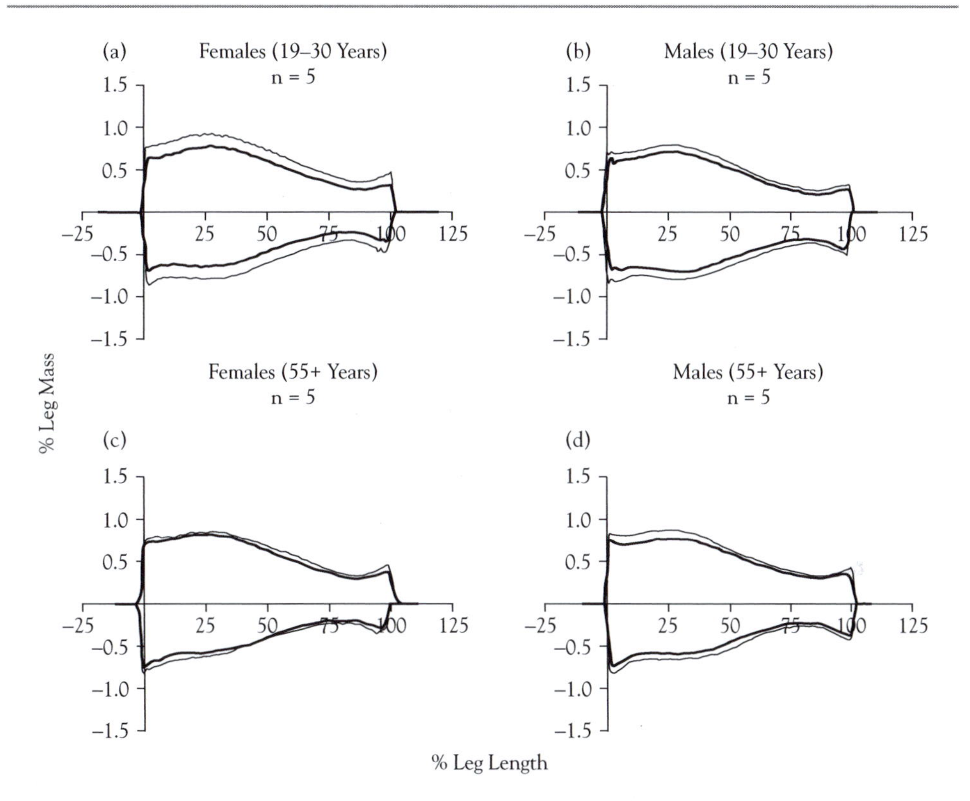

Figure 14.3 Ensemble averages of frontal plane leg mass distribution plots as determined using DEXA. Positive standard deviation values (thin lines) for four populations are also presented. Model mass plots are normalized to 100% DEXA mass. Reprinted with permission from Durkin and Dowling (2006).

measurement techniques, and cadaver methods suffer from many of the same limitations. First, segment parameter measurements are affected by the chosen segment boundaries. Second, it is assumed that these segments are rigid and that segment boundaries do not change with movement. Several other limitations specific to each of the direct and cadaveric methods, such as those discussed in the previous section, must be considered as well. These limitations are in addition to other sources of error that are unaccounted for when using a mathematical model to estimate a given parameter on a group of participants.

With the increased use and noted accuracy of medical imaging techniques, many authors have used these methods to evaluate the error from popular predictive models in the literature. Pearsall *et al.* (1996) compared male and female trunk segment parameters measured using CT to 9 other sources in the literature. The error in mass estimations for the upper trunk, mid-trunk and lower trunk segments ranged from 1.5–10.9 per cent, 0–4.2 per cent, and 0.5–26.3 per cent of whole body mass, respectively. Cheng *et al.* (2000) also examined trunk body segment parameters for eight male Chinese workers. The range of errors for trunk mass was 0.7–4.5 per cent with the most accurate estimates coming from Dempster (1955). Interestingly, Pearsall *et al.* (1994) found a 7.5 per cent difference in mass estimation when compared to Dempster (1955). This discrepancy was attributed to differences in the density of the trunk between cadaveric specimens and living participants. Martin *et al.* (1989) demonstrated that MRI can introduce volume measurement errors when calculating

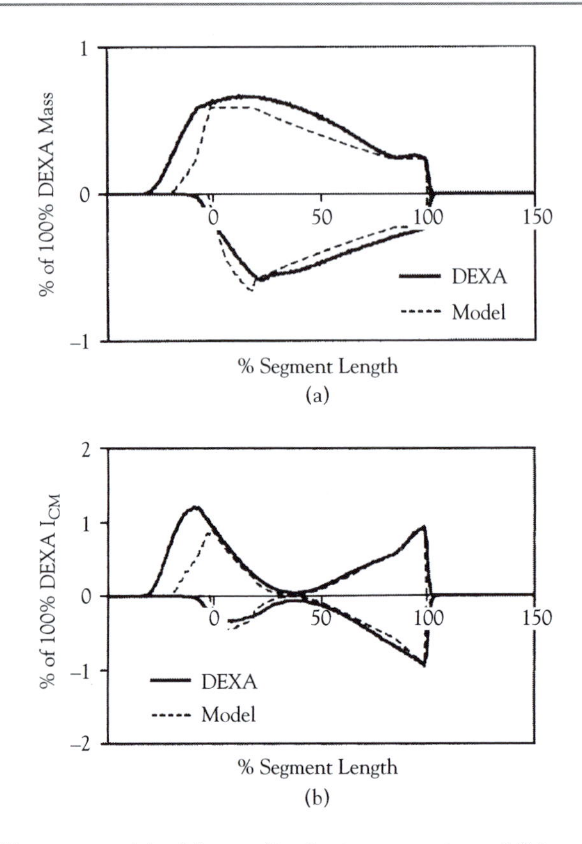

Figure 14.4 Ensemble averages of the (a) mass distribution properties and (b) moment of inertia (I_{CM}) distribution properties estimated by a geometric model of the thigh and measured by DEXA. Mean model estimates are displayed as a percent of DEXA mass to show where the model failed to estimate the desired parameter. The curves represent the mean ensemble averages from a male (19-30 years) group. Reprinted with permission from Durkin et al. (2005).

body segment parameters and that mass estimation errors may result from constant density assumptions (Martin *et al.*, 1989). The differences in the results between Cheng *et al.* (2002) and those of Pearsall *et al.* (1994) may be due to these effects. It is important to note, however, that the error found by Pearsall *et al.* (1994) is similar to the 7.2 per cent error found by Pearsall *et al.* (1996) when comparing trunk mass measured from CT data with estimates from Dempster (1955). Cheng *et al.* (2002) also found errors ranging from 12.2–22.2 per cent for trunk CM, and an error of 1.16 kg•m^2 for moment of inertia estimations when compared to Pearsall *et al.* (1996).

Ganley and Powers (2004a) compared DEXA derived body segment parameter measurements with estimates from Dempster (1955) for males and females aged 20–50 years. Mungiole and Martin (1990) compared MRI measurements on 12 males aged 21–32 years with estimates from several sources. Further, Cheng *et al.* (2000) compared estimates from five sources to MRI-derived data on eight Chinese males in their mid-twenties. For the most part, these studies examined body segment parameter errors of the lower limb only with the exception of Cheng *et al.* (2000) who looked at all the segments of the human body. The ranges of the reported errors for thigh, lower leg and foot segments resulting from these comparisons are listed in Table 14.1. Only a few studies have examined the effects of body segment parameter models when applied to different populations. These include Durkin and Dowling (2003), Durkin *et al.* (2005) and Durkin and Dowling (2006) whereby four populations that varied by age and sex were compared. The results of these studies are presented in Tables 14.2 through 14.4.

How much error is too much?

Although many studies have examined the body segment parameter estimation errors from various models in the literature, the effects of these errors on kinetic calculations remains in question. Several studies have examined these effects with varying results. Pearsall and Costigan (1999) investigated the effects of body segment parameter error on gait. Body segment parameter error from the thigh and leg was varied +/– 40 per cent over 9 levels. The results showed that body segment parameter variation had very little effect on net joint moments with differences amounting to less than 1 per cent body weight during stance. The reasons for the limited effect of body segment parameter error on moment calculations in gait was attributed to the closed loop nature of gait and the low accelerations involved (Pearsall and Costigan, 1999). Ganley and Powers performed similar analyses on young adults (2004a) and children (2004b). Their results supported those of Pearsall and Costigan (1999) where neither study produced differences that were clinically relevant.

The results found in gait analyses are not surprising when one considers that segmental mass and MOI values are small for human body segments. Even a 50 per cent variation in a body segment parameter would have little effect when used in a situation with low accelerations and high GRFs. The effect of body segment parameter error could have more important effects when examining open chain movements or movements with high accelerations. Andrews and Mish (1996) examined the sensitivity of joint resultants to body segment parameter error by simulating oscillations of a leg-foot segment while varying body segment parameter error by +/– 5 per cent. They found that moment errors can be more than double the body segment parameter error present, highlighting the importance of reducing body segment parameter error for open chain movements.

Table 14.1 Range of errors reported in lower limb BSP estimates when compared to criterion values. Criterion measures were made using CT (Cheng et al., 2000), DEXA (Ganley and Powers, 2004a) and MRI (Mungiole and Martin, 1990). Mass errors are reported as a percentage of whole body mass. Centre of mass (CM) errors are reported as a percentage of segment length. Moment of inertia (I_{CM}) errors are reported as a percent difference from the criterion values with the exception of Ganley and Powers (2004a) who reported errors for radius of gyration as a percentage of segment length.

		Cheng et al. (2000)	*Ganley and Powers (2004a)*	*Mungiole and Martin (1990)*
Thigh	Mass	3.0–3.4	1.3	–
	CM	1.4–7.5	2.8	–
	I_{CM}	–	7.2	–
Leg	Mass	0–0.1	0.2	5–21.6
	CM	0.9–7.1	2.1	0.6–2.5
	I_{CM}	88	3.6	0.5–21.2
Foot	Mass	0.5–0.6	0.5	–
	CM	9.1–11.1	0.3	–
	I_{CM}	43.3–55.6	22.1	–

Table 14.2 Percent root mean squared errors (%RMSE) of segment mass estimations from 7 sets of predictive equations. $\%RMSE = [((Calculated\text{-}DEXA)/DEXA)^2/n]^{0.5}$. Reprinted with permission from Durkin and Dowling (2003)*, Durkin et al., (2005) and Durkin and Dowling (2006).

		Durkin and Dowling (2003)*	Durkin et al. (2005)	Durkin and Dowling (2006)	Dempster (1955) via Winter (1990) *	Zatsiorsky et al. (1990) *		Hanavan (1964) *
Segment	*Group*	*Linear[a] Regression*	*Geometric Solid*	*Geometric[d] Solid*	*Linear[a] Regression*	*Multiple[b] Regression*	*Geometric Solid*	*Linear[c] Regression*
Forearm	Males (19–30 Years)	9.37	–	–	13.28	11.99	10.18	10.14
	Females (19–30 Years)	8.11	–	–	10.47	10.34	6.79	21.50
	Males (55+ Years)	8.24	–	–	8.55	8.20	11.03	15.59
	Females (55+ Years)	10.87	–	–	16.53	11.98	8.71	29.67
Hand	Males (19–30 Years)	15.25	–	–	15.57	18.01	–	28.53
	Females (19–30 Years)	9.38	–	–	9.66	9.92	–	34.44
	Males (55+ Years)	16.87	–	–	16.55	16.54	–	22.04
	Females (55+ Years)	17.79	–	–	19.18	16.07	–	37.52
Thigh	Males (19–30 Years)	7.91	18.3	–	21.30	16.08	8.51	21.21
	Females (19–30 Years)	6.65	13.4	–	27.71	13.57	9.97	25.76
	Males (55+ Years)	7.93	24.8	–	16.30	23.52	14.71	17.28
	Females (55+ Years)	10.56	16.1	–	23.57	26.25	15.76	22.38
Leg	Males (19–30 Years)	11.88	–	6.18	17.19	12.84	11.6	21.98
	Females (19–30 Years)	7.04	–	5.76	7.95	15.13	15.62	7.88
	Males (55+ Years)	9.48	–	4.59	22.48	13.83	13.95	30.57
	Females (55+ Years)	12.59	–	10.59	16.27	19.65	18.41	19.47
Foot	Males (19–30 Years)	11.47	–	–	16.81	11.91	–	17.84
	Females (19–30 Years)	9.58	–	–	19.14	15.58	–	29.80
	Males (55+ Years)	14.54	–	–	21.37	10.64	–	17.02
	Females (55+ Years)	14.28	–	–	30.53	26.30	–	36.07

[a]Segment mass predicted from whole body mass (WBM) in kg
[b]Segment mass predicted from WBM and whole body height^2 (WBH^2)
[c]Segment mass predicted from WBM using regression equations from Barter (1957)
[d]Model errors have been converted to per cent RMSE to correspond with other sources

Table 14.3 Percent root mean squared errors (%RMSE) of segment centre of mass estimations from 7 sets of predictive equations. %RMSE = $[((\text{Calculated-DEXA})/\text{DEXA})^2/n]^{0.5}$. Reprinted with permission from Durkin and Dowling (2003)*, Durkin et al., (2005) and Durkin and Dowling (2006).

		*Durkin and Dowling (2003)**	*Durkin et al. (2005)*	*Durkin and Dowling (2006)*	*Dempster (1955) via Winter (1990)* *	*Zatsiorsky et al. (1990)* *		*Hanavan (1964)* *
Segment	*Group*	*Linear[a] Regression*	*Geometric Solid*	*Geometric[c] Solid*	*Linear[a] Regression*	*Multiple[b] Regression*	*Geometric Solid*	*Geometric Solid*
Forearm	Males (19–30 Years)	2.79	–	–	5.84	32.15	20.81	5.10
	Females (19–30 Years)	2.07	–	–	5.37	37.05	20.89	5.21
	Males (55+ Years)	3.51	–	–	6.37	29.80	21.17	5.79
	Females (55+ Years)	2.80	–	–	7.43	41.45	23.37	7.50
Thigh	Males (19–30 Years)	7.06	3.2	–	13.25	36.32	15.30	11.31
	Females (19–30 Years)	10.45	4.4	–	21.00	11.75	17.66	17.47
	Males (55+ Years)	9.64	9.5	–	19.78	48.62	19.16	21.61
	Females (55+ Years)	16.41	7.1	–	34.28	23.78	30.25	32.95
Leg	Males (19–30 Years)	2.44	–	2.94	14.25	4.82	38.83	54.17
	Females (19–30 Years)	1.89	–	3.07	19.88	5.11	36.81	55.22
	Males (55+ Years)	2.50	–	2.93	13.03	6.06	53.62	53.73
	Females (55+ Years)	1.93	–	3.22	20.81	5.60	47.36	56.12

[a]Segment centre of mass (CM) predicted from segment length (SL) in cm
[b]Segment CM predicted from WBM and whole body height^2 (WBH^2)
[c]Model errors have been converted to %RMSE to correspond with other sources

Table 14.4 Percent root mean squared errors (%RMSE) of segment ICM estimations from 7 sets of predictive equations. %RMSE = $[((\text{Calculated-DEXA})/\text{DEXA})^2/n]^{0.5}$. Reprinted with permission from Durkin and Dowling (2003)*, Durkin et al., (2005) and Durkin and Dowling (2006).

		*Durkin and Dowling (2003)**	*Durkin et al. (2005)*	*Durkin and Dowling (2006)*	*Dempster (1955) via Winter (1990)**	*Zatsiorsky et al. (1990)**		*Hanavan (1964)**
Segment	*Group*	*Linear[a] Regression*	*Geometric Solid*	*Geometric[c] Solid*	*Linear[a] Regression*	*Multiple[b] Regression*	*Geometric Solid*	*Geometric Solid*
Forearm	Males (19–30 Years)	3.96	–	–	9.44	5.81	6.37	14.73
	Females (19–30 Years)	3.71	–	–	9.68	6.90	3.54	14.31
	Males (55+ Years)	4.77	–	–	9.69	4.97	6.78	14.39
	Females (55+ Years)	6.15	–	–	11.44	6.87	4.80	13.32
Thigh	Males (19–30 Years)	8.45	14.7	–	12.24	5.34	10.14	17.37
	Females (19–30 Years)	8.04	12.2	–	9.15	6.86	10.89	19.01
	Males (55+ Years)	8.16	25.5	–	7.81	6.25	18.64	24.64
	Females (55+ Years)	11.24	20.5	–	7.77	10.84	17.32	24.12
Leg	Males (19–30 Years)	6.78	–	14.03	16.22	5.54	10.89	9.51
	Females (19–30 Years)	6.57	–	13.07	16.15	3.87	18.64	9.94
	Males (55+ Years)	7.87	–	16.65	16.32	4.74	17.32	9.72
	Females (55+ Years)	6.53	–	24.27	14.70	6.71	11.39	11.10

[a]Segment radius of gyration (K) predicted from SL in cm
[b]Segment K predicted from WBM and whole body height^2 (WBH^2)
[c]Model errors have been converted to per cent RMSE to correspond with other sources

Desjardins *et al.* (1998) and Lariviere and Gagnon (1999) both investigated the effect of body segment parameter models on L5/S1 extension moments in lifting activities. Desjardins *et al.* (1998) varied body segment parameter values by 5 per cent and compared bottom up and top-down link-segment models in asymmetrical lifting movements. It was found that the top-down approach was sensitive to body segment parameter error in segment mass. Lariviere and Gagnon (1999) compared symmetrical and asymmetrical lifting using an upper body model. Five trunk body segment parameter models that varied by: (a) the number of segments used (2 vs. 3); (b) the position of the COM (hip to shoulder line vs. trunk line) and; (c) the type of model (geometric vs. regression), were tested. They found that the effects of body segment parameter error were greater for asymmetric lifts, larger participants and more flexed postures. The use of a three-segment geometric model with a trunk line COG to define anteroposterior COM position was suggested.

So how much error is too much in body segment parameter estimations? Based on the available data, it is reasonable to conclude that the effect of body segment parameter error on kinetic calculations depends on the task being performed, the link-segment model being used, and the characteristics of the participants in the study. If the effects of body segment parameter error on net joint force and moment calculations are not known, then a sensitivity analysis should be performed.

Conclusion

The measurement and estimation of human body segment parameters has been an important biomechanical issue for decades. Of the methods available, each has its own limitations that should be understood before they are used in kinetic calculations of movement. Mathematical models remain the most convenient method for body segment parameter determination; however, the user must be aware of the population on which they are based and the limitations of the data used to generate the equations. Medical imaging techniques have enabled the accurate measurement of human body segment parameters and have permitted an assessment of many of the mathematical models available. Overall, the regression equations of Zatsiorsky *et al.* (1990) are the most complete and provide the greatest accuracy for lower limb segments. Geometric models have the potential to improve on body segment parameter estimates; however, they should be based on the mass-distribution characteristics of segments rather than volume distribution characteristics. Future research examining segmental mass-distribution properties in 3D for many different populations is needed to improve the availability of useful, reliable body segment parameter estimators.

References

Andrews, J.J. and Mish, S.P. (1996) 'Methods for investigating the sensitivity of joint resultants to body segment parameter variations', *Journal of Biomechanics*, 29: 651–654.

Barter, J.T. (1957) *Estimation of the mass of body segments*. TR-57-260. Ohio: Wright-Patterson Air Force Base.

Braune, W. and Fischer, O. (1889) *The centre of gravity of the human body as related to the German infantryman*. ATI 138 452. Leipzig: National Technical Information Service.

Cappozzo, A. and Berme, N. (1990) 'Subject-specific segmental inertia parameter determination', in Berme, N. and Cappozzo, A. (eds), *Biomechanics of Human Movement: Applications in Rehabilitation, Sports and Ergonomics* (pp. 179–185), Washington, D.C.: Bertec Corporation.

Chandler, R.F., Clauser, J.T., McConville, H.M., Reynolds, H.M. and Young, J.W. (1975) *Investigation of inertial properties of the human body*. AMRL-TR-74-137. Ohio: Aerospace Medical Research Laboratory, Wright-Patterson Air Force Base.

Cheng, C-K., Chen, H-H., Chen, C-S., Lee, C-L. and Chen, C-Y. (2000) 'Segment inertial properties of Chinese adults determined from magnetic resonance imaging', *Clinical Biomechanics*, 15(8): 559–566.

Clauser, C.E., McConville, J.T. and Young, J.W. (1969) *Weight, volume, and COM of segments of the human body*. AMRL-TR-69-70. Ohio: Aerospace Medical Research Laboratory, Wright-Patterson Air Force Base.

Dempster, W.T. (1955) *Space requirements of the seated operator*. WADC-TR-55-159. Ohio: Wright Air Development Center, Wright-Patterson Air Force Base.

Desjardins, P., Plamondon, A. and Gagnon, M. (1998) 'Sensitivity analysis of segment models to estimate the net reaction moments at the L5/S1 joint in lifting', *Medical Engineering and Physics*, 20: 153–158.

Dowling, J.J, Durkin, J.L. and Andrews, D.M. (2006) 'The uncertainty of the pendulum method for the determination of the moment of inertia', *Medical Engineering and Physics*, 28: 837–841.

Drillis, R. and Contini, R. (1966) *Body segment parameters*. TR-1166-03. New York, NY: New York University School of Engineering Science.

Durkin, J.L. (1998) *The prediction of body segment parameters using geometric modelling and dual photon absorptiometry*. Masters Thesis. Hamilton, ON: McMaster University.

Durkin, J.L., Dowling, J.J. and Andrews, D.A. (2002) 'The measurement of body segment inertial parameters using dual energy x-ray absorptiometry', *Journal of Biomechanics*, 35(12): 1575–1580.

Durkin, J.L. and Dowling, J.J. (2003) 'Analysis of body segment parameter differences between four human populations and the estimation errors of four popular mathematical models', *Journal of Biomechanical Engineering*, 125(4): 515–522.

Durkin, J.L. and Dowling, J.J. (2006) 'Body segment parameter estimation of the human lower leg using an elliptical model with validation from DEXA', *Annals of Biomedical Engineering*, 34: 1483–1493.

Durkin, J.L., Dowling, J.J. and Scholtes, L. (2005) 'Using mass distribution information to model the human thigh for body segment parameter estimation', *Journal of Biomechanical Engineering*, 127: 455–464.

Erdmann, W.S. (1997) 'Geometric and inertial data of the trunk in adult males', *Journal of Biomechanics*, 30(7): 679–688.

Ganley, K.J. and Powers, C.M. (2004a) 'Determination of lower extremity anthropometric parameters using dual energy x-ray absorptiometry: the influence on net joint moments during gait', *Clinical Biomechanics*, 19: 50–56.

Ganley, K.J. and Powers, C.M. (2004b) 'Anthropometric parameters in children: a comparison of values obtained from dual energy x-ray absorptiometry and cadaver-based estimates', *Gait and Posture*, 19: 133–140.

Hanavan, E.P. (1964) *A mathematical model of the human body*. AMRL-TR-64-102. Ohio: Aerospace Medical Research Laboratories, Wright-Patterson Air Force Base.

Harless, H. (1860) *Die statischen Momente der menschlichen Gliedmassen. Abhandl Mathematische-Physikalischen Classe Konigl Bayerischen Akad Wissenschaft*, 8: 69-96, 257–294.

Hatze, H. (1980) 'A mathematical model for the computational determination of parameter values of anthropomorphic segments', *Journal of Biomechanics*, 13(10): 833–843.

Huang, H.K. and Suarez, F.R. (1983) 'Evaluation of cross-sectional geometry and mass density distributions of humans and laboratory animals using computerized tomography', *Journal of Biomechanics*, 16(10): 821–832.

Jensen, R.K. (1978) 'Estimation of the Biomechanical Properties of Three Body Types Using a Photogrammetric Method', *Journal of Biomechanics*, 11: 349–358.

Jensen, R.K. (1989) 'Changes in Segment Inertia Proportions Between 4 and 20 Years', *Journal of Biomechanics*, 22(6/7): 529–536.

Kim, S.J., Son, K., and Choi, K.H. (2003) 'Construction and evaluation of scaled Korean side impact dummies', *KSME International Journal*, 17(12): 1894–1903.

Lariviere, C. and Gagnon, D. (1999) 'The influence of trunk modelling in 3D biomechanical analysis of simple and complex lifting tasks', *Clinical Biomechanics*, 14: 449–461.

Martin, P.E., Mungiole, M., Marzke, M.W. and Longhill, J.M. (1989) 'The use of magnetic resonance imaging for measuring segment inertial properties', *Journal of Biomechanics*, 22(4): 367–376.

Mungiole, M. and Martin, P.E. (1990) 'Estimating segment inertial properties: Comparison of magnetic resonance imaging with existing methods', *Journal of Biomechanics*, 23(10): 1039–1046.

Pearsall, D.J. and Costigan, P.A. (1999) 'The Effect of Segment Parameter Error on Gait Analysis Results', *Gait and Posture*, 9: 173–183.

Pearsall, D.J. and Reid, J.G. (1994) 'The study of human body segment parameters in biomechanics', *Sports Medicine*, 18(2): 126:140.

Pearsall, D.J., Reid, J.G. and Livingston, L.A. (1996) 'Segmental inertial parameters of the human trunk as determined from computed tomography', *Annals of Biomedical Engineering*, 24(2): 198–210.

Plagenhoef, S., Evans, F.G. and Abdelnour, T. (1983) 'Anatomical data for analyzing human motion', *Research Quarterly for Exercise and Sport*, 52(4): 169–178.

Reid, J.G. and Jensen, R.K. (1990) 'Human body segment inertia parameters: A survey and status report', *Exercise and Sport Science Reviews*, 18: 225–241.

Wei, C. and Jensen, R.K. (1995) 'The application of segment axial density profiles to a human body inertia model' *Journal of Biomechanics*, 28(1): 103–108.

Yeadon, M.R. and Morlock, M. (1989) 'The appropriate use of regression equations for the estimation of segmental inertia parameters', *Journal of Biomechanics*, 22: 683–689.

Young, J.W., Chandler, R.F., and Snow, C.C. (1983) *Anthropometric and Mass Distribution Characteristics of the Adult Female*. FAA-AM-83-16. Oklahoma: FAA Civil Aeromedical Institute.

Zatsiorsky, V. and Seluyanov, V. (1983) 'The mass and inertia characteristics of the main segments of the human body', in H. Matsui and K. Kobayashi (eds) *Biomechanics V-IIIB* (pp. 1152–1159), Champaign, IL: Human Kinetics Publishers Inc.

Zatsiorsky, V.M., Aruin, A.S. and Selujanov, V.N. (1984) *Biomechanik des mensclichen Bewegungsapparates*. Berlin: Sportverlag.

Zatsiorsky, V., Seluyanov, V. and Chugunova, L.G. (1990) 'Methods of determining mass-inertial characteristics of human body segments', in G.G. Chernyi and S.A. Regirer (eds), *Contemporary Problems of Biomechanics* (pp. 272–291), Boston, MA: CRC Press.

15

Use of electromyography in studying human movement

Travis W. Beck[1] *and Terry J. Housh*[2]
[1]*University of Oklahoma, Norman;* [2]*University of Nebraska-Lincoln, Lincoln*

Introduction

Electromyography involves recording the action potentials that activate skeletal muscle fibres. As early as the 1600s, scientists knew that the electromyography (EMG) signal existed (Basmajian and De Luca, 1985), but they were limited by a lack of appropriate equipment to detect the signal. Thus, the science of recording EMG signals probably received its greatest impetus from the related technique of electrocardiography (ECG), which involves detecting the electrical activity of the heart. With the advent of the electrical equipment needed to detect the EMG signal in the early 1900s, physicians began using qualitative interpretation of EMG to diagnose neuropathies and various neuromuscular disorders. Quantitative EMG, however, was not commonly used until the 1950s and 1960s, when the development of electronic integrators allowed researchers to more easily quantify the amplitude of the EMG signal (de Vries, 1966, 1968). With gradual improvements in EMG equipment and signal processing techniques, EMG has evolved to the point where it is now considered to be a very useful biological signal that contains a great deal of information regarding the state of the neuromuscular system (De Luca, 1997).

There are two primary techniques for detecting EMG signals: (a) surface EMG; and (b) indwelling EMG. Surface EMG involves detecting the electrical activity of a muscle with electrodes placed on the surface of the skin. With indwelling EMG, however, the electrodes (needle or wire) are inserted directly into the muscle. In addition, there are two important factors that distinguish surface EMG from indwelling EMG: (a) pick-up area of the recording electrodes; and (b) filtering of the EMG signal by the tissue between the active muscle fibres and recording electrodes. These factors have such a large influence on the EMG signal that valid comparisons cannot be made between surface and indwelling EMG. For example, with indwelling EMG, the signal is generated by just a few muscle fibres that are located within the immediate area of the recording electrodes. Thus, the amplitude of the EMG signal from indwelling electrodes usually ranges from 0.05–5 millivolts root-mean-square (rms), with a bandwidth of 0.1–10,000 Hz (Basmajian and De Luca, 1985). With surface EMG, however, the EMG signal reflects an algebraic summation of the motor unit action potential trains from many muscle fibres (De Luca, 1979), and the tissue between the active fibres and recording electrodes acts as a low pass filter that greatly attenuates the EMG signal. Thus, the amplitude of the EMG signal from surface electrodes usually ranges from 0.01–5 millivolts rms, with a useful bandwidth of 10–500 Hz (Basmajian and De Luca, 1985).

It is important to note that, in most cases, biomechanists are interested in detecting the electrical activity of a large portion of the muscle, rather than just a few muscle fibres. Therefore, surface EMG is more common in biomechanics studies than invasive EMG, and the surface EMG signal is usually used as an indicator of gross muscle activity. As stated previously, however, indwelling EMG can be used for a number of clinical applications, including nerve conduction studies and diagnosis of various neuropathies and neuromuscular disorders.

EMG Equipment

Electrodes

There are two general types of surface electrodes: (a) active electrodes; and (b) passive electrodes. With active electrodes, the input impedance of the electrodes is purposely increased (input impedance is greater than 10^{12} ohms) to make them less sensitive to the quality of the skin-electrode interface. Thus, active EMG electrodes can usually be coupled to the skin surface without the use of abrasive pastes or conducting gels (Gerleman and Cook, 1992). Passive electrodes, however, are designed to have a very low input impedance (e.g., less than 2,000–10,000 ohms). Thus, the skin-electrode interface is a very important factor in determining the quality of the EMG signal that is detected with passive electrodes (Tam and Webster, 1977). In most cases, the very thin dead surface layer of the skin and all skin oils must be removed in the areas where the electrodes will be placed. In addition, conducting gels are usually rubbed into the skin to improve the quality of the skin-electrode interface. The most common passive surface electrodes are pre-gelled silver/silver chloride electrodes.

Unlike surface electrodes, which usually have a fairly large recording surface (e.g., 100–200 mm^2), indwelling EMG electrodes typically consist of very small diameter (e.g., 25–75 micrometre) wires or needles. These small diameter electrodes are necessary because, in most cases, the purpose of using indwelling electrodes is to examine the electrical activities of just a few muscle fibres. The material for needle and wire electrodes usually consists of platinum alloys, silver, and nickel-chromium alloys, with larger diameter electrodes being used in applications where the electrode is susceptible to breakage within the muscle (Basmajian and De Luca, 1985).

With the exception of the electrode array (a specialized surface EMG technique that will be discussed later in this chapter), there are two primary electrode configurations used for surface EMG: (a) a monopolar electrode configuration; and (b) a bipolar electrode configuration. With a monopolar electrode configuration, the EMG signal is detected (with respect to a ground electrode placed in a neutral electrical environment) by a single recording electrode. The primary advantage of the monopolar electrode configuration is that the pick-up area is relatively large, which provides an EMG signal that is generated by many muscle fibres. A disadvantage, however, is that the EMG signal is contaminated by noise from electromagnetic fields (e.g., 50 or 60 Hz power lines and the electrical devices that operate on them, radio signals, television signals and communication signals). Thus, in the early studies that used a monopolar electrode configuration, the EMG signals were detected from subjects in a room/cage lined with copper (i.e., a Faraday cage, see Figure 15.1) to reduce electromagnetic noise (de Vries, 1966). The purpose of the bipolar electrode configuration, however, is to reject electromagnetic noise with differential amplification. Specifically, bipolar EMG involves using two recording electrodes to

Figure 15.1 Example of a copper Faraday cage used to record monopolar electromyography (EMG) signals. The cage reduces electromagnetic noise that can contaminate the EMG signal.

detect two EMG signals from the muscle of interest. The EMG signals are fed into a differential amplifier that subtracts the two signals and amplifies the difference (Soderberg, 1992). Theoretically, this allows the electromagnetic noise present at both electrodes (i.e., common-mode noise) to be removed from the EMG signal. The ability of the amplifier to reject common-mode noise is measured by the common-mode rejection ratio (CMRR), which is a characteristic of EMG amplifiers that will be discussed in the next section. In addition to noise reduction, differential amplification provides an important distinction between monopolar and bipolar EMG signals. Specifically, monopolar EMG signals are generated by the electrical activities of the active muscle fibres within the pick-up area of the recording electrode, plus electromagnetic noise. The bipolar EMG signal, however, is actually a 'difference signal' in the sense that it reflects the difference between the EMG signals detected by the two recording electrodes, rather than the electrical activities of the muscle fibres per se (Soderberg, 1992, p. 30). Figure 15.2 shows an example of a bipolar surface electrode arrangement and two monopolar needle electrodes used for recording indwelling EMG signals.

Amplifiers

The advantage of rejecting electromagnetic noise makes bipolar EMG the method of choice for recording EMG signals. Thus, there are many commercially available differential amplifiers designed specifically for surface and/or indwelling EMG. The primary function of the differential EMG amplifier is to increase the magnitude of the EMG signal detected by the recording electrodes while producing minimal distortion. Table 15.1 shows several recommendations provided by Basmajian and De Luca (1985) regarding optimal characteristics for differential EMG amplifiers.

Analog-to-digital converters

In many of the early EMG studies, the analog EMG signal was visually displayed on an oscilloscope and then stored on magnetic tape with frequency modulated (FM) tape recorders.

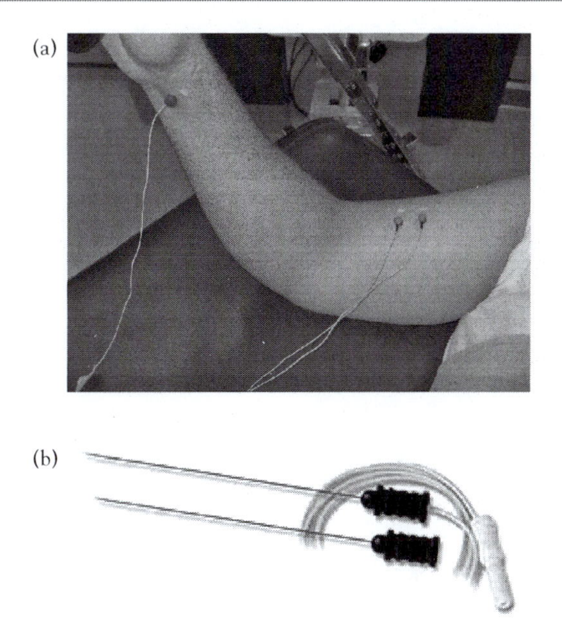

Figure 15.2 Example of two different types of EMG electrodes: (a) A bipolar surface electrode arrangement placed over the biceps brachii muscle. The reference (i.e., ground) electrode is located over the wrist; and (b) two monopolar EMG needle electrodes.

The magnetic tape was then played back at a slow speed such that the EMG signal could be transferred to a strip chart with a mechanical pen or ink jet recorder (Ladegaard, 2002). The advent of commercially available digital computers in the 1980s revolutionized EMG data acquisition. Specifically, instead of storing EMG signals on magnetic tape or strip charts, the EMG signal could be digitized with an analog-to-digital (A/D) converter and stored with a digital computer. The process of converting the EMG signal to digital form involves sampling the signal (i.e., in time) and quantizing the signal (i.e., converting the analog voltages to discrete numbers). The most important factor in determining the sampling rate is the bandwidth of the input signal. Specifically, the sampling rate should be at least twice as high as the highest frequency present in the input signal. Thus, the

Table 15.1 Recommendations regarding optimal characteristics for differential electromyography (EMG) amplifiers.

1. The gain of the amplifier should allow the peak-to-peak amplitude of the output to be ±1.0 volts.
2. The input impedance of the amplifier should be greater than 10^{12} ohms in parallel with 5 picofarads of capacitance.
3. The CMRR should be greater than 100 decibels.
4. The input bias current should be as low as possible (usually less than 50 picoamperes).
5. Baseline noise should be less than 5 microvolts rms.
6. The amplifier should have high- and low-pass filters that achieve 12 decibel/octave rolloff with 3 decibel points at the following frequencies:
   Surface EMG = 20–500 Hz
   Indwelling EMG for general use = 20–1,000 Hz
   Indwelling EMG for signal decomposition or single-fiber studies = 20–10,000 Hz

sampling rate for surface EMG signals should be at least 1,000 samples/second to accommodate a maximum usable frequency of 500 Hz (Hermens *et al.*, 1999). For most general indwelling EMG applications, a sampling rate of at least 2,000 samples/second is sufficient for sampling the EMG signal, but special applications (e.g., signal decomposition or single-fibre studies) may require a sampling rate of at least 20,000 samples/second. In addition, the resolution provided by the quantization procedure is directly influenced by the number of bits used to store each sample. Hermens *et al.* (1999) recommended that for EMG signals, the A/D converter should provide 12 or 16 bits of resolution, which means that the voltages in the EMG signal get curtailed to 4,096 (2^{12}) or 65,536 (2^{16}) possible binary levels.

Collecting EMG data

Electrode placement and interelectrode distance

Two important factors in collecting surface EMG data are electrode placement over the muscle of interest and interelectrode distance. Although many different electrode placement strategies have been used, the three most common locations include: (a) over the motor point of the muscle (the location where the peripheral nerve enters the muscle); (b) over the belly of the muscle, and (c) midway between the innervation zone (IZ = the location where nerve terminations and muscle fibres are connected) and tendonous (proximal or distal) regions. Many of the early EMG studies placed the recording electrodes over the motor point or belly of the muscle for the purpose of maximizing EMG amplitude. Recent investigations (Farina *et al.*, 2001; Rainoldi *et al.*, 2000, 2004), however, have suggested that the most important factor in determining electrode location should be the IZ of the muscle. Specifically, several studies (Farina *et al.*, 2001; Rainoldi *et al.*, 2000, 2004) have found that electrode placement over or near the IZ resulted in decreases in EMG amplitude and increases in EMG frequency. Thus, it was suggested that the IZ should be avoided when recording surface EMG signals (Farina *et al.*, 2001; Rainoldi *et al.*, 2000, 2004).

A recent study from our laboratory (Beck *et al.* in press) examined the influence of electrode location with respect to the IZ for the vastus lateralis muscle on the patterns of responses for absolute and normalized EMG amplitude and mean power frequency (MPF) versus isometric torque. Specifically, a linear electrode array was used to detect 15 channels of bipolar surface EMG signals from the vastus lateralis muscle. The electrode array was oriented such that it was parallel with the long axis of the muscle fibres and the IZ was near the middle of the electrode array. The results indicated that the patterns of responses for absolute and normalized EMG amplitude and MPF versus isometric torque for the vastus lateralis were influenced by electrode location, but this effect was not necessarily due to the IZ. Thus, it was concluded (Beck *et al.*, in press) that electrode location is an important factor affecting the EMG amplitude and MPF versus isometric torque relationships, even when the EMG signals are detected from the same (or nearly the same) muscle fibres. Although a consensus has not been reached, general recommendations for surface EMG electrode placement can be found in Hermens *et al.* (1999) and Zipp (1982).

Selecting the appropriate interelectrode distance is another important aspect of recording bipolar surface EMG signals. The two primary factors that should be considered when choosing interelectrode distance are pick-up area and bandwidth. Specifically, Lynn *et al.* (1978) found that most of the energy in the surface EMG signal comes from muscle fibres that are located within 0.4 length units (1 length unit = interelectrode distance) of the recording electrodes. Thus, smaller interelectrode distances generally result in smaller pick-up areas. In addition, the process of differential amplification eliminates frequencies that are a multiple of the interelectrode distance (De Luca, 1997; Basmajian and De Luca, 1985). Since larger interelectrode distances correspond to longer wavelengths (i.e., lower frequencies), increases in interelectrode distance generally result in lower frequency EMG signals.

Two recent studies from our laboratory (Beck *et al.*, 2005; Malek *et al.*, 2006) examined the influence of interelectrode distance on EMG amplitude and MPF values during concentric isokinetic muscle actions, isometric muscle actions, and cycle ergometry. Specifically, Beck *et al.* (2005) reported that during submaximal to maximal concentric isokinetic and isometric muscle actions of the forearm flexors, a bipolar electrode arrangement with a 60 mm interelectrode distance over the biceps brachii generally resulted in greater EMG amplitude values, and lower EMG MPF values than electrode arrangements with 20 and 40 mm interelectrode distances. Malek *et al.* (2006) reported similar findings for the vastus lateralis muscle during incremental cycle ergometry. In addition, although there were differences among the three interelectrode distances (20, 40, and 60 mm) for absolute EMG amplitude and MPF values, these differences were eliminated with normalization, and differences in interelectrode distance did not affect the patterns of responses for normalized EMG amplitude and MPF versus torque (Beck *et al.*, 2005) or power output (Malek *et al.*, 2006). Thus, it was suggested (Beck *et al.*, 2005; Malek *et al.*, 2006) that normalization is important when comparing EMG amplitude and MPF data from different interelectrode distances. Two general recommendations have been made regarding interelectrode distance. Hermens *et al.* (1999) recommended an interelectrode distance of 20 mm, and Basmajian and De Luca (1985) recommended an interelectrode distance of 10 mm.

Impedance, baseline noise, and preventing aliasing

As stated previously, interelectrode impedance is an important consideration when using passive EMG electrodes. Specifically, the very thin dead surface layer of skin cells and all protective oils must be removed in the areas where the electrodes will be placed to reduce impedance to electrical flow. In most cases, this is done with light abrasion of the skin, cleansing with alcohol, and rubbing conducting gel into the areas where the electrodes will be placed. Although no specific recommendations have been made regarding appropriate interelectrode impedance levels, our laboratory does not collect surface EMG data until interelectrode impedance levels have been reduced to less than 2,000 ohms.

Baseline noise is another important factor when recording surface EMG signals. Generally speaking, noise is any unwanted data that contaminates the signal of interest (Basmajian and De Luca, 1985). Although noise can come from many different sources, the two most important sources in surface EMG are electromagnetic noise and motion artifact. As stated previously, differential amplification usually removes most of the electromagnetic noise in bipolar surface EMG signals. Noise from motion artifact, however, can come from movement of the electrodes on the skin surface or movement of the wire leads connecting the electrodes to the EMG amplifiers. The best strategy for preventing movement of electrodes on the skin surface is to cleanse the skin thoroughly with alcohol to remove dirt and oils, and to use electrodes with very strong adhesives. The adhesives should also be resistant to sweating to prevent electrode movement during sustained exercise, such as cycle ergometry or running. In addition, although movement of the wire leads is less of a concern during isometric muscle actions, it cannot usually be prevented during dynamic muscle actions. In many cases when recording surface EMG signals from large limb muscles, motion artifact does not significantly contaminate the EMG signal because the electrical signal from the muscle is much larger in magnitude than the motion artifact. It can, however, be a problem when recording EMG signals that have very low amplitudes (e.g., detecting EMG signals from very small muscles or during passive movement). In our laboratory, we do not record surface EMG data unless baseline noise levels are less than 5 microvolts rms.

A final consideration that is important when collecting surface EMG data is the use of appropriate sampling rates and analog filters to prevent aliasing. Aliasing occurs when the analog signal contains frequencies that are greater than one-half of the sampling frequency. For example, if an analog signal that contains frequencies from 0–750 Hz is passed through an A/D converter that samples the signal at 1,000 samples/second, then the resulting digitized signal can only contain frequencies between 0 and 500 Hz. Thus, the higher frequency (i.e., 500–750 Hz) information in the analog signal cannot be properly represented in the digitized signal. Instead, this information gets misrepresented (i.e., aliased) to lower frequencies and added to frequencies that legitimately belong in the digitized signal. Therefore, aliasing results both in a loss of information and contamination of the digitized EMG signal. However, an equally important factor for preventing aliasing is to use the appropriate analogue low-pass filter to attenuate all frequencies above one-half of the sampling rate. This procedure is important because it eliminates frequencies that cannot be properly represented in the digitized signal. Thus, these filters are often referred to as 'antialias' filters (Smith, 1997, p. 49). The general guideline for surface EMG is that the antialias filter should have a cutoff frequency of approximately 500 Hz, and the sampling rate should be at least 1,000 samples/second (Hermens *et al.*, 1999).

Signal processing

Digital filtering

The advent of digital computers also changed the way that surface EMG signals are processed. Digital signal processing (DSP) is the mathematics, techniques, and algorithms that are used to manipulate signals once they have been converted into digital form. Perhaps the most common DSP technique is digital filtering. In principle, digital filters are used for the same two purposes that analogue filters are used: (a) separating signals that have been combined; and (b) restoring signals that have been distorted in some way. The performance of digital filters is far superior, however, to analogue filters. In addition, digital filters can be used for many different purposes, which makes them very useful for processing surface EMG signals. For example, one of the most important uses of digital filters in surface EMG research is for removing noise. In most cases, a band-pass filter (i.e., a combination of a low- and a high-pass filter) with cutoff frequencies of 10 and 500 Hz is used to remove both low and high frequency noise. Another common use of digital filters is for data smoothing. For example, with the linear envelope technique (i.e., a signal processing method that will be discussed in the next section), a low-pass filter is used to smooth the EMG signal, thereby providing information regarding slow changes in the amplitude of the signal over time.

Time domain methods

Before digital computers were used to collect and store EMG data, rectification and integration were the two most important techniques involved in quantifying EMG amplitude. Two types of rectification have been used in EMG studies: (a) full-wave rectification; and (b) half-wave rectification. Full-wave rectification is analogous to calculating the absolute value of the EMG signal and involves changing the negative voltages in the signal to positive voltages. With half-wave rectification, however, all negative voltages are ignored, thereby eliminating approximately one-half of the EMG signal. As acknowledged by Basmajian and De Luca (1985), full-wave rectification is used more than half-wave rectification because it preserves all of the energy in the EMG signal. After rectification, two techniques have been used, sometimes interchangeably, to calculate EMG amplitude: (a) the mean rectified value; and (b) integration over a fixed-time period. The mean rectified value is simply the mean, or average voltage of the rectified EMG signal. Integration, however, involves calculating the area under the rectified EMG signal and must be performed over a fixed-time period. Perhaps the most commonly used parameter for quantifying EMG amplitude since the advent of digital computers is the rms voltage of the signal. Since most EMG signals have no direct current (DC) component, the rms value is mathematically equivalent to the standard deviation of the EMG signal. The rms value was recommended by Basmajian and De Luca (1985) as the preferred parameter for quantifying EMG amplitude.

Another important time domain signal processing technique is known as the linear envelope. The linear envelope is calculated by low-pass filtering the rectified EMG signal at a cutoff frequency of approximately 10 Hz (Winter, 1990). The result of the filtering procedure is a smoothed version of the rectified EMG signal that provides information regarding slow changes in the amplitude of the signal over time. The degree of smoothness can be changed by adjusting the cutoff frequency of the low-pass filter (i.e., lower cutoff frequencies result in a smoother linear envelope). It is also important to note that the rectification and filtering procedures are most efficient when performed on the digital EMG signal because digital filters outperform analogue filters. In addition, a zero-phase low-pass filter is usually recommended to prevent time lag in the linear envelope (Hermens *et al.*, 1999).

Frequency analysis methods

Spectral analysis of EMG signals was made possible with the development of the fast Fourier transform (FFT) algorithm by Cooley and Tukey (1965). Although the mathematics of the Fourier transform were developed in the early 1800s (Smith 1997), the technique required far too many calculations than could be performed by the first digital computers. The FFT algorithm, however, reduced computation time by hundreds, thereby introducing frequency analysis to the world of commercial computers and EMG research. It is also important to note that the FFT is mathematically equivalent to the discrete Fourier transform (DFT), with the distinction that the FFT can only be used to process digital signals whose length is a power of two (i.e., 2^N data points). Without addressing the mathematical details of the Fourier transform and frequency analysis, the technique is based on decomposing the input signal onto a series of infinite length (in theory) basis functions (sine waves and cosine waves) that have frequencies ranging from DC to one-half of the sampling rate. The results of the decomposition provide information regarding both the amplitude and phase of the input signal at all frequencies between DC and one-half of the sampling rate. Although the phase provides information regarding the shape of the time domain waveform and is necessary to reconstruct the signal, the magnitude is more commonly used in surface EMG studies. Specifically, the magnitude is usually squared (magnitude squared = power), and the negative frequencies in the spectrum are ignored, which results in a power spectrum (i.e., power as a function of frequency). The frequency resolution of the power spectrum (i.e., distance between spectral lines) is calculated as the sampling frequency divided by the number of data points in the input signal. In addition, the two most common parameters that are calculated from the power spectrum are the median frequency (i.e., the frequency that splits the integral of the power spectrum) and the MPF (i.e., the frequency at mean spectral power). Although the median frequency is less affected by noise, both parameters accurately describe shifts in the power spectrum (i.e., shifts to either higher or lower frequencies) and are frequently used in EMG studies.

Two additional methods that have been used to examine changes in the frequency of the EMG signal are the zero-crossings and turn techniques. The zero-crossings method involves counting the number of times that the time domain EMG signal crosses the baseline (zero volts) in a given time period. The turns method, however, requires counting the number of times that the EMG signal changes direction in a given time period. In many cases, the zero-crossings and turns methods provide similar information. However, several turns can occur in the EMG signal before it crosses the baseline. In addition, the parameters calculated by these techniques (i.e., the number of times the signal crosses the baseline or turns) are influenced by changes in frequency, but they do not measure frequency (i.e., Hz) per se. Both the zero-crossings and turns methods have been used in clinical EMG studies, but they are rarely used in biomechanics.

Time-frequency signal processing

As stated previously, the standard DFT and FFT techniques decompose the input signal onto a set of infinite length sine and cosine waves that have various frequencies. Since the sine and cosine waves are not localized in time, the power spectrum cannot provide information regarding when certain frequencies occurred in the input signal. The most common method for introducing time-dependency into the Fourier transform is to break the EMG signal into epochs (overlapping or not), and then perform the DFT or FFT on each epoch (MacIsaac *et al.*, 2001). This technique is usually referred to as the short-time Fourier transform (STFT) and is the most basic method for tracking time-dependent changes in the frequency of the input signal (Mallat, 1998). An important limitation of the STFT, however, is that the time-frequency resolution of the analysis is greatly influenced by the length of the epoch used (i.e., short windows have good time resolution, but poor frequency resolution, and long windows have good frequency resolution, but poor time resolution).

Another time-frequency procedure that has been used to process EMG signals is the Wigner-Ville transform (Bonato *et al.*, 1996). In a very simplistic sense, the Wigner-Ville transform involves correlating the input signal with itself (i.e., performing autocorrelation) at all time points. The Fourier transform of each autocorrelation function is then calculated, resulting in a power spectrum for every point in time (Cohen, 1995). Although the Wigner-Ville transform has excellent time-frequency localization, it often results in interference terms when it is used to process multicomponent signals. In addition, the interference terms can be reduced with smoothing functions, but only at the expense of spreading out of the primary signal terms (Mallat, 1998).

The wavelet transform, however, is designed to overcome the limitations of the STFT and Wigner-Ville transform. Specifically, the wavelet transform uses short basis functions (i.e., wavelets) at high frequencies and long basis functions at low frequencies. Although previous EMG studies have used both continuous (Karlsson *et al.*, 2000, 2001) and discrete (Constable and Thornhill, 1993, 1994) versions of the wavelet transform, von Tscharner (2000) recently developed a wavelet analysis designed specifically for processing EMG signals. Specifically, the wavelet analysis uses a filter bank of 11 nonlinearly scaled wavelets that maintain a relatively constant relationship between time and frequency resolution across the entire frequency range for surface EMG (approximately 20–400 Hz). Another important advantage of the wavelet analysis is that it is designed to extract the intensity (which is analogous to power) of the events that generate EMG signals. Thus, the intensity of an event that occurs at one frequency can be compared with those that occur at other frequencies. Furthermore, the wavelet spectra generated by the wavelet analysis are very similar in shape to power spectra from the DFT or FFT. Thus, the wavelet-based estimates of MPF or median frequency are comparable to those from the DFT or FFT. Unlike the DFT or FFT, however, a wavelet spectrum can be calculated for each time point, thereby providing information regarding time-dependent changes in the frequency of the input signal.

The most comprehensive form of information provided by the wavelet analysis is an intensity pattern (see Figure 15.3), which is a two-dimensional (2D) matrix of data that gives information regarding time-dependent changes in the intensity and frequency of the input signal. Since each intensity pattern is a two-dimensional matrix of data, it cannot be analyzed with traditional parametric statistics. von Tscharner (2002) and von Tscharner and Goepfert (2003, 2006), however, recently used a pattern recognition technique known as principal components analysis to examine differences in EMG intensity patterns. Although the mathematical details of this procedure will not be discussed, the technique involves decomposing the intensity pattern onto a set of orthogonal eigenvectors. The result of this decomposition is a set of contributing weights that reflect the relative contributions of each eigenvector to the intensity pattern. Thus, this technique is very useful for discriminating between two intensity patterns that have different characteristics (either in time or in frequency). Another procedure that was recently introduced by von Tscharner and Goepfert (2006) involves decomposing intensity patterns onto a set of low- and high-frequency generating spectra. The result of this decomposition is a set of contributing weights that reflect the relative contributions of the low- and high-frequency generating spectra to the intensity pattern. This procedure has been effectively used to decompose EMG intensity patterns from the tibialis anterior and gastrocnemius muscles recorded during the heel-strike

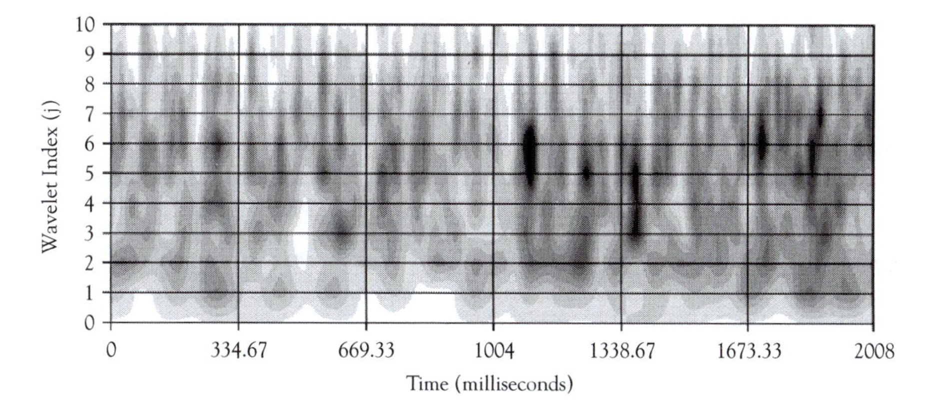

Figure 15.3 Example of an EMG intensity pattern calculated with the EMG wavelet analysis developed by von Tscharner (2000). The intensity pattern was calculated from an EMG signal recorded from the vastus lateralis muscle during a two second isometric MVC of the leg extensors. Increases in the wavelet index (j) represent increases in frequency from $j = 0$ (centre frequency = 6.90 Hz) to $j = 10$ (centre frequency = 395.46 Hz). The dark shaded areas in the intensity pattern reflect high EMG intensity, and the lightly shaded areas reflect low EMG intensity.

and toe-off phases in jogging (von Tscharner and Goepfert, 2006). Thus, both the principal components analysis and generating spectra decomposition are techniques that are designed to extract more information from the EMG signal than could be obtained from Fourier-based estimates of MPF or median frequency.

Data interpretation

Normalization

Normalization is a very important technique in interpreting EMG amplitude and frequency data. The method involves expressing the EMG amplitude and frequency values for each subject relative to a reference value. The reason that this technique is important is because absolute EMG amplitude and frequency values are greatly influenced by factors that are unique for each subject, such as the thickness of the tissue between the muscle and recording electrodes, muscle size, and the locations of the active motor units within the muscle (Farina *et al.*, 2004). These factors have such a large influence on the EMG signal that valid comparisons cannot be made between absolute EMG amplitude and frequency values from different subjects, muscles, and/or studies (Soderberg and Knutson, 2000). It has also been suggested that normalization is necessary when comparisons need to be made between absolute EMG amplitude and/or frequency data obtained with different interelectrode distances, electrode locations, and signal processing techniques (Basmajian and De Luca, 1985; Beck *et al.*, 2005a,b). The most common normalization procedure in surface EMG research is to normalize all EMG amplitude and frequency values relative to the values recorded during an isometric maximum voluntary contraction (MVC). The reason that an MVC is used to generate the reference values is because, theoretically, the amount of electrical activity being generated by the muscle is roughly proportional to force output (LeVeau and Andersson, 1992). Another fairly common technique, however, is to express the EMG amplitude and frequency values for each subject relative to the maximal values recorded during data collection. Our laboratory has successfully used this technique to normalize EMG amplitude and frequency data recorded during isometric and isokinetic (concentric and eccentric) muscle actions, as well as during incremental and constant power output cycle ergometry.

What do EMG amplitude and frequency reflect?

Although the surface EMG signal provides a great deal of valuable information regarding muscle function, several recent studies (Farina *et al.*, 2004; Tucker and Türker, 2005; Keenan *et al.*, 2005) have addressed the importance of proper interpretation of surface EMG amplitude and frequency data. For example, a common misconception is that the amplitude of the surface EMG signal provides quantitative information regarding neural output from the CNS. As acknowledged by Farina *et al.* (2004, p. 1487), however, there are many 'nonphysiological' factors that can affect the surface EMG signal, independent of neural output from the CNS (see Table 15.2). Specifically, factors such as the thickness of the tissue between the muscle and recording electrodes, the locations of the active motor units within the muscle, differential amplification, interelectrode distance, electrode location, and amplitude cancellation can all affect EMG amplitude. In most cases, the relative influence of these factors on the EMG signal is different for each subject and recording condition. Thus, valid comparisons cannot be made between absolute EMG amplitude and/or frequency data obtained from different subjects, muscles, and/or studies.

Table 15.2 Examples of some factors that can affect the amplitude and frequency contents of the surface electromyography (EMG) signal.

1. Thickness of the tissue between the muscle and recording electrodes.
2. Locations of the active motor units within the muscle.
3. Muscle fibres type composition.
4. Number of active motor units.
5. The average muscle fibre action potential conduction velocity.
6. The shapes of the action potentials that propagate along the active muscle fibres.
7. Interelectrode distance.
8. Electrode location.
9. Differential amplification.
10. Amplitude cancellation.
11. Interelectrode impedance.

It is also important to note that the same factors that affect EMG amplitude can also influence EMG frequency. Lindström *et al.* (1970) and Lindström and Magnusson (1977) were among the first to demonstrate that the primary factor influencing the frequency content of the surface EMG signal is muscle fibre action potential conduction velocity. Specifically, decreases in muscle fibre action potential conduction velocity (such as during fatigue) result in lower EMG frequency values (Lindstöm *et al.*, 1970). However, the tissue between the active muscle fibres and recording electrodes acts as a low-pass filter that can also decrease EMG frequency (Basmajian and De Luca, 1985). Thus, individuals with a thick skinfold layer usually demonstrate lower EMG frequency values than those with a thin skinfold layer. This is just one example of a nonphysiological factor that can affect the frequency content of the surface EMG signal, and it is possible that there are other factors that have not been identified yet.

Overall, it is very important for researchers to understand that the surface EMG signal does not simply reflect the activation signal sent from the CNS. Specifically, there are many factors that can affect the signal, which places additional emphasis on normalizing EMG amplitude and frequency values. Furthermore, research is still being conducted to examine the technical aspects of EMG (e.g., electrode location, interelectrode distance, signal processing methodologies, etc.), and it is likely that this research will provide new insight regarding the uses/limitations of the EMG signal. Both surface and indwelling EMG provide useful information regarding muscle function, but their limitations must be understood so that the information they provide can be interpreted properly.

References

1. Basmajian, J.V., De Luca, C.J. (1985) '*Muscles alive: their functions revealed by electromyography*, 5th edn. Baltimore, MD: Williams and Wilkins.
2. Beck, T.W., Housh, T.J., Johnson, G.O., Weir, J.P., Cramer, J.T., Coburn, J.W., Malek, M.H. (2005a) 'The effects of interelectrode distance on electromyographic amplitude and mean power frequency during isokinetic and isometric muscle actions of the biceps brachii', *Journal of Electromyography and Kinesiology*, 15:482–495.
3. Beck, T.W., Housh, T.J., Johnson, G.O., Weir, J.P., Cramer, J.T., Coburn, J.W., Malek, M.H. (2005b) 'Comparison of Fourier and wavelet transform procedures for examining the mechanomyographic and electromyographic frequency domain responses during fatiguing isokinetic muscle actions of the biceps brachii', *Journal of Electromyogaphy and Kinesiology*, 15:190–199.

4. Beck, T.W., Housh, T.J., Cramer, J.T., Weir, J.P. (in press) 'The effects of electrode placement and innervation zone location on the electromyographic amplitude and mean power frequency versus isometric torque relationships for the vastus lateralis muscle', *Journal of Electromyography and Kinesiology*.
5. Bonato, P., Gagliati, G., Knaflitz, M. (1996) 'Analysis of myoelectric signals recorded during dynamic contractions', *IEEE Engineering in Medicine and Biology*, 15:102–111.
6. Cohen, L. (1995) *Time-frequency analysis*. Upper Saddle River, NJ: Prentice Hall.
7. Constable, R., Thornhill, R.J. (1993) 'Using the discrete wavelet transform for time-frequency analysis of the surface EMG signal', *Biomedical Sciences Instrumentation*, 29:121–127.
8. Constable, R., Thornhill, R.J., Carpenter, D.R. (1994) 'Time-frequency analysis of surface EMG during maximum height jumps under altered-G conditions', *Biomedical Sciences Instrumentation*, 30:69–74.
9. Cooley, J.W., Tukey, J.W. (1965) 'An algorithm for the machine calculation of complex Fourier Series', *Mathematics Computation*, 19:297–301.
10. De Luca, C.J. (1979) 'Physiology and mathematics of myoelectric signals', *IEEE Transactions on Biomedical Engineering*, 26:313–325.
11. De Luca, C.J. (1997) 'The use of surface electromyography in biomechanics', *Journal of Applied Biomechanics*, 13:135–163.
12. De Vries, H.A. (1966) 'Quantitative electromyographic investigation of the spasm theory of muscle pain', *American Journal of Physical Medicine*, 45:119–134.
13. De Vries, H.A. (1968) '"Efficiency of electrical activity" as a physiological measure of the functional state of muscle tissue', *Archives of Physical Medicine and Rehabilitation*, 47:10–22.
14. Farina, D., Merletti, R., Enoka, R.M. (2004) 'The extraction of neural strategies from the surface EMG', *Journal of Applied Physiology*, 96:1486–1495.
15. Farina, D., Merletti, R., Nazzaro, M., Caruso, I. (2001) 'Effect of joint angle on EMG variables in leg and thigh muscles', *IEEE Engineering In Medicine and Biology Magazine*, 20:62–71.
16. Gerleman, D.G., Cook, T.M. Instrumentation (1992). In: *Selected Topics in Surface Electromyography For Use In The Occupational Setting: Expert Perspectives*. US Department of Health and Human Services. Center for Disease Control, pp. 43–68.
17. Hermens, H.J., Freriks, B., Merletti, R., Stegeman, D., Blok, J., Rau, G., Disselhorst-Klug, C., Hägg, G. (1999) *SENIAM European Recommendations for Surface ElectroMyoGraphy: Results of the SENIAM project*. Enschede, The Netherlands: Roessingh Research and Development.
18. Karlsson, J.S., Gerdle, B., Akay, M. (2001) 'Analyzing surface myoelectric signals recorded during isokinetic contractions', *IEEE Engineering In Medicine and Biology*, 20:97–105.
19. Karlsson, S., Yu, J., Akay, M. (2000) 'Time-frequency analysis of myoelectric signals during dynamic contractions: a comparative study', *IEEE Transactions On Biomedical Engineering*, 47:228–238.
20. Keenan, K.G., Farina, D., Maluf, K.S., Merletti, R., Enoka, R.M. (2005) 'Influence of amplitude cancellation on the simulated electromyogram', *Journal of Applied Physiology*, 98:120–131.
21. Ladegaard, J. (2002) 'Story of electromyography equipment', *Muscle & Nerve*, 11:S128–S133.
22. Lindström, L., Magnusson, R. (1977) 'Interpretation of myoelectric power spectra: A model and its applications', *Proceedings of the IEEE*, 65:653–662.
23. Lindström, L., Magnusson, R., Petersén, I. (1970) 'Muscular fatigue and action potential conduction velocity changes studied with frequency analysis of EMG signals', *Electromyography*, 4:341–356.
24. Lynn, P.A., Bettles, N.D., Hughes, A.D., Johnson, S.W. (1978) 'Influence of electrode geometry on bipolar recordings of the surface electromyogram', *Medical & Biological Engineering & Computing*, 16:651–660.
25. MacIsaac, D., Parker, P.A., Scott, R.N. (2001) 'The short-time Fourier transform and muscle fatigue assessment in dynamic contractions', *Journal of Electromyography and Kinesiology*, 11:439–449.
26. Malek, M.H., Housh, T.J., Coburn, J.W., Weir, J.P., Schmidt, R.J., Beck, T.W. (2006) 'The effects of interelectrode distance on electromyographic amplitude and mean power frequency during incremental cycle ergometry', *Journal of Neuroscience Methods*, 151:139–147.

27. Mallat, S. (1998) *A Wavelet Tour of Signal Processing*. San Diego, CA: Academic Press.
28. Rainoldi, A., Melchiorri, G., Caruso, I. (2004) 'A method for positioning electrodes during surface EMG recordings in lower limb muscles', *Journal of Neuroscience Methods*, 134:37–43.
29. Rainoldi, A., Nazzaro, M., Merletti, R., Farina, D., Caruso, I., Gaudenti, S. (2000) 'Geometrical factors in surface EMG of the vastus medialis and lateralis muscles', *Journal of Electromyography and Kinesiology*, 10:327–336.
30. Smith, S.W. (1997) *The scientist and engineer's guide to digital signal processing*. San Diego: California Technical Publishing.
31. Soderberg, G.L. Recording Techniques (1992). In: *Selected Topics in Surface Electromyography For Use In The Occupational Setting: Expert Perspectives*. U.S. Department of Health and Human Services. Center for Disease Control, pp. 23–41.
32. Soderberg, G.L., Knutson, L.M. (2000) 'A guide for use and interpretation of kinesiologic and electromyographic data', *Physical Therapy*, 80:485–498.
33. Tam, H., Webster, J.G. (1977) 'Minimizing electrode motion artifact by skin abrasion', *IEEE Transactions on Biomedical Engineering*, 24:134–139.
34. Tucker, K.J., Türker, K.S. (2005) 'A new method to estimate signal cancellation in the human maximal M-wave', *Journal of Neuroscience Methods*, 149:31–41.
35. Von Tscharner, V. (2000) 'Intensity analysis in time-frequency space of surface myoelectric signals by wavelets of specified resolution', *Journal of Electromyography and Kinesiology*, 10:433–445.
36. Von Tscharner, V. (2002) 'Time-frequency and principal-component methods for the analysis of EMGs recorded during mildly fatiguing exercise on a cycle ergometer', *Journal of Electromyography and Kinesiology*, 12:479–492.
37. Von Tscharner, V., Goepfert, B. (2003) 'Gender dependent EMGs of runners resolved by time/frequency and principal pattern analysis', *Journal of Electromyography and Kinesiology*, 13:253–272.
38. Von Tscharner, V., Goepfert, B. (2006) 'Estimation of the interplay between groups of fast and slow muscle fibres of the tibialis anterior and gastrocnemius muscle while running', *Journal of Electromyography and Kinesiology*, 16:188–197.
39. Winter, D.A. (1990) *Biomechanics and motor control of human movement*, 2nd edn. New York, NY: John Wiley & Sons, Inc.
40. Zipp, P. (1982) 'Recommendations for the standardization of lead positions in surface electromyography', *European Journal of Applied Physiology*, 50:41–54.

Section IV

Engineering technology and equipment design

16

Biomechanical aspects of footwear

Ewald M. Hennig
University of Duisburg-Essen, Essen

With the onset of the running boom in the USA in the 1970s many runners experienced overuse injuries and became more and more interested in the quality of their shoes. The magazine *Runners World* started the first surveys of running shoes and soon realized that biomechanical methods are available to quantify the properties of running shoes. Running shoes are without a question the footwear that has been explored the most by scientists in the field of biomechanics. Following the running shoe research peak other products became the focus of interest. In particular, many shoe studies were performed in the field of basketball and other indoor sports. Nowadays, the research extends to all kinds of athletic footwear, ranging from track and field disciplines to tennis, inline skating and footwear for winter sports. Due to differences in foot anatomy (Krauss, 2006) and specific injury patterns for women, gender specific footwear has become an important topic in athletic shoe research. Especially, the much higher incidence of knee injuries in certain sport disciplines requires additional research and efforts by the shoe industry. For example, women have a higher degree of rearfoot motion in running and need more stable shoes that provide an adequate pronation control (Hennig, 2001). Only recently, soccer boots have received a lot of attention and were explored by various research groups. Other than in running shoes, soccer boots have additional tasks to perform. These shoes are used for kicking, they should provide sufficient traction for rapid cutting manoeuvres and assist the players in rapid acceleration and stopping movements. Football shoe design, for example, can influence maximum kicking speed (Hennig and Zulbeck, 1999). Biomechanical soccer shoe research is still in its infancy and many different aspects will have to be considered. Especially, the often conflicting demands of injury prevention and high performance properties remain to be solved. On the following pages measuring and test methods for athletic footwear as well as research results will be presented. This will demonstrate how important biomechanics has become in providing the necessary knowledge for the design of functional footwear. The discussion will be limited to various aspects of running shoe research. This field has received the most attention in the literature and many of the concepts and measuring principles can be transferred to other areas of athletic footwear research.

Running shoe biomechanics

Much effort has been spent in the past to investigate the influence of running shoe construction on the human locomotor system during running. A good summary of research on running shoes during the early years (1970–1990) is given by Nigg (1986) and Cavanagh (1990). From the beginning of biomechanical research on footwear in running overpronation of the foot and high impact shocks have been suspected to be related to overuse injuries in runners. Especially, the limitation of excessive rearfoot motion is believed to reduce the likelihood of injuries.

The etiology of overuse injuries in running

Injury surveys on runners have concluded that the occurrence of overuse injuries is the most common reason for runners to quit running and to continue with other kinds of physical exercise (Koplan *et al.*, 1995). For 2002 distance runners Taunton *et al.* (2002) identified patellofemoral pain syndrome as the most common injury, followed by iliotibial band friction syndrome, plantar fasciitis, meniscal injuries of the knee, and tibial stress syndrome. With 42 per cent, the knee was the most frequently injured anatomical location for these runners. Between genders, women had higher relative incidences of gluteus medius injuries (70 per cent), PFP syndrome (58 per cent), iliotibial band friction syndrome (58 per cent), and tibial stress fractures (55 per cent). The male runners had more frequently meniscal injuries (75 per cent), achilles tendinitis (63 per cent), patellar tendinitis (62 per cent), and plantar fasciitis (58 per cent).

Shock absorption

The high frequency of impacts on the human body during distance running has always been suspected to be important in the etiology of overuse injuries. In animal experiments repetitive impulse loading of joints was found to cause degenerative knee joint changes, as they are also observed in osteoarthritis (Radin *et al.*, 1972; Dekel and Weissman, 1978; Radin *et al.*, 1982). Recently, Milner *et al.* (2006) reported in a thoroughly conducted study a higher incidence of tibial stress fractures for women distance runners, exhibiting a running style that showed increased tibial shock and vertical force loading rates. The authors found that the magnitude of tibial shock predicted injury group membership successfully in 70 per cent of the cases. From space flight and research for the prevention of osteoporosis it has been found that shocks to the body are essential to counteract bone mineral loss and to maintain bone integrity. Astronauts exposed to weightlessness for extended periods experience significant decreases in bone mineral density. Heinonen *et al.* (1996) studied the effects of high-impact loading on determinants of osteoporotic fractures. They found that high-impact exercises with a rapidly rising force profile improved skeletal integrity and muscular performance in the studied premenopausal women. The authors suggest that this type of exercise may help decrease the risk of osteoporotic fractures in later life. In summary, shock to the human body can be positive or of disadvantage as a predisposing factor for overuse injuries in runners. The question still remains to be solved what frequency and intensity of impact loading is positive or negative for the body.

Overpronation

Rearfoot motion during barefoot running has been shown to be lower as compared to shod running (Nigg and Morlock, 1986). Due to the shoe construction the rearfoot is higher above the ground when the lateral heel makes first contact with the ground during running. Thus, the shoe increases the lever arm length and creates a larger turning moment for the calcaneus to rotate. Reducing this lever arm, e.g. by a high compression of the midsole material in the lateral heel area at ground contact, is an excellent method to lower the amount of rearfoot motion towards values that are present in natural barefoot running. The compression of the lateral midsole material also improves shock absorption. Among many other authors Messier *et al.* (1988), Hreljac *et al.* (2000), and Bennett *et al.* (2001) showed that excessive rearfoot motion and increased shock can be identified as risk factors for injuries in runners. Therefore, pronation control and shock absorption are still considered to be the main concepts for injury prevention in running shoe design. Although most researchers have based their studies on these assumptions and have performed shoe testing along these criteria, a few authors deny a relationship between running injuries and overpronation or shock absorption. Based on the review of literature from 25 years, Nigg (2001) comes to the conclusion that overpronation and shock are not related to overuse injuries in runners. However, the contradicting results in the literature are often based on biased studies.

Retrospective and prospective study designs

When it comes to relate biomechanical variables to the etiology of running injuries, the following study concept is the most frequently found in the literature: a comparison of currently or previously injured runners is made towards control groups that are not or never had been injured. Such study designs will have a number of problems, sometimes making the results meaningless. It has to be questioned whether movement and kinetic parameters from acutely injured runners are representing their normal running style. Movement patterns in an injured group are likely to be dominated by pain avoiding strategies and may not represent their natural running style. Even in the more recent literature comparisons of injured vs. noninjured runners are still found (Duffey *et al.*, 2000). Most frequently study designs were chosen that investigate currently pain free runners against runners that never had such overuse injuries. Even these studies are biased for a number of reasons. Runners that experience problems will stop running altogether, reduce training effort, mileage, and they may go through a footwear selection process to reduce the pain. These runners are much more aware of their feet and running style. If these runners do not have a current pain problem they are still the ‘survivors’ that went through a complex adaptation process to reduce their pain.

The following example will highlight the problems with retrospective studies: At the finish line of a marathon select all the runners that are severe overpronators and compare them to a group of runners with normal rearfoot motion. You ask all of them about the incidence of injuries and pain during their training in the last year. Let us assume that from the questionnaires no difference in either the frequency or the severity of pain is present between the groups. Does this mean overpronation is not a predisposing factor for the occurrence of overuse injuries and pain? I suggest alternative explanations. Those overpronators that really suffer from severe pain due to their excessive rearfoot motion never make it to the marathon. As shown by Koplan *et al.* (1995) overuse injuries is the most common reason for quitting running. Those runners that are still able to prepare for and run a marathon may have the right genes, may have altered their running style, footwear, and training regime.

It should be pointed out that retrospective studies, as they have been published almost exclusively until the 1990s, have provided a lot of information and were useful in the interpretation of injury patterns, as well as in providing guidelines for the construction of athletic footwear. In a comparison of a retrospective and prospective study on tibial stress fractures, Hamill and Davis (2006) came to identical conclusions from both studies. Similar parameters from the two studies could differentiate between the injured groups and the control groups. Peak tibial shocks and higher loading rates were related to a greater likelihood of suffering a TSF from running. The authors used the prospective study as a validation for the retrospective study and point out that the matching results of both study approaches may only apply to their specific research question of TSF injuries. Future prospective studies on the etiology of injuries will give us more confidence in the information and concepts that had been gained in the past by retrospective studies. A good example is the study by Willems *et al.* (2007). In a prospective study on 400 physical education students it was found that increased pronation excursion did result in more frequent exercise-related lower leg pain, a common chronic sports injury. According to Macera (1992) weekly distance is the strongest predictor of future injuries in runners. Previous injuries and training habits, especially a rapid build up in mileage, are other important risk factors (Van Mechelen, 1992).

Running shoe testing

For the last 18 years our laboratory has performed ten different running shoe tests for a government supported German consumer product testing agency. In these tests approximately 200 different running shoe models have been investigated biomechanically and were evaluated in field tests. This chapter will describe methods of running shoe testing and present some of our results. Pronation control and shock absorption are considered to be the main concepts for injury prevention in running shoe design even though there is still a lack of conclusive epidemiological and clinical proof. All major running shoe manufacturers follow these concepts. Before going into more detail about the running shoe test procedures a description of the foot loading behaviour of the foot inside of running shoes is presented here. A good picture of the interaction of the foot with the ground can be gained by pressure distribution measurements. A typical development of plantar pressures in a running shoe is shown in Figure 16.1.

Pressures were measured, using eight discrete piezoceramic transducers under well defined anatomical foot structures (Hennig, 1988). Figure 16.1 shows a time sequence of averaged pressure patterns from 12 subjects running in the same shoes. Shortly after ground contact high pressures occur under the rearfoot. An initial higher lateral heel loading is typical for running in almost all shoe models. The higher lateral pressures are caused by the ground contact in a supinated foot position. Lateral mid- and forefoot loading begins earlier than 30 ms after ground contact. This happens before the occurrence of the initial impact peak of the vertical ground reaction force. At 100 ms after heel strike surprisingly high lateral midfoot pressures can be observed. Because of high ground reaction forces the longitudinal arch of the foot collapses under the higher load, thus causing high midfoot pressures. From 100 ms after heel strike until lift off, medial forefoot loading becomes increasingly important. The first metatarsus and hallux are the main load bearing structures during the push-off phase.

Different testing methods have to be used to judge the performance of running shoes. Using technical tests, impacter devices may be used to quantify midsole material properties, thus providing information about production tolerances. Biomechanical measurements are essential in determining the influence of footwear construction on kinematic and kinetic

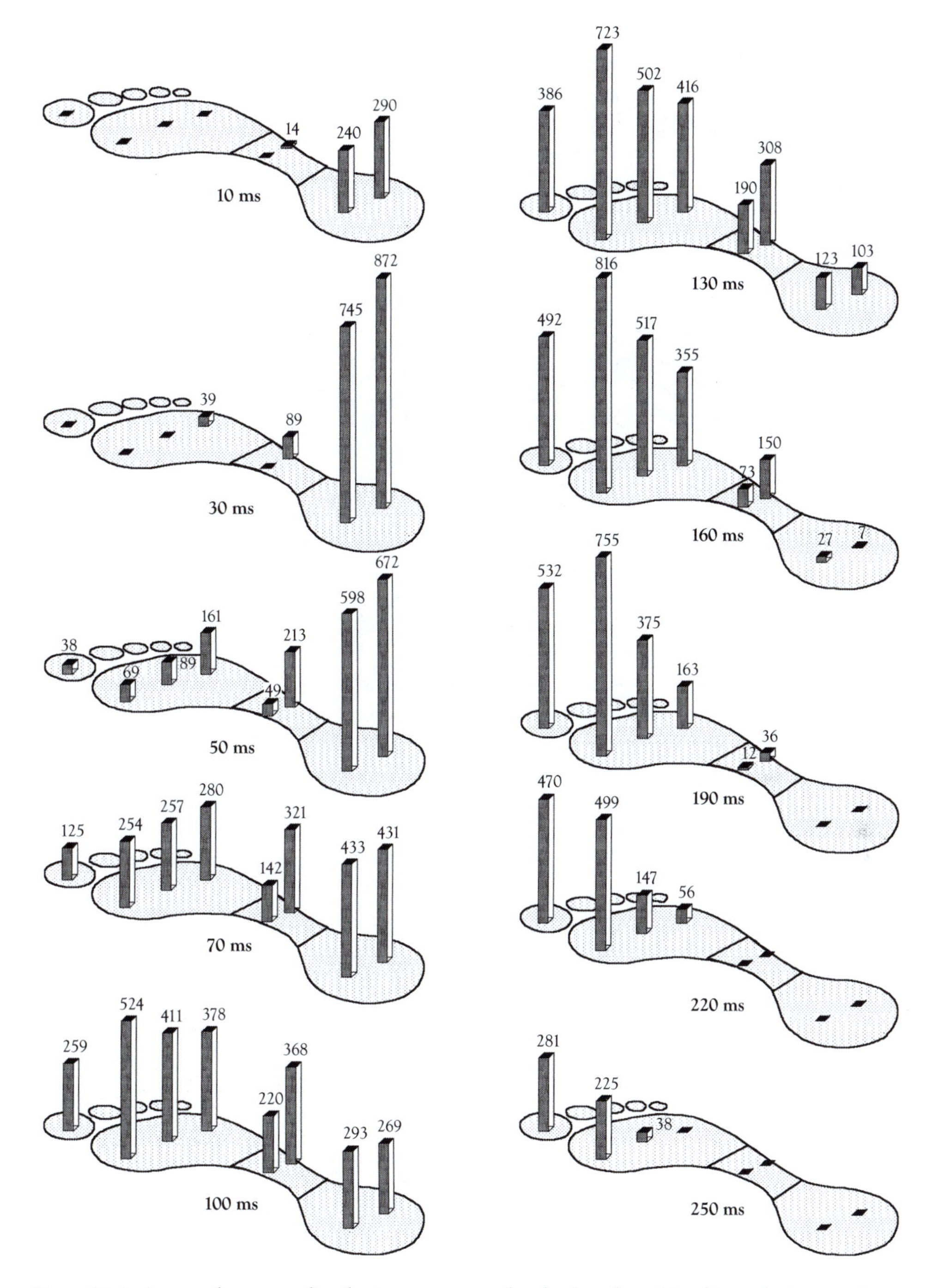

Figure 16.1 Averaged pressure distribution pattern under the foot from 12 subjects during running at 3.3 m/s in the same footwear model (values in kPa).

variables of human performance. Field tests with many subjects are necessary to evaluate shoe fit and comfort during running on varying terrain. Only the combination of a technical test, biomechanical evaluation and subjective ratings by the athletes provides a complete picture for the quality of athletic shoes.

Material testing

Impacter devices are widely distributed in the shoe industry to measure midsole properties of running shoes. Typically, a mass of 8.5 kg falls from a height of 5 cm onto the heel area of running shoes. The mass of 8.5 kg will mimic the effective mass of a runner's leg at initial foot contact during running. The 5 cm falling height will result in an impact velocity of 1 m/s that had been identified as a typical touch down velocity of the foot in running (Cavanagh and Lafortune, 1980). Hennig *et al.* (1993) compared the impacter data from 19 different running shoes to the shocks that were measured on 27 subjects when they were running in these 19 shoes. Peak accelerations of the impacter tests were compared to the peak tibial accelerations during running. The comparison of the two variables resulted in a determination coefficient of only 7 per cent. Impacter peak g values from a more recent study on 13 different shoe models are summarized in Table 16.1. In this table the peak tibial accelerations (PACC) as they were determined from 22 subjects, running in the 13 shoe models, are also included. The determination coefficient between the impacter scores and PACC of our recent study is very low again (10 per cent). Although the impacter scores demonstrate substantial differences in shoe midsole stiffness and elasticity between the shoes, the material properties had only a minor influence on the shock attenuation behaviour of shoes. The mechanical impacter seems to produce an unrealistic impact simulation of the initial foot to ground contact during running. Therefore, shock attenuation is substantially influenced by construction features of the shoe and is probably also influenced by locomotor adaptation of the subjects to the footwear (Hennig *et al.*, 1996). In summary, impacter tests are not able to mimic realistically the impact in running. Biomechanical measurements on the human body have to be performed to evaluate and judge the shock attenuation properties of footwear.

However, mechanical impacter devices are useful in determining the homogeneity of materials and shoe constructions between different pairs of shoes of the same shoe type or even within shoe pairs (left-right comparison). Mechanical impacters are well suited to determine production tolerances.

Table 16.1 Biomechanical variables from running shoe testing with 22 subjects in 13 different shoe models at a running speed of 3.3 m/s

Variable	A	B	C	D	E	F	G	H	I	J	K	L	M
Impacter [g]	9.8	10.3	10.8	10.2	11.5	10.2	10.3	9.7	9.2	9.0	10.2	11.6	10.5
Pron (d)	7.0	7.6	6.3	7.4	7.2	8.2	5.3	10.1	7.8	7.9	6.8	7.7	7.5
DPGN (d/s)	663	646	574	599	681	702	575	735	657	601	585	686	598
PACC (g)	6.0	6.0	7.1	7.0	6.1	6.3	7.0	7.0	5.7	6.0	8.0	7.1	6.3
DPVF (bw/s)	105	109	129	119	110	105	126	125	107	97	137	120	113
PP-Heel (kPa)	785	782	824	782	822	844	894	853	859	722	946	788	845
PP-Midf (kPa)	290	281	256	240	236	285	219	308	296	280	354	199	290
PP-Foref (kPa)	491	494	531	478	510	573	557	515	548	518	502	524	533

Biomechanical testing

As mentioned before, rearfoot motion control and shock absorption are the two main criteria for a biomechanical evaluation of running shoes. Additionally, in-shoe pressure distribution information is useful in providing more insight in the effect of footwear on foot loading and foot motion. The plantar pressure distribution information is like a finger imprint. It provides detailed information about individual features of shoes.

Measurement of rearfoot motion

Traditionally, high-speed films are used to determine rearfoot motion at ground contact on a treadmill or during overground running. Furthermore, high-speed video and opto-electric techniques with the possibilities of automation and digital processing are used in quantifying rearfoot motion. Typically, analyses are performed by using two markers on the heel counter of the shoe and two markers on the leg below the knee in the direction of the Achilles tendon (Nigg, 1986). Film analyses are highly technical and time-consuming. Alternatively, electronic goniometers can be used to measure rearfoot motion. The main advantage of electronic goniometers is that the results are immediately available. Electronic goniometers are typically fixed onto the heel counter of the shoe. A lightweight half-circular metal construction with a potentiometer is fixed at the heel counter with its axis of rotation at the approximate height of the subtalar joint. The movable part of the goniometer is fixed at the lower leg in parallel to the Achilles tendon orientation. The advantage of the goniometric method is a fast and uncomplicated data acquisition that can be performed simultaneously with the measurements of ground reaction forces, tibial accelerations and pressure distributions. Table 16.1 provides information on maximum pronation (PRON) in degrees and peak pronation velocities (DPGN) in degrees per second from 13 different running shoe models that were measured in 2007 in a recent study of our laboratory for a government supported consumer testing agency in Germany. The values were collected from 22 subjects who ran in the 13 different shoe conditions across a force platform. In each shoe five repetitive trials were performed from all subjects at a running speed of 3.3 m/s. Mean values from the five trials were calculated for all parameters and used for further statistical analysis. Data collection of all measuring values was done at a rate of 1 kHz.

Measurement of shock absorption

Using force platforms, several studies provided information about the magnitude and variability of the GRF and the path of the centre of pressure for running at different speeds (Cavanagh and Lafortune, 1980; Bates *et al.*, 1983; Nigg *et al.*, 1987). In a review of ground reaction force (GRF) research in running, Miller (1990) concluded that GRF is not a particularly sensitive measure because its signals reflect the acceleration of the body's centre of mass (COM). Even the first peak of the ground reaction force from heel strike runners is not closely related to the shock absorbing properties of shoes. However, a close relationship exists between the vertical GRF rising rate at initial foot contact and the tibial acceleration, as measured by light weight accelerometers at the lower leg of runners (Hennig *et al.*, 1993). The authors studied the influence of footwear on the peak tibial acceleration of 27 runners in 19 different running shoe models. From force plate measurements they calculated the maximum force rate (DPVF) as the highest differential quotient of adjoining vertical GRF values divided by the time resolution of 1 ms. When comparing this maximum force rate with the peak tibial acceleration values in the different shoe conditions, a simple regression analysis

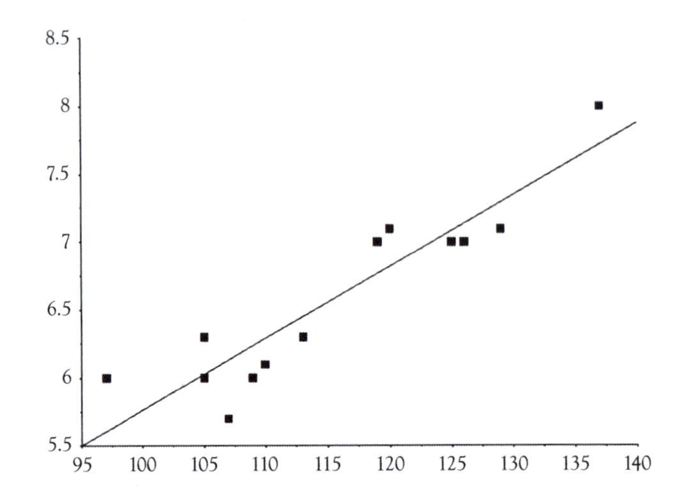

Figure 16.2 Regression between peak tibial acceleration (PACC in g) and vertical ground reaction force rate (DPVF in bw/s) for 13 different running shoe models.

showed a determination coefficient of 95 per cent between these two variables. In the same study Hennig *et al.* also calculated a median power frequency from the Fourier spectrum of the vertical ground reaction force signal. The median power frequency of the vertical force (MPFZ) signal was calculated from a 1024 point FFT power spectrum analysis. For the calculation of MPFZ a low frequency cut-off of 10 Hz was chosen, because only the initial ground reaction force peak was of interest. The median power frequency also showed a very high correlation of $r = +0.94$ with the peak tibial acceleration values from the different shoe models. In conclusion, the following three variables are very well suited to differentiate between the shock absorption properties of footwear in running: peak tibial acceleration (PACC), maximum force rate of the vertical GRF signal (DPVF), and the median power frequency of the vertical GRF. In Figure 16.2 the regression between peak tibial accelerations (PACC) in g and the force rates in bw/s are summarized for the 13 different shoes of our recent study (Table 16.1).

Depending on the shoe model tibial shock (PACC) varies from 5.7 g to 8.0 g. Thus, the best shoe model I provides approximately 30 per cent better shock absorption as compared to shoe model K. The values of the variables PACC and DPVF of the current study (Table 16.1) correlate with a high value of $r = +0.92$.

Plantar in-shoe pressures measurements

Miller (1990) concluded from a literature review that the ground reaction force is not a particularly sensitive measure to detect differences between athletic footwear. Miller also stated that the full potential of GRF analyses in clinical and sport applications will only be achieved when simultaneously miniaturized in-shoe force transducers are applied. Force plate measurements only provide information about the total ground reaction force acting on the body. The loading of individual foot structures which may play an important role in the occurrence of overuse injuries can only be determined using pressure distribution devices. During recent years, the availability of inexpensive force transducers and modern data acquisition systems have made the construction of various pressure distribution measuring systems possible. Transducer technologies for pressure distribution devices are based on capacitive, piezoelectric and resistive principles. Because footwear modifies the foot to ground interaction considerably, in-shoe pressure measurements are of special interest. During ground contact a slight movement of the bones with respect to the plantar skin may be expected. Because plantar pressure measuring systems are placed between the skin and the shoe, sliding in the shoe and/or movement of the transducer with respect to bony landmarks may occur. Nevertheless, bearing this limitation in mind, in-shoe pressure measurements provide a valuable insight into the influence of footwear on foot function. Matrix sensor insoles can easily be placed inside shoes. Because most feet exhibit individual anatomical variations, the exact placement of the sensors under the anatomical sites of interest are not known. Furthermore, depending upon shoe construction peculiarities, the relative positioning of the insole matrix under the foot may vary. To overcome these problems, pressure insoles with a high number of small transducers are necessary. However, pressure insoles with a high density of accurate sensors are expensive. They generate many data but have restricted time resolution. The use of a limited number of discrete rigid transducers offers an alternative for gathering in-shoe pressure information. They are typically fixed with adhesive tape under anatomical foot structures that are manually palpated. The major advantage of this technique is an exact positioning of the sensors under the foot structures of interest, independent of individual foot shape. It also guarantees that the sensor locations remain independent from footwear construction peculiarities. Their major disadvantage is an incomplete mapping of the foot to shoe interaction as pressure knowledge is limited to the chosen sensor locations. Figure 16.3 represents peak pressures, averaged from 20 subjects, who ran in two different running shoes at a speed of 3.3 m/s.

These pressures were recorded with eight individual piezoceramic transducers, fastened under specific anatomical locations of the foot. Large differences in pressure magnitudes but also in the distribution pattern can be observed between the shoes. Especially, the pressure

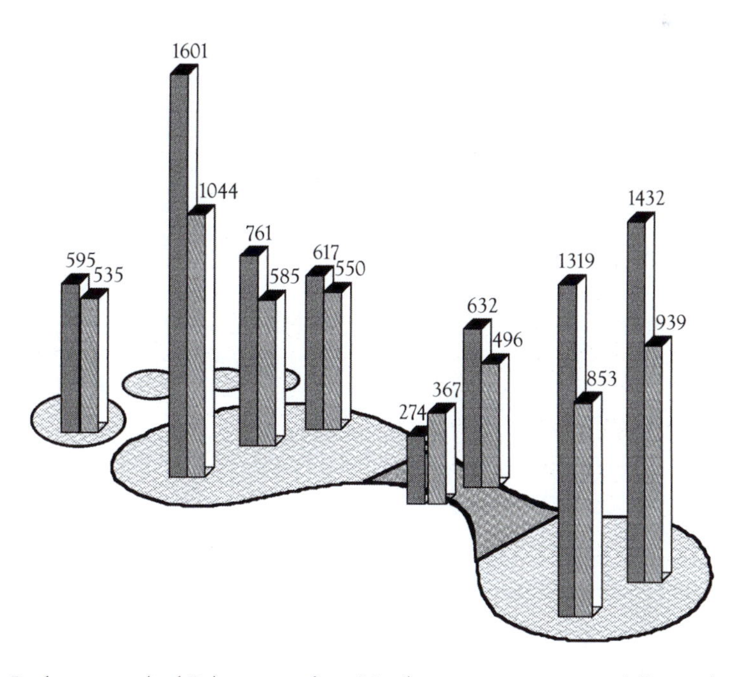

Figure 16.3 Peak pressure (in kPa) patterns from 20 subjects, running in two different shoes at a speed of 3.3 m/s.

differences underneath the first metatarsal head are remarkable. Table 16.1 summarizes the peak pressures under the foot for the 13 different shoe models of our recent study. Data for heel pressures, lateral midfoot pressures under the longitudinal arch, and forefoot pressures are presented. The forefoot pressure values are averages from 4 sensor locations under the 1st, 3rd, and 5th metatarsal heads and the hallux. Especially under the midfoot area, large differences can be observed between shoe model G (219 kPa) and model K (354 kPa). Furthermore, larger differences are present between the shoe models in the heel as compared to the forefoot areas.

Subjective shoe evaluation

Some shoe features are difficult or impossible to measure. To judge fit of the shoe, comfort and foot climate are a few of the criteria that will have to be answered by the runners. It is also important whether the runners will experience pressure sores or blisters during running. In our running shoe tests we typically have between 20 and 30 runners who all run a distance of 10 km in each of the shoe constructions on different days. Assignment of the shoes is randomized and the runners have to fill out a questionnaire at the end of the run. Using a 15-point perception scale the runners will answer questions on the following items. How is your overall liking of the shoe and how do you rate shock absorption, pronation control, fit, handling of the shoe, traction properties, and foot climate. A separate section is included in the questionnaire where subjects report on the occurrence of pressure sores and blisters and at which location of the shoe they occurred. At the end of the field test we have ratings and comments from all runners for each of the shoes. Depending on the number of tested shoes, the field test period will have a duration of six to eight weeks. A major observation that we have made in all of our running shoe tests within the last 18 years is the following. Subjects are not well able to distinguish between the various items. If they like a shoe, all of the properties of the shoe are good and if they do not like a shoe, all shoe features are rated negatively. For a recent study Table 16.2 shows the correlations between the subjective ratings of overall liking, shock absorption, pronation control, shoe fit, and foot climate. All variables correlate with a value above r = +0.9 with the overall liking of the shoes by our subjects.

This observation emphasizes the importance of biomechanical shoe testing. Subjects are not able to differentiate between very different and important shoe features. Their response in perception rating is dominated by the overall liking of the different shoe models. Nevertheless, field tests and subjective evaluations are necessary to find out about deterioration of shoe properties with use, pressure sore and blister problems and the overall liking of the different shoes. In the end, what is the use of a running shoe test, if the biomechanical data are excellent but runners do not like the shoe because of a lack of fit and comfort,

Table 16.2 Correlation matrix of subjective perception ratings on Overall Shoe Liking, Shock Absorption, Fit, and Foot Climate

	O-Liking	*Shock-A*	*Pronation-C*	*Fit*	*Climate*
O-Liking	1				
Shock-A	0.96	1			
Pronation-C	0.93	0.87	1		
Fit	0.96	0.88	0.94	1	
Climate	0.94	0.84	0.87	0.96	1

Table 16.3 Mean change and range of changes in biomechanical properties after 220 km of running with 19 different running shoe models

	Pronation	*Pronation Velocity*	*Peak Tibial Acceleration*	*Peak Pressure (Heel)*	*Peak Pressure (Midfoot)*	*Peak Pressure (Forefoot)*
Mean change in %	+21.5	+4.9	+10.1	+8.5	+7.6	+6.6
Range of changes (%)	–4.3 to + 68.4	–22.9 to +25.6	–5.8 to +27.7	0 to +20.5	–11.4 to 29.2	–8.2 to +14.3

pressure sores, or other features. In conclusion, only the combination of biomechanical data, subjective evaluations from a field study, and material test results provide the necessary information for a good running shoe survey.

Evaluation of used running shoes

During running a deterioration of footwear material occurs when shoes are exposed to many repetitive impacts with high mechanical loads. The protective functions of footwear, such as the restriction of excessive rearfoot motion and shock attenuation, is likely due to changes with the number of miles run with the shoe. A study by Hennig and Milani (1995) identified the changes of footwear properties after wearing different running shoes for a distance of 220 km. In Table 16.3 the changes in biomechanical parameters of 19 used against 19 new shoe models are shown.

Mean changes as well as the range of changes are summarized in this table. The biggest change occurs for rearfoot motion control. In the used shoes maximum pronation across all shoes increased by 21.5 per cent. Softening of the heel counter as well as midsole material aging are likely causes for the deterioration of rearfoot motion control by the shoes. Shock to the body increases after 220 km of use by 10.1 per cent and as much as 27.7 per cent for a specific shoe model. Overall, there are considerable differences in the ageing behaviour of different shoe models. Therefore, the deterioration of shoe properties is an important factor to be considered in the judging of running shoe quality.

Conclusions

With the running boom in the USA in the 1970s a high incidence of running injuries was observed. This was the beginning of biomechanical research for the evaluation and design of athletic footwear, especially running shoes. Pronation control and shock absorption are considered to be the main concepts for injury prevention in running shoe design, even though there is still a lack of conclusive epidemiological and clinical proof. In the last 30 years much biomechanical research has been performed to understand the role of athletic footwear on injury prevention and sports performance. Running and running shoes had been the focus of this research. No other footwear has received a similar attention to date. To evaluate the effect of footwear on the body, biomechanical measurements have to be performed at the body of runners. Material tests are of very limited value for the prediction of the influence of shoes on the human body. Cinematographic recordings, EMG equipment, force platforms, accelerometers and pressure distribution devices are valuable measuring tools for the evaluation of athletic footwear. For the judgment of running shoe quality only the combination of biomechanical data, subjective evaluations from a field test, and material test results provide the necessary information. Even today, there are remarkable differences in the performance of high quality running shoes. Not only the performance of new shoes but also the deterioration of shoe properties in used shoes has to be considered in the judging of running shoe quality. During recent years more research has been performed to design adequate footwear for women and to explore the possibilities of footwear for other sports, especially soccer. Other than in running shoes, soccer boots have additional tasks to perform. The shoes are used for kicking, they should provide sufficient traction for rapid cutting manoeuvres and assist the players in rapid acceleration and stopping movements. It has been shown that soccer boots can improve kicking velocity and sprinting speed. It is also likely that shoes can be modified to provide a better kicking accuracy. This kind of research is still in its infancy but will provide a lot of new information for the future design of better footwear.

References

1. Bates, B. T., Osternig, L. R., Sawhill, J. A. and James, S. L. (1983) 'An assessment of subject variability, subject-shoe interaction and the evaluation of running shoes using ground reaction force data', *Journal of Biomechanics*, 16: 181–191.
2. Bennett, J. E., Reinking, M. F., Pluemer, B., Pentel, A., Seaton, M. and Killian, C. (2001) 'Factors contributing to the development of medial tibial stress syndrome in high school runners', *Journal of Orthopaedic & Sports Physical Therapy*, 31: 504–510.
3. Cavanagh, P. R. (1990) *Biomechanics of distance running*, Champaign, IL, Human Kinetics Books.
4. Cavanagh, P. R. and Lafortune, M. A. (1980) 'Ground reaction forces in distance running', *Journal of Biomechanics*, 13: 397–406.
5. Dekel, S. and Weissman, S. L. (1978) 'Joint changes after overuse and peak overloading of rabbit knees in vivo', *Acta Orthopaedica Scandinavica*, 49: 519–528.
6. Duffey, M. J., Martin, D. F., Cannon, D. W., Craven, T. and Messier, S. P. (2000) 'Etiologic factors associated with anterior knee pain in distance runners', *Medicine and Science in Sports & Exercise*, 32: 1825–1832.
7. Hamill, J. and Davis, I. (2006) 'Can we learn more from prospective or retrospective studies?', *Journal of Biomechanics*, 39: 173.
8. Heinonen, A., Kannus, P., Sievanen, H., Oja, P., Pasanen, M., Rinne, M., Uusi-Rasi, K. and Vuori, I. (1996) 'Randomised controlled trial of effect of high-impact exercise on selected risk factors for osteoporotic fractures', *Lancet*, 348: 1343–1347.
9. Hennig, E. (1988) Piezoelectric sensors. In J. G. Webster (ed.) *Encyclopedia of Medical Devices and Instrumentation* (pp. 2310–2319), New York, J. Wiley & Sons.
10. Hennig, E. M., Milani, T. L. and Lafortune, M. A. (1993) 'Use of ground reaction force parameters in predicting peak tibial accelerations in running', *Journal of Applied Biomechanics*, 9: 306–314.
11. Hennig, E. M. and Milani, T. L. (1995) 'Biomechanical profiles of new against used running shoes'. In K. R. Williams (ed.) *Biomechanical profiles of new against used running shoes* (pp. 43–44), Stanford University, American Society of Biomechanics, Palo Alto.
12. Hennig, E. M., Valiant, G. A. and Liu, Q. (1996) 'Biomechanical variables and the perception of cushioning for running in various types of footwear', *Journal of Applied Biomechanics*, 12: 143–150.
13. Hennig, E. M. and Zulbeck, O. (1999) 'The influence of soccer boot construction on ball velocity and shock to the body'. In E. M. Hennig and D. J. Stefanyshin (eds.) *The influence of soccer boot*

construction on ball velocity and shock to the body (pp. 52–53), Canmore / Canada, University of Calgary.

14. Hennig, E. (2001) 'Gender differences for running in athletic footwear'. In E. Hennig and A. Stacoff (eds.) *Gender differences for running in athletic footwear* (pp. 44–45), Zuerich, Dept. of Materials, ETH Zuerich.
15. Hreljac, A., Marshall, R. N. and Hume, P. A. (2000) 'Evaluation of lower extremity overuse injury potential in runners', *Medicine and Science in Sports & Exercise*, 32: 1635–1641.
16. Koplan, J. P., Rothenberg, R. B. and Jones, E. L. (1995) 'The natural history of exercise: a 10-yr follow-up of a cohort of runners', *Medicine and Science in Sports & Exercise*, 27: 1180–1184.
17. Krauss, I. (2006) 'Frauenspezifische Laufschuhkonzeption (Running Footwear for Women)', *Fakultät für Sozial- und Verhaltenswissenschaften*. Tuebingen, Eberhard-Karls-Universität Tübingen.
18. Macera, C. A. (1992) Lower extremity injuries in runners. Advances in prediction', *Sports Medicine*, 13: 50–57.
19. Messier, S. P. and Pittala, K. A. (1988) 'Etiologic factors associated with selected running injuries', *Medicine and Science in Sports & Exercise*, 20: 501–505.
20. Milgrom, C., Giladi, M., Kashtan, H., Simkin, A., Chisin, R., Margulies, J., Steinberg, R., Aharonson, Z. and Stein, M. (1985) A prospective study of the effect of a shock-absorbing orthotic device on the incidence of stress fractures in military recruits', *Foot & Ankle*, 6: 101–104.
21. Miller, D. I. (1990) 'Ground reaction forces in distance running'. In P. R. Cavanagh (ed.) *Biomechanics of distance running* (pp. 203–224), Champaign, Ilinois, Human Kinetics.
22. Milner, C. E., Ferber, R., Pollard, C. D., Hamill, J. and Davis, I. S. (2006) 'Biomechanical factors associated with tibial stress fracture in female runners', *Medicine and Science in Sports and Exercise*, 38: 323–328.
23. Nigg, B. and Morlock, M. (1986) 'The Influence of Lateral Heel Flare of Running Shoes on Pronation and Impact Forces', *Medicine and Science in Sports and Exercise*, 19: 294–302.
24. Nigg, B. M. (1986) *Biomechanics of running shoes*, Champaign, IL, Human Kinetics Publishers.
25. Nigg, B. M. (2001) 'The role of impact forces and foot pronation: a new paradigm', *Clinical Journal of Sport Medicine*, 11: 2–9.
26. Nigg, B. M., Bahlsen, H. A., Luethi, L. M. and Stokes, S. (1987) 'The influence of running velocity and midsole hardness on external impact forces for heel-toe running', *Journal of Biomechanics*, 20: 951–959.
27. Radin, E. L., Orr, R. B., Kelman, J. L., Paul, I. L. and Rose, R. M. (1982) 'Effect of prolonged walking on concrete on the knees of sheep', *Journal of Biomechanics*, 15: 487–492.
28. Radin, E. L., Paul, I. L. and Rose, R. M. (1972) 'Role of mechanical factors in pathogenesis of primary osteoarthritis', *Lancet*, 1: 519–522.
29. Taunton, J. E., Ryan, M. B., Clement, D. B., Mckenzie, D. C., Lloyd-Smith, D. R. and Zumbo, B. D. (2002) 'A retrospective case-control analysis of 2002 running injuries', *British Journal of Sports Medicine*, 36: 95–101.
30. Van Mechelen, W. (1992) 'Running injuries. A review of the epidemiological literature', *Sports Medicine*, 14: 320–335.
31. Willems, T. M., Witvrouw, E., De Cock, A. and De Clercq, D. (2007) 'Gait-related risk factors for exercise-related lower-leg pain during shod running', *Medicine and Science in Sports and Exercise*, 39: 330–339.

17

Biomechanical aspects of the tennis racket

Duane Knudson
California State University, Chico

Improvements in the engineering, materials, and manufacturing of sports equipment have contributed to the enjoyment of sports all over the world. These improvements in sports equipment must be tailored to the rules of sport and the biomechanical factors of the users. The integration of rules, technology, and the biomechanical characteristics of players have been especially important in how changes in the tennis racket have affected the sport of tennis.

Advances in the tennis racket have had a major impact in the sport. The International Tennis Federation (ITF) has had to change the rules of tennis several times to adjust to changes in racket and stringing technologies. The ITF has also hosted several international meetings on tennis science and technology (Haake and Coe, 2000). It is possible that we may be at the point in history where advances in tennis racket technologies have reached a limit because of biomechanical responses or human factors.

This chapter focuses on how the tennis racket design interacts with the player to affect tennis play. A summary of key design changes in tennis rackets that had significant effects on tennis play is presented. Changes in racket technology and design are then reviewed in the context of the interaction of the racket with the tennis player. The chapter concludes with a summary of the research on racket designs purported to improve performance or reduce risk of injury. It should be noted that there have been several reviews and books on tennis rackets and other tennis equipment (Brody, 1987, 2002; Brody *et al.*, 2002; Cross and Lindsey, 2005; Elliott, 1989).

Key changes in racket technology

Tennis racket design and construction has undergone radical changes in the last 100 years. Early tennis rackets were made of laminated strips of ash or beech wood and were not as long (0.686 m) but had greater mass (about 0.4 kg) than today's rackets. The racket head was roughly oval, about 20 cm wide with a hitting area of about 450 cm^2. These rackets were prone to warping, so wooden presses were often clamped onto the racket head to keep the racket face flat (Figure 17.1).

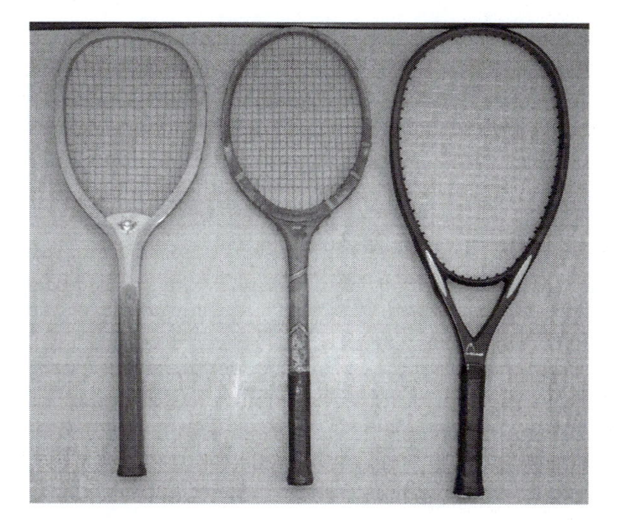

Figure 17.1 A century of change in tennis rackets.

There were several attempts to use other materials in racket construction early in the twentieth century, but it was not until the 1960s that rackets made of steel and aluminium were commercially available to most players. An even more significant change in racket design was the mid-1970s patent of a larger head size racket given to Howard Head. His Prince Pro aluminium racket had over 80 per cent larger hitting area (840 cm^2) than traditional rackets. Larger headed rackets had higher ball rebound speeds, a larger effective hitting zone, and greater resistance to adverse effects of high or low off-centre impacts on the racket face compared to traditional rackets.

The 1980s saw an acceleration in the experimentation with rackets designed with composites of many new materials, including graphite, fiberglass, Kevlar, and boron. Even though the mass of rackets were decreasing, the potential ball rebound speeds were increasing from the larger heads and greater stiffness of the frames.

Two more physical dimension changes have greatly affected racket performance in the past few years. First, some rackets were made longer (up to 0.813 m), so the ITF limited the length to width of rackets to 73.7 by 31.8 cm (Miller, 2006). Second, in the late 1980s wide-body rackets greatly increased frame stiffness and ball rebound speeds. Building rackets with a wider cross-sectional area can double the stiffness of the frame. Engineering rackets with composites of light, stiff materials with a hollow core and a larger head allows the creation of rackets that have significantly higher ball speeds than older wooden rackets.

Improvements in racket design and performance are not, however, without drawbacks. Like other advancements in technology (Tenner, 1996), improvements in tennis racket technology has resulted in unforeseen changes in the nature of the sport. Several articles have documented a variety of changes in performance, match play statistics, and opinions about tennis that are a result of improvements in tennis rackets (e.g. Arthur, 1992; Coe, 2000; Haake *et al.*, 2000; McClusky, 2003; Sheridan, 2006). While the lighter, more powerful rackets make the game easier for new players to learn to play, in the hands of elite players the very high ball speeds has decreased spectator interest in men's professional tennis.

Interactions with the player

Mechanical theory and simulation of the mechanics of impact of the ball with a strung tennis racket have been proposed as useful methods to optimize the design of tennis rackets (Brannigan and Adali, 1981; Brody, 1979). However, the interaction of the racket with the human body has significant effects on the performance and mechanical responses of the racket that are not accounted for in a purely theoretical analysis of a tennis racket-ball impact. The complexity of the interaction of biological and mechanical issues in tennis play is summarized in this section. Predicting how a modification in racket technology will affect the performance and injury risk for a player is a complex biomechanical problem (Kawazoe, 2002). Two specific issues are reviewed to illustrate the complexity of these problems, the effect of hand forces at the grip and the interaction of racket inertial properties with the player's stroke.

Effect of grip forces

The folklore of tennis instruction includes many beliefs that have been tested, and often refuted by biomechanical studies. One such belief from the very origins of the game was that grip forces, specifically the use of a firm grip at impact would increase the speed of ball rebound. The origin of this belief was probably errant, low-speed shots, and the twisting of the racket frame when the ball was struck well off-centre on the racket face of old wooden rackets (450 cm^2).

Early research to examine this issue used measures of ball and racket momentum from cinematography to estimate the 'striking mass' of the racket (Plagenhoef, 1970:87) as a potential measure of this phenomenon. The theory being that grip forces could create enough coupling between the racket and the hand to effectively increase the mass of the racket through forces from the hand and forearm. Unfortunately, the very low frame rate and errors in velocity calculations near impact (Knudson and Bahamonde, 2001) made these calculations inaccurate and unreliable. Several lines of recent research support the opposite hypothesis that hand forces to not significantly affect the dynamics of a tennis impact, so the racket behaves as if it is unrestrained during the impact with the ball.

The first line of research that pointed to this counterintuitive result were studies of ball rebound speed with rackets in various modes of restraint. For balls impacting the centre of the racket face there were no differences in ball rebound velocity between unrestrained rackets and rackets rigidly clamped (Baker and Putnam, 1979; Watanabe *et al.*, 1979). Both modelling and experimental data have since confirmed that hand-held tennis rackets behave at impact as unrestrained bodies (Brody, 1987; Casolo and Ruggieri, 1991; Hatze, 1992, Kawazoe, 1997; Knudson, 1997; Vallatta *et al.*, 1993), so impact studies using rigidly clamped rackets do not represent the mechanics of impact in actual tennis play. This higher inertial effect of firmly clamped conditions (using vices or bolts) could account for some studies that have reported slightly higher ball rebound velocity in firm grip conditions (Elliott, 1982).

Secondly, there are several problems for grip forces to be transmitted into a significant impulse to affect the momentum of the ball. The duration of ball impact on the strings is only 3 to 5 milliseconds (Brody *et al.*, 2002; Kawazoe, 2002; Groppel *et al.*, 1987). The stiffness of hand tissues, frame and strings is also not high enough to transmit significant impulse from the hand to the ball in 5 ms. The essential transit time of the bending wave of collision in a racket is 2–3 ms (Brody *et al.*, 2002; Cross, 1999; Knudson and White, 1989),

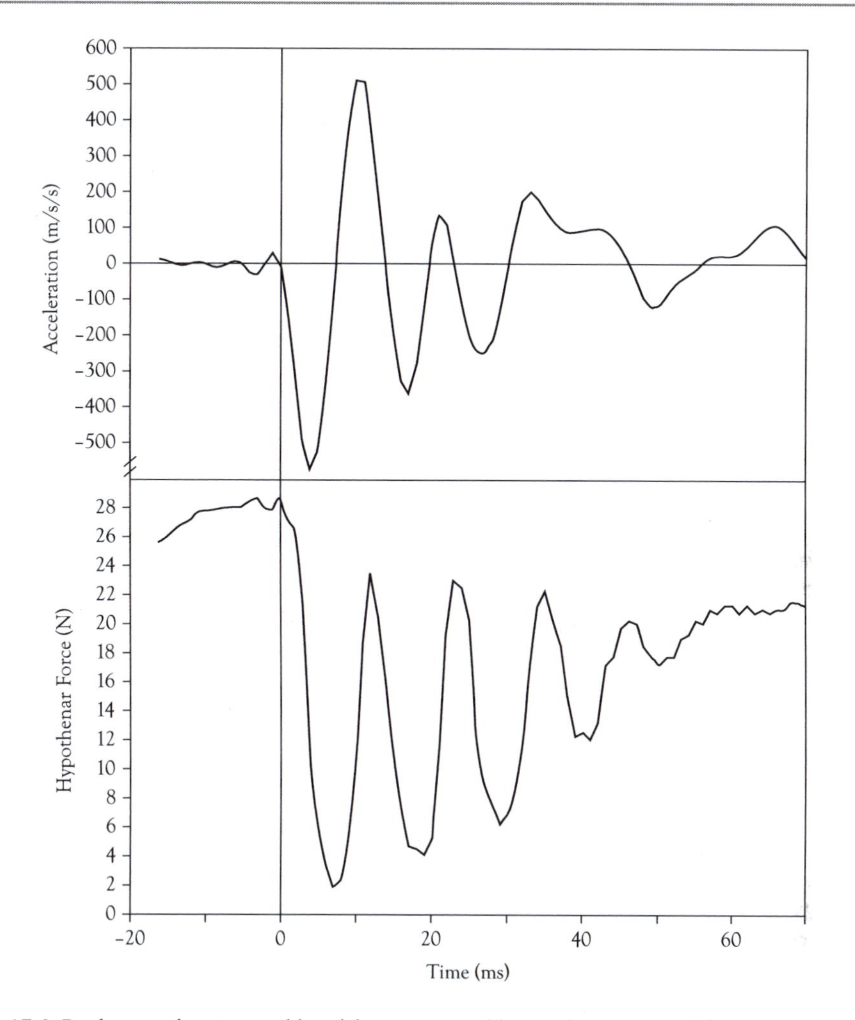

Figure 17.2 Racket acceleration and hand force measured by accelerometer and force sensing resistor in a tennis forehand. Data from Knudson and White (1989).

so hand forces and frame stiffness would have to be exceedingly high to generate and transfer a meaningful impulse to the ball in 2 ms. Figure 17.2 shows a 2 ms delay between racket frame acceleration and force recorded at the grip in a tennis forehand. Rackets would have to be shorter and unusually stiff for grip forces to influence ball rebound speed (Missavage *et al.*, 1984).

Measurements of hand forces during tennis strokes have reported moderate (not near maximal voluntary effort) grip forces up to impact, followed by highly variable forces after impact (for review see Knudson, 2004). The hand forces at impact are correlated with post-impact peak forces (Knudson, 1991), but are not related to ball rebound speed (Knudson, 1989). Ultra high-speed video of a forehand drive (Figure 17.3) shows that the ball has rebounded from the strings before the backward recoil of the frame is completed into the player's hand. All these data support the conclusion that the strength of the human hand is not high enough to significantly increase the resistance of the backward deflection of the racket frame and transfer impulse from the hand through the racket to the strings and ball in this short time interval.

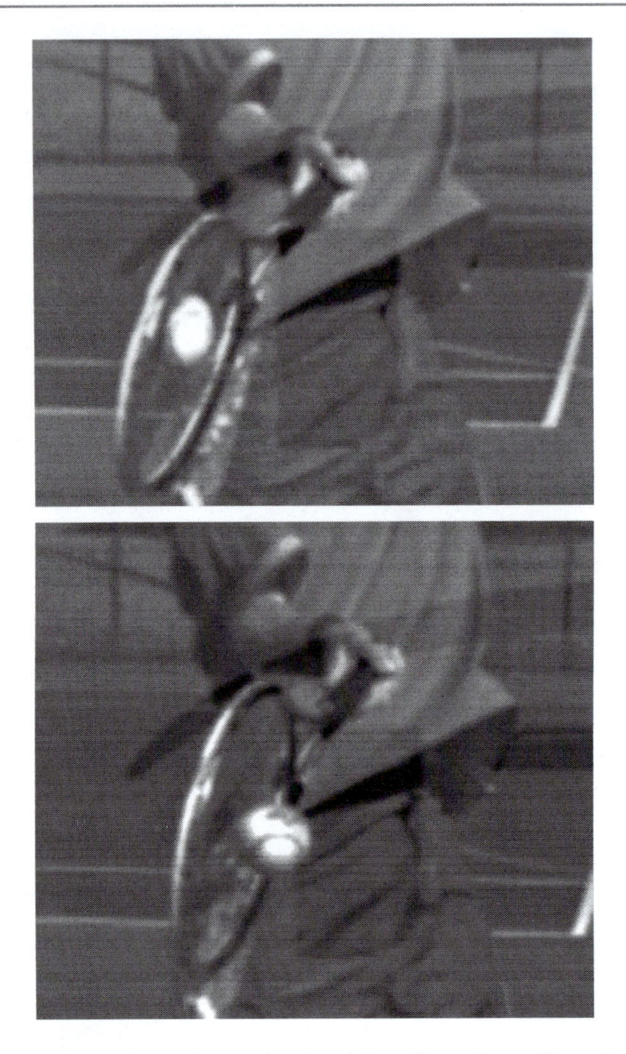

Figure 17.3 Key frames in ultra-high speed video (6000 frames/second) in the forehand of an advanced player. The ball rebounds from the strings before the frame has even stopped recoiling into the player's hand. Notice that the racket face is slightly closed at impact with the ball and the recoil from the off-centre impact squares the racket face and drives the racket backward into the hand. Images courtesy of the ITF.

The unpredictable effects of the large variability in the hand forces and impact positions on the racket face has resulted in several research groups proposing robotic arms for human surrogate testing of tennis rackets (Furusho *et al.*, 2001; Hatze, 1992). Racket manufacturers will likely always use human play testers because tennis player perceptions are very important in predicting sales, but basic evaluation of engineering changes should be systematically evaluated using robots as surrogates for biomechanical testing as well as theoretical modelling (Kawazoe, 2002). The variability of hand held racket performance and the ethical problems of exposing humans to injury risk are also strong arguments for robotic and theoretical testing of racket parameters.

It is clear from the biomechanical research that hand forces are not as influential on impact dynamics as most players think. Player perception of impact severity or sensitivity to the 'feel' and 'comfort' of impacts with a racket means that manufactures must accommodate psychometric and biomechanical factors in racket design (Sol, 1994). Theoretical and robotic testing of new racket technologies must be integrated with player testing to see if player perceptions and stroke accommodations will complement the new design.

Racket inertial properties

Another example of the interaction of biological factors with racket mechanical factors is the effect of racket mass and moment of inertia (MOI) variables on racket performance. All other racket parameters and stroke conditions remaining the same, increasing the mass or MOI of the racket in the plane of the stroke will increase the impulse applied to a ball at impact. The problem is that changing one inertial property of the racket changes other inertial properties, the biomechanics of the swing, and the possible speed and accuracy of the stroke. Table 17.1 shows how the inertial properties of mass and MOI of rotating the racket about the grip has also changed dramatically over the years. The trends are clearly toward lighter rackets.

The recent reduction of racket mass over the past few decades with the adoption of new materials and composite designs, however, has typically not decreased the speed of strokes in tennis play. Mitchell *et al.*, (2000) reported that skilled players could swing the racket faster when serving with lower mass and MOI frames.

The loss of racket mass also does not typically result in a reduction of racket momentum at impact because players tend to swing the lighter rackets faster than older rackets. So the moderate reduction in inertial properties has been more than offset by greater racket speeds and the higher coefficient of restitution afforded by stiff, larger head rackets in modern rackets.

Swing speed is most affected by the weight distribution within a frame which is easily examined by players as the balance point (finding the point near the middle where the frame can be balanced on your finger). Rackets with centre of mass (COMs) closer to the grip (head light) are easier to swing than rackets with the same mass distributed near the head (head heavy). The location of the COM of the racket is a good approximation for the true angular inertia variable of a rigid body: the MOI.

Newton's Second Law of Motion uses the MOI ($I = mr^2$) to represent the angular inertia (resistance to angular acceleration) of a rigid body about some axis of rotation. Since the distribution of mass (r^2) is the most influential variable in determining an object's resistance to rotation, rackets with lower mass and mass concentrated near the handle can be dramatically easier to swing. A tennis racket has the smallest MOIs about its three principal axes through the COM (Figure 17.4).

Table 17.1 Typical inertial properties of tennis rackets over time

Racket Material	*Year*	*Mass(kg)*	*Swing MOI(kg·cm^2)*
Wood	1960	0.41	0.644
Steel	1967	0.38	0.390
Aluminium	1977	0.37	0.394
Titanium	1998	0.21	0.266
Graphite	2000	0.34	0.350
Graphite	2006	0.24	0.300

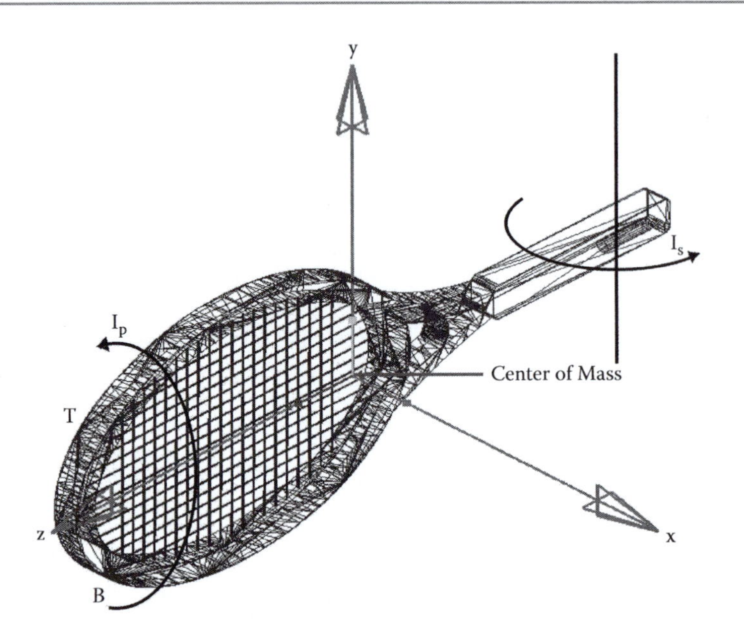

Figure 17.4 The three primary axes of rotation represent the smallest MOIs for a tennis racket about those axes. The swing MOI (I_s) and the polar MOI (I_p) are particularly relevant to tennis strokes. Adding mass (e.g. lead tape) at points T and B will increase both I_s and I_p. Image modified with permission of S. Nesbit, Lafayette College.

So the inertia variable most relevant to tennis strokes is what is commonly called the 'swing weight'. The 'swing weight' of a racket refers to the MOI of the frame at right angles to the typical grip axis of rotation (I_s) used for a stroke (Figure 17.4). The polar MOI (I_p) is the resistant to angular acceleration of the frame about its long axis, and is about 20 times smaller than I. Adding a small amount of mass to the frame with lead tape at points T and B can create dramatic increases both I_s and I_p of the frame. Several chapters in Brody *et al.* (2002) show how to measure MOIs and add mass to a racket to modify them.

Cross and Bower (2006) reported a study of overarm swinging of special weighted rods and found that above 0.2 kg the speed of the swing decreases with increasing mass. They also noted that given the mass of a tennis ball and typical impact position on the racket face, that racket masses between 0.31 and 0.51 kg were near optimal for power transfer to the ball. Since the mass of modern rackets are often at or below 0.3 kg, a limit on lower mass rackets may have been achieved. Many tennis players using very light rackets might benefit in performance and injury risk by adding some mass to the racket.

Players and racket stringers often customize rackets by adding lead tape to the head and other frame locations to adjust balance and swing weight. Adding weight to the sides of the frame increases the polar MOI, which increases the racket stability to resist off-centre impact. Cross (2001) has shown that the best place to increase the speed of ball rebound is to add weight to the tip of the racket. This increases ball speed the most and shifts the maximum power zone closer to the centre of the strings. Skilled players are fairly sensitive to changes in the swing weight of tennis rackets and can detect differences down to 2.5 per cent (Brody, 2000).

The inertial properties of modern tennis rackets have changed dramatically in the last 100 years. The trend is clearly for lighter, head-light frames that are easy to swing, but unfortunately then also provide less resistance to the acceleration from ball impact. It is likely that apparently small differences in racket inertial properties like mass and MOIs can be easily felt by players and would have effects of stroke technique or performance. Highly skilled tennis players are particularly sensitive to these variations in inertial properties and often spend considerable money to customize their rackets.

Rackets to improve performance

There have been a few studies reporting systematic research on racket parameters to improve performance. The major changes in racket head dimensions (Blanksby *et al.*, 1979; Guretter and Davis, 1985) and overall length (Pellett and Lox, 1997) have been shown to have significant effects on tennis skills of students in college tennis classes. Tennis coaches are often less concerned with these standardized tests of tennis skill in beginners than they are with stroke characteristics like ball speed, spin, or accuracy with advanced players.

Research on racket performance usually defines performance as rebound speed of the ball, with less attention paid to ball spin or shot accuracy. Early research confirmed that rackets with oversize heads create significantly greater ball rebound velocity with less frame vibration than traditional sized rackets (Elliott *et al.*, 1980). Much of the early design changes in the size and shape of the racket head focused research on the colloquial ideas of the 'sweet spot' or 'racket power'. The Head patent for the larger head racket, in fact, was based on the larger area of high coefficients of restitution or sweet spot compared to traditional rackets.

The phenomenon of a sweet spot of a tennis racket and string system is really three separate points or regions on the racket face (Brody, 1981; Cross, 1998): the area of maximum rebound coefficient of restitution (e = relative velocity of separation/relative velocity of approach), the point of minimum vibration (node), and the point of no frame reaction (centre of percussion). Static rebound tests of balls bounced off all locations on the racket face creates a map of regions of the coefficient of restitution (e). The area with the highest e is a high power zone that is most relevant to high-level tennis play. This area moves around less that than the other two 'sweet spots' with variations in grip force and stroke pattern (Cross, 2004; Hatze, 1998; Nab *et al.*, 1998).

Another racket design parameter that research has shown to increase ball rebound speed is stiffness. Frame stiffness can be increased by use of stiffer materials in composites or increasing the width of the frame. Greater racket stiffness increases ball rebound velocity (Cross, 2000; Kern and Zimmerman, 1993) and accuracy (Bower and Sinclair, 1999). The stiffness of racket frame, however, does interact with the string tension of the strings (Baker and Wilson, 1978; Elliott, 1982), so it is likely that selecting a racket with greater stiffness should improve player performance if the stringing is also adjusted. The United States Racquet Stringers Association (USRSA) compiles racket and string data in an extensive database and provides a computer model for USRSA members to help players customize their rackets.

Many rackets are marketed with claims about their design features that maximize 'control' or shot accuracy. Unfortunately, this aspect of performance has not been systematically researched in the scientific literature. There is initial support for the belief that increasing frame stiffness through the frame and higher string tensions (Brody and Knudson, 2000), larger racket heads (Brody, 1979), and that stringing the racket at higher tensions will increase the ball rebound accuracy (Knudson, 1997).

The effect of racket inertial properties is likely another design variable that should have a significant effect on stroke accuracy. Unfortunately, there is limited tennis research, and the kinesiology research on how implement mass interacts with other factors. In theory, a lower mass racket is easier to move and accuracy might improve given the more sub-maximal efforts that are used. However, a tennis player is likely to swing the lower mass racket faster; that, in general, means lower accuracy because of the well-known speed-accuracy trade-off. This is another area where the interaction of characteristics of the racket and the player is of great importance. A recent study of tennis stroke speed and accuracy of children (4–10-years-old) using different rackets showed that the best racket for any individual child was unique to a combination of racket size and inertial properties (Gagen *et al.*, 2005). Tennis coaches would likely even take into account the stroke or strategic style of a player when considering a racket change. A racket with greater higher mass and swing weight might help a player with long, erratic strokes and a tendency to over-hit strokes in rallies.

It is likely that racket performance (stroke speed, spin, and accuracy) is a complicated phenomenon that interacts with individual player or stroke characteristics. While there is some basic science research on the general effect of several racket design variables (size, inertial properties, stiffness), their combined effect on the several performance variables is not known. This appears to be a promising area for future research on tennis racket design and tennis performance variables. Racket designs, however, have also been marketed to be effective in reducing the symptoms and risk of the common overuse injury, tennis elbow.

Rackets to reduce risk of overuse injury

Since most middle-aged tennis players will experience elbow pain sometime in their lifetime (Kamien, 1990; Maffulli *et al.*, 2003), many rackets are marketed with claims about material or design properties to relieve symptoms of 'tennis elbow'. Tennis elbow is the common term for any overuse injury of the common wrist flexor or extensor attachments of the elbow. There is very little research on many of the materials or engineering elements in these rackets, and there are virtually no blinded, prospective studies that would be necessary to examine these claims. This section will summarize the biomechanical studies and logic of recent tennis racket designs purported to decrease the risk or pain from tennis elbow.

The first line of evidence is the basic science studies on the mechanics of tennis rackets, similar to the research summarized on hand forces and inertial properties. Since tennis elbow is most likely caused by the rapid elongation of the forearm muscle from the initial shock wave of impact (Knudson, 2004), changes in rackets design and materials to reduce injury risk should work to decrease the peak acceleration after ball impact. It should be noted, however, that the exact mechanism and biomechanical variables that cause tennis elbow are still not conclusively known.

The first design issue would logically be racket inertial properties. Increasing racket mass and swing weight provide greater inertial resistance to the shock wave of impact. This is consistent with the laws of physics and has been verified by experiments and modelling (Brody, 1979; Nesbit *et al.*, 2006). The most effective strategy would be to add mass to the top of the racket head, increasing both mass and swing weight. If a player could not handle the relatively large effect this has on swing weight, the extra mass could be added to the frame closer to the handle and racket COM.

Research has shown that increasing head size decreases the amplitude of post-impact vibration (Elliott *et al.*, 1980; Hennig *et al.*, 1992) compared to regular head rackets. Tennis players wanting to decrease the risk of tennis elbow or current tennis elbow symptoms should strive to use oversize rackets. The larger head allows the strings to absorb more of the energy of ball impact, and their greater polar MOI provides added resistance to off-centre impacts. Off-centre impacts have the largest influence on the peak accelerations transmitted to the player (Knudson, 2004; Plagenhoef, 1982).

Decreasing the stiffness of the racket will also decrease the peak acceleration transmitted to the body. Greater compliance in the frame allows the strings and racket to deflect more and absorb more energy, than a stiffer racket. Remember that frame stiffness interacts with string tension, so players concerned about injury should also tend to string their rackets at the lower end of the range of recommended string tensions for their racket. Besides the mechanical stiffness of the frame, the damping properties of the racket materials and design also affect the energy transfer at impact. Rackets have been designed to essentially isolate the impact energy from the body (e.g. Wilson *Triad*) or damp out the energy of the shock and vibration of impact (e.g. Head *Intelligence*).

Recently, one of these damping designs has begun to be tested with prospective studies with respect to its effects on players suffering with tennis elbow. The Head racket manufacturing company has developed tennis racket frames with piezoelectric ceramic fibres in the shaft that are connected to electronics in the handle. The theory of these designs is that the electricity created by the piezoelectric fibres can be processed by the electronics and sent back to the fibres to more quickly damp out racket vibrations. The first generation or *Intelligence* frames were tested with players with a history of tennis elbow and appeared to reduce the symptoms of the condition (Kotze *et al.*, 2003). A subsequent double-blind prospective 12-week study using a second generation *Protector OS* design seems to support the conclusion that the new damping technology does decrease the symptoms of tennis elbow over play with regular tennis rackets (Cottey *et al.*, 2006). To the author's knowledge no other racket designs (materials, vibration isolation, etc.) have been rigorously tested for efficacy using blinded, prospective research designs.

Selecting a racket to decrease the risk or symptoms of tennis elbow, like trying to optimize performance, is a complicated problem with many factors interacting to determine risk. Attempting to minimize injury risk, although still mostly based on theory, does have the advantage of some basic science and prospective research. There are data to support tentative recommendations of increasing racket mass, decreasing frame stiffness, string tension, and using rackets designed to damp out the energy transmitted to the body. More research is needed on the effectiveness of the various damping technologies, and larger prospective studies with uninjured players to see if these strategies decrease the incidence of tennis elbow.

Conclusion

The materials, design, and construction of rackets has had a major impact on the sport of tennis. Until more tennis-specific basic science and prospective data are available, most players should select rackets that 'feel' good to them during play and do not represent extremes in racket mass, stiffness, or head size. Specifically, racket design characteristics that tend to increase some performance variables (ball speed) without increasing overuse injury risk (peak racket acceleration after impact) may be the best choice for most recreational tennis players: relatively greater racket mass, larger head size, and lower stiffness. Although not a racket design parameter, using a string tension in the lower range of recommended tensions will increase ball rebound speed, but tend to decrease ball spin and rebound accuracy.

Individual player perceptions and preferences should be considered in changing rackets and customizing them. The interaction of biomechanical factors with tennis rackets represents a multifaceted, complex problem that remains a fertile area for future research.

References

1. Arthur, C. (1992) 'Anyone for slower tennis?', *New Scientist*, 134(1819): 24–28.
2. Baker, J.A.W. and Putnam, C.A. (1979) 'Tennis racket and ball responses during impact under clamped and freestanding conditions', *Research Quarterly*, 50: 164–170.
3. Baker, J.A.W. and Wilson, B.D. (1978) 'The effect of tennis racket stiffness and string tension on ball velocity after impact', *Research Quarterly*, 49: 255–259.
4. Blanksby, B.A., Ellis, R. and Elliott, B.C. (1979) 'Selecting the right racquet: Performance characteristics of regular-sized and over-sized tennis racquets', *Australian Journal of Health, Physical Education, and Recreation*, 86: 21–25.
5. Bower, R. and Sinclair, P. (1999) 'Tennis racquet stiffness and string tension effects on rebound velocity and angle for an oblique impact', *Journal of Human Movement Studies*, 37: 271–286.
6. Brannigan, M. and Adali, S. (1981) 'Mathematical modeling and simulation of a tennis racket', *Medicine and Science in Sports and Exercise*, 13: 44–53.
7. Brody, H. (1979) 'Physics of the tennis racket', *American Journal of Physics*, 47: 482–487.
8. Brody, H. (1981) 'Physics of the tennis racket II: The sweet spot', *American Journal of Physics*, 49: 816–819.
9. Brody, H. (1987) 'Models of tennis racket impacts', *International Journal of Sport Biomechanics*, 3: 292–296.
10. Brody, H. (2000) 'Player sensitivity to the moments of inertia of a tennis racket', *Sports Engineering*, 3: 145–148.
11. Brody, H. (2002) 'The tennis racket'. In Renstrom, P.A.F.H. (ed.) *Handbook of Sports Medicine and Science: Tennis* (pp. 29–38). Oxford: Blackwell Science.
12. Brody, H., Cross, R. and Lindsey, C. (2002) *The physics and technology of tennis*. Solana Beach, CA: Racquet Tech Publishing.
13. Brody, H. and Knudson, D. (2000) 'A model of tennis stroke accuracy relative to string tension', *International Sport Journal*, 4: 38–45.
14. Casolo, F. and Ruggieri, G. (1991) 'Dynamic analysis of the ball-racket impact in the game of tennis', *Meccanica*, 26: 67–73.
15. Coe, A. (2000) 'The balance between technology and tradition in tennis. In Haake, S.J. and Coe, A. (eds.) *Tennis Science & Technology* (pp. 3–40). Oxford: Blackwell Science.
16. Cottey, R., Kotze, J., Lammer, H. and Zirngibl, W. (2006) 'An extended study investigating the effects of tennis rackets with active damping technology on the symptoms of tennis elbow'. In Moritz, E. and Haake, S. (eds.) *Engineering of Sport 6* (pp. 391–396). New York: Springer.
17. Cross, R. (1998) 'The sweet spots of a tennis racquet', *Sports Engineering*, 1: 63–78.
18. Cross, R. (1999) 'Impact of a ball with a bat or racket', *American Journal of Physics*, 67: 692–702.
19. Cross, R. (2000) 'Flexible beam analysis of the effects of string tension and frame stiffness on racket performance', *Sports Engineering*, 3: 111–122.
20. Cross, R. (2001) 'Customising a tennis racket by adding weights', *Sports Engineering*, 4: 1–14.
21. Cross, R. (2004) 'Centre of percussion of hand-held implements', *American Journal of Physics*, 72: 622–630.
22. Cross, R. and Bower, R. (2006) 'Effects of swing-weight on swing speed and racket power', *Journal of Sports Sciences*, 24: 23–30.
23. Cross, R. and Lindsey, C. (2005) *Technical Tennis*. Vista, CA: Racquet Tech Publishing.
24. Elliott, B.C. (1989) 'Tennis strokes and equipment'. In Vaughn, C.L. (ed.) *Biomechanics in Sport* (pp. 263–288). Boca Raton, FL: CRC Press.

25. Elliott, B.C. (1982) 'Tennis: the influence of grip tightness on reaction impulse and rebound velocity', *Medicine and Science in Sports and Exercise*, 14: 348–352.
26. Elliott, B.C, Blanksby, B.A and Ellis, R. (1980) 'Vibration and rebound velocity characteristics of conventional and oversized tennis rackets', *Research Quarterly for Exercise and Sport*, 51: 608–615.
27. Furusho, J., Sakaguchi, M., Takesue, N. Sato, F., Naruo, T. and Nagao, H. (2001) 'Development of a robot for evaluating tennis rackets', *Journal of Robotics and Mechatronics*, 13: 74–79.
28. Gagen, L.M., Haywood, K.M. and Spaner, S.D. (2005) 'Predicting the scale of tennis rackets for optimal striking from body dimensions', *Pediatric Exercise Science*, 17: 190–200.
29. Groppel, J.L, Shin, I., Thomas, J. and Welk, G.J. (1987) 'The effects of string type and tension on impact in midsized and oversized tennis racquets', *International Journal of Sport Biomechanics*, 3: 40–46.
30. Gruetter, D. and Davis, T.M. (1985) 'Oversized vs. standard racquets: Does it really make a difference?', *Research Quarterly for Exercise and Sport*, 56: 31–36.
31. Haake, S.J., Chadwidk, S.G., Dignall, R.J., Goodwill, S. and Rose, P. (2000) 'Engineering tennis—slowing the game down', *Sports Engineering*, 3: 131–143.
32. Haake, S.J. and Coe, A. (eds.) (2000) *Tennis Science & Technology*. Oxford: Blackwell Science.
33. Hatze, H. (1998) 'The centre of percussion of tennis rackets: a concept of limited applicability', *Sports Engineering*, 1: 17–25.
34. Hatze, H. (1992) 'Objective biomechanical determination of tennis racket properties', *International Journal of Sport Biomechanics*, 8: 275–287.
35. Hennig, E.M., Rosenbaum, D. and Milani, T.L. (1992) 'Transfer of tennis racket vibrations onto the human forearm', *Medicine and Science in Sports and Exercise*, 24: 1134–1140.
36. Kamien, M. (1990) 'A rational management of tennis elbow', *Sports Medicine*, 9: 173–191.
37. Kawazoe, Y. (1997) 'Experimental identification of a hand-held tennis racket and prediction of rebound ball velocity in an impact', *Theoretical and Applied Mechanics*, 46: 177–188.
38. Kawazoe, Y. (2002) 'Mechanism of high-tech tennis rackets performance', *Theoretical and Applied Mechanics*, 51: 177–187.
39. Kern, J.C. and Zimmerman, W.J. (1993) 'The effect of tennis racquet flexibility on rebound velocity'. In Hamill, J., Derrick, T.R. and Elliot, E.H. (eds.) *Biomechanics in Sports XI* (pp. 193–195). Amherst, MA: University of Massachusetts.
40. Knudson, D. (2006) *Biomechanical Principles of Tennis Technique*. Vista, CA: Racquet Tech Publishing.
41. Knudson, D. (2004) 'Biomechanical studies on the mechanism of tennis elbow'. In Hubbard, M., Mehta, R.D. and Pallis, J.M. *The Engineering of Sport 5: Volume 1* (pp. 135–139). Sheffield: International Sports Engineering Association.
42. Knudson, D. (1997) 'Effect of grip models on rebound accuracy of off-centre tennis impacts'. In Wilkerson, J., Ludwig,K. and Zimmerman, W. *Biomechanics in Sports XV: Proceedings of the 15th International Symposium on Biomechanics in Sports* (pp. 483–487). Denton, TX: Texas Women's University.
43. Knudson, D. (1991) 'Factors affecting force loading in the tennis forehand', *Journal of Sports Medicine and Physical Fitness*, 31: 527–531.
44. Knudson, D. (1989) 'Hand forces and impact effectiveness in the tennis forehand', *Journal of Human Movement Studies*, 17: 1–7.
45. Knudson, D. and Bahamonde, R. (2001) 'Effect of endpoint conditions on position and velocity at impact in tennis', *Journal of Sports Sciences*, 19: 839–844.
46. Knudson, D. and Elliott, B. (2004) 'Biomechanics of tennis strokes'. In Hung, G.K. and Pallis, J.M (eds.) *Biomedical Engineering Principles in Sports* (pp. 153–181). New York: Kluwer Academic/ Plenum Publishers.
47. Knudson, D. and White, S. (1989) 'Forces on the hand in the tennis forehand drive:Application of force sensing resistors', *International Journal of Sport Biomechanics*, 5: 324–331.
48. Kotze, J., Lammer, H., Cottey, R. and Zirngibl, W. (2003) 'The effects of active piezo fibre rackets on tennis elbow'. In Miller, S. (ed.) *Tennis Science and Technology 2* (pp. 55–60). London: ITF.

49. Maffulli, N., Wong, J. and Almekinders, L.C. (2003) 'Types and epidemiology of tendinopathy', *Clinics in Sports Medicine*, 22: 675–692.
50. McClusky, M. (2003). Tennis swaps grace for strength. *Wired News*. Online. Available HTTP: http://www.wired.com/news/technology/0,60177-0.html (accessed 21 November 2006).
51. Miller, S. (2006) 'Modern tennis rackets, balls, and surfaces', *British Journal of Sports Medicine*, 40: 401–405.
52. Missavage, R.J., Baker, J. and Putnam, C. (1984) 'Theoretical modeling of grip firmness during ball-racket impact', *Research Quarterly for Exercise and Sport*, 55: 254–260.
53. Mitchell, S.R., Jones, R. and King, M. (2000) 'Head speed vs racket inertia in the tennis serve', *Sports Engineering*, 3: 99–110.
54. Nab, D., Hennig, E.M. and Schnabel, G. (1998) 'Ball impact location on a tennis racket head and its influence on ball speed, arm shock, and vibration'. In Riehle, H.J. and Vieten, M.M. *Proceedings II of the XVI International Society of Biomechanics in Sports Symposium* (pp. 229–232). Konstanz: Universitatsverlag Konstanz.
55. Pellett, T.L. and Lox, C.L. (1997) 'Tennis racket length comparisons and their effect on beginning college player' playing success and achievement', *Journal of Teaching in Physical Education*, 16: 490–499.
56. Plagenhoef, S. (1970) *Fundamentals of Tennis*. Englewood Cliffs, NJ: Prentice-Hall.
57. Plagehnoef, S. (1982) 'Tennis racket testing'. In Terauds, J. (ed.) *Biomechanics in Sports* (pp. 411–421). Del Mar, CA: Research Centre for Sports.
58. Sheridan, H. (2006) 'Tennis technologies: de-skilling and re-skilling players and the implications for the game', *Sport in Society*, 9: 32–50.
59. Sol, H. (1994) 'Computer aided design of rackets'. In Reilly, T., Hughes, M. and Lees, A. (eds.). *Science and Racket Sports* (pp. 125–133). London: E & FN Spon.
60. Tenner, E. (1996) *Why Things Bite Back: Technology and the Revenge of Unintended Consequences*. New York: Knopf.
61. Vallatta, A., Casolo, F. and Caffi, M. (1993) 'On the coefficients of restitution of tennis racquets'. In Hamill, J., Derrick, T.R. and Elliott, E.H. (eds.) *Biomechanics in Sports XI* (pp. 196–200). Amherst, MA: University of Massachusetts.
62. Watanabe, T.Y., Ikegami, Y. and Miyashita, M. (1979) 'Tennis: the effects of grip firmness on ball velocity after impact', *Medicine in Science in Sports*, 11: 359–361.

18

Sports equipment – energy and performance

Darren J. Stefanyshyn and Jay T. Worobets
The University of Calgary, Calgary

Introduction

Energy is defined as the ability to do work. In sports, the more work you can do, the farther you can throw a javelin, the faster you can skate or run and the higher you can jump. The more energy you have, the greater your potential to do work and, therefore, increase your performance.

Van Ingen Schenau and Cavanagh (1990) stated that the mechanical energy the muscles produce for human movement can flow along one of three main paths (Figure 18.1). The first is a conservative path where the energy can be stored in a piece of equipment and reutilized by the athlete. The second is a non-conservative path where the energy is dissipated as heat, sound, vibration, etc. The final is the application of energy directly to performance to move the athlete or equipment.

To maximize athletic performance, it is necessary to maximize the output energy (E_o). When considering athletic equipment, this can be achieved through three main principles (Nigg *et al.*, 2000). The first is to increase the mechanical energy produced by the muscles (E_m) by optimizing the intrinsic characteristics of the musculoskeletal system. The second is to maximize the conservative energy (E_{CG}) which is stored and returned by a piece of equipment. The third is to utilize a piece of equipment that minimizes the non-conservative energy (E_{nc}) that is lost.

This chapter will focus on three main aspects associated with how sports equipment can influence energy of athletic movement and sport performance:

- optimizing the musculoskeletal system;
- maximizing conservative energy storage and return; and
- minimizing non-conservative energy loss.

Optimizing the musculo-skeletal system

Optimizing muscular output is dependent on several factors. The stronger a muscle is the more force it can produce and the greater the amount of work it can do. Thus, stronger

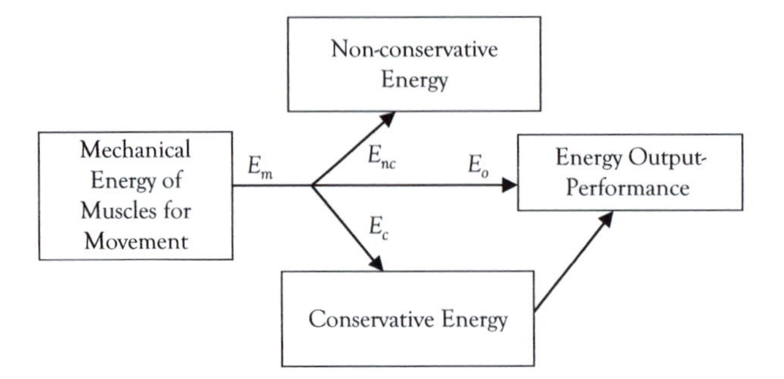

Figure 18.1 Paths of mechanical energy flow from the muscles for sport performance. E_m is the muscle energy, E_{nc} is non-conservative energy dissipated in equipment, E_c is conservative energy stored and returned in equipment, E_o is output energy used for performance. (Adapted from Van Ingen Schenau and Cavanagh, 1990.)

muscles should result in increased athletic performance. This is a physiological characteristic of an athlete and is not influenced by the particular piece of equipment that they are using during their athletic performance. However, equipment can influence dynamic force production of muscles by manipulating the intrinsic musculoskeletal characteristics of the force-length and force-velocity relationships. In simple terms a piece of equipment can have a large influence on the technique that the athlete is using, which can have a substantial influence on performance.

The force that a muscle can produce is severely decreased at very short lengths and at very long lengths. There is an optimal length where the maximal amount of force can be produced (Figure 18.2). Therefore, to optimize the musculoskeletal system during an athletic activity, sports equipment can be manipulated to adjust the operating lengths of different muscles to maximize force production.

Another intrinsic characteristic of muscle is how much force it can produce at a given velocity of contraction. At high shortening velocities, the amount of force that a muscle produces is very low whereas the most force that it can produce is during an eccentric contraction, when the muscle is lengthening. Many sports are classified as power sports where work must be performed quickly. Since power is the time rate of change of energy or work this is very important with respect to athletic performance. Power is also the product of

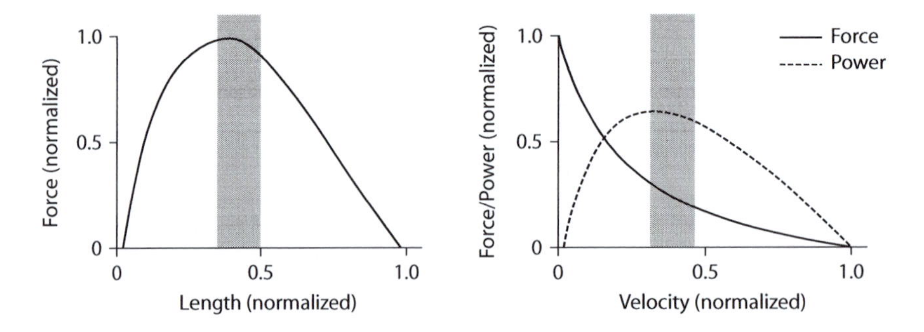

Figure 18.2 Force-length and force-velocity relationships of skeletal muscle, the shaded regions show optimal conditions for maximal force and power production.

force and velocity. Therefore, the force-velocity relationship indicates that there is an optimal shortening velocity where power output is maximized (Figure 18.2). Maximal muscular power is produced at a shortening speed that is approximately 31 per cent of the peak velocity of contraction. Based on this information, sports equipment can be designed to manipulate the contraction velocity of different muscles to maximize power production.

The following paragraphs will provide some examples of how athletic equipment has been utilized to optimize the musculoskeletal system by manipulating either the force-length or force-velocity relationship.

Bicycles

Altering bicycle seat height changes the lower leg kinematics during cycling. By changing the lower leg kinematics, the muscles of the lower extremity can operate at different lengths. So, ultimately the seat height has an influence on the force-length characteristics of the lower extremity muscles. In fact several studies have shown that an optimal bike seat height can be selected for optimizing performance. Hamley and Thomas (1967) investigated how bicycle seat height influences the amount of time necessary to complete a preset amount of work. They found that a seat height of approximately 108 per cent of leg length resulted in the minimum time required to complete the required amount of work. Nordeen-Snyder (1977) showed that oxygen consumption as a function of bicycle seat height was minimized at about 100 per cent of leg length. Lower seat positions lead to much more oxygen consumption while pedalling as did higher seat heights. Seat position also affects aspects such as trunk angle during cycling. Savelberg *et al.* (2003) recently showed that changes in trunk angle during cycling can have a large influence on muscle recruitment due to changes in muscle length. Thus, by manipulating seat height and position you can have a large influence on performance by changing the force-length characteristics of the lower extremities.

A simulation study performed by Yoshihuku and Herzog (1990) showed that pedalling rate could also be optimized. The pedalling rate is directly related to the force velocity characteristics of the lower extremity. Groups of lower extremity muscles were modelled as contractile systems and it was found that the power output of each of the lower extremity muscle groups depended on pedalling rate. When power output of all the individual muscles was summed to quantify total power, an optimal pedalling rate was found. Peak power was achieved at 155 revolutions per minute and slower pedalling rates led to decreases in power output as did higher pedalling rates. By simply manipulating the pedalling rate of a bicycle an athlete can have a large influence on the force-velocity characteristics of the lower leg muscles and ultimately on power output and performance.

Klap skates

The introduction of the klap skate at the elite standard in 1998 led to instantaneous and dramatic increases in speed skating performance. Initially it was believed that the increased work output during speed skating that occurs with the klap skate was due to propulsion at the ankle joint. In actual fact, research has shown that the pivot point mechanism that was introduced in the klap skate allows increased energy production at the knee joint (Houdijk *et al.*, 2000). Houdijk and coworkers showed that the klap skate led to an increase in gross efficiency and an ultimate increase in performance of about 5 per cent.

The klap skate influences where speed skaters operate on the force-velocity and force-length relationships. The fact that the largest influence occurred at the knee joint shows how important manipulating a piece of equipment can be in increasing performance. Moving the pivot point in an anterior direction has the influence of increasing the lever arm for the plantar flexor muscles. By increasing the lever arm the resultant moment and amount of work that the plantar flexor muscles must produce will have to increase. However, because of the position of the speed skater, a more anterior klap skate pivot point position will decrease the amount of work that is necessary from the knee extensor muscles. So an optimization procedure is required to determine exactly where the optimal pivot point position occurs.

Van Horne and Stefanyshyn (2005) showed that by systematically moving the pivot point more anterior, more total energy was produced per push at the ankle joint by the calf muscles. However, there was a systematic decrease in the amount of energy that was output by the quadriceps muscles crossing the knee joint. The total amount of energy at the hip joint remained relatively unchanged because the line of action of the GRF must go through the COM, which is relatively close to the hip joint centre and thus did not have a large influence on the energy produced at that joint. Overall they found that a more posterior pivot point position resulted in a larger amount of total energy produced when all of the joints (the ankle, knee and hip joints) were summed together (Figure 18.3). There was a significant increase in total energy per push at the more posterior positions, compared to the more anterior positions. They concluded that systematically moving the pivot point anteriorly decreases the work output at the knee joint by a greater amount than the increase at the ankle joint. The overall result, therefore, was a decrease in performance. Van Horne (2004) also showed that the optimal pivot point position was athlete specific. Some skaters for example produced more energy during each push at a more posterior position while some skaters produced it at a more anterior position. He further went on to show that this position was correlated to the work that the athlete was capable of producing on an isokinetic dynamometer. Those skaters that had very strong calf muscles were the athletes that had their optimal performance with a more anterior position, while the skaters that had weaker calf muscles had a better performance with a more posterior position. The klap skate is an excellent example of: (1) how the general characteristics of a piece of equipment can be prescribed to improve performance of a given movement; and (2) how the detailed characteristics of a

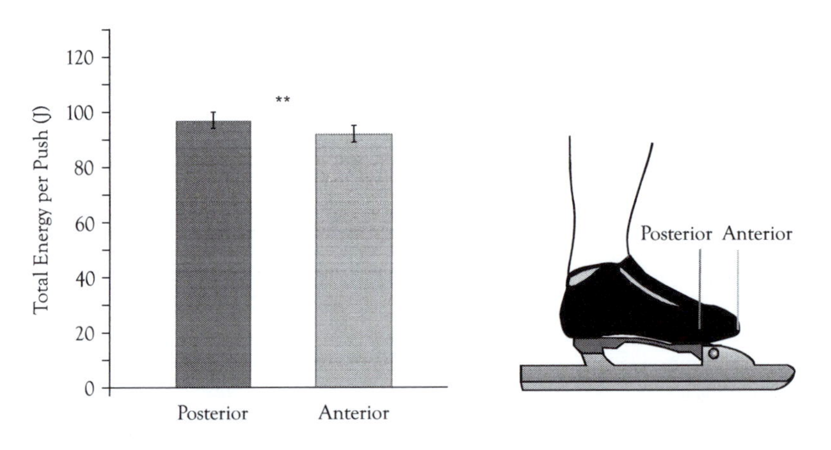

Figure 18.3 Total work produced by the ankle, knee and hip joints during speed skating pushes for two different klap skate pivot point positions. Data are averages of 16 elite skaters and ** indicates a significant difference.

piece of equipment require individual tuning to improve the specific performance of individual athletes.

Sport shoes

Stefanyshyn and Nigg (1998) showed that the metatarsophalangeal joint is a net energy absorber. As athletes roll onto the ball of their foot, the joint bends and absorbs energy in the foot and shoe structures. However, during takeoff, the joint only extends minimally and the energy is not returned. As a result, energy is being lost at that particular joint. For vertical jumping they found this amount of energy to be about 23 Joules per jump, an amount which is relatively large. They concluded that this energy loss likely did not have a positive influence on performance and, therefore, designed some subsequent studies to investigate the influence of reducing the energy lost.

Stefanyshyn and Nigg (2000) found that by placing thin carbon fibre insoles into sport shoes, the energy lost at the metatarsophalangeal joint during jumping decreased. The overall result was a significant increase in jumping performance of about 2.5 per cent when jumping with a stiff shoe compared to a control shoe. Similar improvements were found by Stefanyshyn and Fusco (2004) for sprinting with a stiff plate placed inside of sprint shoes. Overall, average sprint times were approximately 1 per cent faster when sprinting in stiff shoes compared to control shoes. Furthermore, Roy and Stefanyshyn (2006) showed improved running economy with stiff shoes compared to control shoes.

All of the above studies found that continued increases in stiffness started to lead to decrements in performance and that there was an optimal situation where the performance was maximized. The general theory is that the stiff plates act similar to the klap skate mechanism by manipulating the point of application of the GRF. This changes the lever arm with respect to the lower extremity joints. For example, increasing the stiffness increases the lever arm of the GRF about the ankle joint and increases the work that is required by the plantar flexor muscles. If the shoe is too flexible, the point of application of the GRF is too far back, similar to a very low gear, and performance is hindered. However, if the shoe is too stiff, the point of application of the GRF is too far forward, similar to a very high gear, where an athlete may not be able to generate enough power to propel themselves. Thus, an optimal situation is achieved by having a shoe bending stiffness that optimizes the movement with respect to the force-length characteristics of the lower extremities.

Maximizing conservative energy storage and return

Energy storage and return in sports equipment is made possible through the appropriate manipulation of conservative forces. To illustrate the advantage of conservative forces, imagine a force applied to a perfectly elastic linear spring. While the force causes the spring to deform, the work done by this force is stored in the spring as elastic potential energy. This deformed spring now has the capacity to do work, and so if it is released from this deformed state, it would exert force as it returns to its original undeformed position. The work done by the spring, in fact, would be exactly the same in magnitude as the work done to the spring. In this scenario, these forces are considered to be conservative. As such, when applied to an elastic element, the work done by a conservative force is stored during a loading cycle, and can then be later returned during an unloading cycle. It is this ability which can make conservative forces tremendously useful in certain sports settings.

For example, in archery contests, as the athlete pulls back the bow string, they are applying a conservative force to do work deforming the bow. As the bow has elastic properties, this work is stored in the bow as potential energy. When the archer releases the string, this stored energy is returned, resulting in the bow exerting force on the arrow which causes the arrow to accelerate significantly. The advantage of using the bow is clear; no athlete could propel an arrow nearly the same with just their bare hands. The reason for this advantage is that, while the athlete does at least the same amount of work on the bow during the loading cycle as the bow does on the arrow during the unloading cycle, the unloading cycle takes only a fraction of the amount of time of the loading cycle; the bow is capable of doing work at a much higher rate than the athlete.

This archery example illustrates how using sports equipment to store and return energy (i.e. the work done by conservative forces) can result in enhanced performance. The performance benefit here, which is specific to archery, is realized due to differences in the timing aspects of the loading and unloading periods. Although the exact mechanism by which athletes reap the benefits of storing and returning energy in sports equipment varies from sport to sport, there are basic performance enhancement strategies that can be applied to any situation where energy is stored and returned. These basic strategies, which will be commented on in the following sections, are:

- maximizing the amount of energy stored by the equipment;
- minimizing the amount of energy loss;
- returning the energy at the right time;
- returning the energy at the right frequency; and
- returning the energy at the right location.

Energy storage

Any sports equipment with elastic properties has the ability to store and return energy. Energy is stored in the equipment as it deforms, and released as the equipment returns to its original configuration. The amount of energy stored in the equipment is a function of the stiffness of the equipment, and the extent to which it is deformed (Equation 1). Stored energy increases linearly with increasing stiffness, and quadratically with increasing deformation (assuming the elastic properties of the equipment are represented as a linear spring). So for a given deforming force, a more flexible piece of equipment will, therefore, store more energy than a stiffer piece of equipment.

$$\text{Energy Stored} = \tfrac{1}{2} \cdot \text{Stiffness} \cdot \text{Deformation}^2 \qquad (1)$$

The implication of the mechanical relationship between stored energy, stiffness, and deformation for sports equipment design is obvious: in order to store a maximal amount of potential energy, an athlete should be given the stiffest piece of equipment which they are still able to maximally deform. In the archery example, assuming for the moment that stiffness is not an issue, the maximal bow deformation for any athlete is dictated by his or her arm lengths: an archer can only pull back the bowstring so far before becoming anatomically constrained. Based on the strength of the individual, any archer should realize maximal energy storage with the stiffest bow that they can still attain their maximal bow deformation, which would be considered their 'optimal' stiffness. Using a bow that is less stiff than this optimal stiffness will result in decreased energy storage, as there can be no additional deformation if the archer is already stretched to his or her full capability. Increasing bow stiffness beyond the optimal stiffness, on the other hand, will result in decreased deformation, as the athlete will not be strong enough to pull the string to the full extent. Although stiffness is increased, the associated decrease in deformation results in decreased stored energy. This theoretical example works well from a mechanical standpoint; for a given force and a constraint on maximum deformation, the resulting energy stored in bows of different stiffness can easily be calculated with the equations governing this theoretical mechanical system. However, as would be expected when working with athletes, biomechanical systems do not behave so predictably. As it turns out, theoretical novelties such as 'given force' and 'constrained maximum deformation' do not always occur in reality.

An example of how biomechanical systems can disrupt the predictive ability of this basic mechanical theory concerning energy storage can be found in ice hockey. Shooting in ice hockey is characterized by the athlete first imparting a deformation to the hockey stick (Figure 18.4) and then propelling the puck forwards as the stick recoils, which is clearly a

Figure 18.4 Photograph of an ice hockey player storing elastic energy in a hockey stick during shooting.

case of energy storage and return. In a study examining the influence of hockey stick stiffness on puck speed (Worobets *et al.*, 2006), researchers gave hockey players a set of sticks of varying stiffness which they used to shoot pucks using two different types of shooting techniques: a wrist shot and a slap shot. It was predicted that, as outlined above, the more flexible sticks would allow for more deformation, resulting in increased energy storage. Since more energy would be available to be returned to the puck, these sticks would be associated with higher puck speeds. This hypothesis is exactly what the results of the study supported; but only for one of the two types of shots. For the wrist shot, the players stored the most amount of energy in the more flexible sticks and realized higher puck speeds; maximizing the amount of energy storage resulted in enhanced performance. However, for the slap shot, when presented with the different sticks, the players reacted to the varying stiffnesses; they changed the forces they applied to the different sticks. Thus, increased flexibility did not result in increased deformation, and so despite the different stiffnesses, there were no discernible trends in the amount of energy stored in the different sticks. As we can see, designing equipment for athletes to achieve increased energy storage can be much more complex than elementary theoretical mechanics would suggest, as athletes sometimes respond to modified equipment with modified technique.

Energy return

Engineering sports equipment with the express purpose of storing maximal amounts of potential energy is not enough on its own to induce an increase in performance. Even if a vast amount of energy is stored during a loading cycle, it is essentially useless if it is not returned properly during the unloading cycle. While the exact details of how unloading must occur to be beneficial varies from sport to sport, in general the stored energy must be returned; without severe dissipation, at the right time, at the right frequency, and to the correct location.

Any sports equipment capable of storing and returning energy does so due to its elastic properties. These elastic elements, however, are not perfectly elastic; they lose some energy as it is absorbed or dissipated by the material during the unloading cycle (Equation 2). This loss of conservative energy is known as hysteresis. The practical impact of hysteresis is that no matter how much work an athlete does storing energy in a piece of equipment, the amount of work they get out of that equipment will always be less. How much less will be dependent on the properties of the materials from which the equipment is built. From an equipment design standpoint, addressing this issue is for the most part a matter for materials engineering. For example, newer composite materials such as carbon fibre can be produced to have a very low hysteresis. If an athlete's older piece of equipment loses 30 per cent of the energy stored in it, replacing it with a newer piece having a hysteresis of only 10 per cent provides opportunity for increases in performance.

$$\text{Energy return} = \text{Energy stored} - \text{Energy lost} \tag{2}$$

The energy stored in any equipment must be returned at the right time in order to maximize its usefulness. For example, if a golf club was to be clamped at the grip and a deformation imposed upon the shaft, potential energy would be stored in the shaft. When released from this bent orientation, the potential energy would be increasingly converted to club head kinetic energy as the club straightens. The club head has its maximum kinetic energy, therefore, when it has returned to it's fully straightened configuration. Maximum kinetic energy only occurs for a brief moment, as the kinetic energy is again converted to potential energy as the club head continues upon its path to a bent forward position. During the actual golf swing energy is certainly stored and returned by the club as there is substantial shaft deformation, however, for many golfers the shaft is in a bent forward position at the time of ball impact (Wishon and Grundner, 2005). This means that the portion of the club head speed contributed by the returning energy from the shaft was not at its maximum at the time of ball impact. Thus, the timing of the energy return was not optimally synchronized with the timing of the critical event (ball impact, in this case). By changing certain parameters of the golf club, such as shaft stiffness or swing weight, the time at which the club recoils to its straightened position may be delayed to precisely coincide with ball impact. In this case, the energy return would then occur at the right time.

During an unloading cycle, work is being done by the sports equipment: the equipment is applying a force over a certain distance. This force, combined with the time over which the unloading cycle occurs, dictates an impulse curve. This impulse curve can be looked at as a portion of a sine wave, and so the frequency of this impulse can be determined. In order for the energy returned by sports equipment to be effective, the frequency at which the energy is returned should match the frequency of the event which the energy is contributing to. For example, divers use springboards to improve jump height by storing energy in the springboard by deflecting it, then pushing off the board while it returns to its straightened position. In this way, the athlete's total jump height results from a combination of the work of his or her push off, plus the work that the springboard does on them. This cumulative effect is optimized if the impulses exerted by the athlete and the springboard are of the same matched frequency. If the frequency of the diver's push off was much higher than the oscillating springboard, then the diver would leave contact with the board before it had a chance to return the stored energy. If the frequency of the diver's push off was much lower than the oscillating springboard, then the diver would be riding several cycles of oscillations before they completed push off. Either way, jump performance would be significantly diminished compared to the case where the energy return was at the right frequency.

Stored energy must be returned at the right location if it is to effectively add to performance. An example illustrating this is in running footwear. During heel-toe running, the technique adopted by most joggers, the athlete lands on and compresses the heel, storing energy in this part of the shoe. As stance phase progresses, the entire foot comes into contact with the ground. Then, as the shoe continues to rotate forward, the athlete pushes off the ground using the forefoot only. Since the energy stored in the shoe during the landing phase was stored in the heel, it is essentially useless to the athlete, as the push off occurs from the forefoot location. This is not to say that energy must be stored and returned to the same site to be useful. Energy storage and return can take place in completely different sites, as long as the athlete can transfer the stored energy to the correct location upon return. Going back to the ice hockey example, we can see that although the energy is stored in the shaft of the stick, it is transferred by the athlete to the correct location: the distal end of the stick where blade and puck are in contact.

Minimizing non-conservative energy loss

Not all of the total work done by an athlete during a sporting event is directed entirely to achieving his or her goal. Some energy is spent doing non-performance-oriented work. Examples of these include: energy lost to heat and sound, energy required to overcome drag and friction, work done to stabilize, and work done to dampen vibrations. Each of these phenomena requires energy from the athlete, yet none of them directly aid the performance of the athlete. From this viewpoint, these are considered to be sources of non-conservative energy loss. Since this 'wasted' energy could have been utilized by the athlete to increase performance, minimizing these losses can have a profound effect. Designing sport equipment with this strategy in mind can, therefore, be particularly useful to athletes.

Athletes of any timed sport involving moderately high velocities are negatively affected by drag forces. Drag forces are resistive forces acting opposite to the direction of motion of any object travelling through a viscous medium. Since these forces are, by definition, resistive, the work they do reduces the kinetic energy of the athlete. It is, therefore, quite obvious why athletes of timed events would wish to minimize these drag forces, which is something that sport equipment can accomplish. For example, Thompson *et al.* (2001) used wind tunnel testing to appropriately modify helmets, leg fairings, poles, and suits into more aerodynamic designs, and demonstrated decreases of up to 16 per cent in the amount of drag experienced by speed skiers using this modified equipment. Also, Pendergast *et al.* (2006) were able to reduce drag on swimmers by up to 16 per cent by adding 'trip wires' to the backs of swimsuits in order to disrupt laminar flow. In both of these examples, investigators were able to substantially minimize energy loss by implementing equipment designed to reduce drag forces.

One of the functions of human skeletal muscles is the stabilization of joints. This is where agonist and antagonist muscle groups co-contract to increase joint stiffness. This is desirable in scenarios where unwanted joint motions would either disrupt performance or lead to injury, and thus these unwanted motions are avoided by using muscles to stabilize the joint. However, using muscles in this manner requires energy which the athlete may no longer direct towards their performance. Therefore, using sport equipment as an alternate means of increasing joint stiffness serves to minimize this loss of energy. An example where sport equipment is used as a joint stabilizer is in ice hockey and figure skates. Ankle stability is of particular concern when wearing skates. The added lever arm of the skate blade magnifies the moments about the ankle joint caused by the ice reaction forces during the sudden stops of ice hockey and jump landings of figure skating. Controlling these moments with ankle stabilizer muscles would cost the athlete tremendous amounts of energy were it not for the extremely stiff high-top design of hockey and figure skates. Since these stiff skates maintain stabilization for the athlete, they can instead use this saved energy in the pursuit of performance.

Vibrations travelling through the human body can have deleterious effects on performance. These effects can include reduced visual acuity, decreased fine motor control, reduced joint mobility, and general discomfort, depending on the amplitude and frequency of the vibrations (Levy & Smith, 2005). When the soft tissues of the body are subject to vibrations, muscular forces are recruited to dissipate this energy. This is another instance of muscular forces being used for a function which is not directly performance-related. In this setting, any sport equipment capable of damping vibrations would, therefore, minimize the energy wasted by this non-performance-oriented muscular work. For example, the rugged terrain traversed by mountain bikers can subject these athletes to large magnitude vibrations in two distinct frequency regions; 0–100 Hz and 300–400 Hz (Levy & Smith, 2005). Vibrations of this type would elicit an overwhelming damping response by the muscles of the cyclist. However, with the arrival of newer mountain bike suspension systems, these lower frequency vibrations can be reduced by up to 60 per cent and the higher frequency vibrations even more strongly attenuated. With less vibration, less energy is wasted damping soft tissue vibrations, and this energy is now available for performance-oriented tasks.

Conclusion

Sports equipment can be designed to improve performance. From an energy perspective, equipment can increase energy output of an athlete by optimizing the musculoskeletal system, store and return elastic strain energy, and help decrease the amount of energy an athlete loses.

Bicycles, the klap skate, and athletic footwear are some examples of sports equipment that can be utilized to optimize the musculoskeletal system. Specific characteristics of this equipment are designed to exploit the intrinsic force-length and force-velocity characteristics of muscle to maximize force and power production. Hockey sticks, springboards and archery bows are designed to store elastic strain energy. This strain energy is used to increase performance by supplementing the work performed by an athlete or changing the rate at which work is done. In designing these pieces of equipment, special attention has to be given to minimizing the energy loss in the equipment and ensuring the energy is returned at the right time, frequency and location. Aerodynamic apparel, hockey skates and bicycle shocks all aid in decreasing the energy that an athlete loses to drag, joint stabilization and vibrations. This allows an athlete to utilize more energy directly for performance by reducing the work the athlete performs for secondary tasks.

References

1. Hamley, E.J. and Thomas, V. (1967) 'Physiological and postural factors in the calibration of the bicycle ergometer', *Physiological Society*, 14–15: 55P–57P.
2. Houdijk, H., de Koning, J.J., de Groot, G., Bobbert, M.F. and van Ingen Schenau, G.J. (2000a) 'Push-off mechanics in speed skating with conventional skates and klapskates', *Medicine and Science in Sports and Exercise*, 32: 635–641.
3. Levy, M. and Smith, G.A. (2005) 'Effectiveness of vibration damping with bicycle suspension systems', *Sports Engineering*, 8 (2): 99–106.
4. Nordeen-Snyder, K.S. (1977) 'The effect of bicycle seat height variation upon oxygen consumption and lower limb kinematics', *Medicine and Science in Sports*, Summer 9 (2): 113–117.
5. Nigg, B.M., Stefanyshyn, D.J. and Denoth, J. (2000) 'Work and energy – mechanical considerations'. In B.M. Nigg, B.R. MacIntosh and J. Mester (eds) *Biomechanics and Biology of Human Movement* (pp. 5–18). Champaign, IL: Human Kinetics.
6. Pendergast, D.R., Mollendorf, R., Cuviello, R. and Termin II, A.C. (2006) 'Application of theoretical principles to swimsuit drag reduction', *Sports Engineering*, 9 (2): 65–76.
7. Roy, J.P. and Stefanyshyn, D.J. (2006) 'Influence of shoe midsole bending stiffness on running economy, joint energy and EMG', *Medicine and Science in Sports and Exercise*, 38 (3): 562–569.
8. Savelberg, H.H.C.M., Van de Port, I.G.L. and Willems, P.J.B. (2003) 'Body configuration in cycling affects muscle recruitment and movement pattern', *Journal of Applied Biomechanics*, 19: 310–324.
9. Stefanyshyn, D.J. and Fusco, C. (2004) 'Increased shoe bending stiffness increases sprint performance', *Sports Biomechanics*, 3 (1): 55–66.
10. Stefanyshyn, D.J. and Nigg, B.M. (2000) 'Influence of midsole bending stiffness on joint energy and jump height performance', *Medicine and Science in Sports and Exercise*, 32 (2): 471–476.
11. Stefanyshyn, D.J. and Nigg, B.M. (1998) 'Contribution of the lower extremity joints to mechanical energy in running vertical jumps and running long jumps', *Journal of Sports Sciences*, 16: 177–186.
12. Thompson, B.E., Friess, W.A. and Knapp II, K.N. (2001) 'Aerodynamics of speed skiers', *Sports Engineering*, 4 (2): 103–112.

13. Van Horne, S. and Stefanyshyn, D.J. (2005) 'Potential method of optimizing the klapskate hinge position in speed skating', *Journal of Applied Biomechanics*, 21 (3): 211–222.
14. Van Horne, S. (2004) *Mechanical and Performance Effects of a Modified Point of Foot Rotation during the Speed Skating Push*, M.Sc. University of Calgary.
15. Van Ingen Schenau, G.J. and Cavanagh, P.R. (1990) 'Power equations in endurance sports', *Journal of Biomechanics*, 23 (9): 865–881.
16. Yoshihuku, Y. and Herzog, W. (1990) 'Optimal design parameters of the bicycle-rider system for maximal muscle power output,' *Journal of Biomechanics*, 23 (10): 1069–1079.
17. Wishon, T. and Grundner, T. (2005) *The Search for the Perfect Golf Club*. Ann Arbor: Sports Media Group.
18. Worobets, J.T, Fairbairn, J.C. and Stefanyshyn, D.J. (2006) 'The influence of shaft stiffness on potential energy and puck speed during wrist and slap shots in ice hockey', *Sports Engineering*, 9 (4): 191–200.

19

Approaches to the study of artificial sports surfaces

Sharon J. Dixon
University of Exeter, Exeter

Introduction

Artificial sports surfaces have increasingly become the surface of choice in the majority of sports. Suggested advantages of these surfaces compared with natural playing surfaces include year round and repeated daily use, longevity, consistent surface properties and reduced maintenance (Kolitzus, 2003). Technical developments have led to materials being developed for use in specific sports surfaces and claims that performance is improved compared with natural playing surfaces.

For some sports, such as field hockey, playing characteristics have altered dramatically in response to the introduction of artificial playing surfaces. For others, such as rugby and cricket, natural turf remains the surface of choice. For sports such as tennis, artificial surfaces have become part of the game, but natural turf still plays its part. For others still, most notably soccer, there remains much resistance to the acceptance of artificial surfaces for everyday use, despite claims that these surfaces reproduce the properties of natural turf.

Historically sports surfaces have developed from materials such as turf, cinder and concrete into a multimillion pound business, with surfaces produced in a large range of artificial materials. This has led to many surface options being available commercially. A customer selecting a sports surface for a specific application, therefore, requires information on the performance of the surface. For example, selection of surfaces for the running track, cycling velodrome, hockey pitches, soccer pitches, etc., for the London 2012 Olympics should ideally to be based on sound scientific evidence. A key problem for those selecting artificial sports surfaces is to know on which criteria their selection should be based. It seems that both injury and performance considerations are important.

Whilst artificial playing surfaces have allowed sports participation around the clock and at any time of year, they have unfortunately also been associated with an increased incidence of injury and a change in injury patterns compared with participation on natural playing surfaces (Andreasson and Olofsson, 1983; James, Bates and Osternig, 1978). Although evidence linking specific playing surfaces with injury is inconclusive, it is evident that movement patterns and thus loading on the structures of the human body differ for different playing surfaces. Test methods have, therefore, been developed to study the influence of sports surfaces on factors associated with injury.

As well as influencing injury incidence, the development of artificial sports surfaces has also influenced performance in different sports. An extreme example is field hockey which, now predominantly played on artificial pitches, has become a totally different game to the one played on natural turf. Improvements in track and field performances have been contributed to greatly by developments in the track surface. Tennis also provides a good example of the extent to which a playing surface influences performance, with players who excel on one surface being relatively mediocre on another. These specific examples highlight the large extent to which the introduction of artificial playing surfaces has influenced performance across different sports. Performance considerations in sports surface design include the minimization of energy cost when performing on the surface and the optimization of movements such as turns and stops. Test methods have therefore also been developed to study the influence of surface characteristics on these variables.

When injury and performance requirements are considered, it seems that there are some key characteristics that differentiate one sports surface from another. These include cushioning, traction and energy cost on the surface. One approach to the study of artificial sports surfaces has been to utilize mechanical test procedures to assess aspects such as surface cushioning and traction and, where appropriate, ball bounce. This approach involves the replication of typical loads applied to the surface during human or ball interaction and the measurement of surface performance. Standard mechanical tests have been developed for the assessment of aspects such as force reduction (cushioning), traction and energy return. Such test procedures have been criticized for their inability to provide information that is relevant to the surface behaviour during human interaction.

An alternative approach is to quantify surface performance during human interaction, using biomechanical test procedures. Biomechanical testing of sports playing surfaces involves the measurement of loads and kinematics during the performance of typical sports movements on the surface. Biomechanical studies may be performed in the laboratory or in the field and may focus on injury or performance aspects. Biomechanical studies can identify differences in loading or movement patterns for different surfaces. However, relating these study results to injury or performance has proved difficult since our understanding of the forces and movements contributing to injury is still developing. An approach to identifying surface characteristics associated with injury is to monitor injury incidence for different playing surfaces. More of these medical surveillance epidemiological studies are beginning to appear in the literature.

This chapter evaluates the different approaches to sports surface study and summarizes current understanding of sports surface influences on injury and performance. Suggested 'ideal' approaches to increase our understanding of the influence of sports surfaces on injury and playing performance are discussed.

Mechanical approaches to the study of sports surfaces

Quantification of cushioning

Since playing surface characteristics result in them providing differing amounts of cushioning, impact test methods have been developed to quantify force reduction or peak acceleration in different playing surface materials. The earliest approach to this surface testing was

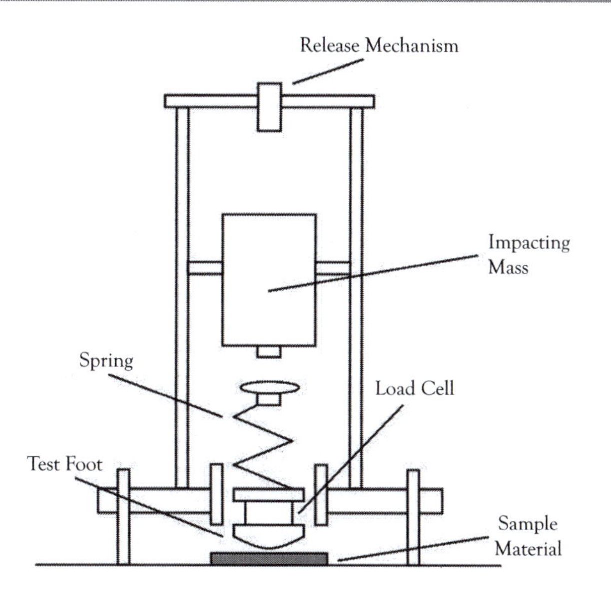

Figure 19.1 Artificial Athlete Berlin.

the development of mechanical test procedures to simulate loads applied during sports participation and the use of these tests to differentiate between playing surfaces.

One of the first documented studies quantifying sports surface cushioning (or resilience) characteristics was documented by Kolitzus (1984). Measurement of typical forces applied by humans during sports movements highlighted impact and active forces that required simulation by the mechanical test equipment. Thus mechanical devices, termed 'Artificial Athletes', were developed to simulate the impact of the human heel with the ground and the active propulsion phase of the step (Figure 19.1). For example, the test developed to simulate heel impact, termed the 'Artificial Athlete Berlin', involved the release of a 20 kg mass from a specified height onto a spring (Kolitzus, 1984). The spring was mounted on a metal foot and introduced compliance to provide a resulting impact force that is comparable in magnitude and timing to that measured for a running human. Using this test, the peak force is measured and compared with the force when impacting a concrete surface. The potential ability of the surface to reduce impact force is then presented as a 'force reduction' relative to concrete. The study presented by Kolitzus (1984) highlighted the possibility of simulating human interaction using a mechanical test device and provided a test procedure still used widely today to differentiate between playing surfaces in terms of their potential to cushion impacts (ITF, International Tennis Federation, 1997).

Several sports governing bodies have adopted these early impact test procedures to quantify the cushioning of different surfaces. For example, the ITF categorizes surfaces with a force reduction of 5–10 per cent as providing low force reduction, 10–20 per cent as medium and >20 per cent as high (ITF, International Tennis Federation, 1997). Example results for sample tennis surfaces tested using the Artificial Athlete Berlin are presented in Table 19.1 (from Dixon and Stiles, 2003). The range from 9.6 per cent to 33.5 per cent force reduction demonstrates the ability of these test procedures to detect differences between surfaces. Using the same test procedures, natural grass surfaces have been reported to provide a force reduction of around 50 per cent (Lees and Nolan, 1998). Whilst these procedures provide

Table 19.1 Force reduction for typical tennis surfaces (from Dixon and Stiles, 2003)

Playing surface	*Force reduction (per cent)*
Concrete	0
Cushioned acrylic hardcourt with concrete base layer	16.1
Polyurethane hardcourt with concrete base layer, 4 mm polyurethane	9.6
Polyurethane hardcourt with concrete base layer, 7 mm polyurethane	26.2
Sand-filled artificial turf surface with shockpad and concrete base layer	33.5

an indication of the relative potential of playing surfaces to reduce mechanical impacts, there is no evidence for stating a safe force reduction.

Test equipment such as the Artificial Athlete Berlin has been demonstrated to clearly differentiate between different sports surfaces. Particularly with recent developments in methods for filtering data, these test procedures provide an accurate and reliable means for comparing surface cushioning characteristics. However, the comparison of results from these mechanical devices with results of biomechanical studies quantifying ground reaction force (GRF) for different surfaces has highlighted discrepancies in results and concerns over the use of mechanical tests to study sports surface behaviour (Kaelin *et al.*, 1985).

A drawback of these impact test procedures is that they are too simplistic to comprehensively determine the mechanical behaviour of sports surface materials. As indicated by Miller, Baroud and Nigg (2000), the force-deformation properties measured by a mechanical impact test represent a structural property rather than a material property. As such, these tests do not account for boundary conditions such as loading rate, loading magnitude, indenter mass and contact area and specimen geometry. Miller *et al.* (2000) describe a set of measures to quantify the elastic behaviour of sports shoe and surface materials which take these factors into account. These include the measurement of unconfined compression and tension tests, and a confined compression test, providing information to fully represent the bimodal stress-stretch and compressible behaviour of the surface materials tested. It is suggested that this comprehensive approach to sports surface mechanical characterization is more likely to reveal relationships between surface material properties and human behaviour on the surface than the sole use of traditional impact test devices.

Quantification of traction

A shoe-surface interface has both translational (sliding) and rotational friction characteristics. Translational friction is present when movement between the two contacting surfaces tends to occur in a straight line, for example, when attempting to stop. The amount of twisting permitted between the shoe and surface is influenced by rotational friction. For example, if a player turns quickly and twists with their weight on the ball of their foot, rotational friction will oppose the ease of rotational sliding. If the rotational friction is high, this may stop the foot rotating easily and may result in high forces occurring at the ankle or knee.

The force opposing sliding movement between two bodies is influenced by the properties of both contacting surfaces. The frictional characteristics of the combined surfaces are represented by a constant value, known as the coefficient of friction. Thus, both the shoe and the sports surface properties are influential in determining the amount of sliding that occurs. The higher the coefficient of friction between the shoe and the surface, the higher the force needed to be applied to cause sliding, and thus the less likely it is that sliding will occur. This highlights the requirement for inclusion of footwear in the testing of sports surfaces when traction is to be assessed.

In order to systematically test the traction properties between sports surfaces and footwear, tests need to cover the range of actual situations that occur during the sports activity and the tests should also be repeatable. Human testing using biomechanical test methods can cover the range of situations that occur in sport. However, if slippage does not occur during the sport movement, this testing method is unable to determine the maximum traction capabilities between the surface and shoe. Mechanical testing therefore has the advantage that slippage can be enforced, thereby ensuring that the maximum traction between two bodies is measured.

Existing mechanical test devices used to quantify sliding friction tend to measure the resistance to horizontal forces to quantify frictional characteristics (ASTM, 1994). Similarly, resistance to rotational motion is quantified by measuring the torque required to rotate two parallel contacting surfaces during the application of a specified normal force. These tests provide reliable methods for comparison of different surfaces, but typically do not reproduce specific conditions occurring during human movement. For example, a mechanical test method used to measure translational friction is the British Transport and Road Research Laboratory portable skid resistance tester (Figure 19.2). A standard rubber foot is attached to the end of a pendulum. The pendulum is released from a horizontal position to slide over the sports surface. A friction coefficient is determined by the maximum height attained by the foot following sliding over the surface. By using a standard rubber foot rather than samples of typical shoe soles, this test does not provide specific friction values for the actual conditions occurring on the sports surface during play. It does, however, provide a standard test procedure for comparison of different surfaces. Such mechanical tests are not sufficient for the testing of shoe-surface properties under required loading conditions.

To overcome the limitation of tests that do not include shoe characteristics, efforts have been made to develop mechanical tests to include the shoe. For example, for frictional characteristics, a weighted shoe may be dragged across a playing surface. The vertical and horizontal force components can be measured using a force platform (Schlaepfer, Unold and

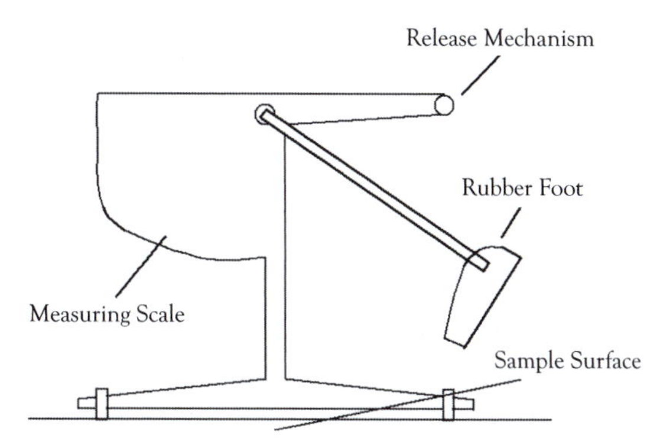

Figure 19.2 British Transport and Road Research skid resistance tester.

Nigg, 1983) or a strain gauge system (Wojcieszak, Jiang and Frederick, 1997), allowing the calculation of friction values for different shoe-surface combinations.

Rotational friction has been quantified using similar procedures, with measurement of torque using a force plate (van Gheluwe, Deporte and Hebbelinck, 1983) or instrumented mechanical leg (Bonstingl, Moorhouse and Nichol, 1975). Using an instrumented leg to simulate human interaction with playing surfaces, Bonstingl and co-workers demonstrated that the torque required for foot rotation is influenced by the characteristics of the shoe sole and playing surface, the weight being supported and the stance adopted by the player. Such findings have led to the suggestion that the most appropriate approach for the assessment of the frictional characteristics of playing surfaces is the development of standardized test movements to be performed mechanically on the test surface using typical shoe sole materials (Dixon, Collop and Batt, 1999). Livesay and colleagues (2006) recently compared artificial turf surfaces and natural turf, employing two footwear types – one designed for play on artificial turf and one for play on natural turf. The test device was constructed to measure the torque in response to an applied rotation, using the two footwear models. Both shoe and surface were found to influence the measured torque. Recent developments of robot systems that attempt to impact the shoe with the surface at typical velocities and angles occurring during sports movements (Cole *et al.*, 2003) may allow the development of an increased understanding of the factors influencing friction between a sports shoe and playing surfaces.

Frequently, studies incorporating shoe and surface properties to study traction involve the development of test equipment specific to the sports application, thus including appropriate footwear characteristics and loading patterns for the sport in question. In this way, limitations of standard mechanical test procedures are avoided. However, the specificity of the approach makes it difficult to compare results between studies. For example, Livesay, Reda and Nauman (2006) reported results of a study comparing the torque for five sports turf playing surfaces. This approach to the mechanical testing of sports surfaces has external validity, by replicating the conditions occurring on the sports surface, but has limitations regarding comparison with other test results where different procedures and/or conditions have been used.

Unfortunately it has been found that conflicting results are obtained when frictional properties of surfaces are quantified using mechanical test procedures compared with human tests (Nigg and Segesser, 1988; van Gheluwe and Deporte, 1992). Evidence that human kinematics are influenced by changes in surface friction (Dura, Hoyes, Martinez *et al.*, 1999), indicates that reliance on mechanical testing alone is not appropriate for the description of traction characteristics.

Using the current methodologies for characterization of traction properties of different playing surfaces means that surface properties cannot be described in isolation. This makes it difficult to uniquely specify surface characteristics. The coefficient of friction is dependent on the properties of both the shoe and surface materials, and therefore cannot be used to present the properties for the surface. An alternative method to uniquely define surface characteristics is by the presentation of surface roughness parameters (Chang *et al.*, 2001), since it has been demonstrated that surface roughness is highly correlated with slipperiness on a surface. Various stylus profilometers are available for the assessment of surface roughness. In addition, the recent development of laser scanning microscopes provides a non-contact method for quantification of surface roughness (Chang *et al.*, 2001). The increased use of such methods to uniquely define surface characteristics should improve understanding of the specific surface properties that influence slip on playing surfaces and the features of shoes that are desirable on a particular surface.

Biomechanical approaches to the study of sports surfaces

Injury considerations – surface cushioning

Chronic injuries, such as stress fractures and Achilles tendinopathy, have been associated with playing surface hardness or stiffness. Some researchers and practitioners have attributed injury occurrence on artificial surfaces to the impact loads experienced by the lower extremity (James, Bates and Osternig, 1978). An alternative suggestion has been that human adaptations in response to increased stiffness, such as increased eccentric muscle activity (Richie, Devries and Endo, 1993) and changes in movement patterns at the ankle and knee (Hamill, Bates and Holt, 1992; Stergiou and Bates, 1997; Dixon, Collop and Batt, 2000), may lead to increased injury incidence.

Biomechanical studies of surface stiffness have tended to measure GRF using a force platform beneath the sports surface (Figure 19.3). The cushioning provided by the surface has been described as the effectiveness to reduce the magnitude of the impact peak of the GRF (Nigg, Cole and Brüggemann, 1995). Several studies have demonstrated that similar impact forces occur during running on different surfaces, despite clear differences in force reduction values in mechanical tests (Feehery, 1986; Nigg and Yeadon, 1987; Dixon, Collop and Batt, 2000; Stiles and Dixon, 2006). The maintenance of similar impact peak of GRF on surfaces with differing mechanical cushioning ability has been suggested to be the result of adaptations in lower limb stiffness and kinematics (Ferris, Liang, Farley, 1999; Dixon *et al.*, 2000). It is also possible that, as a measure of the acceleration of the whole body centre of gravity, GRF is not sensitive enough to changes in surface and thus is not an appropriate variable for quantifying the cushioning effectiveness of sports surfaces. The development of models of the lower extremity has provided methods for estimating internal loads experienced by body structures (Morlock and Nigg, 1991). Using these methods, Cole, Nigg, Fick and Morlock (1995) demonstrated that the loads experienced at the ankle joint were not influenced by changes in the cushioning provided at the shoe-surface interface. In contrast, Krabbe, Farkas and Baumann (1992) reported that the estimation of internal loads highlighted differences in ankle joint loading on different surfaces, with frictional properties of the surfaces and running style of the person being identified as influential factors. It therefore seems that our understanding of the force reducing ability of sports surfaces during human

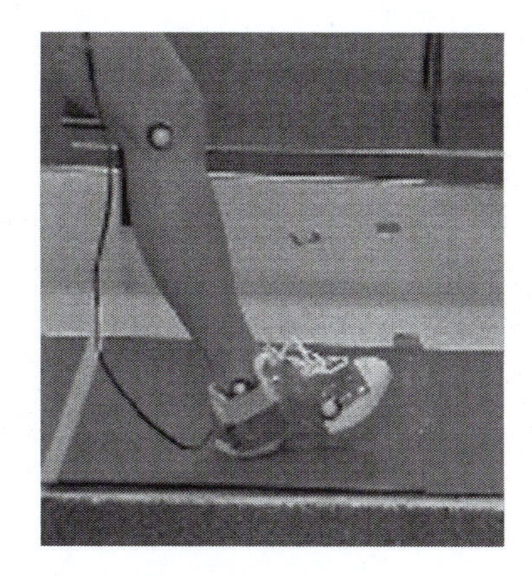

Figure 19.3 Collection of biomechanical data for a tennis shoe-surface combination.

interaction needs to be improved if specific shoe-surface characteristics that are desirable are to be identified.

Recent technological developments have allowed the study of surface cushioning using pressure insoles placed within the shoe. The measurement of pressure distribution for different surfaces allows the detection of differences in contact area and distribution of resultant force, which cannot be obtained using a traditional force plate system. However, these systems have only recently been sufficiently robust and provided a high enough sampling frequency for application in sports. The few studies utilizing pressure insoles to compare the cushioning of sports surfaces have demonstrated the potential of this approach to study surfaces with different mechanical cushioning (Dixon and Stiles, 2001; Ford, Manson, Evans *et al.*, 2006). Dixon and Stiles (2003) demonstrated that tennis surfaces are ranked in the same order using pressure insole data as with mechanical impact testing. Ford *et al.* (2006) detected significantly higher forefoot pressures when a cutting movement was performed on artificial turf compared with natural turf. These examples highlight the potential of studies using pressure insoles to increase our understanding of surface cushioning, suggesting that the future biomechanical study of sports surfaces should ideally include the use of this technology. In addition, it has been suggested that the measurement of pressure distribution may highlight specific differences between shoes that influence the comfort felt by the player (Llane, Brizuala, Dura and Garcia, 2002).

Injury considerations – surface traction

Biomechanical tests have also been used to investigate the friction characteristics associated with typical sports movements, using sample sports shoe-surface combinations. For example, Stucke, Baudzus and Baumann (1984) used a force plate system to measure typical vertical and horizontal forces and movement patterns during starting, stopping and turning movements on three different surfaces. A video system was used to detect the start of sliding between the shoe and the surface, for differentiation of static and dynamic friction. It was found that runners adapted to the change in surface by varying their movement patterns. For example, when performing a stopping movement on a cinder (clay) surface, the amount of knee flexion was found to remain reasonably constant throughout the movement, with sliding occurring between the shoe and surface. However, when performing the same task on an artificial sports surface, the high static friction resulted in negligible sliding occurring, and knee flexion was found to occur throughout the stopping movement. This adaptation to a change in surface by adjustment of movement may increase the energy cost of performing the task, it may reduce the effectiveness of the performance and it may also influence the chances of injury.

The ability of humans to adapt to different amounts of friction provided between the shoe and surface has also been demonstrated in a study by Dura, Hoyes, Martinez and Lozano (1999). However, these authors suggest that performance is not influenced adversely. The performance of a 180° turn was compared on five different playing surfaces. It was found that, despite differences in coefficient of friction indicated by the mechanical test procedures, the time taken to change direction on the different surfaces was comparable. For surfaces with relatively high coefficients of friction, the time taken in the braking phase of the turn was highest. This allowed time for more knee flexion than that observed for the surfaces with lower coefficients of friction. The authors suggested that this increase in knee flexion is a protective mechanism against potential high loads that may result from the limited sliding on these surfaces. For the surfaces with high coefficients of friction, the propulsion phase time was reduced, resulting in comparable total contact times for each of the five surfaces. It was also found that the maximum moment during turning was higher in surfaces with a higher coefficient of friction. The authors suggested that humans adapt to maintain forces and moment values below acceptable limits.

The relative contribution of the shoe and the surface to friction characteristics in tennis has been investigated using an open stance forehand (van Gheluwe and Deporte, 1992). Three typical tennis shoe models and four tennis surfaces with distinct traction characteristics were tested. It was found that, for the shoe and surface conditions tested, the surface characteristics were more influential than the shoe outsole on the friction force. In particular, the surfaces described as having a 'fluid' top layer of rubber granules generally resulted in the lowest frictional forces and torques. Future research should identify the specific design characteristics of playing surfaces that influence the coefficient of friction during interaction with different shoes and for a range of sports movements.

Performance considerations

Performance considerations in sports surface design include the minimization of energy cost when performing on the surface and the optimization of movements such as turns and stops. In an early study of surface effects on performance, McMahon and Greene (1979) demonstrated that the energy cost of running and the time taken to cover a specified distance, could be influenced by track stiffness. However, this research involved an area elastic indoor running track surface, which was able to deform over a large area beyond the area of foot contact, and thus to experience relatively large deformations. The authors suggested that, during deformation of the surface, energy was stored that was then returned to the runner during the propulsion phase of the step. In contrast, many surfaces for sports are point elastic in nature, deforming predominantly below the area of foot contact. These surfaces experience relatively small deformation during foot contact and thus do not have the same potential as area elastic surfaces to influence energy cost by storing and returning energy.

Although the storage and return of energy from artificial surfaces for many sports does not appear to be a realistic concept, it is possible that some surfaces have properties that reduce the energy cost of performing on them. To illustrate this possibility, Nigg and Anton (1995) demonstrated, using a mechanical model, that the work done during ground contact was higher for a more elastic shoe-surface combination than for a relatively viscous shoe-surface interface. This suggestion was subsequently supported by the demonstration of increased oxygen consumption when running in shoes with conventional elastic midsoles compared to those with increased viscoelastic properties (Stefanyshyn and Nigg, 1998). Nigg and Wakeling (2001) have suggested that this finding may be the result of a reduced energy cost of controlling tissue vibrations following ground impact when performing on a more viscous shoe-surface interface.

The design characteristics of a sports surface most likely to influence performance in sports involving stopping and turning are those features influencing traction on the surface. High shoe-surface friction is required for optimum acceleration and changes in direction, but friction should be sufficiently low that loads acting on the body do not exceed thresholds for damage. Since the ideal coefficient of friction will vary for different applications, and the majority of sports are characterized by a range of different movements, it is a difficult task to specify the most appropriate friction properties of a shoe-surface combination. However, Dura *et al.* (1999) conclude in their study of turning, that a coefficient of friction of around 0.4 is desirable. These authors suggest that this relatively low coefficient provides sufficient sliding to minimize the chances of injury, without sacrificing performance time in turning. This value of 0.4 is comparable to the value of 0.5 suggested by previous authors (Nigg and Segesser, 1986).

Modelling approaches to the study of sports surfaces

Modelling methods have been utilized as an alternative approach to the investigation of the effect of defined changes in material properties of sports surfaces. For example, McCullagh and Graham (1985) developed rheological models of materials used in sports shoes and surfaces. The authors claimed reasonable accuracy of the models, confirmed using the results of mechanical tests. Using this approach, it was demonstrated that the behaviour of typical sports surface materials could be modelled mathematically. The authors, therefore, suggested that such methods could potentially be used to develop shoe-surface combinations that are tuned for specific individuals. However, the modelling of materials used in playing surfaces is not as straightforward as McCullagh and Graham (1985) suggest. Shorten and Himmelsbach (2002) described how the non-linear behaviour of materials increases the complexities of models needed to accurately represent surface materials. These authors described how a power-law force displacement model could be used to provide an adequate representation of the impact reducing performance of surfaces with non-linearities.

Finite element models have also been used to investigate the influence of changes in surface construction on sports surface behaviour. For example, Baroud, Nigg and Stefanyshyn (1999) described a 3D model developed using parameters from existing sports surface samples. Finite element analysis was used to investigate the energy input, return and dissipation, using GRF data from a runner as input. The potential influence of controlled changes in surface properties could then be investigated. The development of models of increasing complexity allows the accurate modelling of surface materials and structures. This is likely to result in an increased use of such models to investigate the influence of defined changes on the combined behaviour of shoes and surfaces when used for sports.

Epidemiological approaches to the study of sports surfaces

Injury incidence on artificial sports surfaces has been reported for a range of sports. In tennis, the introduction of relatively hard court surfaces has been associated with an increased incidence of overuse injuries such as 'shin splints', Achilles tendinitis and patello-femoral pain (PFP) (Bocchi, 1984; Chard and Lachmann, 1987). Bastholt (2000) reported injury data in relation to surface for three year's of play in the men's ATP tour, with four surfaces utilized: indoor carpet, clay, hard court and grass. It was found that the distribution of injury location was similar for the four surfaces, with up to 50 per cent of injuries being to the lower extremity. Of these lower extremity injuries, the average number of treatments per match on grass was found to be more than double that on clay. The relative risk of lower extremity injury was also found to be significantly lower on hard court than on grass, but higher on hard court than on clay. Based on these data, it was concluded that playing on grass or hard courts resulted in more lower extremity injuries than playing on clay. Bastholt (2000) discussed differences in style of play on different surfaces possibly contributing to these differences in injury treatments and also highlighted the relative hardness and roughness of hard court surfaces. In support of the suggested relationship between tennis playing surface and injury occurrence, Nigg and Segesser (1988) reported a lower incidence of injury on clay surfaces than on artificial tennis court surfaces. These authors reported results of a questionnaire returned by 1,003 players performing on a range of surfaces including clay, synthetic sand, a synthetic surface and asphalt. The synthetic sand surface included a loose granular topping, whilst the synthetic surface had no granular topping. It was found that there were significantly more injuries reported by players performing on the synthetic surface than for those playing on clay or synthetic sand.

Descriptive epidemiological studies allow speculation of the risk factors associated with injury, but do not allow confirmation of causal relationships. For example, since the synthetic surface and synthetic sand surface studied by Nigg and Segesser (1988) had similar cushioning properties, the authors speculated that differences in cushioning of the surfaces were not the main cause of greater pain or injury on the synthetic surface. The main difference reported between the clay and synthetic sand surfaces and the synthetic and asphalt surfaces was in their frictional behaviour, with the clay and synthetic sand having a relatively low coefficient of sliding friction. Specifically, it was found that there was a lower incidence of injury to the lower limb when playing on surfaces that allow sliding. This was suggested to be the result of relatively high forces acting on the human body when a movement is stopped abruptly, such as when the front foot in a running forehand is fixed on the surface and no sliding occurs. If the surface allows a degree of sliding, lower forces are applied over longer time periods, which appears to reduce the chances of injury. The authors therefore speculated that tennis surfaces with relatively low frictional resistance were associated with lower injury incidence, and suggested an optimal coefficient of friction in the region of 0.5. The speculative nature of associations between sports surface specific characteristics and injury is a feature of these descriptive studies.

Brooks and Fuller (2006) describe how analytical studies, which include case-control, cohort and intervention studies, allow causal relationships between risk factors and injuries to be investigated. For example, a prospective two-cohort study has recently demonstrated few differences between injury risk on artificial turf compared with natural turf (Ekstrand, Timpka and Hagglund, 2006). A key feature of this study was the identification of two study cohorts exposed to different risk and the monitoring of these forward in time. This study was described as the first to compare injury risk on artificial turf and natural grass, providing valuable data for the football community and researchers in this area. Similar approaches to the study of surfaces in other sports should contribute greatly to our understanding of factors associated with the development of injury.

Overall conclusions

This chapter has reviewed literature examples describing the study of artificial sports surfaces using mechanical, biomechanical and epidemiological approaches. Each of these approaches has its merits and its limitations. It is suggested that combining the different approaches may be the most appropriate approach to increasing our understanding of sports surfaces.

Despite there appearing to be a large range of cushioning abilities and traction characteristics for different sports surfaces, the majority of information relating surface characteristics with injury and performance has been anecdotal. The continued use of analytical studies, in the form of case-control, cohort and intervention studies, for a range of different sports, should increase our understanding of sports surfaces linked with the development of injuries. Understanding the development of these injuries requires the subsequent use of mechanical approaches to quantify specific surface material properties and biomechanical approaches to investigate injury mechanisms.

In general, mechanical test procedures provide repeatable tests that are relatively straightforward to perform. However, the relevance of the tests depends on whether the results correlate with human interaction with the sports surface. In fact, several studies have demonstrated that similar impact forces occur during running on different surfaces, despite the clear differences in force reduction values in mechanical tests (Feehery, 1986; Nigg and Yeadon, 1987; Dixon, Collop and Batt, 2000). Similarly, the results of biomechanical testing of traction has tended not be correlate with mechanical test results. The approach suggested by Miller *et al.* (2000) of fully characterizing the material properties of a surface is suggested to be the most appropriate approach to the mechanical testing of sport surface materials. Unique characterization of surface roughness, as described by Chang *et al.* (2001), provides a more comprehensive approach to the mechanical study of traction.

Biomechanical studies utilize humans and thus allow the detection of human adaptation to surface properties. An increased use of sports-specific movements in biomechanical studies should reveal more regarding differences between existing sports surfaces. The development of models to adequately estimate internal loading should also improve our understanding of injury mechanisms. Ideally, future studies should combine mechanical and biomechanical approaches to determine specific material characteristics that influence human biomechanics. This should allow the design of sports surfaces with properties appropriate for the demands of the sport.

References

1. Andreasson, G. and Oloffson, B. (1983) 'Surface and shoe deformation in sports activities and injuries', in *Biomechanical Aspects of Sport Shoes and Playing Surfaces*, B. M. Nigg and B. A. Kerr, eds., University of Calgary, Calgary, Alberta, Canada, pp. 55–61.
2. American Society for Testing and Material Standards (1995) *Test ASTM- F1614–95*.
3. Baroud, G., Nigg, B. M. and Stephanyshyn, D. (1999) 'Energy storage and return in sport surfaces', *Sports Engineering*, 2: 173–180.
4. Bastholt, P. (2000). 'Professional tennis (ATP Tour) and number of medical treatments in relation to type of surface', *Medicine and Science in Tennis*, 5.
5. Bonstingl, R. W., Morehouse, C. A. and Nichol, B. (1975) 'Torques developed by different types of shoes on various playing surfaces', *Medicine and Science in Sports*, 7: 127–131.
6. Brooks, J.H.M. and Fuller, C.W. (2006) 'The influence of methodological issues on the results and conclusions form epidemiological studies of sports injuries', *Sports Medicine*, 36, 459–472.
7. Brown, R. P. (1987) 'Performance tests for artificial sports surfaces', *Polymer Testing*, 7: 279–292.
8. British Standard 7044 (1990) 'Artificial Sports Surfaces. Part 2, Methods of Test. Section 2.2. Methods for determination of person/surface interaction', pp. 1–7.
9. Chang, W. R., Kim, I. J., Manning, D. P. and Bunterngchit, Y. (2001) 'The role of surface roughness in the measurement of slipperiness', *Ergonomics*, 44: 1200–1216.
10. Chard, M. D. and Lachmann, S. M. (1987) 'Racquet sports – patterns of injury presenting to a sports injury clinic', *British Journal of Sports Medicine*, 21: 150–153.
11. Cole, G. K., Nigg, B. M., Fick, G. H. and Morlock, M. M. (1995) 'Internal loading of the foot and ankle during impact in running', *Journal of Applied Biomechanics*, 11: 25–46.
12. Dixon, S. J., Batt, M. E. and Collop, A. C. (1999) 'Artificial playing surfaces research: a review of medical, engineering and biomechanical aspects', *International Journal of Sports Medicine*, 20: 1–10.
13. Dixon, S. J., Collop, A. C. and Batt, M. E. (2000) 'Surface effects on GRF and lower extremity kinematics in running', *Medicine and Science in Sports and Exercise*, 32: 1919–1926.
14. Dixon, S. J. and Stiles V. H. (2003) 'Impact absorption of tennis shoe-surface combinations', *Sports Engineering*, 6: 1–9.
15. Dura, J. V., Hoyos, J. V., Martinez, A. and Lozano, L. (1999) 'The influence of friction on sports surfaces and turning movements', *Sports Engineering*, 2: 97–102.

16. Ekstrand, J., Timpka, T. and Hagglund, M. (2006) 'Risk of injury in elite football played on artificial turf versus natural grass: a prospective two-cohort study', *British Journal of Sports Medicine*, 40: 975–980.
17. Feehery, R. V. (1986) 'The biomechanics of running on different surfaces', *Sports Medicine*, 3: 649–659.
18. Ferris D.P., Liang K., Farley C.T. (1999) 'Runners adjust leg stiffness for their first step on a new running surface', *Journal of Biomechanics*, 32: 787–794.
19. Van Gheluwe, B. and Deporte, E. (1992) 'Frictional measurement in tennis on the field and in the laboratory', *International Journal of Sports Biomechanics*, 8: 48–61.
20. International Tennis Federation (1997) *An Initial Study on Performance Standards for Tennis Court Surfaces*, ITF, London, UK.
21. James, S. L., Bates, B. T. and Osternig, L. R. (1978) 'Injuries to runners', *American Journal of Sports Medicine*, 6: 40–50.
22. Kaelin, X., Denoth, J., Stacoff, A. and Stuessi, E. (1985) 'Cushioning during running – material tests versus subject tests', in *Biomechanics: Principles and Applications*, S. Perren, ed., Martinus Nijhoff Publishers, London, pp. 651–656
23. Krabbe, B., Farkas, R. and Baumann, W. (1992) 'Stress on the upper ankle joint in tennis-specific forms of movement', *Sportverletz Sportschaden*, 6: 50–57.
24. Kolitzus, H. J. (1984) 'Functional standards for playing surfaces', in *Sport Shoes and Playing Surfaces: Biomechanical Properties*, E.C. Frederick, ed., Human Kinetics Publishers Inc., Champaign, IL, pp. 99–117.
25. Kulund, D. N., McCue, F. C., Rockwell, D. A. and Gieck, J. H. (1979) 'Tennis injuries: prevention and treatment. A review', *American Journal of Sports Medicine*, 7: 249–253.
26. Lawn Tennis Association website. www.lta.co.uk/ (retrieved October 2002).
27. Lees, A. and Nolan, L. (1998) 'The biomechanics of soccer: A review', *Journal of Sports Sciences*, 16: 1–14.
28. Llane, S., Brizuela, G., Dura, J. V. and Garcia, A. C. (2002) 'A study of discomfort associated with tennis shoes', *Journal of Sports Sciences*, 20: 671–679.
29. McCullagh, P. J. J. and Graham, I. D. (1985) 'A preliminary investigation into the nature of shock absorbency in synthetic sports materials', *Journal of Sports Sciences*, 3: 103–114.
30. McMahon, T. A. and Greene, P. R. (1979) 'The influence of track compliance on running', *Journal of Biomechanics*, 12: 893–904.
31. Miller, J. E., Baroud, G. and Nigg, B. M. (2000) 'Elastic behaviour of sport surface materials', *Sports Engineering*, 3: 177–184.
32. Morlock, M. M. and Nigg, B. M. (1991) 'Theoretical considerations and practical results on the influence of the representation of the foot for estimation of internal forces with models', *Clinical Biomechanics*, 6: 3–13.
33. Nigg, B. M. and Segesser, B. (1988) 'The influence of playing surfaces on the load on the locomotor system and on football and tennis injuries', *Sports Medicine*, 5: 375–385.
34. Nigg, B. M. and Yeadon, M. R. (1987) 'Biomechanical aspects of playing surfaces', *Journal of Sports Sciences*, 5: 117–145.
35. Nigg, B. M. (1990) 'The validity and relevance of tests for the assessment of sports surfaces', *Medicine and Science in Sports and Exercise*, 22: 131–139.
36. Nigg, B. M. and Anton, M. (1995) 'Energy aspects for elastic and viscous shoe soles and playing surfaces', *Medicine and Science in Sports and Exercise*, 27: 92–97.
37. Nigg, B. M. and Wakeling, J. M. (2001) 'Impact forces and muscle tuning: a new paradigm', *Exercise and Sports Sciences Reviews*, 37–41.
38. SATRA Footwear Technology Centre Bulletin (1990) SATRA, Northants, UK.
39. Schlaepfer, F., Unold, E. and Nigg, B. M. (1983) 'The frictional characteristics of tennis shoes', in *Biomechanical Aspects of Sport Shoes and Playing Surfaces*, B. M. Nigg and B. A. Kerr (eds.), University of Calgary, Calgary, AB, Canada, pp. 153–160.

40. Seaward, E. (2000) 'Construction and maintenance of championship grass courts', *Proc. 1st International Congress of Tennis Science and Technology, Blackwell Science*, Oxford, pp. 201–208.
41. Shorten, M. R. and Himmelsbach, J. A. (2002) 'Shock attenuation of sports surfaces', *Proc. 4th International Conference of Engineering of Sport*, Blackwell Science, Oxford.
42. Stefanyshyn, D. J. and Nigg, B. M. (1998) 'Influence of viscoelastic midsole components on the biomechanics of running', *Proc. 3rd World Congress of Biomechanics*, Sapporo, Japan, p. 376b.
43. Stiles, V. H. and Dixon, (2002) 'Biomechanical response to different surfaces during a running forehand', *Journal of Applied Biomechanics*, 22: 14–24.
44. Stucke, H., Baudzus, W. and Baumann, W. (1984) 'On friction characteristics of playing surfaces', in *Sport Shoes and Playing Surfaces: Biomechanical Properties*, E.C. Frederick, ed., Human Kinetics Publishers Inc., Champaign, IL, pp. 87–97.
45. Wojcieszak, C., Jiang, P. and Frederick, E. C. (1997) 'A comparison of two friction measuring methods', *Proc. 3rd Symposium on Functional Footwear*, Tokyo, pp.32–33.

Section V

Biomechanics in sports

20

Biomechanics of throwing

Roger Bartlett[1] and Matthew Robins[2]
[1]University of Otago, Dunedin; [2]Nottingham Trent University, Nottingham

Introduction

This chapter focuses on those sports or events in which the participant throws, passes, bowls or shoots an object from the hand. The similarities between these activities and striking skills – such as the tennis serve – make much of the research into the latter also relevant to applied work in throwing, but these skills will not be covered explicitly in this chapter.

Throwing movements are often classified as underarm, overarm or sidearm. This chapter will concentrate mainly on overarm throws; much of the material presented can be extrapolated to underarm or sidearm throws. Overarm throws are characterized by lateral rotation of the humerus in the preparation phase and its medial rotation in the action phase. This movement is one of the fastest joint rotations in the human body. The sequence of movements in the preparation phase of a baseball pitch, for example, include, for a right-handed pitcher, pelvic and trunk rotation to the right, horizontal extension and lateral rotation at the shoulder, elbow flexion and wrist hyperextension. These movements are followed, sequentially, by their anatomical opposite at each of the joints mentioned plus radio-ulnar pronation. As Marshall (2000) showed, the long-axis rotations of the arm do not fit easily into the assumed proximal-to-distal sequence of the other joint movements.

The mass and dimensions of the thrown object plus the size of the target area and the rules of the particular sport are constraints on the movement pattern of any throw. Bowling in cricket differs from other similar movement patterns, as the rules do not allow the elbow to extend during the delivery stride. The interpretation of this rule is fraught with difficulty. If the umpires consider that this law has been breached, they can 'call the bowler for throwing' but umpires can err in calling throws. One reason for this is that they only have a two-dimensional (2D) view of the three-dimensional (3D) movements of the arm.

The goal of a throwing movement will generally be distance, accuracy or some combination of the two. In throws for distance, the release speed – and, therefore, the force applied to the thrown object – is crucial. In some throws, the objective is not to achieve maximal distance; instead, it may be accuracy or minimal time in the air. In accuracy dominated skills, such as dart throwing, some passes and free throws in basketball, the release of the object needs to achieve accuracy within the distance constraints of the skill. The interaction of speed and accuracy in these skills is often expressed as the speed–accuracy trade-off. This has been investigated particularly thoroughly for basketball shooting. The shooter has to release the ball with speed and accuracy to pass through the basket.

Optimizing performance

In many throws, the objective is to maximize, within certain constraints, the range achieved. Any increase in release speed (v_0) or release height (y_0) is always accompanied by an increase in the range. If the objective of the throw is to maximize range, it is important to ascertain the best (optimum) release angle to achieve this. The optimum release angle (θ), ignoring air resistance, can be found from: $\cos 2\theta = g\, y_0 / (v_0^2 + g\, y_0)$, where g is the acceleration due to gravity. For a good shot putter, this would give a value around 42°. Although optimum release angles for given release speeds and heights can easily be determined mathematically, they do not always correspond to those recorded from the best performers in sporting events. This is even true for the shot put (Tsirakos *et al.*, 1995) in which the object's flight is the closest to a parabola of all sports objects. The reason is that the calculation of an optimum release angle assumes, implicitly, that release speed and release angle are independent of one another. For a shot putter, the release speed and angle are, however, not independent, because of the arrangement and mechanics of the muscles used to generate the release speed of the shot. A greater release speed, and hence range, can be achieved at an angle (about 35°) that is less than the optimum release angle for the shot's flight phase. If the shot putter seeks to increase the release angle to a value closer to the optimum angle for the shot's flight phase, the release speed decreases and so does the range.

In javelin throwing, some research has assessed the interdependence of the various release parameters. The two for which an interrelationship is known are release speed and angle. Two groups of researchers have investigated this relationship, one using a 1 kg ball (Red and Zogaib, 1977) and the other using an instrumented javelin (Viitasalo and Korjus, 1988). Surprisingly, they obtained very similar relationships over the relevant range, expressed by the equation: release speed (m/s) = nominal release speed (m/s) – 0.13 (release angle (°) – 35°). The nominal release speed is defined as the maximum speed at which a thrower is capable of throwing for a release angle of 35°. In the javelin throw, the aerodynamic characteristics of the projectile can significantly influence its trajectory. It may travel a greater or lesser distance than it would have done if projected in a vacuum. Under such circumstances, the calculations of range and of optimal release parameters need to be modified considerably to take account of the aerodynamic forces acting on the javelin. Furthermore, more release parameters are then important. These include the angular velocities of the javelin at release, such as the pitching and yawing angular velocities, and the 'aerodynamic' angles – the angles of pitch and yaw. A unique combination of these release parameters exists that will maximize the distance thrown (Best *et al.*, 1995). Away from this optimum, many different combinations of release parameters will produce the same distance for sub-optimal throws. The implications of this sub-optimal variability and the different 'steepnesses' of the approaches to the optimal conditions have yet to be fully established.

Another complication arises when accuracy becomes crucial to successful throwing, as in shooting skills in basketball. A relationship between release speed and release angle is then found that will satisfy the speed-accuracy trade-off. For a given height of release and distance from the basket, a unique release angle exists for the ball to pass through the centre of the basket for any realistic release speed. Margins of error for both speed and angle exist about this pair of values. The margin of error in the release speed increases with the release angle, but only slowly. However, the margin of error in the release angle reaches a sharp peak for release angles within a few degrees of the minimum-speed angle (the angle for which the release speed is the minimum to score a basket). This latter consideration dominates the former, particularly as a shot at the minimum speed requires the minimum force from the shooter. The minimum-speed angle is, therefore, the best one (Brancazio, 1992). The role of movement variability – both intra-individual and inter-individual (Hore *et al.*, 1996) – in distance – and accuracy-dominated throws has not been fully explained to date, and is discussed in the next section.

The co-ordination of joint and muscle actions is often considered to be crucial to the successful execution of throwing movements. For example, in kicking a proximal-to-distal sequence has been identified. As kicking has much in common with throwing, we might expect similar distal-to-proximal behaviour for the arm segments in throwing. This is not the case when the movement sequence includes long-axis rotations (Marshall, 2000).

Movement variability in selected throwing skills

Morriss *et al.* (1997) reported the results of a study of the men's javelin final in the 1995 World Athletics Championships, with a focus on arm contributions to release speed. The very large shoulder angular velocity for the silver medallist suggested a reliance on shoulder horizontal flexion and extension to accelerate the javelin, which would suit his linear throwing style. In contrast, the gold medallist used medial rotation of the shoulder as a major method of accelerating the javelin. This movement, combined with an elbow extension angular velocity that was at least 18 per cent larger than for any other of the 12 finalists, was an important part of the reason he was able to achieve the greatest release speed. The other finalists used various combinations of these three arm movements to generate release speed. Such differences between throwers hardly support the idea of a common optimal motor pattern or technique, and question the approach of trying to copy the most successful performers. These differences may be the result of the individual-specific self-organization process (see Clark, 1995) such that performers find unique solutions to a task, although some of these solutions may be sub-optimal.

Morriss *et al.* (1997) speculated that these differences in the movements of the upper arm and forearm between throwers have important implications for their physical training. The training exercises performed by each thrower should be done in a way that replicates their individual movement patterns such that the gold medallist, for example, when ball throwing should emphasize shoulder medial rotation and elbow extension to ensure movement specificity (Enoka, 1994).

Further evidence against the idea of a common optimal technique has been provided, for example, from self-organizing Kohonen maps for javelin throwing by Schöllhorn and Bauer (1998) and for discus throwing by Bauer and Schöllhorn (1997). Kohonen maps are artificial neural networks that can be used to reduce complexity by, for example, mapping multiple time-series data on to simple 2D outputs. Although this approach is still novel in sports biomechanics, the results of these theoretical studies support empirical throwing research.

Bauer and Schöllhorn (1997) used 53 discus throws (45 of a decathlete, 8 of a specialist) recorded using semi-automated marker tracking over a one-year training period. There were 34 kinematic time series for each throw, for 51 normalized times; these complex,

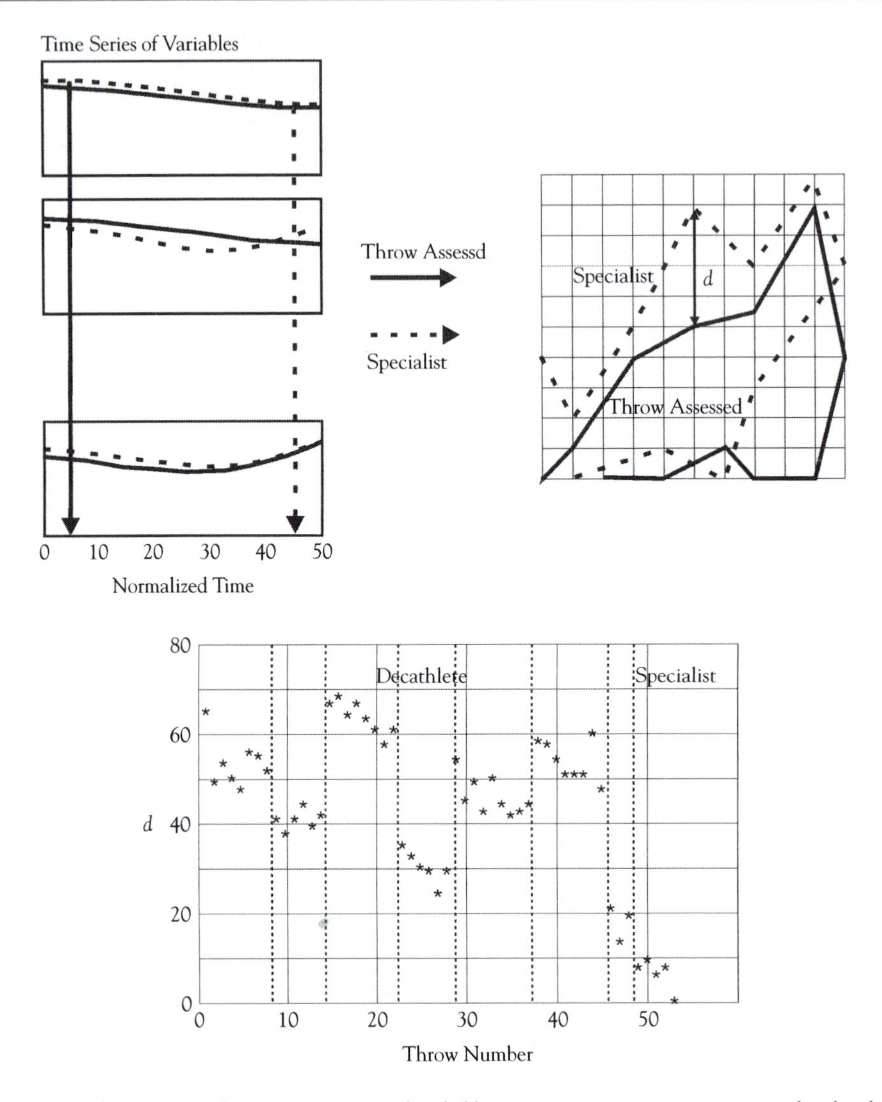

Figure 20.1 (a) Mapping of input times series (on left) onto an 11 × 11 output matrix for the discus throw to be assessed (solid line) against the reference throw from the specialist thrower; (b) values of the mean distance; (d) for various throws show grouping within training or competition sessions, between the vertical lines (adapted from Bauer and Schöllhorn, 1997).

multi-dimensional time series were mapped on to a simple 11 × 11 neuron output space (Figure 20.1a). Each sequence was then expressed as the mean deviation (*d* in the figure) of the output map – the continuous line – from the output map of one of the throws by the specialist thrower, shown by the dashed line. It should be noted here that this distance between network nodes has no Euclidean significance, although the authors' calculations imply that it has; this might have affected their findings.

The deviations for the eight specialist throws are shown on the right of Figure 20.1b and the decathlete's 45 throws on the left. The 'distances' are less for the specialist thrower as the comparator was one of his throws. Note the clustering of groups of throws, between the vertical lines, within training or competition sessions. There was more variability between than within sessions – for five groups of five trials, the authors computed inter- and intra-cluster variances, giving an inter-to-intra variance ratio of 3.3 ± 0.6. This shows that even elite throwers cannot reproduce invariant movement patterns between sessions. The supposed existence of such invariant patterns – which arose mainly from the motor program concept of cognitive motor control – has often been used, explicitly or implicitly, to justify the use of a 'representative trial' in sports biomechanics; such trials clearly do not exist.

Schöllhorn and Bauer (1998) reported a similar approach to analyze 49 javelin throws from eight elite males, nine elite females and ten heptathletes. This time, manual digitizing of estimated joint centre locations was used. Clustering was found for the male throwers – as a group – and for the two females for whom multiple trials were recorded. Variations in the cluster for international male athletes were held again to contradict any existence of an 'optimal movement pattern', even for 'good' efforts. Intra-individual coordination variability was reported for elite throwers in an empirical study by Morriss *et al.* (unpublished data). They studied four throws, all for maximum range, from the men's gold medallist at the 1996 Olympic Games, and presented the results as cross-correlation coefficients. The cross-correlations between the right shoulder and elbow joint angles of the throwing arm (Figure 20.2), for example, showed very similar patterns for rounds 2 and 6, and for rounds 4 and 5, within the limits of experimental error, as outlined below. The same was not true between the 2–6 and 4–5 pairs, which had substantial amplitude and phase differences (Figure 20.2). Bartlett *et al.* (1996) reported intra-individual differences in novice, club and elite javelin throwers; although not reported explicitly in that paper, intra-individual differences were greater for the novice and the elite throwers than for the club throwers. Even throwers striving for maximum distance cannot generate identical coordination patterns.

In the context of some of the above studies (Schöllhorn and Bauer, 1998; Morriss *et al.*, 1997), the results of the study of Bartlett *et al.* (2006) merit attention. They reported the results of a study of five trials of treadmill running in a laboratory to ascertain the reliability of manual digitizing of body coordinates with and without markers; four experienced operators digitized the trials on each of five consecutive days, with the no-markers trials

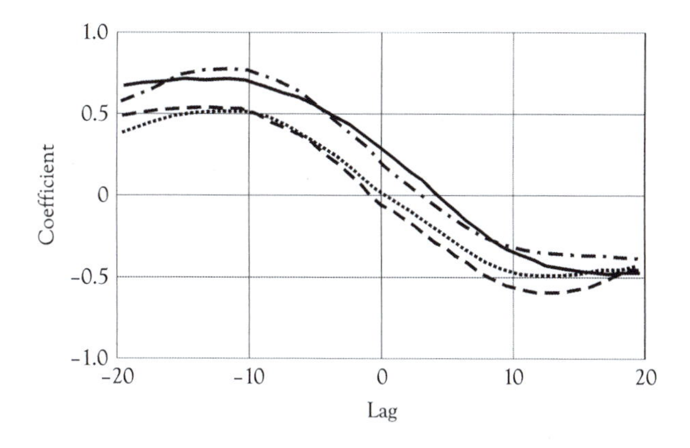

Figure 20.2 Cross-correlation functions at various phase lags between the throwing arm shoulder and elbow joints for four throws by the 1996 men's Olympic gold medallist (Rounds (R) as follows: continuous line R2; dashed R4; chained dashed R5; dotted R6).

being digitized before the ones with markers, to inhibit learning of marker positions. In the marker trials, the intra- and inter-operator reliability was good, with the former similar to autotracking; movement (trial-to-trial variability) dominated the other sources of variance. However, this was not true for the no-marker condition, in which movement variability was often swamped by the other sources of variance. They concluded that movement variability could not be determined reliably without the use of markers and speculated that this would be even worse for 3D studies in competition in which the positions of some 'joint centres' have to be estimated from invisible landmarks. It is with those results in mind that the four trials in the study of Morriss *et al.* (unpublished data) have been divided into two pairs – the differences within each pair would fall inside the limits of such experimental errors.

A computer simulation model that could be considered to have predicted variability in sports movements was seen in Best *et al.* (1995), who presented their results as contour maps of two variables. Figure 20.3 provides an example for the release angle of attack of the javelin against release angle, with other release parameters kept constant, as it is difficult to represent *n*-dimensional space in two dimensions. The contour lines are lines of equal distance thrown. The peak of the 'hill', shown by the black circle, represents the maximum distance that a given thrower could throw a particular make of javelin. It should be noted that only one combination of release parameters gave the maximum throw. However, on any 2D contour map, any pair of release parameters on a constant range line will produce that sub-optimal throw, even when the sub-optimal range is only slightly less than maximal, as for contour line 29 on Figure 20.3 indicated by the arrow; this generalizes to *n*-dimensional representations of the release parameters. These results show that infinite combinations of

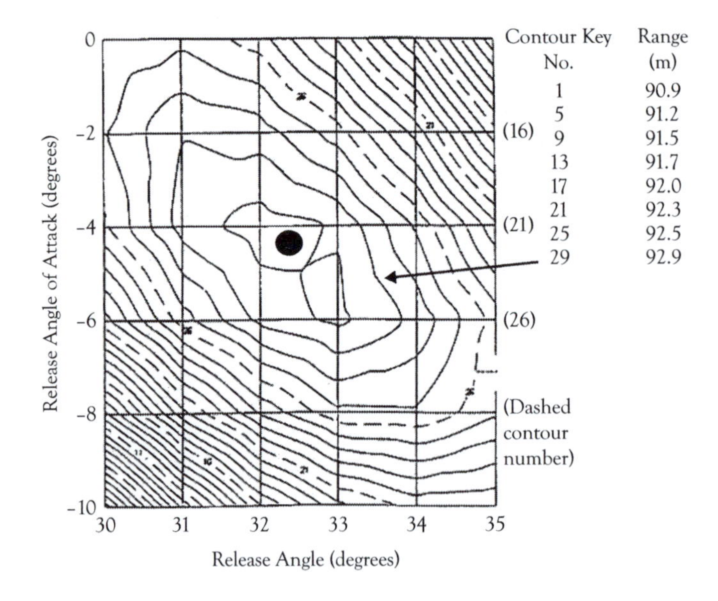

Figure 20.3 Contour map of simulated range thrown for different combinations of release angle and release angle of attack for the maximum release speed at which that thrower could throw and for a given model of javelin; contours are lines of constant range (adapted from Best *et al.*, 1995).

release parameters will result in the same sub-optimal range; each of these combinations could have arisen from kinematically different movements of the thrower. Furthermore, the unique maximal throw combination of release parameters could also have arisen from kinematically different motions that generated the optimal release parameter values (see Kudo *et al.*, 2000). Outcome consistency does not require movement consistency.

These computer simulation results appear to predict 'sub-optimal' movement variability. This variability in javelin throwing could be functional in allowing adaptations to environmental changes, such as wind conditions, or to distribute maximal loads on each throw among different tissues. However, the functionality of such variability is still somewhat speculative.

Shooting is the most important skill in the sport of basketball and many researchers have, consequently, studied this aspect of the game. Changes in basketball shooting kinematics have been examined as a function of distance (Elliott and White, 1989; Miller and Bartlett, 1996; Miller, 2002; Robins *et al.*, 2006), sex (Elliott, 1992), ability (Penrose and Blanksby, 1976; Hudson, 1985; Button *et al.*, 2003) or shooting accuracy (Miller, 1998). One of the general trends to emerge is that shooting is a compromise between the allowable margin for error and energy expenditure (see, for example, Miller and Bartlett, 1996). The joint configurations and release parameters used by players are, therefore, tailored to a particular shooting distance.

For example, players use a shallower shooting trajectory at greater distances to reduce the required ball release speed (the minimum speed principle, see Miller and Bartlett 1996). However, this produces a correspondingly lower margin for error because the basketball ring has a smaller elliptical area through which the ball can travel. The impact of these changing accuracy demands on measures of movement variability has rarely been addressed.

Movement variability in basketball shooting has received little attention until recently (Miller, 2002; Button *et al.*, 2003; Robins *et al.*, 2006). In a study by Miller (1998, 2002), 12 experienced basketball players each completed five successful and five not-deliberately unsuccessful shots at distances of 2.74, 4.23 (free-throw line) and 6.4 m. Measures of movement variability included coefficients of variation for ball release parameters, standard deviations for ranges of motion for the wrist, elbow and shoulder joints, and absolute and relative variability of segment end-point linear speeds at ball release. No evidence was found to suggest that the players could generate identical movements from shot to shot. There was increased absolute variability, expressed as standard deviation, in segment end-point speeds, from the longest to the shortest distances, which was reversed for relative variability, expressed by the coefficient of variation. No significant difference in absolute variability of joint position at release was found between successful and unsuccessful throws. For example, the standard deviation for range of movement (ROM) in free throws for the wrist and elbow joints was 7.5 and 6.3° for accurate shots and 7.5 and 5.7° for inaccurate shots. An increasing trend in absolute variability of segment end-point speed along the segment chain was also apparent, but a decreasing trend in relative variability.

Button *et al.* (2003) examined how movement variability in the basketball free-throw was affected by differing abilities among female players, ranging from a senior national team captain and two under-18 national team players to a player of very little experience. Skilled performers were characterized by increased inter-trial consistency from the elbow and wrist joints, but no clear reduction in trajectory variability occurred as skill increased. Trajectory variability referred to the standard deviation of linear elbow displacement at discrete points during the throwing action. Of particular interest, the observed increase in the standard deviation of joint angle at ball release from the elbow to the wrist joint was proposed to act as a compensatory mechanism. This compensatory interaction of joints along the kinematic chain has been suggested to minimize the variability of release parameters. Compensatory variability has also been reported in other dynamic throwing tasks (McDonald *et al.*, 1989; Muller and Loosch, 1999; Kudo *et al.*, 2000). Finally, skilled players exhibited a greater range of wrist motion but the authors did not allude to the potential importance of this finding from a coaching perspective.

Robins and co-workers (Robins, 2003; Robins *et al.*, 2006) analyzed five standardized, successful jump shots from each of six experienced basketball players. Shots were taken at distances of 4.25 (free-throw line), 5.25 and 6.25 m (three-point line). Their findings suggested that all participants were capable of replicating the desired movement pattern at all three distances, and showed a narrow bandwidth of movement variability (see Figure 20.4). This narrow bandwidth of movement variability corroborates earlier research demonstrating a

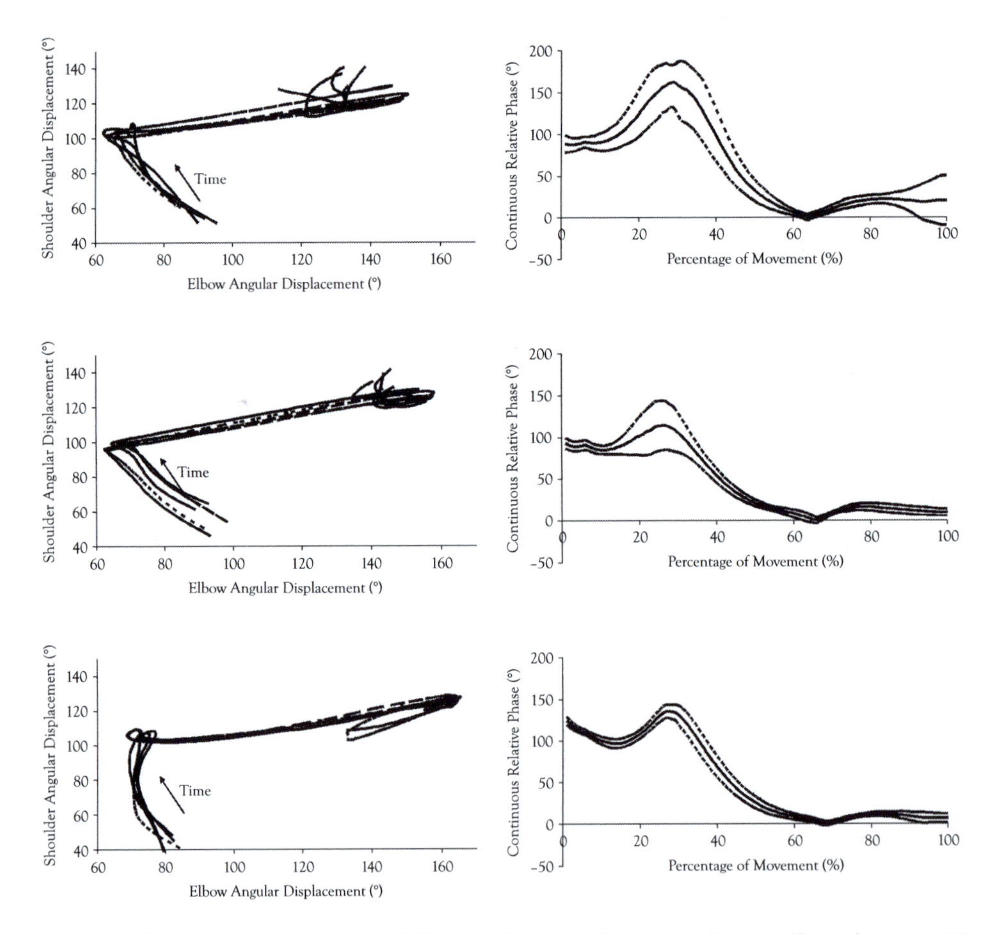

Figure 20.4 Changes in movement variability as a function of shooting distance (from the top: 4.25, 5.25 and 6.25 m) for five trials for a typical player expressed as angle-angle diagrams for the sagittal plane coupling of the shoulder and elbow (left-hand figures) and as continuous relative phase (CRP; shown as mean – continuous line – ± SD as dashed lines) between the same angles (right-hand figures) (from Robins, 2003).

reduction in movement variability with practice in basketball (Button *et al.*, 2003) and dart throwing (McDonald *et al.*, 1989). Although, considering the premise of the paper, a reduction in movement variability as a function of practice appears paradoxical, there is evidence to suggest that experts can exploit the available movement variability functionally to satisfy the constraints of the task. Robins *et al.* (2006) found that the movement variability at the shoulder and elbow did not increase as a function of distance; the wrist was the only joint with an increased variability of joint position at release. Standard deviations of 3.5°, 3.2° and 2.2° were reported at ball release for the shoulder joint, for example, at distances of 4.25, 5.25 and 6.25 m, respectively. This finding counters those reported by Miller (1998, 2002) and opposes the views proposed by impulse-variability theory (see Schmidt *et al.*, 1979).

Robins and co-workers (Robins 2003; Robins *et al.*, 2006) also observed an increase in the standard deviation of joint angle at ball release from the shoulder to the wrist joint. Furthermore, the variability of joint angles at release did not adversely affect the height, angle or speed of release, suggesting that compensatory mechanisms were present at the wrist and elbow joints to minimize the variability of projectile release parameters, thus implying a more functional role for movement variability. Values for height, angle and speed of release were delimited within 50 mm, 2° and 0.7 m/s, respectively, regardless of shooting distance. This finding agrees with those reported by Button *et al.* (2003), and offers additional support for the theory of compensatory variability. This compensatory strategy might explain why the variability values for the height of release reported by both Button *et al.* (2003) and Robins *et al.* (2006) were relatively unchanged. Within a practical context, a reduction in shoulder extension could be corrected for by increased elbow extension. An elevated height of release, a commonly reinforced coaching point (Wissel, 1994), would, therefore, have the benefit of requiring a lower speed of release for a given release angle (Miller and Bartlett, 1996).

The freeing of biomechanical degrees of freedom (DoF), which occurs with motor learning (Vereijken *et al.*, 1992), may assist with the apparent compensatory strategy exhibited by skilled performers. Skilled performers have been shown to display more than twice the wrist flexion (82°) of any other performer (Button *et al.*, 2003). An increase in wrist amplitude may serve several purposes. First, a larger ROM at the wrist may be used in conjunction with an increase in vertical and horizontal displacement of the jump (Elliott, 1992) to assist with impulse generation. Secondly, exploring a fuller ROM may enable a more effective compensatory mechanism to ensure end-point accuracy.

Although the analysis of discrete variables of interest yields important information at key instants, such as ball release, it is often necessary to supplement discrete measurements with continuous variables. The use of coordination profiling and, in particular, coordination variability, is often favoured by dynamical systems theorists and provides an invaluable insight into how individuals adapt to changing constraints. When assessing the effect of an increased shooting distance on coordination variability, Robins *et al.* (2006) reported a significant reduction in continuous coordination variability for the joint couplings of the wrist-elbow ($p = 0.01$) and wrist-shoulder ($p = 0.01$). Significant reductions were also observed for the variability of the joint couplings of the wrist-elbow ($p = 0.0001$), elbow-shoulder ($p = 0.07$) and wrist-shoulder ($p = 0.001$) at ball release. The decrease in both discrete and continuous measures of movement variability with distance can be attributed to the reduction in margin for error. A smaller margin for error at longer shooting distances requires a more constrained movement pattern, one that is characterized by lower movement variability. Therefore, it is feasible that the magnitude of variability is dependent on the constraints of the task (Newell and Vaillancourt, 2001).

The availability of equally functional movement patterns is important because it offers greater flexibility to adapt to potential perturbations and environmental uncertainty. This is particularly important during basketball competition because the extent of defender interference or pressure increases as players move closer to the basket. However, this flexibility is not available at greater distances, because the margin for error demands that the coordination pattern is more closely constrained. Coaches are, therefore, advised to devise strategies and play patterns that provide free scoring opportunities when shooting from the perimeter. This will minimize defender interference and prevent the shooter from having to manipulate his or her technique to any great extent.

Conclusions

In seeking to maximize the performance of an athlete in a throwing event, we need the correct optimal release model against which to evaluate throwers' performances. In throwing activities, long-axis rotations complicate proximal-to-distal sequencing, which needs to be addressed in, for example, devising training schedules.

The implications of sub-optimal variability in throwing and the different approaches to the optimal conditions have yet to be fully established. If movement variability is ubiquitous across throwing skills, as we have proposed in this chapter for javelin throwing, a speed-dominated skill, and basketball shooting, in which there is a speed-accuracy trade-off, and across stages of skill acquisition, what does this imply for the movement scientist and the sports practitioner? As different athletes perform the same task, such as a javelin throw, in different ways, there is no optimal movement pattern to achieve that task for athletes as a whole. Therefore, it makes no sense to try to copy specific details of a successful athlete's technique. These differences in movements between athletes have important implications for their physical training. The training exercises performed by each athlete should be done in a way that replicates their individual movement patterns, to ensure movement specificity. Because athletes do not replicate a movement exactly from trial to trial, for example, in basketball shooting, then the use of many trials in training needs to be carefully weighed against potential risks of overuse injury, particularly in activities in which loads on tissues are large, as in javelin throwing. As there is probably no unique movement that optimizes the performance of a given sports task, it makes much sense to allow athletes, particularly in the early stages of learning, to explore possible solutions, rather then for the coach to impose too many unnecessary constraints upon them. We mentioned in the previous section when discussing basketball shooting that the availability of equally functional outcomes is important because it offers greater flexibility to adapt to environmental uncertainty. This is clearly important in competition in many invasive team sports, because the defender interference or pressure often increases as players move closer to the 'target', be that the goal, try line, basket or whatever. However, this flexibility is not available at greater distances, because the margin for error demands that the coordination pattern is more closely constrained. Coaches are, therefore, advised to devise training strategies that provide free scoring opportunities from different distances from the target.

References

1. Bartlett, R. M., Bussey, M., and Flyger, N. (2006) 'Movement variability cannot be determined reliably in no marker conditions', *Journal of Biomechanics*, 39: 3076–3079.

2. Bartlett, R. M., Müller, E., Lindinger, S., Brunner, F., and Morriss, C. (1996) 'Three-dimensional evaluation of the kinematic release parameters for javelin throwers of different skill levels', *Journal of Applied Biomechanics*, *12*: 58–71.
3. Bauer, H. U. and Schöllhorn, W. I. (1997) 'Self-organizing maps for the analysis of complex movement patterns', *Neural Processing Letters*, 5: 193–199.
4. Best, R. J., Bartlett, R. M., and Sawyer, R. A. (1995) 'Optimal javelin release', *Journal of Applied Biomechanics*, *11*: 371–394.
5. Brancazio, P. J. (1992) 'Physics of basketball'. In *The Physics of Sports—Volume I* (pp. 86–95). New York: American Institute of Physics.
6. Button, C., MacLeod, M., Sanders, R., and Coleman, S. (2003) 'Examining movement variability in the basketball free-throw action at different skill levels', *Research Quarterly for Exercise and Sport*, *74*: 257–269.
7. Clark, J. E. (1995) 'On becoming skilful: Patterns and constraints', *Research Quarterly for Exercise and Sport*, 66: 173–183.
8. Elliott, B. (1992) 'A kinematic comparison of the male and female two-point and three-point jump shots in basketball', *Australian Journal of Science and Medicine in Sport*, *24*: 111–118.
9. Elliott, B. and White, E. (1989) 'A kinematic and kinetic analysis of the female two point and three point jump shots in basketball', *The Australian Journal of Science and Medicine in Sport*, *21(2)*: 7–11.
10. Enoka, R. M. (1994) *Neuromechanical Basis of Kinesiology*. Champaign: Human Kinetics.
11. Hudson, J. L. (1985) 'Prediction of basketball skill using biomechanical variables', *Research Quarterly*, 56: 115–121.
12. Hore, J., Watts, S., and Tweed, D. (1996) 'Errors in the control of joint rotations associated with inaccuracies in overarm throws', *Journal of Neurophysiology*, *75*: 1013–1025.
13. Kudo, K., Ito, T., Tautsui, S., Yamamoto, Y., and Ishikura, T. (2000) 'Compensatory coordination of release parameters in a throwing task', *Journal of Motor Behavior*, *32*: 337–345
14. Marshall, R.N. (2000) 'Applications to throwing of recent research on proximal-to-distal sequencing'. In Y. Hong and D. Johns (eds), *Proceedings of XVIII International Symposium on Biomechanics in Sport* (pp. 878–881). Hong Kong: CUHK Press.
15. McDonald, P. V., van Emmerik, R. E. A., and Newell, K. M. (1989) 'The effects of practice on limb kinematics in a throwing task', *Journal of Motor Behaviour*, *21*: 245–264.
16. Miller, S. A. (1998) 'The kinematics of inaccuracy in basketball shooting'. In H. J. Riehle and M. M. Vieten (eds), *Proceedings of XVI International Symposium on Biomechanics in Sports* (pp. 188–191). Konstanz: Universitatsverlag Konstanz.
17. Miller, S. A. (2002) 'Variability in basketball shooting: Practical implications'. In Y. Hong (ed.), *International Research in Sports Biomechanics* (pp. 27–34). London: Routledge.
18. Miller, S. A., and Bartlett, R. M. (1996) 'The relationship between basketball shooting kinematics, distance and playing position', *Journal of Sports Sciences*, *14*: 243–253.
19. Morriss, C., Bartlett, R. M., and Fowler, N. (1997) 'Biomechanical analysis of the men's javelin throw at the 1995 World Championships in Athletics', *New Studies in Athletics*, *12*: 31–41.
20. Muller, H. and Loosch, E. (1999) 'Functional variability and an equifinal path of movement during targeted throwing', *Journal of Human Movement Studies*, 36: 103–126.
21. Newell, K. M. and Vaillancourt, D. (2001) 'Dimensional change in motor learning', *Human Movement Science*, *20*: 695–715.
22. Penrose, T. and Blanksby, B. (1976) 'Film analysis: Two methods of basketball jump shooting techniques by two groups of different ability levels', *The Australian Journal for Health, Physical Education and Recreation*, March, 14–21.
23. Red, W.E. and Zogaib, A.J. (1977) 'Javelin dynamics including body interaction', *Journal of Applied Mechanics*, *44*: 496–497.
24. Robins, M. T. (2003). An investigation into the variability of segment limb coordination of the shooting arm in basketball. Unpublished master's thesis. University of Wales Institute Cardiff. UK.
25. Robins, M., Wheat, J.S., Irwin, G., and Bartlett, R. (2006) 'The effect of shooting distance on movement variability in basketball' *Journal of Human Movement Studies*, *20*: 218–238.

26. Schmidt, R. A., Zelaznik, H., Hawkins, B., Frank, J. S., and Quinn, J. T. (1979) 'Motor-output variability: A theory for the accuracy of rapid motor tasks', *Psychological Reviews*, 86: 415–451.
27. Schöllhorn, W. I. and Bauer, H. U. (1998) 'Identifying individual movement styles in high performance sports by means of self-organizing Kohonen maps'. In H. J. Riehle and M. Vieten (eds), *Proceedings of the XVI Congress of the ISBS, 1998* (pp. 754–577). Konstanz: Konstanz University Press.
28. Tsirakos, D.T., Bartlett, R.M., and Kollias, I.A. (1995) 'A comparative study of the release and temporal characteristics of shot put', *Journal of Human Movement Studies*, 28: 227–242.
29. Vereijken, B., Whiting, H. T. A., Newell, K. M., and van Emmerik, R. E. A. (1992) 'Freezing degrees of freedom in skill acquisition', *Journal of Motor Behavior*, 24: 133–142.
30. Viitasalo, J.T., and Korjus, T. (1988) 'On-line measurement of kinematic characteristics in javelin throwing'. In G. de Groot, A. P. Hollander, P. A. Huijing and G. J. van Ingen Schenau (pp. 583–587). Amsterdam: Free University Press.
31. Wissel, H. (1994) *Basketball: Steps to Success*. Champaign: Human Kinetics.

21

The biomechanics of snowboarding

Greg Woolman
Sportsmed, Christchurch

Introduction

Once banned from many ski areas, snowboarding has achieved dramatic growth and has emerged as the world's fastest growing winter sport. The endorsement of snowboarding as an Olympic sport in 1998 highlighted the sport's popularity. During 2004 snowboard participation in the USA was estimated to be 6.3 million (National Ski and Snowboard Retailer Association, 2004). Recent figures from the Canadian Snow Industry highlight that snowboarders occupied more than 30 per cent of the ski areas during 2005 (Canadian Ski Council, 2006).

However, as expected with greater participation, snowboarding injuries have increased steadily. One report highlighted a dramatic 42 per cent increase in US snowboarding injuries from 1993 to 1994 (US Consumer Product Safety Commission, 1995). Epidemiological studies have revealed injury patterns in snowboarding not previously seen in alpine sports. Wrist injuries have been prevalent since the sport first began and have continued to plague beginner snowboarders in particular (Abu-Laban, 1991; Chow *et al.* 1996; Sutherland *et al.* 1996; Idzikowski *et al.*, 2000; Rønning *et al.* 2001; Matsumoto *et al.*, 2004). As well as the escalating cost of this injury, it also threatened the healthy image of the sport. The use of wrist guards has since been introduced and promoted to help reduce this injury's incidence.

More recent studies are reporting an increase of injuries to regions such as the head, abdomen and spine, highlighting a changing injury trend associated with more aerial snowboarding maneuvers (Dohjima *et al.*, 2001; Fukuda *et al.*, 2001; Nakaguchi, 2002; Rajan and Zellweger, 2005). Terrain parks and half pipes offer the snowboarder an aerial incentive and are found in nearly every ski field. Even the ski industry has acknowledged the popularity of these areas now reflected in new ski and boot designs. Shorter twin tip skis and soft boots now allow the skier to also frequent such areas and perform jumps more easily.

In the largest study of foot and ankle injuries in over 3,000 snowboarders, a high incidence of fractures to the lateral process of the talus, known as snowboarders' ankle, was reported (Kirkpatrick, 1998). Laboratory research has subsequently identified several biomechanical causes of snowboarders' ankle (Boon, 2002; Funk, 2005) and, perhaps, also highlighted the failure of equipment to protect the snowboarder from this injury. As a result the snowboard boot has received the most criticism with many authors claiming a direct relationship between boot and injury (Ganong *et al.*, 1992; Shealy, 1993; Bladin *et al.*, 1993; Janes and Fincken, 1993; Kirkpatrick *et al.*, 1998). An accurate classification of boot type was not provided in these studies. Various descriptions of either soft, hard, ski or hybrid boot types are often published, with no real differentiation between these being made. In order that this information is made more useful and can be linked with injury, a more accurate equipment reporting system is required.

Epidemiological studies on snowboarding injuries are increasing; however, there are very few biomechanical studies on the sport itself. The injuries sustained while snowboarding highlight the biomechanical uniqueness of this sport and the equipment used.

Snowboarding equipment

Snowboard boots

Snowboard boots are designed to secure the rider's feet to the snowboard as well as help facilitate edge control and assist steering. They must protect the foot and ankle not only from the environment but also from potentially harmful motion. The snowboard boot must also dampen shock and vibrations. When compared with alpine ski boots, snowboard boots are designed specifically to allow for more freedom of ankle joint motion. The boots should allow for the rider to naturally perform snowboarding specific movements.

Snowboard boot models vary depending upon the style of snowboarding. Boots are generally classified by the range of motion that they allow at the ankle joint although this is primarily in their ability to allow ankle joint dorsiflexion. Softer boots allow greater ankle flexibility which is important for manoeuvrability and tricks usually seen in half pipe riding and jumping. This soft boot type is favoured by the majority of snowboarders with reports of 70–90 per cent usage (Ganong *et al.*, 1992; Bladin and McCrory, 1995; Chow *et al.*, 1996; Davidson and Laliotis, 1996; Kirkpatrick *et al.*, 1998). Stiffer boots reduce ankle motion and are used generally for faster speeds where linked slalom-like turns are performed. This boot rigidity means that movements of the foot and shank are transmitted to the snowboard edge more rapidly.

Boots designed specifically for 'step-in' bindings are more rigid in their shell construction given that they have no binding frame for additional support. Two studies have highlighted that step–in boot types are more restrictive of ankle joint motion when compared to strap in boots (Torres *et al.*, 2000; Delorme *et al.*, 2005).

Soft snowboard boots have been implicated as contributing to ankle injuries by several authors (Pino and Colville, 1989; Abu-Laban, 1991; Ganong *et al.*, 1992; Shealy, 1993; Bladin *et al.*, 1993; Paul and Janes, 1996; Ferrera *et al.*, 1999; Machold *et al.*, 2000). From a safety point of view the boot should not allow any joint motion beyond some critical point, although these critical points have yet to be defined.

Snowboard bindings

Bindings are used to couple both feet to the snowboard and are designed specifically not to release during a fall. These assist sagittal plane motion about the ankle joints so that forces can be transferred efficiently to either edge of the snowboard. Bindings primarily restrict movement of ankle joint plantar flexion, however do also reduce medial and lateral ankle movements. Most bindings are designed with a relatively low profile to maintain a rider's low center of mass (COM). Snowboard bindings are primarily of two distinct types with these being the strap binding, and the step-in binding.

The strap binding is where the boot is simply strapped into the binding, usually via several ratcheted closures. This is by far the most commonly used system. The step-in binding is a system where the sole of the boot integrates directly with the binding through a locking system. The step-in binding's popularity has diminished steadily in recent years.

Snowboards

Snowboards are manufactured like alpine skis using a composite sandwich type construction. Snowboards generally range in length from 140–170 cm with widths of 20–30 cm. Shorter snowboards are used for aerial manoeuvres and sharp turns as in freestyle, whereas longer snowboards are used for greater speeds and larger turns when freeriding. They comprise of a base, a core, and a top sheet with most snowboards being 10 to 12 mm thick. The larger surface area of the snowboard allows for less penetration and better gliding on softer snow compared to skis. A significant difference between snowboards and skis is the side-cut radius. Their shorter length and greater side-cut radius than skis allows snowboarders to perform tighter turns. Radius is expressed in metres with snowboards having radii below 10 m and skis more commonly between 14 m and 45 m (Lind, 1997; Federation Internationale de Ski, 2006).

The snowboard itself also offers an important means of deceleration. Maximal braking occurs between 3.6 m/s^2 and 8.7 m/s^2, comparing favourably to skiing, which highlights longer braking distances of between 5.0 m/s^2 and 14.1 m/s^2 (Nachbauer and Paps, 2004).

Stance position

The stance position (Figure 21.1) is very dependent upon the individual snowboarding style or discipline. The snowboarder will ride either a left foot forward (natural) or right foot forward (goofy) position. The effect of having both feet attached to the snowboard is considered protective for rotational injuries to the knee as the snowboard cannot rotate freely as a ski can in alpine skiing. There are several key stance positions that can vary with each snowboarder; the front foot angle, the back foot angle, and the stance width. More forward-orientated foot angles are usually associated with faster speeds for freeriding with turns performed through a greater arc. An alpine stance position where the front foot is angled +18° to +21° forward and back foot angled +3° to +6° forward is an example of a common freeriding stance (Gibbins, 1996; Leijen, 2007). A 'duck stance' position where the front foot is angled 15° to 21° forwards and the back foot angled 15° to 21° backwards is favoured by many snowboarders for freestyling (Gibbins, 1996; Leijen, 2007). This position enables riders to perform aerial rotations and is commonly used for half pipe and terrain parks. They can also select which end of the board they lead with.

The stance width is largely dependent upon the snowboarder's height; the taller the rider the wider the stance. With the stance being slightly wider than the hips there is an immediate triangulation effect of the lower limbs that contributes to additional stability.

The stance position places the snowboarder's hip, knee, and ankle joint axes perpendicular to the long axis of the snowboard. Flexion and extension about these joints will allow regulation of the rider's COM, as this can be raised or lowered, or shifted toward either edge

Figure 21.1 A typical alpine stance position with the back and front foot both angled forwards by 6° and 21°.

of the snowboard. Lower limb rotations occur naturally while turning and initiate subtle motions through the foot and ankle. As the foot is fixed to the snowboard through the boot and binding coupling it cannot move freely. As a result there is likely to be additional torque applied to the subtalar and ankle joints and may have significant implications for ankle joint injuries.

Snowboard injuries

Snowboarding research has identified unique trends and injury patterns quite different to those injuries of alpine skiing. These injuries provide an insight into the biomechanical forces involved with this sport. In snowboarding lower limb injuries, the left leg is injured 62–75 per cent of the time (Pino & Colville, 1989; Abu-Laban, 1991; Davidson and Laliotis, 1996; Kirkpatrick *et al.*, 1998). This is the leading leg in the more commonly adopted position of a right leg dominant rider. The mechanism of injury has been reported rather loosely in the literature and it is difficult to conclude such precise pathomechanics. Most injuries are defined as impact with the snow, with most knee and ankle injuries occurring when falling forward or to the toeside of the board (Shealy, 1993; Davidson and Laliotis, 1996).

A steady annual increase in the number of snowboarding knee injuries has been reported by Janes and Abbott (1999). Anterior cruciate ligament (ACL) injuries in expert snowboarders constituted 59 per cent in this study. Torjussen and Bahr (2005), in their review of elite snowboarding injuries, found that knee injuries constituted between 16–21 per cent of all snowboarding injuries. They also considered the nature of higher impacts such as those in elite snowboarding to be a cause of this twofold increase when compared to recreational snowboarders.

Ankle injuries

Unlike skiing, injuries to the ankle amongst snowboarders are more frequent. Ankle injuries in snowboarding have been reported with an incidence of between 5–38 per cent (Pino and Colville, 1989; Abu-Laban, 1991; Ganong *et al.*, 1992; Shealy, 1993; Nicholas *et al.*, 1994; Bladin and McCrory, 1995; Davidson and Laliotis, 1996; Sutherland *et al.*, 1996; Kirkpatrick *et al.*, 1998; Janes and Abbott, 1999; Shealy, 2007). By comparison, skiing ankle injuries constitute approximately 5–8 per cent of all skiing injuries (Shealy, 1993; Langran and Selvaraj, 2002; Fong *et al.*, 2007).

More recently a 50 per cent reduction in the incidence of snowboarding ankle injuries has been identified in 14 studies from 1985 to 2005 using a 'mean days between injury' meta-analysis (Shealy, 2007). This review highlights ankle injuries as constituting approximately 5 per cent of all injuries. Advances in boot design were suggested as the reason for this reduction. Fractures to the lateral process of the talus were rare and not commonly seen before the advent of snowboarding. In fact only 60 cases were reported in the literature before this injury was seen in snowboarders (Mills and Horne, 1987). Kirkpatrick *et al.* (1998) found in a study of over 3,000 snowboarding injuries that snowboarder's ankle represented a significant 2.3 per cent of these injuries. Hyperdorsiflexion and simultaneous inversion had been postulated as the mechanism of this injury by several authors (Hawkins, 1965; Nicholas *et al.*, 1994; McCrory and Bladin, 1996).

An *in vitro* study was performed to reproduce fractures to the lateral process of the talus, with measured ankle joint motion under axial loading (Boon *et al.*, 2001). This study highlighted that under axial loading external rotation of the tibia as well as ankle dorsiflexion and inversion reproduced this fracture type. Using similar methods, Funk *et al.* (2005) attempted to reproduce such fractures *in vitro* using different ankle positions. Each specimen subjected to dorsiflexion and eversion sustained a fracture to the lateral process of the talus. Both studies highlight unique biomechanical mechanisms that may occur during a snowboarding fall. Hyperdorsiflexion is a movement that is likely with toeside landings owing to the increased lever arm of the snowboard. In alpine skiing a significant reduction of ankle injuries was achieved with the advent of stiffer boot types. The compromise has, however, been an increase of serious knee injuries (Nantri *et al.*, 1999). It would, therefore, appear that the use of rigid snowboard boots is likely to also place the knee joint at greater risk of injury.

Snowboard biomechanics research

Despite the sport's huge popularity and rapid growth, biomechanical research related to snowboarding is scarce. Loads transmitted during snowboarding were measured by Bally and Taverney (1996). Ten participants were tested performing a series of linked turns down a marked course using a binding prototype that incorporated force transducers. Vertical forces of up to 1540 N and 1740 N and inversion-eversion forces of up to 1070 N and 780 N were recorded for the back and front foot. Flexion forces were recorded up to 310 N and 320 N for the back and front foot. Moments of rotation around the ankle were also recorded with 71 N·m and 62 N·m for the back and front ankle. The researchers found no significant correlation between the weight of the riders and the loads measured.

A releasable binding prototype was tested on ten snowboarders of varying ability (Bally and Shneegans, 1999). A ski-type binding that could be preset to a selection of minimal retention moments was utilized. Minimum retention moments for internal and external torsion were determined to be between 40 and 60 N·m with lateral release moments measured between 110 and 220 N·m. The authors concluded that the development of a releasable binding for snowboarding was at least feasible.

Knünz *et al.* (2001) measured both forces and moments during snowboarding. Load cells mounted between the snowboard and bindings were used. Two participants wore hard racing type boots, for whom data on carving turns were obtained for a series of both toe-side and heel-side turns. Peak vertical forces for the back boot were consistently greater for the heel side and toeside turns (1350 N, 1270 N) compared to the front boot (750 N, 300 N). Moments around the transverse axis were also found to be greater for the back boot

(170 N·m, 10 N·m) compared to the front boot (80 N·m, 10 N·m) for heel side and toe side turns. Moments around the boots longitudinal axis were again greater for the back boot (–110 N·m, 70 N·m) than the front boot (–40 N·m, 40 N·m) for heel side and toe side turns. The authors highlighted the fact that the loads through the back foot are consistently greater than the front foot, highlighting that the back foot is used much more for holding an edge and steering than was previously thought. It had been previously taught that the snowboarder should load more through their front foot while riding.

Knutz (2001) also tested one participant who wore softer freestyle boots where the rider was asked to perform turns generating high impact forces. The peak vertical forces were significantly higher than those recorded for the harder racing boots above. Maximum values of up to 2700 N and 2750 N were measured for the back and front boot. Transverse rotational moments were also measured with these highlighting moments of up to 140 N·m. The authors compared these to the lower values in skiing of up to 25 N·m. Flexion forces were measured up to 370 N and 390 N and inversion forces recorded up to 360 N and 890N for the back and front foot respectively.

Estes *et al.* (1999) calculated ankle deflection during a mathematically modelled forward fall in snowboarding. The equation assumed the rider of average height and weight would drop vertically and land on the front of the board at an angle of 45° to the horizontal snow surface at this velocity. Their modelling highlighted that maximum ankle deflection had the greatest sensitivity to boot stiffness. This sensitivity was more than 1.5 times greater than for snow damping, two times greater for snow stiffness, and more than four times greater for board stiffness and mass/inertia. These results highlighted the significance of snowboard boot stiffness over the other tested variables on ankle flexion.

The specific stiffness properties of snowboard boots have also been investigated (Torres *et al.*, 2000). An articulated and a rigid prosthesis were both tested to flex a snowboard boot mounted on a rotating disc through 15° intervals. The leg was loaded monotonically between 0 N and 1112 N using a tension cable system. Both a strap-in binding and a step-in binding were tested. Using the articulating model, noticeable differences were recorded between the two systems with the step-in binding having a greater overall stiffness, particularly for forward and backward lean (1.16 N·m/°, 3.02 N·m/°) compared to the strap-in (0.77 N·m/°, 2.07 N·m/°). Significant increases in stiffness values were obtained using the rigid model, where forward and backward lean values of 5.25 N·m/° and 10.38 N·m/° compared to 0.77 N·m/°, and 2.07 N·m/° for the articulating model. These results are the first published to quantify boot stiffness.

Woolman *et al.* (2002) studied 3D ankle joint motion of the front foot during simulated snowboarding landings (Figure 21.2). Twelve participants were tested in a laboratory using a Polhemus Fastrak® electromagnetic tracking system (Polhemus, Colchester, Vermont, USA). Three different boot types were tested with the participants falling at two different angles of 0° and 30° to the falling plane.

Of interest during this study were the findings of high ranges of external ankle rotation measured (Figure 21.3). Large mean external rotation values of 14.8° were measured in the 0° fall position. This was theorized to be based upon the greater transverse rotation moment placed on the ankle joint when the foot is positioned more perpendicular to the falling plane. Unusually, the timing of peak ankle flexion, external rotation, and inversion were identified to all occur nearly simultaneously during the falls. This was reported to be an unnatural sequence resulting in an abnormal foot and ankle position. Differences were also found between the boots with the softer boot type allowing greater range of motion (ROM) for ankle internal rotation, ankle flexion, and ankle inversion than the harder boot types. External rotation, however, was recorded as being greater in the harder boot types.

Figure 21.2 The laboratory drop tester; used to simulate snowboard landings at the University of Otago, Department of Physical Education.

Ankle joint motion while snowboarding was measured by Delorme *et al.* (2005). Five snowboarders were tested on a ski slope in both strap (soft) and step-in (stiff) binding systems using a portable Polhemus Fastrak® system. A total of 105 heel side and 109 toe side turns were recorded. The snowboarders assumed a common stance position of +21° front foot and +6° back foot angles. The authors highlighted the most noticeable differences being decreased dorsiflexion, and increased eversion in the front, compared to the back ankle. The stiffer boot also restricted ankle joint motion in dorsiflexion, eversion, and external rotation when compared to the soft boot type.

Tests in soft boots for heel side and toe side turns (in parentheses) highlighted maximum dorsiflexion of 3.6° (13.6°) for the front ankle, and 13.9° (25.4°) for the back ankle. Using hard boots maximum dorsiflexion values of 1.0° (8.6°) for the front ankle and 11.8° (21.2°) for the back ankle were recorded. Ankle dorsiflexion was consistently greater for both the back foot position and also the soft boot type, possibly indicative of a greater level of control through the back foot during turns. Using soft boots maximum values of ankle joint eversion were measured during heel side turns of 14.4° (15.1°) for the front and 1.6° (2.3°) for the back ankle. In hard boots maximum eversion values of 9.6° (10.9°) for the front and 0.1° (–0.5°) for the back ankle were recorded. Ankle joint eversion between boot types was statistically significant for both maximum and ranges of eversion with soft boots allowing greater eversion. Asymmetrical ankle rotations were observed for both heel side and toe side

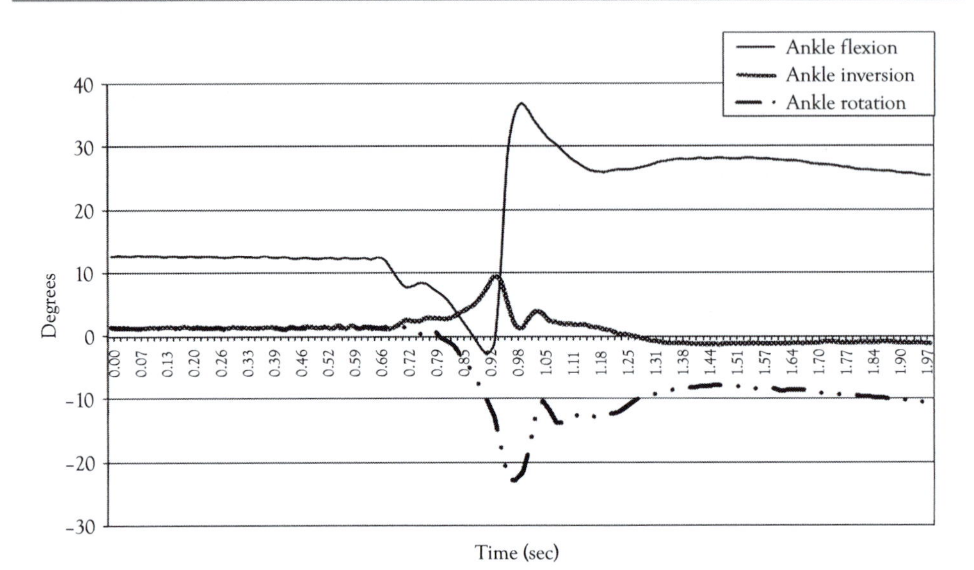

Figure 21.3 A single trial example of a simulated snowboard landing highlighting the synchronized peak values for ankle dorsiflexion, inversion and external rotation.

turns with the front ankle being everted while the back ankle was inverted. Movement of the snowboarder's weight toward the front foot was speculated as the reason for this finding, however, ankle joint eversion is not a position commonly described in the literature.

Woolman and Wilson (2004) performed a similar field study of ankle joint motion of the front ankle in nine snowboarders. A portable Polhemus Fastrak® system was used in a similar set-up to that used by Delorme (2005). Stiff and soft boot types were compared by participants who performed a series of linked turns down a 20° slope. Strap bindings were used and set at 0°, 0° angle. Interestingly no statistically significant differences were identified between both types of boots for any of the variables tested, despite these being classified as being soft and hard by the manufacturers.

Of interest in this study, however, were the high angles of maximum external tibial rotation recorded in both boot types (Table 21.1: 14.1°, 14.8°). Studies of normal gait have measured external tibial rotation as having relatively low values of between 2.5° to 7.3° with external rotation always occurring during ankle plantar flexion (Ramsey *et al.*, 1999; Levinger *et al.*, 2005; Levinger and Gilleard, 2007). Both Funk (2003) and Boon (2001) have reported that external rotation is a contributing movement to fractures to the lateral process of the talus in snowboarders. Another significant finding was the foot position, with the front foot maintaining an everted position relative to the snowboard itself. Maximum foot eversion values of –14° and –12.7° were recorded for the hard and soft boot types. In fact it would seem that this front foot maintains an everted (or pronated) position throughout both turns. Intuitively it would seem that being on a 20° sloped surface the front foot would assume a more inverted position, however, this was not observed.

Another interesting finding was the high angle of ankle eversion measured. Maximum values of –14.2° and –13.1° for the hard and soft boots were recorded. In this position both

Table 21.1 Foot and ankle joint motion during snowboarding. Maximum values of the leading foot for both hard and soft boot types.

Boot Type	*Ankle Flexion*	*Std Error*	*Ankle Extension*	*Std Error*	*Ankle Inversion*	*Std Error*	*Ankle Eversion*	*Std Error*
Hard	17.5°	.787	–10.3°	.867	11.5°	2.025	–14.2°	1.180
Soft	15.9°	.787	–13.1°	.867	12.0°	2.025	–13.1°	1.180
	Ankle Int Rot	*Std Error*	*Ankle Ext Rot*	*Std Error*	*Foot Inversion*	*Std Error*	*Foot Eversion*	*Std Error*
Hard	–8.7°	1.161	14.8°	.443	2.7°	.906	–14.0°	.923
Soft	–8.4°	1.161	14.1°	.443	4.4°	.942	–12.7°	.960

the ankle and subtalar joints are compressed laterally. Funk *et al.* (2003) has speculated that ankle eversion is one of the components of fractures of the lateral process of the talus when the snowboarder falls toward the tip of the board.

Woolman and McNair (2006) performed a field study on the effect of binding position on foot and ankle joint motion (Figure 21.4). Using a Polhemus Patriot® system (Polhemus, Colchester, Vermont, USA), ankle joint motion for the front foot was measured in 3D for three common stance angles. Eleven snowboarders performed a series of linked turns on an indoor slalom course. Each participant performed two trials using three different stance positions – alpine (21°, 6°), neutral (0°, 0°), and duck (18°, 18°) angles of the front and back feet, respectively.

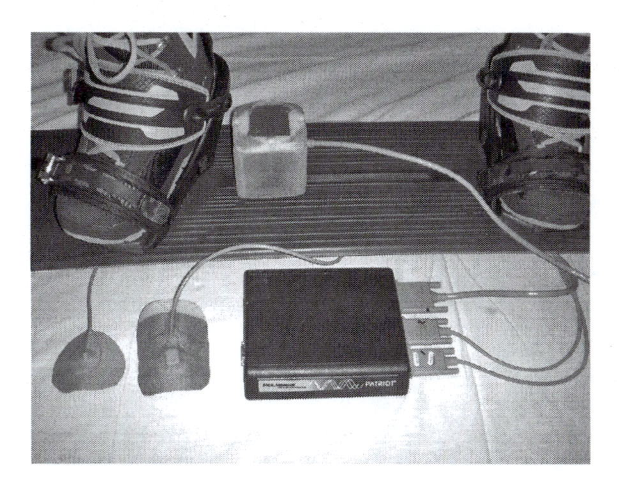

Figure 21.4 The Polhemus Patriot® 3D electromagnetic tracking system. The transmitter mounted to the snowboard with both sensors attached to thermoplastic moulds.

Table 21.2 Foot and ankle joint motion during snowboarding. Maximum values of the leading foot for alpine, neutral, and duck stance positions.

Stance Position	*Ankle Flexion*	*Std Error*	*Ankle Extension*	*Std Error*	*Ankle Eversion*	*Std Error*	*Ankle Inversion*	*Std Error*
Neutral	14.70°	.652	–5.90°	.751	–9.10°	.472	2.44°	.627
Alpine	17.61°	.652	–2.89°	.751	–12.12°	.472	–0.21°	.627
Duck	15.84°	.652	–3.90°	.751	–10.17°	.472	1.94°	.627
	Ankle Int Rot	*Std Error*	*Ankle Ext Rot*	*Std Error*	*Foot Inversion*	*Std Error*	*Foot Eversion*	*Std Error*
Neutral	5.06°	.541	–5.60°	.539	–1.74°	.317	–8.96°	.431
Alpine	1.32°	.541	–11.58°	.539	–2.33°	.317	–9.42°	.431
Duck	5.60°	.541	–5.60°	.539	–1.86°	.317	–9.51°	.431

With the exception of foot eversion significant differences were recorded for all other variables (Table 21.2). The greatest differences were highlighted for the alpine stance with this position recording higher values for ankle external rotation (–11.6°), ankle eversion (–12.1°), and ankle flexion (17.6°) compared to either the neutral, or duck stance positions. Alpine stance recorded more than twice the amount of external rotation than the other stance positions.

Ankle flexion angles for the front foot were lower than expected (14.7° to 17.6°) with foot eversion being greater than expected for all three conditions (–9.0° to –9.5°) with the foot maintaining an everted position throughout testing. Flexion, eversion, and external rotation of the ankle joint have all been implicated in snowboarding ankle injuries. It would appear that the more commonly adopted alpine stance position may place the ankle at greater injury risk.

Conclusion

Prevention of excessive movements known to be harmful is important in reducing snowboarding injuries. It seems intuitive that increasing ankle stability should reduce ankle injuries, however, reducing ankle ROM is likely to predispose the snowboarder to more serious knee injuries and thus a careful compromise is required.

Future equipment design needs to focus upon allowing movements to occur within normal ROMs, as well as controlling the velocities at which these movements occur. Thus if the boot or binding could decelerate movements so as to allow more time to react to a potentially abnormal position then this could also reduce injury risk.

Movements through the front and back limbs during snowboarding are asymmetrical, and the development of customizable boot and binding systems that allow for these differences will no doubt occur. In addition this equipment could also be constructed specifically for gender differences, and differing snowboarding disciplines. Anisotropic materials that have different properties in different directions may provide a significant breakthrough in equipment customization. The development of equipment where the boot and binding have quite separate and independent features, yet become integrated when combined together, also seems appropriate.

The shortage of current biomechanical information on snowboarding highlights the ongoing need for further research in this popular sport.

References

1. Abu-Laban, R. B. (1991) 'Snowboarding injuries: an analysis and comparison with alpine skiing injuries'. *Journal of the Canadian Medical Association*, 145 (9): 1097–1103.
2. Bally, A., Boreiko, M. Bonjour, F. and Brown, C. A. (1989) 'Modelling forces on the anterior cruciate knee ligament during backward falls while skiing'. In M. H. Binet, C. D. Mote, and R Johnson (eds) *Skiing Trauma and Safety* (pp. 267–275), Seventh Volume. American Society for Testing and Materials, West Conshohocken, PA.
3. Bally, A. and Schneegans, F. (1999) 'A release binding for snowboards'. In R. J. Johnson (ed) *Skiing Trauma and Safety* (pp. 132–137), Twelfth Volume. American Society for Testing and Materials, West Conshohocken, PA.
4. Bally, A. and Taverney, O. (1996) 'Loads transmitted in the practice of snowboarding'. In C. D. Mote, R. J. Johnson and P. S. Schaff (eds) *Skiing Trauma and Safety* (pp. 196–205), Tenth Volume. American Society for Testing and Materials, Philadelphia.
5. Bladin, C. Giddings, P. and Robinson, M. (1993) 'Australian injury database study – a four year prospective study', *American Journal of Sports Medicine*, 21: 701–704.
6. Bladin, C. and McCrory, P. (1995) 'Snowboarding injuries. An overview', *Sports Medicine*, 19 (5): 358–64.
7. Boon, A. J., Smith, J. Zobitz, M. E. and Amrami, K. (2001) 'Snowboarder's talus fracture'. Mechanism of injury. *American Journal of Sports Medicine*, 29 (3): 333–8.
8. Boone, D. and Azen, S. (1979) 'Normal range of motion of joints in male subjects'. *The Journal of Bone and Joint Surgery*, 61A (5): 756–759.
9. Canadian Ski Council (2006) *Canadian Snow Industry in Review: 2005—2006*. September 2006. Mississauga, ON L4W4Y4, Canada.
10. Chissell, H. R., Feagin, J. A., Lambert, W. K., King, P. and Johnson, L. (1996) 'Trends in ski and snowboard injuries'. *Sports Med.* 22 (3): 141–145.
11. Chow, T. K., Corbett, S. W. and Farstad, D. J. (1996) 'Spectrum of injuries from snowboarding'. *Journal of Trauma*, 41 (2): 321–325.
12. Davidson, T. M. and Laliotis, A. T. (1996) 'Snowboarding injuries, a four-year study with comparison with alpine ski injuries'. *Western Journal of Medicine*, 164 (3): 231–237.
13. Delorme, S., Tavoularis, S. and Lamontagne, M. (2005) 'Kinematics of the ankle joint complex in snowboarding'. *Journal of Applied Biomechanics*, 21: 394–403.
14. Dohjima, T., Sumi, Y. Ohno, T. Sumi, H. and Shimizu, K. (2001) 'The dangers of snowboarding. A 9-year prospective comparison of snowboarding and skiing injuries'. *Acta Orthop Scand*, 72 (6): 657–660.
15. Estes, M., Wang, E. and Hull, M. (1999) 'Analysis of ankle deflection during a forward fall in snowboarding', *Journal of Biomechanical Engineering*, 121 (2): 243–248.
16. Federation Internationale de Ski (2006) Specifications for competition equipment and commercial markings 2006/2007. Available Online: http://www.fis-ski.com/uk/rulesandpublications/fisgeneralrules/equipment.html (16 January 2007).
17. Ferrera, P. C., McKenna, D. P. and Gilman, E. A. (1999) 'Injury patterns with snowboarding'. *American Journal of Emergency Medicine*, 17 (6) 575–577.
18. Fong, D. T., Hong, Y., Chan, L., Yung, P. S. and Chan, K. (2007) 'A systematic review on ankle injury and ankle sprain in sports'. *Sports Med*, 37 (1): 73–94.
19. Fukuda, O., Takaba, M., Saito, T. and Endo, S. (2001) 'Head injuries in snowboarders compared with head injuries in skiers'. *The American Journal of Sports Medicine*, 29 (4): 437–440.

20. Funk, J. R., Srinivasan, S. C. and Crandall, J. R. (2003) 'Snowboarders talus fractures experimentally produced by eversion and dorsiflexion'. *The American Journal of Sports Medicine*, 31 (6): 921–928.
21. Ganong, R., Heneveld, E., Beranek, S. R. and Fry, P. (1992) 'Snowboarding injuries'. *The Physician and Sports Medicine*, 20 (12): 114–122.
22. Gibbins, J. (1996) *Snowboarding*. Avonmouth, Bristol BS11 9QD, United Kingdom, Parragon Books Ltd, p 36–37.
23. Hagel, B. E., Goulet, C., Platt, R. W. and Pless, I. B. (2004) 'Injuries amongst skiers and snowboarders in Quebec'. *Epidemiology*, 15 (3): 279–286.
24. Hawkins, L. (1965) 'Fracture of the lateral process of the talus'. *The Journal of Bone and Joint Surgery (American)*, 47: 1170–1175.
25. Idzikowski, J. R., Janes, P. C. and Abbott, P. J. (2000) 'Upper extremity snowboarding injuries: Ten-year results from the Colorado Snowboard Injury Survey'. *The American Journal of Sports Medicine*, 28 (6): 825–832.
26. Janes, P. C. and Abbott, P. (1999) 'The Colorado Snowboarding Injury Study: Eight year results'. In R. J. Johnson (ed) *Skiing Trauma and Safety* (pp. 141–149), Twelfth Volume. American Society for Testing and Materials, West Conshohocken, PA.
27. Janes, P. C. and Fincken G. T. (1993) 'Snowboarding injuries'. In R. J. Johnson, C. D. Mote and J. Zelcer (eds), *Skiing trauma and safety*: Ninth international symposium (pp. 255–261). American Society for Testing and Materials, Philadelphia.
28. Kirkpatrick, D. P., Hunter, R. E., Janes, P. C. Mastrangelo, J. and Nicholas, R. A. (1998) 'The snowboarder's foot and ankle'. *American Journal of Sports Medicine*, 26 (2): 271–277.
29. Knunz, B., Nachbauer, W., Schindelwig, K. and Brunner, F. (2001) 'Forces and moments at the boot sole during snowboarding'. In E. Muller, H. Schwameder, C. Raschner, S. Lindinger, and E. Kornexl (eds) *Science and Skiing II*. (pp. 242–249). Hamburg. Proceedings of II International Congress on Skiing and Science, St Christoph, Austria.
30. Langran, M., and Selvaraj, S. (2002) 'Snow sports injuries in Scotland: a case-control study. British'. *Journal of Sports Medicine*, 36: 135–140.
31. Langran, M., and Selvaraj, S. (2004) 'Increased Injury Risk Among First-Day Skiers, Snowboarders, and Skiboarders'. *The American Journal of Sports Medicine*, 32 (1): 96–103.
32. Leijen, D. (2007) Everything about snowboard stance. [Online] Available: http://www.cs.uu.nl/~daan/snow/stance.html (16 January 2007).
33. Levinger, P., Gilleard, W. and Coleman, C. (2005) 'Reliability of an individually molded shank shell for measuring tibial transverse rotations during the stance phase of walking'. *Journal of Applied Biomechanics*, 21: 198–205.
34. Levinger, P. and Gilleard, W. (2007) 'Tibia and rearfoot motion and ground reaction forces in subjects with patellofemoral pain syndrome during walking'. *Gait and Posture*, 25 (1): 2–8.
35. Lind, D., and Sanders, S. (1997) *The Physics of Skiing: Skiing at the Triple Point*. Woodbury, NY: American Institute of Physics, pp. 255–258.
36. McCrory, P. and Bladin, C. (1996) 'Fractures of the lateral process of the talus: a clinical review', Snowboarder's ankle. *Clinical Journal of Sport Medicine*, 6 (2): 124–128.
37. Machold, W., Kwasny, O., Gäßler, P., Kolanja, A., Reddy, B. Bauer, E. and Lehr, S. (2000) 'Risk of injury through snowboarding'. *The journal of trauma: injury, infection and critical care*, 48 (6): 1109–1114.
38. Matsumoto, K., Sumi, H., Sumi, Y. and Shimizu, K. (2004) 'Wrist fractures from snowboarding; A prospective study for 3 seasons from 1998 to 2001'. *Clinical Journal of Sports Medicine*, 14 (2): 64–71.
39. Mills, K. and Horne, G. (1987) 'Fractures of the lateral process of the talus'. *Australian and New Zealand Journal of Surgery*, 57: 643–646.
40. Nachbauer, W., and Kaps, P. (2001) 'Maximal braking deceleration for alpine skiers and snowboarders'. In *Knee Surgery, Sports Traumatology and Arthroscopy*, 12: 169–177. Proceedings of the XV International Conference on Skiing Trauma and Safety.

41. Nakaguchi, H. and Tsutsumi, K. (2002) 'Mechanisms of snowboarding-related severe head injury: shear strain induced by the opposite-edge phenomenon'. *Journal of Neurosurgery*, 97: 542–548.
42. Nantri, A., Beynnon, B., Ettlinger, C. F., Johnson, R. J. and Shealy, J. E. (1999) 'Alpine ski bindings and injuries'. *Sports Medicine*, 28 (1): 35–48.
43. National Ski & Snowboard Retailers Association (2004). Available online: http://www.nsaa.org/nsaa/press/0506/nsga-snbd-part-2004.pdf. (16 January 2007).
44. Nicholas, R., Hadley, J., Paul, C. and Janes, P. (1994) 'Snowboarder's fracture: fracture of the lateral process of the talus'. *Journal of the American Board of Family Practice*, 7 (2): 130–133.
45. O'Neill, D. F. and McGlone, M. R. (1999) 'Injury risk in first-time snowboarders versus first-time skiers'. *American Journal of Sports Medicine*, 27 (1): 94–97.
46. Paul, C. C. and Janes, P. C. (1996) 'The Snowboarder's Talus Fracture'. In C. D. Mote, R. J. Johnson, and P. S. Schaff (eds) *Skiing Trauma and Safety* (pp. 388–393), Tenth volume. American Society for Testing and Materials, Philadelphia.
47. Pino, E. C. and Colville M. R. (1989) 'Snowboard injuries'. *American Journal of Sports Medicine*, 17 (6): 778–781.
48. Polhemus Inc., P.O. Box 560, Colchester, Vermont, USA.
49. Rajan, G. P. and Zellweger, R. (2005) 'Half pipe snowboarding: an (un)forgettable experience or an increasing risk for head injury?'. *British Journal of Sports Medicine*, 38: 753.
50. Ramsey, D. K. and Wretenberg, P. F. (1999) 'Biomechanics of the knee: methodological considerations in the in vivo kinematic analysis of the tibiofemoral and patellofemoral joint'. *Clinical Biomechanics*, 14: 595–611.
51. Rønning, R., Rønning, I., Gerner, T. Engebretsen, L. (2001) 'The efficacy of wrist protectors in preventing snowboarding injuries'. *The American Journal of Sports Medicine*, 29 (5): 581–585.
52. Sacco, D. E., Sartorelli, D. H. and Vane, D.W. (1998) 'Evaluation of alpine skiing and snowboarding injury in a Northeastern state'. *Journal of Trauma*, 44 (4): 654–659.
53. Shealy, J. E., and Miller, D. A. (1989) 'Dorsiflexion of the human ankle as it relates to ski boot design in downhill skiing'. In R. J. Johnson, C. D. Mote and M. Binet (eds) *American Society for Testing and Materials* (pp. 146–152), Philadelphia.
54. Shealy, J. (1993) 'Snowboard vs. downhill skiing injuries'. In R. J. Johnson, C. D. Mote and J. Zelcer (eds) *Skiing Trauma and Safety* (pp. 241–254), IX International Symposium American Society for Testing and Materials, Philadelphia.
55. Shealy J. E., Ettlinger E. F. and Johnson R. J. (2004) 'How fast do people go on Alpine Slopes?'. *Skiing Trauma and Safety*, Fifteenth Volume. Proceedings of XV International Conference on Skiing Trauma and Safety.
56. Shealy, J. E. (2007) Unpublished review of snowboarding injuries. Personal communication (10 January 2007)
57. Sutherland, A. G., Holmes, J. D. and Myers, S. (1996) 'Differing injury patterns in snowboarding and alpine skiing'. *Injury*, 27 (6): 423–425.
58. Torjussen, J., and Bahr, R. (2005) 'Injuries Among Competitive Snowboarders at the National Elite Level'. *The American Journal of Sports Medicine*, 33 (3): 370–377.
59. Torres, K., Crisco, J. and Greenwald, R. (2000) 'Development and validation of an apparatus for determining snowboard boot stiffness'. In R. J. Johnson, P. Zucco and J. E. Shealy (eds) *Skiing Trauma and Safety* (pp. 68–83), Thirteenth Volume. American Society for Testing and Materials, West Conshohocken, PA.
60. U.S. Consumer Product Safety Commission (1995) "CPSC Says Snowboarding Boom Leads To More Injuries". [Online] Available: http://cpsc.gov/cpscpub/prerel/prhtml95/95067.html. Washington, DC: 1. [10 January 2007].
61. Woolman, G., Wilson, B. and Milburn, P. (2002) 'Ankle joint motion during simulated snowboard landings'. A thesis submitted for the degree of Master of Health Sciences of the University of Otago, Dunedin, New Zealand. 30 October 2002. *Proceedings of XIV International Conference on Skiing Trauma and Safety*.

62. Woolman, G. and Wilson, B. (2004) 'Three–dimensional foot and ankle joint motion during snowboarding', A study undertaken for the Accident Compensation Corporation, Injury Prevention Programme, Seabridge House, Wellington, New Zealand. *Proceedings of XV International Conference on Skiing Trauma and Safety.*
63. Woolman, G. and McNair, P. (2006) 'The effect of binding stance position upon foot and ankle joint motion during snowboarding'. *Proceedings of XVII International Conference on Skiing Trauma and Safety.*

22

Biomechanics of striking and kicking

Bruce Elliott[1], Jacqueline Alderson[1] and Machar Reid[2]
[1]University of Western Australia, Crawley, Western Australia;
[2]Tennis Australia, Melbourne

Introduction

Many sports involve striking and kicking manoeuvres where energy flows from proximal to more distal segments, such that the endpoint segment, for example, the foot in a kick, or the hand, racquet or stick in a hit, impact the ball at the highest possible speed (summation of speed principle). This chapter comprises a discussion of the special problems that researchers face in performing analyses of upper limb striking-based and lower limb kicking-based skills, and concurrently outlines the key mechanical considerations in these movements.

Technical and practical considerations in three-dimensional analyses of striking and kicking skills

Lower versus upper limb motion

Research of the upper and lower limbs has been driven by biomechanics teams with varying interests. Lower limb models have been developed primarily to answer clinical questions involving gait, whereas the development of upper limb models has largely been at the behest of the sports biomechanist interested in hitting and throwing activities.

The development of a true three-dimensional (3D) approach to motion analysis first occurred in the lower limbs as a result of clinical necessity. Consequently, scapular movement, full ranges of shoulder circumduction and internal-external rotation, all difficulties associated with upper limb motion, were not encountered during the development of lower limb biomechanical models.

The reconstruction of human motion requires the definition of a biomechanical model that is intended to accurately represent the motion of the underlying skeletal system. As each body segment is deemed to be a rigid body connected by a relevant joint type, it is necessary to determine the instantaneous orientation and position of each segment in three dimensions via the definition of orthogonal sets of axes, commonly termed local or technical coordinate systems (Cappozzo *et al.*, 1995). While generally defined by external markers on the body, these local coordinate systems are considered embedded in each rigid bony segment of the biomechanical model (Cappozzo *et al.*, 1996). The orientation and position of each rigid segment's local coordinate system is determined ideally by three reconstructed marker positions per segment (Della Croce *et al.*, 2003) (see Figure 22.1 for athlete with a representative upper body marker set). The position of these markers should be non-collinear but they may be placed anywhere on the segment. However, this results in movement data that has little anatomical relevance and simply provides the researcher with the relative motion between segments. To the coach, clinician or biomechanist, such an approach has little functional value, in that the interpretation and, by extension, practical application of the data, is limited to within subject, relative segment comparisons.

To make these data comparable and functionally meaningful, the kinematic and kinetic scalar information must be referred to anatomical systems of reference (Della Croce *et al.*, 2003). This requires anatomical landmarks to be defined and the position of identifiable external landmarks to be determined with respect to each segment's local coordinate system (see Figure 22.2, where the pointer method is used to record the position of the lateral epicondyle with reference to two local coordinate systems positioned on the upper arm).

Provided that sufficient anatomical landmarks are defined for each bony segment (commonly joint centres and bony prominences such as epicondyles), anatomical coordinate systems (ACS) can be defined and their position and orientation determined (Della Croce *et al.*, 1997). When the position and orientation of the ACSs of two adjacent segments (e.g. forearm and upper arm) are accurately defined, joint kinematics and kinetics may then be determined within and between subjects (Grood and Suntay, 1983). However, the accuracy of the resulting joint kinematics and kinetics is largely dependent on two key issues in this process. How closely do the external markers used to define a segment's local coordinate system couple with the underlying bony motion, and secondly, how accurately and repeatedly can the anatomical landmarks used to define the crucial ACSs be identified.

The need for external markers to couple closely with underlying skeletal motion is essential for accurate and valid measurement of upper arm motion in throwing and striking activities. Primarily due to the large amount of soft tissue in the upper arm and extensive muscle activity that occurs during throwing and striking motions, correct placement of external markers on this segment is paramount. The approach whereby markers are affixed to clusters may allow for the implementation of methods to remove marker shift attributed to

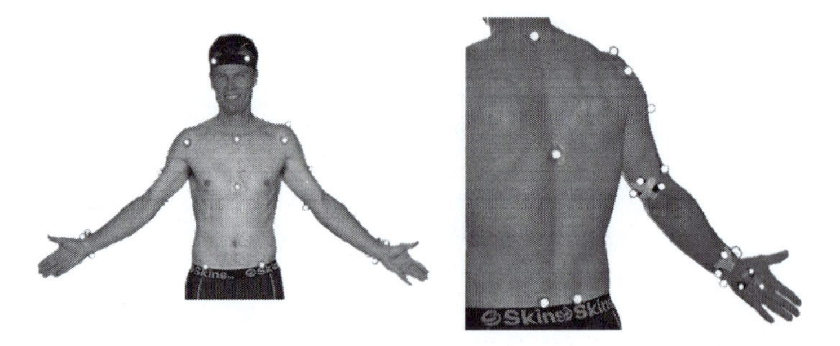

Figure 22.1 Subject with upper body marker set.

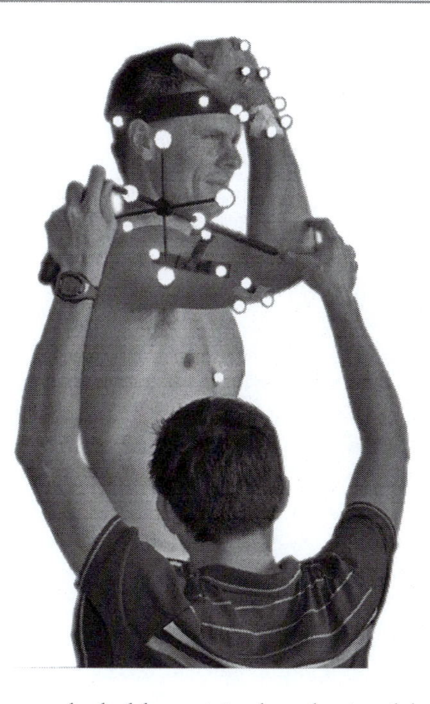

Figure 22.2 Pointer method of determining lateral epicondyle of the humerus.

excessive skin movement (Manal *et al.*, 2002). However, if accurate shoulder internal/external rotation is to be obtained, these clusters should not be placed mid-segment (over the bulk of the soft tissue). Conversely, due to the excessive marker motion that occurs when individual markers are placed directly over anatomical landmarks (e.g. medial and lateral epicondyles) this approach is also unworkable. More research is clearly needed to address this problem; however, it is likely that multiple individual markers or multiple clusters placed across the length of the segment may go some way to addressing these issues.

Much research has recently been directed at accurately determining the joint centre locations and relevant axes of rotations for the ankle, knee and hip joints. Approaches based on regression equations using cadaver anthropometry (Shea *et al.*, 1997) and, more recently, the introduction of numerical techniques, such as the functional method for defining the hip joint centre (Piazza *et al.*, 2001; Leardini *et al.*, 1999) and 'optimal' axes or mean instantaneous helical axes of rotation for the knee (Besier *et al.*, 2003; Churchill *et al.*, 1998) are beyond the scope of this chapter. However, given the complexity of upper limb modelling, the use of these approaches may well be methods that once implemented will advance upper limb modelling protocols beyond the current position of a poor cousin to lower limb modelling procedures. While analyses around the wrist and elbow joints have been relatively constructive thus far (Chéze *et al.*, 1998; Stokdijk *et al.*, 1999), more work is needed to accurately and repeatedly estimate the shoulder joint centre location (Stokdijk *et al.*, 2000), particularly during dynamic overhead motion.

Three dimensional analysis of segments in striking and kicking typically require a minimum of three markers per segment, and a repeatable and accurate method to determine joint centres and their respective axes of rotation.

Laboratory versus match testing

A conundrum faces the sport biomechanist with reference to the laboratory versus field-testing scenario. First, what analysis system should be selected to collect the objective striking or kicking data? Do we select video-based systems where images of the performance, without markers attached to the athlete, can be collected during a match, then digitized and relevant data extracted? Or do we select marker identification systems that more accurately track the body during the motion in question (typically passive where athletes are not required to wear transmitters and light is reflected back from markers to the cameras)?

Another consideration in this debate is the level of error acceptable to the coach or clinician. The root mean squared (RMS) error of 0.6° for an opto-reflective (passive) system (Vicon – 612) has been shown to be more accurate in reconstructing a known angle than a video-based system (Peak Motus) (RMS error 2.3°) in the laboratory, when the same mathematical procedure (model) was applied to calculate an elbow flexion-extension angle. When different models were applied to the raw marker trajectories collected using the video-based system, the RMS errors increased (Elliott *et al.*, 2007).

If the magnitude of error is a primary concern in testing (research or servicing) then laboratory testing using a passive opto-reflective system should be used and ecological validity of the testing environment maximized. However, in all circumstances, whether using a video or opto-reflective system, ideally three markers per segment should generally be adopted (e.g. hitting – Agkutagawa and Kojima, 2005; kicking – Nunome *et al.*, 2002).

The influence of impact on data smoothing

Valid and reliable mechanical descriptors of segments or joint movements necessitate that an optimal level of filtering be applied to remove signal noise. This chapter will not address the issue of filtering (e.g. spline versus digital filter and so on) or methods procedures used to determine the appropriate filtering method (e.g. residual analysis and so on). However, it is imperative that the biomechanist does consider the influence of smoothing data through impact, as this is a characteristic of both striking and kicking movements. While various impact smoothing approaches (extrapolation or interpolation) have been used to remove the influence of the impact on the resulting displacement curve (e.g. Knudson and Bahamonde, 2001), the final decision should be made following consideration of the momentum of the impacting bodies and the duration of the impact.

It would appear that impact affects endpoint velocity in different ways, depending on the momentum of the two colliding bodies. When a peak value about impact is required, your decision to deal with the influence of impact from a smoothing perspective should then be guided by consideration of: conservation of momentum, the time of impact, the need for interpolated or extrapolated data and the best type of smoothing algorithm to deal with impact.

Number and repeatability of trials

While mechanical factors involved in upper limb striking have been shown to vary, particularly velocity and acceleration measures as compared with displacement (Knudson, 1990), many biomechanical studies have used one superior performance to characterize a player's movement coordination. However, anomalies underlying this assumption are well documented (Mullineaux *et al.*, 2001; Salo *et al.*, 1997). Recent investigations of the tennis serve (continuous data of selected variables considered important in the forward swing of the racquet to ball impact) and punt kicks (discrete data at selected points in the motion) in our laboratory have used coefficients of multiple correlations (CMCs) and Cronbach alpha statistics to evaluate inter-trial repeatability. The basic premise behind such approaches is to identify how many trials are needed to provide a representative sample of performance.

In the service analysis it was found that five trials did not add significantly to the compilation of data from three trials. Typically, CMC values of ≥0.9 were recorded from three trials, irrespective of the type of serve hit (first and second) during the phases up to impact (Reid, 2006). In pilot work on the punt kick, it was shown that the repeatability over three trials was generally high (≥0.8) for logically selected discrete two dimensions (digital video – SiliconCOACH software) and three dimensions (Vicon opto-reflective system) kinematic parameters.

Consideration of three trials would appear to be representative of most striking and kicking skills that are primarily closed such as the tennis serve and football penalty or punt kick.

Key mechanical considerations in striking and kicking movements

Several review articles have been published in the general areas of hitting (Elliott, 2000), major racquet sports (Lees, 2003) and kicking (soccer: Lees and Nolan, 1998) and readers are directed to those for in-depth critiques of these topic areas.

The first consideration in all of striking and kicking movements is preparation and this involves: visual perception, movement to the ball (balance) and the ability to appropriately position the body for the swing of the stick, foot or arm to impact.

Preparation

Visual perception

While awaiting the ball, the batter in baseball and footballer preparing to kick a ball face two problems, namely a sensory-perceptual issue related to the ball or the position of players on the field, followed by the selection of an appropriate motor response. The sensory-perceptual problems consist of visually tracking a ball such that a decision can be made as to where and when to swing, or alternatively, where to kick given the position of opposition players. The correct timing and sequencing of movements must occur so as to ensure the desired impact situation. Research clearly shows that continuous time-to-contact information is critical to successful striking performances (Peper *et al.*, 1994). Actions are continuously geared to source information. Harrison (1978) wrote that visual dynamics would be greatly improved if baseball hitters adopted the below sequence of visual foci while watching the pitcher.

Soft focus – watch the whole body-general area of mound.
Fine focus – focus on something in the plane of the ball at release.
Specific fine focus – focus on the area of release, i.e. the hand and ball.

Minor modifications to this approach could see it used in many striking-based skills.

Higher performance players are able to process critical information earlier in the opponent's action, thus permitting more time to move to the ball (Goulet *et al.*, 1989) and a more accurate prediction of ball flight (Paul and Glencross, 1997). These players appear to be able to 'look' at the right cues and 'see' the information these cues provide, yet this should not be seen to detract from the purposeful practice of cueing (Abernethy, 1996). Accurate processing of information from the hitting-throwing action of the opponent, together with tracking information, is obviously critical in sports when the ball approaches at high speed.

A further logical expectation with respect to striking or kicking sports is that the expert performer possesses a more precise knowledge of event probabilities than a novice. This experience will guide their selective attention, with respect to realistic outcomes, and ultimately their skilled performance.

Equally, alignment of the body, and more specifically the head, is hypothesized to accommodate preferences of eye dominance, such that task preparation and execution is enhanced. While a source of growing investigative interest in sport, early indications are that eye dominance may exert some influence on swinging-hitting stance and set-up (Sugiyama and Lee, 2005). Several researchers have revealed no differences in the hitting, fielding and pitching skills among professional baseball players who possessed either matched (ipsilateral) or crossed (contralateral) hand-eye dominance (e.g. Laby *et al.*, 1998), yet it may be that more dichotomous performances present among athletes with more variable preparations or set-ups.

Movement to the ball

Perhaps one of the most important aspects of movement is the way that it is initiated. This involves the concepts of 'unweighting' and first foot movement. What is this unweighting and how does it relate to rapid movement about a court or field? Before an athlete prepares to move, the force applied to the ground, and therefore the equal and opposite force applied by the ground to the feet, is equal to body weight. Once the decision to move is made, the knees flex and the body is accelerated towards the ground. Mean peak speeds of the hip for high-performance tennis players during this preparation were approximately 0.5 m s^{-1} for ground strokes (e.g. Elliott *et al.*, 1989). The reaction force from the court is, therefore, lowered as the body accelerates downward.

The deceleration of the downward movement is controlled by eccentric contraction of the quadriceps muscle group; such that tension is developed in these muscles and elastic energy is stored. The preloading of the quadriceps, in combination with the stored energy, is used to assist the quadriceps' concentric contraction (knee extension) during the subsequent lower limb drive to the ball. The key to these movements is to coordinate the knee flexion-extension action with the opponent's stroke or movement.

The need for athletes to possess an 'explosive first step' forms part of modern sports coaching lore. A certain amount of conjecture, however, surrounds just what type of 'first step' athletes should take. Kraan *et al.* (2001) demonstrated that subjects who took a paradoxical first step (i.e. away from the intended direction of travel) backwards generated the highest GRFs and greatest forward acceleration. Indeed, the movement of the limb backwards (unilateral hip extension) manipulates the position of a person's centre of gravity with respect to their base of support to facilitate explosive forward progression. This finding contrasts with many time-honoured coaching practices that encourage athletes to always take their first step toward the oncoming ball or opponent.

Segment-implement displacement

Body segment rotations position the foot in kicking and the hand-implement in striking such that appropriate speed, trajectory and orientation of the endpoint can be achieved at impact. In general terms, an increased displacement arc permits this speed to be 'developed' under more control during the forward swing than would occur with a smaller backward movement that would require the forward swing to be performed very quickly to achieve the same velocity.

In striking skills, while accepting that 'drive' begins from the ground, a key factor in trunk rotation is the concept of the separation angle. Separation angle contrasts the shoulder-thorax

Figure 22.3 Separation angle in a tennis backhand (picture published with permission of the ITF).

alignment to pelvic rotation (hips). In the backswing for golf, tennis, baseball hits, and so on, the 'shoulders' are rotated more than the 'hips' to assist in backward displacement of the hand-implement and to put the muscles of the trunk on-stretch (Figure 22.3). Typically this angle is 20–30° in tennis ground strokes (Reid and Elliott, 2002) and the golf drive (McTeigue *et al.*, 1994). The hand with the implement is then positioned away from and posterior to the back-shoulder joint. In major league baseball, a professional player rotated the bat by approximately 130° (Galinas and Hoshizaki, 1988) and in tennis, rotations of the racquet from 135°–270° have been recorded in preparation for the forward swing (Reid and Elliott, 2002). The lesser rotation in baseball may be related to the smaller movement time to respond to the pitched ball compared with that for a tennis groundstroke. In kicking, the kicking-limb is positioned posterior to the trunk to achieve this displacement. This is obviously created by thigh extension at the hip, increased pelvic tilt and rotation and lower leg flexion at the knee joint during the backswing.

Forward swing

Movement sequencing

Bunn (1955) was the first to popularize the summation of speed principle or kinematic chain, where speed is developed by a succession of movements of members of the body such that the speed of each successive member should be faster than that of its predecessor. The movement of each member should start at the moment of greatest velocity but least acceleration of the preceding member. However, it was not until the 1990s that Putnam (1991, 1993) described segment interaction with a more causal approach. She proposed that in a two-segment system, such as the thigh and lower leg in kicking, that a force is applied across the knee joint (internal flexor moment) actively to restrain extension of the leg in the early swing phase. Once this force has been removed, the leg will extend at the knee and thus increase the knee extension velocity and the linear velocity of the ankle (segment endpoint). The intriguing dilemma that still faces biomechanists is whether the extension of the lower leg slows the thigh (Dunn and Putnam, 1988) or if the deceleration of the thigh increases the velocity of the lower leg (Luhtanen, 1988). Nunome *et al.* (2002) in a soccer kick presented data to suggest that an eccentric thigh extensor moment was likely involved in thigh deceleration. Certainly research is still needed across sporting movements to address the quandary of whether the distal segment rotation is active or passive from a mechanical joint power perspective.

Irrespective of the dilemma above, research has shown that a kinematic chain does exist in the upper limb, although humeral internal rotation has been shown to occur very late in the forward swing of many striking movements (e.g. Elliott *et al.*, 1995). It can be generalized that in striking manoeuvres the complete kinematic chain involves:

- drive from the ground and extension of the joints of the lower limbs;
- multi-planar trunk rotation, where the primary axis of rotation will be dependent on the type of striking movement; and
- upper limb rotation, ending with an optimal speed of the hand or striking implement.

In kicking skills, while pelvic rotation and forward movement of the body all play a role, the primary source of velocity generation comes from the hip and knee joints. Papers by Nunome and colleagues (2006) and Dörge *et al.* (2002) summarized the kinematic chain in soccer kicking under varying conditions. Dörge *et al.* (2002) reported that deceleration of the thigh did not enhance the extension of the leg in kicking.

A chain, in which the more distal segment endpoint increases in speed, characterizes striking and kicking skills, however, if all the degrees of freedom are considered for each segment, traditional sequencing approaches that have largely focused on flexion-extension motion may not hold true. That is, in striking skills significant contributions may come from internal-external rotation, which to date has only occasionally been reported in the movement sequencing research.

Impact

Obviously the velocity profile and orientation of the foot in a kick, the hand in a punch, or the stick in a hit are critical to a successful impact. However, other mechanisms will influence the interaction between the two impacting bodies.

Short period impacts

Impacts in golf (0.5 ms – Gobush, 1990) and tennis (4-5 ms – Brody, 1987) typify short-period impacts. During these impacts both bodies deform in an imperfectly elastic collision. As segment rotation plays such an important role in modern striking and kicking movements, players must practice 'flattening the arc of the swing' in an endeavour to optimize the likelihood of hitting or kicking the ball in the desired direction. This is a challenge to coaches and young players alike, as the desire for 'power' must be matched with the need for control and accuracy.

The key to these short period impacts is that whatever you want to occur at impact must be happening prior to impact. For example, to hit a forehand topspin groundstroke the necessary racquet speed, trajectory and orientation must all be in place at impact.

Long period impacts

These are characterized by impacts of approximately 15 ms and commonly occur in football codes (Tsaousidis and Zatsiorsky, 1996). During this period the ball travels a considerable distance forward on the foot (15–26 cm). In these impacts mechanical work applied to the ball during contact may have a greater influence on the final mechanical characteristics of ball flight than occurs for the short duration impact.

Movements of the foot during these impacts may have serious effects on accuracy and distance. While eccentric forces applied to the ball to create a 'curved flight path' in soccer may be desirable, the same may not be the case in a punt kick for distance and accuracy.

Follow through

The follow through in striking and kicking skills has two broad functions; to dissipate force in a controlled manner and to prepare the player for the tactical considerations of the game. The influence of the variety of tactical requirements for striking and kicking sports on subsequent movement precluded their discussion in this chapter, and the dissipation of energy built during the forward swing will be dealt with separately for the upper and lower limb movements.

Upper limb striking skills

The role of internal rotation in the forward swing of so many striking manoeuvres is a major consideration in the conditioning of athletes. The very large internal rotator muscles of the shoulder, when compared with the size of the external rotators, place a great deal of load on the external rotators, as they act eccentrically to slow the rapidly internally rotating upper arm. A typical sequence of muscular contractions in many striking movements involves:

> *Eccentric contraction* of the internal rotators (Figure 22.4A – this creates pre-stretch and stores elastic energy in the internal rotators).
> *Concentric contraction* of the internal rotators (Figure 22.4B).
> *Eccentric contraction* of the external rotators to slow internal rotation (Figure 22.4C).

Research has shown that muscles involved with shoulder internal rotation are very well developed in striking athletes (Cook *et al.*, 1987) and that injury to the external rotator group or capsule of the shoulder is common (Blevins *et al.*, 1996). Specific eccentric training of the shoulder joint external rotator musculature, in isolation to the humeral internal rotators is then essential to protect this area of the body from injury.

Lower limb kicking

While the sequence of muscle contractions mentioned above is the same for the lower limb, the problem is in a way exaggerated by the fact that the hamstrings are a two-joint muscle.

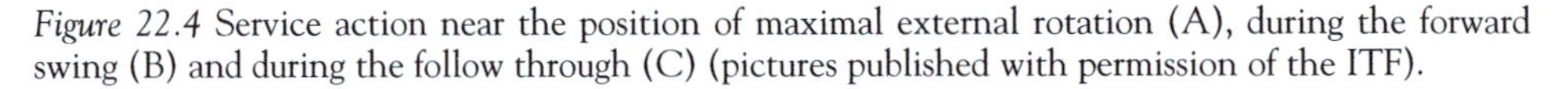
Figure 22.4 Service action near the position of maximal external rotation (A), during the forward swing (B) and during the follow through (C) (pictures published with permission of the ITF).

That is, while they are required to eccentrically contract and slow a rapidly extending lower leg at the knee joint during the follow through of a kick, they are also required to slow the rate of thigh flexion toward end range of motion at the hip. This certainly makes the hamstring group susceptible to injury during the follow through phase of kicking.

Eccentric training of the appropriate muscle groups plays an integral role in preparing athletes to perform striking and kicking manoeuvres.

Conclusion

While athlete skill, including flair, will always be the cornerstone of high performance striking and kicking, sports science is today playing an ever-increasing role in athlete development. There is little doubt that the optimal servicing of athletes requires the use of research to underpin applied science and so sport biomechanists must perform research based on appropriate practical questions. The application of the findings resulting from such research must be applied in a robust clinical manner if the benefits are to be realized by the coach, athlete and community at large.

References

1. Abernethy, B. (1996) 'Training the visual perceptual skills of athletes: Insights from the study of motor expertise', *American Journal of Sports Medicine*, 24: S89–92
2. Akutagawa, S. and Kojima, T. (2005) 'Trunk rotation torques through the hip joints during the one- and two-handed backhand tennis strokes', *Journal of Sports Sciences*, 23: 781–793.
3. Besier, T. F., Sturnieks, D. L., Alderson, J. A. and Lloyd, D. G. (2003) 'Repeatability of gait data using a functional hip joint centre and a mean helical knee axis', *Journal of Biomechanics*, 36: 1159–1168.
4. Brody, H. (1987) *Tennis Science for Tennis Players*. Philadelphia, Pennsylvania: University of Pennsylvania Press.
5. Blevins, F. T., Hayes, W. M., and Warren, R. F. (1996) 'Rotator cuff injury in contact athletes', *American Journal of Sports Medicine*, 24: 263–267.

6. Bunn, J. (1955) *Scientific Principles of Coaching*. New Jersey: Englewood Cliffs.
7. Cappozzo, A., Catani, F., Croce, U. D. and Leardini, A. (1995) 'Position and orientation in space of bones during movement: anatomical frame definition and determination', *Clinical Biomechanics*, 10: 171–178.
8. Cappozzo, A., Catani, F., Leardini, A., Benedetti, M. and Croce, U. (1996) 'Position and orientation in space of bones during movement: experimental artefacts', *Clinical Biomechanics*, 11: 90–100.
9. Chèze, L., Fregly, B.J. and Dimnet, J. (1998) 'Determination of joint functional axes from noisy marker data using the finite helical axis', *Human Movement Science*, 17: 1–15.
10. Churchill, D. L., Incavo, S. J., Johnson, C. C. and Beynnon, B. D. (1998) 'The transepicondylar axis approximates the optimal flexion axis of the knee', *Clinical Orthpaedics and Related Research*, 356: 111–118.
11. Cook, E. E., Gray, V. L., Savinar–Nogue, E. and Medeiros, J. (1987) 'Shoulder antagonistic strength ratios: A comparison between college-level baseball pitchers and non-pitchers', *Journal of Orthopaedic Sports Physical Therapy*, 8: 451–461.
12. Della Croce, U., Cappozzo, A., Kerrigan, D. C. and Lucchetti, L. (1997) 'Bone position and orientation errors: pelvis and lower limb anatomical landmark identification reliability', *Gait and Posture*, 5: 156–157.
13. Della Croce, U., Camomilla, V., Leardini, A. and Cappozzo, A. (2003) 'Femoral anatomical frame: assessment of various definitions', *Medical Engineering and Physics*, 25: 425–431.
14. Dörge, H., Andersen, T., Sorensen, H. and Simonsen, E. (2002) 'Biomechanical differences in soccer kicking with the preferred and non-preferred leg', *Journal of Sports Sciences*, 20: 293–299.
15. Dunn, E. and Putnam, C. (1988) 'The influence of lower leg motion on thigh deceleration in kicking'. In G. de Groot, A. Hollander, P. Huijing and G. van Ingen Schenau (eds), *Biomechanics X1-B* (pp. 787–790). Amsterdam: Free University Press.
16. Elliott, B. (2000) 'Hitting and kicking'. In V. Zatsiorsky (ed) *Biomechanics of Sport* (pp. 487–506). Oxford: Blackwell Science.
17. Elliott, B., Marsh, T. and Overheu, P. (1989) 'A biomechanical comparison of the multisegment and single unit topspin forehand drives in tennis', *International Journal of Sport Biomechanics*, 5: 350–364.
18. Elliott, B., Marshall, R. and Noffal, G. (1995) 'Contributions of upper limb segment rotations during the power serve in tennis', *Journal of Applied Biomechanics*, 11: 433–442.
19. Elliott, B., Alderson, J. and Denver, E. (2007) 'System and modelling errors in motion analysis: implications for the measurement of the elbow angle in cricket bowling', accepted *Journal of Biomechanics*.
20. Galinas, M. and Hoshizaki, T. (1988) 'Kinematic characteristics of opposite field hitting. In K. Kreighbaum and A. McNeill (eds), *Biomechanics of Sports V1* (pp. 519–530). Bozeman: Montana State University Press.
21. Gobush, W. (1990) 'Impact force measurements on golf balls'. In A. Cochran (ed), *First World Scientific Congress on Golf* (pp. 219–224). London: E & FN Spon.
22. Goulet, C., Bard, C. and Fleury, M. (1989) 'Expertise differences in preparing to return a tennis serve: a visual information processing approach', *Journal of Sport and Exercise*. Psychology, 11: 382–398.
23. Grood, E. S. and Suntay, W. J. (1983) 'A joint coordinate system for the clinical description of three-dimensional motions: application to the knee', *Journal of Biomechanical Engineering*, 105: 136–144.
24. Harrison, W. (1978) 'Visual dynamics', *Scholastic Coach*, 47: 38–40.
25. Kraan, G. A., van Veen, J., Snijders, C. J. and Storm, J. (2001) 'Starting from standing; why step backwards', *Journal of Biomechanics*, 3: 211–215.
26. Knudson, D. (1990) 'Intrasubject variability of upper extremity angular kinematics in the tennis forehand drive', *International Journal of Sport Biomechechanics*, 6: 415–421.
27. Knudson, D. and Bahamonde, R. (2001) 'Effect of endpoint conditions on position and velocity near impact in tennis', *Journal of Sports Sciences*, 19: 839–844.
28. Laby, D. M., Kirschen, D. G., Rosenbaum, A. L., and Mellman, M. F. (1998) 'The effect of ocular dominance on the performance of professional baseball players', *Ophthalmology*, 105: 864–846.

29. Leardini, A., Cappozzo, A., Catani, F., Toksvig-Larsen, S., Petitto, A., Sforza, V., Cassanelli, G. and Giannini, S. (1999) 'Validation of a functional method for the estimation of hip joint centre location', *Journal of Biomechanics*, 32: 99–103.
30. Lees, A. and Nolan, L. (1998) 'The biomechanics of soccer: A review', *Journal of Sports Sciences*, 16: 211–234.
31. Lees, A. (2003) 'Science and major racquet sports: A review', *Journal of Sports Sciences*, 21: 707–732.
32. Luhtanen, P. (1988) 'Kinematics and kinetics of maximal instep kicking in junior soccer players'. In T. Reilly, A. Lees, K. Davids and W. Murphy (eds), *Science and Football* (pp. 441–448). London: E&FN Spon.
33. Manal, K., McClay, I., Richards, J., Galinat, B. and Stanhope, S. (2002) 'Knee moment profiles during walking: errors due to soft tissue movement of the shank and the influence of the reference coordinate system', *Gait and Posture*, 15: 10–17.
34. McTeigue, M., Lamb, S. and Mottram, R. (1994) 'Spine and hip motion analysis during the golf swing', *Science and Golf II*. Proceedings of the 1994 World Scientific Congress of Golf, St Andrews, 50–57.
35. Mullineaux, D. R., Bartlett, R. M. and Bennett, S. (2001) 'Research design and statistics in biomechanics and motor control', *Journal of Sports Sciences*, 19: 739–760.
36. Nunome, H., Asai, T., Ikegami, Y. and Sakurai, S. (2002) 'Three-dimensional kinetic analysis of side-foot and instep soccer kicks', *Medicine and Science in Sports and Exercise*, 34: 2028–2036.
37. Nunome, H., Lake, M., Georgakis A. and Stergioulas, K. (2006) 'Impact phase kinematics of instep kicking in soccer, *Journal of Sports Sciences*, 24: 11–12.
38. Paul, G. and Glencross, D. (1997) 'Expert perception and decision making in baseball', *International Journal of Sports Psychology, 28: 35—56.*
39. Peper, L., Bootsma, R. J., Mestre, D. R. and Bakker, F. C. (1994) 'Catching balls: how to get the hand to the right place at the right time', *Journal of Experimental Psychology Human Perceptual Performance*, 20:591–612.
40. Piazza, S. J., Okita, N. and Cavanagh, P. R. (2001) 'Accuracy of the functional method of hip joint center location: effects of limited motion and varied implementation', *Journal of Biomechanics*, 34: 967–973.
41. Putnam, C. (1991) A segment interaction analysis of proximal-to-distal sequential motion patterns, *Medicine and Science in Sports and Exercise*, 23: 130–144.
42. Putnam, C. (1993) 'Sequential motions of body segments in striking and throwing skills – descriptions and explanations', *Journal of Biomechanics*, 26: 125–135.
43. Reid, M. and Elliott, B. (2002) 'The one- and two-handed backhands in tennis', *Sports Biomechanics*, 1: 47–68.
44. Reid, M. (2006) 'Loading and velocity generation in the high performance tennis', unpublished PhD, The University of Western Australia, Australia.
45. Salo, A., Grimshaw, P. N. and Viitasalo, J. T. (1997) 'Reliability of variables in the kinematic analysis of sprint hurdles', *Medicine and Science in Sports and Exercise*, 29: 383–389.
46. Shea, K. M., Lenhoff, M. W., Otis, J. C. and Backus, S. I. (1997) 'Validation of a method for location of the hip joint centre', *Gait & Posture*, 5: 157–158.
47. Sugiyama, Y. and Lee, M.S. (2005) 'Relation of eye dominance with performance and subjective ratings in golf putting', *Perceptual Motor Skills*, 100: 761–766.
48. Stokdijk, M., Meskers, C. G., Veeger, H. E., de Boer, Y. A. and Rozing, P. M. (1999) 'Determination of the optimal elbow axis for evaluation of placement of prostheses', *Clinical Biomechanics*, 14: 177–184.
49. Stokdijk, M., Biegstraaten, M., Ormel, W., de Boer, Y. A., Veeger, H. E. and Rozing, P. M. (2000) 'Determining the optimal flexion–extension axis of the elbow in vivo – a study of interobserver and intraobserver reliability', *Journal of Biomechanics*, 33: 1139–1145.
50. Tsaosidis, N. and Zatsiorsky, V. (1996) 'Two types of ball–effector interaction and their relative contribution to soccer kicking', *Human Movement Science*, 15: 861–876.

23

Swimming

Ross H. Sanders, Stelios Psycharakis, Roozbeh Naemi, Carla McCabe and Georgios Machtsiras
The University of Edinburgh, Edinburgh

Introduction

Analysis in swimming has been, and continues to be, limited by the technical ability to quantify motion. Swimming motion occurs across air and water and results from forces which are both difficult to measure and unpredictable. While many authors have risen to the challenge of analysing swimming technique, much remains unknown. This chapter provides some insight into how knowledge has emerged and forecasts future research given the increasing ability to analyse aquatic motion afforded by methodological and technological advances.

Propulsion in swimming

Several mechanisms of propulsion have been proposed with regard to human swimming (for a review see McCabe and Sanders, 2006).

Propulsion from lift and drag

Many researchers have acknowledged the interaction of forces in the direction of fluid flow (drag forces) and forces that are perpendicular to the fluid flow (lift forces). Counsilman (1968) 'borrowed' the Bernoulli Principle from aerodynamics to explain why swimmers used curved hand paths. It was proposed that by using curved hand paths in a 'sculling' motion and continuously adjusting the pitch of the hand, lift forces were produced with a strong component in the desired direction of travel. However, it is now widely accepted that the Bernoulli Principle is an inappropriate explanation for propulsion in swimming.

Propulsion may be obtained by accelerating a mass of water opposite the swimming direction, in accordance with Newton's second and third laws of motion. (See Sprigings and Koehler, 1990, for a detailed description). In this mechanism of propulsion the magnitude of the drag and lift components (D and L) depends on the density of the fluid (ρ), the velocity of the limb relative to the fluid (V), the surface area of the limb (S) and a coefficient for drag (C_D) or lift (C_L) that varies according to the shape of the limb and its orientation to the flow.

$$L = \frac{1}{2} \rho V^2 C_L S$$

$$D = \frac{1}{2} \rho V^2 C_D S$$

Several authors have estimated the drag and lift coefficients across a range of hand orientations using immersed hand/forearm models in fluid channels (Schleihauf, 1979; Schleihauf *et al.*, 1983; Berger *et al.*, 1995; Berger *et al.*, 1999; Sanders, 1999; Rouboa *et al.*, 2006). By combining lift and drag coefficients with hand velocity and hand-orientation data from digitised video the lift and drag forces are estimated.

From ensuing studies it was observed that the pitch of the hands during the arm stroke is within the range of angles that maximise drag force rather than lift. The drag forces are particularly dominant during the propulsive phase of the front crawl, back crawl and butterfly strokes (Rushall *et al.*, 1994; Sanders, 1999; Bixler, 1999; Maglischo 1989; Maglischo, 2003). Maglischo (1989) commented that not every movement in the stroke cycle is propulsive and that lift forces are minor and overshadowed by drag forces when propulsion and acceleration of the swimmer exists.

Through experimentation and simulation, Hay *et al.* (1993) and Lui *et al.* (1993) demonstrated that the curved path of the hand in freestyle was actually due to the body roll action and not to the mediolateral motion of the arm. In fact, it was suggested that the arm motion acts to flatten the curved path (Hay *et al.*, 1993). This further dispelled the idea that good swimmers are 'scullers' relying strongly on lift forces.

Nevertheless, based on the above equations, some curve or directional change in hand path offers advantages. If the water and body segment are moving in the same direction, the relative velocity is reduced as the water starts to 'move with the hand' and thus the propulsive forces diminish even when the hand is moving with the same speed and orientation relative to an external frame of reference (see, for example, Sanders, 1999). Manipulating the direction of the hand throughout the stroke cycle allows the swimmer to 'find still water' to accelerate large masses of water over a longer period of time.

Unfortunately, the coefficients obtained for the hand and forearm models yield only 'ball park' estimates of the forces produced in swimming. They assume that the flow under steady (constant velocity, angle of attack and sweepback angle) is comparable to the flow during 'real' swimming. Further, vortex shedding, limb orientation, rotation and added mass effects are all ignored and assumed negligible (Toussaint *et al.*, 2002; Lauder and Dabnichki, 2005; Rouboa *et al.*, 2006; Gardano and Dabnichki, 2005). Additionally, the accuracy of quantifying the three-dimensional (3D) orientation of the limbs throughout a stroke cycle is limited by the accuracy and reliability of digitising which, in turn, is limited by the size and clarity of the image that can be obtained for a small limb moving rapidly in a fluid medium through a large space.

There is considerable scope for advancement in methods to calculate lift and drag forces according to the formulae presented above. Application of the rapidly emerging capabilities of computational fluid dynamics (CFD) (Bixler and Riewald, 2002; Rouboa *et al.*, 2006) to scanned limbs of individual swimmers combined with improving 3D data collection techniques and digitising software could provide better estimates of forces produced. However, there is growing evidence that other mechanisms may contribute greatly to propulsion.

Propulsion from vortices

'Vortex theory', used extensively to quantify characteristics of propulsion executed by marine animals, birds and insects, even in unsteady conditions, is now being intensively investigated with respect to human swimming. A 'shed' vortex indicates the end of a propulsive impulse and occurs during major transitions of the underwater stroke. It is, therefore, possible to speculate that the curved hand path of the underwater stroke cycle continuously sheds vortices to assist propulsion (Colwin, 2002; Ungerechts *et al.*, 1999; Arellano, 2006). A shed vortex represents a transfer of momentum from the water to the swimmer, similar to that observed in studies of aquatic animals, whereby backward momentum of the vortex rings corresponds to a forward momentum gained by the aquatic animal (Lighthill, 1970; Ungerechts *et al.*, 1999; Müller *et al.*, 1997; Arellano *et al.*, 2002).

Incorporation of basic flow visualisation techniques such as tufts, colour dyes, injection of air bubbles, air bubble wall and small buoyant particles, have provided the researcher with tools to qualitatively analyse water flow behaviour (Colwin, 1985; Hay and Thayer, 1989; Persyn, *et al.*, 1998; Ungerechts *et al.*, 1999; Colman *et al.*, 1999; Arellano, 1999). Particle image velocimetry (PIV) (Videler *et al.*, 1999; Müller *et al.*, 2001) and digital particle velocimetry (DPIV) (Wilga and Lauder, 2000; Wilga and Lauder, 2002; Birch and Dickinson, 2001; Lu *et al.*, 2006, Lehmann *et al.*, 2005, Arellano, 2006) techniques allow quantification of vortex properties by tracking neutrally buoyant particles, allowing the generation of velocity vector graphs of the vortices in response to water flow movements.

Propulsion from axial flow

Toussaint *et al.* (2002) have proffered 'axial flow' as an explanation for propulsion from the upper limbs in front crawl. In accordance with Bernoulli's Law a pressure gradient develops where local pressure close to the limb decreases in the direction of the fingertips. Thus, rotating the limb 'pumps' fluid along the arm toward the hand on both the front and back sides of the arm. Arm translation magnifies the pressure on the palmar (leading) side of the hand and a lower pressure on the dorsal (trailing) side of the hand creates propulsive force.

Clearly, much further research is required to fully understand how swimmers propel themselves through the water and how effectiveness of propulsion might be improved.

Resistance in swimming

Characteristics of resistive forces in water

Just as movement of a swimmer's limbs through the water can produce forces in the desired direction of travel, motion of the body parts in water also produce forces that are not in the desired direction, that is, resistive forces.

Total resistance is equal to the sum of 'frictional' (F_f), 'pressure' (F_p) and 'wave' (F_w) contributions when swimming at a constant speed (Toussaint, 2000).

$$F_{tot} = F_{frictional} + F_{pressure} + F_{wave}$$

Frictional resistance

Frictional resistance can be estimated according to the equation:

$$F_{fr} = \mu \, (dV / dZ) \, S_{fr}$$

where μ is coefficient of dynamic viscosity, dV is the difference between velocity of water layers, dZ the difference in thickness of water layers and S_{fr} the wetted body surface area (Vorontsov and Rumyantsev, 2000).

The magnitude of frictional forces depends on: (i) total immersed surface area of the swimmer; and; (ii) the flow conditions within the boundary layer (Webb, 1975). Boundary flow can be: (i) laminar; (ii) turbulent; or (iii) transitional. The state of the flow depends on the velocity of the flow, the size and shape of the swimmer, as well as the density and viscosity of the water. In laminar flow, fluid flows in parallel layers without any disturbances or disruption between the layers. As the velocity increases flow 'separates' and becomes turbulent. Separation and turbulent flow occur when friction within the boundary layer increases. The formation of eddies after the point at which the flow separates results in variation of pressure and velocity of the water as well as the formation of unsteady vortices.

For a given shape, the point of separation is related to the 'Reynolds number', a dimensionless ratio of inertial to viscous forces:

$$R_e = \rho . V L/\mu$$

where ρ is water density, V is flow velocity, L is body length and μ is the coefficient of dynamic viscosity.

For competitive swimming R_e is of the order of 2×10^5 to 2.5×10^6 (Clarys, 1979). The size and shape of swimmers is such that at these Reynolds numbers the flow is predominantly turbulent. This being the case, resistance due to friction drag is small relative to resistance due to pressure drag and wave drag.

Pressure resistance

Pressure resistance, otherwise termed 'form drag' or 'pressure drag' is the result of differential pressure between the front and rear of a swimmer's body or individual limbs. Boundary separation between the water layer and the surface of the swimmer causes turbulent flow along the body. The rotational movement of the water forms an area of low pressure called 'wake'.

Pressure resistance can be estimated according to the equation:

$$F_p = \frac{1}{2} Cp \, \rho . AV^2$$

Where C_P is the pressure resistance coefficient and is influenced by shape and orientation to the flow, ρ is fluid density, A is the projected area to the flow and V is the velocity of the moving body.

A swimmer does not want to reduce average velocity. However, because the resistance is related to velocity squared, it is better having a close to constant velocity than having periods of very high velocity and periods of very low velocity. The swimmer can also minimise resistance by orienting the body to have a small cross-sectional area and to 'streamline' their shape as much as possible.

Even though a strong relationship between hydrodynamic resistance and human morphology has been confirmed (Huijing *et al.*, 1988), it was argued that differences in drag between subjects depend also on different technique characteristics (Clarys, 1979). Being comprised of many body segments, it is difficult for swimmers to maintain a streamlined position during active swimming; consequently, local areas of pressure difference develop.

Huijing *et al.* (1988) identified body cross-sectional area as the factor that most accurately distinguishes swimmers in terms of pressure drag. However, the resistive drag is influenced also by the swimming technique.

Wave resistance

Waves generated when swimming on the surface or at small depths contribute significantly to overall resistance when speed is close to maximum (Miller, 1975). Part of a swimmer's kinetic energy is lost in displacing water and consequently forming waves. The energy wasted to form these waves is proportional to the energy contained in the prime wave (Rumyantsev, 1982).

Wave resistance can be calculated according to the equation:

$$F_w = \rho \,(A^3/\lambda^2)\,(V_w \sin \alpha)^3 \cos \alpha \,\Delta t$$

where ρ is water density, A is wave amplitude, λ is wave length, V_w is wave velocity and α is the angle between the direction of general centre of mass (COM) motion and the front of the prime wave (Rumyantsev, 1982).

When increasing speed both wave length and wave amplitude increase. For competitive swimmers maximum speed corresponds to 'hull speed' (Kolmogorov and Duplishcheva, 1992) which occurs when the bow wave length becomes equal to water line length of the swimmer (swimmer's height). Any further increase in speed is inhibited as the swimmer is 'trapped' in the interim space between crests of waves.

The relationship between relative velocity and body height is defined by a dimensionless 'Froude number' determining the magnitude of resistance and can be calculated according to the equation:

$$F_r = \frac{v}{\sqrt{gL}}$$

Where V is swimming velocity, L is the swimmer's height and g is gravitational acceleration (9.81 m/s^2).

Wave resistance is significantly related to cruising depth (Vennell *et al.*, 2006) becoming negligible at a depth greater than 0.6m (Lyttle *et al.*, 1998). Thus, wave drag, a major contributor to resistance at high speed, can be minimised by travelling underwater. Since swimmers reach high speed after starts and turns at a depth where wave drag is small, gliding, followed by underwater kicking when the glide has slowed to the speed sustainable by underwater kicking, can be fast and economical compared to surface swimming, provided that the swimmer maintains a streamlined position.

Interaction between propulsive and resistive forces

Propulsive and resistive forces change continuously within and between stroke cycles. Unfortunately, we cannot readily determine the magnitude of propulsive and resistive forces separately. However, we can observe the effect of the difference between them, that is, the net force acting on a swimmer.

Swimming velocity of the whole body COM fluctuates during a stroke cycle due to the variation in the magnitude and direction of propulsive and resistive forces. It has been proposed that elite swimmers have smaller velocity fluctuations than sub-elite swimmers, since the former are believed to be more effective in minimising the resistive forces than the latter (e.g. Takagi *et al.*, 2004). This had been supported by evidence of early two-dimensional (2D) studies that reported lower fluctuations for the hip or whole body COM speeds of faster swimmers (e.g. Togashi and Nomura, 1992; Sanders, 1996a, 1996b). While a linear relationship between the magnitude of velocity fluctuations and performance is intuitively appealing it does not take into account that resistive forces during the active phases of swimming (active drag) are proportional to the square of velocity (e.g. Toussaint and Beek, 1992) rather than merely being proportional to velocity. Therefore, the higher the maximum intracycle speed reached by a swimmer the larger the magnitude of the resistive forces that the swimmer experiences. Recent 3D data have shown that faster freestyle swimmers actually had a tendency towards greater velocity fluctuations than slower swimmers (Psycharakis, 2006).

Small local minima in the intracycle velocity profiles indicate that resistive forces have been considerably larger than propulsive forces during the preceding period in the stroke cycle. Biomechanists, coaches and swimmers can use this information to develop technique strategies for reducing the difference between propulsive and resistive impulses during those periods. These may include improving the postures and alignment of the body and reducing the period of time that the resistive forces are larger than the propulsive forces.

Direct (e.g. Toussaint *et al.*, 1988) and indirect methods (e.g. Di Prampero *et al.*, 1974) have been developed to measure active drag forces. Researchers have also attempted to calculate hand propulsive forces with the use of purpose-built equipment such as pressure transducers on the hands (e.g. Takagi and Wilson, 1999). However, the methods developed are not completely free of perturbations to the normal swimming technique. The identification of the net forces experienced during a stroke cycle and their influence on swimming performance could be facilitated in future research by the combination of 3D biomechanical analysis and a bio-energetic/physiological assessment of energy expenditure. To do so requires quantifying the whole body COM using 3D analysis methods, including digitising a full body model. Above and below water views must be digitised from at least two camera positions and preferably more for the underwater view. Given that automatic digitising of underwater views is currently problematic due to the interference of bubbles and the 'blue filter' effect of the water, the task is extremely labour intensive.

Rotations around longitudinal and transverse axes

The combined effect of the rolling actions of shoulders and hips plays an important role in swimming performance in freestyle and backstroke. Computer simulation studies indicated that body roll might assist in the generation of propulsive lift forces (e.g. Payton *et al.*, 1997). Yanai (2004) found that the primary source of body roll was the buoyancy force and not the hydrodynamic forces, and that good swimmers used the former effectively in driving the body roll.

The vast majority of investigations of body roll have considered the trunk as a whole and calculated body roll based on 2D data. Cappaert *et al.* (1995) investigated hip and shoulder roll as separate variables using 3D analysis methods and reported that the shoulders and hips of some sub-elite swimmers, despite having similar roll ranges to elite swimmers, were rolling in opposite directions. The investigators stated that the opposite roll between the shoulders and the hips of the sub-elite group might have increased resistance by increasing frontal surface area. Interestingly, the 3D analysis of Psycharakis (2006) showed that faster swimmers had a tendency to roll their shoulders and hips less than slower swimmers throughout a maximum 200 m freestyle swim.

The above evidence suggests that a more complete understanding of the inter-relationships between forces and rolling actions should be sought by separate examination of shoulder and hip roll. Future research needs to focus on the effect of shoulder and hip roll on swimming performance by identifying the contributions of the rolling actions to propulsive and resistive forces. Biomechanists could also investigate the influence of different roll magnitudes and timing on the positioning of the arms and legs for their consequent application of propulsive forces.

Sanders *et al.* (1995) established that rotations about transverse axes in butterfly swimming are sequenced to produce a 'body wave' that contributes to propulsion. While intracyclic velocity fluctuations of the whole body COM are known to be large in butterfly swimming, it is likely that efficiencies are gained by transfer of energy by the body wave culminating in large propulsive forces during the kick. Similarly, in breaststroke (Sanders, 1995; Sanders *et al.*, 1998) and freestyle kicking (Sanders, 2007), body waves produced by learned sequencing of joint actions is related to effective technique. Swimming is clearly an activity in which rhythm and timing are very important for efficient swimming. There is great scope for further investigations to establish how underlying rhythms of the rotations about different axes are related to performance.

Starts and turns

Freestyle, breaststroke and butterfly starts

The importance of starts is indicated by the significant correlation between start time and performance in the 50, 100 and 200 m freestyle events during the 1992 Olympic Games (Arellano *et al.*, 1994) and in most events of the 1999 Pan Pacific Championships (Mason and Cossor, 2000) regardless of the stroke. Cossor and Mason (2001) found that the start time, defined as the time between the starting signal until the centre of head reached the 15 m mark, represented between 0.8 per cent (in 1500 m freestyle) to 26.1 per cent (in 50m sprints) of the total race time depending on the event.

A traditional deterministic model (Hay, 1987) lends itself to identification of variables that are important during start phases (Sanders, 2002).

The conventional starting technique in which the swimmer swung the arms (Lewis, 1980) has been replaced by variants of 'grab' starts in which the swimmer dives from a position holding onto the block. Zatsiorsky *et al.* (1979) and Bowers and Cavanagh (1975) found that the grab start provides a faster block time and a higher ground reaction force (GRF) than the conventional swing arm start. Lewis (1980) showed that the grab technique is faster in leaving the block but does not allow the swimmer to travel as far in flight as the conventional technique.

In general, the grab start is advantageous because the forces are in a desirable direction from the beginning of the dive, time for an arm swing and counter movement are minimised, and less time is required for the body to lean to a suitable position for the main push than in the conventional start.

The 'track' start is basically a variant of the grab start that was introduced by Fitzgerald (1973). In the track start one foot is further forward than the other. Within track starts two variants have been identified. These are the forward weighted track start known as 'bunch start' (Ayalon, *et al.* 1975) and the rear weighted 'slingshot' track start.

Vilas-Boas *et al.* (2000) compared the two variants of the track start technique. Although the rear weighted track start resulted in greater horizontal impulses and thereby greater horizontal velocity than the start with the forward weighted track start, this advantage was offset by an increased block time. Vilas-Boas *et al.* (2003) found that the whole body COM had a greater horizontal displacement during the block time for the rear weighted track start than in the forward weighted track start or grab start techniques. Also the grab start had a longer movement time and a faster reaction time than the two variants of track start.

Pearson *et al.* (1998) found that performances to 7 m for the 'handle' start were equal to those of the grab start, which was the subjects' preferred technique. Thus, it was recognised that with practice, the blocks with handles may yield improved performances. To further test this possibility Blanksby *et al.* (2002) conducted an intervention study to determine whether any of the grab, track or handle swimming starts yielded a faster time. While there were no significant differences between dive groups in time to 10 m pre- or post-training, time to 10m, reaction, movement, block and flight times all improved irrespective of the technique used. The implication for swimmers and coaches, supported also by the work of Arellano *et al.* (2000), is that intensive practice of a technique variant is more important than the choice of technique variant.

The force development characteristics during the grab and track start have been investigated recently by numerous researchers (Arellano *et al.*, 2000; Kruger *et al.*, 2003; Vilas-Boas *et al.*, 2003; Benjanuvatra *et al.*, 2004). Results indicate differences in the patterns of force development in both the vertical and horizontal directions. Modelling and simulation by Takeda and Nomura (2006) indicated that the track start increases the ability to generate rotational motion due to the large moment about the COM that can be created by the reaction force from the rear foot, whereas the grab start affords the ability to extend the joints rapidly to yield a high resultant velocity at take-off. However, the track start could have a shorter block time than the grab start due to the small range of joint extension (Takeda and Nomura, 2006). Benjanuvatra *et al.* (2004) found that the front and rear legs differed in their contributions to impulse in the track start with the front leg, producing more impulse due to the longer period of contact as well as higher peak forces in both the vertical and horizontal directions.

Thus, there is a complex of factors interacting to affect time on the block and the release characteristics at takeoff. There is a trade-off between time on the blocks and speed of release with a small range of movement (ROM) allowing a small block time at the expense of speed of release.

While increases in angle of release up to about 40° will increase the range for any given speed, there is also an increase in time in the air, a reduction in the horizontal speed, and a steeper entry angle. Take-off angles between –5 to 10° have been found in recent studies (Arellano *et al.*, 2000; Holthe and McLean, 2001; Miller *et al.*, 2003). Thus, angles of release among competitive swimmers do not approach the angles that would maximise the flight distance.

The time below the surface includes the period of immersion between first contact with the water and complete immersion, the period of the glide in which the swimmer maintains a passive motion under water, and the post-glide action, including the kick (long pull and glide in breaststroke) until resumption of stroking at the surface.

Hay and Guimaeres (1983) found that the underwater glide time accounted for 95 per cent of the variance in starting time, indicating that the glide phase is crucially important. Vilas-Boas *et al.* (2003) observed that any advantage in horizontal velocity at entry was quickly lost during immersion and gliding. Thus, the ability to minimise resistance during the glide has a major influence on overall dive performance. This is supported by race analysis data, indicating that good start performance is associated with a long underwater distance (Cossor and Mason, 2001). Clearly, swimmers who can glide well glide further before surfacing and gain an advantage from doing so.

The depth affects the amount of wave drag (Hertel, 1966). This has implications for the desired trajectory path. A shallow trajectory increases wave drag. However, a deep trajectory means that the motion is not entirely in the direction of desired horizontal travel. Thus, additional research to supplement the pioneering work of Lyttle and Blanksby (2000) is required to establish the optimal glide paths taking into account speeds and angles of entry.

The time of initiating the kick is also important. Initiating the kick after the time corresponding to that when the swimmer is slower than can be sustained by kicking reduces performance (Blanksby *et al.*, 1996). Initiating the kick when the speed is still greater than that sustainable by kicking causes excess active drag and the net force to be higher than the passive drag force during a glide (Lyttle and Blanksby, 2000). Preliminary research (Sanders and Byatt-Smith, 2001) indicates that many swimmers kick too soon and that this is more common than kicking too late.

Assuming the time of commencement of the kick is adjusted appropriately, a swimmer can improve performance in the glide phase by having an optimal body orientation and posture during immersion to maximise initial speed, having a streamlined body position during the glide, and by improving the effectiveness of the underwater kick.

Backstroke starts

Despite the relatively large contribution of start time to backstroke performance, limited research has been conducted on the backstroke start. Kruger *et al.* (2006) tested nine German elite backstroke sprinters, finding a strong correlation between the magnitude of the resultant reaction forces at take-off and the time to reach 7.5 m distance from the wall ($r = -0.83$). Currently, there is a paucity of data on the entry and the glide phase.

Turns

The turn time becomes increasingly important with increasing swimming distance and, of course, is more important in short course events than in long course events. In 200 m breaststroke more than one third of the race time is spent from 2 m into the wall to 9 m out of the wall (Thayer and Hay, 1984). A significant correlation was found between the turn time (7.5 m round trip time) and the event time for the 50 m, 100 m and 200 m freestyle events during the 1992 Olympic Games (Arellano *et al.*, 1994). However, there was no significant relationship between turn times and race times in the Sydney 2000 Olympic Games (Mason and Cossor, 2001) using the same 7.5 m round trip time criterion.

Swimming turns consist of two types of turn, the tumble turn used in freestyle and backstroke, and the open or pivot turn used in breaststroke and butterfly. This review focuses predominantly on the tumble turn (freestyle and backstroke). However, many of the points can be directly related to the breaststroke and butterfly turns.

The contact period of a turn may be likened to a counter-movement jump. That is, there is a period of flexion after initial contact in which the major extensors of the hip, knee, and ankle are working eccentrically to absorb the energy possessed by virtue of the swimmer's movement towards the wall. This is then followed by concentric work of muscles to extend the hip, knee, and ankle to generate speed away from the wall. The object is to minimise the time of contact while also maximising the exit speed.

As in jumping, there is an optimal amount of counter movement or joint flexion that yields a maximum exit speed. Unlike jumping, the amount of initial flexion and the amount of flexion that occurs following initial contact are important in their influence on the time of contact as well as the time prior to contact (time in). Therefore, the appropriate magnitude of initial flexion and counter movement flexion depends on a complex mix of variables (Sanders, 2002).

Researchers have used force plates to determine the wall contact and the push-off characteristics (Nicol and Kruger, 1979; Blanksby *et al.*, 1998; Blanksby *et al.*, 2004; Lyttle *et al.*, 1999). Lyttle *et al.* (1999) showed that in a push-off it is important to flex prior to contact, minimise flexion after contact, and to minimise the wall contact time. The initial speed at contact should be as large as possible as it reflects a fast swimming speed and little loss during the tumble. The change in speed results from resistive impulse, that is, from the forces and the time over which they act. Time is minimised by using the wall, rather than hydrodynamic resistive forces, to arrest the motion. Thus, the swimmer seeks a rapid tumble from a distance that allows early contact with the wall. At this time the distance should be such that the hips and knees flex to a position that allows the combination of speed generated by extension and time of contact to be optimised. To date, there has been little research to establish optimum amounts of joint flexion at contact. Based on knowledge of elastic energy restitution (see, for example, Wilson, 1994), it is likely that small distances of flexion (counter movement) are desirable. A long period of flexion resulting from a slow initial speed and a large amplitude of counter movement is undesirable, as this would reduce the restitution of elastic energy as well as contributing to the overall time of contact. A fast initial speed, combined with a small amplitude of counter movement, would have the advantage of ensuring that joint torques with an extension influence are large at the commencement of extension.

Blanksby *et al.* (1996) defined a 'tuck index' as the ratio of the distance between the hip and the wall and the standing trochanteric height. They found a correlation between tuck index and the round trip time, indicating that the larger the tuck index the faster the time. The ideal knee angle was found to be between 110–120°. A smaller knee angle would cause the quadriceps muscles to be at a less than optimal length to produce large forces quickly (Lyttle and Benjanuvatra, 2006; Pereira *et al.*, 2006).

It has been found that the effectiveness of the push off is not dependent only on the impulse generated during contact but also on the streamlining during the push-off (Lyttle, *et al.*, 1999). Thus the maximum forces should be achieved when the swimmer's body, particularly the trunk, head, and arms, have attained a streamlined position oriented in the swimming direction.

In the freestyle tumble turn it is possible to complete the somersault in a supine position and then rotate to a prone position during the glide. While this may enable a reduced contact time, the rotation of the body may increase drag due to the body being in a less than optimal streamlined posture, thereby affecting the 'time out' (Counsilman, 1955).

The interplay of propulsive and resistive forces that determines the change in speed during the kicking phase is linked to the amplitude and frequency of the kick. There is a natural tendency for swimmers to simply 'kick as hard as they can' but this may not be conducive to optimising speed and may also waste physiological energy. In comparing the kicking styles, Clothier *et al.* (2000) demonstrated that the dolphin kick was superior to the flutter kick as it enables the swimmer to maintain a higher speed.

Resistance to progression in the swimming direction during periods when the swimmer is not attempting to generate propulsion through body actions is known as 'passive drag'. These periods are referred to as 'glide phases'. In addition to being a significant part of starts and turns, glides occur during transitional phases of all strokes (Vorontsov and Rumyantsev, 2000), and are particularly important in breaststroke, in which the gliding phase corresponds to more than 40 per cent of the swimming time at all race distances. Thus, minimising passive drag is an important factor distinguishing the turn performance of swimmers (D'Acquisto *et al.*, 1988; Chatard *et al.*, 1990). The ability to minimise resistance in glide phases depends on a swimmer's body shape, orientation of the swimmer with respect to the swimming direction and the postures adopted.

Resistive forces in glides are commonly measured by transducing cables (Karpovich, 1933; Alley, 1952; Counsilman, 1955; Hairabedian, 1964; Chatard, *et al.*, 1990; Sheehan and Laughrin, 1992; Klauck, 1999; Lyttle *et al.*, 1998; Benjanuvatra *et. al.*, 2001) or rods attached to devices that tow a swimmer, either with the apparatus moving, for example, in towing tanks (Van Manen and Rijken, 1975; Jiskoot and Clarys, 1975; Clarys and Jiskoot, 1975; Clarys 1979), or with the water moving, for example, in a flume (Miyashita and Tsunoda, 1978; Maiello *et al.*, 1998; Roberts *et al.*, 2003; Chatard and Wilson, 2003; Vennell *et al.*, 2006). Unfortunately these methods affect the swimmer's posture, and tend to apply turning forces that affect a swimmer's alignment. Further, they do not account for the inertial effects of 'added mass' entrained by a decelerating swimmer in actual glides in swimming.

Indirect methods such as digitised video have had problems dealing with the inherent unsteady effects that are difficult to separate from the noise amplified from small measurement errors when deriving velocity and force data from displacement data (Klauck and Daniel, 1976; Oppenheim, 1997). Recently, Naemi (2007) applied a curve fitting approach to digitised displacement data. The method is sensitive enough to detect small differences in glide efficiency within and among swimmers in actual non-tethered glides while also taking into account the effects on glide coefficients of varying Reynolds number. The method shows great promise to improve understanding of how individual swimmers can 'fine-tune' their gliding techniques to maximise performance.

References

1. Alley, L. E. (1952) 'An analysis of water resistance and propulsion in swimming the crawl stroke'. *Research Quarterly*, 23: 257–270.
2. Arellano, R., Brown, P., Cappaert, J. and Nelson, R. C. (1994) 'Analysis of 50-100-, and 200-m freestyle swimmers at the 1992 Olympic Games'. *Journal of Applied Biomechanics*, 10: 189–199.
3. Arellano, R. (1999) Vortices and Propulsion. In R. Sanders and J. Linsten (eds.), *Swimming: Applied Proceedings of the XVII International Symposium on Biomechanics in Sports* (Vol. 1, pp. 53–66). Perth, Western Australia: School of Biomedical and Sports Science.
4. Arellano, R., Pardillo, S., De La Fuente, B. and Garcia, F. (2000). A system to improve the swimming start technique using force recording, timing and kinematic analyses. In R. Sanders and Y. Hong (eds.), *Proceedings of XVIII Symposium on Biomechanics in Sports: Applied Program: Application of Biomechanical Study in Swimming* (pp. 609–613). Hong Kong: International Society of Biomechanics in Sports.
5. Arellano, R., Pardillo, S. and Gavilán, A. (2002) Underwater undulatory swimming: kinematic characteristics, vortex generation and application during the start, turn and swimming strokes. In K. E. Gianikelis, B. R. Mason, H. M. Toussaint, R. Arellano and R.H. Sanders (eds.), *Applied Proceeding – Swimming – XXth International Symposium of Biomechanics in Sports Proceeding*. Caceres: International Society of Biomechanics in Sports.
6. Arellano, R. *(2006) Understanding swimming propulsion based on new technologies* [In press].
7. Ayalon, A., Gheluwe, B. and Kanitz, M. (1975) 'A comparison of four styles of racing in swimming'. In L. Lewillie and J. P. Clarys (eds.), *Swimming II* (pp. 233–240). Baltimore: University Park Press.
8. Benjanuvatra, N., Blanksby, B. A. and Elliott, B. C. (2001) 'Morphology and hydrodynamic resistance in young swimmers'. *Paediatric Exercise Science*, 13: 246–255.

9. Benjanuvatra, N., Lyttle, A., Blanksby, B. A. B. and Larkin, D. (2004). Force development profile of the lower limbs in the grab and track start. In M. Lamontagne, D. Gordon, E. Roberstson, and H. Sveistrup (eds.),*Proceedings of XXII International Symposium on Biomechanics in Sports* (pp. 430–433). Ottawa: International Society of Biomechanics in Sports.
10. Berger, M., de Groot, G. and Hollander, A. P. (1995) 'Hydrodynamic drag and lift forces on human hand/arm models'. *Journal of Biomechanics*, 28(2): 125–133.
11. Berger, M., Hollander, A. P. and de Groot, G. (1999) 'Determining propulsive force in front crawl swimming: a comparison of two methods'. *Journal of Sports Sciences*, 17: 97–105.
12. Birch, J. M. and Dickinson, M. H. (2001) 'Spanwise flow and the attachment of the leading-edge vortex on insect wings'. *Nature*, 412: 729–733
13. Bixler, B. and Riewald, S. (2002) 'Analysis of a swimmer's hand and arm in steady flow conditions using computational fluid dynamics'. *Journal of Biomechanics*, 35(5): 713–717.
14. Bixler, B. S. (1999) 'The bombastic Bernoulli bandwagon: it's time to step off'. *American Swimming Magazine*, 4: 4–10.
15. Blanksby, B. A., Gathercole, D. G. and Marshall, R. N. (1996) 'Force plate and video analysis of the tumble turn by age-group swimmers'. *Journal of Swimming Research*, 11: 40–45.
16. Blanksby, B. A., Simpson, J. R., Elliott, B. C. and McElroy, G. K. (1998). 'Biomechanical factors influencing breaststroke turns by age-group swimmers'. *Journal of Applied Biomechanics*, 14: 180–189.
17. Blanksby, B., Nicholson, L. and Elliott, B. (2002). 'Biomechanical analysis of the grab, track and handles starts: an intervention study'. *Sports Biomechanics*, 1(1): 11–24.
18. Blanksby, B. A., Skender, S., Elliott, B. C., McElroy, G. K. and Landers, G. (2004). 'An analysis of rollover backstroke turns by age-group swimmers'. *Sports Biomechanics*, 3(1): 1–14.
19. Bowers, J. E. and Cavanagh, P. R. (1975) 'A biomechanical comparison of the grab and conventional sprint starts in competitive swimming'. In L. Lewillie and J. P. Clarys (eds.), *Swimming II*, (pp. 225–232). Baltimore: University Park Press.
20. Cappaert, J. M., Pease, D. L. and Troup, J. P. (1995) '3D analysis of the men's 100-m freestyle during the 1992 Olympic Games'. *Journal of Applied Biomechanics*, 11: 103–112.
21. Chatard, J. C., Lavoie, J. M., Bourgoin, B. and Lacour, J. R. (1990) 'The contribution of passive drag as a determinant of swimming performance'. *International Journal of Sports Medicine*, 11: 367–372.
22. Chatard, J. C. and Wilson, B. (2003) 'Drafting distance in swimming'. *Medicine and Science in Sports and Exercise*, 35: 1176–1181.
23. Clarys, J. P. and Jiskoot, J. (1975) 'Total resistance of selected body positions in the front crawl'. In L. Lewillie and J. P. Clarys (eds.), *International Series on Sport Sciences, Volume 2; Swimming II* (pp. 110–117). Baltimore: University Park Press.
24. Clarys, J. P. (1979) 'Human morphology and hydrodynamics'. In J. Terauds and E. W. Bedingfield (eds.), *International Series on Sports Science, Volume 8; Swimming III* (pp. 3–41). Baltimore: University Park Press.
25. Clothier, P. J., McElroy, G. K., Blanksby, B.A. and Payne, W. R. (2000) 'Traditional and modified exits following freestyle tumble turns by skilled swimmers'. *South African Journal for Research in Sport, Physical Education and Recreation*, 22(1): 41–55.
26. Colman, V., Persyn, U., and Ungerechts, B. E. (1999) 'A mass of water added to the swimmer's mass to estimate the velocity in dolphin-like swimming bellow the water surface'. In K. L. Keskinen, P. V. Komi, and A. P. Hollander (eds.), *Biomechanics and Medicine in Swimming VIII* (1st edn., pp. 89–94). Jyväskylä, Finland: Department of Biology of Physical Activity, University of Jyväskylä, Gummerus Printing House
27. Colwin, C. M. (2002) *Breakthrough Swimming*. Human Kinetics.
28. Cossor, J., and Mason, B. (2001) 'Swim start performances at the Sydney 2000 Olympic Games' In J. Blackwell and R. Sanders (eds.), *Proceedings of XIX Symposium on Biomechanics in Sports*,(pp. 70–74). San Francisco: University of California at San Francisco.
29. Counsilman, J. E. (1955) 'Forces in swimming .two types of crawl stroke'. *Research Quarterly*, 26: 127–139.

30. Counsilman, J. E. (1968). *The science of swimming* Englewood Cliffs, N.J.: Prentice-Hall.
31. D'Acquisto, L. J., Costill, D. L., Gehlsen, G. M., Young, W. T. and Lee, G. (1988) 'Breaststroke economy skill and performance: study of breaststroke mechanics using a computer based "velocity video"'. *Journal of Swimming Research*, 4: 9–14.
32. Di Prampero, P. E., Pendergast, D. R., Wilson, C. W., and Renny, D. W. (1974) 'Energetics of swimming in man'. *Journal of Applied Physiology*, 37: 1–5.
33. Fitzgerald, J. (1973). 'The track start in swimming'. *Swimming Technique*, 10(3): 89–94.
34. Gardano, P. and Dabnichki, P. (2005) 'On hydrodynamics of drag and lift of the human arm'. *Journal of Biomechanics* [In press].
35. Hairabedian, A. (1964). *Kinetic Resistance Factors Related to Body Position in Swimming*. Unpublished Ph.D. thesis, Stanford University, Stanford.
36. Hay, J. G. (1987). The development of deterministic models for qualitative analysis. In R. Shapiro and J. R. Marett (eds.), *Proceedings of the Second National Symposium on Teaching Kinesiology and Biomechanics in Sports*.
37. Hay, J. G. and Guimares, A. C. S. (1983) 'A quantitative look at the swimming biomechanics'. *Swimming Technique*, Aug–Oct, 11–17.
38. Hay, J.G. and Thayer, A. M. (1989) 'Flow visualization of competitive swimming techniques: the tufts method'. *Journal of Biomechanics*, 22(1): 11–19.
39. Hay, J. G., Lui, Q. and Andrews, J. G. (1993) 'Body roll and handpath in freestyle swimming: a computer simulation study'. *Journal of Applied Biomechanics*, 9, 227–237.
40. Hertel, H. (1966). *Structure-Form-Movement*. New York: Reinhold Publishing Corporation.
41. Holthe, M. J. and McLean, S. P. (2001). Kinematic comparison of grab and track starts in swimming. In J. Blackwell and R. H. Sanders (eds.), *Proceedings of Swim Sessions XIX International Symposium on Biomechanics in Sports* (pp. 31–34). San Francisco: International Society of Biomechanics in Sports.
42. Huijing, P. A., Toussaint, H. M., Clarys, J. P., *et al.* (1988). *Active drag related to body dimensions*. In Ungerechts *et al.* (eds), *Swimming Science V*, Human Kinetics Book, Champaign, III.
43. Jiskoot, J. and Clarys, J. P. (1975). Body resistance on and under the water surface. In L. Lewillie and J. P. Clarys (eds.), *International Series of Sport Sciences, Volume 2; Swimming II* (pp. 105–109). Baltimore: University Park Press.
44. Karpovich, P. V. (1933) 'Water resistance in swimming'. *Research Quarterly*, 4: 21– 28.
45. Klauck, J. (1999). Man's water resistance in accelerated motion: An experimental evaluation of the added mass concept. In K. L. Keskinen, P. V. Komi and A. P. Hollander (eds.), *Proceedings of the VIII International Symposium on Biomechanics and Medicine in Swimming; Biomechanics and Medicine in Swimming VIII* (pp. 83–88). Saarijavi: University of Jyväskylä.
46. Klauck, J. and Daniel, K. (1976). Determination of man's drag coefficient and effective propelling forces in swimming by means of chronocyclography. In P. V. Komi (ed.), *Biomechanics VB, Intentional Series on Biomechanics*. (pp. 250–257) Vol. 1 B.
47. Kolmogorov, S. V. and Duplishcheva O. A. (1992) 'Active drag, useful mechanical power output and hydrodynamic force coefficient in different swimming strokes at maximal velocity', *Journal of Biomechanics*, 25: 311–318.
48. Kruger, T., Wick, D., Hohmann, A., El-Bahrawi, M. and Koth, A. (2003). Biomechanics of the grab and track start technique. In J. C. Chatard (ed.), *Biomechanics and Medicine in Swimming IX. Proceeding of the IX International Symposium on Biomechanics and Medicine in Swimming* (pp. 219–223). University of Saint-Etienne, France.
49. Kruger, T., Hohmann, A., Kirsten, R. and Wick, D. (2006). Kinematics and kinetics of the backstroke start technique. In J. P. Vilas-Boas, F. Alves and A. Marques (eds.) *Biomechanics and Medicine in Swimming X. Proceeding of the X International Symposium on Biomechanics and Medicine in Swimming* (pp. 58–60). Portugal: University of Porto.
50. Lauder, M. A. and Dabnichki, P. (2005) 'Estimating propulsive forces – sink or swim?' *Journal of Biomechanics*, 38: 1984–1990.
51. Lehmann, F. O., Sane, S. P. and Dickinson, M. (2005) 'The aerodynamic effects of wing-wing interaction in flapping insect wings'. *Journal of Experimental Biology*, 208: 3075–92.

52. Lewis, S. (1980) 'Comparison of five swimming starting techniques'. *Swimming Technique*, 16(4): 124–128.
53. Lighthill, M. J. (1970) 'Aquatic animal propulsion of high hydromechanical efficiency'. *Journal of Fluid Mechanics*, 44(2): 265–301.
54. Lu, Y., Shen, G. X. and Lai, G. J. (2006) 'Dual leading-edge vortices on flapping wings'. *Journal of Experimental Biology* 209: 5005–5016.
55. Lui, Q., Hay, J. G. and Andrews, J. G. (1993) 'Body roll and handpath in freestyle swimming: an experimental study'. *Journal of Applied Biomechanics*, 9: 238–253.
56. Lyttle, A. D., Blanksby, B. A., Elliott, B. C. and Lloyd, D. G. (1998) 'The effect of depth and velocity on drag during the streamlined glide'. *Journal of Swimming Research*, 13: 15–22.
57. Lyttle, A. D., Blanksby, B. A. B., Elliott, B. C., and Lloyd, D. G. (1999) 'Investigating kinetics in the freestyle flip turn push-off'. *Journal of Applied Biomechanics*, 15(3): 242–252.
58. Lyttle, A. and Blanksby, B. (2000). 'A look at gliding and underwater kicking in the swim turn'. In R. Sanders and Y. Hong (eds.), *Proceedings of XVIII Symposium on Biomechanics in Sports: Applied Program: Application of Biomechanical Study in Swimming* (pp. 56–63). Hong Kong: International Society of Biomechanics in Sports.
59. Lyttle, A. and Benjanuvatra, N. (2006). Optimising swim turn performance. http://www.coacesinfo.com/category/swimming/281/, Date accessed 04/07/07.
60. Maglischo, E.W. (1989) 'The basic propulsive sweeps in competitive swimming'. *Proceedings of the VIIth International Symposium of Biomechanics in Sports* (pp. 151–162). Melbourne, Victoria, Australia.
61. Maglischo, E.W. (2003) *Swimming Fastest*. Human Kinetics.
62. Maiello, D., Sabatini, A., Demarie, S., Sardella, F. and Dal Monte, A. (1998) 'Passive drag on and under the water surface'. *Journal of Sports Sciences*, 16(5): 420–421.
63. Mason, B. and Cossor, J. (2000). 'What can we learn from competition analysis at the 1999 Pan Pacific Swimming Championships?', In R. Sanders and Y. Hong (eds.), *Proceedings of XVIII Symposium on Biomechanics in Sports: Applied Program: Application of Biomechanical Study in Swimming* (pp. 75–82). Hong Kong: International Society of Biomechanics in Sports.
64. Mason, B. and Cossor, J. (2001) 'Swim turn performance at the Sydney 2000 Olympic Games' In J. Blackwell and R. H. Sanders (eds.), *Proceedings of Swim Sessions XIX International Symposium on Biomechanics in Sports* (pp. 65–69). San Francisco: International Society of Biomechanics in Sports.
65. McCabe, C. and Sanders, R (2006) Propulsion in swimming. http://www.coachesinfo.com/category/swimming/323/, Date accessed 04/04/07.
66. Miller, D. I. (1975) '*Biomechanics of Swimming*'. In J. H. Wilmore and J. F. Keogh (eds.), Exercise and Sport Sciences Reviews, pp. 219–248. New York, USA: Academic Press.
67. Miller, M., Allen, D. and Pein, R. (2003) 'A kinetic and kinematic comparison of the grab and track starts in swimming'. In J. C. Chatard (ed.), *Biomechanics and Medicine in Swimming IX. Proceedings of the IX International Symposium on Biomechanics and Medicine in Swimming* (pp. 231–235). University of Saint-Etienne, France.
68. Miyashita, M. and Tsunoda, R. (1978) 'Water resistance in relation to body size'. In B. Eriksson and B. Furberg (eds.), *International Series of Sport Sciences, Volume 6; Swimming Medicine IV* (pp. 395–401). Baltimore: University Park Press.
69. Müller, U. K., Van den Heuvel, B. L. E., Stamhuis, E. J. and Videler, J. J. (1997) 'Fish foot prints: morphology and energetics of the wake behind a continuously swimming mullet (chelon labrosus risso)'. *Journal of Experimental Biology*, 200: 2893–2906.
70. Müller, U. K., Smit, J., Stamhuis, E. J. and Videler, J. J. (2001). 'How the body contributes to the wake in undulatory fish swimming: flow fields of a swimming eel'. *Journal of Experimental Biology*, 204(16): 2751–2762.
71. Naemi, R. (2007) *A Hydro-kinematic Method for Quantifying Glide Efficiency of Swimmers*. Doctoral Thesis, The University of Edinburgh, Edinburgh.
72. Nicol, K., and Kruger, F. (1979) 'Impulse exerted in performing several kinds of swimming turns'. In J. Terauds and E. W. Bedingfield (eds.), *International Series of Sport Sciences, volume 8; Swimming III* (pp. 222–232). Baltimore: University Park Press.

73. Oppenheim, E. (1997) *Model Parameter and Drag Coefficient Estimation from Swimmer Velocity Measurements*. Unpublished MSc dissertation, University of Buffalo, Buffalo.
74. Payton, C. J., Hay, J. G. and Mullineaux, D. R. (1997) 'The effect of body roll on hand speed and hand path in front crawl swimming – a simulation study'. *Journal of Applied Biomechanics*, 13: 300–315.
75. Pearson, C. T., McElroy, G. K., Blitvich, J. D., Subic, A. and Blanksby, B. A. (1998) 'A comparison of the swimming start using traditional and modified starting blocks'. *Journal of Human Movement Studies*, 34: 49–66.
76. Pereira, S., Araujo, L., Freita, E., Gatti, R., Silveira, G. and Roesler, H. (2006). 'Biomechanical analysis of turn in front crawl swimming'. In J. P. Vilas-Boas, F. Alves and A. Marques (eds.), *Biomechanics and Medicine in Swimming X, Proceeding of the X International Symposium on Biomechanics and Medicine in Swimming* (pp. 77–79). University of Porto, Portugal.
77. Persyn, U., Colman, V. and Soons, B. (1998) 'New Combinations of Hydrodynamic Concepts in Swimming'. In K. Keskinen, P. V. Komi, and A. P. Hollander (eds.), *Biomechanics and Medicine in Swimming VIII, Proceedings of the VIII International Symposium on Biomechanics and Medicine in Swimming* (pp. 7–14). Jyväskylä: Gummerus Printing House
78. Psycharakis, S. (2006) 'A *3D analysis of intra-cycle kinematics during 200m freestyle swimming*', Edinburgh. Doctoral Thesis, The University of Edinburgh, Edinburgh.
79. Roberts, B. S., Kamel, K. S., Hedrick, C. E., McLean, S. P. and Sharp, R. L. (2003) 'Effect of a FastSkin suit on sub maximal freestyle swimming'. *Medicine and Science in Sports and Exercise*, 35: 519–524.
80. Rouboa, A., Sliva, A., Leal, A., Rocha, J. and Alves, F. (2006) 'The effect of swimmer's hand/forearm acceleration on propulsive forces generation using computational fluid dynamics'. *Journal of Biomechanics*, 39: 1239–1248.
81. Rumyantsev, V. A. (1982) *Biomechanics of Sport Swimming*. Central State Institute of Physical Culture. Moscow.
82. Rushall, B. S, Holt, L. E, Sprigings, E. J, and Cappaert, J.M. (1994) 'A Re-evaluation of Forces in Swimming'. *Journal of Swimming Research*, 10: 6–30.
83. Sanders, R. H. (1995) 'Can skilled performers readily change technique? An example, conventional to wave action breaststroke'. *Human Movement Science*, 14: 665–679.
84. Sanders, R. H., Cappaert, J. and Devlini, R. K. (1995) 'Wave characteristics of butterfly swimming'. *Journal of Biomechanics*, 28: 9–16.
85. Sanders, R. H. (1996a). Breaststroke technique variations among New Zealand Pan Pacific squad swimmers. In J. P. Troup, A. P. Hollander, D. Strasse, S. W. Trappe, J. M. Cappaert and T. A. Trappe (eds.), *Biomechanics and Medicine in Swimming VII* (pp. 64–69). London: E&FN Spon.
86. Sanders, R. H. (1996b) Some aspects of butterfly technique of New Zealand Pan Pacific squad swimmers. In J. P. Troup, A. P. Hollander, D. Strasse, S. W. Trappe, J. M. Cappaert and T. A. Trappe (eds.), *Biomechanics and Medicine in Swimming VII* (pp. 23–28). London: E&FN Spon.
87. Sanders, R. H., Cappaert, J. and Pease, D. L. (1998) 'Wave characteristics of Olympic breaststroke swimmers'. *Journal of Applied Biomechanics*, 14: 40–51.
88. Sanders, R. (1999) 'Hydrodynamic characteristics of a swimmer's hand'. *Journal of Applied Biomechanics*, 15: 3–26
89. Sanders, R. H., and Byatt-Smith, J. (2001). Improving feedback on swimming turns and starts exponentially. In J. Blackwell and R. H. Sanders (eds.), *Proceedings of Swim Sessions XIX International Symposium on Biomechanics in Sports* (pp. 91–94). San Francisco: International Society of Biomechanics in Sports.
90. Sanders, R. H. (2002). New analysis procedure for giving feedback to swimming coaches and swimmers. In K. E. Gianikelis, B. R. Mason, H. M. Toussaint, R. Arellano and R.H. Sanders (eds.), *Applied Proceeding – Swimming– XXth International Symposium of Biomechanics in Sports Proceeding*. Caceres: International Society of Biomechanics in Sports.
91. Sanders, R. H. (2003). Start Technique – recent findings. http://coachesinfo.com/category/swimming/88/, Date accessed 04/07/07.

92. Sanders, R. H. (2007) 'Kinematics, coordination, variability, and biological noise in the prone flutter kick at different levels of a "learn-to-swim" programme'. *Journal of Sports Sciences*, 25: 213–227.
93. Schleihauf, R. E. (1979). A hydrodynamic analysis of swimming propulsion. In J. Terauds and E.W. Bedingfield (eds.), *Swimming III* (pp. 70–109). Baltimore: University Park Press.
94. Schleihauf, R. E., Gray, L. and DeRose, J. (1983). Three dimensional analysis of swimming propulsion in sprint front crawl stroke. In A. P. Hollander, P. A. Huying and G. de Groot (eds.) *Swimming IV)* (pp. 173–183). Baltimore: University Park Press.
95. Sheehan, D. P. and Laughrin, D. M. (1992) 'Device for qualitative measurements of hydrodynamic drag on swimmers'. *Journal of Swimming Research*, 8: 30–34.
96. Sprigings, E. J, and Koehler, J. A. (1990) 'The choice between Bernoulli's or Newton's model in predicting dynamic lift'. *International Journal of Sports Biomechanics*, 6: 235–245.
97. Takagi, H. and Wilson, B. (1999). Calculating hydrodynamic force by using pressure differences in swimming. In K. L. Keskinen, P. V. Komi and A. P. Hollander (eds.), *Proceedings of the Biomechanics and Medicine in Swimming VIII* (pp. 101–106). Jyväskylä, Finland: University of Jyväskylä, Gummerus Printing House.
98. Takagi, H., Sugimoto, S., Nishijima, N. and Wilson, B. (2004) 'Differences in stroke phases, arm-leg coordination and velocity fluctuation due to event, gender and performance level in breaststroke'. *Sports Biomechanics*, 3: 15–27.
99. Takeda, T. and Nomura, T. (2006). What are the differences between grab and track start. In J. P. Vilas-Boas, F. Alves and A. Marques (eds.), *Biomechanics and Medicine in Swimming X, Proceedings of the X International Symposium on Biomechanics and Medicine in Swimming* (pp. 102–105). University of Porto, Portugal.
100. Thayer, A. L. and Hay, J. C. (1984) 'Motivating start and turn improvement'. *Swimming Technique*, Feb/Apr, 17–20.
101. Togashi, T. and Nomura, T. (1992). A biomechanical analysis of the novice swimmer using the butterfly stroke. In D. MacLaren, A. Lees and T. Reilly (eds.), *Biomechanics and Medicine in Swimming. Swimming science VI* (pp. 87–90). London: E & FN Spon.
102. Toussaint, H. M., Beelen, A., Rodenburg, A., Sargeant, A. J., DeGroot, G., Hollander, A. P., and Ingen Schenau, G. J. (1988) 'Propelling efficiency of front-crawl swimming'. *Journal of applied physiology*, 65: 2506–2512.
103. Toussaint, H. M., and Beek, P. J. (1992) 'Biomechanics of competitive front crawl swimming'. *Sports Medicine*, 13: 8–24.
104. Toussaint, H. M. (2000). An alternative fluid dynamic explanation for propulsion in front crawl swimming. In R. Sanders and Y. Hong (eds.) *Proceedings of the XVIII International Symposium on Biomechanics in Sports* (pp. 96–103), Chinese University of Hong Kong, China,
105. Toussaint, H. M., Van den Berg, C. and Beek, W. J. (2002) '"Pumped-Up Propulsion" during front crawl swimming'. *Medicine and Science in Sports and Exercise*, 34(2): 314–319.
106. Ungerechts, B. E., Persyn, U. and Colman, V. (1999) Application of vortex flow formation to self-propulsion in water. In K. L. Keskinen, P. Komi and A. P. Hollander (eds.) *Biomechanics and Medicine in Swimming VIII* (pp. 95–100) Jyväskylä, Finland: Gummerus Printing House.
107. Van Manen, J. D. and Rijken, H. (1975). Dynamic measurement techniques on swimming bodies at the Netherlands ship model basin. In L. Lewillie and J. P. Clarys (eds.), *International Series on Sport Sciences, Volume 2; Swimming II* (pp. 70–79). Baltimore: University Park Press.
108. Vennell, R., Pease, D. and Wilson, B. (2006) 'Wave drag on human swimmers'. *Journal of Biomechanics*, 39: 664–671.
109. Videler, J. J., Müller, U. K., and Stamhuis, E. J. (1999) 'Aquatic vertebrate locomotion: wakes from body waves'. *The Journal of Experimental Biology*, 202(23): 3423–3430.
110. Vilas-Boas, J. P. Cruz, M. J., Sousa, F., Conceicao, F. and Carcvalho, J.M. (2000). 'Integrated kinematic and dynamic analysis of two track-start techniques'. In R. Sanders and Y. Hong (eds.), *Proceedings of XVIII Symposium on Biomechanics in Sports: Applied Program: Application of Biomechanical Study in Swimming* (pp. 75–82). Hong Kong: International Society of Biomechanics in Sports.

111. Vilas-Boas, J. P., Cruz, J., Sousa, F., Conceicao, F., Fernandes, R. and Carvalho, J. (2003) Biomechanical analysis of ventral swimming starts: comparison of the grab start with two track-start techniques'. In J. C. Chatard (ed.), *Biomechanics and Medicine in Swimming IX. Proceedings of the IX International Symposium on Biomechanics and Medicine in Swimming.* (pp. 249–253). Saint-Etienne.
112. Vorontsov, A. R. and Rumyantsev, V. A. (2000) 'Resistive forces in swimming'. In V. Zatsiorsky (ed.), *Biomechanics in Sport* (Vol. 1, pp. 184–204). Oxford: Blackwell Science Ltd.
113. Webb, P. W. (1975) *Hydrodynamics and energetics of fish propulsion.* Bulletin of the Fisheries Research Board of Canada, 190: 1–159.
114. Wilga, C. D. and Lauder, G. V. (2000) '3D kinematics and wake structure of the pectoral fins during locomotion in leopard sharks Triakis Semifasciata'. *The Journal of Experimental Biology*, 203: 2261–2278.
115. Wilga, C. D. and Lauder, G. V. (2002) 'Function of the heterocercal tail in sharks: quantitative wake dynamics during steady horizontal swimming and vertical maneuvering'. *The Journal of Experimental Biology*, 205: 2365–2374.
116. Wilson, G. J. (1994). Strength and power in sport: Elastic properties of muscles and tendons. In J. Bloomfield, T. R. Ackland, B. C. Elliott (eds.), *Applied Anatomy and Biomechanics in Sport* (pp. 123–124). Melbourne. Blackwell Scientific Publications.
117. Yanai, T. (2004) 'Buoyancy is the primary source of generating body roll in front-crawl swimming'. *Journal of Biomechanics*, 37: 605–612.
118. Zatsiorsky, V., Bulgakova, N. and Chaplinsky, N. (1979) 'Biomechanical analysis of starting techniques in swimming'. In J. Terauds and E. W. Bedingfield (eds.), *Swimming III* (pp. 199–206). Baltimore: University Park Press.

24

Biomechanics of the long jump

Nicholas P. Linthorne
Brunel University, Uxbridge

Introduction

The basic technique used in long jumping has remained unchanged since the beginning of modern athletics in the mid-nineteenth century. The athlete sprints down a runway, jumps up from a wooden take-off board, and flies through the air before landing in a pit of sand. A successful long jumper must, therefore, be a fast sprinter, have strong legs for jumping, and be sufficiently coordinated to perform the moderately complex take-off, flight, and landing maneuvers. The best women long jumpers achieve distances of about 6.5–7.5 m, whereas the best men (who are faster and stronger) reach about 8.0–9.0 m.

The objectives in each phase of the jump are the same regardless of the athlete's gender or ability. To produce the greatest possible jump distance the athlete must reach the end of the run-up with a large horizontal velocity and with the take-off foot placed accurately on the take-off board. During take-off the athlete attempts to generate a large vertical velocity while minimizing any loss of horizontal velocity, and in the flight phase the athlete must control the forward rotation that is produced at take-off and place their body in a suitable position for landing. During the landing the athlete should pass forward of the mark made by their feet without sitting back or otherwise decreasing the distance of the jump.

This chapter presents a review of the most important biomechanical factors influencing technique and performance in the long jump. The biomechanical principles behind the successful execution of the run-up, take-off, flight, and landing phases of the jump are explained. The effects of changes in run-up velocity on the athlete's take-off technique are also examined, as are the design principles of long jump shoes and the techniques used by disabled athletes.

Typical values of selected long jump parameters are presented in Table 24.1. The values in this table are based on studies of elite long jumpers at major international championships (Arampatzis, Brüggemann, and Walsch, 1999; Hay, Miller, and Canterna, 1986; Lees, Fowler, and Derby, 1993; Lees, Graham-Smith, and Fowler, 1994; Nixdorf and Brüggemann, 1990). The table will be a useful reference while reading this chapter.

Table 24.1 Typical values of selected parameters for elite long jumpers

Parameter	*Men*	*Women*
Athlete's height (m)	1.82	1.75
Athlete's body mass (kg)	76	62
Jump distance (m)	8.00	6.80
Run-up length (m)	48	40
Horizontal velocity at touchdown (m/s)	10.6	9.5
Vertical velocity at touchdown (m/s)	–0.1	–0.1
Horizontal velocity at take-off (m/s)	8.8	8.0
Vertical velocity at take-off (m/s)	3.4	3.1
Take-off velocity (m/s)	9.4	8.6
Take-off angle (°)	21	21
Change in horizontal velocity during take-off (m/s)	–1.8	–1.5
Change in vertical velocity during take-off (m/s)	3.5	3.2
Leg angle at touchdown (°)	61	63
Knee angle at touchdown (°)	166	161
Take-off duration (s)	0.11	0.11
Touchdown height (m)	1.03	0.96
Take-off height (m)	1.29	1.20
Landing height (m)	0.65	0.60
Height difference between touchdown and take-off (m)	0.26	0.24
Height difference between take-off and landing (m)	–0.64	–0.60
Height at the peak of the jump (m)	1.88	1.69

Run-up

The run-up phase is crucial in long jumping; it is impossible to produce a good performance without a fast and accurate run-up. The three main tasks of the athlete during the run-up are: to accelerate to near-maximum speed, lower the body during the final few steps and bring it into position for take-off, and place the take-off foot accurately on the take-off board.

Run-up velocity

In long jumping, the distance achieved is strongly determined by the athlete's horizontal velocity at the end of the run-up. To produce a fast run-up, most long jumpers use 16–24 running strides performed over a distance of about 35–55 m. By the end of the run-up the athlete reaches about 95–99 per cent of their maximum sprinting speed. Long jumpers do not use a longer run-up length that gives 100 per cent sprinting speed because the advantage of a faster run-up speed is outweighed by the increased difficulty in accurately hitting the take-off board (Hay, 1986). Faster athletes tend to use a longer run-up because it takes them longer to build up to their maximum sprinting speed. Most long jumpers start their run from a standing position with one foot forward of the other. Some athletes prefer to take several walking strides onto a check mark before accelerating. However, this technique is believed to produce a less consistent velocity profile and hence a less accurate run-up.

Studies of competition jumps have consistently found high correlations between run-up velocity and jump distance. Figure 24.1 shows an example of this association (Hay, 1993). The data in the figure are from 306 jumps by men and women with a wide range of ability, from high school athletes through to elite athletes. However, one must recognize that the

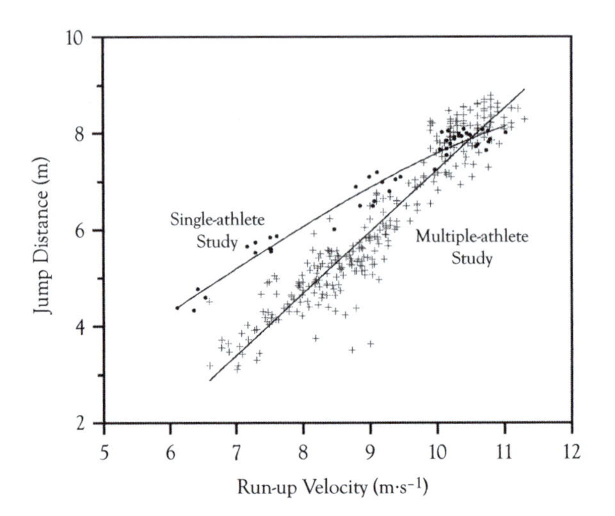

Figure 24.1 Increase in jump distance with run-up velocity in a multiple-athlete study (+) and in a single-athlete study (·). Data for the single-athlete study are for an elite male long jumper. Adapted with permission from Bridgett and Linthorne (2006).

slope of a regression line from a multiple-athlete study does not indicate the expected rate of improvement in jump distance for an individual athlete. The main cause of variations in jump distance among athletes is probably differences in muscular strength. The slope of the regression line in a multiple-athlete study therefore indicates how an individual's jump distance changes in response to a change in muscular strength, rather than how the jump distance changes with run-up velocity. For the individual athlete the relation between jump distance and run-up velocity is not quite linear (Bridgett and Linthorne, 2006). Figure 24.1 shows an example of the relation for an elite male long jumper.

Run-up accuracy

To produce the best possible jump distance, a long jumper must place their take-off foot close to, but not over, the take-off line that is marked by the front edge of the take-off board. The long jump run-up has two main phases; an acceleration phase during which the athlete produces a stereotyped stride pattern; and a 'zeroing-in' phase during which the athlete adjusts their stride pattern to eliminate the spatial errors that have accrued during the first phase (Hay, 1988). During the last few strides before take-off, the athlete uses their visual perception of how far away they are from the board as a basis for adjusting the length of their strides. Top long jumpers start using a visual control strategy at about five strides before the board and are able to perform the stride adjustments with only a small loss of horizontal velocity. Athletes of lesser ability tend to have a greater accumulated error and anticipate their stride adjustment later than highly-skilled jumpers. Many long jumpers use a check-mark at 4–6 strides before the board so that their coach can monitor the accumulated error in the first phase of the run-up.

Transition from run-up to take-off

Skilled long jumpers maintain their normal sprinting action up until about two to three strides before take-off (Hay and Nohara, 1990). The athlete then begins to lower their centre of mass (COM) in preparation for the take-off. A low position into the take-off is necessary to give a large vertical range of motion (ROM) over which to generate upwards velocity. The athlete lowers their COM to the required height and tries to keep a flat trajectory in the last stride before take-off. This ensures that the athlete's COM has minimal downward vertical velocity at the instant of touchdown and so the upwards vertical impulse exerted by the athlete during the take-off produces the highest possible vertical velocity at the instant of take-off. Most long jumpers spend a lot of time practicing to lower their COM while minimizing any reduction in run-up velocity.

The entry into the take-off is usually performed using a 'pawing' action, where the take-off leg is swept down and back towards the athlete (Koh and Hay, 1990). The take-off foot has a negative velocity relative to the athlete's COM, but the velocity of the foot relative to the ground is not quite reduced to zero (about 4–5 m/s). This 'active' landing technique is believed to reduce the braking force experienced by the athlete during the initial stages of the take-off.

Take-off

Although long jump performance is determined primarily by the athlete's ability to attain a fast horizontal velocity at the end of the run-up, the athlete must also use an appropriate take-off technique to make best use of this run-up velocity. Long jumpers place their take-off foot well ahead of their COM at touchdown to produce the necessary low position at the start of the take-off. The jumper's body then pivots up and over the take-off foot, during which time the take-off leg rapidly flexes and extends. Long jumping is essentially a projectile event, and the athlete wishes to maximize the flight distance of the human projectile by launching it at the optimum take-off velocity and take-off angle. In launching the body into the air, the athlete desires a large horizontal velocity at take-off to travel forward and a large vertical velocity to give time in the air before landing back on the ground. A fast run-up produces a large horizontal take-off velocity, but it also shortens the duration of the ground contact and hence the ability of the athlete to generate a vertical impulse (force integrated over time). To increase the duration of the foot contact, the athlete plants their foot ahead of the COM at touchdown. However, the resulting increase in vertical propulsive impulse is accompanied by an undesirable increase in horizontal braking impulse. Therefore, there is an optimum leg angle at touchdown which offers the best compromise between vertical propulsive impulse and horizontal braking impulse. In the long jump, the optimum take-off technique is to run up as fast as possible and plant the take-off leg at about 60–65° to the horizontal (Bridgett and Linthorne, 2006; Seyfarth, Blickhan, and Van Leeuwen, 2000).

Take-off mechanism

Just before touchdown the athlete pre-tenses the muscles of the take-off leg. The subsequent bending of the leg during the take-off is due to the force of landing, and is not a deliberate yielding of the ankle, knee, and hip joints. Flexion of the take-off leg is unavoidable and is limited by the eccentric strength of the athlete's leg muscles. Maximally activating the muscles of the take-off leg keeps the leg as straight as possible during the take-off. This enables the athlete's COM to pivot up over the foot, generating vertical velocity via a purely mechanical mechanism. Over 60 per cent of the athlete's final vertical velocity is achieved by the instant of maximum knee flexion, which indicates that the pivot mechanism is the single most important mechanism acting to create vertical velocity during the take-off (Lees, Fowler, and Derby, 1993; Lees, Graham-Smith, and Fowler, 1994). The knee extension phase makes only a minor contribution to the generation of vertical velocity, and the rapid plantar flexion of the ankle joint towards the end of the take-off contributes very little to upward velocity. Long jumpers spend a lot of time on exercises to strengthen the muscles of their take-off leg. Greater eccentric muscular leg strength gives the athlete a greater ability to resist flexion of the take-off leg, which enhances the mechanical pivot mechanism during the take-off and hence produces a greater take-off velocity.

The stretch-shorten cycle, where the concentric phase of a muscle contraction is facilitated by a rapid eccentric phase, does not play a significant role in the long jump take-off (Hay, Thorson, and Kippenhan, 1999). Rather, fast eccentric actions early in the take-off enable the muscles to exert large forces and thus generate large gains in vertical velocity. In the long jump take-off the instant of maximum knee flexion is a poor indicator of when the extensor muscles of the take-off leg change from eccentric activity to concentric activity. In long jumping, the *gluteus maximus* is active isometrically at first and then concentrically; the hamstrings are active concentrically throughout the take-off; rectus femoris acts either isometrically at first then eccentrically or eccentrically throughout the take-off; and the vasti, soleus, and gastrocnemius act eccentrically at first and then concentrically.

The explosive extension of the hip, knee, and ankle joints during the last half of the take-off is accompanied by a vigorous swinging of the arms and free leg. These actions place the athlete's COM higher and farther ahead of the take-off line at the instant of take-off, and are also believed to enhance the athlete's take-off velocity. Some athletes use a double-arm swing to increase the take-off velocity, but it is difficult to switch smoothly without loss of running velocity from a normal asynchronous sprint arm action during the run-up to a double-arm swing at take-off.

Take-off angle

It is well known that take-off angles in the long jump are substantially less than the 45° angle that is usually proposed as the optimum for a projectile in free flight. Video measurements of world-class long jumpers consistently give take-off angles of around 21°. The notion that the optimum take-off angle is 45° is based on the assumption that the take-off velocity is constant for all choices of take-off angle. However, in the long jump, as in most other sports projectile events, this assumption is not valid. The take-off velocity that a long jumper is able to generate is substantially greater at low take-off angles than at high take-off angles and so the optimum take-off angle is shifted to below 45° (Linthorne, Guzman, and Bridgett, 2005).

From a mathematical perspective the athlete's take-off velocity is the vector sum of the horizontal and vertical component velocities, and the take-off angle is calculated from the ratio of the component velocities. A take-off angle of 45° requires that the horizontal and vertical take-off velocities are equal in magnitude. The maximum vertical velocity an athlete can produce is about 3–4 m/s (when performing a running high jump), but an athlete can produce a horizontal take-off velocity of about 8–10 m/s through using a fast run-up. By deciding to jump from a fast run-up, the athlete produces a high take-off velocity at a low take-off angle. In long jumping, generating a higher take-off velocity gives a much greater performance advantage than jumping at closer to 45°.

Take-off forces

During the take-off the athlete experiences a ground reaction force (GRF) that tends to change the speed and direction of the athlete's COM. The horizontal force during the take-off is predominantly a backwards braking force, and only for a very short time at the end of the take-off does it switch over to become a forwards propulsive force. Because the braking impulse is much greater than the propulsive impulse, the athlete's forward horizontal velocity is reduced during the take-off (by about 1–3 m/s). The vertical GRF exerted on the athlete produces the athlete's vertical take-off velocity. The vertical force initially acts to reverse the downward velocity possessed by the athlete at touchdown, and then accelerates the athlete upwards. The athlete always experiences a slight reduction in upwards velocity in the last instants before take-off. This decrease occurs because the vertical force must drop down to zero at the instant of take-off. For a short time before take-off the vertical GRF is less than body weight and is therefore not enough to overcome the gravitational force on the athlete. Both the horizontal and vertical components of the GRF display a sharp impact peak at touchdown when the take-off leg strikes the ground and is rapidly reduced to near zero velocity.

As well as changing the speed and direction of the athlete's COM, the GRF tends to produce angular acceleration of the athlete's body about its somersaulting axis. The GRF produces a forward or backward torque about the athlete's COM depending on whether the line of action of the force passes behind or ahead of the COM (Hay, 1993). In the initial stages of the take-off the torque acts to produce backwards acceleration, but it soon changes to produce forwards acceleration. Overall, the athlete experiences a large forwards rotational impulse, and so the athlete leaves the take-off board with a large amount of forward-somersaulting angular momentum. Forward angular momentum is consistently a source of difficulty for the athlete. Unless the jumper takes appropriate steps to control the angular momentum during the flight, excessive rotation of the body will reduce the distance of the jump by producing a landing with the feet beneath the body rather than extended well in front of the body.

Flight and landing

During the flight phase, most long jumpers either adopt a 'hang' position or perform a 'hitch-kick' movement (a modified running-in-the-air action). In both techniques the athlete's actions are designed to control the forward rotation that is imparted to the body at take-off and hence allow the athlete to attain an effective landing position (Hay, 1993). The hang and hitch-kick techniques deal with the angular momentum during the flight phase in different ways. In the hang technique the athlete attempts to minimize the forward rotation of the body, whereas in the hitch-kick technique the athlete performs movements that actively counter the forward rotation.

In the hang technique the athlete reaches up with their arms and extends their legs downwards just after take-off. This extended body position gives the athlete a large moment of inertia (MOI) about their somersaulting axis and hence reduces the athlete's forward angular velocity. The athlete maintains the hang position for as long as possible during the flight so as to minimize the amount of forward rotation.

The hitch-kick technique involves using the motions of the arms and legs to evoke a contrary reaction of the trunk, thereby maintaining the athlete's upright posture in the air. The athlete rotates their arms and legs forward in a movement that is similar to running.

Because the athlete is in free flight the athlete's total angular momentum must be conserved. The forward angular momentum generated by circling the arms and legs forward is, therefore, countered by an equal backwards angular momentum in the trunk. The athlete is thus able to counter the forward rotation of the body that was developed in the take-off.

Long jumpers choose their flight technique according to the amount of angular momentum they generate during the take-off and the time they have available before landing. Many coaches recommend the hang technique for athletes of lesser ability, who usually generate a lower angular momentum during the take-off and spend less time in the air. The hitch-kick is recommended for better athletes, who usually generate a higher angular momentum during the take-off and have a longer flight time. The hitch-kick technique has two main variants, '2½-step' and '3½-step', named according to the number of steps the athlete executes during the flight. Contrary to what is sometimes thought, the long jumper's forward circling actions in the air do not increase the distance of the jump by propelling the athlete through the air.

Towards the end of the flight phase the athlete prepares for landing by lifting their legs up and extending them in front of the body. The goal of the landing is to create the greatest possible horizontal distance between the take-off line and the mark made by the heels in the sand. The landing technique should not result in the athlete falling backwards into the pit or otherwise producing a mark that is closer to the take-off board than that made by the heels. There are several basic variations on landing technique, including the 'orthodox', 'slide-out' and 'swivel-out' techniques, but there is currently no consensus on the optimum technique that produces the longest jump distance (Hay, 1986).

Flight distance equation

The athlete's jump distance is measured from the take-off line to the nearest mark made by the athlete in the landing area. Jump distance may be considered as the sum of the take-off distance, the flight distance, and the landing distance (Figure 24.2):

$$d_{\text{jump}} = d_{\text{take-off}} + d_{\text{flight}} + d_{\text{landing}}.$$

In most jumps the flight distance is about 90 per cent of the total jump distance. Therefore, the biomechanical factors that determine the athlete's flight distance are very important in long jumping.

During the flight phase of the jump the effects of gravity are much greater than those of aerodynamic forces and so the jumper may be considered as a projectile in free flight. The trajectory of the athlete's COM is determined by the conditions at take-off, and the flight distance is given by

$$d_{\text{flight}} = \frac{v^2 \sin 2\theta}{2g}\left[1+\left(1+\frac{2gh}{v^2\sin^2\theta}\right)^{1/2}\right], \quad (1)$$

where v is the take-off velocity, θ is the take-off angle, and g is the acceleration due to gravity. Here, the relative take-off height, h, is given by:

$$h = h_{\text{take-off}} - h_{\text{landing}},$$

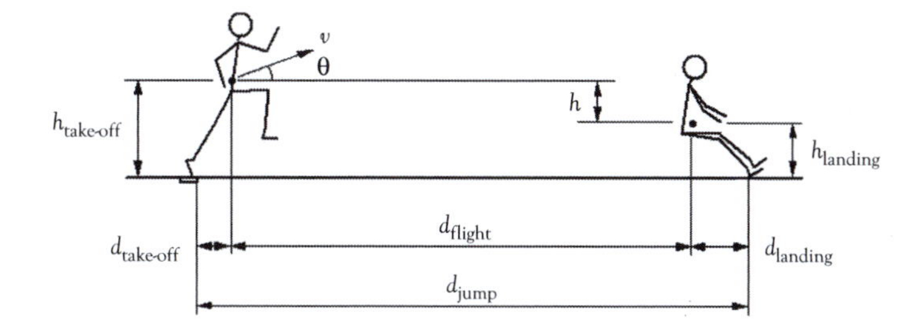

Figure 24.2 The total jump distance is the sum of the take-off, flight, and landing distances. The flight distance is determined by the take-off velocity (v), the take-off angle (θ), and the relative height difference between take-off and landing (h). Adapted with permission from Linthorne, Guzman, and Bridgett (2005).

where $h_{take\text{-}off}$ is the take-off height and $h_{landing}$ is the landing height (Figure 24.2). When $h = 0$, equation (1) reduces to the familiar expression for the range of a projectile launched from ground level over a horizontal plane, $d_{flight} = (v^2 \sin 2\theta)/g$.

An examination of equation (1) reveals how the athlete can maximize the flight distance of the jump. By far the most important variable is the take-off velocity. The flight distance is proportional to the square of the take-off velocity, and, therefore, the athlete should strive for a high velocity at take-off. The athlete should also aim to maximize the height difference between take-off and landing by having a high body position at take-off and a low body position at landing. However, any actions to achieve a large height difference should not come at the expense of a fast take-off velocity. At first glance, equation (1) suggests that the athlete should jump with a take-off angle of about 45° so as to maximize the 'sin 2θ' term. However, it is important to recognize that the take-off velocity (v) and relative take-off height (h) are not constants, but are functions of the take-off angle (θ). These relations must be determined and inserted into equation (1) in order to determine the athlete's optimum take-off angle.

Optimum take-off angle

A long jumper's optimum take-off angle may be determined by using high-speed video to measure the athlete's relations between take-off velocity and take-off angle, $v(\theta)$, and between relative take-off height and take-off angle, $h(\theta)$ (Linthorne, Guzman, and Bridgett, 2005). To obtain reliable measures of these relations the athlete must jump many times using a wide range of take-off angles (0–90°). The highest take-off velocities are obtained when the jumper uses a fast run-up and then attempts to jump up as much as possible. However, long jumpers cannot attain take-off angles greater than about 25° using this technique (Figure 24.3a). To achieve greater take-off angles the athlete must use a slower run-up and so the take-off velocity is reduced. In the extreme case of a near-vertical take-off angle, the run-up velocity must be reduced to walking pace and so the take-off velocity is at its lowest. The take-off height and landing height are determined by the athlete's body configuration. Although the take-off and landing heights both increase with increasing take-off angle, the height difference between the two remains approximately constant (Figure 24.3b).

To find the athlete's optimum take-off angle, the mathematical expressions for $v(\theta)$ and $h(\theta)$ are inserted into the equation for the flight distance (equation 1). The flight distance is then

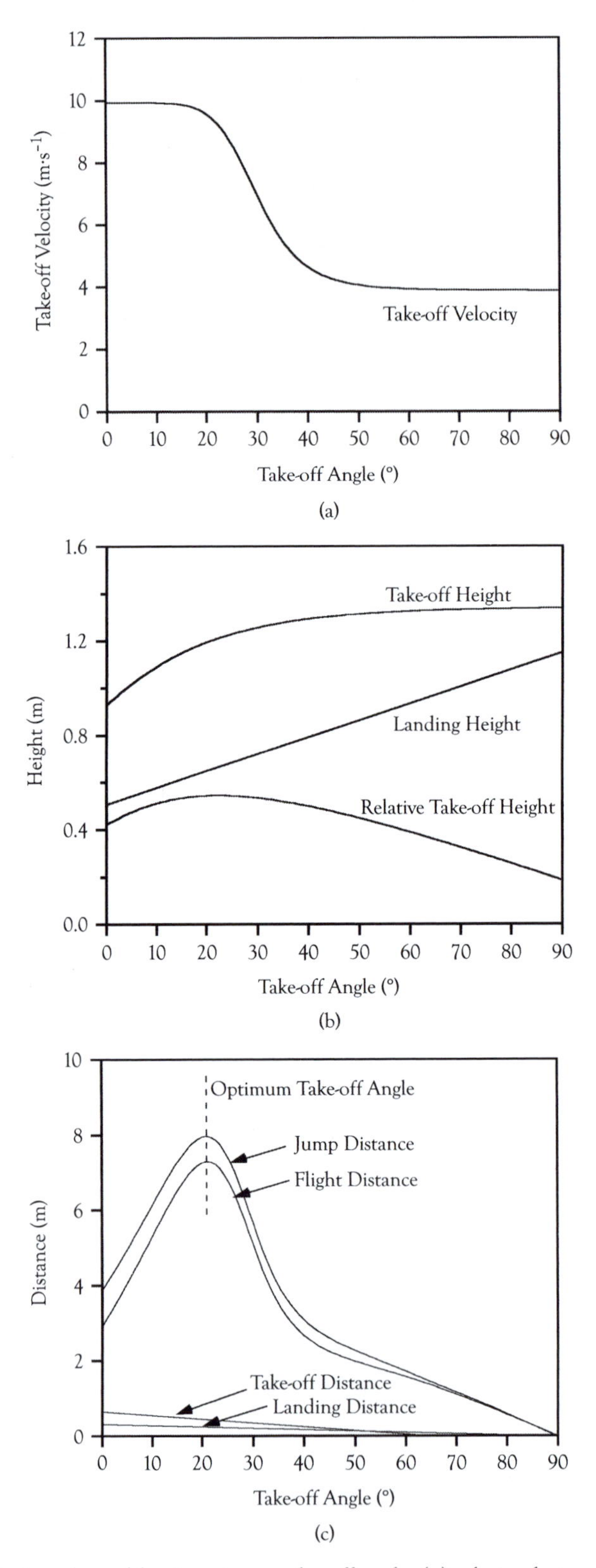

Figure 24.3 Calculation of an athlete's optimum take-off angle: (a) relation between take-off velocity and take-off angle; (b) relation between relative take-off height and take-off angle; and (c) component distances and total jump distance. Curves are for an elite male long jumper, and the optimum take-off angle is about 21°. Adapted with permission from Linthorne, Guzman, and Bridgett (2005).

plotted as a function of the take-off angle, and the optimum take-off angle is the point on the curve at which the flight distance is greatest (Figure 24.3c). In the long jump, $v(\theta)$ has a strong influence on the optimum take-off angle, whereas $h(\theta)$ is relatively unimportant. In long jumping, the total jump distance is slightly more than the flight distance (Figure 24.2). However, the take-off and landing distances make relatively small contributions to the total jump distance and have little effect on the optimum take-off angle (Figure 24.3c). Launching the body at close to the optimum take-off angle is essential for a successful long jump. The distance achieved by the jumper is sensitive to take-off angle, and so large deviations from the optimum take-off angle cannot be tolerated (Figure 24.3c).

Run-up velocity and take-off technique

The most important determinant of success in long jumping is the athlete's ability to produce a fast run-up velocity and hence a fast take-off velocity. Long jumpers, therefore, spend a lot of time developing their maximum sprint speed and improving their ability to get close to this maximum speed in the last stride before take-off. However, an athlete must make adjustments to their take-off technique in order to benefit from a faster run-up velocity. Bridgett and Linthorne (2006) determined the relations between run-up velocity and take-off technique by manipulating the run-up length used by an elite male long jumper. The improvement in jump distance with increasing run-up velocity is shown in Figure 24.1, and the effects of run-up velocity on the athlete's take-off technique are shown in Figure 24.4.

A long jumper must use a straighter knee at touchdown in order to benefit from a faster run-up velocity (Figure 24.4a). A straighter take-off leg has a smaller moment arm about the knee for the GRF and is, therefore, more resistant to flexion. By preventing excessive flexion, less energy is dissipated by the leg muscles in eccentric contraction. A faster run-up speed also requires the athlete to use a slightly lower leg angle at touchdown (Figure 24.4b). The lower leg angle arises because a faster run-up velocity requires the athlete to increase the duration of the foot contact in order to maintain a high vertical take-off velocity. The athlete therefore plants the foot farther ahead of the COM at touchdown and hence has a lower leg angle.

Even though a long jumper performs actions during the take-off that are aimed at generating vertical velocity, the athlete is still able to transfer much of his run-up velocity through to horizontal take-off velocity. The resultant take-off velocity therefore steadily increases with increasing run-up velocity (Figure 24.4c). In the long jump, the jumping action results in a reduction in horizontal velocity, and this loss becomes greater as the run-up velocity is increased.

A long jumper produces a lower take-off angle at faster run-up velocities (Figure 24.4d). The take-off angle is determined by the ratio of the vertical velocity and the horizontal velocity. At all run-up velocities the optimum take-off strategy that produces the greatest jump distance is to generate close to the maximum possible vertical velocity. Changes in the take-off angle are therefore determined by changes in the horizontal velocity. Because the athlete's horizontal take-off velocity increases with increasing run-up velocity, the angle of the take-off velocity vector to the horizontal steadily decreases.

In jumps from a full-speed run-up the take-off foot is in contact with the ground for about 0.12 s. The duration of the take-off decreases in proportion to $1/v^{0.6}$, where v is the run-up velocity (Figure 24.4d). A simplistic model of the long jump take-off, in which the rotational ROM of the take-off leg is the same at all run-up velocities, suggests that the take-off

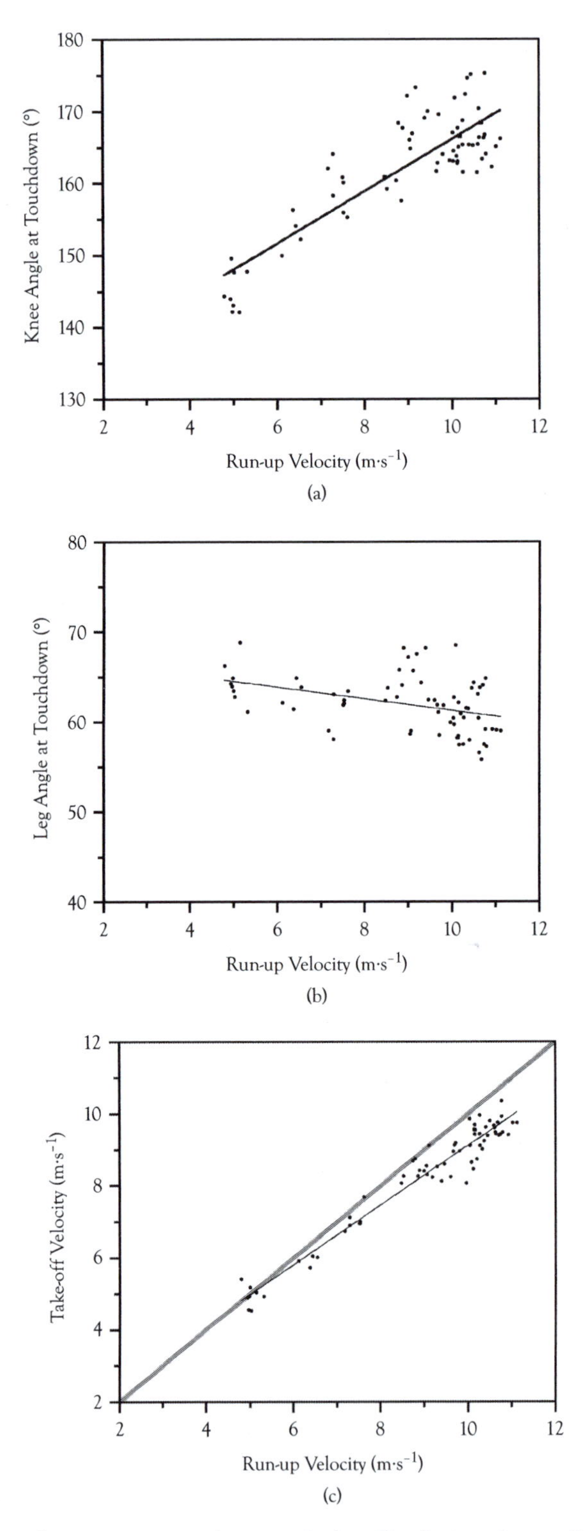

Figure 24.4 Relations between run-up velocity and take-off technique for an elite male long jumper: (a) knee angle at touchdown; (b) leg angle at touchdown; (c) take-off velocity; (d) take-off angle; and (e) take-off duration. Adapted with permission from Bridgett and Linthorne (2006).

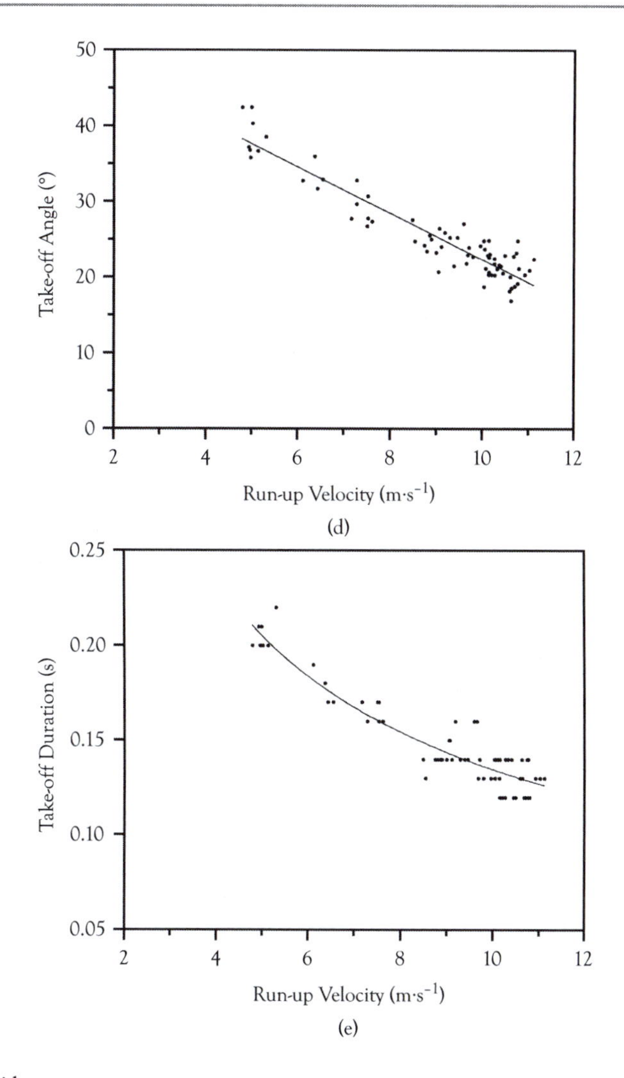

Figure 24.4—Cont'd

duration should vary in proportion to $1/v$. However, long jumpers tend to use a greater rotational ROM of their take-off leg at faster run-up velocities.

Long jump shoes

Long jumpers use shoes that are designed specifically for the event. The shoes have several spikes positioned under the ball of the foot to give a firm grip on the board at take-off, and a beveled toe to help reduce the chances of marking the plasticine indicator board at take-off. Some companies manufacture long jump shoes with stiff soles under the balls of the feet. Studies of the foot during the stance phase of running and jumping have shown that the metatarsophalangeal joint flexes as the athlete rolls onto the forefoot and does not extend until after take-off. The metatarsophalangeal joint is, therefore, believed to be a large energy absorber (Stefanyshyn and Nigg, 1998). In shoes with stiff soles, the energy that is normally dissipated during the bending of the metatarsophalangeal joints at touchdown is stored in the sole and returned to the athlete during the take-off phase of the stride.

Shoes with stiff soles are expected to benefit the athlete in both the run-up and take-off phases of the jump. Experiments on sprinters with stiffening plates in their shoes showed an improvement in sprinting speed of just over 1 per cent (Stefanyshyn and Fusco, 2004). In the long jump, a 1 per cent increase in run-up velocity produces a 6 cm increase in jump distance (Bridgett and Linthorne, 2006). About 60–100 J is believed to be absorbed in the metatarsophalangeal joint during a long jump take-off. A sole that stores and returns a large fraction of this energy to the athlete (say, 40 J) will increase the jump distance by about 10 cm.

Some shoe companies manufacture long jump shoes that have a tapered sole. The sole is thickest under the ball of the foot (13 mm; the maximum allowed by the rules) and thinnest under the heel. These shoes give the athlete an effectively longer leg during the take-off. The advantage is maximized by making the difference between the length of the leg at touchdown (with the landing on the heel) and at the instant of take-off (on the ball of the foot) as great as possible. A longer take-off leg may give a biomechanical advantage by allowing the athlete to generate a greater take-off velocity, and hence produce a longer jump.

Disabled athletes

Disabled athletes are classified according to their functional ability, and for lower limb amputee athletes the relevant classifications are 'below-knee amputee' and 'above-knee amputee'. Most single-leg amputee athletes jump from their intact leg and use the same basic jumping technique as able-bodied athletes (Nolan, Patritti, and Simpson, 2006). The jump distances achieved by able-bodied athletes are usually greater than those achieved by amputee athletes, and below-knee amputees generally jump farther than above-knee amputees.

Performances achieved by amputee athletes are limited by their asymmetrical gait and its detrimental effect on the run-up velocity that the athlete can attain. Amputee athletes spend longer in stance and have a shorter swing phase on their intact limb than on their prosthetic limb. They also spend longer stepping into their prosthetic limb than into their intact limb. These differences probably arise because of the lower mass and different inertial characteristics of the prosthetic limb.

Even at equal run-up velocities, amputee athletes do not produce jump distances as great as those produced by able-bodied athletes because they find it more difficult to get into the correct position at touchdown. Amputee athletes have difficulty in lowering their COM during the support phase of the last stride before take-off because their prosthetic knee needs to be locked to support their body weight during stance. Amputee athletes therefore have a greater downward vertical velocity into the take-off stride than able-bodied athletes, and so are less effective at generating upwards vertical velocity. Above-knee amputee athletes are less able than below-knee amputees to lower their COM into the take-off. Their jump distances are therefore usually not a great as those by below-knee amputees. Both above-knee and below-knee amputees attempt to compensate for the partial loss of a lower limb by using a greater ROM at the hip joint of the intact limb during the take-off.

References

1. Arampatzis, A., Brüggemann, G.P. and Walsch, M. (1999) 'Long jump'. in G.P. Brüggemann, D. Koszewski and H. Müller (eds.) *Biomechanical Research Project Athens 1997 Final Report*, Oxford: Meyer & Meyer Sport.
2. Bridgett, L.A. and Linthorne, N.P. (2006) 'Changes in long jump take-off technique with increasing run-up speed'. *Journal of Sports Sciences*, 24: 889–97.
3. Hay, J.G. (1986) 'The biomechanics of the long jump'. *Exercise and Sport Sciences Reviews*, 14: 401–46.
4. Hay, J.G. (1988) 'Approach strategies in the long jump'. *International Journal of Sport Biomechanics*, 4: 114–29.
5. Hay, J.G. (1993) 'Citius, altius, longius (faster, higher, longer): The biomechanics of jumping for distance'. *Journal of Biomechanics*, 26 (Suppl 1): 7–21.
6. Hay, J.G. and Koh, T.J. (1988) 'Evaluating the approach in the horizontal jumps'. *International Journal of Sport Biomechanics*, 4: 372–92.
7. Hay, J.G. and Nohara, H. (1990) 'Techniques used by elite long jumpers in preparation for take-off'. *Journal of Biomechanics*, 23: 229–39.
8. Hay, J.G., Miller, J.A. and Canterna, R.W. (1986) 'The techniques of elite male long jumpers'. *Journal of Biomechanics*, 19: 855–66.
9. Hay, J.G., Thorson, E.M. and Kippenhan, B.C. (1999) 'Changes in muscle-tendon length during the take-off of a running long jump'. *Journal of Sports Sciences*, 17: 159–72.
10. Koh, T.J. and Hay, J.G. (1990) 'Landing leg motion and performance in the horizontal jumps I: The long jump'. *International Journal of Sport Biomechanics*, 6: 343–60.
11. Lees, A., Fowler, N. and Derby, D. (1993) 'A biomechanical analysis of the last stride, touch-down and take-off characteristics of the women's long jump'. *Journal of Sports Sciences*, 11: 303–14.
12. Lees, A., Graham-Smith, P. and Fowler, N. (1994) 'A biomechanical analysis of the last stride, touchdown, and takeoff characteristics of the men's long jump'. *Journal of Applied Biomechanics*, 10: 61–78.
13. Linthorne, N.P., Guzman, M.S. and Bridgett, L.A. (2005) 'Optimum take-off angle in the long jump'. *Journal of Sports Sciences*, 23: 703–12.
14. Nixdorf, E. and Brüggemann, G-P. (1990) 'Biomechanical analysis of the long jump'. in G.P. Brüggemann and B. Glad (eds.) *Scientific Research Project at the Games of the XXIVth Olympiad — Seoul 1988 Final Report*. Monaco: International Athletic Foundation
15. Nolan, L., Patritti, B.L. and Simpson, K.J. (2006) 'A biomechanical analysis of the long jump technique of elite female amputee athletes'. *Medicine & Science in Sports & Exercise*, 38: 1829–35.
16. Seyfarth, A, Blickhan, R. and Van Leeuwen, J.L. (2000) 'Optimum take-off techniques and muscle design for long jump'. *Journal of Experimental Biology*, 203: 741–50.
17. Stefanyshyn, D. and Fusco, C. (2004) 'Increased shoe bending stiffness increases sprint performance'. *Sports Biomechanics*, 3: 55–66.
18. Stefanyshyn, D.J. and Nigg, B.M. (1998) 'Contribution of the lower extremity joints to mechanical energy in running vertical jumps and running long jumps'. *Journal of Sports Sciences*, 16: 177–86.

25

External and internal forces in sprint running

Joseph P. Hunter[1], Robert N. Marshall[2] and Peter J. McNair[3]
[1]Sports Science Center, Auckland; [2]Eastern Institute of Technology, Hawkes Bay;
[3]Auckland University of Technology, Auckland

Introduction

When running as fast as possible, a sprinter will produce and experience many forces. These forces can be divided into two main categories: external forces and internal forces. External forces are those forces that act on the sprinter from outside of the body. For unobstructed sprint-running under normal circumstances, there are only three external forces: a ground reaction force (GRF), a gravitational force, and wind resistance (see Figure 25.1). Internal forces, in contrast, are those forces that act within the body. Some examples of internal forces are muscle, ligament, and joint forces.

This chapter focuses on unobstructed sprint-running in a straight line. First, we review the three external forces acting on a sprinter. We make particular reference to the magnitudes and effects of these forces. Then, we review the internal forces and moments and the relationships they have with kinematics, muscle activity, changes in muscle length, and segment interactions.

External forces

The sum effect of the GRF, gravitational force, and wind resistance acting on a sprinter of given body mass, ultimately determines how fast the sprinter will run. Of the three external forces, the GRF is by far the largest and is the force over which the athlete has by far the most control. Nonetheless, the forces due to gravity and wind are significant and should not be ignored.

Ground reaction force

Components of the ground reaction force

During ground contact, a sprinter exerts a force against the ground. An equal and opposite force is exerted by the ground against the sprinter. This force acting on the sprinter is

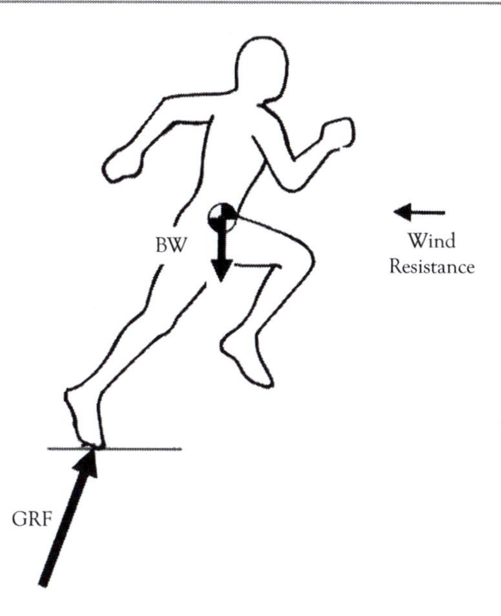

Figure 25.1 The three external forces that determine the acceleration of the centre of mass (COM) of a sprinter: ground reaction force (GRF), gravitational force equivalent to body weight (BW), and wind resistance.

referred to as the GRF. For analysis purposes, the GRF is commonly divided into its vertical, anterior-posterior, and medial-lateral components, with the latter two typically based on the intended direction of progression of the athlete.

The GRF components provide information concerning the vertical and horizontal acceleration of the sprinter's centre of mass (COM). The vertical GRF is related to the athlete's ability to halt the downward vertical velocity of the COM at touchdown, and reverse it to an upward vertical velocity at take-off (Miller, 1990).

The anterior-posterior GRF is 'biphasic'(Miller, 1990) and can be classified depending on the direction in which it acts. That is, a braking GRF usually occurs early in the stance phase and acts posteriorly. The braking ground reaction impulse causes a decrease in horizontal velocity of the sprinter. In contrast, a propulsive GRF usually occurs later in the stance phase and acts anteriorly (Kyrolainen *et al.*, 1999). The propulsive ground reaction impulse causes an increase in the horizontal velocity of the sprinter (Mero *et al.*, 1983).

The medial-lateral GRF is typically the smallest of the three GRF components (Kyrolainen *et al.*, 1999), and can be, at least in the case of sub-maximal running, highly variable among athletes, and for each foot of the same athlete (Miller, 1990). Regardless of its variability, however, one would expect that if an athlete is running in a straight line, the net medial-lateral ground reaction impulse would be zero, at least if averaged over a number of consecutive steps (Miller, 1990).

Quantification and comparison

The GRF during sprint running has previously been quantified by a number of measures, including: the maximum and average values of each component (Hunter *et al.*, 2004b; Mero *et al.*, 1987; Weyand *et al.*, 2000); the impulse associated with each component (Hunter *et al.*, 2004a; Hunter *et al.*, 2004b; Hunter *et al.*, 2005; Mero, 1988; Mero and Komi, 1994);

and the magnitude of the resultant force (Kyrolainen *et al.*, 1999; Mero and Komi, 1986). To allow meaningful comparison among individuals, the magnitude of the GRF relative to body weight is often reported. However, the anterior-posterior and medial-lateral ground reaction impulses can be expressed relative to body mass. Then, according to the impulse-momentum relationship, the results would reflect the change in velocity experienced by the athlete (ignoring wind resistance) during the respective period over which the impulse was measured. For the vertical ground reaction impulse, the impulse due to body weight should first be deducted before dividing by body mass (Hunter *et al.*, 2005). The result would then indicate the change in vertical velocity experienced by the athlete during the stance phase. An example of these methods of expressing the GRF variables is shown in Table 25.1.

Changes during a sprint race

The GRF components change noticeably during a sprint race. Table 25.1 shows group mean values of time, average force, and impulse for the vertical, braking, and propulsive GRF components during the early and late acceleration phases of a sprint. The data are from two separate studies (Mero, 1988; Mero *et al.*, 1987) in which the average forces and the impulses were not always specifically provided, but could be derived.

As shown in Table 25.1 the transition from the early to late acceleration phase involves a decrease in stance time, decrease in propulsive time, and an increase in braking time. In addition, the average propulsive GRF decreases, but the average vertical and braking GRF values increase in magnitude. Consequently, as shown by the relative ground reaction impulses, during the early acceleration phase, the loss in horizontal velocity during braking is small and the gain during propulsion is large. In contrast, during the late acceleration phase, the loss in horizontal velocity during braking is relatively large and the gain during propulsion is barely greater than this loss.

Relationships with sprint performance

Investigation into the relationships between sprint performance and GRF variables has resulted in researchers suggesting hypotheses regarding the relative importance of the various GRF components. For example, Mero and Komi (1986) reported a significant correlation between maximal sprint-velocity and average net-resultant-force during propulsion ($r = 0.84$, $P < 0.001$). However, when the resultant force was expressed relative to body mass, the strength of the correlation decreased ($r = 0.65$, $P < 0.01$). In the same study, there

Table 25.1 GRF components for the early and late acceleration phases

	Vertical component			*Braking component*			*Propulsive component*		
	Stance time (ms)	*Average vertical GRF (BW)*	*Relative vertical GRI (m/s)*	*Braking time (ms)*	*Average braking GRF (BW)*	*Relative braking GRI (m/s)*	*Propulsive time (ms)*	*Average propulsive GRF (BW)*	*Relative propulsive GRI (m/s)*
[a]Early	193	1.6	1.0	22	−0.2	−0.05	171	0.7	1.2
[b]Late	96	2.3	1.2	44	−0.5	−0.2	52	0.5	0.3

GRI, Ground reaction impluse; GRF, Ground reaction force; BW, Body weight

[a]The early acceleration phase of a sprint. The times were obtained, and the force and impulse values were derived, from data provided by Mero (1988). The data were measured from the first stance after a block start.

[b]The late acceleration phase of a sprint. The times were obtained, and the force and impulse values were derived, from data provided by Mero et al. (1987). The data were measured from a stance beyond the 35 m mark of a sprint.

was a significant correlation between maximal sprint-velocity and average net-resultant-force during braking ($r = 0.65$, $P < 0.01$). This indicates that the faster sprinters had greater braking forces; possibly a result of inclusion of males and females in the same sample. In the discussion, however, the authors still suggested the possibility that: 'the braking force ... should be as small as possible to avoid loss of velocity during the impact phase.'

In another study, Mero (1988) reported significant correlations between sprint-running velocity at the end of the first contact after a block start and various anterior-posterior GRF variables ($r = 0.62$ to 0.71, $P < 0.05$). The authors suggested that the braking and propulsive forces that occur during the ground contact after a block start 'strongly affect running velocity in the sprint start.' This study, however, included only eight participants.

In contrast to the two previously discussed studies, Weyand *et al.* (2000) suggested that the vertical GRF may play a crucial role in sprint running, at least during maximal velocity. Weyand *et al.* reported that for a mixed-gender, heterogeneous group of sprinters, all athletes produced similar vertical ground reaction impulse when sprinting on a treadmill; however, the faster sprinters produced this impulse with a shorter contact time and greater average vertical GRF (expressed relative to body weight). The authors concluded that faster sprinters achieve their maximal velocity not by repositioning their limbs more quickly, but rather by applying greater vertical forces during stance.

From the discussion above, it is apparent that differing opinions exist regarding the relative importance of the various GRF components in sprint running. Some researchers have warned against accepting any such theories until further investigation has been conducted. Putnam and Kozey (1989), for example, highlighted that it is unknown if the braking GRF is related to other mechanical factors which also affect performance, such as the propulsive and vertical GRF components, and step length and step rate.

More recently, Hunter *et al.* (2005) used stepwise, multiple linear regression to investigate the relationship of braking, propulsive, and vertical ground reaction impulse (expressed relative to body mass) with sprint velocity at the 16 m mark of a sprint. Propulsive ground reaction impulse explained 57 per cent of the variance in sprint velocity, braking ground reaction impulse explained a further 7 per cent, but vertical ground reaction impulse did not explain any further variance. These and other results from the study showed a trend for the faster sprinters to produce greater propulsive ground reaction impulse, lower braking ground reaction impulse, and only moderate vertical ground reaction impulse. However, it was also suggested that the faster sprinters may not have actually *minimized* the braking ground reaction impulse and that the possibility that some braking ground reaction impulse is of advantage could not be ruled out.

The lack of agreement regarding the relative importance of the GRF components has not stopped various hypotheses being advanced regarding how the GRF components might be optimized. The first group of theories are based on the belief that it is of advantage to minimize the braking GRF during sprinting. The braking GRF is thought to be minimized by:

- Minimizing the forward horizontal velocity of the foot relative to the ground immediately before touchdown (Hay, 1994: 407–8; Mann and Sprague, 1983; Wood, 1987). This technique is sometimes referred to as using a highly 'active' touchdown (Knoedel, 1984).
- Use of a high extension velocity of the hip joint and high flexion velocity of the knee joint at the instant of touchdown (Mann and Sprague, 1983; Mann *et al.*, 1984).
- Minimizing the touchdown distance (i.e., the distance the foot is placed in front of the COM at the moment of touchdown) (Mero *et al.*, 1992).

- Related to the above three theories is the notion that a high knee lift is beneficial because it would allow easier achievement of the above three techniques (Mann and Herman, 1985).

The second group of hypotheses are based on the belief that it is of advantage to maximise the propulsive GRF. The propulsive GRF is thought to be maximized by:

- Ensuring a high angular velocity of the stance-limb hip joint throughout the stance phase (Mann and Sprague, 1983; Mann *et al.*, 1984; Wiemann and Tidow, 1995).
- Fully extending the stance-limb hip, knee, and ankle joints at take-off (see Hay, 1994: 409, for discussion).

Some of these 'optimization techniques,' however, have received criticism. For example, it has been suggested:

- Use of a high knee lift may slow leg recovery (Wood *et al.*, 1987), and therefore, decrease step rate.
- Use of a highly active touchdown may result in increased risk of injury (Mann and Sprague, 1980).
- Fully extending the stance limb at take-off might be unproductive in increasing propulsion, and might have a negative effect on step rate (Hay, 1994: 409; Mann and Herman, 1985).

Hunter *et al.* (2005) investigated some of the above theories regarding methods of minimizing braking and maximizing propulsion. Horizontal ground reaction impulse expressed relative to body mass was used as the GRF variable of interest. The results showed that lower braking ground reaction impulse was associated with an active touchdown and with a smaller touchdown distance. There was only partial support for the theory that a high hip extension velocity during stance is associated with a high propulsive ground reaction impulse. However, due to the cross-sectional design of the study, causation cannot be assumed from these results. There is a great need for research of longitudinal design to assess all the theories in this section more completely.

Gravitational force

The magnitude of the gravitational force experienced by a sprinter is equivalent to the sprinter's body weight and proportional to the sprinter's body mass. Unlike the other external forces acting on a sprinter (i.e., GRF and wind resistance), the body weight of any given athlete, for any given race, is constant in magnitude and direction (i.e., downward).

Elite sprinters are generally the heaviest of all elite track athletes (Carter, 1984). Nonetheless, the high 'power to mass ratio' requirement of sprint running (Baker and Nance, 1999; Young *et al.*, 1995) means that elite sprinters are generally lean (Norton *et al.*, 2000), muscular, but not heavy by general population standards (Uth, 2005). The mean and standard deviation of body mass of elite sprinters are shown in Table 25.2.

Wind resistance

Of the three external forces acting on a sprinter, wind resistance is the smallest. Nonetheless, wind resistance does have a significant effect on sprint performance.

Table 25.2 Body mass of elite sprinters

Sample	*Reference*	*Males*		*Females*	
		Mean	*(SD)*	*Mean*	*(SD)*
World class sprinters from the IAAF all-time 100 m top 50 list	(Uth, 2005)	77.0	(6.6)	58.1	(5.2)
Elite sprinters and hurdlers (100–400 m) from the Australian Institute of Sport	(Smith *et al.*, 2000)	75.8	(6.2)	57.6	(4.3)
100 m sprinters from the 1976 Olympic Games	(Carter, 1984)	71.5	not available	56.0	not available

Stance time is equal to the sum of braking time and propulsive time. Average GRF values are expressed in multiples of body weight (BW). Relative ground reaction impulse (GRI) values are expressed in m/s and reflect the change in velocity experienced by the athlete.

The magnitude of the aerodynamic drag force experienced by an athlete is determined, in part, by the velocity of the wind relative to the athlete. Kyle and Caiozzo (1986) estimated that for a male, 1.75 m tall, 71 kg body mass, and running with a relative wind velocity of 9 m/s, the drag force would be about 26 N. This may seem small, relative to the large horizontal GRF that can be produced by a sprinter. However, the drag force acts continuously throughout a race, whereas GRF is produced intermittently. Consequently, the drag force will, at various times throughout a stride (particularly during the flight phase) cause the horizontal velocity of a sprinter to decrease.

Researchers have used mathematical models to predict the effects of wind on 100 m race times of elite sprinters. Linthorne (1994) predicted that a 2 m/s tailwind, when compared to no wind, would result in a 0.10 s decrease in 100 m time for males and a 0.12 s decrease for females. In a more recent study, Quinn (2003) predicted very similar results: a 0.10 s decrease for males and a 0.11 s decrease for females. In contrast, a 2 m/s head wind, when compared to no wind, would result in a 0.12 s increase in 100 m time for males and 0.13 s increase for females (Quinn, 2003).

At high altitude, the air is less dense. Consequently, at altitude the aerodynamic drag force experienced by an athlete is less. The IAAF does not have a restriction on altitude, but scientists generally consider a race performed higher than 1,000 m above sea level to be 'altitude assisted' (Linthorne, 1994; Quinn, 2003) Quinn (2003) calculated that in no wind, an altitude of 1,000 m would provide a 0.03 s advantage in a 100 m race. In Mexico City (2,250 m above sea level), the advantage would be 0.07 s (Linthorne, 1994) or 0.08 s (Kyle and Caizzo, 1986).

Various steps can be taken to minimize wind resistance. Kyle and Caiozzo (1986) calculated that a 2 per cent decrease in aerodynamic drag force might improve 100 m time by 0.01 s, which is equivalent to about a 0.10 m lead at the end of the race. Their results from wind-tunnel testing indicated that a significant reduction in drag force, possibly greater than 2 per cent, could be achieved by shaving the head or wearing a tight smooth cap, shaving the limbs or covering them with sleeves or tights of appropriate material, and 'smoothing' the shoes and laces. Socks and shoes were thought to be critical because during sprinting, the legs and feet reach velocities much higher than the COM of the athlete.

Internal forces and moments

In this section we discuss the internal forces and moments and how they are coordinated to generate the typical kinematics seen in sprinters. In addition, reference has at times been made to muscle activity, changes in muscle length, segment interactions, as well as evidence and theories regarding how sprint technique might be optimized. In sprinting, the coordination of the entire body is vital; however, for descriptive purposes, the lower limbs, upper limbs, and trunk are discussed separately. Most research on the topics in this section of the review has focused on the maximal-velocity phase or thereabouts. Consequently, the following descriptions are most applicable to a stride during that phase.

The lower limbs

In the case of the lower limbs, a description of a single stride for one limb is provided. The stride is divided into the swing phase and the stance phase. The swing phase refers to the period from the moment of take-off to the moment of touchdown for the same foot. The stance phase refers to the period the lower limb of interest is in contact with the ground. Furthermore, the swing phase is divided into three sub-phases: early swing, mid-swing, and late swing. The stance phase is divided into two sub-phases: early stance and late stance (see Figure 25.2).

Early swing

The early swing phase starts at the instant of take-off and continues until the swinging thigh is approximately vertical (see Figure 25.3a). Immediately after take-off, the hip joint continues to extend a few degrees (Mann and Herman, 1985), but soon starts to flex (Johnson and Buckley, 2001). This is due to a flexion resultant joint moment at the hip, caused by an eccentric, then concentric action of the hip flexor musculature (Novacheck, 1998). Also, during early swing, the knee undergoes flexion, which continues until a minimum knee-angle is obtained during the mid-swing phase (Novacheck, 1998). This results in the foot being positioned close to the buttocks, and the moment of inertia (MOI) of the entire swing-limb being decreased (Hay, 1994: 409). This knee flexion, at least during sub-maximal running, is initiated by a motion-dependent s egment interaction with the thigh (Putnam, 1991). Throughout most of the early swing phase, the resultant joint moment at the knee acts in opposition to the knee flexion (Mann, 1981; Novacheck, 1998).

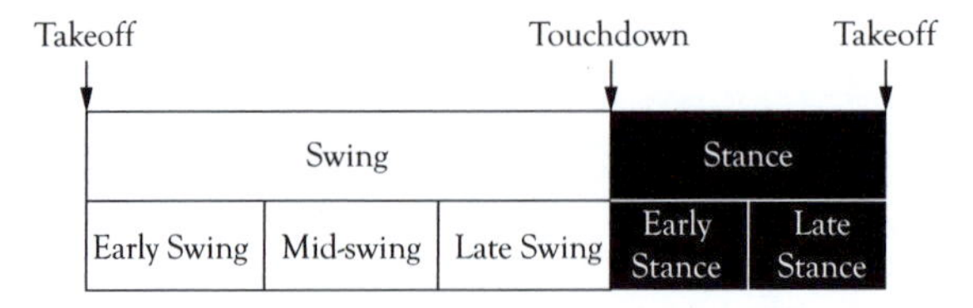

Figure 25.2 Subdivisions of a stride, based on events of a single lower-limb.

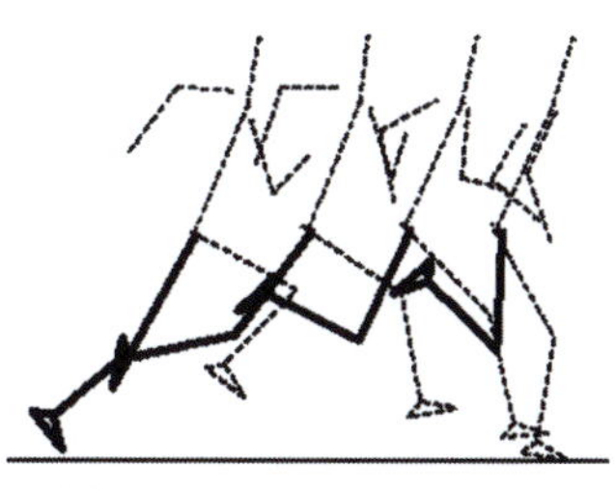

(a) Early Swing of Lower Limb

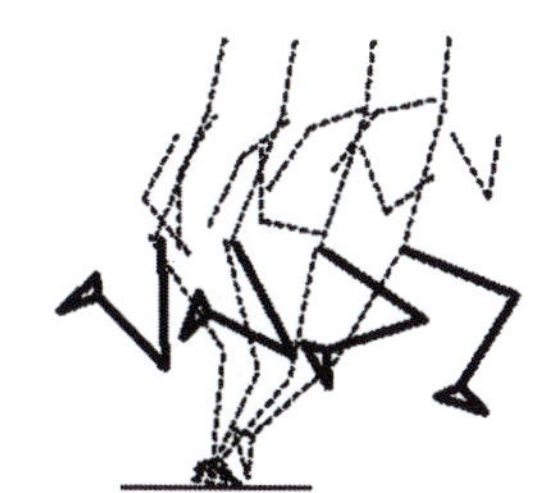

(b) Mid-swing of Lower Limb

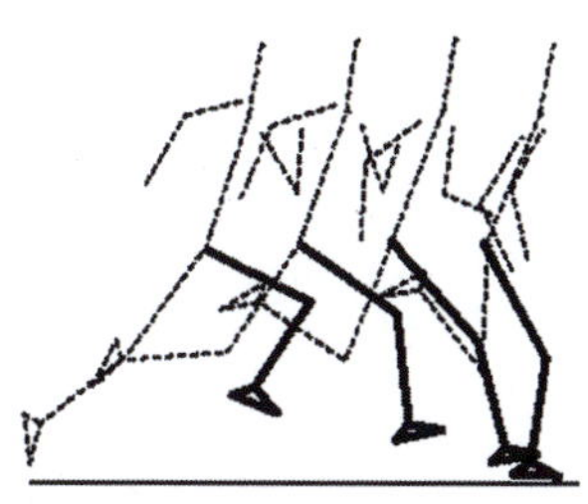

(c) Late Swing of Lower Limb

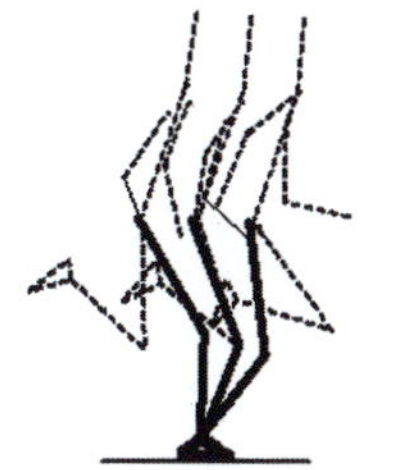

(d) Early Stance of Lower Limb

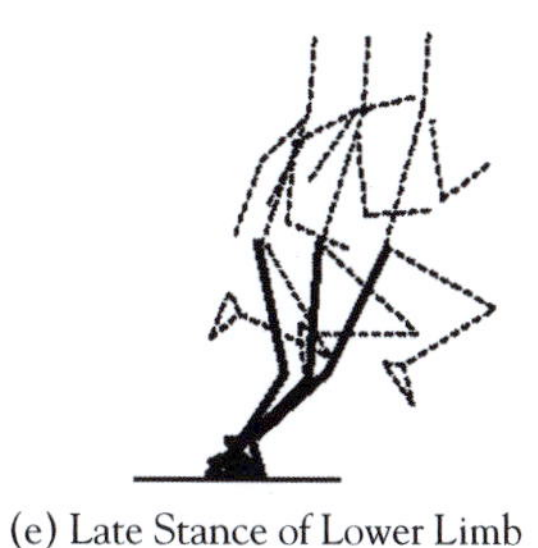

(e) Late Stance of Lower Limb

(f) Mid-swing Position of Upper Limbs

Figure 25.3 Key phases and positions during a sprinting stride.

Mid-swing

The mid-swing phase starts from when the swinging thigh is vertical and continues until the instant of maximum knee lift (see Figure 25.3b). Close to the time that the knee of the swing limb passes besides the knee of the stance limb, the hip resultant joint moment changes from flexion to extension (Mann, 1981; Novacheck, 1998). As the swing progresses, the knee extends and the foot comes away from the buttocks. The gluteal and hamstrings musculature act eccentrically to slow the flexing hip, and the hamstrings also act in opposition to the extending knee (Novacheck, 1998; Simonsen *et al.*, 1985). That is, the hip resultant joint moment is one of extension and the knee resultant joint moment is one of flexion. During this time, the hamstrings, which cross both the hip and knee joints, undergo extensive and rapid lengthening while under tension (Simonsen *et al.*, 1985; Wood, 1987). This is possibly related to the risk of hamstring injury that seems so common in sprinters.

The mid-swing phase concludes when the hip obtains its maximum flexion and the knee is in an elevated position. Some researchers consider a high knee-lift to be an ideal sprinting technique (Deshon and Nelson, 1964). A high knee-lift might also be one way of reducing injury risk to the hamstrings. That is, the forward swing of the lower limb could be slowed over a greater angular range, thereby reducing the forces and moments that must be generated by the hamstrings. Nonetheless, a high knee-lift might also have a negative effect on the quickness of leg recovery and, therefore, step rate (Wood, 1987).

Late swing

The late-swing phase starts with the swing limb in the 'high-knee position' and continues with the hip and knee extending in preparation for the upcoming touchdown (see Figure 25.3c).

During this time, an extension resultant joint moment acts at the hip and contributes to the increasing rearward angular velocity of the thigh (Johnson and Buckley, 2001; Novacheck, 1998). The knee resultant joint moment is one of flexion, and acts to slow the extending knee (Johnson and Buckley, 2001; Novacheck, 1998). The path of the foot during this phase has been likened to a 'pawing' action. That is, the foot travels rearward relative to the COM of the body and is thought to play a role in minimizing the braking GRF; (Hay, 1994: 407–8; Hunter *et al.*, 2005) but may also increase the risk of injury (Mann and Sprague, 1980).

Early stance

The early stance phase starts at the instant of touchdown and continues for approximately half the stance phase (see Figure 25.3d). Touchdown occurs with the stance foot typically placed in front of the COM (Hay, 1994: 408). A small touchdown distance (i.e., a small horizontal distance between the foot and the COM at the instant of touchdown) is thought to be a means by which the braking GRF can be minimized (Hunter *et al.*, 2005; Mero *et al.*, 1992). During approximately the first-third or first-half of stance, a large extension resultant joint moment acts at the hip (Mann, 1981; Novacheck, 1998). Some researchers believe the large extension resultant joint moment at the hip is the main cause of the increasing angular velocity of the thigh during stance, which, in turn, is thought to play a major role in minimizing the braking GRF and maximizing the propulsive GRF (Mann and Sprague, 1983; Mann *et al.*, 1984; Wiemann and Tidow, 1995). However, a more recent study (Hunter *et al.*, 2004c) has suggested that the increasing angular velocity of the thigh during stance is not only generated from the hip resultant joint moment, but also from the knee resultant joint moment and segment interaction effects.

During the early stance phase, the ankle and knee joints flex. Throughout virtually the entire stance phase the resultant joint moment at ankle and knee are extension moments (Johnson and Buckley, 2001; Novacheck, 1998). One important exception to this is a very brief period after touchdown during which the knee resultant joint moment is sometimes shown to be a flexion moment (Mann, 1981; Putnam and Kozey, 1989). Some researchers (Mann and Sprague, 1980; Mann and Sprague, 1983; Mann, 1981) have interpreted this as a means by which the braking GRF can be minimized. Very soon after touchdown occurs, however, the resultant joint moment at the knee is clearly one of extension and the knee joint continues to flex (Mann, 1981; Putnam and Kozey, 1989).

Late stance

The late stance phase starts approximately halfway through stance and continues until the instant of take-off (see Figure 25.3e). When running at maximal velocity, the horizontal GRF changes from braking to propulsion approximately halfway through stance (Kyrolainen *et al.*, 1999). During the second half of stance, the thigh continues to be accelerated until peak extension velocity occurs approximately two-thirds through stance (Johnson and Buckley, 2001; Putnam and Kozey, 1989). This increasing extension velocity of the thigh has often been attributed to an extension resultant joint moment at the hip, produced by the gluteal and hamstring musculature (Mann and Sprague, 1980; Mann and Sprague, 1983; Wiemann and Tidow, 1995). However, some researchers have reported that the hip resultant joint moment changes to flexion even before halfway through stance (Mann, 1981; Novacheck, 1998); that is, well before maximum hip extension velocity is expected to occur. This suggests that there may be other sources, such as the knee resultant joint moment and segment interactions, that contribute significantly to the increasing angular velocity of the thigh (Hunter *et al.*, 2004c).

Peak propulsive GRF occurs approximately two-thirds through stance (Kyrolainen *et al.*, 1999). By this time, the hip resultant joint moment is one of flexion; however, the resultant joint moments at the knee and ankle joints generally remain extension moments through to the end of stance (Novacheck, 1998). Peak extension velocity of the knee joint, then the ankle joint, occurs in the final third of the stance phase (Johnson and Buckley, 2001; Putnam and Kozey, 1989). That is, a proximal-to-distal sequence is evident with regards to the stance-limb hip, knee, and ankle joint angular velocities and joint powers (Jacobs and van Ingen Schenau, 1992; Johnson and Buckley, 2001).

During the last third of stance, due to a large flexion resultant joint moment at the hip, there is a rapid slowing of the rearward swinging thigh (Johnson and Buckley, 2001; Putnam and Kozey, 1989). This is in preparation for the upcoming swing phase. At the instant of take-off, the hip joint has extended beyond 180° (Novacheck, 1998) and after take-off, it continues to extend by a few degrees (Mann and Herman, 1985). In contrast, late in stance, the knee joint obtains maximum extension of about 160°, but then begins to flex a little, immediately before take-off (Novacheck, 1998). The ankle joint is similar to the hip joint in that it continues to extend during take-off (Novacheck, 1998). Although a definitive study has not been conducted on the topic, it has been suggested that any additional propulsion gained from excessively extending the stance limb might be more than offset with an increase in stance time, and therefore, a decrease in step rate (Hay, 1994: 409; Mann *et al.*, 1984).

The upper limbs

When running, the arms rotate backwards and forwards. As the knee of the swing leg passes in close proximity to the knee of the stance leg, both the left and right upper arms are approximately aligned with the trunk (when viewed in the sagittal plane). We will refer to this position as the mid-swing position of the arms (see Figure 25.3f). As explained in the following paragraph, the mid-swing position is an important transition point for the resultant joint moment at the shoulder.

The resultant joint moment acting at the shoulder during sprint running has been documented by Mann (1981). As the upper arm is swung forward of the mid-swing position, the shoulder resultant joint moment is one of extension, and acts to slow the forward swinging limb and then initiate the backward swing. Then, as the upper arm rotates backward past the mid-swing position, the resultant joint moment at the shoulder changes to flexion and acts to slow the rearward motion of the limb and initiate a forward swing.

Apart from the research of Mann and colleagues, the role of the arms in sprinting has not received a great deal of research attention. However, the role of the arms in sub-maximal running (up to 5.4 m/s pace) has been documented by Hinrichs (1990). Hinrichs concluded that the main role of the arms during sub-maximal running was to balance the angular momentum of the body about the vertical axis. That is, '... the arms and upper trunk provide angular impulses to the lower body needed for the legs to alternate through stance and swing phases'. In addition, the arms were thought to make a small contribution to lift (i.e., vertical impulse), but did not contribute to propulsion (horizontal impulse). Finally, Hinrichs noted that the arms were found to assist in maintaining a more constant horizontal velocity (both anterior-posterior and medial-lateral velocity).

The trunk

The trunk (including the pelvis) is thought to play a major role in controlling the amount of rotation of the body about its transverse axis (Hay, 1994: 411–2). When an athlete is in contact with the ground, the resultant vector of the vertical and anterior-posterior GRF components, if not acting directly through the COM, will create a moment tending to rotate the athlete about the transverse axis. However, instinctively, an athlete will adjust the position of his or her trunk, which has a large influence on the position of the COM. As a result, the moment arm for the resultant GRF is modified, as is the tendency to rotate.

The involvement of the trunk in assisting with control of rotation of the body about its transverse axis is very noticeable in the transition from the early acceleration phase to the maximal-velocity phase. During this transition, the magnitude of the vertical, braking, and propulsive GRF components change notably. As a result, the angle at which the resultant GRF acts also changes. In response, the athlete alters the forward lean of the body. During the first stance after leaving the blocks, the trunk, when viewed in the sagittal plane, has a pronounced lean, but becomes more upright with each step. For example, for eight male, elite sprinters, the mean trunk-lean (measured from horizontal) at the moment of take-off from the first, second, third, and fourth stance phases (from a block start) was 24°, 30°, 37°, and 44°, respectively (Atwater, 1982). Mean trunk-lean for the same athletes at the maximal-velocity phase of the sprint was 81° at take-off.

Another role of the trunk is to assist the arms in balancing the angular momentum of the body about the vertical axis. At sub-maximal running velocity, the upper trunk rotates in the same direction as the arms, and the lower trunk (including the pelvis) rotates in the same direction as the legs (Hinrichs, 1990). As a result, the upper body (head, upper trunk, and arms) produces an angular momentum about the vertical axis that is about equal and opposite to that produced by the lower body (lower trunk and legs). Without this balance, an athlete would have great difficulty swinging the lower limbs backward and forward, particularly during the flight phase (Hinrichs, 1990).

Conclusion

Much research about the internal and external forces in sprint running exists. It is clear that the GRF and that the internal forces and moments produced are extremely important in sprint running. However, our understanding and interpretation of these kinetics is still incomplete. For example, there is still disagreement among researchers as to what the optimal GRF is and how it can be produced. Furthermore, what are the most important internal kinetics? How can these internal kinetics be influenced by training? What contribution do segment interactions of the stance limb and the actions of the swinging limbs make to the GRF? While the research conducted to date has been very informative, more research is required that can establish causation among interventions, the GRF, and the resulting sprint performance.

References

1. Atwater, A. E. (1982) 'Kinematic analyses of sprinting'. *Track and Field Quarterly Review*, 82: 12–16.
2. Baker, D. and Nance, S. (1999) 'The relationship between running speed and measures of strength and power in professional rugby league players'. *Journal of Strength and Conditioning Research*, 13: 230–235.

3. Carter, J. (1984) 'Age and body size of Olympic athletes'. In Carter, J. (ed.) *Medicine and Sport Science: Physical Structure of Olympic Athletes, Part II* (pp. 53–79), Basel, Karger.
4. Deshon, D. E. and Nelson, R. C. (1964) 'A cinematographical analysis of sprint running'. *Research Quarterly*, 35: 453–454.
5. Hay, J. G. (1994) *The Biomechanics of Sports Techniques*. London, Prentice Hall International.
6. Hinrichs, R. (1990) 'Upper extremity function in distance running'. In Cavanagh, P. (ed.) *Biomechanics of distance running* (pp. 107–133), Champaign, IL, Human Kinetics.
7. Hunter, J., Marshall, R. and McNair, P. (2004a) 'Interaction of step length and step rate during sprint running'. *Medicine and Science in Sports and Exercise*, 36: 261–271.
8. Hunter, J., Marshall, R. N. and McNair, P. (2004b) 'Reliability of biomechanical variables of sprint running'. *Medicine and Science in Sports and Exercise*, 36: 850–861.
9. Hunter, J. P., Marshall, R. N. and McNair, P. J. (2005) 'Relationships between GRF impulse and kinematics of sprint-running acceleration'. *Journal of Applied Biomechanics*, 21.
10. Hunter, J. P., Marshall, R. N. and McNair, P. J. (2004c) 'Segment-interaction analysis of the stance limb in sprint running'. *Journal of Biomechanics*, 37: 1439–1446.
11. Jacobs, R. and van Ingen Schenau, G. (1992) 'Intermuscular coordination in a sprint push-off'. *Journal of Biomechanics*, 25: 953–965.
12. Johnson, M. D. and Buckley, J. G. (2001) 'Muscle power patterns in the mid-acceleration phase of sprinting'. *Journal of Sports Sciences*, 19: 263–272.
13. Knoedel, J. (1984) 'Active landing in the triple jump'. *Track Technique*, 88: 2814–2816.
14. Kyle, C. and Caizzo, V. (1986) 'The effect of athletic clothing aerodynamics upon running speed'. *Medicine and Science in Sports and Exercise*, 18: 509–515.
15. Kyrolainen, H., Komi, P. and Belli, A. (1999) 'Changes in muscle activity patterns and kinetics with increasing running speed'. *Journal of Strength and Conditioning Research*, 13: 400–406.
16. Linthorne, N. (1994) 'The effect of wind on 100 m sprint times'. *Journal of Applied Biomechanics*, 10: 110–131.
17. Mann, R. and Herman, J. (1985) 'Kinematic analysis of Olympic sprint performance: men's 200 meters'. *International Journal of Sport Biomechanics*, 1: 151–162.
18. Mann, R. and Sprague, P. (1980) 'A kinetic analysis of the ground leg during sprint running'. *Research Quarterly for Exercise and Sport*, 51: 334–348.
19. Mann, R. and Sprague, P. (1983) 'Kinetics of sprinting'. *Track and Field Quarterly Review*, 83: 4–9.
20. Mann, R. V. (1981) 'A kinetic analysis of sprinting'. *Medicine and Science in Sports and Exercise*, 13: 325–328.
21. Mann, R. V., Kotmel, J., Herman, J., Johnson, B. and Schultz, C. (1984) 'Kinematic trends in elite sprinters'. *Proceedings of the International Symposium of Biomechanics in Sports* (pp. 17–33), Del Mar, California, Academic Publishers.
22. Mero, A. (1988) 'Force-time characteristics and running velocity of male sprinters during the acceleration phase of sprinting'. *Research Quarterly for Exercise and Sport*, 59: 94–98.
23. Mero, A. and Komi, P. (1994) 'EMG, force, and power analysis of sprint-specific strength exercises'. *Journal of Applied Biomechanics*, 10: 1–13.
24. Mero, A., Komi, P., Rusko, H. and Hirvonen, J. (1987) 'Neuromuscular and anaerobic performance of sprinters at maximal and supramaximal speed'. *International Journal of Sports Medicine*, 8: 55–60, Supplement.
25. Mero, A. and Komi, P. V. (1986) 'Force-, EMG-, and elasticity-velocity relationships at submaximal, maximal and supramaximal running speeds in sprinters'. *European Journal of Applied Physiology*, 55: 553–561.
26. Mero, A., Komi, P. V. and Gregor, R. J. (1992) 'Biomechanics of sprint running'. *Sports Medicine*, 13: 376–392.
27. Mero, A., Luhtanen, P. and Komi, P. (1983) 'A biomechanical study of the sprint start'. *Scandinavian Journal of Sports Science*, 5: 20–28.
28. Miller, D. (1990) 'GRFs in distance running'. In Cavanagh, P. (ed.) *Biomechanics of Distance Running* (pp. 203–224), Champaign, IL, Human Kinetics.

29. Norton, K., Marfell-Jones, M., Whittingham, N., Kerr, D., Carter, L., Saddington, K. and Gore, C. (2000) 'Anthropometric Assessment Protocols'. In Gore, C. (ed.) *Physiological Tests for Elite Athletes* (pp. 66–85), Champaign, IL, Human Kinetics.
30. Novacheck, T. (1998) 'The biomechanics of running'. *Gait and Posture*, 7: 77–95.
31. Putnam, C. A. (1991) 'A segment interaction analysis of proximal-to-distal sequential segment motion patterns'. *Medicine and Science in Sports and Exercise*, 23: 130–144.
32. Putnam, C. A. and Kozey, J. W. (1989) 'Substantive issues in running'. In Vaughan, C. L. (ed.) *Biomechanics of Sport*. (pp. 1–33), Boca Raton, Florida, CRC Press.
33. Quinn, M. (2003) 'The effects of wind and altitude in the 200 m sprint'. *Journal of Applied Biomechanics*, 19: 49–59.
34. Simonsen, E. B., Thomsen, L. and Klausen, K. (1985) 'Activity of mono- and biarticular leg muscles during sprint running'. *European Journal of Applied Physiology*, 54: 524–532.
35. Smith, D., Telford, R., Peltola, E. and Tumilty, D. (2000) 'Protocols for the physiological assessment of high-performance runners'. In Gore, C. (ed.) *Physiological Tests for Elite Athletes* (pp. 334–344), Champaign, IL, Human Kinetics.
36. Uth, N. (2005) 'Anthropometric comparison of world-class sprinters and normal populations'. *Journal of Sports Science and Medicine*, 4: 608–616.
37. Weyand, P., Sternlight, D., Bellizzi, M. J. and Wright, S. (2000) 'Faster top running speeds are achieved with greater ground forces not more rapid leg movements'. *Journal of Applied Physiology*, 89: 1991–1999.
38. Wiemann, K. and Tidow, G. (1995) 'Relative activity of hip and knee extensors in sprinting – implications for training'. *New Studies in Athletics*, 10: 29–49.
39. Wood, G. (1987) 'Biomechanical limitations to sprint running'. In van Gheluwe, B. and Atha, J. (eds.) *Medicine and Sport Science* (pp. 58–71), Basel, Karger.
40. Wood, G. A., Marshall, R. N. and Jennings, L. S. (1987) 'Optimal requirements and injury propensity of lower limb mechanics in sprint running'. *Biomechanics X* (pp. 869–874), Champaign, Ill, Human Kinetics.
41. Young, W., McLean, B. and Ardagna, J. (1995) 'Relationship between strength qualities and sprinting performance'. *Journal of Sports Medicine and Physical Fitness*, 35: 13–19.

26

Biomechanical simulation models of sports activities

M. R. Yeadon and M. A. King
Loughborough University, Loughborough

Introduction

Experimental science aims to answer research questions by investigating the relationships between variables using quantitative data obtained in an experiment and assessing the significance of the results statistically (Yeadon and Challis, 1994). In an ideal experiment the effects of changing just one variable are determined. While it may be possible to change just one variable in a carefully controlled laboratory experiment in the natural sciences, this is problematic in the sports sciences in general and in sports biomechanics in particular.

Theoretical approaches to answering a research question typically use a model that gives a simplified representation of the physical system under study. The main advantage of such a model is that ideal experiments can be carried out since it is possible to change just one variable. This chapter will describe theoretical models used in sports biomechanics, detailing their various components and discussing their strengths and weaknesses.

Models may be used to address the *Forward Dynamics* problem and the *Inverse Dynamics* problem. In the forward dynamics problem the driving forces are specified and the problem is to determine the resulting motion. In the inverse dynamics problem the motion is specified and the problem is to determine the driving forces that produced the motion (Zatsiorsky, 2002). The focus of this chapter will be forward dynamics modelling since the range of research questions that can be addressed is much greater.

This chapter will first describe the process of building a mathematical model using rigid bodies and elastic structures to represent body segments and various ways of representing the force generating capabilities of muscle. Direct and indirect methods of determining the physical parameters associated with these elements will be described. Before using a model to answer a research question it is first necessary to establish that the model is an adequate representation of the real physical system. This process of model evaluation by comparing model output with real data will be described. Finally, various issues in model design will be discussed.

The forward dynamics problem

In forward dynamics the driving forces are specified and the problem is to determine the resulting motion. Muscle forces or joint torques may be used as the drivers in which case the joint angle time histories will be part of the resulting motion. If joint angle time histories are used as drivers for the model then the resulting motion will be specified by the whole body mass centre movement and whole body orientation time history. When a model is used to solve the forward dynamics problem it is known as a simulation model.

Model building

The human body is very complex with over 200 bones and 500 muscles and, therefore, any human body model will be a simplification of reality. The degree of simplification of a simulation model will depend on the activity being simulated and the purpose of the study. For example a one-segment model of the human body may adequately represent the aerial phase of a straight dive but a model with two or three segments would be required for a piked dive to give an adequate representation. As a consequence a single model cannot be used to simulate all activities and so specific simulation models are built for particular tasks. As a general rule the model should be as simple as possible, while being sufficiently complex to address the questions of interest. This simple rule of thumb can be quite difficult to implement since the level of complexity needed is not always obvious.

Model components

The following section will discuss the various components that are used to build a typical simulation model.

Linked segment models

Most of the whole body simulation models in sports biomechanics are based on a collection of rigid bodies (segments) linked together, and are generically called 'linked segment systems'. The rigid bodies are the principal building blocks of simulation models and can be thought of as representing the basic structure and inertia of the human body. For each rigid segment in a planar model four parameters are usually required: length, mass, mass centre location and moment of inertia (MOI). The number of segments used depends on the aim of the study and the activity being modelled. For example, Alexander (1990) used a two-segment model to determine optimum approach speeds in jumps for height and distance, Neptune and Kautz (2000) used a planar two-legged bicycle-rider model to look at muscle contributions in forward and backward pedalling, and King and Yeadon (2003) used a planar five-segment model to investigate take-offs in tumbling.

Wobbling masses

Although linked rigid body models have been used extensively to model many activities, a recent development has been to modify some of the rigid segments in the model by incorporating wobbling mass elements (Gruber *et al.*, 1998). This type of representation allows some of the mass (soft tissue) in a segment to move relative to the bone, the fixed part. For impacts the inclusion of wobbling masses within the model is crucial as the loading on the system can be up to nearly 50 per cent lower for a wobbling mass model compared to the equivalent rigid segment model (Pain and Challis, 2006). The most common way to model wobbling masses is to attach a second rigid element to the first fixed rigid element, representing the bone, within a segment using non-linear damped passive springs with spring force $F = kx^3 - d\dot{x}$ where x is displacement, $\dot{x}$ is velocity and k and d are constants (Pain and Challis, 2001a).

The disadvantage of including wobbling mass elements within a simulation model is that there are more parameter values to determine and the equations of motion are more complex, leading to longer simulation times. Wobbling mass segments should, therefore, only be included when necessary. Whether to include wobbling masses depends on the activity being modelled, although it is not always obvious whether they are needed. For example, a simulation model of springboard diving (Yeadon, Kong and King, 2006b) included wobbling mass segments, but when the springs were made 500 times stiffer the resulting simulations were almost identical.

Connection between rigid links

Typically, the rigid links in the simulation model are joined together by frictionless joints, whereby adjacent segments share a common line or a common point. For example, Neptune and Kautz (2000) used a hinge joint to allow for flexion-extension at the knee while Hatze (1981) used a universal joint at the hip with three degrees of freedom to allow for flexion-extension, abduction-adduction and internal-external rotation. The assumption that adjacent segments share a common point or line is a simplification of reality and, although reasonable for most joints, it is questionable at the shoulder where motion occurs at four different joints. Models of the shoulder joint have ranged in complexity from a one degree of freedom pin joint (Yeadon and King, 2002) to relatively simple viscoelastic representations (Hiley and Yeadon, 2003) and complex finite element models (van der Helm, 1994). The complexity to be used depends on the requirements of the study. Simple viscoelastic representations have been used successfully in whole body models, where the overall movement is of interest, whereas complex models have been used to address issues such as the contribution of individual muscles to movement at the shoulder joint.

Interface with external surface

The simplest way to model contact between a human body model and an external surface, such as the ground or sports equipment, is to use a 'joint' so that the model rotates about a fixed point on the external surface (Bobbert *et al.*, 2002). The disadvantage of this method is that it does not allow the model to translate relative to the point of contact or allow for a collision with the external surface, since for an impact to occur the velocity of the point contacting the surface has to be non-zero initially. Alternatively, forces can be applied at a finite number of locations using viscoelastic elements at the interface, with the forces determined by the displacements and velocities of the points in contact. The viscoelastic elements can be used to represent specific elastic structures within the body such as the heel pad (Pain and Challis, 2001b) or sports equipment such as the high bar (Hiley and Yeadon, 2003) or tumble-track/foot interface (King and Yeadon, 2004). The equations used for the viscoelastic elements have varied in complexity from simple damped linear representations (King and Yeadon, 2004) through to highly non-linear equations (Wright *et al.*, 1998). The number of points of contact varies but it is typically less than three (Yeadon and King, 2002) although 66 points of contact were used to simulate heel-toe running (Wright *et al.*, 1998).

The horizontal forces acting while in contact with an external surface can be calculated using a friction model (Gerritsen *et al.*, 1995) where the horizontal force is expressed as a function of the vertical force and the horizontal velocity of the point in contact or by using viscoelastic springs (Yeadon and King, 2002). If viscoelastic springs are used the horizontal force should be expressed as a function of the vertical force so that the horizontal force falls to zero at the same time as the vertical force (Wilson *et al.*, 2006).

Muscle models

Muscle models in sports biomechanics are typically based upon the work of A.V. Hill, where the force-producing capabilities of muscle are divided into contractile and elastic elements (lumped parameter models) with the most commonly used version being the three-component Hill model (Caldwell, 2004). The model consists of a contractile element and two elastic elements: the series elastic element and the parallel elastic element. Mathematical relationships are required for each element in the muscle model so that the forces exerted by a muscle in the simulation model can be defined throughout a simulation.

Contractile element

The force that a contractile element produces can be expressed as a function of three factors; muscle length, muscle velocity and muscle activation. The force-length relationship for a muscle is well documented as being bell-shaped with small tensions at extremes of length and maximal tension in between (Edman, 1992). As a consequence, the force-length relationship is often modelled as a simple quadratic function.

The force-velocity relationship for a muscle can be split into two parts: the concentric phase and the eccentric phase. In the concentric phase tetanic muscle force decreases hyperbolically with increasing speed of shortening to approach zero at maximum shortening speed (Hill, 1938). In the eccentric phase maximum tetanic muscle force increases rapidly to around 1.4–1.5 times the isometric value, with increasing speed of lengthening and then plateaus for higher speeds (Dudley *et al.*, 1990). Maximum voluntary muscle force shows a similar force-velocity relationship in the concentric phase, but plateaus at 1.1–1.2 times the isometric value in the eccentric phase (Westing *et al.*, 1988; Yeadon, King and Wilson, 2006a).

The voluntary activation level of a muscle ranges from 0 (no activation) to 1 (maximum voluntary activation) during a simulation and is defined as a function of time. This function is multiplied by the maximum voluntary force given by the force-length and force-velocity relationships to give the muscle force exerted. Ideally the function used to define the activation time history of a muscle should have only a few parameters. One way of doing this is to define a simple activation profile for each muscle (basic shape) using a few parameters (Yeadon and King, 2002). For example, in jumping the activations of the extensors rise up from a low initial value to a maximum and then drop off towards the end of the simulation, while the flexor activations drop from an initial to a low value and then rise towards the end of the simulation (King *et al.*, 2006). These parameters are varied within realistic limits to define the activation-time history used for each muscle during a specific simulation.

Series and parallel elastic elements

The series elastic element represents the connective tissue (the tendon and aponeurosis) in series with the contractile element. The force produced by the series elastic element is typically expressed as an increasing function of its length with a slack length below which no force can be generated. The effect of the parallel elastic element is often ignored in models of sports movements as this element does not produce high forces for the normal working ranges of joints.

Torque generators vs. individual muscle representations

All simulation models that include individual muscle models have the disadvantage that it is very difficult to determine individual parameters for each element of each muscle, as it is impossible to measure all the parameters required non-invasively. As a consequence, researchers rely on data from the literature for their muscle models and so the models are not specific to an individual. An alternative approach is to use torque generators to represent the net effect of all the muscles crossing a joint (e.g. King and Yeadon, 2002) as the net torque produced by a group of muscles can be measured on a constant velocity dynamometer. More recently the extensor and flexor muscle groups around a joint have been represented using separate torque generators (King *et al.*, 2006). In both cases each torque generator consists of rotational elastic and contractile elements. Using torque generators instead of individual muscles gives similar mathematical relationships, with the maximum voluntary torque produced by the contractile element being expressed as a function of the muscle angle and muscle angular velocity (Yeadon *et al.*, 2006a).

Model construction

The following sections will discuss the process of building a simulation model and running simulations using the components described in the previous section.

Generating the equations of motion

The equations of motion for a mechanical system can be generated from first principles using Newton's Second Law for relatively simple models with only a few segments (e.g. Hiley and Yeadon, 2003). For a planar link model, three equations of motion are available for each segment using Newton's Second Law (force = mass × acceleration) in two perpendicular directions and taking moments about the mass centre of the segment (moment = MOI × angular acceleration).

For more complex models a computer package is recommended, as it can take a long time to generate the equations of motion by hand and the likelihood of making errors is high. There are several commercially available software packages, for example DADS, ADAMS, AUTOLEV and SD Fast, that can generate equations of motion for a user-defined system of rigid and elastic elements. Each package allows the user to input a relatively simple description of the model and the equations of motion are then automatically generated, solved and integrated. Note that with all packages that automatically generate equations of motion it is important to learn how to use the specific software by building simple models and performing checks to ensure that the results are correct.

Model input and output

Two sets of input are required for a simulation to run. Firstly, there are the initial kinematics which comprises the mass centre velocity, and the orientation and angular velocity of each segment. The initial kinematics can be obtained from recordings of actual performances, although it may be difficult to obtain accurate velocity estimates (Hubbard and Alaways, 1989). Secondly, there is information required during the simulation. A kinematically-driven model requires joint angle-time histories (Yeadon, 1990a) while a kinetically-driven model requires activation histories for each actuator (muscle or torque generator) in the model (Alexander, 1990).

The output from both types of simulation model comprises time histories of all the variables calculated in the simulation model. For a kinematically-driven model this is the whole body orientation, linear and angular momentum and joint torques, while for a kinetically-driven model it comprises the whole body orientation, linear and angular momentum and joint angle-time histories and may include joint reaction forces.

Integration

Running a simulation to calculate how a model moves requires a method for integrating the equations of motion over time. The simplest method to increment a set of equations of motion (ordinary differential equations) through a time interval *dt* is to use derivative information from the beginning of the interval. This is known as the 'Euler method' (Press *et al.*, 1988):

$$x_{n+1} = x_x + \dot{x}_n dt + \tfrac{1}{2}\ddot{x}_n dt^2$$

where x, $\dot{x}$, $\ddot{x}$ are, respectively, the displacement, velocity and acceleration and suffices n and n+1 denote the nth and (n+1) step separated by a fixed step length *dt*. A better method is to use a fourth order Runge-Kutta in which four evaluations of the function are calculated per step size (Press *et al.*, 1988).

A kinetically-driven model requires the force or torque produced by each actuator to be input to the model at each time step. The force or torque produced is a function of the actuator's activation, length and velocity. The movement of the contractile element or series elastic element must, therefore, be calculated. Caldwell (2004) gives an in-depth account of this procedure, but essentially at each time step the total length of the actuator is split between the contractile element and series elastic element in such a way that the force or torque in each element are equal.

Optimisation

Simulation models can be used to find the optimum technique for a specific task by running many simulations with different inputs. To perform an optimisation is a three-stage process. Firstly, an objective function, or performance score, must be formulated which can be maximised or minimised by varying inputs to the model within realistic limits. For jumping simulations the objective function can simply be the jump height or jump distance, but for movements in which rotation is also important a more complex function incorporating both mass centre movement and rotation is required. The challenge for formulating such an objective function is to determine appropriate weightings for each variable in the function since the weightings affect the solution.

Secondly, realistic limits need to be established for each of the variables, typically activation parameters to each muscle and initial conditions. Additionally the activation patterns of each muscle need to be defined using only a few parameters to keep the optimisation run time reasonably low and increase the likelihood of finding a global optimum.

Thirdly, an algorithm capable of finding the global optimum rather than a local optimum is needed. Of the many algorithms available the Simplex algorithm (Nelder and Mead, 1965), the Simulated Annealing algorithm (Goffe *et al.*, 1994) and Genetic algorithms (van Soest and Cassius, 2003) have proved popular. The Simplex algorithm typically finds a solution quickly but can terminate at a local optimum as it only accepts downhill solutions, whereas the Simulated Annealing and Genetic algorithms are better at finding the global optimum as they can escape from local optima.

Parameter determination

Determining parameters for a simulation model is difficult but vital as the values chosen can have a large influence on the resulting simulations. Parameters are needed for the fixed and wobbling mass elements within a segment, muscle-tendon complexes, and viscoelastic elements in the model. Fundamentally there are two different ways to approach this, either to estimate values from the literature, or take measurements on an individual to determine individual-specific parameters. There is a clear advantage to determining individual-specific parameters as it allows a model to be evaluated by comparing simulation output with performance data on the same individual.

Inertia parameters

Accurate segmental inertia values are needed for each segment in the simulation model. For a rigid segment the inertia parameters consist of the segmental mass, length, mass centre location and MOI: one MOI value is needed for a planar model, while three MOI values are needed for a 3D model. For a wobbling mass segment there are twice as many inertia parameters needed since a wobbling mass segment comprises two rigid bodies connected by viscoelastic springs.

There are two methods of obtaining rigid segmental inertia parameters. The first is to use regression equations (Hinrichs, 1985) based upon anthropometric measurements and inertia parameters determined from cadaver segments (Chandler, 1975). The disadvantage of this method is that the accuracy is dependent on how well the morphology of the participant compares with the cadavers used in the study. A better method, which only requires density values from cadaver studies, is to take anthropometric measurements on the participant and use a geometric model (Hatze, 1980; Yeadon, 1990b) to determine the segmental inertia parameters. Although it is difficult to establish the accuracy of these geometric models for determining segmental inertia parameters, error values of around two per cent have been reported for total body mass (Yeadon, 1990b).

Including wobbling mass segments within the model increases the number of unknown parameters that are needed for each segment. The combined segmental inertia parameters can be calculated using a geometric model or regression equations. However, the calculation of the inertia parameters of the separate fixed and wobbling masses requires additional information on the ratio of bone to soft tissue, which is typically obtained from cadaver dissection studies (Clarys and Marfell-Jones, 1986). This ratio data can then be scaled to the specific individual using total body mass and percentage body fat (Pain and Challis, 2006; Wilson *et al.*, 2006).

Strength parameters

Determining accurate individual-specific strength parameters for muscle-tendon complexes is a major challenge in sports biomechanics, which has resulted in two different ways to represent the forces produced by muscles. The first is to include all the major muscles that cross a joint in the simulation model as individual muscle-tendon complexes with the parameters for the individual muscles obtained mainly from animal experiments (e.g. Gerritsen *et al.*, 1995). The alternative approach is to use torque generators at each joint in the model to represent the effect of all the muscles around a joint (flexors and extensors represented by separate torque generators). The advantage of this approach is that the net torque at a given joint can be measured on a constant velocity dynamometer over a range of joint angular velocities and joint angles for the participant and so individual-specific parameters can be determined that define maximal voluntary torque as a function of muscle angle and velocity (King and Yeadon, 2002; Yeadon *et al.*, 2006a). With this approach it is still necessary to use data from the literature to determine the parameters for the series elastic element for each torque generator. In recent studies (King *et al.*, 2006) it has been assumed that the series elastic element stretches by 5 per cent of its resting length during isometric contractions (Muramatsu *et al.*, 2001).

Viscoelastic parameters

Viscoelastic parameters are required for springs that are included within a simulation model: connection of wobbling masses, shoulder joint, foot- or hand-ground interface and equipment. Sometimes these springs represent specific elements where it is possible to determine viscoelastic properties from measurements (Pain and Challis, 2001b) while in other models the springs represent more than one viscoelastic element and so make it much harder to determine the parameters from experiments (e.g. Yeadon and King, 2002). Viscoelastic parameters should ideally be determined from independent tests and then fixed within the model for all simulations (Gerritsen *et al.*, 1995; Pain and Challis, 2001a). If this is not possible the viscoelastic parameters can be determined through an optimisation procedure by choosing initial values and then allowing the parameters to vary within realistic bounds until an optimum match between simulation and performance is found.

Model evaluation

Model evaluation is an essential step in the process of developing a simulation model and should be carried out before a model is used in applications. Although this step was identified as an important part of the process over 25 years ago (Panjabi, 1979) the weakness of many simulation models is still that their accuracy is unknown (Yeadon and Challis, 1994). While a number of models have been evaluated to some extent, such as those of Hatze (1981), Yeadon *et al.*, (1990), Neptune and Hull (1998), Brewin *et al.* (2000), Hiley and Yeadon (2003), Yeadon and King (2002), and King *et al.* (2006), many have not been evaluated at all.

The complexity of the model and its intended use should be taken into account when evaluating a model. For a simple model (e.g. Alexander, 1990), which is used to make general predictions, it may be sufficient to show that results are of the correct magnitude. In contrast, if a model is being used to investigate the factors that determine optimum performance in jumping, the model should be evaluated quantitatively so that the accuracy of the model is known (e.g. King *et al.*, 2006). Ideally the model evaluation should encompass the range of initial conditions and activities that the model is used for with little extrapolation of the model to cases in which the accuracy is unknown (Panjabi, 1979). For example, if a simulation model of springboard diving is evaluated successfully for forward dives, the model may not work for reverse dives and so it should be also evaluated using reverse dives.

The purpose of model evaluation is to determine the accuracy, which can then be borne in mind when considering the results of simulations. Furthermore, a successful evaluation gives confidence that the model assumptions are not erroneous and that there are no gross modelling defects or simulation software errors. Ideally the evaluation process should include all aspects of the model that are going to be used to make predictions. If a model is going to be used to investigate the effect of initial conditions on maximum jump height then the model should be evaluated quantitatively to show that for a given set of initial conditions the model can perform the movement in a similar way and produce a similar jump height. If a model is to be used to examine how the knee flexor and extensor muscles are used in jumping, the model should be evaluated to show that for a given jump the model uses muscle activations similar to the actual performance.

To evaluate a simulation model is challenging and may require several iterations of model development before the model is evaluated satisfactorily. Initially, data must be collected on an actual performance by the sports participant. Ideally this should be an elite performer who is able to work maximally throughout the testing and produce a performance that is close to optimal. Time histories of kinematic variables, from video or an automatic system, kinetic variables from force plate or force transducers, and electromyograms (EMG) histories, if possible, should be obtained. Individual-specific model parameter values, such as anthropometry and strength, are then determined from the measurements taken on the individual, with as little reliance on data from the literature as possible (Wilson *et al.*, 2006; Yeadon *et al.*, 2006a). The initial kinematic conditions – positions and velocities – for the model are then determined from the performance data and input to the model along with any other time histories that are required for the model to run a single simulation. If the model is kinetically-driven this will consist of the activation-time history for each actuator (Yeadon and King, 2002), while if the model is kinematically-driven the time history of each joint angle will be required (Hiley and Yeadon, 2003). Once a single simulation has been run, a difference score should be calculated by quantitatively comparing the simulation with the actual performance. The formulation of the score depends on the activity being simulated, but it should include all features of the performance that the model should match, such as joint angle changes, linear and angular momentum, and floor movement. The input to the model is then varied until the best comparison is found (score minimised) using an optimisation routine. If the comparison between performance and simulation is close (Figure 26.1) then the model can be used to run simulations. If not then the model complexity or model parameters need to be modified and the model re-evaluated. If the comparison gives a percentage difference of less than 10 per cent this is often sufficient for applications in sports biomechanics.

Issues in model design

The design of a particular model should be driven by the intended use and the questions to be answered. For example, if the aim is to determine the forces that act within the human body during running then an inverse dynamics model may be more appropriate than a forward dynamics model. If the aim is to demonstrate some general mechanical principles for a type of movement then a simple model may be adequate. The issue of model complexity is not straightforward, however. While it is evident that simple models

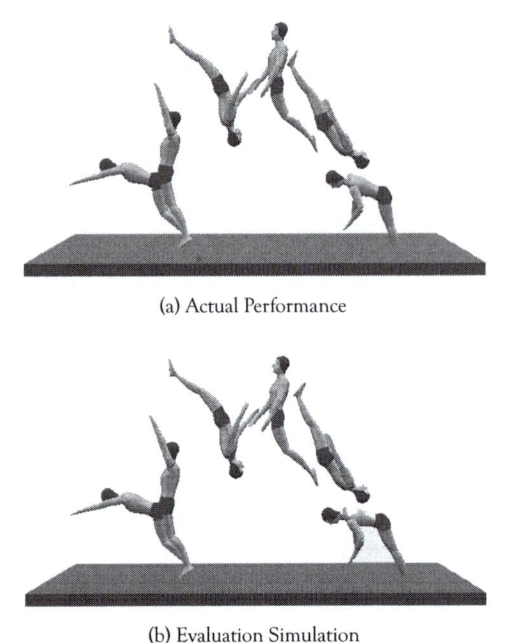

(a) Actual Performance

(b) Evaluation Simulation

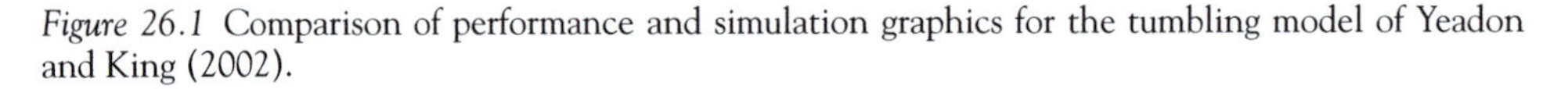

Figure 26.1 Comparison of performance and simulation graphics for the tumbling model of Yeadon and King (2002).

such as Alexander's (1990) model of jumping can give insight into the mechanics of technique, there is often a tendency to rely on the quantitative results without recourse to model evaluation. The issue of model evaluation for a simple model is problematic since all that can be realistically expected is ballpark or order of magnitude accuracy. To achieve anything approaching 10 per cent accuracy when compared with actual performance a model of some complexity is usually required, comprising several segments, realistic joint drivers and elastic elements. The development of such a model is a non-trivial endeavour. Sprigings and Miller (2004) argue the case for 'the use of the simplest possible model capable of capturing the essence of the task being studied', citing Alexander (1990) and Hubbard (1993) in support. The problem here is deciding at what point a model is too simple. If a model is so simple that it is 30 per cent inaccurate then it is difficult to justify conclusions indicated by the model results unless they are robust to a 30 per cent inaccuracy. It is evident that some measure of model accuracy is needed to reach conclusions.

Simple models of throwing in which the implement is modelled as an aerodynamic rigid body (Hubbard and Alaways, 1987) need to be complemented by a representation of the ability of the thrower to impart velocity in a given direction (Hubbard *et al.*, 2001) so that realistic simulations may be carried out. The same considerations apply to other models that do not include the human participant.

While a rigid body may be adequate for a model of equipment it is likely to be too simple for a model of an activity such as high jumping (Hubbard and Trinkle, 1985) although a rigid body model has been used to give insight into the two general modes of rotational aerial motion (Yeadon, 1993).

Joint angle time-histories are sometimes used as drivers for a simulation model. In the case of aerial movement (van Gheluwe, 1981; Yeadon *et al.*, 1990) it can be argued that this is a reasonable approach so long as the joint angular velocities are limited to achievable values. In activities where there are large contact forces with the external surroundings this approach is more problematic since steps need to be taken to ensure that the corresponding joint torques are achievable. Hiley and Yeadon (2003) and Brewin *et al.* (2000) used angle-driven models to simulate swinging on the high bar and on the rings and eliminated simulations that required larger torques than were achieved by the participant on a constant velocity dynamometer. Another approach is to use joint torques as drivers where the maximum voluntary joint torque (MVJT) is a function of angular velocity (Alexander, 1990) and, possibly, of joint angle (King and Yeadon, 2003). This approach leads to more realistic simulations than the use of angle-driven models but there is a corresponding loss of the simple control of joint angles. Finally, there are models that use representations of individual muscles or muscle groups crossing a joint (Hatze, 1981; Neptune and Hull, 1998) and these have the potential to provide even more accurate representations but pose the problem of determining appropriate muscle parameter values.

Conclusion

The use of simulation models in sport can give insight into what is happening or, in the case of a failing model, what is not happening (Niklas, 1992). Models also provide a means for testing hypotheses generated from observations or measurements of performance. It should be remembered, however, that all models are simplifications and will not reflect all aspects of the real system. The strength of computer simulation modelling of sport is that it can provide general research results for the understanding of elite performance although there is also the potential to investigate individual characteristics of technique using personalised models.

Acknowledgement

The content of this chapter is largely based on Chapter 9 on 'Computer Simulation Modelling in Sport' in the book: *Biomechanical Evaluation of Movement in Sport and Exercise: The British Association of Sport and Exercise Guide* edited by C. Payton and R.M. Bartlett, 2007, published by Routledge and is reproduced here by kind permission of the publishers.

References

1. Alexander, R.M. (1990) 'Optimum take-off techniques for high and long jumps', *Philosophical Transactions of the Royal Society of London – Series B*, 329: 3–10.
2. Bobbert, M.F., Houdijk, J.H.P., Koning, J.J. de and Groot, G. de (2002) 'From a one-legged vertical jump to the speed-skating push-off: A simulation study', *Journal of Applied Biomechanics*, 18: 28–45.
3. Brewin, M.A., Yeadon, M.R. and Kerwin, D.G. (2000) 'Minimising peak forces at the shoulders during backward longswings on rings', *Human Movement Science*, 19: 717–736.
4. Caldwell, G.E. (2004) 'Muscle modeling', in G.E. Robertson, G.E. Caldwell, J. Hamill, G. Kamen, S.N. and Whittlesey (eds) *Research Methods in Biomechanics*, Champaign, IL: Human Kinetics.

5. Chandler, R.F., Clauser, C.E., McConville, J.T., Reynolds, H.M. and Young, J.W. (1975) 'Investigation of inertial properties of the human body', *AMRL Technical Report* 74–137, Wright-Patterson Air Force Base, Drayton, OH, USA.
6. Clarys, J.P. and Marfell-Jones, M.J. (1986) 'Anthropometric prediction of component tissue masses in the minor limb segments of the human body', *Human Biology*, 58: 761–769.
7. Dudley, G.A., Harris, R.T., Duvoisin, M.R., Hather, B.M. and Buchanan, P. (1990) 'Effect of voluntary vs. artificial activation on the relationship of muscle torque to speed', *Journal of Applied Physiology*, 69: 2215–2221.
8. Edman, K.A.P. (1992) 'Contractile performance of skeletal muscle fibres', in P.V. Komi (ed.) *Strength and power in sport. Vol. III of the Encyclopaedia of Sports Medicine*, Oxford: Blackwell Scientific.
9. Gerritsen, K.G.M., van den Bogert, A.J. and Nigg, B.M. (1995) 'Direct dynamics simulation of the impact phase in heel-toe running', *Journal of Biomechanics*, 28: 661–668.
10. Greig, M.P. and Yeadon, M.R. (2000) 'The influence of touchdown parameters on the performance of a high jumper', *Journal of Applied Biomechanics*, 16: 367–378.
11. Gruber, K., Ruder, H., Denoth, J. and Schneider, K. (1998) 'A comparative study of impact dynamics: wobbling mass model versus rigid body models', *Journal of Biomechanics*, 31: 439–444.
12. Goffe, W.L., Ferrier, G.D. and Rogers, J. (1994) 'Global optimisation of statistical functions with simulated annealing', *Journal of Econometrics*, 60: 65–99.
13. Hatze, H. (1980) 'A mathematical model for the computational determination of parameter values of anthropomorphic segments', *Journal of Biomechanics*, 13: 833–843.
14. Hatze, H. (1981) 'A comprehensive model for human motion simulation and its application to the take-off phase of the long jump', *Journal of Biomechanics*, 14: 135–142.
15. Hiley, M.J. and Yeadon, M.R. (2003) 'Optimum technique for generating angular momentum in accelerated backward giant circles prior to a dismount', *Journal of Applied Biomechanics*, 19: 119–130.
16. Hill, A.V. (1938) 'The heat of shortening and the dynamic constants of muscle', *Proceedings of the Royal Society Series B*, 126: 136–195.
17. Hinrichs, R.N. (1985) 'Regression equations to predict segmental MOI from anthropometric measurements: An extension of the data of Chandler *et al.* (1975)', *Journal of Biomechanics*, 18: 621–624.
18. Hubbard, M. (1993) 'Computer Simulation in Sport and Industry', *Journal of Biomechanics*, 26, Supplement 1: 53–61.
19. Hubbard, M. and Alaways, L.W. (1987) 'Optimum release conditions for the new rules javelin', *International Journal of Sport Biomechanics*, 3: 207–221.
20. Hubbard, M. and Alaways, L.W. (1989) 'Rapid and accurate estimation of release conditions in the javelin throw', *Journal of Biomechanics*, 22: 583–596.
21. Hubbard, M., de Mestre, N.J. and Scott, J. (2001) 'Dependence of release variables in the shot put', *Journal of Biomechanics*, 34: 449–456.
22. Hubbard, M. and Trinkle, J.C. (1985) 'Clearing maximum height with constrained kinetic energy', *ASME Journal of Applied Mechanics*, 52: 179–184.
23. King, M.A. and Yeadon, M.R. (2005) 'Factors influencing performance in the Hecht vault and implications for modelling', *Journal of Biomechanics*, 38: 145–151.
24. King, M.A., Wilson, C. and Yeadon, M.R. (2006) 'Evaluation of a torque-driven computer simulation model of jumping for height', *Journal of Applied Biomechanics*, 22: 264–274.
25. King, M.A. and Yeadon, M.R. (2002) 'Determining subject specific torque parameters for use in a torque driven simulation model of dynamic jumping', *Journal of Applied Biomechanics*, 18: 207–217.
26. King, M.A. and Yeadon, M.R. (2003) 'Coping with perturbations to a layout somersault in tumbling', *Journal of Biomechanics*, 36: 921–927.
27. King, M.A. and Yeadon, M.R. (2004) 'Maximising somersault rotation in tumbling', *Journal of Biomechanics*, 37: 471–477.
28. Muramatsu, T., Muraoka, T., Takeshita, D., Kawakami, Y., Hirano, Y. and Fukunaga, T. (2001) 'Mechanical properties of tendon and aponeurosis of human gastrocnemius muscle in vivo', *Journal of Applied Physiology*, 90: 1671–1678.

29. Nelder, J.A. and Mead, R. (1965) 'A simplex method for function minimisation', *Computer Journal*, 7: 308–313.
30. Neptune, R.R. and Hull, M.L. (1998) 'Evaluation of performance criteria for simulation of submaximal steady-state cycling using a forward dynamic model', *Journal of Biomechanical Engineering*, 120: 334–341.
31. Neptune, R.R. and Kautz, S.A. (2000) 'Knee joint loading in forward versus backward pedaling: implications for rehabilitation strategies', *Clinical Biomechanics*, 15: 528–535.
32. Niklas, K.J. (1992) *Plant Biomechanics: an Engineering Approach to Plant Form and Function*, Chicago, IL: University of Chicago Press.
33. Pain, M.T.G. and Challis, J.H. (2001a) 'High resolution determination of body segment inertial parameters and their variation due to soft tissue motion', *Journal of Applied Biomechanics*, 17: 326–334.
34. Pain, M.T.G. and Challis, J.H. (2001b) 'The role of the heel pad and shank soft tissue during impacts: a further resolution of a paradox', *Journal of Biomechanics*, 34: 327–333.
35. Pain, M.T.G. and Challis, J.H. (2006) 'The influence of soft tissue movement on ground reaction forces, joint torques and joint reaction forces in drop landings', *Journal of Biomechanics*, 39: 119–124.
36. Panjabi, M. (1979) 'Validation of mathematical models', *Journal of Biomechanics*, 12: 238.
37. Press, W.H., Flannery, B.P., Teukolsky, S.A. and Vetterling, W.T. (1988) *Numerical Recipes. The Art of Scientific Computing*, Cambridge: Cambridge University Press.
38. Sprigings, E.J. and Miller, D.I. (2004) 'Optimal knee extension timing in springboard and platform dives from the reverse group', *Journal of Applied Biomechanics*, 20: 244–252.
39. Van der Helm, F.C.T. (1994) 'A finite element musculoskeletal model of the shoulder mechanism', *Journal of Biomechanics*, 27: 551–569.
40. Van Gheluwe, B. (1981) 'A biomechanical simulation model for airborne twist in backward somersault', *Journal of Human Movement Studies*, 7: 1–22.
41. Van Soest A.J. and Casius, L.J.R. (2003) 'The merits of a parallel genetic algorithm in solving hard optimization problems', *Journal of Biomechanical Engineering*, 125: 141–146.
42. Westing, S.H., Seger, J.Y., Karlson, E. and Ekblom, B. (1988) 'Eccentric and concentric torque-velocity characteristics of the quadriceps femoris in man', *European Journal of Applied Physiology*, 58: 100–104.
43. Wilson, C., King, M.A. and Yeadon, M.R. (2006) 'Determination of subject-specific model parameter visco-elastic elements', *Journal of Biomechanics*, 39: 1883–1890.
44. Wright, I.C., Neptune, R.R., van den Bogert, A.J. and Nigg, B.M. (1998) 'Passive regulation of impact forces in heel-toe running', *Clinical Biomechanics*, 13: 521–531.
45. Yeadon, M.R. (1990a) 'The simulation of aerial movement – I: The determination of orientation angles from film data', *Journal of Biomechanics*, 23: 59–66.
46. Yeadon, M.R. (1990b) 'The simulation of aerial movement – II: A mathematical inertia model of the human body', *Journal of Biomechanics*, 23: 67–74.
47. Yeadon, M.R. (1993) 'The biomechanics of twisting somersaults. Part I: Rigid body motions', *Journal of Sports Sciences*, 11: 187–198.
48. Yeadon, M.R., Atha, J. and Hales, F.D. (1990) 'The simulation of aerial movement – IV: A computer simulation model', *Journal of Biomechanics*, 23: 85–89.
49. Yeadon, M.R. and Challis, J.H. (1994) 'The future of performance related sports biomechanics research', *Journal of Sports Sciences*, 12: 3–32.
50. Yeadon, M.R. and King, M.A. (2002) 'Evaluation of a torque driven simulation model of tumbling', *Journal of Applied Biomechanics*, 18: 195–206.
51. Yeadon, M.R., King, M.A. and Wilson, C. (2006a) 'Modelling the maximum voluntary joint torque/angular velocity relationship in human movement', *Journal of Biomechanics*, 39: 476–482.
52. Yeadon, M.R., Kong, P.W. and King, M.A. (2006b) 'Parameter determination for a computer simulation model of a diver and a springboard', *Journal of Applied Biomechanics*, 22: 167–176.
53. Zatsiorsky, V.M. (2002) *Kinetics of Human Motion*, Champaign, IL: Human Kinetics.

Section VI

Injury, orthopedics and rehabilitation

27

Lower extremity injuries

William C. Whiting[1] and Ronald F. Zernicke[2]
[1]California State University, Northridge; [2]University of Michigan, Ann Arbor

Introduction

Lower-extremity injuries are significant, given the important role of the lower extremities in walking, running, and jumping. Representative injuries are presented in this chapter for the major lower-extremity joints (hip, knee, and ankle) and the segments spanning those joints (thigh, lower leg, and foot), with emphasis on mechanisms of injury.

Hip injuries

Hip fracture: Bone fractures in the hip and pelvis arise from high-energy trauma such as with vehicle crashes and falls. Pelvic fractures, while not as common as femoral fractures, constitute a significant problem. More than 65,000 pelvic fractures happen in the US annually. Nearly 25 per cent of pelvic fractures result from motor vehicle crashes (Moffatt *et al.*, 1990). In motor vehicle crashes the direction of force determines the injury patterns, with pelvic fractures significantly more frequent in side-impact collisions (Gokcen *et al.*, 1994).

With more than 350,000 fractures occurring annually in the US, proximal hip (femoral) fractures are a major health concern (AAOS, 2007a). Worldwide, hip fracture affects millions and is a significant cause of morbidity and mortality (Johnell and Kanis, 2004). The situation becomes more serious because many hip fracture victims, especially older persons, die within a year of injury, usually due to complications and chronic conditions that worsen after the injury. (Davidson *et al.*, 2001), for example, reported a 12-month mortality rate after hip fracture of 26 per cent.

Risk factors for hip fracture include decreased bone mass and bone mineral density, falls, small stature, decreased muscular strength, physical inactivity, chronic illnesses, drugs, and cognitive, visual, or perceptual impairment (Marks *et al.*, 2003).

Hip fracture risk increases dramatically with advancing age. Hip fractures in the elderly are associated with falls caused by unsteady gait or tripping. In most cases, the impact force on landing from the fall causes the fracture; rarely does spontaneous hip fracture cause the fall. Fall risk is multidimensional, with common risk factors including chronic illnesses, compromised strength, balance, coordination and reflexes, dizziness, postural hypotension, fainting, history of falls, environmental factors, and neurological, cerebrovascular, cardiovascular and cognitive disorders (Cummings *et al.*, 1985; Marks *et al.*, 2003; Rubenstein and Josephson, 2002).

Hip osteoarthritis: Osteoarthritis (OA), also termed degenerative joint disease, is the most common form of arthritis. Osteoarthritis of the hip is a major cause of disability, particularly in older persons. As a major load-bearing joint, the hip is exposed to high mechanical loads and consequent risk for OA.

Hip OA affects an estimated 1.5 per cent of adults in the US, and up to 5 per cent of those between the ages of 55–74 (AAOS, 2007b). While the exact mechanism remains unclear, age is greatest risk factor for hip OA. One possible explanation involves a molecular mechanism in which advanced glycation end products (AGEs) accumulate in the cartilage collagen. AGE cross-linking increases collagen network stiffness and may reduce the network's ability to resist damage (Verzijl *et al.*, 2003; Verzijl *et al.*, 2002). Other risk factors for OA include higher-than-normal bone mineral density, developmental deformities, ethnicity, gender, genetic predisposition, joint loading during physical activity, nutrition, obesity, smoking and traumatic injury (AAOS, 2007b).

Thigh injuries

Femoral fracture: Femoral neck fractures are an urgent health concern in older people. Fracture to other areas of the femur is also of consequence. Femoral shaft fractures, for example, number nearly 30,000 in the US annually (DeCoster and Swenson, 2002). Most femoral shaft fractures result from high-energy trauma. Winquist *et al.*, 1984, reported that nearly 78 per cent of 520 femoral fractures studies arose from automobile, motorcycle, or automobile-pedestrian accidents.

Skiing provides an example of femoral fracture mechanisms. Femoral fractures depend on skier ability, snow conditions, physical conditioning, age, and injury mechanism. Sterett and Krissoff (1994) examined 85 cases of femoral fractures in alpine skiing with respect to injury mechanisms as a function of skier age. In the youngest age group (3–18 years), femoral fracture tended to result from torsional loading of the femoral shaft, usually while skiing fast and catching a ski in heavy or wet snow. In young adults (18–45 years), fracture mostly resulted from high-energy, direct-impact collisions (e.g., with a rock or tree) and resulted in bone-shattering comminuted fractures. In older skiers (>45 years), most fractures were localized in the hip area (e.g., femoral neck) and were caused by low-energy falls on firm snow. Femoral fracture is one of the few injuries more common in advanced skiers than in beginners. This is due primarily to the high energy required for femoral fracture. Advanced skiers typically ski at higher speeds and consequently have higher kinetic energy that may be transferred to the femur at impact.

Hamstring strain: Muscle strain involves injury to a musculotendinous complex and typically occurs during eccentric muscle action used to decelerate high-velocity movements (e.g., sprinting). At the muscle level, animal studies implicate excessive sarcomere strain as the primary cause of injury (Lieber and Friden, 2002). Many risk factors for muscle strain have been identified. These include muscle imbalance, lack of flexibility, fatigue, insufficient warm-up, age, history of injury, muscle weakness, poor training, use of inappropriate drugs, presence of scar tissue, and incomplete or aggressive rehabilitation (Croisier, 2004; Garrett *et al.*, 1987; Verrall *et al.*, 2003; Worrell, 1994).

Certain muscles are more prone to strain injury than others. The hamstrings, for example, are particularly susceptible to muscle strain. In a study of strain injuries in Australian football, 69 per cent involved the hamstring group (Orchard, 2001). Why? The hamstring group muscles (with the exception of the short head of the biceps femoris) are biarticular. This structural arrangement results in muscle length being determined by the combined action of the knee and hip joints. Knee extension and hip flexion both act to lengthen the semitendinosus, semimembranosus, and biceps femoris (long head). Simultaneous knee extension and hip flexion lengthen the hamstrings and contribute to the muscles' risk of injury.

Hamstring strains commonly recur, often due to premature return to action, weakened tissues, strength deficits, and altered mechanical characteristics. Recent studies suggest that previously injured hamstring muscles reach their peak torque at significantly shorter lengths than do uninjured muscles (Brockett *et al.*, 2004; Proske *et al.*, 2004). This shift in torque-length profile may predispose the hamstring muscles to recurrent injury.

Knee injuries

Cruciate ligament sprain: Growing participation in physical exercise and sports in recent years, especially by girls and women, has been accompanied by increased incidence of cruciate injuries. Anterior cruciate ligament (ACL) injuries are more common than posterior cruciate ligament (PCL) injuries. Due to the multifactorial nature of ACL pathology, ACL injury rates are both sport and gender specific (e.g., Agel *et al.*, 2005; Bjordal *et al.*, 1997; Bradley *et al.*, 2002).

Commonly, injury to the ACL happens in response to valgus loading combined with external tibial rotation (valgus-rotation mechanism) or to knee hyperextension combined with internal tibial rotation. Such combined loading places the ACL at particular risk (Markolf *et al.*, 1995). *Non-contact* ACL injuries are due to valgus-rotation, which can happen when the foot is planted on the ground, with the tibia externally rotated, knee near full-extension, and subsequent knee collapse into further valgus (Myer *et al.*, 2005). The situation worsens if, while the foot is in ground contact, an external force is applied to the knee in what is termed a *contact* ACL injury. Contact ACL injuries are common in sports such as soccer, rugby, and American football. ACL failure resulting from knee hyperextension-internal tibial rotation is less common, but may predominate (e.g., gymnasts and basketball players) when the knee violently hyperextends upon landing from a jump.

Downhill skiing provides two mechanisms for ACL injury. The first involves a backward fall in which the skis and boots accelerate forward relative to the body and pull the tibia forward. This creates an anterior drawer (i.e., anterior tibial translation) and possible ACL failure (a boot-induced ACL injury).

The second mechanism of ACL injury unique to skiing is described as the phantom foot in which the rear section of the ski effectively forms a 'foot' in the posterior direction. In a backward fall, this phantom foot levers the flexed knee into internal tibial rotation and amplifies stress in the ACL.

Injury to the ACL may be an isolated injury or may happen with damage to other structures. An example of such a combination injury is one known as the 'unhappy triad', in which the ACL, medial collateral ligament (MCL), and medial meniscus are conjointly damaged during valgus-rotation.

Though injured less frequently than the ACL, the PCL also is of clinical significance. The causes of PCL injury vary but about half of the cases are due to trauma resulting from motor vehicle accidents (MVA). Most of the remaining cases happen during sporting activities (40 per cent) and industrial accidents (10 per cent). MVA and industrial accidents typically involve high-energy dynamics, while sport-related injuries are considered low-energy.

The mechanisms of PCL failure are related to its role in limiting posterior tibial movement relative to the femur, and secondarily in restricting hyperflexion and hyperextension, and stabilizing the femur during weight bearing on a flexed knee. Some PCL injury mechanisms (Figure 27.1) are common, while others are rare (Andrews *et al.*, 1994).

Most PCL injuries involve posteriorly directed tibial forces. The level of posterior tibial force transmitted to the PCL depends on knee flexion angle, with greater knee flexion angles associated with higher posterior tibial force in the PCL (Markolf *et al.*, 1997). Adding a valgus torque to the posterior tibial forces increases PCL forces and further increases injury risk.

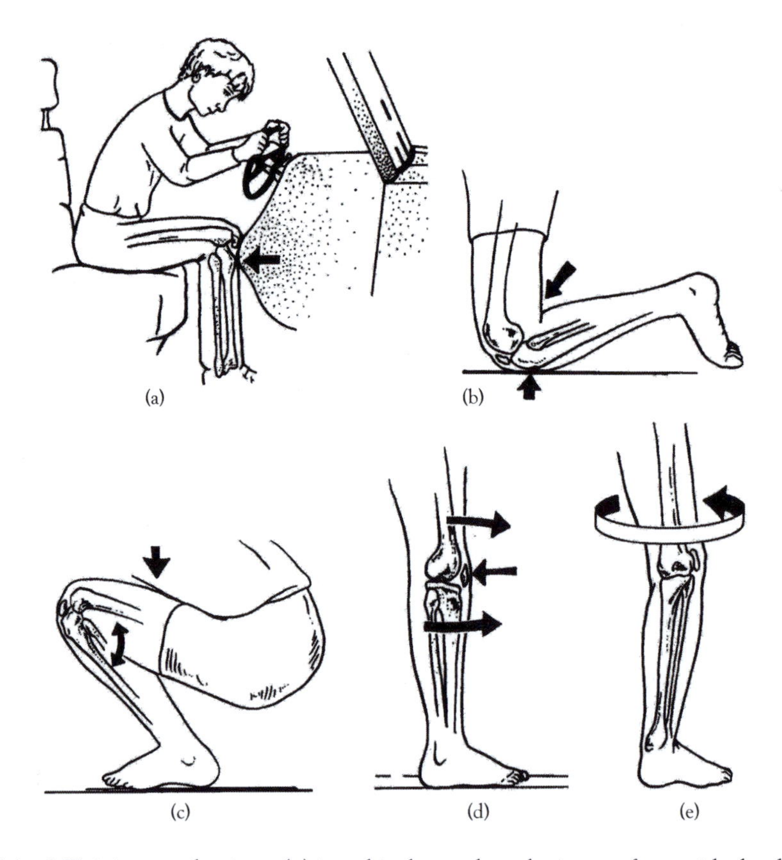

Figure 27.1 PCL injury mechanisms: (a) in vehicular crashes, the impact force with the dashboard pushes the tibia posteriorly causing PCL rupture; (b) falling on a flexed knee with impact on the tibial tuberosity driving the proximal tibia posteriorly; (c) forcible knee flexion; (d) violent knee hyperextension; and (e) rapid weight transfer from one foot to another while rotating quickly on a minimally flexed knee. From Whiting and Zernicke (1998). Reprinted by permission.

Meniscus injury: The menisci play an essential role in maintaining normal knee function. Physiologically, meniscal compression helps distribute nutrients. Mechanically, the menisci are involved in weight bearing, shock absorption, stabilization, and rotational facilitation.

The magnitude of joint compressive forces accepted by the menisci depends on joint position. In full extension the menisci accept 45–50 per cent of the load, and in 90° of flexion they accept 85 per cent of the load (Ahmed and Burke, 1983; Ahmed *et al.*, 1983). Load distribution also differs between the medial and lateral menisci. The lateral meniscus assumes 70 per cent of load transmission, while the medial meniscus shares the compressive load equally with the articular cartilage (Seedhom and Wright, 1974; Walker and Erkman, 1975).

Meniscal injury is either traumatic or degenerative. Traumatic injuries happen from acute insult to the meniscus and are usually seen in young and active persons. Degenerative meniscal tears develop chronically, usually in older persons, and typically result from simple movements (e.g., deep knee bends) that load weakened meniscal tissue.

Meniscal damage usually happens when the meniscus is subjected to combined flexion-rotation or extension-rotation during weight bearing. These movements create injury-causing shear forces between the tibial and femoral condyles (Siliski, 2003; Silbey and Fu, 2001).

Due to structural considerations, the medial meniscus is five times more likely to be injured than the lateral meniscus. At times, loads during rapid knee extension are large enough to cause a vertical longitudinal tear that extends into the anterior horn (i.e., bucket-handle tear). More frequently, however, a bucket-handle tear results from repeated insult to a partial tear that propagates to span a substantial portion of the meniscus (Figure 27.2).

Meniscal tears are associated with certain sports. In soccer, for example, players frequently collide with opponents and change direction with their cleats embedded in the turf. Meniscal injuries also are common in track and field (e.g., knee torsion in the discus throw) and skiing (e.g., sudden twisting of the knee). Occupations involving repeated squatting also can lead to meniscus injury, often due to the degenerative processes that accompany prolonged and repeated knee flexion.

Collateral ligament sprain: The knee, because of its poor bony fit, relies on ligaments for structural support. The primary ligaments are the cruciates (ACL and PCL) and the collateral ligaments, specifically the medial collateral ligament (MCL) and lateral collateral ligament (LCL). The MCL is a capsular ligament that connects the medial femoral epicondyle with the superomedial surface of the tibia. As a capsular ligament, the MCL has direct connection to the joint capsule and the medial meniscus. The LCL, in contrast, is extra-capsular and connects the lateral epicondyle of the femur with the lateral surface of the fibular head.

Injury to both ligaments usually results from sudden, violent loading and is more common in the MCL. Medial collateral ligament damage usually is due to impact on the lateral side of the knee that causes knee valgus and consequent tensile loading of the medial structures. The MCL most effectively resists valgus loading when the knee is flexed 25–30° (Swenson and Harner, 1995). At full knee extension other structures play a greater role in knee stability. Many studies (e.g., Grood *et al.*, 1981; Piziali *et al.*, 1980; Seering *et al.*, 1980) have confirmed the role of the MCL in resisting valgus loading. The MCL acts as the primary valgus restraint, with only secondary involvement provided by the cruciate ligaments. In cases of isolated MCL failure, however, residual structures such as the ACL resist valgus loading (Inoue *et al.*, 1987).

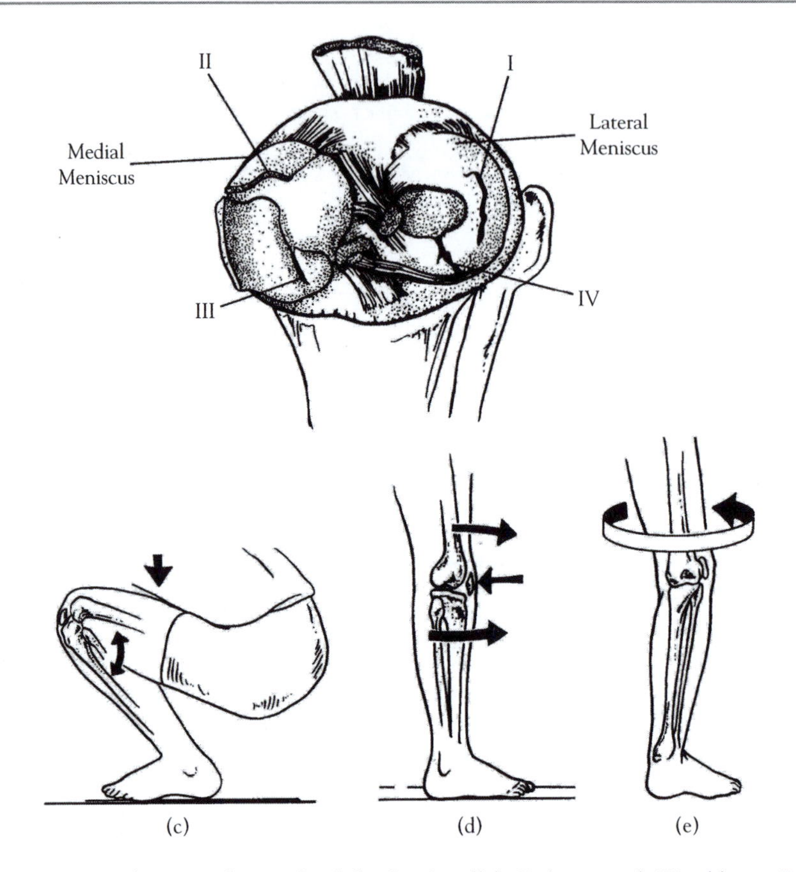

Figure 27.2 Meniscal tears: I, longitudinal (bucket-handle); II, horizontal; III, oblique; IV, radial. From Whiting and Zernicke (1998). Reprinted by permission.

LCL injury usually is due to varus loading, often in combination with knee hyperextension. Varus loading from impact to the knee's medial aspect while the foot is planted on the ground creates tensile forces in the lateral structures, including the LCL. Because of its extracapsular location, the LCL is more likely than the MCL to sustain isolated injury. Conjoint injury to the LCL and one or both of the cruciate ligaments is not uncommon.

Knee extensor disorders

The knee joint complex provides the middle link in the lower extremity's kinetic chain. An important component of the knee complex is the knee extensor mechanism (KEM), which consists of the quadriceps muscle group, PF joint, and tendons (quadriceps and patellar) connecting these structures. The patella serves as the central structure in the KEM, acting as a fulcrum to enhance the mechanical advantage of the quadriceps and displace the tendon line of action away from the knee joint axis, thereby increasing the moment arm.

Patellofemoral disorders: Patellofemoral pain is a common lower extremity pathology, but controversy persists regarding its causation, evaluation and treatment. Effective evaluation of PFP should not be limited solely to consideration of knee structure and function, but also should include actions of the hip/pelvis and foot/ankle complexes (Powers, 2003).

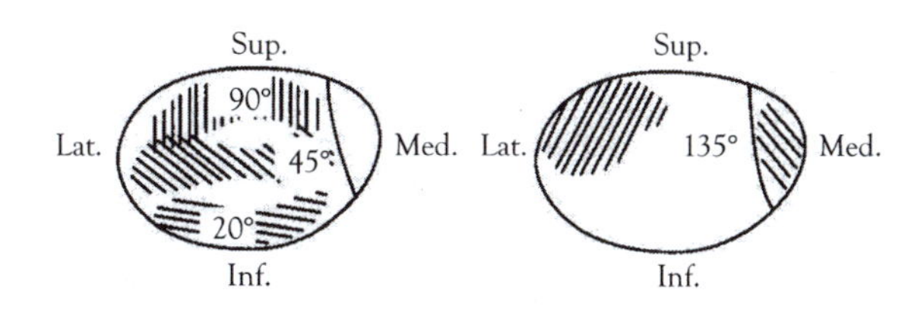

Figure 27.3 Patellofemoral contact areas as a function of knee flexion angle. From Whiting and Zernicke (1998). Reprinted by permission.

When the knee is fully extended the patella rides high on the femur. As the knee flexes, the patella slides in the intercondylar groove in a movement termed patellar tracking. Effective patellar tracking depends on structural congruence between the patella and femur, the net effect of muscle forces (i.e., vastus medialis, vastus lateralis, vastus intermedius, and rectus femoris) and other structural considerations (e.g., Q-angle). As the patella tracks, contact pressures develop between the patella and femur. These pressures, determined by the contact forces and areas, vary with knee flexion (Figure 27.3). As the knee flexes, PF joint contact area increases and the knee flexes from full-extension, with most of the increase happening in the first 45–60° of flexion (Salsich *et al.*, 2003). In addition to increasing in size, the contact area migrates superiorly on the retropatellar surface as the patella slides deeper into the intercondylar groove (Figure 27.3).

Patellofemoral contact area and joint stress are clinically relevant. For example, higher PF joint stress (due to smaller contact area) has been reported in persons with PFP compared to subjects without PFP (Brechter and Powers, 2002). In addition, persons with patella alta have significantly less contact area and greater PF stress than controls (Ward and Powers, 2004). Despite extensive study, the precise relation remains elusive between patellofemoral movement and patellofemoral pain.

Quadriceps tendon and patellar tendon ruptures: Extreme acute forces or chronic trauma to a weakened KEM may lead to tendon rupture. Quadriceps tendon rupture usually happens to people over 40 years old as a focal lesion at the osteotendinous junction at the superior pole of the patella. In contrast, patellar tendon ruptures tend to happen before the age of 40 years, and most often rupture at the inferior patellar pole.

The injury mechanism typically involves violent quadriceps muscle action against resistance at the knee. Some tendon injuries have identifiable mechanisms (e.g., jumping or tripping) and predisposing risk factors (e.g., systemic disease or steroid use), while others are of unknown origin.

In rare instances, circumstances permit evaluation of an actual injury. One such case, reported by (Zernicke *et al.*, 1977), involved rupture of a human patellar tendon during sports competition. A world-class weight lifter was attempting to clean-and-jerk 3770 N (385 lb). At the beginning of the second phase of the lift, his right patellar tendon fully ruptured at the distal patellar pole and the patellar-tendon insertion on the tibia. Using a biomechanical analysis, they calculated a tensile force in the tendon at the instant of rupture of 14.5 kN (3258 lb) – more than 17.5 times the lifter's body weight.

Lower-leg injuries

Compartment syndrome: Acute injury or chronic exertion can induce fluid accumulation (hemorrhage, edema, or both) within muscle compartments. The excess fluid increases intracompartmental pressure and creates a compartment syndrome (CS). The mechanism of CS may be either acute or, more commonly, chronic. Specific causal agents include soft tissue contusion, blunt force trauma, bleeding disorders, venous blockage, arterial occlusion, burn, prolonged compression after drug overdose, surgery, apparel, or exercise.

Any of the lower leg's four muscle compartments (i.e., anterior, lateral, superficial posterior, or deep posterior) may be affected. Most commonly, the anterior compartment is involved with all muscles in the compartment (i.e., tibialis anterior, extensor hallucis longus, extensor digitorum longus, and peroneus tertius) affected (Schepsis *et al.*, 2005).

Increased compartmental pressure affects vascular and neural function and may presage ischemia and a negative feedback cycle of fluid accumulation and restricted flow. The situation is worsened by the surrounding fascia's mechanical properties, as fascia becomes thicker and stiffer in response to CS (Hurschler *et al.*, 1994). The situation is exacerbated by a decrease in compartment volume that may be caused by wearing tight clothing or compression wraps. Sufficiently high compartment pressures result in vessel closure and consequent catastrophic physiological consequences. Venous collapse dramatically reduces blood return and leads to capillary congestion and decreased tissue perfusion. Resulting tissue hypoperfusion may lead to ischemia and eventual necrosis.

Tibial stress reaction and stress fracture: Bone normally responds to repetitive loading by structurally adapting consistent with Wolff's law. This adaptive remodelling process results in net bone resorption under low-loading conditions and net deposition in regions of high- and repeated-loading where added bone is needed to sustain the mechanical loads. However, if load magnitude and frequency exceed bone's ability to adapt, injury may occur. The most common form of bone injury is fracture. Bone fracture may happen acutely as with traumatic fractures, or in response to chronic loading (e.g., stress fracture). Chronic fractures are either caused by a sudden increase in activity, as seen in military recruits (fatigue fractures), or in persons with no increase in activity, but with decreased bone density (insufficiency fractures).

Detectable bone stress fractures are typically preceded by increased activity in the bone. The term *stress reaction* describes areas of bone with such evidence of remodelling but with an absence of radiological evidence of fracture (e.g., radiographs, bone scans, and magnetic resonance imaging).

The tibia accounts for up to 50 per cent of all stress fractures. Most long-bone stress fractures are oriented transversely to the long axis of the bone. Longitudinally-oriented stress fractures are less common and when present usually are found on the anterior cortex of the distal tibia (Tearse *et al.*, 2002).

Fracture site appears to be activity dependent with the mechanical demands of specific movements playing an important role in determining fracture location. Runners most often exhibit fractures between the middle and distal thirds of the tibia. Athletes in jumping sports (e.g., volleyball and basketball) tend to have proximal fractures. Dancers, in contrast, sustain more mid-diaphyseal stress fractures.

Ankle and foot injuries

The foot and ankle's many bones, ligaments, and articulations, together with its load-bearing function, make this area particularly susceptible to injury. The determining factors in ankle-foot injuries include joint position, the magnitude, direction, and rate of applied forces, and the resistance provided by joint structures. The mechanisms involved in ankle-foot injuries include walking on uneven surfaces, stepping in holes, rolling the ankle during a cutting manoeuvre, or landing on another player's foot when descending from a jump in sporting events. Resulting injuries range from fracture-dislocation to ligamentous damage (sprain).

Ankle sprain: The ankle joint's instability and load-bearing function contribute to its being the most frequently injured joint (e.g., basketball).

The ankle joint comprises the tibia, fibula, and talus, and is reinforced by numerous ligaments. The body of the talus is wedge-shaped, with its anterior portion wider than its posterior. This asymmetry directly contributes to the ankle's position-dependent stability. In extreme dorsiflexion, the wider part of the talus tightly wedges between the malleoli and stabilizes the joint. As the ankle plantar flexes, the narrower portion of the talus rotates between the malleoli. The looser talar fit in extreme plantar flexion permits talar translation and tilt that contributes to lateral ankle instability in this position.

An 'ankle sprain' actually involves both the ankle and subtalar (talocalcaneal) joints. These two joints move synchronously to execute combined ankle-foot movement. The majority (~85 per cent) of ankle sprains result from foot-ankle supination (i.e., combined ankle plantar flexion, subtalar inversion, and internal rotation of the foot). Though often referred to as *inversion sprains*, these injuries are more appropriately termed *lateral ankle sprains.*

Lateral ankle sprains exhibit sequential ligament failure. The anterior talofibular ligament (ATFL) fails first because of its orientation at the instant of loading and its inherent weakness (Siegler *et al.*, 1988). When the ankle is plantar flexed (as in ankle-foot supination), the ATFL aligns with the fibula and acts as a collateral ligament (Carr, 2003). This alignment and the ATFL's structural weakness predispose the ATFL to injury. The calcaneofibular ligament (CFL) typically tears next, followed rarely by failure of the posterior talofibular ligament (PTFL).

The opposite movement pattern creates an *eversion sprain* (actually *pronation*, or combined ankle dorsiflexion, subtalar eversion, and lateral rotation of the foot), which is more accurately termed a *medial ankle sprain*. Given the deltoid ligament's inherent strength, medial ankle sprains are both less frequent and less severe than lateral sprains. Deltoid ligament rupture is rare and always happens in conjunction with other ligament tears. In medial sprains, the talus is forced against the lateral malleolus, which may result in malleolar fracture.

Applied forces can wedge the fibula apart from the tibia with sufficient force to tear the interosseous membrane and tibiofibular ligaments in what is termed a high ankle sprain. The most likely mechanisms for high ankle sprain are talar torsion and forced ankle dorsiflexion.

Calcaneal tendon pathologies: The calcaneal (Achilles) tendon is formed by confluence of the distal tendons of the gastrocnemius and soleus about 5–6 cm proximal to its insertion site on the posterior aspect of calcaneus. The calcaneal tendon transmits substantial forces developed by the triceps surae muscle group (i.e., gastrocnemius and soleus). For example, Fukashiro *et al.* (1995) reported peak Achilles tendon force of 2233 N (502 lb) in the squat jump, 1895 N (426 lb) in the countermovement jump, and 3786 N (851 lb) in hopping. Giddings *et al.* (2000) predicted maximal Achilles tendon force 3.9 BW (body weight) for walking and 7.7 BW for running, with the peak loads at 70 per cent of the stance phase for walking and 60 per cent of stance for running. Pourcelot *et al.* (2005) found peak tendon forces of about 850N (191 lb) during the stance phase of walking.

Despite being the longest and strongest tendon in the body, the calcaneal tendon is predisposed to overuse pathologies due to its frequent and repeated loading. These pathologies include peritenonitis, bursitis, myotendinous junction injury, or tendonopathies (Kvist, 1994).

Tendon degeneration may lead to tendon rupture. Calcaneal tendon ruptures most often are seen in sedentary, 30–40-year-old men who suddenly exert themselves in a sporting task that involves running, jumping, or rapid change of direction (Jarvinen *et al.*, 2001; Schepsis *et al.*, 2002; Yinger *et al.*, 2002). While such tendon ruptures may happen spontaneously, tendon rupture often is secondary to degenerative processes. Four primary mechanisms have been implicated in calcaneal tendon rupture (Mahan and Carter, 1992): (1) sudden dorsiflexion of a plantar flexed foot; (2) pushing off the weight-bearing foot while extending the ipsilateral knee joint; (3) sudden excess tension on an already taut tendon; and (4) a taut tendon struck by a blunt object (Figure 27.4).

Calcaneal tendon ruptures about 2–6 cm proximal to the calcaneal insertion. An avascular region lies close to the calcaneal insertion site, and regional differences in vascular density (Zantop *et al.*, 2003) exist along the tendon length. The middle part of the calcaneal tendon has a much lower vascular density than either the proximal or distal regions. This regional hypovascularity (and resulting hypoxia) may be a predisposing factor for calcaneal tendon degeneration and rupture.

Plantar fasciitis: The plantar fascia is a tough, fibrous band of connective tissue running along the plantar surface of the foot, parallel to the longitudinal arch. Compressive forces associated with load-bearing activities (e.g., running) flatten the longitudinal arch of the foot and stretch the plantar fascia. Repeated loading can cause microtears or partial rupture of fascial fibres (in the midfoot or at the insertion site on the medial tuberosity of the calcaneus) with inflammation – plantar fasciitis (PF). Plantar fascia forces during running have been estimated at 1.3–2.9 times body weight (Scott and Winter, 1990). Lack of flexibility exacerbates plantar fasciitis. Calcaneal tendon tightness limits ankle dorsiflexion and results in greater plantar fascial stress. Ankle strength and flexibility deficits have been reported in the symptomatic feet (Kibler *et al.*, 1991).

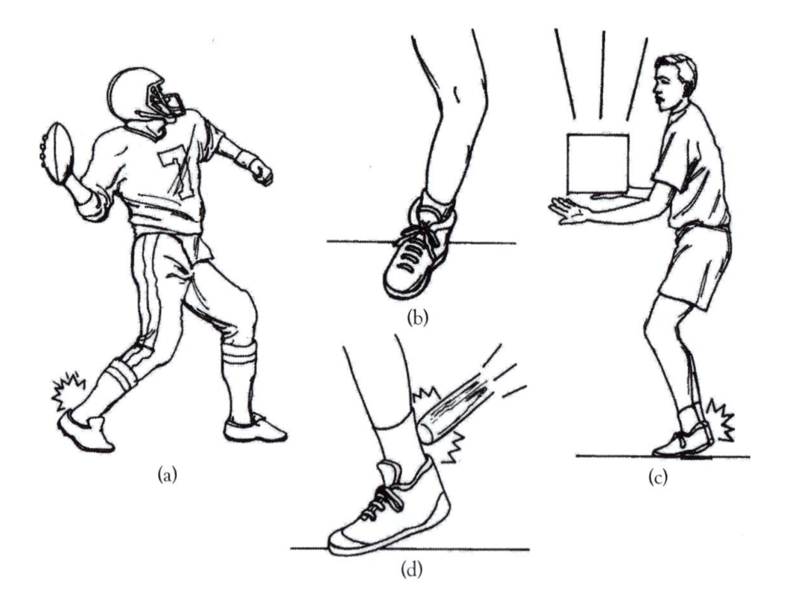

Figure 27.4 Mechanisms of calcaneal tendon rupture. (a) rapid dorsiflexion of the ankle; (b) cutting manoeuvre with quick change of direction; (c) catching a falling object; and (d) blunt force trauma to a taut calcaneal tendon. From Whiting and Zernicke (1998). Reprinted by permission.

Other factors associated with PF include overtraining, limb length discrepancies, fatigue, fascial inextensibility, and poor movement mechanics. Excessive pronation during running can contribute to PF. During pronation the subtalar joint everts and results in stretching of the plantar fascia. This stretching increases tissue stress. Repeated stretching can lead to microdamage and inflammation.

Conclusion

Injuries to the lower extremities are common and affect posture and many loading-bearing tasks. Greater participation in physical activities, particularly competitive and recreational sports, has contributed to the increased incidence of lower extremity injuries. The increase has been especially profound in girls and women. Demographic and sociological changes (e.g., an ageing population) portend future increases in certain lower extremity injuries (e.g., hip fractures in older persons).

References

1. Agel, J., Arendt, E. and Bershadsky, B. (2005) 'Anterior cruciate ligament injury in National Collegiate Athletic Association basketball and soccer', *American Journal of Sports Medicine*, 33: 524–31.
2. Ahmed, A.M. and Burke, D.L. (1983) 'In-vitro measurement of static pressure distribution in synovial joints: Part I. Tibial surface of the knee', *Journal of Biomechanical Engineering*, 105: 216–25.
3. Ahmed, A.M., Burke, D.L. and Yu, A. (1983) 'In-vitro measurement of static pressure distribution in synovial joints: Part II. Retropatellar surface', *Journal of Biomechanical Engineering*, 105: 226–36.
4. American Academy of Orthopaedic Surgeons. (2007a) Online. Available HTTP: <http://orthoinfo.aaos.org/fact/thr_report.cfm?Thread_ID=77&topcategory=Hip> (accessed 29 May 2007).
5. American Academy of Orthopaedic Surgeons. (2007b) Available HTTP: <http://www.aaos.org/Research/documents/OAinfo_hip_state.pdf> (accessed 29 May 2007).
6. Bjordal, J., Arnly, F., Hannestad, B. and Strand,T. (1997) 'Epidemiology of anterior cruciate ligament injuries in soccer', *American Journal of Sports Medicine*, 25: 341–5.
7. Bradley, J., Klimkiewicz, J., Rytel, M. and Powell, J. (2002) 'Anterior cruciate ligament injuries in the National Football League: epidemiology and current treatment trends among team physicians', *Arthroscopy*, 18: 502–9.
8. Brechter, J.H., Powers, C.M., Terk, M.R., Ward, S.R. and Lee, T.Q. (2003) 'Quantification of patellofemoral joint contact area using magnetic resonance imaging', *Magnetic Resonance Imaging*, 21: 955–9.
9. Brockett, C.L., Morgan, D.L. and Proske, U. (2004) 'Predicting hamstring strain injury in elite athletes', *Medicine and Science in Sports and Exercise*, 36: 379–87.
10. Carr, J.B. (2003) 'Malleolar fractures and soft tissue injuries of the ankle'. in B.D. Browner, J.B. Jupiter, A.M. Levine and P.T. Trafton (eds.) *Skeletal Trauma: Basic Science, Management, and Reconstruction*, Philadelphia: Saunders.
11. Croisier, J.L., Forthomme, B., Namurois, M.H., Vanderthommen, M. and Crielaard, J.M. (2002) 'Hamstring muscle strain recurrence and strength performance disorders', *American Journal of Sports Medicine*, 30: 199–203.
12. Cummings, S.R., Kelsey, J.L., Nevitt, M.C. and O'Dowd, K.J. (1985) 'Epidemiology of osteoporosis and osteoporotic fractures', *Epidemiologic Review*, 7: 178–208.
13. Davidson, C.W., Meriles, M.J., Wilkinson, T.J., McKie, J.S. and Gilchrist, N.L. (2001) 'Hip facture mortality and morbidity—can we do better?', *New Zealand Medical Journal*, 114: 329–32.

14. DeCoster, T.A. and Swenson, D.R. (2002) 'Femur shaft fractures', in C. Bulstrode, J. Buckwalter, A. Carr, L. Marsh, J. Fairbank, J. Wilson-MacDonald and G. Bowden (eds.) *Oxford Textbook of Orthopedics and Trauma*, Oxford, UK: Oxford University Press.
15. Felson, D.T., Anderson, J.J., Naimark, A., Walker, A.M. and Meenan, R.F. (1988) 'Obesity and knee osteoarthritis: The Framingham study', *Annals of Internal Medicine*, 109: 18–24.
16. Fukashiro, S., Komi, P.V., Jarvinen, M. and Miyashita, M. (1995) 'In vivo Achilles tendon loading during jumping in humans', *European Journal of Applied Physiology and Occupational Physiology*, 71: 453–8.
17. Garrett, W.E., Jr., Safran, M.R., Seaber, A.V., Glisson, R.R. and Ribbeck, B.M. (1987) 'Biomechanical comparison of stimulated and nonstimulated skeletal muscle pulled to failure', *American Journal of Sports Medicine*, 15: 448–54.
18. Giddings, V.L., Beaupre, G.S., Whalen, R.T. and Carter, D.R. (2000) 'Calcaneal loading during walking and running', *Medicine & Science in Sports & Exercise*, 32: 627–34.
19. Gokcen, E.C., Burgess, A.R., Siegel, J.H., Mason-Gonzalez, S., Dischinger, P.C. and Ho, S.M. (1994) 'Pelvic fracture mechanism of injury in vehicular trauma patients', *Journal of Trauma*, 36: 789–96.
20. Grood, E.S., Noyes, F.R., Butler, D.L. and Suntary, W.J. (1981) 'Ligamentous and capsular restraints preventing straight medial and lateral laxity in intact human cadaver knees', *Journal of Bone and Joint Surgery*, 63A: 1257–69.
21. Inoue, M., McGurk-Burleson, E., Hollis, J.M. and Woo, S.L.Y. (1987) 'Treatment of the medial collateral ligament injury', *American Journal of Sports Medicine*, 15: 15–21.
22. Jarvinen, T.A., Kannus, P., Paavola, M., Jarvinen, T.L., Jozsa, L. and Jarvinen, M. (2001) 'Achilles tendon injuries', *Current Opinions in Rheumatology*, 13: 150–5.
23. Johnell, O. and Kanis, J.A. (2004) 'An estimate of the worldwide prevalence, mortality and disability associated with hip fracture', *Osteoporosis International*, 15: 897–902.
24. Kibler, W.B., Goldberg, C. and Chandler, T.J. (1991) 'Functional biomechanical deficits in running athletes with plantar fasciitis', *American Journal of Sports Medicine*, 19: 66–71.
25. Kvist, M. (1994) 'Achilles tendon injuries in athletes', *Sports Medicine*, 18: 173–201.
26. Lieber, R.L. and Friden, J. (2002) 'Mechanisms of muscle injury gleaned from animal models', *American Journal of Physical Medicine and Rehabilitation*, 81(11 Suppl): S70–9.
27. Mahan, K.T. and Carter, S.R. (1992) 'Multiple ruptures of the tendo Achillis', *Journal of Foot Surgery*, 31: 548–59.
28. Markolf, K.L., Burchfield, D.M., Shapiro, M.M., Shepard, M.F., Finerman, G.A. and Slauterbeck, J.L. (1995) 'Combined knee loading states that generate high anterior cruciate ligament forces', *Journal of Orthopaedic Research*, 13: 930–5.
29. Markolf, K.L., Slauterbeck, J.R., Armstrong, K.L., Shapiro, M.S. and Finerman, G.A. (1997) 'A biomechanical study of replacement of the posterior cruciate ligament with a graft. Part II: Forces in the graft compared with forces in the intact ligament', *Journal of Bone and Joint Surgery*, 79A: 381–6.
30. Marks, R., Allegrante, J.P., MacKenzie, C.R. and Lane, J.M. (2003) 'Hip fractures among the elderly: causes, consequences and control', *Ageing Research Reviews*, 2: 57–93.
31. Moffatt, C.A., Mitter, E.L. and Martinez, R. (1990) 'Pelvic fractures crash vehicle indicators', *Accident Analysis and Prevention*, 22: 561–9.
32. Myer, G., Ford, K., and Hewett, T. (2005) 'The effects of gender on quadriceps muscle activation strategies during a maneuver that mimics a high ACL injury risk position', *Journal of Electromyographical Kinesiology*, 115: 181–9.
33. Orchard, J.W. (2001) 'Intrinsic and extrinsic risk factors for muscle strains in Australian football', *American Journal of Sports Medicine*, 29: 300–3.
34. Piziali, R.L., Rastegar, J., Nagel, D.A. and Schurman, D.J. (1980) 'The contribution of the cruciate ligaments to the load-displacement characteristics of the human knee joint', *Journal of Biomechanical Engineering*, 102: 277–83.
35. Pourcelot, P., Defontaine, M., Ravary, B., Lemâtre, M. and Crevier-Denoix, N. (2005) 'A noninvasive method of tendon force measurement', *Journal of Biomechanics*, 38: 2124–9.

36. Powers, C.M. (2003) 'The influence of altered lower-extremity kinematics on patellofemoral joint dysfunction: a theoretical approach', *Journal of Orthopaedic & Sports Physical Therapy*, 33: 639–46.
37. Proske, U., Morgan, D.L., Brockett, C.L. and Percival, P. (2004) 'Identifying athletes at risk of hamstring strains and how to protect them', *Clinical Experiments in Pharmacology and Physiology*, 31: 546–50.
38. Rubenstein, L.Z. and Josephson, K.R. (2002) 'The epidemiology of falls and syncope', *Clinics in Geriatric Medicine*, 18: 141–58.
39. Salsich, G.B., Ward, S.R., Terk, M.R. and Powers, C.M. (2003) 'In vivo assessment of patellofemoral joint contact area in individuals who are pain free', *Clinical Orthopaedics and Related Research*, 417: 277–84.
40. Schepsis, A.A., Fitzgerald, M. and Nicoletta, R. (2005) 'Revision surgery for exertional anterior compartment syndrome of the lower leg', *American Journal of Sports Medicine*, 33: 1040–7.
41. Schepsis, A.A., Jones, H. and Haas, A.L. (2002) 'Achilles tendon disorders in athletes', *American Journal of Sports Medicine*, 30: 287–305.
42. Scott, S.H. and Winter, D.A. (1990) 'Internal forces at chronic running injury sites', *Medicine and Science in Sports and Exercise*, 22: 357–69.
43. Seedhom, B.B. and Wright, V. (1974) 'Functions of the menisci: a preliminary study', *Journal of Bone and Joint Surgery*, 56B: 381–2.
44. Seering, W.P., Piziali, R.L., Nagel, D.A. and Schurman, D.J. (1980) 'The function of the primary ligaments of the knee in varus-valgus and axial rotation', *Journal of Biomechanics*, 13: 785–94.
45. Sibley, M.B. and Fu, F.H. (2001) 'Knee injuries', in F.H. Fu and D.A. Stone (eds.) *Sports Injuries: Mechanisms, Prevention, Treatment*, Philadelphia: Lippincott Williams & Wilkins.
46. Siegler, S., Block, J. and Schneck, C.D. (1988) 'The mechanical characteristics of the collateral ligaments of the human ankle joint', *Foot & Ankle*, 8: 234–42.
47. Siliski, J.M. (2003) 'Dislocations and soft tissue injuries of the knee', in B.D. Browner, J.B. Jupiter, A.M. Levine, and P.T. Trafton (eds.) *Skeletal Trauma: Basic Science, Management, and Reconstruction*, Philadelphia: Saunders.
48. Sterett, W.I. and Krissoff, W.B. (1994) 'Femur fractures in alpine skiing: Classification and mechanisms of injury in 85 cases', *Journal of Orthopaedic Trauma*, 8: 310–14.
49. Swenson, T.M. and Harner, C.D. (1995) 'Knee ligament and meniscal injuries: Current concepts', *Orthopaedic Clinics of North America*, 26: 529–46.
50. Tearse, D., Buckwalter, J.A., Marsh, J.L. and Brandser, E.A. (2002) 'Stress fractures', in C. Bulstrode, J. Buckwalter, A. Carr, L. Marsh, J. Fairbank, J. Wilson-MacDonald, and G. Bowden (eds.) *Oxford Textbook of Orthopedics and Trauma*, Oxford: Oxford University Press.
51. Verrall, G.M., Slavotinek, J.P., Barnes, P.G., Fon, G.T. and Spriggins A.J. (2001) 'Clinical risk factors for hamstring muscle strain injury: a prospective study with correlation of injury by magnetic resonance imaging', *British Journal of Sports Medicine*, 35: 435–9.
52. Verzijl, N., Bank, R.A., TeKoppele, J.M. and DeGroot, J. (2003) 'AGEing and osteoarthritis: a different perspective', *Current Opinions in Rheumatology*, 15: 616–22.
53. Verzijl, N., DeGroot, J., Ben, Z.C., Brau-Benjamin, O., Maroudas, A., Bank, R.A., Mizrahi, J., Schalkwijk, C.G., Thorpe, S.R., Baynes, J.W., Bijlsma, J.W., Lafeber, F.P. and TeKoppele, J.M. (2002) 'Crosslinking by advanced glycation end products increases the stiffness of the collagen network in human articular cartilage: a possible mechanism through which age is a risk factor for osteoarthritis', *Arthritis and Rheumatology*, 46: 114–23.
54. Walker, P. and Erkman, M. (1975) 'The role of the menisci in force transmission across the knee', *Clinical Orthopaedics*, 109: 184–92.
55. Ward, S.R. and Powers, C.M. (2004) 'The influence of patella alta on patellofemoral joint stress during normal and fast walking', *Clinical Biomechanics*, 19: 1040–7.
56. Whiting, W.C. and Zernicke, R.F. (1998) *Biomechanics of Musculoskeletal Injury*, Champaign, IL: Human Kinetics.
57. Worrell, T.W. (1994) 'Factors associated with hamstring injuries: An approach to treatment and preventative measures', *Sports Medicine*, 17: 338–45.

58. Yinger, K., Mandelbaum, B.R. and Almekinders, L.C. (2002) 'Achilles rupture in the athlete: current science and treatment', *Clinics in Podiatric Medicine and Surgery*, 19: 231–50.
59. Zantop, T., Tillmann, B. and Petersen, W. (2003) 'Quantitative assessment of blood vessels of the human Achilles tendon: an immunohistochemical cadaver study', *Archives of Orthopaedic Trauma Surgery*, 123: 501–4.
60. Zernicke, R.F., Garhammer, J. and Jobe, F.W. (1977) 'Human patellar-tendon rupture: A kinetic analysis', *Journal of Bone and Joint Surgery*, 59A: 179–83.

28

Upper extremity injuries

Ronald F. Zernicke[1], William C. Whiting[2] and Sarah L. Manske[3]
[1]University of Michigan, Ann Arbor; [2]California State University, Northridge; [3]University of Calgary, Calgary

Introduction

Upper extremity injuries pervade sports and recreational activities, particularly those that involve overhead activities such as throwing. Likewise, injuries to the upper extremity frequently happen in contact sports and with falls on an outstretched hand. Here, we narrow the focus to biomechanical concepts related to overhead activities (e.g., throwing and swimming) and link those concepts to risk of upper extremity injury. We elaborate on the common mechanisms of injury in the shoulder, elbow, wrist, hand, humerus, radius, and ulna – including acute and chronic/overuse injuries.

Mechanisms of upper extremity movements and injuries

Overhead throwing: The throwing motion is similar in many sports (e.g., baseball, cricket, football, tennis serve, and water polo). One of the critical moments in throwing occurs when there is reversal in shoulder motion between cocking and acceleration (zero velocity; backward-to-forward motion reversal). At this point in the motion, significant glenohumeral joint loads and significant elbow valgus loads are created as the rotator cuff musculature (e.g., supraspinatus, infraspinatus, and teres minor) reach peak activity. Significant joint and muscle-tendon loads again happen during deceleration after ball release due to the eccentric contractions of humeral external rotator and horizontal abductor muscle groups firing at significant percentages of their maximum to slow joint rotation and translation (Meister, 2000).

One of the common findings in individuals involved in intense throwing activities, such as baseball pitching, is an exaggerated ratio of the strength of the internal to external rotators. In addition, repeated stresses within the anterior capsule, stretching of the anterior stabilizers, and exaggerated external rotator ROM can happen with repeated throwing and lead to anterior instability and secondary impingement – when pressure increases within a confined anatomical space, and the enclosed tissues are deleteriously affected (Whiting and Zernicke, 1998). The anterior joint instability arises from the inability of the anterior part of the inferior glenohumeral ligament (IGL) to prevent translation (Baeyens *et al.*, 2001).

Another common problem in throwing is abnormal scapular biomechanics, which can result in the loss of a link in the kinetic chain that leads to inefficient energy and angular momentum transfers. For example, inability to sufficiently elevate the acromion leads to impingement during the cocking and follow-through stages. Lack of scapular stability impairs the ability to develop maximal torque and can lead to muscular imbalance.

Swimming: Ninety per cent of propulsion in swimming comes from the upper limbs (Deschodt *et al.*, 1999), and thus upper extremity injuries are common, with 80 per cent of swimmers reporting a history of shoulder pain (McMaster and Troup, 1993). Weldon and Richardson (2001) suggest that most shoulder pain in swimming results from instability due to increased shoulder ROM, imbalance in internal rotator and adductor muscle strength, and excessive shoulder-intensive training (e.g., 20–30 hr/wk).

Swimmers strive to have a long stroke to increase efficiency, but the prolonged shoulder adduction and internal rotation can lead to hypovascularity of the supraspinatus muscle and tendon, which can increase the risk of tendinopathy (Brukner and Khan, 2007). Because the rotator cuff muscles, particularly subscapularis, are active throughout the stroke cycle, they are especially prone to fatigue (Pink *et al.*, 1991). The scapular stabilizers, particularly the serratus anterior, fatigue before the internal rotators fatigue and thereby render the humeral head less stable.

Specific aspects of the swimming stroke can affect susceptibility to injury. For example, using a straight arm during recovery and/or insufficient body roll can lead to shoulder impingement. Body roll allows the shoulder to function in a more neutral position relative to the coronal plane, which improves efficiency and requires less force to achieve the same forward propulsion. In addition, elite swimmers have greater shoulder laxity (i.e., hypermobility) than recreational swimmers (Zemek and Magee, 1996), which can effectively increase ROM but lead to joint instability. The increased flexibility that this laxity creates at the glenohumeral joint increases the swimmer's ability to generate power throughout the full pull-through phase (Troup, 1999). Use of hand paddles as a training tool increases the resistance provided by water, and thus increases the force production on internal rotation, which can further reduce joint stability (Weldon and Richardson, 2001).

Shoulder injuries

Rotator cuff: The rotator cuff undergoes considerable tension, particularly in throwing and swimming and is a frequent site of injury. The rotator cuff comprises the supraspinatus, infraspinatus, subscapularis, and teres minor muscles. The etiology for many injuries in the rotator cuff region is multifactorial, and frequently there is more than one problem that leads to diagnosis. There is a strong relation among glenohumeral impingement, joint instability, and rotator cuff pathology. Here, we present some of the signs and diagnoses related to the rotator cuff.

Shoulder impingement happens when rotator cuff tendons are pinched as they pass through the subacromial space created between the acromion, coracoacromial ligament, and acromioclavicular joint above, and the humeral head below. These tendons experience irritation due to the impingement that can lead to soft-tissue swelling and damage. The etiology for the change in the tendon or space that leads to the impingement, however, has not been confirmed. Clinically, rotator cuff impingement is considered a *sign* rather than a diagnosis, as it is a sign associated with a host of different diagnoses (e.g., subacromial bone spurs and/or bursal hypertrophy, acromioclavicular joint arthrosis and/or bone spurs, rotator cuff disease, superior labral injury, glenohumeral internal rotation deficit, glenohumeral instability, biceps tendinopathy, scapular dyskinesis, glenohumeral instability or superior labral injury, and cervical radiculopathy (Brukner and Khan, 2007)). Thus, we use the collective term impingement, generically, rather than impingement syndrome.

There are two subtypes of impingement: external impingement and internal impingement. Internal impingement or glenoid impingement happens when there is impingement (pinching) of the under surface of the rotator cuff against the posterior-superior surface of the glenoid (Whiting and Zernicke, 1998). This can happen, for example, during the late cocking stage of overhead throwing.

There are two potential mechanisms for the development of external impingement or subacromial impingement, or alternatively, both mechanisms may be combined. The first is intrinsic impingement that suggests that long-term degenerative changes that happen as a result of overuse, tension overload, or trauma of the tendons (Budoff *et al.*, 1998) lead to osteophyte formation, muscle imbalance, acromical changes, and altered shoulder kinematics that produce impingement (Michener *et al.*, 2003). The alternative explanation is that impingement happens via extrinsic mechanisms, such as mechanical compression by a structure external to the tendon (Neer, 1972). There is often no evidence of extrinsic causes in rotator cuff pathology (Nirschl, 1989), but there is evidence that shoulder pain due to impingement is related to multiple factors. These include altered posture (Michener *et al.*, 2003), altered kinematics (Yamaguchi *et al.*, 2000), and rotator cuff muscular weaknesses (Reddy *et al.*, 2000) in the shoulder. It is unclear, however, whether muscle weakness results from the impingement, or if weakened musculature causes the impingement.

Pathologies that were traditionally termed tendonitis are now preferentially termed tendinopathies, to reflect the absence of inflammation in later stages of chronic pathology. The early stages of rotator cuff tendinopathies are characterized by inflammation, hypercellularity, disorganization of the collagen matrix, and tendon weakness (Brukner and Khan, 2007). Yuan and colleagues (2003) demonstrated that mechanical loading of the tendon activates protein kinases that can trigger tenocytes to undergo apoptosis (programmed cell death) if activated continually without recovery. Increased apoptosis results in less collagen synthesis and a weaker matrix that is prone to overall tendon tears (Murrell, 2002). In chronic tendinopathies there may be more degenerative disease (Matthews *et al.*, 2006).

The supraspinatus tendon is the most commonly injured tendon in the rotator cuff (Li *et al.*, 1999; Figure 28.1). It was previously believed that the high prevalence of tendinopathies in the supraspinatus tendon is due to a hypovascular region found in the distal 1.0–1.5 cm of the tendon (Moseley and Goldie, 1963). More recently, however, researchers found that the vascular pattern may be sufficient to meet the metabolic needs of the muscle (Malcarney and Murrell, 2003).

Evidence suggests that most torn rotator cuff tendons are preceded by degenerative changes (Kannus and Jozsa, 1991). However, studies have also found rotator cuff degeneration and tears in asymptomatic individuals (Milgrom *et al.*, 1995). Thus, rotator cuff tears – to some extent – should be considered a part of normal aging and do not necessarily lead to pain and dysfunction sufficient to seek medical attention.

Glenohumeral instability and dislocation: The substantial mobility in the shoulder is a trade-off for stability. The shoulder is more prone to dislocation than the hip because of the relatively poorer fit between the bones and the limited supporting musculature. The glenoid fossa is shallow, and there is limited area for contact between the humeral head and fossa. The glenoid labrum deepens the fossa, which increases contact area, but the glenohumeral joint nonetheless is the least stable joint in the body. The factors that contribute to

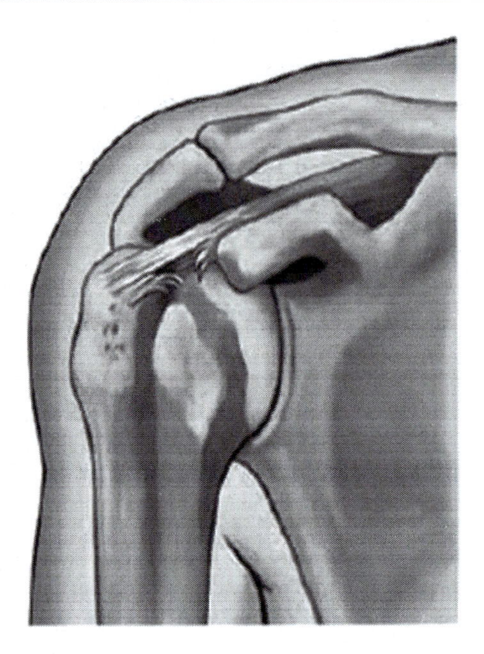

Figure 28.1 Rotator cuff tendinopathy. Pathology often begins on the inferior surface of the supraspinatus tendon. From Brukner and Khan (2007). Permission pending.

glenohumeral stability help to explain the mechanisms involved in glenohumeral luxation (dislocation).

Both active and passive elements assist in stabilizing the glenohumeral joint. Active stabilization occurs dynamically through the action of muscles surrounding the joint, including the deltoid, trapezius, latissimus dorsi, pectoralis major, and the rotator cuff muscles (i.e., subscapularis, supraspinatus, infraspinatus, and teres minor). The joint capsule and supporting ligaments provide passive stabilization in the extremes of joint motion. As there is considerable laxity in these structures through the normal ROM, a small negative intracapsular pressure helps to stabilize the joint (Speer, 1995), along with concavity compression and scapulohumeral balance. Concavity compression contributes to stability by having a convex object (humeral head) press against a concave surface (glenoid fossa), providing some resistance to translational movement between the surfaces (Lippitt and Matsen, 1993).

Most glenohumeral joint luxations happen as a result of trauma, and most luxate anteriorly. The most common mechanism for anterior luxation occurs when indirect forces are applied to an abducted, extended, and externally rotated arm (Figure 28.2). Anterior luxation can also occur via direct forces to the posterior aspect of the humerus.

In contrast, posterior luxations typically occur as a result of indirect forces transmitted through a flexed, adducted, and internally rotated arm, which drive the humerus posteriorly. Posterior luxations due to direct trauma have also been reported. Rarely, inferior luxations happen through hyperabduction, where a fulcrum is created between the humeral neck and acromion, and the humeral head gets pushed out inferiorly.

Acromioclavicular joint injuries: A glenohumeral luxation is distinct from an acromioclavicular sprain – often referred to as a separated shoulder or shoulder separation. An acromioclavicular sprain can happen when applied forces displace the acromion process of

Figure 28.2 Anterior glenohumeral luxation (dislocation) from indirect force applied through the arm in an extended, abducted, and externally rotated position. From W.C. Whiting and R.F. Zernicke, 1998, *Biomechanics of Musculoskeletal Injury*, p. 182. © 1998 by William C. Whiting and Ronald F. Zernicke. Reprinted with permission from Human Kinetics (Champaign, IL).

the scapula from the distal end of the clavicle. The AC joint is stabilized by the joint capsule, acromioclavicular ligaments (horizontally), and coracoclavicular ligaments (vertically). The acromioclavicular joint is stabilized passively through the extracapsular (coracoclavicular) and capsular acromioclavicular ligaments (Buss and Watts, 2003). The deltoid and trapezius muscles act as dynamic stabilizers (Galatz and Williams, 2006).

Acromioclavicular injury results from directly or indirectly applied forces. Direct impacts to the acromion (when the humerus is adducted) can force the acromion inferiorly and medially (Buss and Watts, 2003). Injury from indirect forces typically involves a fall onto an outstretched arm or elbow with a superiorly directed force. Either mechanism can lead to ligamentous injury and/or fracture of the distal one-third of the clavicle. The severity of the injury increases with the magnitude of the applied impact force: lower forces produce a mild sprain of the acromioclavicular ligament, greater forces produce moderate acromioclavicular ligament sprain and involvement of the coracoclavicular ligament, while maximal forces result in complete acromioclavicular dislocation with tearing of the clavicular attachments to the deltoid and trapezius muscle and rupture of the coracoclavicular ligament (Figure 28.3).

Clavicular fracture: Clavicular fractures are one of the most common fractures that occur in sporting and recreational activities. They are typically caused by a fall on the shoulder. Most clavicular fractures occur in the middle third of the bone, where the distal fragment displaces inferiorly and the proximal fragment displaces superiorly. Fractures to the distal one third of the clavicle usually result from a direct force applied to the shoulder, often in athletes throwing in a contact sport (Buss and Watts, 2003).

Upper-arm injuries

Humeral fracture: Humeral fractures are common, accounting for about 8 per cent of all fractures (Praemer *et al.*, 1999). The majority (65 per cent) of humeral fractures happen in the middle third of the diaphysis (shaft), while 25 per cent happen in the proximal humerus and 10 per cent in the distal humerus (Tytherleigh-Strong *et al.*, 1998).

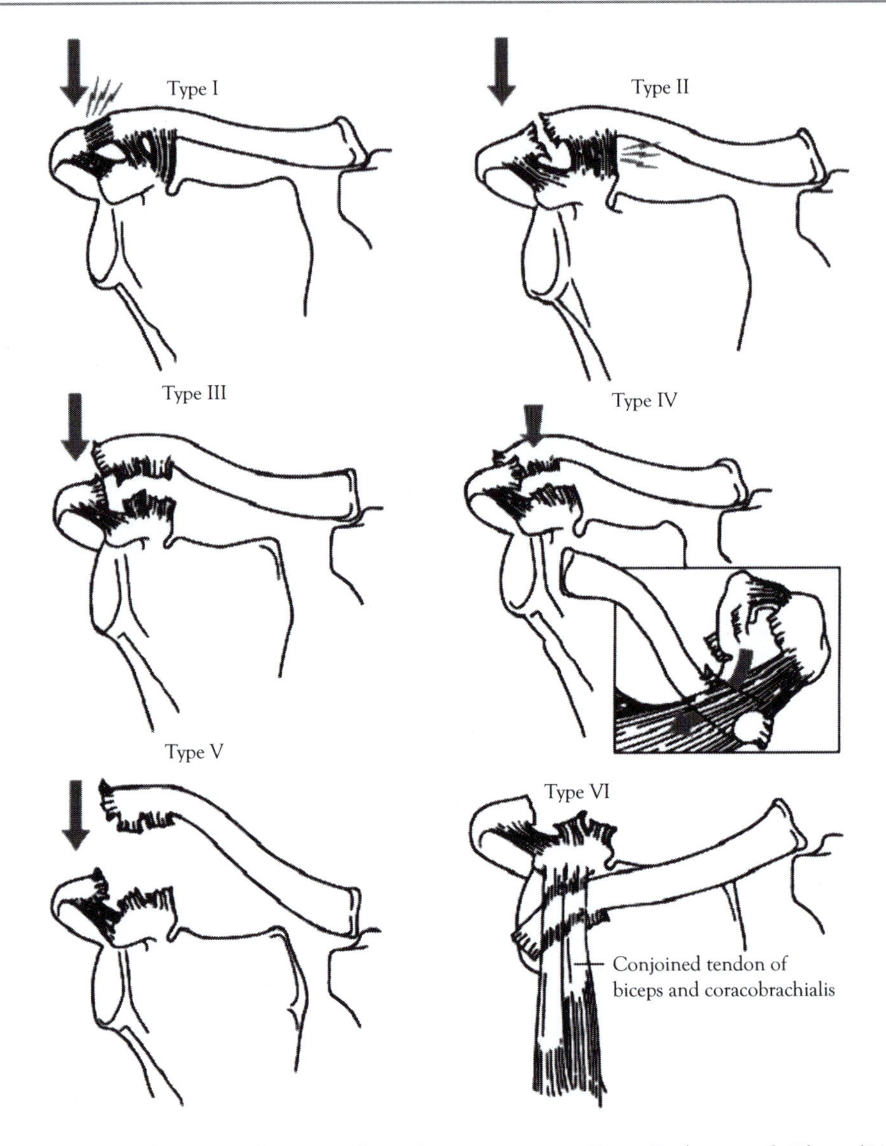

Figure 28.3 Mechanisms of acromioclavicular joint injury. From Brukner and Khan (2007). Permission pending.

Like many injuries, humeral fractures result from direct or indirect trauma, but the majority are due to direct trauma (e.g., from a fall on an outstretched arm or motor vehicle crash). Fractures due to direct trauma usually involve considerable comminution (i.e., bone fragmentation) and soft tissue disruption due to the high energy input. In contrast, less energy and less comminution and displacement happen from indirect trauma such as violent muscular contractions.

As with most fractures, the fracture pattern varies depending on the magnitude, location, and direction of the applied forces. Compressive forces lead to proximal and distal fractures, bending leads to transverse shaft fractures, and torsional loads lead to spiral fractures.

The bony displacement also depends on the muscle activity in the area near the applied forces. For example, if fracture occurs proximal to the pectoralis major attachment site, the distal segment is displaced medially by the pectoralis major, while the proximal segment is abducted and rotated internally by the rotator cuff musculature.

The nature and extent of the damage in a humeral fracture also depends in part on age. In older persons, humeral fractures are more likely a result of a fall (Tytherleigh-Strong *et al.*, 1998). In younger individuals, fracture usually results from direct impact or throwing objects such as a baseball (Ogawa and Yoshida, 1998). In middle age, disuse osteopenia may play a role as individuals returning to baseball pitching after years of disuse can experience spontaneous humeral fractures (Branch *et al.*, 1992). Thus an individual's physical condition and activity patterns play an important role in determining injury risk.

Elbow injuries

Elbow injuries are common in many sports and recreational activities, especially racquet sports and sports involving overhead throwing. Elbow injuries can significantly affect everyday activities. The causative mechanisms have been researched extensively in recent years.

Elbow fractures: Elbow fractures can involve any bone within the elbow joint: humerus (supracondylar fracture), ulna (olecranon fracture), or radius (radial head fracture). Each of these fractures can result from a fall on an outstretched arm. The involvement of each bone depends on the nature, magnitude, location, and direction of the applied force. Supracondylar fractures are common in children around 12 years of age, frequently due to a fall from a bicycle. Olecranon fractures can also happen with direct impact to the posterior aspect of the elbow. Radial head fractures can occur with elbow dislocation, which results in a very unstable fracture.

Lateral epicondylitis or **extensor tendinopathy**: Lateral epicondylitis typically presents as lateral elbow pain arising from pathology (tendinosis) of the extensor carpi radialis brevis (ECRB) tendon, close to the extensor origin at the lateral epicondyle. As inflammation tends to be present only in the early stages of development, some prefer to refer to the condition as epicondylosis. As the site of pathology is usually below the lateral epicondyle in the ECRB tendon, and the primary pathology is collagen disorganization, it has also been called extensor tendinopathy (Whaley and Baker, 2004). Significantly, epicondylitis is a degenerative process rather than an inflammatory process (Whaley and Baker, 2004). With continued use and microdamage, partial tears of macroscopic size may appear (Regan *et al.*, 1992).

Lateral epicondylitis is common among tennis players, as 10 per cent to 50 per cent of players experience it over the course of their years playing (Nirschl, 1992). The backhand stroke has been implicated as a causative factor in lateral epicondylitis, as the wrist extensor activity, especially in the ECRB, is high (Giangarra *et al.*, 1993). Novice tennis players tend to use different kinematics (e.g., greater eccentric contraction of wrist extensor muscles (Blackwell and Cole, 1994)) than experts, which predisposes them to injury. Use of a two-handed backhand decreases the risk of lateral epicondylitis as there is a lower demand on the dominant arm extensors.

The injury, however, is not exclusive to tennis playing. Other striking sports such as racquetball, squash, and golf can be causative, as well as occupations involving repetitive motions of the wrist and elbow, particularly with overuse of the extensors in pinching and grasping (e.g., chronic work with hand tools as in surgery, dentistry, carpentry, or writing) (Snijders *et al.*, 1987).

Medial epicondylitis or **flexor-pronator tendinopathy**: Medial elbow pain tends to originate from one of two sources. The first is what is traditionally termed medial epicondylitis, which is a similar pathological process to that which happens in the extensor tendons of the lateral epicondyle, but is usually a tendinosis in the tendons of the pronator teres and the wrist flexor group. The pain is due to excessive activity of the wrist flexors. This is seen especially in golf (e.g., golfer's elbow) and in tennis when players apply substantial top spin to their forehand shots (Brukner and Khan, 2007). Medial epicondylitis is less prevalent than lateral epicondylitis.

Ulnar or **medial collateral ligament sprain — valgus extension loading injuries**: The second source of medial elbow pain is related to a sprain of the ulnar (medial) collateral ligament complex. The lateral collateral ligaments and ulnar collateral ligaments are often referred to as 'complexes' to emphasize that their contributions to elbow stability are enhanced by adjacent capsuloligamentous, fascial, and musculotendinous structures (Cohen and Hastings, 1997).

The coupling of valgus torque and elbow extension in throwing can produce a valgus-extension overload that strains the ulnar collateral ligament and can lead to other medial elbow injuries, including ulnar collateral ligament sprain or rupture and avulsion fracture. The largest internal varus torque developed in the medial tissues to resist the external valgus torque occurs near the end of the cocking phase of throwing (Fleisig *et al.*, 1995). In baseball pitching, the ulnar collateral ligament approaches its maximal capacity, making it vulnerable to injury (Fleisig *et al.*, 1995). In the cocking and acceleration phase, as substantial internal varus torque is generated, the elbow extends from 85° to 20° (Fleisig *et al.*, 1995). The combination of internal varus torque and elbow extension supports the 'valgus extension loading' mechanism for medial elbow injury – a common injury to baseball pitchers and javelin throwers.

Elbow dislocation: Due to the elbow's relative stability, elbow dislocations are three times less common than shoulder dislocations (Praemer *et al.*, 1999). The elbow is inherently stable due to its bony structure, primarily because of the close interaction between the trochlear notch of the ulna and the trochlea of the humerus, as well as the olecranon process of the ulna and the olecranon fossa of the humerus (Ring, 2006). On the radius, the coronoid process and radial head contact the coronoid and radial fossae of the humerus. The ulnar collateral ligament and lateral collateral ligament complexes, with the anterior capsule, account for the majority of the elbow's stability. Nevertheless, elbow dislocations do occur, and when they do happen, considerable soft tissue damage occurs, frequently with rupture of the ulnar and lateral collateral ligaments.

Because of the strong anterior stability in the elbow, most dislocations occur posteriorly (Rettig, 2002). The elbow commonly dislocates via an axial force applied to an extended or hyperextended elbow. That axial force levers the ulna out of the trochlea, which causes ligament and capsular ruptures (Hotchkiss, 1996). If the elbow is forcibly hyperextended, there are also valgus stresses that can lead to rupture of the ulnar collateral ligament and sometimes rupture at the medial origin of the flexor-prontator muscle group.

Forearm injuries

Diaphyseal fractures of the radius and ulna: Radial and ulnar fractures happen singularly or in combination. Identification of the causal mechanism typically suggests the location and type of fracture. When both radius and ulna are injured, it is usually due to a direct, high-energy trauma such as a motor vehicle accident or gunshot wound. An isolated proximal radius fracture rarely happens because of the protection provided by overlying musculature. If there is sufficient force to fracture the proximal radius, the ulna will also likely fracture. Fractures in the middle and distal thirds of the radius occur more frequently. Isolated fractures of the distal third of the radius are named Galeazzi fractures and are often accompanied by dislocation of the distal radioulnar joint. Galeazzi fractures typically happen after a direct trauma on the dorsolateral side of the wrist or a fall on an outstretched arm that axially loads a hyperpronated forearm.

Ulnar variance: Ulnar variance refers to the length discrepancy between the radius and ulna. This discrepancy or variance influences forearm and wrist mechanics, particularly when the wrist acts as a load-bearing joint, as in gymnastics (e.g., pommel horse). There is no ulnar variance when the radius and ulna are the same length. With positive ulnar variance, the ulna is longer than the radius, and with negative ulnar variance, the radius is longer than the ulna. Typical ulnar variance varies depending on age, ethnicity, genetics, elbow pathology, and loading history and possibly gender.

Loading history may be the most relevant factor to the discussion of ulnar variance and wrist injury mechanisms (De Smet *et al.*, 1994). In gymnastics, for example, with pommel horse or tumbling, the wrist is subject to considerable compressive loads. The radius receives approximately 80 per cent of the load. In a skeletally immature individual, the repetitive compressive loading can cause premature closure of the radial growth plate. Continued ulnar growth would create positive ulnar variance, with potential ulnar impact syndrome and degeneration of the triangular fibrocartilage and ulnar carpus. While wrist pain is common in gymnasts, ulnar variance itself is not associated with wrist pain or radiographically defined injury in the distal radius (DiFiori *et al.*, 2002).

Wrist and hand injuries

Fracture of the distal radius: Distal radial fractures tend to occur in three different populations: children aged 5 to 14, males under age 50, and females older than age 40 (Ruch, 2006). There are many types of distal radius fractures. We will refer to the clinical descriptions and newer classification systems, with eponyms (named fractures) noted parenthetically. One classification system groups the fractures according to their mechanism of injury rather than by radiological characteristics (Jupiter and Fernandez, 1997). The five types described are: type I – bending fractures; type II – shearing fractures of the articular surface; type III – compression of the articular surface; type IV – avulsion fractures or radiocarpal fracture dislocations; and type V – combined fractures with significant soft tissue involvement.

Type I fractures are caused by the axial compressive forces that occur in landing on an outstretched arm and cause bending of the radius (Figure 28.4). The fracture pattern (Colles' fracture) demonstrates failure of the anterior metaphyseal cortex in tension and varying degrees of comminution on the posterior surface. Most distal radius fractures are type I fractures. If a fall happens on a flexed wrist or an outstretched and supinated arm, the radius bends in the opposite direction. As a result, the compressive loading causes tensile failure on the posterior surface of the metaphysis, with comminution on the anterior surface (Smith's fracture).

With higher-energy loading (typically observed in younger individuals), type II shearing fractures happen with the anterior (volar) lip of the radial articular surface sheared off (Barton's fracture). Type III fractures occur when high compressive loads (e.g., from a fall

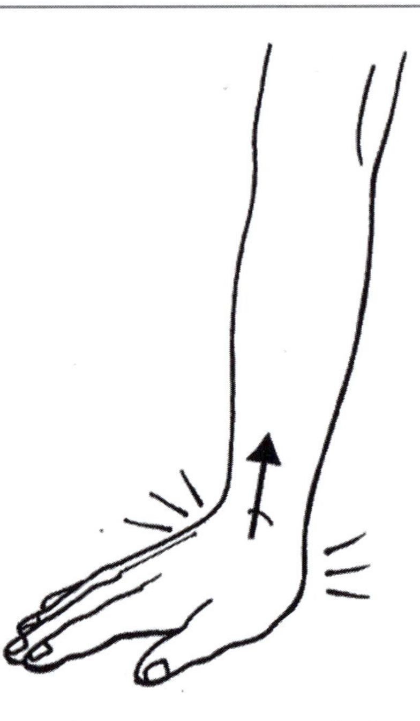

Figure 28.4 Common mechanism of injury from compressive load applied to a hyperextended wrist at impact. From W.C. Whiting and R.F. Zernicke, 1998, *Biomechanics of Musculoskeletal Injury*, p. 202. © 1998 by William C. Whiting and Ronald F. Zernicke. Reprinted with permission from Human Kinetics (Champaign, IL).

from a high height) cause intra-articular fractures of the articular surface and disruption of the subchondral and cancellous bone.

When high stresses are created on the osteoligamentous attachments (e.g., in exaggerated torsion), type IV avulsion fractures can occur. Type V fractures usually result from high-energy trauma that combines bending, compression, shear, and/or avulsion.

Carpal fractures: Most carpal fractures occur as a result of an axial compressive load applied to a hyperextended wrist, as when a person trips and falls on an outstretched upper extremity (Figure 28.4). When the wrist is hyperextended (dorsiflexed), compressive forces are transmitted through the carpals to the distal radioulnar complex. Less commonly, hyperflexion and torsional loading can produce carpal fracture. Which carpal fractures depends on the degree and direction of ulnar or radial deviation, the energy absorbed during the load application, the location and direction of the applied load, and the relative strength of the bones and ligaments. The scaphoid and lunate are the most likely to be fractured because they articulate with the radius, and the distal radius absorbs 80 per cent of the transmitted load in a fall (compared with 20 per cent for the ulna). These fractures typically occur when the wrist is hyperextended past 90–95°, and the radius accepts most of the load on the wrist (Weber and Chao, 1978). Tensile failure can occur on the palmar side of the scaphoid, while the dorsal aspect fails in compression (Ruby and Cassidy, 2003).

Thumb injuries: The most common hand sprain is the ulnar collateral ligament of the first metacarpophalangeal joint. The sprain is colloquially termed gamekeeper's thumb or skier's thumb as it commonly occurs when a skier falls onto an outstretched hand with the thumb in an abducted position (Figure 28.5). The handle of the ski pole holds the thumb

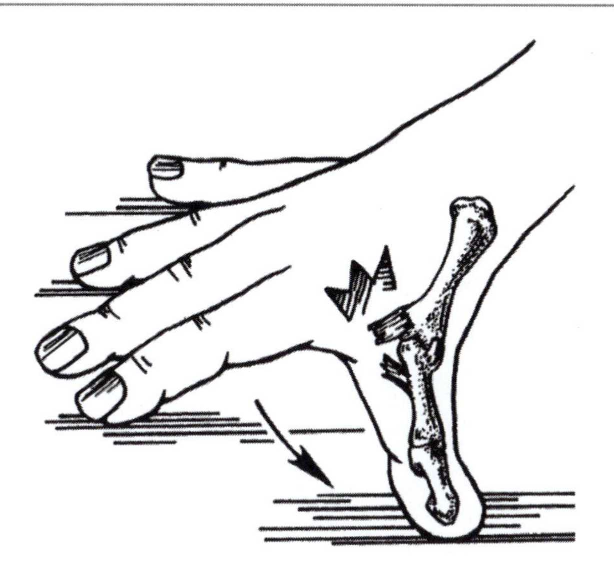

Figure 28.5 Mechanism of first metacarpophalangeal sprain, where a fall occurs on an outstretched hand in an abducted position. From W.C. Whiting and R.F. Zernicke, 1998, *Biomechanics of Musculoskeletal Injury*, p. 203. © 1998 by William C. Whiting and Ronald F. Zernicke. Reprinted with permission from Human Kinetics (Champaign, IL).

in abduction as the load of the fall is absorbed by the hand, placing excessive tensile forces on the ulnar collateral ligament. The other common mechanism for ulnar collateral ligament sprain is hyperextension of the first metacarpophalangeal joint. This commonly occurs in a collision between two athletes, such as the hand of a softball player tagging an opponent sliding into a base.

Conclusion

Upper extremity injuries are common in many persons participating in sports and recreational activities. While the specific injuries vary in each joint, causal mechanisms are often similar. Overuse can lead to a loss of stability in a joint that makes the person more susceptible to both chronic overuse and acute injuries. Most injuries are multifactorial and involve combined disorders and causes. The upper extremity is designed for mobility, and acute or chronic injuries to any of the upper-extremity joints can significantly reduce a person's quality of life – involving activities of daily living, during work, or in sport and recreational activities (Whiting and Zernicke, 1998). It is fundamental to understand the links among normal joint structure, joint biomechanics, and injury mechanics to understand fully the *what*, *why*, and *how* of joint injuries.

References

1. Baeyens, J. P., Van Roy, P., De Schepper, A., Declercq, G. and Clarijs, J. P. (2001) 'Glenohumeral Joint Kinematics Related to Minor Anterior Instability of the Shoulder at the End of the Late Preparatory Phase of Throwing'. *Clin Biomech* (Bristol, Avon) 16(9): 752–7.

2. Blackwell, J. R. and Cole, K. J. (1994) 'Wrist Kinematics Differ in Expert and Novice Tennis Players Performing the Backhand Stroke: Implications for Tennis Elbow'. *J Biomech* 27(5): 509–16.
3. Branch, T., Partin, C., Chamberland, P., Emeterio, E. and Sabetelle, M. (1992) 'Spontaneous Fractures of the Humerus During Pitching. A Series of 12 Cases'. *Am J Sports Med* 20(4): 468–70.
4. Brukner, P. and Khan, K. (2007) *Clinical Sports Medicine*. Sydney, McGraw-Hill.
5. Budoff, J. E., Nirschl, R. P. and Guidi, E. J. (1998) 'Debridement of Partial-Thickness Tears of the Rotator Cuff without Acromioplasty. Long-Term Follow-up and Review of the Literature'. *J Bone Joint Surg Am* 80(5): 733–48.
6. Buss, D. D. and Watts, J. D. (2003) 'Acromioclavicular Injuries in the Throwing Athlete'. *Clin Sports Med* 22(2): 327–41, vii.
7. Cohen, M. S. and Hastings, H., 2nd (1997) 'Rotatory Instability of the Elbow. The Anatomy and Role of the Lateral Stabilizers'. *J Bone Joint Surg Am* 79(2): 225–33.
8. De Smet, L., Claessens, A., Lefevre, J. and Beunen, G. (1994) 'Gymnast Wrist: An Epidemiologic Survey of Ulnar Variance and Stress Changes of the Radial Physis in Elite Female Gymnasts'. *Am J Sports Med* 22(6): 846–50.
9. Deschodt, V. J., Arsac, L. M. and Rouard, A. H. (1999) 'Relative Contribution of Arms and Legs in Humans to Propulsion in 25-M Sprint Front-Crawl Swimming'. *Eur J Appl Physiol Occup Physiol* 80(3): 192–9.
10. DiFiori, J. P., Puffer, J. C., Aish, B. and Dorey, F. (2002) 'Wrist Pain, Distal Radial Physeal Injury, and Ulnar Variance in Young Gymnasts: Does a Relationship Exist?' *Am J Sports Med* 30(6): 879–85.
11. Fleisig, G. S., Andrews, J. R., Dillman, C. J. and Escamilla, R. F. (1995) 'Kinetics of Baseball Pitching with Implications About Injury Mechanisms'. *Am J Sports Med* 23(2): 233–9.
12. Galatz, L. M. and Williams, G. R. J. (2006) Acromioclavicular Injuries. *Rockwood and Green's Fractures in Adults*. R. W. Bucholz, J. D. Heckman and C. M. Court-Brown. Philadelphia, Lippincott Williams & Wilkins. 2.
13. Giangarra, C. E., Conroy, B., Jobe, F. W., Pink, M. and Perry, J. (1993) 'Electromyographic and Cinematographic Analysis of Elbow Function in Tennis Players Using Single- and Double-Handed Backhand Strokes'. *Am J Sports Med* 21(3): 394–9.
14. Hotchkiss, R. N. (1996) 'Fractures and Dislocations of the Elbow'. *Rockwood and Green's Fractures in Adults*. C. A. Rockwood, D. P. Green, R. W. Bucholz and J. D. Heckman. Philadelphia, Lippincott-Raven: 929–1024.
15. Jupiter, J. B. and Fernandez, D. L. (1997) 'Comparative Classification for Fractures of the Distal End of the Radius'. *J Hand Surg [Am]* 22(4): 563–71.
16. Kannus, P. and Jozsa, L. (1991) 'Histopathological Changes Preceding Spontaneous Rupture of a Tendon. A Controlled Study of 891 Patients'. *J Bone Joint Surg Am* 73(10): 1507–25.
17. Li, X. X., Schweitzer, M. E., Bifano, J. A., Lerman, J., Manton, G. L. and El-Noueam, K. I. (1999) 'Mr Evaluation of Subscapularis Tears'. *J Comput Assist Tomogr* 23(5): 713–17.
18. Lippitt, S. and Matsen, F. (1993) 'Mechanisms of Glenohumeral Joint Stability'. *Clin Orthop Relat Res* (291): 20–8.
19. Malcarney, H. L. and Murrell, G. A. (2003) 'The Rotator Cuff: Biological Adaptations to Its Environment'. *Sports Med* 33(13): 993–1002.
20. Matthews, T. J., Hand, G. C., Rees, J. L., Athanasou, N. A. and Carr, A. J. (2006) 'Pathology of the Torn Rotator Cuff Tendon. Reduction in Potential for Repair as Tear Size Increases'. *J Bone Joint Surg Br* 88(4): 489–95.
21. McMaster, W. C. and Troup, J. (1993) 'A Survey of Interfering Shoulder Pain in United States Competitive Swimmers'. *Am J Sports Med* 21(1): 67–70.
22. Meister, K. (2000) 'Injuries to the Shoulder in the Throwing Athlete. Part One: Biomechanics/Pathophysiology/Classification of Injury'. *Am J Sports Med* 28(2): 265–75.

23. Michener, L. A., Mcclure, P. W. and Karduna, A. R. (2003) 'Anatomical and Biomechanical Mechanisms of Subacromial Impingement Syndrome'. *Clin Biomech (Bristol, Avon)* 18(5): 369–79.
24. Milgrom, C., Schaffler, M., Gilbert, S. and Van Holsbeeck, M. (1995) 'Rotator-Cuff Changes in Asymptomatic Adults. The Effect of Age, Hand Dominance and Gender'. *J Bone Joint Surg Br* 77(2): 296–8.
25. Moseley, H. F. and Goldie, I. (1963) 'The Arterial Pattern of the Rotator Cuff of the Shoulder'. *J Bone Joint Surg Br* 45: 780–9.
26. Murrell, G. A. (2002) 'Understanding Tendinopathies'. *Br J Sports Med* 36(6): 392–3.
27. Neer, C. S., 2nd (1972) 'Anterior Acromioplasty for the Chronic Impingement Syndrome in the Shoulder: A Preliminary Report'. *J Bone Joint Surg Am* 54(1): 41–50.
28. Nirschl, R. P. (1989) 'Rotator Cuff Tendinitis: Basic Concepts of Pathoetiology'. *Instr Course Lect* 38: 439–45.
29. Nirschl, R. P. (1992) 'Elbow Tendinosis/Tennis Elbow'. *Clin Sports Med* 11(4): 851–70.
30. Ogawa, K. and Yoshida, A. (1998) 'Throwing Fracture of the Humeral Shaft. An Analysis of 90 Patients'. *Am J Sports Med* 26(2): 242–6.
31. Pink, M., Perry, J., Browne, A., Scovazzo, M. L. and Kerrigan, J. (1991) 'The Normal Shoulder During Freestyle Swimming. An Electromyographic and Cinematographic Analysis of Twelve Muscles'. *Am J Sports Med* 19(6): 569–76.
32. Praemer, A., Furner, S. and Rice, D. P. (1999) *Muscoskeletal Conditions in the United States*. Park Ridge, IL, American Academy of Orthopaedic Surgeons.
33. Reddy, A. S., Mohr, K. J., Pink, M. M. and Jobe, F. W. (2000) 'Electromyographic Analysis of the Deltoid and Rotator Cuff Muscles in Persons with Subacromial Impingement'. *J Shoulder Elbow Surg* 9(6): 519–23.
34. Regan, W., Wold, L. E., Coonrad, R. and Morrey, B. F. (1992) 'Microscopic Histopathology of Chronic Refractory Lateral Epicondylitis'. *Am J Sports Med* 20(6): 746–9.
35. Rettig, A. C. (2002) 'Traumatic Elbow Injuries in the Athlete'. *Orthop Clin North Am* 33(3): 509–22, v.
36. Ring, D. (2006) 'Fractures and Dislocations of the Elbows'. *Rockwood and Green's Fractures in Adults*. R. W. Bucholz, J. D. Heckman and C. M. Court-Brown. Philadelphia, Lippincott Williams & Wilkins. 1.
37. Ruby, L. K. and Cassidy, C. (2003) 'Fractures and Dislocations of the Carpus'. *Skeletal Trauma*. B. D. Browner, J. B. Jupiter, A. M. Levine and P. G. Trafton. Philadelphia, Saunders.
38. Ruch, D. S. (2006) 'Fractures of the Distal Radius and Ulna'. *Rockwood and Green's Fractures in Adults*. R. W. Bucholz, J. D. Heckman and C. M. Court-Brown. Philadelphia, Lippincott Williams & Wilkins. 1.
39. Snijders, C. J., Volkers, A. C., Mechelse, K. and Vleeming, A. (1987) 'Provocation of Epicondylalgia Lateralis (Tennis Elbow) by Power Grip or Pinching'. *Med Sci Sports Exerc* 19(5): 518-23.
40. Speer, K. P. (1995) 'Anatomy and Pathomechanics of Shoulder Instability'. *Clin Sports Med* 14(4): 751–60.
41. Troup, J. P. (1999) 'The Physiology and Biomechanics of Competitive Swimming'. *Clin Sports Med* 18(2): 267–85.
42. Tytherleigh-Strong, G., Walls, N. and Mcqueen, M. M. (1998) 'The Epidemiology of Humeral Shaft Fractures'. *J Bone Joint Surg Br* 80(2): 249–53.
43. Weber, E. R. and Chao, E. Y. (1978) 'An Experimental Approach to the Mechanism of Scaphoid Waist Fractures'. *J Hand Surg [Am]* 3(2): 142–8.
44. Weldon, E. J., 3rd and Richardson, A. B. (2001) 'Upper Extremity Overuse Injuries in Swimming. A Discussion of Swimmer's Shoulder'. *Clin Sports Med* 20(3): 423–38.
45. Whaley, A. L. and Baker, C. L. (2004) 'Lateral Epicondylitis'. *Clin Sports Med* 23(4): 677–91, x.
46. Whiting, W. C. and Zernicke, R. F. (1998) *Biomechanics of Musculoskeletal Injury*. Champaign, IL, Human Kinetics.

47. Yamaguchi, K., Sher, J. S., Andersen, W. K., Garretson, R., Uribe, J. W., Hechtman, K. and Neviaser, R. J. (2000) 'Glenohumeral Motion in Patients with Rotator Cuff Tears: A Comparison of Asymptomatic and Symptomatic Shoulders'. *J Shoulder Elbow Surg* 9(1): 6–11.
48. Yuan, J., Wang, M. X. and Murrell, G. A. (2003) 'Cell Death and Tendinopathy'. *Clin Sports Med* 22(4): 693–701.
49. Zemek, M. J. and Magee, D. J. (1996) 'Comparison of Glenohumeral Joint Laxity in Elite and Recreational Swimmers'. *Clin J Sport Med* 6(1): 40–7.

29

Biomechanics of spinal trauma

Brian D. Stemper and Narayan Yoganandan
Medical College of Wisconsin, Milwaukee

Introduction

The human vertebral column sustains mechanical loads during physiological situations and responds through deformations of its components. The main physiologic function of the column is to protect the spinal cord and maintain normal interrelationships between the various intervertebral components. During traumatic loadings, due to excessive deformations, injuries can occur in the form of fractures to bony regions and/or disruption of the integrity of the soft tissues surrounding the vertebral complex. Spinal injuries can vary from minor anatomic abnormalities with no long-term or neurological involvement to severe consequences. This chapter presents fundamental biomechanics from component, segmental, and vertebral column perspectives, leading to discussions on trauma mechanisms focused on acute loading. A brief introduction is presented on the biomechanical aspects of spine anatomy.

Biomechanical anatomy

The human spinal column is divided into cervical, thoracic, lumbar, and sacral regions consisting of seven, twelve, five, and five vertebrae. The cervical region is in the neck, thoracic region is in the thorax, and lumbar and sacral regions are in the abdomen. In the adult spine, cervical and lumbar regions have convex or lordotic, and thoracic and sacral regions have concave or kyphotic curvatures. Intervertebral discs and ligaments, termed as soft tissues, connect the vertebrae. The connecting components are flexible while vertebrae are relatively rigid. External forces applied during physiologic or traumatic activities are transmitted through the medium of vertebrae and interconnecting structures.

Vertebrae increase in size inferiorly from cervical to lumbar regions and demonstrate region-dependent anatomical characteristics. Cervical vertebrae, C1 to C7, are the smallest of the spinal column. Lower cervical vertebrae, C3 to C7, demonstrate approximately consistent anatomy. The most distinct feature is its anteriorly-oriented vertebral body. The bodies are oval-shaped in the horizontal plane. The saddle-shape of the bodies in the coronal plane are due to the bilateral uncinate processes. All vertebral bodies consist of trabecular bone and cortical shell. Superior and inferior surfaces have endplates, consisting of a thin shell of horizontally-oriented cortical bone. Postero-laterally oriented pedicles connect the vertebral body to the articular pillars, known as lateral masses. The two articular pillars are the second most massive portions of the vertebrae. Superior and inferior surfaces are flat and oriented at approximately 45° in the sagittal plane. Postero-medially oriented laminae connect the pillars to the spinous process. The posterior edge of the vertebral body, pedicles, pillars, and laminae enclose the vertebral foramen, through which the spinal cord traverses. Spinous processes extend posteriorly and are approximately one-half of the anterior-posterior vertebral length. The posterior-most extent of spinous processes C3 to C6 bifurcates into two tubercles. C7 spinous process does not bifurcate and is more prominent than other vertebrae, a feature evident on lateral x-rays.

The superior-most vertebrae are C1 and C2, known as the atlas and the axis. C2 consists of an enlarged vertebral body with superiorly oriented prominence known as the dens or odontoid process. As C1 has no vertebral body, it is commonly theorized that C2 odontoid process was once the vertebral body of C1. Other C2 features are similar to C3 to C6 vertebrae, with the exception of large horizontally oriented superior surfaces of the articular processes. C2 odontoid process is oriented posteriorly to the anterior arch of C1. The anterior arch is connected to C1 articular processes, with flat and horizontal inferior surfaces. Superior surfaces demonstrate an anterior-posteriorly oriented concave curvature and articulate with the occipital condyles of the cranium. C1 posterior arches enclose the vertebral foramen and terminate at the posterior tubercle, without the spinous process.

Features of the thoracic and lumbar vertebrae are similar C3 to C7. Vertebral bodies are anterior-to-posteriorly elongated in the horizontal plane and thicker posteriorly. Articular pillar surfaces are oriented almost vertically. The superior surface faces posteriorly and the inferior surface faces anteriorly. In contrast to the horizontal cervical spinous processes, thoracic spinous processes are angled inferiorly, approximately 60°. Another unique feature is that pedicles extend postero-laterally from the articular pillars to form the transverse costal facet. Coupled with superior costal facets on the postero-lateral corners of the vertebral body, the transverse costal facets form the interface for the ribs, which articulate with all thoracic vertebrae.

Lumbar vertebrae are the most massive. Kidney-shaped bodies in the horizontal plane are characteristic of these vertebrae. Superior and inferior surfaces are flat and oriented horizontally. Articular pillar surfaces are oriented almost vertically. Superior surfaces face postero-medially, and inferior surfaces face antero-laterally. Horizontally-oriented spinous processes are relatively shorter and vertically thicker than cervical and thoracic regions.

The sacral and coccyx regions are the inferior ends of the spinal column. These regions are distinct from cervical, thoracic, and lumbar regions. The sacrum consists of five fused vertebrae that form the dorsal wall of the pelvis. The wide triangular shaped sacrum articulates with the ilium of the pelvis. Superiorly, the sacrum is attached to L5 vertebra through L5-S1 disc and posterior articulations. The coccyx is the inferior projection of the sacrum and consists of four small independent bones.

As indicated, soft tissues interconnect vertebrae and consist of intervertebral discs and ligaments. Discs are located between C2 and S1 bodies and consist of annulus fibrosus and nucleus pulposus. Disc height varies by spinal region. The annulus fibrosus, the outer portion of the disc, is made up of concentric rings of fibrous tissue. Adjacent rings are approximately perpendicular. Fibres are oriented at 45° relative to the vertebral body in one ring and at 135° relative to the vertebral body in the adjacent ring. The nucleus pulposus residing at approximately the center of the disc, consists of gelatinous material. The size of the nucleus varies with the spinal region. While more prominent in cervical and lumbar regions, the nucleus is smaller in the thoracic region.

Several ligaments interconnect each vertebral segment. Ligaments are elastic bands made up of collagen and elastin fibres arranged in parallel (Nachemson *et al.*, 1968). Five primary ligaments interconnect each segmental level: anterior and posterior longitudinal ligaments, ligamentum flavum, interspinous ligament, and facet joint capsular ligament. The longitudinal ligaments run continuously along anterior and posterior vertebral body surfaces from cervical to lumbosacral regions. The anterior longitudinal ligament is a broad ligament covering most of the vertebral body anterior surface. The posterior longitudinal ligament is similar to its anterior counterpart, although about one-third as wide. Individual deep fibers of these liagments traverse a single disc and superficial fibers may traverse several segments. The ligamentum flavum, or yellow ligament, spans between adjacent laminae. The thickness of the ligamentum flavum increases inferiorly. The interspinous ligament spans between spinous processes of adjacent levels. The ligamentum flavum and interspinous ligaments are not continuous as fibers connect opposing surfaces of adjacent vertebrae. Capsular ligaments along with synovial membranes and synovial fluid form the facet joints. These synovial joints consist of fluid between the adjacent articular processes. The synovial membrane contains the fluid, and the capsular ligament forms a band around the membrane. The membrane and ligament are collectively known as the joint capsule. Posterior to the spinous processes is the supraspinous ligament in thoracic and lumbar regions and the ligamentum nuchae in the cervical region. The nuchae extends from the occipital bone of the cranium to the posterior extents of cervical spinous processes. The thin supraspinous ligament runs continuously along the posterior extents of spinous processes from C7 to the sacral region.

Fundamental biomechanics

Physiologic activities induce mechanical loading on the spine, resulting in internal deformations. Due to the segmented nature, the biomechanical behavior of the vertebral column is governed by the segmental response. Translations and rotations form the primary response components, and under physiologic loads, soft and hard tissues respond differently. Ligaments are tension-only elements and pre-stressed in vivo. The pre-stress induces nominal resting stiffness to the spine. In contrast, intervertebral discs provide compressive resistance, are responsible for supporting the body weight, and offer resistance in other loading modes. The cervical spine supports the weight of the head and neck, thoracic spine supports the weight of the head, neck, and thorax, and lumbar spine supports the weight of the head, neck, thorax, and abdomen. Facet joints have high compressive stiffness and very low shear stiffness. This permits opposing articular surfaces to slide, resisted primarily by joint capsule tension.

The application of an external force results in deformations and the force-deformation characteristics of a spinal structure are typically sigmodial. Based on the stiffness of the structure, Yoganandan *et al.* characterized the force-deformation response of intervertebral joints into the physiologic phase, traumatic phase, and failure, or the post traumatic phase (Figure 29.1) (Yoganandan *et al.*, 1985). In the physiologic phase, the structure acts as an integral unit; stiffness increases gradually with increasing resistance. This region represents the highest mechanical efficiency domain with no trauma. Although not precisely quantified, the toe region,

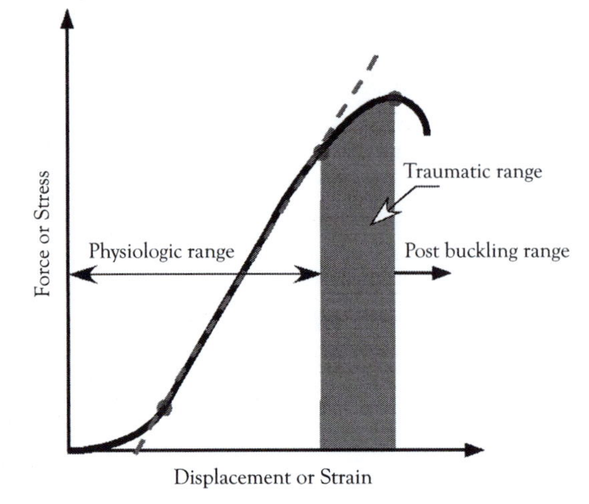

Figure 29.1 Force-deformation response of intervertebral joints demonstrating the physiologic phase, traumatic phase, and the post-traumatic phase (post buckling range).

a terminology frequently used in spine biomechanics literature, refers to the initial portion in the described physiologic phase wherein the structure has very low resistance to applied load. The structure yields with increase in loading and is identified by the onset of decreased stiffness. Microfailure occurs during this phase (Yoganandan *et al.*, 1988). After reaching its peak during the physiologic phase, stiffness decreases to zero at the end of the traumatic phase, indicating that the structure has reached its ultimate load carrying capacity. In the subsequent post traumatic phase, the structure responds with negative resistance. Spinal trauma has been identified on radiographs when the load reaches this level. This concept has been used to describe other structures (White *et al.*, 1990). It has been reported that continued deformation of the anterior longitudinal ligaments leads to failure of individual fibers, commonly referred to as yield or subfailure (Neumann *et al.*, 1994). Other ligament studies have also shown similar nonlinear responses (Chang *et al.*, 1992; Chazal *et al.*, 1985; Myklebust *et al.*, 1988; Pintar *et al.*, 1992; Winkelstein *et al.*, 1999; Yoganandan *et al.*, 1989, 2000).

Segmental motions in response to external loads are categorized into primary and coupled components. Primary motions occur in the direction or plane of the applied load or moment and coupled motions occur in other directions or planes. Translations describe motion of one intervertebral joint (IVJ) relative to the adjacent IVJ, and angular displacements describe rotation of one IVJ relative to the adjacent IVJ. Linear motions occur along the anterior-posterior (AP shear), lateral (side shear), and inferior-superior (compression-tension) directions. Angular motion occurs in the sagittal (flexion/extension), coronal (lateral bending), and horizontal (axial rotation) planes. Complex motions involve combinations of two or more linear or angular motions. Due to complex spinal geometry, motions are typically three-dimensional (3D).

Axial (tension or compression) and shear forces applied to the spine result in linear motions. Primary motions resulting from axial forces are distraction and compression (Brinckmann *et al.*, 1983; Pintar *et al.*, 1995; Yoganandan *et al.*, 1989). Due to facet joint orientation, particularly in the cervical spine, compressive forces result in coupled motions including posterior rotation (extension) of the superior relative to the inferior vertebra (Panjabi *et al.*, 1986). Tensile forces lead to approximately pure distraction on a segmental basis due to minimal bony interaction (Pintar *et al.*, 1986; Yoganandan *et al.*, 1996).

Primary motions from shear forces applied along the anterior-posterior and lateral directions result in anterior-posterior or lateral displacement of the superior relative to the inferior vertebra (Ordway *et al.*, 1999; Panjabi *et al.*, 1986; Yingling *et al.*, 1999). Due to the orientation of facet joints in the cervical region, anterior-posterior shear forces result in coupled motions, including flexion and extension rotation for anteriorly and posteriorly directed shear forces (Panjabi *et al.*, 1986). The almost vertical orientation of facet joints in thoracic and lumbar regions contributes to high anterior-posterior shear stiffness. Under physiologic loading, anterior-posterior shear displacements are small (Stokes *et al.*, 1987). Lateral shear leads to lateral displacement coupled with axial rotation of the superior relative to the inferior vertebra (Panjabi *et al.*, 1986). Uncinate processes in the cervical region and axially-rotated facet joints in the lumbar region lead to coupled axial rotation of the superior vertebra.

Primary motions from bending moments applied in the sagittal, coronal, and horizontal planes result in flexion/extension, lateral bending, and axial rotations. Flexion and extension moments induce sagittal plane rotations of the superior relative to the inferior vertebra in the anterior and posterior directions (Dimnet *et al.*, 1982, 1991, 1993; Frobin *et al.*, 2002; Goel *et al.*, 1984; Lysell *et al.*, 1972; Miura *et al.*, 2002; Ordway *et al.*, 1999; Penning, 1978; Richter *et al.*, 2000; Wen *et al.*, 1993). Lateral bending moment induces coronal plane rotation of the superior relative to the inferior vertebra (Goel *et al.*, 1984; Lysell *et al.*, 1972; Miura *et al.* 2002; Richter *et al.*, 2000; Wen *et al.*, 1993). Axial twisting moment induces horizontal plane rotation of the superior relative to the inferior vertebra (Goel *et al.*, 1984; Lysell *et al.*, 1972; Miura *et al.*, 2002; Richter *et al.*, 2000; Sonoda *et al.*, 1962; Wen *et al.*, 1993). Coupled motions during flexion and extension are limited to anteriorly and posteriorly-directed displacement of the superior relative to the inferior vertebra. Coupled motions are greatest in the cervical and less in thoracic and lumbar regions. Lateral bending and axial twisting moments are linked in the cervical region. Under lateral bending moment, coronal rotation is the primary and axial rotation is the coupled motion, and vice versa for the axial twisting moment. This relationship is attributed to the characteristic anatomy of the cervical region.

Trauma mechanisms

Spine trauma can occur during acute loading such as that resulting from falls, diving, athletic-related activities, and motor vehicle and helicopter crashes. Trauma due to repetitive or cyclic loading can occur during manual material handling tasks, albeit the magnitude of the external force lies below the acute loading injury tolerance. Acute loading event is important in the design and development of biofidelic anthropomorphic test devices, commonly termed dummies, geared toward improvements in crashworthiness of motor vehicles and user-friendly devices such as helmets and playground surfaces. Federal Motor Vehicle Safety Standards, Snell Standards and others focus on this area. Cyclic loading is relevant in manual material handling tasks. The National Institute of Occupational Safety and Health focuses on this area. Spinal injuries from acute loadings can be classified according to impact mode: bending, i.e., sagittal (flexion, extension) and coronal planes; axial loading, i.e., compression and tension; shear; and torsion-related groups; other classifications also exist (Maiman *et al.*, 1991; Myers *et al.*, 1995; Pike *et al.*, 2002; Sances *et al.*, 1984; Yoganandan *et al.*, 1987,1990). Biomechanical studies confirm hypotheses of trauma mechanisms based on field data, epidemiological results, and clinical experience. Because of the complex anatomy and heterogeneity, mechanisms of injury often involve combined vectors.

Flexion-related injuries

These injuries occur due to forward bending of the spine in the sagittal plane, commonly with compressive forces applied through the head (cervical), shoulder (thoracic), or buttocks (lumbar). When the line of action of the compressive force lies anterior to the spine, injuries occur under the compression-flexion mechanism. Wedge fractures, fracture-dislocations, facet dislocations, and spinous process avulsions, especially at the lower cervical spine (clay shoveler), belong to this group.

Vertebral body fractures associated with the disruption of posterior longitudinal ligament integrity can occur with injuries related to the compression-flexion mechanism. Although wedge fractures can occur to any region of the column, the lower cervical spine and the thoracolumbar junction are the two most common regions (Sances *et al.*, 1981, 1984, 1986; Yoganandan *et al.*, 1989, 1990). This is because of the characteristic anatomy and mechanisms of load transfer in the two regions. The degree of wedging depends of the severity of flexion. Greater magnitude of flexion moment is needed to induce a wedge fracture with posterior ligament involvement than a burst fracture. Minor wedge fractures do not often result in neurological deficit. Anterior subluxation of the rostral with respect to the caudal vertebra is a more severe injury than a simple minor wedge fracture.

Facet dislocation represents forward movement of the rostral over the caudal facet, causing the superior vertebral body to sublux anteriorly with respect to the inferior body/segment. If both facets override, dislocation is bilateral, else unilateral. Bilateral facet dislocations often result in anterior shift of the rostral body by more than one-half of its width and involve neurological (Yoganandan *et al.*, 1990) consequences. Perched facets represent a special type of dislocation wherein the inferior tip of the rostral facet lies anatomically over the superior facet of the inferior vertebra. Because of the anatomy and function of the cervical spine and the eccentricity of the head mass, the cervical spine is more susceptible to facet dislocations than thoracic and lumbar regions. These injuries are hypothesized to be due to compression-flexion with muscular involvement due to the flexion component (Pintar *et al.*, 1998). Inverted post mortem human subject (PMHS) osteoligamentous cervical spine-head drops have produced bilateral facet dislocations (Nightingale *et al.*, 1997).

Spinous process avulsions, less critical from a clinical perspective, are hypothesized to occur due to flexion inducing tension in the posterior ligament complex and distracting the tips of the two adjacent processes. This injury is essentially confined to the lower cervical spine. Hyperflexion injuries are an extreme case wherein loading exceeds flexion limits and, without vertebral body fracture, this mechanism results in pure subluxation with ligamentous disruption or sprains with less involvement of the ligamentous complex.

Motor vehicle- and diving-related epidemiological studies indicate that the lower cervical spine is the most susceptible region for injuries related to the compression-flexion mechanism (Portnoy *et al.*, 1972; Sances *et al.*, 1986; Yoganandan *et al.*, 1990).

Laboratory studies have confirmed the mechanism with single cycle anterior eccentric impacts to the cranium of PMHS intact head-neck complexes. According to Pintar *et al.*, 25% probability of flexion-induced cervical spine injury occurs at 62 Nm flexion bending moment and 1.9 kN compressive force (Pintar *et al.*, 1998). Thoracic spine injuries related to the compression-flexion mechanism are minimal due to ribcage, although the thoracolumbar junction is more prone (Sances *et al.*, 1984, 1986). These injuries involve anterior and posterior columns and, from a clinical perspective, ligament status affects spine instability (Denis *et al.*, 1984).

Extension-related injuries

These injuries occur due to rearward bending of the cervical spine in the sagittal plane, with compressive or tensile forces applied through the head or chin. Such injuries are less common in the other two regions of the vertebral column. The kyphotic curvature and ribcage provide structural stability to the thoracic spine, and external loads inducing thoracic or lumbar extension are rare. When the line of action of the compressive force lies posterior to the spine, or the tensile force lies anterior, injuries result from the compression/tension-extension mechanism. Compression vector induces fractures of the posterior bony complex, including spinous process, lamina, pedicle, and vertebral arch (Harris *et al.*, 1996). Isolated posterior arch fracture of the atlas is an example. Similar to flexion, the degree of extension at the segmental level dictates component(s) involved in trauma. Because the outer annulus and anterior longitudinal ligament resist local tension due to posterior eccentric compression, disc-related pathologies can occur (Yoganandan *et al.*, 1997). Laboratory studies have confirmed the compression-extension mechanism using inverted PMHS osteoligamentous cervical spine-head specimens subjected to drop-induced impact (Nightingale *et al.*, 1996). Airbag-related injuries in vehicular environments have been attributed to the tension-extension mechanism. Stretch-related trauma occurs to the spine, essentially confined to the upper cervical regions and soft tissues. This is because of the lack of discs in the occipt-C1-C2 complex. Early field experiences with airbag loading resulting in atlanto-occipital/axial dislocations have implicated the tension-extension mechanism. Injuries can be fatal because of vital function involvement at these levels. Using intact PMHS specimens, Pintar *et al.* reported 50% probability of upper cervical injuries at an extension moment of 75 Nm and tensile force of 3.2 kN (Pintar *et al.*, 2005). Whiplash traumas, classified under the hyperextension mechanism, often occur without direct head contact loading in rear end impacts, are discussed later.

Axial loading-related injuries

These injuries occur due to axial compression or tension loads with the acknowledgement that no loading paradigm is pure. As indicated, impact-induced compressive force application occurs commonly in athletic- and motor vehicle-related incidents (Huelke *et al.*, 1973, 1986; Portnoy *et al.*, 1972; Torg *et al.*, 1987, 1990; Yoganandan *et al.*, 1986, 1998). Burst fractures form the fundamental type of axial loading injury to the spine, most common in the mid-lower cervical spine and thoracolumbar junction. By definition, axial compressive forces predominate and local bending is minimal. Thus, typical fracture-dislocations are not included. With minimal eccentricities, 'pure' burst fractures can occur. Neurological functions remain intact if fractured fragments do not compromise cord integrity. Although the cervical spine has normal lordosis in head impact environments, preflexed head-neck removes the curvature, resulting in a straightened column, aligns the cervical vertebrae, places the neck, resulting in the 'stiffest axis' alignment, and renders the structure susceptible for injuries related to the axial loading mechanism (Liu *et al.*, 1989; Pintar *et al.*, 1995; Torg *et al.*, 1990). Burst fractures occur early, within the first 20 milliseconds during the impact loading event and to the least resistive mid-lower cervical segment (Pintar *et al.*, 1995). PMHS studies have shown that vertical compression leads to disc bulge and endplate tears in association with body fracture (Yoganandan *et al.*, 1986). Compressive impacts to the anterior edge of the vertebral body producing burst fractures confirm the axial loading (with minimal bending) mechanism.

Injuries due to tensile loads involve soft tissue component(s) and bony abnormalities are due to avulsions in contrast to compressive fractures. As discussed, airbag deploying under the chin induces tension and depending on the position of the occupant and airbag interaction, if extension moments are minimal, injuries can occur under this mechanism. Laboratory studies have confirmed this mechanism by reproducing basilar, ring, and dens fractures using intact PMHS subjected to uniaxial tensile loading (Yoganandan *et al.*, 1996). Upper cervical dislocations, as described, have a tensile component. Because of the developing anatomy of the human, distractive forces applied inertially via the head to the child spine can result in a unique category of trauma, termed in clinical literature as spinal cord injuries without radiographic abnormalities.

Lateral bending injuries

Although uncommon, these injuries are attributed to the bending of the neck in the coronal plane and side impact loading in vehicular environments is an example (Allen *et al.*, 1982). Thoracic and lumbar columns are even less involved. Because the cervical spine demonstrates coupling, these injuries are associated with bending in the axial plane (Sherk *et al.*, 1989; Yoganandan *et al.*, 2007). Because the weight of the head always acts on the neck and its center of gravity/mass is eccentrically located with respect to the cervical column, compression and flexion are also associated. Odontoid fractures stemming from the combined lateral bending and compression mechanism have been identified in modern motor vehicle environments and supported by laboratory studies (Mouradian *et al.*, 1978; Yoganandan *et al.*, 1986, 2005). Although rare, unilateral facet trauma in the form of pillar fractures can occur because of ipsilateral compression; diastasis of the contralateral facet joint is also possible.

Torsion-related injuries

These injuries are attributed to rotation in the axial plane. Similar to lateral bending, cervical spine injuries due to torsion are uncommon. Because of the presence of intervertebral discs in the subaxial cervical spine, the role of pure torsion is minimal. However, torsion may play a role in the upper head-neck complex as it is devoid of discs. Laboratory studies using PMHS spines have produced torsion-related trauma to the atlanto-axial complex (Myers *et al.*, 1991). Under in vivo situations, mechanisms are coupled. Although torsion has been postulated as a mechanism for unilateral facet dislocations, laboratory studies have been unable to confirm and, as described, compression combined with bending has produced this trauma. Role of torsion is minimal in the thoracic spine due to the added stability by the ribcage. Because of the orientation of the facet joints and other anatomical characteristics, lower lumbar and lumbosacral regions are susceptible to injury in this mode in situations such as materials handling (Farfan, 1970). Torsion can occur in vehicular environments due to occupant preposition/out-of-position with respect to the seat.

Shear-related injuries

These injuries are specific to the lumbar region and can occur as horizontal fractures of the vertebra due with minimal to no soft tissue involvement. The trauma, termed as Chance fracture, occurs due to the application of the anteroposterior shear force through the lumbar vertebra from lap belt loading in frontal impacts. These injuries are uncommon in current motor vehicle environments because the diagonal shoulder belt restrains torso kinematics. Injuries due to pure shear loading of the thoracic spine are also uncommon because of rib structures. Although lateral impacts may predispose the human cervical spine to lateral shear, as described earlier, injuries are rare. Laboratory studies have been unsuccessful in producing pure shear-related cervical spine trauma under impact-induced lateral accelerations using intact PMHS (Yoganandan *et al.*, 2006).

Whiplash injuries

Due to the occult nature of injuries sustained during low-speed automotive rear impacts, several mechanisms have been proposed. Head-neck hyperextension was an original mechanism. It was based on the large posterior head rotation in simulated rear impacts with human volunteers and PMHS, and field studies with automobiles with no head restraints (Macnab *et al.*, 1964, 1971; Mertz *et al.*, 1967; Severy *et al.*, 1955). Head restraints were, therefore, implemented in all passenger vehicles in the US since 1970. However, whiplash injury rates did not markedly decrease (Kahane *et al.*, 1982; O'Neill *et al.*, 1972; States *et al.*, 1972). Nerve root injury due to pressure gradients in the spinal canal and anterior neck muscle injury due to eccentric contraction are other mechanisms (Aldman *et al.*, 1986; Bostrom *et al.*, 1996; Brault *et al.*, 2000; Svensson *et al.*, 1993). Nerve root injury leads to radicular pain in the upper extremities and muscular injuries typically heal. Consequently, the most common complaints, i.e., suboccipital headache and neck pain, are not supported by these mechanisms (Deng *et al.*, 2000). Recent experimental studies have identified a shear mechanism (Cusick *et al.*, 2001; Deng *et al.*, 2000; Pearson *et al.*, 2004; Stemper *et al.*, 2004, 2005; Yang *et al.*, 1996). It is based on the development of large anteroposterior shear force in the lower cervical spine during the initial stages rear impact loading, as the head remains stationary due to its inertia. The anterior displacement of the inferior relative to the superior vertebra results in facet capsule stretch (Stemper *et al.*, 2005). Depending on the stretch magnitude, subfailure may occur and allodynia may follow (Lee *et al.*, 2004).

Trauma variables

The human spine is dynamic in terms of anatomy, physiology, biomechanics, and trauma. Early developmental changes gradually lead to the ossified column, age-related changes in the adult spine often lead to degenerative changes such as spondylolsis, some functional adaptations occur due to anatomical changes, and these affect biomechanics, i.e., mechanism of load transfer within the components, and hence, resistance to traumatic loads. Factors such as alignment, age, gender, and loading rate play a role in trauma.

Maiman *et al.* showed that injuries and injury mechanisms depend on initial cervical alignment; spines aligned along the stiffest axis produce injuries related to the axial loading mechanism, aligned posterior to the stiffest axis result in injuries related to the compression-extension mechanism, and aligned anteriorly induce injuries related to the compression-flexion mechanism (Maiman *et al.*, 2002). Significant differences in vertebral body architecture between lumbar and cervical spines contribute to variations in kinetic (and kinematic) responses between the two regions (Yoganandan *et al.*, 2006).

Females are more susceptible than males to acute trauma and chronic symptoms. Biomechanical, psychological, sociological, or anthropometry-related factors may influnce injury rates. Simulated automotive rear impacts with human volunteer and PMHS have shown greater segmental angulations and linear facet joint shear motions in females (Siegmund *et al.*, 1997, 1999; Stemper *et al.*, 2203, 2004; van den Kroonenberg *et al.*, 1998).

Gender dependence has been attributed to a more slender neck and cervical column (Stemper *et al.*, 2008), decreased cartilaginous cover on the cervical facet joint surface (Yoganandan *et al.*, 2003), differing ligamentous material properties (Chandrashekar *et al.*, 2006), differing ligamentous collagen and elastin content (Osakabe *et al.*, 2001), increased segmental ROM (Gore *et al.*, 1986; Kuhlman *et al.*, 1993; Lind *et al.*, 1989; Youdas *et al.*, 1992), and early signs of degeneration (Gore *et al.*, 2001; Harrison *et al.*, 2002).

Advancing age has an effect on spine trauma. Clinical whiplash injury studies indicate that older age has a significant correlation with higher pain scores and neck disability index (Crouch *et al.*, 2006). Another study of 80 individuals with neck pain following automotive impact showed that age is a significant predictor of persistent moderate to severe symptoms (Sterling *et al.*, 2005). At one-year follow-up, patients older than 60 years reported more impaired conditions than at the time of initial injury.

Older age is independently associated with longer recovery time (Suissa *et al.*, 2003). A study of 1,147 occupants in rear impacts found age to be an independent predictor of trauma (Pobereskin *et al.*, 2005).

From a biomechanical perspective, vertebral compressive strength decreases with age (Hansson *et al.*, 1980, 1981; Riggs *et al.*, 1981; Ross *et al.*, 1987), compressive breaking load decreases by 50% for the 60–79 year group compared to the 20–39 year group (Yamada *et al.*, 1973), and disc degeneration affects spine biomechanics (Rohlmann *et al.*, 2006). Sagittal plane ROM is greater in younger than older PMHS cervical spine (Miura *et al.*, 2002; Richter *et al.*, 2000; Schulte *et al.*, 1989; Wen *et al.*, 1993). Spinal curvature alteration is a sign of degeneration. A computational study modeled different spinal curvatures and measured cervical facet joint capsular ligament elongations due to rear impact loading (Stemper *et al.*, 2005). C2 to C3 and C4 to C7 facet joint ligament elongations were greater in straightened and kyphotic than normal lordotic curvatures. These results support clinical observations, i.e., patients with abnormal curvatures have poorer long-term outcomes than patients with normal lordosis at the time of rear impact (Hohl *et al.*, 1990).

Loading rate studies in intervertebral discs, spine and anterior cruciate ligaments, patellar tendons, temporomandibular joint discs, and arterial vessels have shown that stiffness increases with increasing rate (Chin *et al.*, 1996; Collins *et al.*, 1972; Danto *et al.*, 1993; Haut *et al.*, 1997; Lawton *et al.*, 1955; Race *et al.*, 2000; Stemper *et al.*, 2007; Yoganandan *et al.*, 1989). Vertebral compressive strength increases by approximately two-fold between 0.53 and 5,334 cm/minute loading rates (Kazarian *et al.*, 1977). Ligament tensile strength increases by approximately four-fold between 0.7 and 2,500 mm/sec loading rates. Thus, as a first step, it behoves to examine the role of age, gender, and loading rate on trauma biomechanics. Pintar *et al.*, determined the tolerance of the cervical spinal column to be 18 mm mean axial compression and 3.3 KN of mean compressive force delivered to the head (Pintar *et al.*, 1995). All specimens sustained at least one vertebral body fracture, such as burst and wedge fractures related to the axial loading mechanism (Pintar *et al.*, 1990). Pintar *et al.* also conducted statistical analysis of their experimental data using the Cox proportional hazards model and derived injury probability curves (Figure 29.2) as a function of age, loading rate, and gender (Pintar *et al.*, 1998). Briefly, age and rate had an interactive effect. Increasing age reduced the rate effect and, approximately at 80 years, rate was insignificant. Males were consistently 600 N stronger than females. Decreased trabecular lattice density with advancing age and marrow content alterations are factors for the interactive effect.

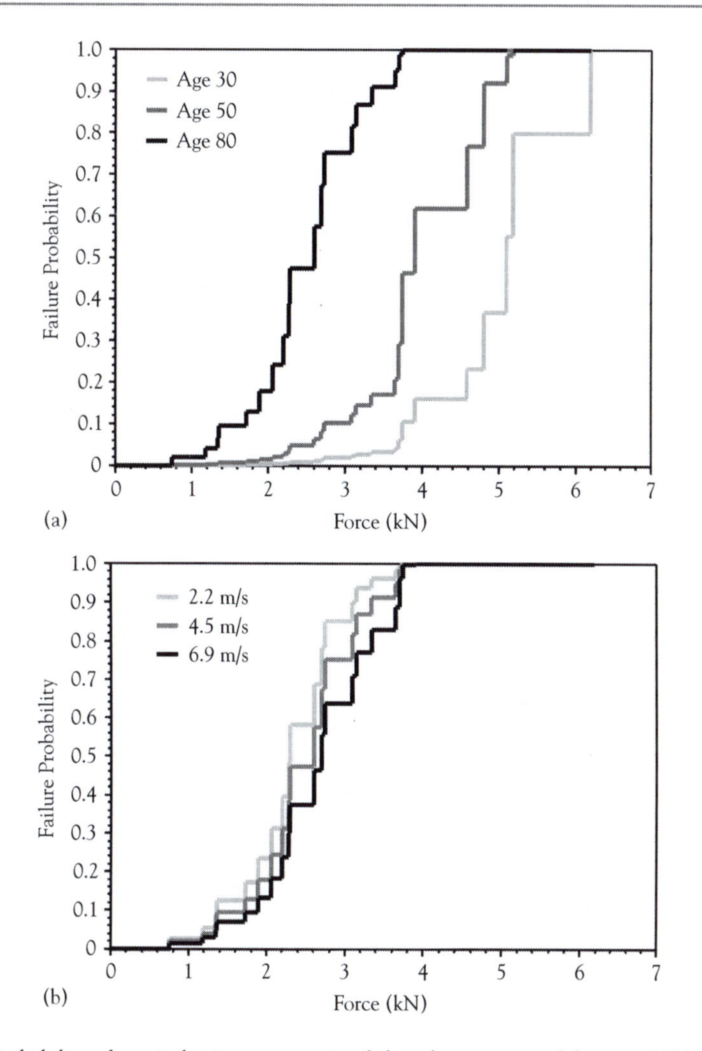

Figure 29.2 Probability of cervical spine compressive failure for increasing (a) age and (b) loading rates.

These results underscore the need for clinicians to incorporate rate in addition to demographics during trauma assessment and treatment. In addition, these biomechanical data are valuable in the development of biofidelic dummies for different adult age groups.

Summary

An attempt has been made in this chapter to provide basic insights into the biomechanics of spine trauma through a description of the anatomy and basic mechanics of spine components and segments, leading to a discussion of injuries and injury mechanisms. Information is focused on acute traumatic loads such as those occurring in vehicular environments. The cervical spine is covered more than the thoracic and lumbar spines because of its vital function and injury frequency. However, the detailed bibliography should assist the reader in additional information. Understanding spine trauma is complex owing to its 3D anatomy, heterogeneous components, loading rate dependent responses of different tissues, gender-related variations, developmental and physiological changes resulting in altered load carrying capacities of components and mechanisms of load transfer, and variability in the specificity of the external load vector. Studies have begun to crystallize on tolerances in areas such as injuries associated with the axial loading mechanism. As applied to current motor vehicle environments, with continuing advances in technology such as frontal and side airbags and seatbelt pretensioners and load limiters, and public awareness for safety, it is important to pursue this area to further mitigate trauma. Since the cervical spine controls head kinematics and plays a role in brain injury mechanics, it is also important to pursue this research for head trauma mitigation.

Acknowledgements

This research was supported by VA Medical Research.

References

1. Aldman, B. (1986) 'An analytical approach to the impact biomechanics of head and neck injury'. 30th Annual Conference of the Association for the Advancement of Automotive Medicine. Montreal, Quebec, 439–54.
2. Allen, B.L., Ferguson, R.L., Lehmann, T.R., *et al.* (1982) 'A mechanistic classification of closed, indirect fractures and dislocations of the lower cervical spine'. *Spine* 7:1–27.
3. Bostrom, O., Svensson, M., Aldman, B., *et al.* (1996) 'A new neck injury criterion candidate-based on injury findings in the cervical spinal ganglia after experimental sagittal whiplash'. International Research Council on the Biomechanics of Impact (IRCOBI). Dublin, Ireland, 123–6.
4. Brault, J., Siegmund, G., Wheeler, J., (2000) 'Cervical muscle response during whiplash: Evidence of a lengthening muscle contraction'. *Clin Biomech* 15:426–35.
5. Brinckmann, P., Frobin, W., Hierholzer, E., *et al.* (1983) 'Deformation of the vertebral end-plate under axial loading of the spine'. *Spine* 8:851–6.
6. Chandrashekar, N., Mansouri, H., Slauterbeck, J., *et al.* (2006) 'Sex-based differences in the tensile properties of the human anterior cruciate ligament'. *J Biomech* 39:2943–50.
7. Chang, H., Gilbertson, L.G., Goel, V.K., *et al.* (1992) 'Dynamic response of the occipito-atlanto-axial (C0-C1-C2) complex in right axial rotation'. *J Orthop Res* 10:446–53.
8. Chazal, J., Tanguy, A., Bourges, M., *et al.* (1985) 'Biomechanical properties of spinal ligaments and a histological study of the supraspinal ligament in traction'. *J Biomech* 18:167–76.
9. Chin, L.P.Y., Aker, F.D., Zarrinnia, K. (1996) 'The viscoelastic properties of the human temporomandibular joint disc'. *J Oral Maxillofac Surg* 54:315–18.
10. Collins, R., Hu, W.C.L. (1972) 'Dynamic deformation experiments on aortic tissue'. *J Biomech* 5:333–7.
11. Crouch, R., Whitewick, R., Clancy M, *et al.* (2006) 'Whiplash associated disorder: incidence and natural history over the first month for patients presenting to a UK emergency department'. *Emerg Med J.* 23:114–18.
12. Cusick, J.F., Pintar, F.A., Yoganandan, N. (2001) 'Whiplash syndrome: Kinematic factors influencing pain patterns'. *Spine* 26:1252–8.
13. Danto, M.I., Woo, S.L. (1993) 'The mechanical properties of skeletally mature rabbit anterior cruciate ligament and patellar tendon over a range of strain rates'. *J Orthop Res* 11:58–67.
14. Deng, B., Begeman, P.C., Yang, K.Y., *et al.* (2000) 'Kinematics of human cadaver cervical spine during low speed rear-end impacts'. *Stapp Car Crash J* 44:171–88.

15. Denis, F. (1984) 'Spinal instability as defined by the three-column spine concept in acute spinal trauma'. *Clin Orthop Relat Res* 189:65–76.
16. Dimnet, J., Pasquet, A., Krag, M.H., *et al.* (1982) 'Cervical spine motion in the sagittal plane: kinematic and geometric parameters'. *J Biomech* 15:959–69.
17. Dvorak, J., Panjabi, M.M., Grob, D., *et al.* (1993) 'Clinical validation of functional flexion/extension radiographs of the cervical spine'. *Spine* 18:120–7.
18. Dvorak, J., Panjabi, M.M., Novotny, J.E., *et al.* (1991) 'Clinical validation of functional flexion-extension roentgenograms of the lumbar spine'. *Spine* 16:943–50.
19. Farfan. 'Torsion of the lumbar intervertebral joint'. *JBJS* 1970:109–15.
20. Frobin, W., Leivseth, G., Biggemann, M., *et al.* (2002) 'Sagittal plane segmental motion of the cervical spine'. A new precision measurement protocol and normal motion data of healthy adults. *Clin Biomech* (Bristol, Avon), 17:21–31.
21. Goel, V.K., Clark, C.R., McGowan, D., *et al.* (1984) 'An in-vitro study of the kinematics of the normal, injured and stabilized cervical spine'. *J Biomech* 17:363–76.
22. Gore, D.R. (2001) 'Roentgenographic findings in the cervical spine in asymptomatic persons: a ten-year follow-up'. *Spine* 26:2463–6.
23. Gore, D.R., Sepic, S.B., Gardner, G.M. (1986) 'Roentgenographic findings of the cervical spine in asymptomatic people'. *Spine* 11:521–4.
24. Hansson, T., Roos, B. (1981) 'The relation between bone mineral content, experimental compression fractures, and disc degeneration in lumbar vertebrae'. *Spine* 6:147–53.
25. Hansson, T., Roos, T., Nachemson, A. (1980) 'The bone mineral content and ultimate compressive strength of lumbar vertebrae'. *Spine* 5:46–55.
26. Harris, J.H., Jr, Mirvis, S.E. (1996) *The Radiology of Acute Cervical Spine Trauma.* Third ed. Philadelphia: Williams & Wilkins.
27. Harrison, D.E., Bula, J.M., Gore, D.R. (2002) 'Roentgenographic findings in the cervical spine in asymptomatic persons: A 10-year follow-up'. *Spine* 27:1249–50.
28. Haut, T.L., Haut, R.C. (1997) 'The state of tissue hydration determines the strain-rate-sensitive stiffness of human patellar tendon'. *J Biomech*, 30:79–81.
29. Hohl, M. (1990) 'Soft tissue neck injuries – a review'. *Rev Chir Orthop*, 76:15–25.
30. Huelke, D.F., Marsh, J.C., DiMento, L., *et al.* (1973) 'Injury causation in rollover accidents'. 17th Annual AAAM Conf. Morton Grove, IL, 87–115.
31. Huelke, D.F., Nusholtz, G.S. (1986) 'Cervical spine biomechanics: a review of the literature'. *J Orthop Res*. 4:232–45.
32. Kahane, C. (1982) 'An evaluation of head restraints-federal motor vehicle safety standard 202'. Springfield, VA: National Technical Information Service, NHTSA, 1982.
33. Kazarian, L., Graves, G.A., Jr. (1977) 'Compressive strength characteristics of the human vertebral centrum'. *Spine* 2:1–14.
34. Kuhlman, K.A. (1993) 'Cervical ROM in the elderly'. *Arch Phys Med Rehabil* 74:1071–9.
35. Lawton, R.W. (1955) 'Measurements on the elasticity and damping of isolated aortic strips in the dog'. *Circ Res* 3:403–8.
36. Lee, K.E., Thinnes, J.H., Gokhin, D.S., *et al.* (2004) 'A novel rodent neck pain model of facet-mediated behavioural hypersensitivity: implications for persistent pain and whiplash injury'. *J Neurosci Methods* 137:151–9.
37. Lind, B., Sihlbom, H., Nordwall, A., *et al.* (1989) 'Normal ROM of the cervical spine'. *Arch Phys Med Rehabil* 70:392–695.
38. Liu, Y.K., Dai, Q.G. (1989) 'The second stiffest axis of a beam-column: implications for cervical spine trauma'. *J Biomech Eng*, 111:122–7.
39. Lysell, E. (1972) 'The pattern of motion in the cervical spine'. In Zotterman, C, Hirsch, Y., eds. Cervical pain. Oxford, UK: Pergamon Press, 53–8.
40. Macnab, I. (1964) 'Acceleration injuries of the cervical spine'. *J Bone Joint Surg Am* 46–A:1797–800.
41. Macnab, I. (1971). 'The "Whiplash Syndrome"'. *Orthop Clin North Am* 2:389–403.

42. Maiman, D.J., Yoganandan, N. (1991) 'Biomechanics of cervical spine trauma'. *Clin Neurosurg* 37:543–70.
43. Maiman, D.J., Yoganandan, N., Pintar, F.A. (2002) 'Preinjury cervical alignment affecting spinal trauma'. *J Neurosurg* 97:57–62.
44. Mertz, H.J. (1967) 'The kinematics and kinetics of whiplash'. *Engineering Mechanics*. Detroit, MI: Wayne State University, 370.
45. Mertz, H.J., Patrick, L.M. (1967) 'Investigation of the kinematics and kinetics of whiplash'. 11th Stapp Car Crash Conf. Anaheim, CA: Society of Automotive Engineers, Inc., 267–317.
46. Miettinen, T., Airaksinen, O., Lindgren, K.A., *et al.* (2004) 'Whiplash injuries in Finland – the possibility of some sociodemographic and psychosocial factors to predict the outcome after one year'. *Disabil Rehabil* 26:1367–72.
47. Miura, T., Panjabi, M.M., Cripton, P.A. (2002) 'A method to simulate in vivo cervical spine kinematics using a compressive preload'. *Spine* 27:43–8.
48. Mouradian, W., Fietti, V.J., Cochran, G., *et al.* (1978) 'Fractures of the odontoid: a laboratory and clinical study of mechanisms'. *Orthop Clin North Am* 9:985–1001.
49. Myers, B., McElhaney, J.H., Doherty, B.J., *et al.* (1991) 'The role of torsion in cervical spine trauma'. *Spine* 16:870–4.
50. Myers, B.S., Winkelstein, B.A. (1995) 'Epidemiology, classification, mechanism, and tolerance of human cervical spine injuries'. *Crit Rev Biomed Eng* 23:307–409.
51. Myklebust, J.B., Pintar, F.A., Yoganandan, N., *et al.* (1988) 'Tensile strength of spinal ligaments'. *Spine* 13:526–31.
52. Nachemson, A.L., Evans, J.H. (1968) 'Some mechanical properties of the third lumbar interlaminar ligament (ligamentum flavum). *J Biomech* 1:211–20.
53. Neumann, P., Keller, T.S., Ekstrom, L. *et al.* (1994) 'Effect of strain rate and bone mineral on the structural properties of the human anterior longitudinal ligament'. *Spine* 19:205–11.
54. Nightingale, R., McElhaney, J., Camacho, D., *et al.* (1997) 'The dynamic responses of the cervical spine: Buckling, end conditions, and tolerance in compressive impacts'. 41st Stapp Car Crash Conf. Lake Buena Vista, FL,451–71.
55. Nightingale, R.W., McElhaney, J.H., Richardson, W.J., *et al.* (1996) 'Dynamic responses of the head and cervical spine to axial impact loading'. *J Biomech* 29:307–18.
56. O'Neill, B., Haddon, W., Kelley, A., *et al.* (1972) 'Automobile head restraints: Frequency of neck injuries insurance claims in relation to the presence of head restraints'. *Am J Public Health* 62:569–73.
57. Ordway, N.R., Seymour, R.J., Donelson, R.G., *et al.* (1999) 'Cervical flexion, extension, protrusion, and retraction: a radiographic segmental analysis'. *Spine* 24:240–7.
58. Osakabe, T., Hayashi, M., Hasegawa, K., *et al.* (2001) 'Age- and gender-related changes in ligament components'. *Ann Clin Biochem* 38:527–32.
59. Panjabi, M.M., Summers, D.J., Pelker, R.R., *et al.* (1986) 'Three-dimensional load–displacement curves due to forces on the cervical spine'. *J Orthop Res* 4:152–62.
60. Pearson, A.M., Ivancic, P.C., Ito, S., *et al.* (2004) 'Facet joint kinematics and injury mechanisms during simulated whiplash'. *Spine* 29:390–7.
61. Penning, L. (1978) 'Normal movements of the cervical spine'. 130:317–26.
62. Pike, J., Pintar, F., Yoganandan, N., *et al.* (2002) 'The use of x-rays, CT and MRI to study crash-related injury mechanisms'. In JA P ed. Neck injury. Warrendale, PA: SAE, 95–151.
63. Pintar, F., Voo, L., Yoganandan, N. (1998) 'The mechanisms of hyperflexion cervical spine injury'. *IRCOBI*. Goteborg, Sweden, 349–63.
64. Pintar, F.A., Myklebust, J., Sances, Jr. A., *et al.* (1986) 'Biomechanical properties of the human intervertebral disk in tension'. *ASME Adv Bioeng*. New York, NY, 38–9.
65. Pintar, F.A., Sances, Jr. A., Yoganandan, N., *et al.* (1990) 'Biodynamics of the total human cadaveric cervical spine'. 34th Stapp Car Crash Conf. Orlando, FL.
66. Pintar, F.A., Yoganandan, N., Baisden, J. (2005) 'Characterizing occipital condyle loads under high-speed head rotation'. *Stapp Car Crash J* 49:33–47.

67. Pintar, F.A., Yoganandan, N., Myers, T., *et al.* (1992) 'Biomechanical properties of human lumbar spine ligaments'. *J Biomech* 25:1351–6.
68. Pintar, F.A., Yoganandan, N., Pesigan, M., *et al.* (1995) 'Cervical vertebral strain measurements under axial and eccentric loading'. *J Biomech Eng* 117:474–8.
69. Pintar, F.A., Yoganandan, N., Voo, L. (1998) 'Effect of age and loading rate on human cervical spine injury threshold'. *Spine* 23:1957–62.
70. Pintar, F.A., Yoganandan, N., Voo, L.M., *et al.* (1995) 'Dynamic characteristics of the human cervical spine'. *SAE Transactions* 104:3087–94.
71. Pobereskin, L.H. (2005) 'Whiplash following rear end collisions: a prospective cohort study. *J Neurol Neurosurg Psychiatry*' 76:1146–51.
72. Portnoy, H.J., McElhaney, J.H., Melvin, J., *et al.* (1972) 'Mechanism of cervical spine injury in auto accidents'. *15th Annual Conference of the Association for the Advancement of Automotive Medicine*, 58–83.
73. Race, A., Broom, N.D., Robertson, P. (2000) 'Effect of loading rate and hydration on the mechanical properties of the disc'. *Spine* 25: 662–9.
74. Richter, M., Wilke, H.J., Kluger, P., *et al.* (2000) 'Load-displacement properties of the normal and injured lower cervical spine in vitro'. *Eur Spine J*; 9:104–8.
75. Riggs, B.L., Wahner, H.W., Dunn, W.L., *et al.* (1981) 'Differential changes in bone mineral density of the appendicular and axial skeleton with aging'. *J Clin Invest* 67:328–35.
76. Rohlmann, A., Zander, T., Schmidt, H., *et al.* (2006) 'Analysis of the influence of disc degeneration on the mechanical behaviour of a lumbar motion segment using the finite element method'. *J Biomech* 39:2484–90.
77. Ross, P.D., Wasnich, R.D., Heilbrun, L.K., *et al.* (1987) 'Definition of a spine fracture threshold based upon prospective fracture risk'. *Bone* 8:271–8.
78. Sances, A., Jr, Myklebust, J.B., Maiman, D.J., *et al.* (1984) 'The Biomechanics of Spinal Injuries'. *CRC Critical Reviews in Biomedical Engineering* 11:1–76.
79. Sances, A., Jr, Thomas, D.J., Ewing, C.L., *et al.* eds. (1986) *Mechanisms of Head and Spine Traumaed.* Goshen, NY: Aloray
80. Sances, A., Jr., Weber, R.C., Larson, S.J., *et al.* (1981) 'Bioengineering analysis of head and spine injuries'. *Crit Rev Bioeng* 5:79–122.
81. Schulte, K.R., Clark, C.R., Goel, V.K. (1989) 'Kinematics of the cervical spine following discectomy and stabilization'. *Spine* 14:1116–21.
82. Severy, D.M., Mathewson, J.H., Bechtol, C.O. (1955) 'Controlled automobile rear-end collisions, an investigation of related engineering and medical phenomena'. *Can Serv Med J* 11:727–59.
83. Sherk, H. (1989) *The Cervical Spine: Second Edition.* Philadelphia: J.B. Lippincott Company
84. Siegmund, G.P., Heinrichs, B.E., Wheeler, J.B. (1999) 'The influence of head restraint position and occupant factors on peak head/neck kinematics in low-speed rear-end collisions'. *Accid Anal Prev* 31:393–407.
85. Siegmund, G.P., King, D.J., Lawrence, J.M., *et al.* (1997) 'Head/neck kinematic response of human subjects in low-speed rear-end collisions'. *41st Stapp Car Crash Conf.* Lake Buena Vista, FL, 357–85.
86. Sonoda, T. (1962) 'Studies on the strength for compression, tension and torsion of the human vertebral column'. *J Kyoto Prefectural Medical University* 71:659–62.
87. States, J.D., Balcerak, J.C., Williams, J.S. (1972) 'Injury frequency and head restraint effecttiveness in rear–end impact accidents'. *16th Stapp Car Crash Conf.* New York: Society of Automotive Engineers, Inc., 228–57.
88. Stemper, B.D., Yoganandan, N., Gennarelli, T.A., *et al.* (2005) 'Localized cervical facet joint kinematics under physiological and whiplash loading'. *J Neurosurg Spine* 3:471–6.
89. Stemper, B.D., Yoganandan, N., Pintar, F.A. (2003) 'Gender dependent cervical spine segmental kinematics during whiplash'. *J Biomech* 36:1281–9.
90. Stemper, B.D., Yoganandan, N., Pintar, F.A. (2004) 'Gender- and region-dependent local facet joint kinematics in rear impact: implications in whiplash injury'. *Spine* 29:1764–71.

91. Stemper, B.D., Yoganandan, N., Pintar, F.A. (2005) 'Effects of abnormal posture on capsular ligament elongations in a computational model subjected to whiplash loading'. *J Biomech* 38:1313–23.
92. Stemper, B.D., Yoganandan, N., Pintar, F.A. (2007) Mechanics of arterial subfailure with increasing loading rate. *J Biomech*, 40(8): 1806–1812.
93. Stemper B.D., Yoganandan N., Pintar F.A., *et al.* (2008) Anatomical gender differences in cervical vertebrae of size-matched volunteers. *Spine* 33(2): E44–E49.
94. Sterling, M., Jull, G., Vicenzino, B., *et al.* (2005) 'Physical and psychological factors predict outcome following whiplash injury'. *Pain* 114:141–8.
95. Stokes, I.A., Frymoyer, J.W. (1987) 'Segmental motion and instability'. *Spine* 12:688–91.
96. Suissa, S. (2003) 'Risk factors of poor prognosis after whiplash injury'. *Pain Res Manag* 8:69–75.
97. Svensson, M.Y., Aldman, B., Hansson, H.A., *et al.* (1993) 'Pressure effects in the spinal canal during whiplash extension motion'. *International Research Council on the Biomechanics of Impact (IRCOBI)*. Eindhoven, Netherlands, 189–200.
98. Torg, J., Sennett, B., Vegso, J. (1987) 'Spinal injury at the level of third and fourth cervical vertebrae resulting from the axial loading mechanism: an analysis and classification'. *Clin Sports Med* 6:159–83.
99. Torg, J., Vegso, J., O'Neil, M., *et al.* (1990) 'The epidemiologic, pathologic, biomechanical, and cinematographic analysis of football-induced cervical spine trauma'. *Am J Sports Med* 18:50–7.
100. van den Kroonenberg, A., Philippens, M., Cappon, H., *et al.* (1998) 'Human head-neck response during low-speed rear end impacts'. *42nd Stapp Car Crash Conf.* Tempe, AZ, 207–21.
101. Wen, N., Lavaste, F., Santin, J.J, *et al.* (1993) 'Three-dimensional biomechanical properties of the human cervical spine in vitro'. *Eur Spine J* 2:2–11.
102. White, A.A., Panjabi, M.M. (1990) *Clinical Biomechanics of the Spine*. Philadelphia: JB Lippincott.
103. Winkelstein, B.A., Nightingale, R.W., Richardson, W.J., *et al.* (1999) 'Cervical facet joint mechanics: Its application to whiplash injury'. *43rd Stapp Car Crash Conf.* San Diego, CA: Society of Automotive Engineers, 243–52.
104. Yamada, H. (1973) *Strength of Biological Materialised.* Baltimore, MD: Williams & Wilkins Company,.
105. Yang, K.J., Begeman, P.C. (1996) 'A proposed role for facet joints in neck pain after low to moderate speed rear end impacts'. 6th Injury Prevention Through Biomechanics Symposium, 59–63.
106. Yingling, V.R., McGill, S.M. (1999) 'Anterior shear of spinal motion segments. Kinematics, kinetics, and resultant injuries observed in a porcine model'. *Spine* 24:1882–9.
107. Yoganandan, N., Ray, G., Sances, A., *et al.* (1985) 'Assessment of traumatic failure load and microfailure load in an intervertebral disc segment'. *Adv Bioeng* 130–1.
108. Yoganandan, N., Baisden, J.L., Maiman, D.J., *et al.* (2005) 'Type II odontoid fracture from frontal impact: case report and biomechanical mechanism of injury'. *J Neurosurg Spine* 2:481–5.
109. Yoganandan, N., Haffner, M., Maiman, D.J., *et al.* (1990) 'Epidemiology and injury biomechanics of motor vehicle related trauma to the human spine'. *SAE Transactions* 98:1790–807.
110. Yoganandan, N., Knowles, S.A., Maiman, D.J., *et al.* (2003) 'Anatomic study of the morphology of human cervical facet joint'. *Spine* 28:2317–23.
111. Yoganandan, N., Kumaresan, S., Pintar, F.A. (2000) 'Geometric and mechanical properties of human cervical spine ligaments'. *J Biomech Eng* 122:623–9.
112. Yoganandan, N., Maiman, D.J., Pintar, F., *et al.* (1988) 'Microtrauma in the lumbar spine: a cause of low back pain'. *Neurosurgery* 23:162–8.
113. Yoganandan, N., Myklebust, J.B., Ray, G., *et al.* (1987) 'Mathematical and finite element analysis of spine injuries'. *Crit Rev Biomed Eng* 15:29–93.
114. Yoganandan, N., Pintar, F., Butler, J., *et al.* (1989) 'Dynamic response of human cervical spine ligaments'. *Spine* 14:1102–10.
115. Yoganandan, N., Pintar, F., Cusick, J.F. (1997) 'Biomechanics of compression-extension injuries to the cervical spine'. *41st Association for the Advancement of Automotive Engineering.* Orlando, FL, 331–44.
116. Yoganandan, N., Pintar, F.A., eds. (1998) 'Frontiers in Head and Neck Trauma: Clinical and Biomechanical'. The Netherlands: IOS Press.

117. Yoganandan, N., Pintar, F.A., Butler, J., *et al.* (1989) 'Dynamic response of human cervical spine ligaments'. *Spine* 14:1102–10.
118. Yoganandan, N., Pintar, F.A., Gennarelli, T., *et al.* (2006) 'Head linear and rotational accelerations and craniocervical loads in lateral impact'. IRCOBI. Madrid, Spain,127–42.
119. Yoganandan, N., Pintar, F.A., Maiman, D.J., *et al.* (1996) 'Human head-neck biomechanics under axial tension'. *Med Eng Phys* 18:289–94.
120. Yoganandan, N., Pintar, F.A., Maiman, D.J., *et al.* (1996) 'Human head-neck biomechanics under axial tension'. *Med Eng Physics* 18:289–94.
121. Yoganandan, N., Pintar, F.A., Sances, A., Jr, *et al.* (1989) 'Biomechanical investigations of the human thoracolumbar spine'. *SAE Transactions* 97:676–81.
122. Yoganandan, N., Pintar, F.A., Stemper, B.D., *et al.* (2006) 'Bone mineral density of human female cervical and lumbar spines from quantitative computed tomography'. *Spine* 31:73–6.
123. Yoganandan, N., Pintar, F.A., Stemper, B.D., *et al.* (2006) 'Trabecular bone density of male human cervical and lumbar vertebrae'. *Bone* 39:336–44.
124. Yoganandan, N., Pintar, F.A., Stemper, B.D., *et al.* (2007) 'Level-dependent coronal and axial moment-rotation corridors of degeneration-free cervical spines in lateral flexion'. JBJS, in press.
125. Yoganandan, N., Ray, G., Pintar, F.A., *et al.* (1989) 'Stiffness and strain energy criteria to evaluate the threshold of injury to an intervertebral joint'. *J Biomech* 22:135–42.
126. Yoganandan, N., Sances, A., Jr., Maiman, D.J., *et al.* (1986) 'Experimental spinal injuries with vertical impact'. *Spine* 11:855–60.
127. Yoganandan, N., Sances, Jr, A., Maiman, D.J., *et al.* (1986) 'Experimental spinal injuries with vertical impact'. *Spine* 11:855–60.
128. Youdas, J.W., Garrett, T.R., Suman, V.J., *et al.* (1992) 'Normal ROM of the cervical spine: An initial goniometric study'. *Phys Ther* 72:770–80.

30

In vivo biomechanical study for injury prevention

Mario Lamontagne[1], D. L. Benoit[1], D. K. Ramsey[1], A. Caraffa[1] and G. Cerulli[2]
[1]University of Ottawa, Ottawa; [2]University at Buffalo, Buffalo

Introduction

Accurate and precise measurements of the kinematics and kinetics of human joints are necessary to understand the normal and pathological function of the musculoskeletal system during movement performance [1-6]. Yet knowledge about joint kinematics and kinetics are limited by the accuracy of measuring system used [7]. Usually joint kinematics are obtained by attaching reflective markers to the segment on the skin. Based upon rigid body mechanics, three-dimensional kinematics assumes that markers placed on the skin represent the position of bony landmarks of the segment. However, skin markers move in relation to these bony landmarks resulting in relative errors [8]. Consequently, considerable questions remain concerning the accuracy of joint kinematics [9]. Reinschmidt *et al.* [10] have reported a segmental error due to skin movement artefact of approximately 5 degrees. *In vivo* measurements of the skeletal motion by typically using markers fixed on bone pins represent an accurate technique but ethically questionable [11-13]. This provides one of the most accurate means for determining bone movements [14]. Since *in vivo* and invasive methods for joint kinematics are not suitable for routine analyses, surface marker optimisation methods have been proposed to correct skin movement artefacts through the application of clusters of markers [15-17]. Some methods reported reduced kinematics errors between 25 and 33%, however, no studies have validated their techniques with *in vivo* kinematics data.

Recently, much advancement has been committed to the measurement of in vivo kinematics using various imaging techniques [5, 18-20]. These techniques include using sagittal plane fluoroscopy combined with computed tomography (CT) image-models to examine the position of joint contact points [21-23], using cine phase-contrast MRI techniques to determine the motion of the bone [24] and using biplanar radiograph techniques to determine *in vivo* joint motion. Dynamic imaging techniques are very promising for measuring 3D joint kinematics [20], however, these techniques have important limitations: small field of view limiting range of motion of the joint, not able to capture cyclic motion like gait under weight-bearing and low sampling rate. Given the aforementioned limitations, bone pins combined with high-speed stereophotogrammetry remains one of the most accurate and valid means of measuring physiological knee motions during normal activity [12, 13, 25].

Despite recent technological progress in the study of *in vivo* measurements, the *in vivo* function of tissues is not well understood. The investigations in tissue loading, in particular in bone, ligaments and tendons, are essential from a clinical perspective to gain insight into injury mechanism [26] and to prevent catastrophic damage to the tissues. Several *in vivo* strain studies in ligaments [6, 26-34] and in bone [35-37] have provided significant insight into biomechanical function but have offered limited information when investigating dynamic motion. Few investigations have been carried out in order to study the *in vivo* ACL mechanical behaviour during dynamics motions such as the jump, quick stop, and cut [6, 31, 38, 39]. However, the direct relationship between the ACL elongation and the neuromuscular control of the flexor and extensor muscles has not been frequently investigated.

In this chapter, we present two in vivo biomechanical methods to address: skin movement artefacts on 3D kinematics of the tibio-femoral joint and anterior cruciate ligament (ACL) mechanical behaviour during various dynamic motions. Finally, this chapter will attempt to answer the usefulness of *in vivo* direct experimental data.

Skin movement artefacts and 3D kinematics of the tibio-femoral joint

The first objective was to quantify the error caused by skin movement artefact when reporting the kinematics of the tibio-femoral joint during movements that incorporate sagittal and non-sagittal plane rotations.

Participants: Eight healthy male participants with no history of knee injury or prior surgical treatment of the lower limbs were selected by an orthopaedic surgeon to participate in the study. All participants were informed of the risks involved with all procedures and all foreseeable complications. A consent form was accepted and signed by all participants and the study was conducted with the approval of the Ethics Committee of the Karolinska Hospital, Stockholm, Sweden.

Surgical procedure: A complete description of the surgical technique, as well as associated limitations and methodological concerns can be found in [13]. In short, stainless steel Apex self-drilling/self-tapping pins (Stryker Howmedica AB, Sweden, 3.0 mm diameter, #5038–2-110) were inserted into the distal femur and proximal tibia of the right leg under local anaesthetic. Following surgery, participants were then transported by wheelchair to the motion analysis laboratory (Astrid Lindgren Hospital-Stockholm, Sweden) for data collection. The pins remained inserted for the duration of the test. Upon completion of the experiments (approximately 2 hours), participants returned to the operating room to have the pins removed.

Motion recordings: Triads consisting of four non-collinear 7 mm reflective markers (pin-markers) were affixed to the bone pins. Additional clusters, comprised of four 10 mm surface markers (skin-markers), were affixed onto the lateral and frontal aspects of both the right thigh and shank. Skin markers were spaced 10-15 cm from adjacent markers within their respective cluster and their arrangement was chosen to ensure they remained non-coplanar in at least two camera views throughout the range of motion. Other reflective markers were also placed to define the segmental anatomical coordinate system and they were recorded and removed prior to the movement trials (Figure 30.1). Motion recordings and the force platform signal were synchronously collected.

Bone pin-marker and skin-marker trajectories were simultaneously tracked within a $0.8m^3$ measurement volume ($1.1m \times 0.8m \times 0.9m$) using four infrared cameras (ProReflex, Qualisys AB, Sweden), sampling at a frequency of 120 Hz. Marker coordinates were transformed using the direct linear transform (DLT) and the raw 3D coordinates exported and saved to a local computer for later analysis.

Participants walked along a 12 m walkway at a self selected pace. Five successful walking trials (contact with the force plate and no evidence of targeting) were recorded for each participant. Prior to performing the lateral cutting manoeuvre, participants jumped for maximal horizontal distance. Their longest measurement was recorded and marked on the floor to determine the proper takeoff distance to the force platform. From an initial standing position the participant pushed off using the left leg and, upon landing onto their right foot, immediately pushing off the platform, cutting to the left at an angle of approximately 45°. Five measurement trials were recorded for each movement task followed by a second standing reference trial. The orientation of the target clusters from the first reference trial

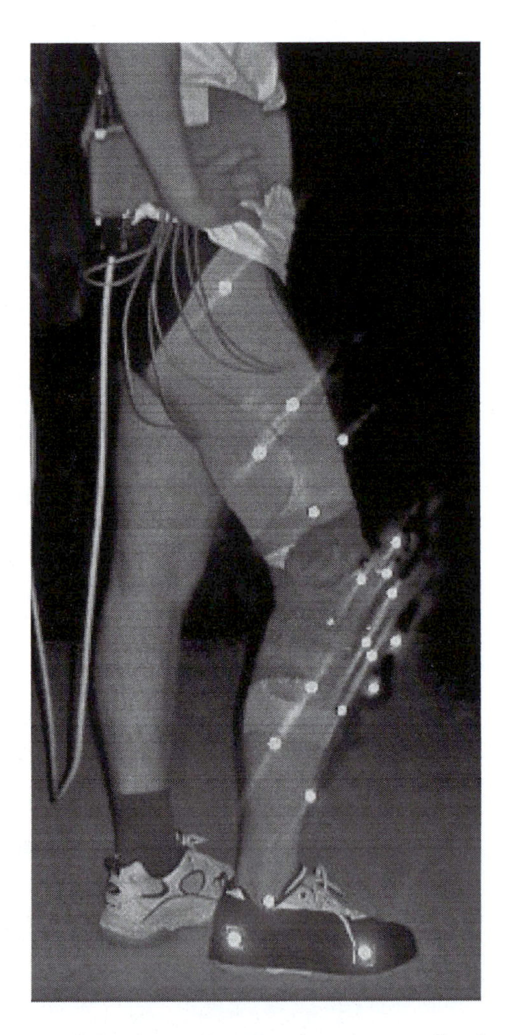

Figure 30.1. Configuration of the skin and pin marker clusters affixed on the right lower-limb. Cluster marker arrangement was chosen to ensure they remained non-coplanar in at least two camera views throughout the range of motion (adapted from: Benoit *et al.* 2006 [44]).

was matched against the second to verify the pins did not bend and the triad did not rotate during testing.

Kinematic analysis: Custom-made software (Matlab, Mathworks Inc, Natick, USA) was developed and validated to process the 3D kinematic information derived from the bone pins and surface markers respectively [40]. The kinematic patterns are described using the terminology and the ordered operations of the Joint Coordinate System (JCS) [41]. Kinematic data (movement of the tibia relative to the femur) for both the pin and skin markers were computed and low-passed filtered at 12 Hz using a 20th order FIR digital filter (Matlab). The kinematic data was normalised to 100% stance phase (foot-strike to toe-off). Pre-foot-strike was expressed as a function of the normalised stance phase and ranges from –10% (or the longest duration of pre-foot strike for that given participant) to 0% (foot strike).

Statistical analysis: Three points of interest during the stance phase of the walking and cutting cycle were chosen for statistical analysis: heel strike (HS); mid-stance point (corresponding with maximum knee flexion angle during the first 60% of stance) (MS); and toe-off (TO). The kinematic data derived from the bone-pins was considered the 'Gold Standard' of measurement. Paired, two-tailed Student's T-tests were used to determine if skin derived kinematics at the three time-points differed from those derived from the bone-pins.

Results and discussion: skin movement artefacts

Of the eight participants, six participants had usable data due to technical problems. No participants experienced significant pain and/or discomfort during the experiments and all reported being able to move their knee freely despite pin implantation. From the 3D kinematics, data in rotation and translation for walking showed that in the primary axis of rotation (flexion-extension), bone-pin markers and skin markers are similar, whereas in the secondary axes (adduction-abduction or internal and external rotation) the differences between bone-pin and skin markers are relatively much larger (Figure 30.2a and Figure 30.2b). In translation, data showed large divergence between bone-pin and skin markers for both types of motion. More specifically, anterior-posterior (A-P) translation calculated with bone-pin and skin markers presented a close relationship in mid-stance whereas this A-P translation showed divergence at heel strike and toe-off.

Absolute error between the skin-marker and pin-marker kinematics at heel strike, mid-stance and toe-off during the walking and cutting motions are noted in Table 30.1. A significant difference in reporting skin-marker derived kinematics with respect to actual tibio-femoral kinematics is evidenced at heel strike, mid-stance and toe-off for both walking and cutting rotations and translations. In the stance phase of walking, the average rotational absolute error ranged from 2.1° to 4.4° while translational errors ranged from 3.3 to 13.2 mm. In the cutting movement, the range of absolute errors and maximum absolute errors were higher for both rotations (3.3° to 13.1°) and translations (5.6 to 16.1 mm), respectively.

While the absolute error is the absolute difference between the skin-marker and pin-marker derived kinematics, the average standard error of the estimate (S) describes the error associated with predicting pin-marker based tibio-femoral kinematics from skin-marker derived kinematics. The average S for walking and cutting movements is found in Table 30.2. These error values were higher in the cutting movement for all measured rotations and translations. This data was calculated by comparing the pin- and skin-marker data across all participants for each trial and at every time point, with the average calculated across time points (n = 110 walking, n = 105 cutting due to a shorter pre-foot-strike phase).

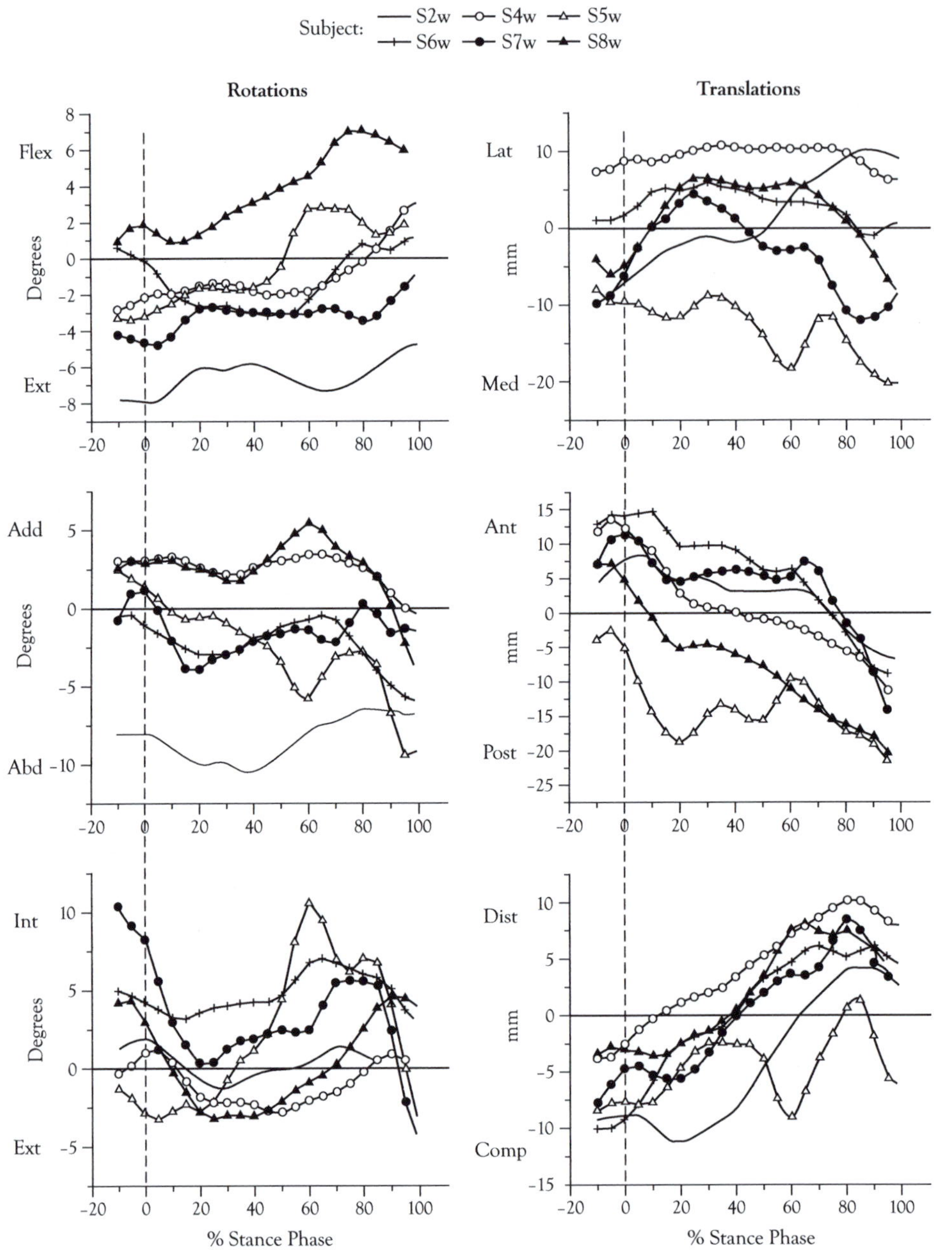

Figure 30.2a. Three-dimensional kinematics of walking trials of a representative subject, subject-4: Rotation data is presented in the left column while translation data are in the right column for the stance phase. Pin-marker labels are unfilled while skin-marker labels are in bold (adapted from: Benoit *et al.* 2006 [44]).

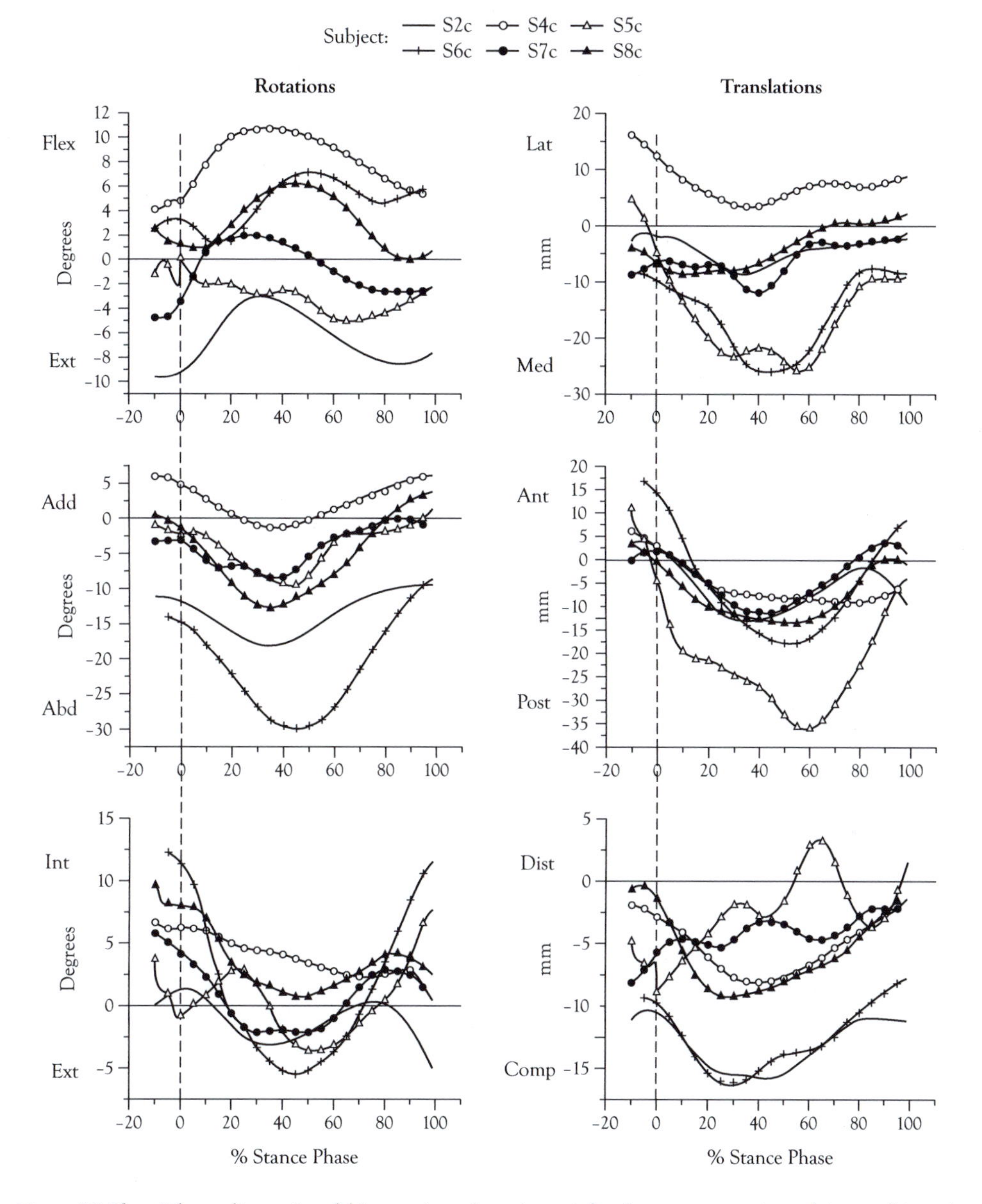

Figure 30.2b. Three-dimensional kinematics of cutting trials of a representative subject, subject-4: Rotation data is presented in the left column while translation data are in the right column for the stance phase. Pin-marker labels are unfilled while skin-marker labels are in bold (adapted from: Benoit *et al.* 2006 [44]).

This work aimed to quantify the error caused by skin movement artefact when reporting the kinematics of the tibio-femoral joint during movements that incorporate sagittal and non-sagittal plane rotations. We found within participant data to be repeatable when using either the skin or surface mounted markers for both the walk and cut. This was encouraging, however, the error associated with skin movement artefact differed widely across participants. Unfortunately, skin movement of the thigh and shank may be large enough to

Table 30.1 Absolute error values of skin-marker derived kinematics at three time points during walking and cutting of knee rotations and translations: flexion extension (Flex/ext), adduction-abduction (Add/abd), internal-external rotation (Int/Ext); medio-lateral

		Rotations (degrees +/– StDev)						*Translations (mm +/– StDev)*					
		Flex/ext		*Add/abd*		*Int/ext*		*Med/lat*		*Ant/post*		*Dist/comp*	
Walk	Foot-strike	2.8†	(2.6)	2.6	(2.8)	2.7†	(2.0)	5.1	(2.6)	7.7†	(4.4)	5.0†	(2.9)
	Mid-stance	2.4†	(2.0)	3.1	(3.3)	2.3	(1.0)	5.5	(3.2)	6.1	(5.4)	3.3†	(2.4)
	Toe-off	2.8	(2.4)	4.4†	(3.3)	2.1	(2.1)	8.3	(5.5)	13.2†	(5.0)	5.0†	(2.5)
Cut	Foot-strike	3.9†	(2.9)	6.7†	(5.4)	5.4†	(4.2)	7.3†	(4.4)	5.6	(5.1)	6.3†	(4.0)
	Mid-stance	4.0	(2.5)	5.9†	(3.1)	5.4†	(4.0)	5.9†	(4.5)	6.7†	(4.4)	5.6†	(3.8)
	Toe-off	4.1	(2.7)	13.1	(9.8)	3.3†	(1.8)	13.9†	(10.1)	16.1†	(8.9)	8.3†	(6.2)

Note: † denotes significant difference between skin and pin-marker data (two tailed paired Students T-test, $p < 0.05$) (adapted from: Benoit *et al.* 2006 [44]).

Table 30.2 Average standard error of the estimate (S) describing the error associated with predicting tibio-femoral kinematics from skin-marker derived kinematics.

	Rotations (degrees)			*Translations (mm)*		
	Flex/ext	*Add/abd*	*Int/ext*	*Med/lat*	*Ant/post*	*Dist/comp*
Walk	2.5	3.6	2.9	6.0	6.8	2.8
Cut	6.3†	4.5	3.0	8.0†	5.5†	7.1††

Note: Average calculated for each data point of the stance phase (average of 110 data points) based on the estimated prediction of all walking (n = 25) and cutting (n = 28) trials.
†: Cut motion has at least twice as much error associated with predicting tibio-femoral kinematics from skin markers.
††: Cut motion has at least three times as much error associated with predicting tibio-femoral kinematics from skin markers (adapted from: Benoit *et al.*, 2006 [44]).

mask the actual movements of the underlying bones, thus making reporting of knee joint kinematics potentially uncertain. The data from this study suggested that the use of skin-markers to describe knee joint motion must be presented with an envelope of accuracy that describes the artefact imparted by skin movement of the markers. Although this error varies throughout the stance phases of gait and cutting, we propose the use of the average standard error of the estimate (see Table 30.2) when reporting the accuracy of skin-marker derived kinematics. This estimate of the error associated with predicting tibio-femoral kinematics from skin-markers would allow for the reporting of non-sagittal plane kinematics within approximately 65% confidence interval (for 95% confidence interval use $1.96 \times S$) that may be relevant in situations where large differences between populations may be detected.

With recent motion analysis systems, the 3D kinematic data are relatively easily obtained and the non-sagittal plane measurements are generally calculated. This work indicated the need for caution when evaluating kinematic patterns of the knee using surface markers. A very important observation from this work is that the surface marker derived kinematics can present repeatable patterns within a subject for various movements. These repeatable patterns must not be misinterpreted as accurately representing skeletal kinematics, at least beyond the sagittal plane of movement where the error is small relative to the total movement [45].

Anterior cruciate ligament (ACL) mechanical behaviour during various dynamic motions

The second objective consisted of applying similar *in vivo* methods of ACL strain measurement during a backward fall in skiing and during rapid deceleration, stop and cut motions, all movements that have been previously shown to precede injuries to the ACL. The research has been carried out over three periods distributed over three years.

Participants: Overall, five healthy males from the Medical School at the University of Perugia (mean: age, 25 yrs, height, 167 cm; weight, 71.5 kg) with no previous knee joint injuries volunteered for the study. Prior to the Differential Variable Reluctance Transducer (DVRT, MicroStrain Inc., Burlington, USA) implantation, the participant was informed of the surgical procedure and the laboratory testing protocol. After reading and signing the informed consent, the participant went to the biomechanics laboratory the day before data collection to practice the various tasks. The first movement tested consisted of a simulated backward fall in downhill skiing. The instrumented leg of the participant was placed in a ski boot fixed to a force plate (Bertec Corp., Model 4060, Ohio, USA). The subjects then forcefully propelled themselves backward in an attempt to cause release of the boot rear-spoiler. The second task consisted of standing on the right leg with the left leg slightly flexed at the knee. This position would prevent the DVRT to impinge with the upper medial intercondylar femoral notch. The participant jumped 1.5 m to the target, an X taped at the centre of a force plate (Bertec Corp., Model 4060), landing with the instrumented left leg, stopping in the landing position and maintaining this position for two seconds. The third task required hopping and reaching the target as quickly as possible, landing with the instrumented leg and stopping in the landing position, without touching the right foot to the ground for at least two seconds. The fourth task, a simulation of a cut manoeuvre, consisted of a hop on the force plate and a change of direction with the instrument leg. At all times the participant was asked to maintain partial knee flexion in the instrumented leg (at least 10°) to prevent impingement of the DVRT on the condylar notch. The task was repeated until the participant was able to perform it comfortably without going into full extension.

Surgical procedure: The following day, the DVRT was implanted on the antero-medial band of the intact ACL. The participant consciousness is necessary during the surgical procedures to control muscular contraction which affects ACL strain. All procedures are therefore performed under local intra-articular anaesthetic. The DVRT is inserted via standard arthroscopic knee portals: the antero-medial portal is used for the arthroscopic optic device and the antero-lateral portal is used to insert the DVRT. The DVRT is then inserted through a 10 mm cannulae and trocar inserted through the antero-medial portal. The trocar/cannular is inserted into the joint space and pressed against the antero-medial bundle of the ACL. The trocar is then removed and the insertion tool inserted into the cannulae. The DVRT is thus introduced into the joint space and aligned with the ligament fibres (Figure 30.3) The barbed ends of the DVRT are then inserted into the ligament bundle and fixed in place. The sutures holding the DVRT onto the tool are then released and removed, thus leaving the DVRT implanted into the ligament.

The surgical instruments were removed and the wounds were closed around the exiting instrument wire and removal sutures. Before closing the wounds, the zero strain position of the ACL was determined using the technique previously described by Fleming [42]. If the DVRT signal was responding well, then the wounds were sutured and covered with sterile bandages and the limb wrapped in a sterile elastic bandage.

The participant was then transported to the biomechanics laboratory for data collection. The zero strain position of the ACL was determined using a technique previously described with an anterior translation force of 140 N during the instrumented Lachman test. Four high-speed

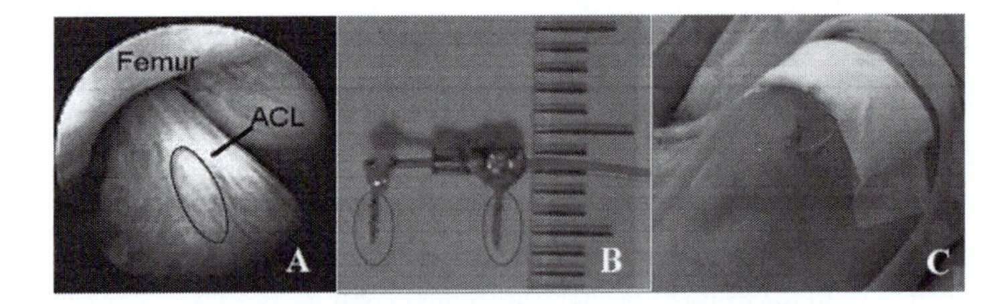

Figure 30.3. DVRT arthroscopically implanted on the antero-medial band (A) of the intact ACL under local anesthesia. The barbed ends of the DVRT (B) are inserted into the ligament bundle and fixed in place and the wounds (C) were closed and sutured around the exiting instrument wire.

digital video cameras (JVC GR-DVL9600) connected to a PC computer equipped with the SIMI* Motion system (SIMI* Reality Motion Systems GmbH) were positioned on the same side of the participants' instrumented leg to record all trials. The cameras recorded at a speed of 50 Hz and were zoomed to include only the instrumented leg in the field of view. The calibrated volume was approximately 1.5 m × 1.0 m × 0.75 m. The entire collection window was 8 seconds at 1000 Hz for the electromyography, force plate, and DVRT signals and at 50 Hz for the kinematics data. A total of three trials per movements were collected. The zero-strain test was then repeated to ensure proper operation of the DVRT.

Data processing: The rectified EMG signals recorded during the three motions were synchronised to match the time of DVRT, ground reaction force, and kinematics data. The rectified EMG signals were normalised by peak amplitude for the dynamic contractions of the three manoeuvres using the stopping motion EMG data as normalisation basis. All data were processed using SIMI Motion system. Manoeuvres were analysed five frames before heel strike until the participant moved outside of the force plate. The data from all three trials was ensemble averaged over the cycle and the data reported corresponds to the average over the three trials.

Results and discussion: *in-vivo* ACL strain

During the backward falling movement, the strain of the antero-medial band of the ACL was directly related to the modelled shear force at the knee joint (r2 = 0.9239). This indicates that the backward falling motion caused a strain to the antero-medial band of the ACL up to 230% of the relative strain recorded during the instrumented Lachman tests. In all trials, maximal strain corresponded to the maximum tibial shear force prior to the subject placing the trailing leg on the ground. These findings concur that anterior tibial translation caused by the boot rear spoiler during the backward fall induces increased strain on the ACL.

During the stopping movement task, the average peak strain of the ACL was 5.47 ± 0.28%. It is noted that the strain in the ACL rises during the flight phase prior to impact. Peak strain occurs at the point of impact of the left foot (instrumented leg) and the strain in the ACL was maintained relatively high for the duration of the hop cycle. The ACL strain at the impact with ground after the flight phase was more than 2.75 times higher than the instrumented Lachman test. Qualitative video analysis of the movement trials indicated that the knee never reached full extension during the stopping tasks. It was further observed that the leg was most extended at the time of impact during the landing phase. During the flight phase, there is an extension of the knee joint and muscle activation of the quadriceps, hamstring and gastrocnemius muscles in preparation for landing. Although there is a continuous increase in ACL strain until reaching maximum strain at the impact, immediately prior to impact, ACL strain is already at 5%. Consequently, the bone geometry and muscle contraction induce important ACL strain.

As shown in the left column of Figure 30.4, the knee angle (a), in vivo ACL elongation (b), vertical ground reaction force (c) and in the right column, EMG of the vastii (d), hamstrings (e), and both gastrocnemii (f) are depicted for the cutting manoeuvres.

During the cutting task, the maximum knee angle coincides with the maximum contraction of the hamstrings and gastrocnemii. On the other hand, the maximum peak elongation of the ACL coincided with the maximum vertical ground reaction force and at a less contracted state of hamstrings and gastrocnemii (dash line). This shows for the cutting motion that ACL could be in a vulnerable state for injury.

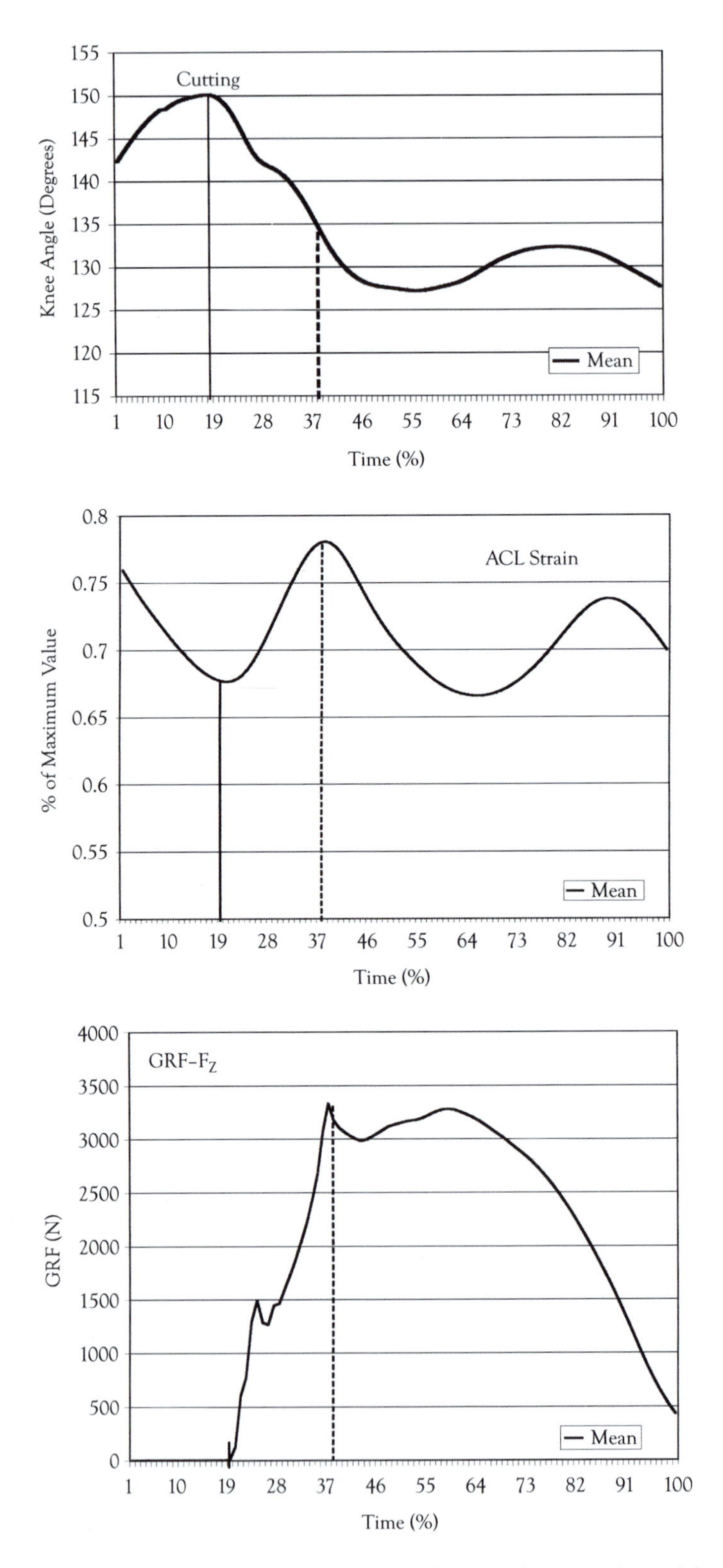

Figure 30.4. Knee angle (a), ACL in vivo elongation (b), ground reaction forces (c) with the peak normalised EMG of the vastii.

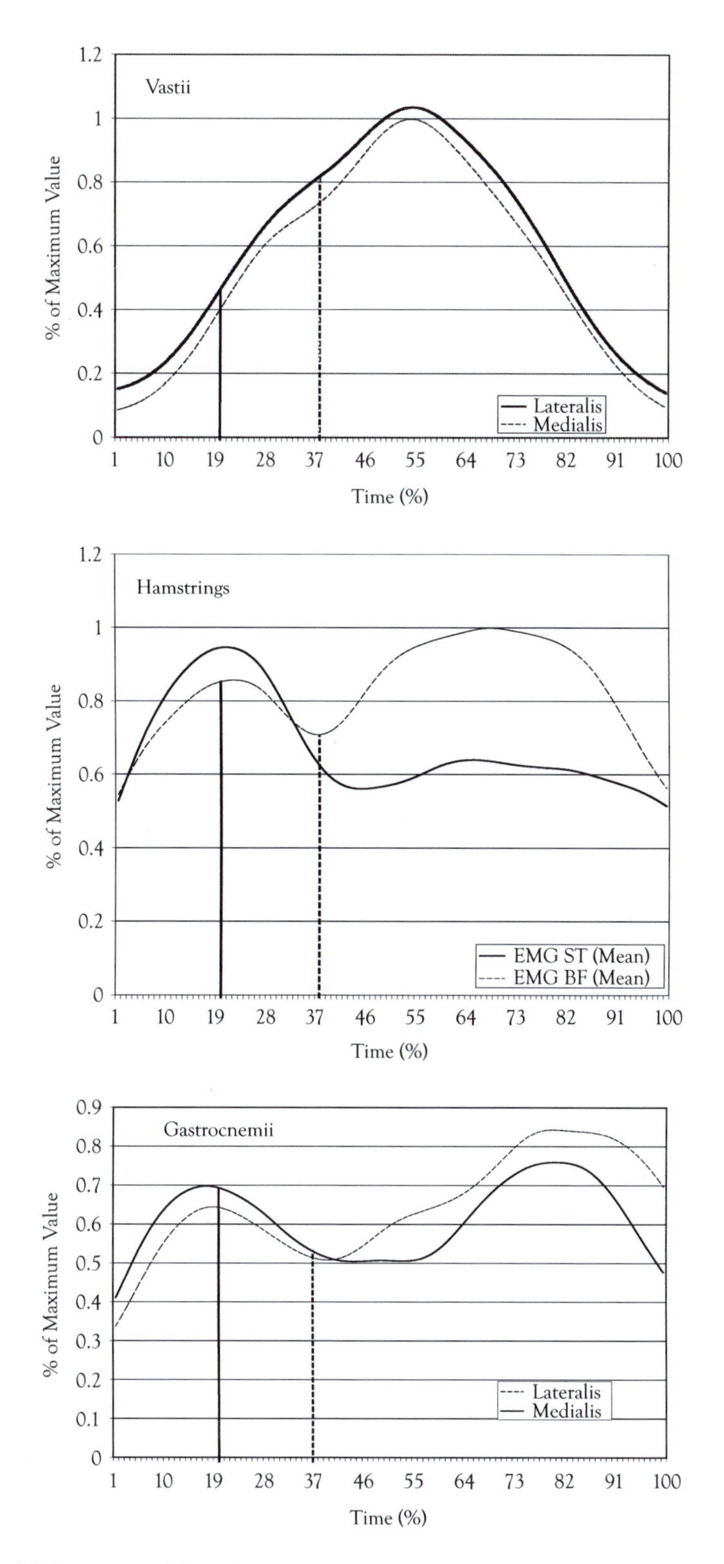

Figure 30.4 (d) hamstrings (e), and gastrocnemii (f) during the cutting task (Note: the dashed line indicates the maximum ligament strain).

As shown in Figure 30.5, knee angle (a), *in vivo* ACL elongation (b), vertical ground reaction force (c) and EMG of the vastii (d), medial and lateral hamstrings (e) and both gastrocnemii (f) are depicted for the stopping task. During this task, the maximum peak elongation of the ACL coincided with the maximum knee angle and the maximum vertical ground reaction force occurred later in the cycle. The maximum normalised EMG of the semitendinosus and both gastrocnemii took place before the maximum ACL elongation. On the other hand, the quadriceps muscle reached their maximum after the peak ground reaction force (dashed line).

In the two conditions tested with less extent for the cutting, the participant's neuromuscular strategy did anticipate the impact by contracting the hamstrings and gastrocnemius muscles with high intensity, whereas the quadriceps muscles contracted right at or after the impact with the ground. It showed that the hamstring and gastrocnemius muscles are used as protective mechanisms to the ACL elongation as in stopping conditions. The quadriceps muscles played their anti-gravitational role to avoid collapsing of the knee when the foot impacts the ground. During the cutting motion, it seems that the hamstrings and gastrocnemius muscles do not have as much a stabilisation effect than in the stopping motion. Similar muscular patterns of the hamstrings and gastrocnemius have been reported for unanticipated cutting manoeuvres [43]. It seems that maximum hamstrings contraction occurred much before foot strike in order to protect against overloading the anterior cruciate ligament.

The instrumented portion of the ligament in this study also corresponds to the portion of the ligament that may first be damaged during ACL injuries [26]. The fact that the strain increased in the stressful movement in this portion of the ACL indicated that the movement might cause an increased load across the fibres.

Conclusion

Our studies highlight the need for caution when evaluating kinematic parameters of the knee joint using surface markers. A very important observation was that the surface marker derived kinematics can present repeatable profiles within a participant for various movements. These repeatable patterns must not be misinterpreted as accurately representing skeletal kinematics, at least beyond the sagittal plane of movement where the error is small relative to the total movement. The absolute errors presented in this study offer a guideline to which conclusions may be drawn from 3D knee joint kinematics under similar testing conditions. And also the study of skin movement artefact proposed potential guidelines when discussing findings.

In the second half of this chapter, the findings have confirmed that the stopping and cutting manoeuvres generate a relatively high level of ACL elongation that initiates at or just before foot contact, when the leg is most extended. The muscle contraction anticipatory of the hamstring and gastrocnemius play an important role of protecting excessive ACL elongation, whereas the quadriceps muscle prevents the collapsing of the knee joint after the foot impact with the ground. Specific neuromuscular, proprioceptive, and motor control factors associated with ACL injury must be investigated in depth to better understand the risk factors. *In vivo* experimentation yielded useful information, which could not have been found without direct measurement. These *in vivo* findings can be used to validate musculoskeletal models or to establish the real function of the anatomical structures in human movement performance.

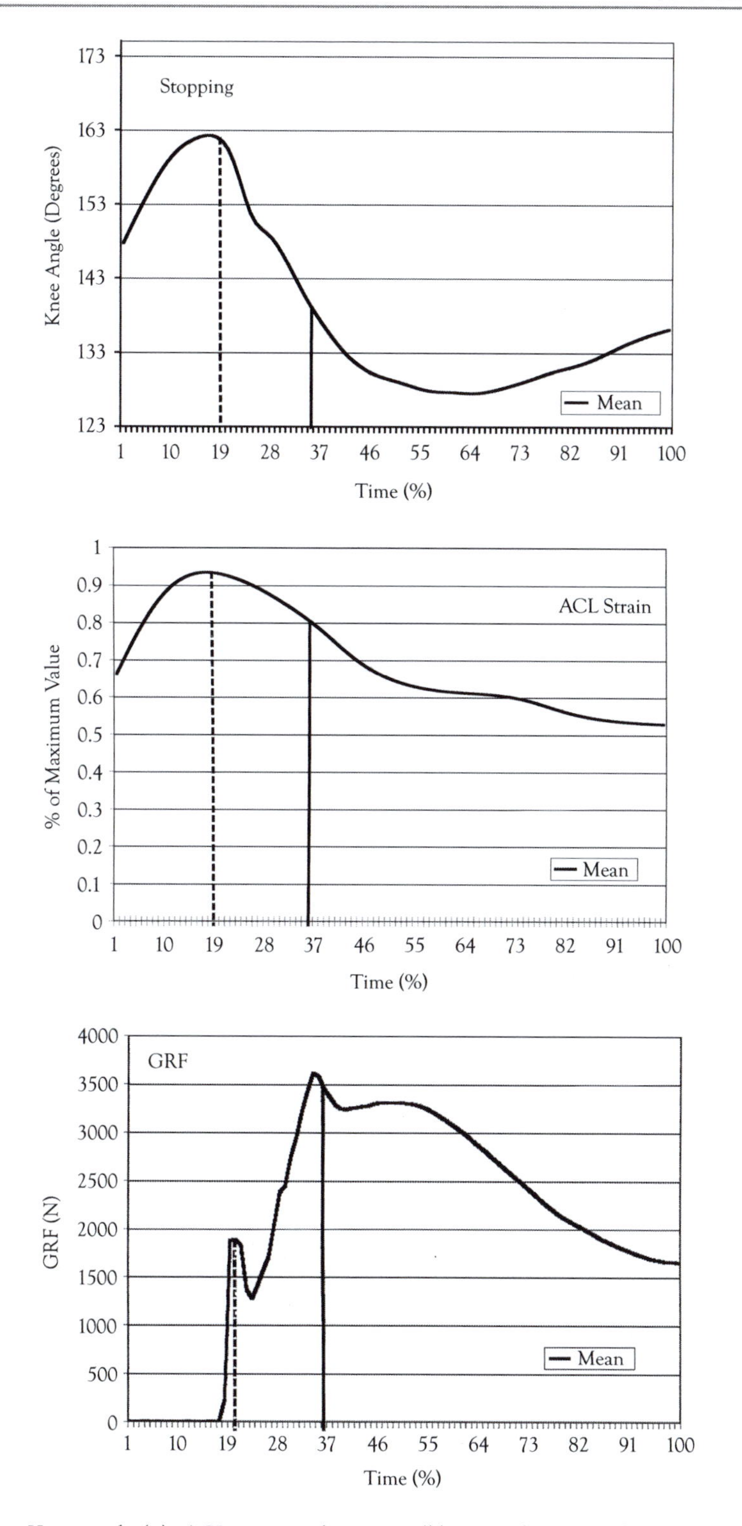

Figure 30.5. Knee angle (a), ACL in vivo elongation (b), ground reaction forces (c) with the peak normalised EMG of the vastii. *(Continued)*

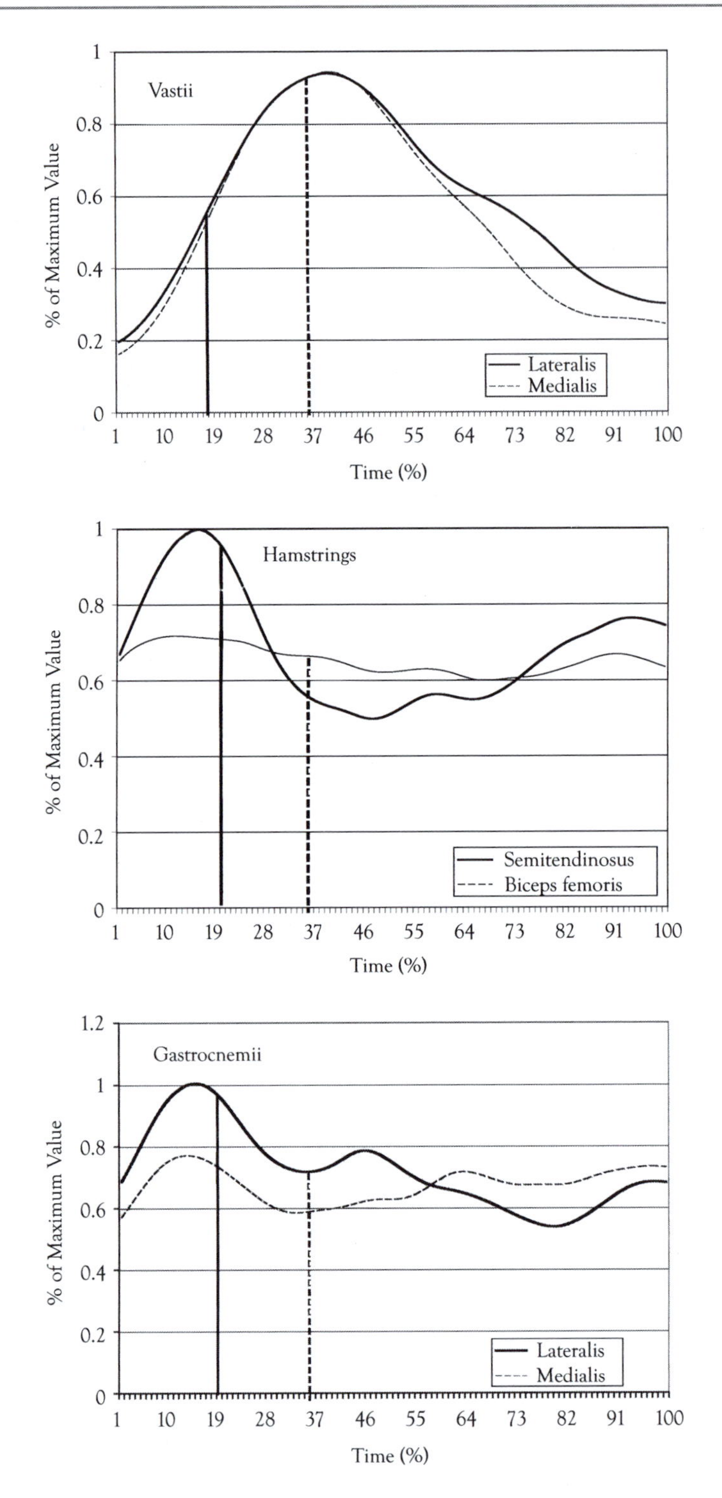

Figure 30.5 (Continued) (d) hamstrings (e), and gastrocnemii (f) during the stopping task (Note: the dashed line indicates the maximum ligament strain).

Acknowledgments

These studies have been partly funded by Natural Sciences and Engineering Research Council of Canada and Let People Move (Perugia, Italy) as well as Centrum fõr Iddrottsforskning (Sweden) . The *in vivo* experimental data collections have been carried out with the researchers and residents at the 'laboratorio di biomeccanica' of Let People Move and Karolinska University Hospital, Stockholm, Sweden.

References

1. Li G, Defrate LE, Rubash HE, Gill TJ. (2005) 'In vivo kinematics of the ACL during weight-bearing knee flexion'. *J Orthop Res.* 23(2):340–4.
2. Komistek RD, Kane TR, Mahfouz M, Ochoa JA, Dennis DA. (2005) 'Knee mechanics: a review of past and present techniques to determine in vivo loads'. *Journal of Biomechanics.* 38(2):215.
3. Ravary B, Pourcelot P, Bortolussi C, Konieczka S, Crevier-Denoix N (2004) 'Strain and force transducers used in human and veterinary tendon and ligament biomechanical studies'. *Clin Biomech* (Bristol, Avon) 19(5):433–47.
4. Fleming BC, Beynnon BD. (2004) 'In vivo measurement of ligament/tendon strains and forces: a review'. *Ann Biomed Eng.* 32(3):318–28.
5. DeFrate LE, Sun H, Gill TJ, Rubash HE, Li G. (2004) 'In vivo tibiofemoral contact analysis using 3D MRI-based knee models'. *J Biomech.* 37(10):1499–504.
6. Cerulli G, Benoit DL, Lamontagne M, Caraffa A, Liti A. (2003) 'In vivo anterior cruciate ligament strain behaviour during a rapid deceleration movement: case report'. *Knee Surg Sports Traumatol Arthrosc.* 11(5):307–11.
7. Ramsey DK, Wretenberg PF. (1999) 'Biomechanics of the knee: methodological considerations in the in vivo kinematic analysis of the tibiofemoral and patellofemoral joint'. *Clin Biomech* (Bristol, Avon) 14(9):595–611.
8. Ishii Y, Terajima K, Terashima S, Koga Y. (1997) 'Three-dimensional kinematics of the human knee with intracortical pin fixation'. *Clin Orthop Relat Res.* October 1997; 343:144–50.
9. Reinschmidt C, van Den Bogert AJ, Murphy N, Lundberg A, Nigg BM. (1997) 'Tibiocalcaneal motion during running, measured with external and bone markers'. *Clin Biomech* (Bristol, Avon) 12(1):8–16.
10. Reinschmidt C, van den Bogert AJ, Nigg BM, Lundberg A, Murphy N. (1997) 'Effect of skin movement on the analysis of skeletal knee joint motion during running'. *J Biomech.* 30(7):729–32.
11. Lafortune MA, Cavanagh PR, Sommer HJ, 3rd, Kalenak A. (1992) Three-dimensional kinematics of the human knee during walking. *J Biomech.* 1992 Apr;25(4):347–57.
12. Ramsey DK, Lamontagne M, Wretenberg PF, Valentin A, Engstrom B, Nemeth G. (2001) 'Assessment of functional knee bracing: an in vivo three-dimensional kinematic analysis of the anterior cruciate deficient knee'. *Clin Biomech* (Bristol, Avon) 16(1):61–70.
13. Ramsey DK, Wretenberg PF, Benoit DL, Lamontagne M, Nemeth G. (2003) 'Methodological concerns using intra-cortical pins to measure tibiofemoral kinematics'. *Knee Surg Sports Traumatol Arthrosc* 11(5):344–9.
14. Cappozzo A. (1991) 'Three dimensional analysis of human walking: experimental methods associated artifacts'. *Human Movement Science* 10(5):589–602.
15. Andriacchi TP, Alexander EJ, Toney MK, Dyrby C, Sum J. (1998) 'A point cluster method for in vivo motion analysis: applied to a study of knee kinematics'. *J Biomech Eng.* 120(6):743–9.
16. Cheze L, Fregly BJ, Dimnet J. (1995) 'A solidification procedure to facilitate kinematic analyses based on video system data'. *J Biomech.* 28(7):879–84.
17. Lucchetti L, Cappozzo A, Cappello A, Della Croce U. (1998) 'Skin movement artefact assessment and compensation in the estimation of knee-joint kinematics'. *J Biomech.* 31(11):977–84.

18. Li G, DeFrate LE, Park SE, Gill TJ, Rubash HE. (2005) 'In vivo articular cartilage contact kinematics of the knee: an investigation using dual-orthogonal fluoroscopy and magnetic resonance image-based computer models'. *Am J Sports Med* 33(1):102–7.
19. Komistek RD, Stiehl JB, Buechel FF, Northcut EJ, Hajner ME. (2000) 'A determination of ankle kinematics using fluoroscopy'. *Foot Ankle Int.* 21(4):343–50.
20. Sheehan FT, Seisler AR, Siegel KL. (2007) 'In vivo talocrural and subtalar kinematics: a non-invasive 3D dynamic MRI study'. *Foot Ankle Int.* 28(3):323–35.
21. Stiehl JB, Komistek R, Dennis DA. (2001) 'A novel approach to knee kinematics'. *Am J Orthop.* 30(4):287–93.
22. Marai GE, Laidlaw DH, Demiralp C, Andrews S, Grimm CM, Crisco JJ.(2004) 'Estimating joint contact areas and ligament lengths from bone kinematics and surfaces'. *IEEE Trans Biomed Eng.* 51(5):790–9.
23. Wretenberg P, Ramsey DK, Nemeth G. (2002) 'Tibiofemoral contact points relative to flexion angle measured with MRI'. *Clin Biomech* (Bristol, Avon) 17(6):477–85.
24. Gilles B, Perrin R, Magnenat-Thalmann N, Vallee J-P. (2005) 'Bone Motion Analysis From Dynamic MRI: Acquisition and Tracking'. *Academic Radiology* 12(10):1285–92.
25. Ramsey DK, Wretenberg PF, Lamontagne M, Nemeth G. (2003) 'Electromyographic and biomechanic analysis of anterior cruciate ligament deficiency and functional knee bracing'. *Clin Biomech* (Bristol, Avon) 18(1):28–34.
26. Zavatsky AB, Wright HJ. (2001) 'Injury initiation and progression in the anterior cruciate ligament'. *Clin Biomech* (Bristol, Avon) 16(1):47–53.
27. Arms SW, Pope MH, Johnson RJ, Fischer RA, Arvidsson I, Eriksson E. (1984) 'The biomechanics of anterior cruciate ligament rehabilitation and reconstruction'. *Am J Sports Med.* 12(1):8–18.
28. Beynnon BD, Fleming BC. (1998) 'Anterior cruciate ligament strain in-vivo: a review of previous work'. *J Biomech.* 31(6):519–25.
29. Beynnon BD, Johnson RJ, Fleming BC, Peura GD, Renstrom PA, Nichols CE, *et al.* (1997) 'The effect of functional knee bracing on the anterior cruciate ligament in the weightbearing and non-weightbearing knee'. *Am J Sports Med.* 25(3):353–9.
30. Beynnon BD, Johnson RJ, Fleming BC, Stankewich CJ, Renstrom PA, Nichols CE. (1997) 'The strain behavior of the anterior cruciate ligament during squatting and active flexion-extension. A comparison of an open and a closed kinetic chain exercise'. *Am J Sports Med.* 25(6):823–9.
31. Fleming BC, Beynnon BD, Renstrom PA, Johnson RJ, Nichols CE, Peura GD, *et al.* (1999) 'The strain behavior of the anterior cruciate ligament during stair climbing: an in vivo study'. *Arthroscopy* 15(2):185–91.
32. Fleming BC, Renstrom PA, Ohlen G, Johnson RJ, Peura GD, Beynnon BD, *et al.* (2001) 'The gastrocnemius muscle is an antagonist of the anterior cruciate ligament'. *J Orthop Res.* 19(6):1178–84.
33. Li G, DeFrate LE, Sun H, Gill TJ. (2004) 'In vivo elongation of the anterior cruciate ligament and posterior cruciate ligament during knee flexion'. *Am J Sports Med.* 32(6):1415–20.
34. Yoo JD, Papannagari R, Park SE, DeFrate LE, Gill TJ, Li G. (2005) 'The Effect of Anterior Cruciate Ligament Reconstruction on Knee Joint Kinematics Under Simulated Muscle Loads'. *Am J Sports Med.* 33(2):240–6.
35. Demes B. (2007) 'In vivo bone strain and bone functional adaptation'. *Am J Phys Anthropol.* 2007, Mar 1.
36. Milgrom C, Finestone A, Ekenman I, Simkin A, Nyska M. (2001) 'The effect of shoe sole composition on in vivo tibial strains during walking'. *Foot Ankle Int.* 22(7):598–602.
37. Rolf C, Westblad P, Ekenman I, Lundberg A, Murphy N, Lamontagne M, *et al.* (1997) 'An experimental in vivo method for analysis of local deformation on tibia, with simultaneous measures of ground reaction forces, lower extremity muscle activity and joint motion'. *Scand J Med Sci Sports.* 7(3):144–51.
38. Benoit D, Cerulli G, Caraffa A, Lamontagne M, Liti A, Brue S. (2000) 'In-Vivo Anterior Cruciate Ligament Strain Behaviour During a Rapid Deceleration Movement (Abstract)'. *Archives of Physiology and Biochemistry* 18(1/2): p. 100.

39. Fleming BC, Beynnon BD, Renstrom PA, Peura GD, Nichols CE, Johnson RJ. (1998) 'The strain behavior of the anterior cruciate ligament during bicycling. An in vivo study'. *Am J Sports Med.* 26(1):109–18.
40. Lafontaine D, Benoit D, Xu L, Lamontagne M. (2003) 'Validation of Joint Rotation and Translation Calculations using a Joint Coordinate System Approach'. Twelfth Annual Meeting of European Society of Movement Analysis for Adults and Children. Marseille, France, 2003.
41. Grood ES, Suntay WJ. (1983) 'A Joint Coordinate System for the Clinical Description of Three-Dimensional Motions: Application to the Knee'. *Journal of Biomechanical Engineering* 105:136–44.
42. Fleming BC, Beynnon BD, Tohyama H, Johnson RJ, Nichols CE, Renstrom P, *et al.* (1994) 'Determination of a zero strain reference for the anteromedial band of the anterior cruciate ligament'. *J Orthop Res.* 12(6):789–95.
43. Beaulieu M, Lamontagne M. (2006) 'EMG and 3D kinematics of the lower limb of male and female elite soccer players performing an unanticipated cutting task'. *XXIV International Symposium on Biomechanics in Sports.* Salzburg, Austria: Department of Sport Science and Kinesiology, University of Salzburg, Austria, 2006.
44. Benoit DL, Ramsey DK, Lamontagne M, Xu L, Wretenberg P, Renstrom P. Effect of skin movement artifact on knee kinematics during gait and cutting motions measured in vivo. Gait & Posture. 2006;24(2):152–164.
45. Benoit DL, Ramsey DK, Lamontagne M, Xu L, Wretenberg PF, Renström P. (2007) In Vivo Knee Kinematics during Gait Reveals New Rotation Profiles and Smaller Translations. Clinical Orthopedics and Related Research. Jan;454:81–88.

31

Impact attenuation, performance and injury related to artificial turf

Rosanne S. Naunheim
University of Perugia, Perugia

A history of artificial turf: introduction

In the 1950s US inventors at Monsanto designed artificial turf in hopes of improving fitness in urban dwellers. It had been noticed that children who lived in cities were less physically fit than their rural counterparts and it was thought that a low maintainance surface on which to play would promote more recreational activity. In 1965, Monsanto filed a patent for Astroturf, originally called Chemgrass.

Astroturf was thought to have advantages over natural grass in addition to its low maintainence properties. It could be used in a domed stadium and did not deteriorate during bad weather; avoiding the costly resodding of a field during the season.

American professional football teams were the first to embrace the new turf as multipurpose-domed stadiums became popular. The Houston Astros placed the turf in their domed stadium in 1965. The original Astroturf was had blades of polyethylene-polypropylene woven into a mat backing. The PVC grass fiber is described as 6,6 which refers to the fact that it is two monomers with six carbon chains. The grass mat was placed on top of a foam pad of variable thickness and could be installed over concrete for use in a domed stadium. Newer forms of artificial turf were developed using a shredded rubber base instead of foam and improvements were made to the artificial blades of grass by coating them with silicone and combining them with curled blades to decrease friction. While artificial turf was popular for some time, with 68 of 245 games being played on the surface in American football from 68–85; however, by 2002, only 9 of the NFL's 32 teams still had artificial turf. Some teams, such as the team in Phoenix AZ, have gone to great lengths to maintain a natural grass field with an entire grass field that can be moved out into the sunshine when not in use. While artificial turf is no longer popular with professional teams, it continues to be used at the high school and college level where it is an easily maintained all-weather surface. The reasons for this decrease in popularity among professional teams are related to the properties of artificial turf.

Characteristics of artificial turf

Resilience

Several characteristics of sports surfaces are important in causing injuries. One of the first characteristics is resilience. This is the measure of the energy absorbed by the surface that is returned to the object striking it. Concrete is very resilient with most of the energy of impact being transferred back to the object striking the surface. This rebound resilience is the kinetic energy of the object after impact divided by the kinetic energy before impact. Resilience is very important in ball sports since the ball will bounce higher on a surface that is more resilient. In cricket, for example, resilience is 7.8 per cent on a slow field and 15.6 per cent on a fast field. In other sports resilience differs, with 20–40 per cent for hockey, 20–45 per cent for soccer, 42 per cent for grass tennis and 60 per cent for tennis played on a synthetic court (Bell, 1985; Bartlett, 1999). Playing on a surface that is resilient makes the game faster, but also will be responsible for more injuries, since a player's head striking the more resilient field will sustain a higher transfer of energy, potentially causing brain injury. It becomes important for fast moving sports like track and field for the resilience of the surface to be high.

Comparisons have been made of the resilience of artificial turf. (Naunheim, 2005). In this study, older Astroturf used by a professional football team was replaced by a newer turf, called Field Turf. The older turf at the practice field consisted of an AstroTurf carpet on top of 5/8 inches of padding over concrete. The newer field had a base of shredded rubber with a grass carpet on top of concrete. The study showed the newer surfaced tended to be compacted at areas of the field which sustained high traffic, the 50-yard line and the near the end zones. The new surface behaves much the same as playground surfaces which are covered with wood chips where the surface becomes stiffer as the the subsurface is compacted.

Stiffness: is another property of a sports surface that is assumed to influence the frequency of injuries. It is assumed that a 'hard' surface will cause more injuries than a 'soft' surface. Stiffness is defined as the ratio of applied force to deflection and varies with the magnitude of the force applied. Stiffness may be particularly important in causing chronic overuse injuries. Increased stiffness also transmits more of the impact to the bones, joints, ligaments and tendons of players. Compliance is the inverse of stiffness and refers to how much the surface deforms under a load. Many of the standard 'hardness' tests cannot tell the difference between playing surfaces of different thickness. Hardness on natural grass surfaces correlates with moisture content and temperature.

Friction: is an important characteristic of the playing surface. In running the coefficient of friction should exceed 1.1 (Bartlett, 1999). Friction is 10–40 per cent greater on artificial turf than grass (Figure 31.1).

Friction on natural grass is closely correlated to the amount of grass cover. Ryegrass is associated with less shoe-surface traction than Kentucky bluegrass or Bermuda grass.

Frictional forces are both linear (in the direction of movement) and rotational, (Nigg and Yeadon, 1987) enabling the player to start and stop and to make rapid changes in motion. It depends on the material properties of the shoe/surface interface, considering their relative velocities. For rotational friction, the amount of area making contact and the pressure distribution determine the amount of friction.The ideal shoe would be one with high translational friction to allow rapid starting and stopping, but low rotational friction, so if a player planted his foot and was hit, the rotational stress would not tear ligaments in the

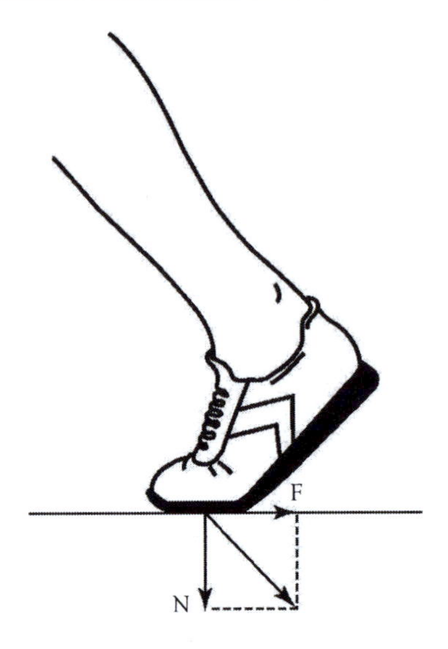

Figure 31.1 Friction coefficient.

knee or ankle. It is thought that these rotational forces are more important for soccer where dribbling and cutting are more prevalent than in American football. While knee injuries are still the most prevalent injury in American football, the injuries are often the result of collisions (Stanford Research Institute, 1975).

Figure of friction

A study by Torg (1971) showed that cleats which increase the amount of friction, and therefore the release coefficient, are associated with increased injuries. The release coefficient is defined as $r = F/w$ where F is the rotational force and w is the axial load. The release coefficient also increases as the temperature of the field increases, suggesting that shorter cleats, or no cleats should be used on warmer days. The average release coefficient of all the shoes tested in the Torg study increased 19.4 per cent when the artificial turf temperatures increased from 52–110° F. A release coefficient of 0.50 was calculated as a level at which knee injury was possible. This study and others (Meyers, 2004) were responsible for the change in shoe type for play on artificial surfaces, short cleats or no cleats at all. Force measurements of players in actual games show that coefficients of friction in the range of 1.2–1.5 are not uncommon and that if the coefficient of friction is less than 0.8, slips become frequent (Valiant, 1990). The coefficient of friction can be direction dependent in the case of artificial turf with woven carpets.

Force reduction: is defined by the German Standards Institute (DIN) It compares the impact on concrete with the surface being tested. A playing surface would be expected to be better than concrete in terms of impact attenuation. The International Athletes Federation (IAF) recommends a force reduction of 35–50 per cent for athletic tracks. In competitive situations however, such as the 1996 Olympic Games, the force reduction of the track was minimal, only 36 per cent, presumably to make the track faster.

While all the tests for turf are good for their intended purposes, none of them duplicates the force-time relationship that occurs when an athlete runs or falls on a surface. A runner's foot has a velocity of several feet per second when it makes contact with the playing surface. Standard compression testing (ASTM D-575) moves at 0.5 inches per minute. This is important in measuring resilience since some surfaces are shaped by slow deformation but become brittle with rapid deformation.

Standardization of surfaces

For the assessment of vertical loads, a drop test device may be used. A weight is dropped on to the surface mounted on a rigid base or on a force platform. The artificial athlete Stuttgart is a drop test device which has a duration of impact of 100–200 ms. This is similar to the contact time for the athlete in many sports. In this test a weight is dropped on to a spring that compresses a piston indenting the surface to be tested. Displacement of the surface is measured by an inductive displacement transducer and the piston pressure is measured by a pressure transducer. Energy is calculated from a force-deformation curve (Kolitzus, 1984). Similar devices are used in many countries. In the US, the American Society of Testing and Materials (ASTM) evaluates turf with the F-355 dynamic shock test. The basic shock absorbency test is performed with a free-falling missile containing an accelerometer and devices to measure velocity at impact. The ASTM has published standards for this test and recommends that fields be replaced if the test exceeds 200 Gs. (ASTM, 1994). A second test used on artificial turf is the Clegg hammer test. It is also conducted with an accelerometer falling on to turf, but it is dropped from a different height and has a different mass resulting in a different g force. A third accelerometer used in field testing is the IS-100 (Techmark, Lansing, MI) which is a self-contained recording device designed to be transported to the surface being tested (Figure 31.2).

Injuries on turf

Estimates range from 30 injuries/1000 hrs on artificial turf (Engebretsen, 1987) to 5 injuries/1000 hrs (Hagel, 2003). Many factors determine the number of injuries sustained on turf. The time of the year may be important. An early season bias was noted by Orchard (2002) in which he related to the drier earlier season in Australian rugby. During the pre-season training, there are often more overuse injuries while during the regular season there are more traumatic injuries.

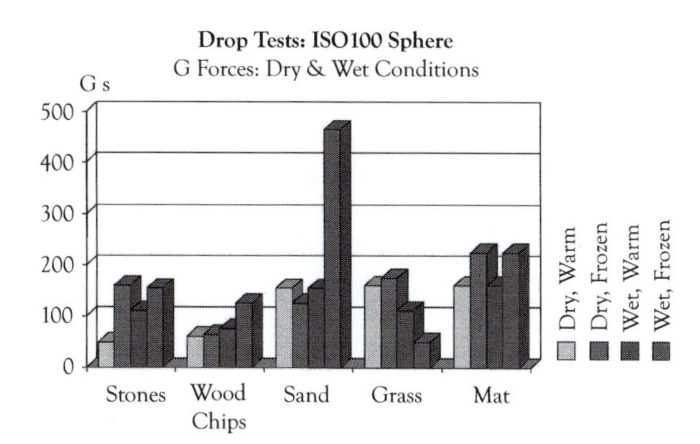

Figure 31.2 Drop tests: ISO100 Sphere.

There is a definite bias for injury if players have had a previous injury. Annegers (1980) reported that if you have one concussion you are three times more likely to have a second and eight times more likely to have a third. Previous injury predisposes to knee and ankle injuries as well.

Injuries may be more likely depending on the position played. For example, special punt return teams in American football tend to consist of smaller, faster players which are often hit by larger heavier ones.

Injuries are more common in experienced players. Many studies note that seniors in high school have more injuries than less experienced players, but this may be related to the fact that they have more playing time.

Both the surface of play and protective equipment like helmets must be considered in injury prevention. A softer surface for example will increase the duration of impact and decrease the impact to the head during a fall, but it may also increase friction and therefore cause more flexion injuries to the spine. The next section of the chapter will focus on sports injuries which may be related to artificial turf.

Cervical spine injuries

It is clear that the position of the neck on impact is the single most important determinant of cervical cord injury with axial loading causing many cervical injuries (Gosch, 1972).

In the 1975 football season, 12 football players in Pennsylvania and New Jersey sustained severe cervical spine injuries with 8 of them resulting in quadriplegia. The majority of these injuries were caused by axial loading (Figure 31.3), where the player put his head down going into a block straightening the cervical spine and causing an injury to the cord. Because of this, the NCAA implemented rule changes banning spearing, using the top of the helmet to strike an opponent. With this change in rules came a significant reduction in the number of cases of cervical spine injury. The yearly incidence of quadriplegia decreased from a high of 34 in 1976 to 18 the following year with the implementation of the spearing rule change.

The position of the neck is also important as a player hits the turf (Torg, 1990). If the player contacts the turf with the face mask, it can become lodged in the surface holding the head stationary while the body rotates forward over the neck, causing hyperflexion injury

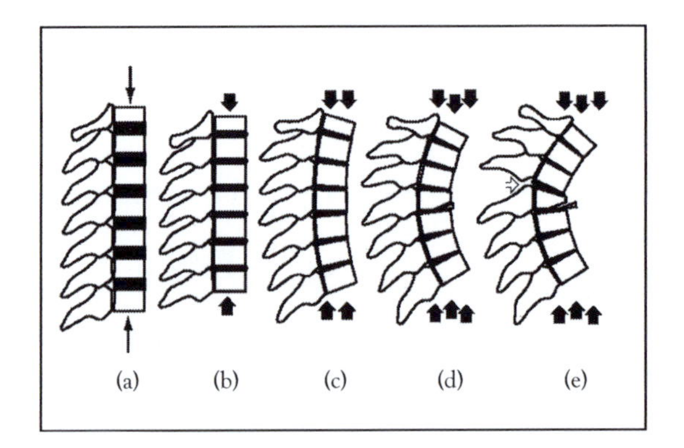

Figure 31.3 Cervical injuries.

to the neck. The friction of the surface is important in this injury. Hyperflexion upon contacting the turf can also cause cervical spine fractures and ligamentous injuries, resulting in atlantoaxial dislocation, anterior teardrop fracture with cord compression and odontoid fractures. The face mask can also be a cause of hyperextension of the cervical spine either by contact with an opposing player or contacting the turf. Hyperextension can result in vertebral artery injury with resultant quadriplegia or rupture of the anterior longitudinal ligament and neck instability (Schneider, 1973).

Serious head injuries – subdurals and skull fractures

Because of frequent collisions, head injuries have been part of the game of football since its inception, long before the advent of artificial turf. The original game of American football nearly went out of existence because the brutality of the game with mass tackling and flying wedges resulted in 18 deaths on college campuses in 1905. Theodore Roosevelt is credited with saving the game by enforcing rules changes that led to the formation of the National Collegiate Athletic Association (NCAA). The first football helmet was worn in the Army/Navy game of 1893 by a player who was told he risked 'instant insanity' if he were to be kicked in the head again. Helmets were not mandated for play until 1930.

Helmets protect the player from head injury for three reasons. First, energy will be dissipated by the material within the helmet; second, the mass of the helmet will reduce the velocity imparted to the person's head, thus reducing the acceleration his or her head endures; and third, the compliant soft layer within the helmet increases the time involved in the momentum transfer, further reducing head acceleration. Increasing the area over which the force is absorbed, and increasing the duration of time over which the impact occurs, decrease the stresses experienced by the brain. Based on Newton's s law, F = ma, where a is the change in velocity over change in time, if the time of impact increases, the acceleration will decrease. Attempts have been made to look at the duration of impact with the Head Injury Criterion and Gadd Severity Score. The HIC (Versace, 1971) and the Gadd Severity Score (Gadd, 1966) are attempts to incorporate both acceleration and duration of impact to predict the potential for brain injury.

One of the most interesting studies done to establish a tolerance level for brain injury was done by Gurdjian (Gurdjian, 1962). In this study cadavers were dropped down an elevator shaft to impact a car windshield below. The study was done two ways; in one the cadavers were denuded of skin and scalp while in the other group the skin and scalp was intact. It took 20 times more force to cause a skull fracture in the skulls with skin and hair intact. The implication here is that even a small bit of protection is useful in preventing brain injury caused by linear acceleration. From this original work done by Gurdjian, testing linear acceleration, a level of 200 g was thought to be the level at which skull fracture, and, therefore, brain injury, occurs.

A helmet can also have a detrimental effect, however, by increasing angular rotation, If the helmet's size increases the moment arm of rotation. Angular acceleration may be responsible for increasing head injury since rotation causes shear stresses at areas of the brain where there is a change in density, like the gray-white junction or the brainstem (Gennarelli, 1982; Ropper, 2007). This angular rotation may be increased if there is increased friction with the playing surface.

Comparison of head injury data from 1959 to 1963 to data from 1971–1975 show that both intracranial hemorrhages and intracranial deaths decreased. This is probably due to improvements in helmets. Intracranial hemorrhage appeared to increase from 1976 to 1982 but this apparent increase may be due to the advent of CT scanning, which allowed the hemorrhages to be more readily identified. Intracranial hemorrhages resulting in death among high school and college athletes remained constant from 1975–1987 at about ten per year (Torg, 1990).

American football helmets have a stiff outer shell and an inner layer of padding which may consist of a pneumatic bladder or foam. Helmets in American football were initially worn to prevent subdural hematomas caused by skull fractures in head to head contact with other players or hitting the ground. In these cases two stiff objects strike each other and the helmet attenuates the impact by absorbing the energy of the blow. The helmet is less effective in striking a softer surface, i.e., falling on to soft ground. Most of the impact in this case will be absorbed by the softer surface and the helmet will have little effect. In a similar fashion, if the helmet hits artificial turf, and the turf is relatively stiff, causing energy to be transmitted back into the players head, the helmet will be helpful in attenuating the impact.

Minor head injuries – concussions

A concussion is defined by the Congress of Neurological Surgeons as 'any transient change in neurological function including change of vision, equilibrium or other similar symptoms' (Congress of Neurological Surgeons, 1966). From the work of Gurdijian (1966), and Ommaya (1966, 1974) it would seem that decreasing the acceleration of impact should help protect against concussion. However, Ommaya and colleagues have shown that concussion and traumatic unconsciousness may also occur from high accelerational impulse, a mechanism that may not be amenable to protection by a helmet since the helmet cannot prevent the brain from rotating within the skull. This may be the explanation for the fact that concussions in sports are increasing. Concussions affect 50 people per 100,000 populations in the US each year. A well constructed helmet can decrease intracranial acceleration by about 30 per cent, but if artificial turf is 6–20 per cent stiffer than an optimal grass field, a helmet may just barely make up for the change in surface.

Lower extremity injury

Kicking is, without doubt, the most widely studied skill in soccer and has been the subject of numerous biomechanical analyses. The speed obtained by the kicker can be an indication of his/her proficiency. The foot is plantar flexed before contact with the ball, and the toes of the foot reach greater speed than the centre of mass (COM) of the foot. For instep kicks, ball to foot speed ratios range from 1.06 (Asami, 1983) to 1.5 (Aitcheson,1983). If the foot contacts the artificial turf, increased friction from the artificial surface may result in injury.

'Turf toe' is a condition that results from hyperextension of the great toe made possible by the more flexible shoes worn on artificial turf. The type of shoes that are worn tend to be more flexible than the cleats used on grass. While a player's foot is planted and the first metatarsal phalangeal joint hyperdorsiflexes and the heel is raised up off the ground. During an applied downward force, the great toe is dorsiflexed to an extreme, resulting in a tear of the capsule around the first metatarsal phalangeal joint. This injury is normally treated with rest and often can become a chronic disabling problem. Operative repair can sometimes be successful.

A higher incidence of muscle-tendon overload was noted on Fieldturf (Meyers, 2004). The artificial turf is more consistent and leads to a faster game, which may lead to a greater potential for injury because of greater fatigue potential for the players. Players performing at greater speed, acceleration and torque have potential for greater injury when they rapidly decelerate.

Achilles tendonitis, and bursitis are reported with artificial turf, perhaps because of the stiffness of the surface.

Upper extremity injuries

Falls directly on the shoulder cause more injuries if the surface is hard. Clavicle fractures were shown to increase in soccer on dry, hard surfaces. These usually fall on an outstretched hand.

Knees and ankles

The reason for more injuries on artificial turf is that increased friction on artificial surfaces leads to knee and ankle injuries. Meyers (2004) found that more injuries occur on field turf in warm weather than on natural grass possibly because of increased friction at higher temperatures. A study by Torg (1971) found that the frequency of injuries was higher for players using conventional football shoes with 7 3/4 inch cleats than other players using shoes with 14 3/8 inch cleats. This was confirmed in a study by Blyth (1974). Shoe surface traction has a linear association with the bottom cleat surface area. This suggests higher friction between the surface and shoe is an important cause of injury.

Increasing axial load on a shoe increased frictional resistance but it is not a linear relationship. The response to a 400 per cent increase in axial load varied between 30 per cent increases to over 1500 per cent in a study by (Cawley, 2003). They also described the tendency of shoes designed for use on artificial turf to load to their peak limit and then break loose suddenly. A similar phenomenon occurs with cleats on natural grass. In play, when an athlete plants his foot and decelerates prior to making a cut, axial loads on the shoe may be as high as six times body weight. The inertia of the body causes high compressive forces between the sole and surface which locks the foot. As the player then pivots, causing axial rotation, valgus injury can occur at the knee.

Skin abrasions

Although the newer artificial surfaces like field turf claim to be less abrasive, studies have shown they are still responsible for more abrasions than natural grass (Meyers, 2004). Meyers reported a higher incidence of abrasions on field turf than natural grass with 5.8 per cent versus 0.8 per cent. The incidence of abrasions is important since subsequent skin infections have plagued several teams. The St Louis Rams reported an outbreak of methicillin resistant staph aureus (MRSA) in 2003 that they directly associated with skin abrasions.

Summary

Grass fields are not all equal with industry reports of muddy grass fields having G values of 65–70 while frozen grass fields have G values as high as 275 (Lewis, 1993). Yet grass fields are preferred by professional athletes.

Artificial turf, although it certainly is a more consistent surface than natural grass, is unfortunately associated with more injuries (Arnason, 1996). Many of them are minor, relating to the increased friction of the artificial surface, causing problems like turf toe or abrasions but others, like the increased risk of concussion in American football reported by Guskiewicz (Guskiewicz, 2000) have more long-term sequellae.

In addition, because of its increased stiffness, artificial turf is likely to be responsible for chronic damage to the joints of the knee and ankle.

In soccer, as well, because of the differences between artificial turf and natural grass, increased ball roll distance and greater peak deceleration for high energy impacts on artificial turf, grass is preferred. Artificial surfaces are thought to be suitable for lower-level play where the economic advantages of a maintenance-free field become important.

References

1. Aitcheson, I., Lees, A. (1983) 'A biomechanical analysis of place kicking in Rugby Union Football'. *Journal of Sports Sciences*, 1: 136–7.
2. Annegers, J.F. (1980) 'The incidence, causes and secular trends of head trauma in Olmstead County, MN. 1935–1974'. *Neurology*, 30: 912–19.
3. Arnason, A., Gudmundsson, A., Dahl, H.A., Hohannsson, E. (1996) 'Soccer injuries in Iceland'. *Scand J Med Sci Sports*, 6: 40–5.
4. Asami, T., Nolte, V., 'Analysis of powerful ball kicking'. In Biomechanics VIII–B (edited by H. Matsui and K Kobayashi), pp. 695–700. Champaign, IL: Human Kinetics.
5. ASTM D575 Specification for compression. Annual book of ASTM Standards. West Conshohocken, PA: American Society for Testing and Materials, 1994; 15: 07.
6. ASTM F355 Test Method for Shock Absorbing Properties of Playing Surface Systems and Materials. Annual book of ASTM Standards. West Conshohocken, PA: American Society for Testing and Materials, 1994 15: 07.
7. Bartlett, R. (1999). 'Sports Biomechanics. The effect of sports equipment and technique on injury'. Spon Press, London, 69.
8. Barzarian, J.J., McClung, J., Shah, M.N., Cheng, Y.T., Flesher, W., Graus, J. (2003) 'Mild traumatic brain injury in the United States, 1998–2000'. *Brain Injury*, 19: 85–91.
9. Bell, M.J., Baker, S.W., Canaway, P.W. (1985). 'Playing quality of sports surfaces: a review'. *Journal of the Sports Turf Research Institute*, 61: 26–45.
10. Blyth, C.S., Mueller, F.D. (1974) 'Football injury survey'. *Physician and Sportsmedicine*, 2(10): 71–8.
11. Cawley, P.W., Heidt, R.S., Scranton, P.E., Losse, G.M., Howard, M.E. (2003) 'Physiologic axial load, frictional resistance, and the football shoe-surface interface'. *Foot and Ankle International*, 24(7): 551–6.
12. Gadd, C.W. (1966) 'Use of weighted-impulse criterion for estimating injury hazard'. *Proceedings of the 10th Stapp Car Crash Conference*. New York, NY: Society of Automotive Engineers, P–12, paper 660793.
13. Gennarelli, T., Thibault, L. (1982) 'Biomechanics of acute subdural hematom'. *Journal of Trauma*, Vol 22(8): 680–6.
14. Gosch, H.H., Gooding, E., Schneider, R.C. (1972) 'An experimental study of cervical spine and cord injuries'. *Journal of Trauma*; 12: 570–6.
15. Gurdjian, M.D., Lisner, M.S., Patrick, M.S. (1962) 'Protection of the head and neck in sports'. *JAMA* 182(5): 509–12.
16. Gurdijian, E.S., Roberts, V.L., Thomas, L.M. (1966) 'Tolerance curves of acceleration and intracranial pressure and protective index in experimental head injury', *J of Trauma*. 6: 600–5.
17. Guskiewicz, K.M., Weaver, M.L., Padua, D.A., Garrett, W.E. (2000) 'Epidemiology of concussion in collegiate and high school football players'. *Am J Sports Med*, 28: 43–650.
18. Hagel, B.E., Fick, G.H., Meeuwisse, W.H. (2003) 'Injury risk in men's Canada West University football'. *Am. J. Epidemiol*, 157: 825–33.

19. Kolitzus, H.F. (1984) Functional standards for playing surfaces in Sports Shoes and Playing Surfaces (ed. E.C. Frederick), *Human Kinetics*. Champaign, IL, USA, pp. 98–118.
20. Lewis, L.M., Naunheim, R.S., Standeven, J.S., Naunheim, K.S. (1993) 'Quantitation of impact attenuation of different playground surfaces under various environmental conditions using a tri-axial accelerometer'. *Journal of Trauma*, 35(6): 932–5.
21. Meyers, M.C., Barnhill, B.X. (2004) 'Incidence, causes, and severity of high school football injuries on FieldTurf versus natural grass: a 5 yr prospective study'. *American Journal of Sports Medicine*, 32(7): 1626–38.
22. Menlo, P.C. (1975) *National football league 1974 injury study*. Stanford Research Institute.
23. Naunheim, R.S., Parrot, H., Standeven, J.S. (2005) 'A comparison of artificial turf'. *Journal of Trauma*, 57(6): 1311–14.
24. Nigg, B.M., Yeadon, M.R. (1987) 'Biomechanical aspects of playing surfaces'. *Journal of Sports Sciences*, 5: 117–145.
25. Ommaya, A.K., Flamm, E.S., Mahone, R.H. (1966) 'Cerebral concussion in the monkey: an experimental model'. *Science*, 153: 211–12.
26. Ommaya, A.K., Gennarelli, T.A. (1974) 'Cerebral concussion and traumatic unconsciousness; correlation of experimental and clinical observations on blunt head injuries'. *Brain*, 97: 633–54.
27. Orchard, J. (2002) 'Is there a relationship between ground and climatic conditions and injuries in football?' *Sports medicine*, 32(7): 419–32.
28. Ropper, A.H., Gorson, K.C. Concussion. *NEJM* 2007 356(2): 166–71.
29. Schneider, R.C. (1973) *Head and Neck Injuries in Football*. Williams & Wilkins, Baltimore.
30. Stanford Research Institute (1975) National Football League 1974 Injury Study. Menlo Park, California.
31. Torg, J.S., Quedenfeld,T. (1971) 'Effect of shoe type and cleat length on incidence and severity of knee injuries among high school football players'. *Research Quarterly*, 42: 203–11.
32. Torg, J.S., Vegso, J.J., O'Neill M.J., Sennett B. (1990). 'The epidemiologic, pathologic, biomechanical, and cinematographic analysis of football-induced cervical spine trauma'. *Am Journal of Sports Med*, 18(1): 50–7.
33. Valiant, G.A. 'Traction characteristics of outsoles for use on artificial playing surfaces'. *Natural and artificial playing fields: characteristics and safety features*, ASTM STP 1073. (edited by) R.C. Schmidt, E.F. Hoerner, E.M. Milner and C.A. Morehouse, pp. 61–88. Philadelphia: The American Society for Testing and Materials.
34. Versace, J. (1971) 'A review of the severity index'. *Proceedings of the 15th Stapp Car Crash Conference*. New York, NY: Society of Automotive Engineers.

Section VII

Health promotion

32

Influence of backpack weight on biomechanical and physiological responses of children during treadmill walking

Youlian Hong[1], Jing Xian Li[2] and Gert-Peter Brüggemann[3]
[1]The Chinese University of Hong Kong, Hong Kong; [2]University of Ottawa, Ottawa; [3]German Sports University Cologne, Cologne

Abstract

This study examined the biomechanical and physiological responses of children to carrying backpack of 0 per cent (as control), 10 per cent, 15 per cent and 20 per cent of their own body weight during treadmill walking. When compared to 0 per cent load condition, 20 per cent body weight load induced significant increase in trunk forward lean, double support duration, stance duration and metabolic cost, and decrease in trunk angular motion and swing duration with a prolonged blood pressure recovery time. Fifteen per cent load condition induced significant increase in trunk forward lean and prolonged blood pressure recovery time. No significant difference was found in the measured parameters between 10 per cent and 0 per cent load conditions.

Introduction

The overloading of school bags has roused increasing concern in communities in many countries. Reports from different countries found that most students carry school bags that are more than 10 per cent of their body weight, with many bags being heavier than 20 per cent of the child's body weight (Sander, 1979; HKSCHD, 1988; Pascoe *et al.*, 1997; Negrini and Carabalona, 2002). However, compared with the extensive studies on physiological and biomechanical responses to load carriage in adults, investigations for children are limited. Malhotra and Sen Gupta (1965) examined the metabolic cost associated with different ways of carrying schoolbags by children during walking. The bags were 2.6 kg which approximated 10–12 per cent of the subject's own body weight and carried in four different positions: rucksack on the back, low on the back, across the shoulders, and hand held. Pascoe *et al.* (1997) studied the static postures and gait kinematics of children in four different conditions: without load and carrying a book bag with a load of 17 per cent of mean body weight in one-strap backpack, two-strap backpack and one-strap athletic bag. The one-strap book bag induced significant elevation of the strap supporting the shoulder and concomitant lateral bending of the spine. More recently, Chansirinukor *et al.* (2001) reported the effects of weight of a backpack, its position on the spine and time of carrying on adolescents' cervical and shoulder posture. They revealed that both backpack weight and carrying duration influenced cervical and shoulder posture. Carrying a backpack weighing 15 per cent of body weight appeared to be too heavy to maintain standing posture for adolescents. To our knowledge, information about the concomitant biomechanical and physiological stress of children induced by carrying heavy backpack has not ben reported. The purpose of this study was to investigate the effects of carrying different weights during treadmill walking on children by simultaneous measurement of gait, trunk posture, expired air, heart rate and blood pressure. It was hoped that this study would provide useful information to the community in suggesting an appropriate weight of school bags to children.

Methods

A survey of body height and weight was made for all ten-years-old students in a local primary school. The body mass index (BMI), which is a relatively good indicator of total body composition and is related to health outcomes, was then calculated for each student. The 15 students with the most representative BMI for this group were recruited to serve as subjects in this investigation. Before the load carrying testing was implemented, the subjects and their parents were informed about the purpose, procedures and applications of the study and parental consents were obtained. This study was approved by the Clinical Research Ethics Committee, the Chinese University of Hong Kong.

The subjects had a mean body weight of 33.53 ± 2.64 kg and a mean body height of 141.86 ± 3.77 cm. Each subject participated in four treadmill walking trials: without load (0 per cent of body weight) and carrying a school bag of 10 per cent, 15 per cent and 20 per cent of their body weight (BW). For each subject, the different loads carried were randomly assigned on each of four different testing days. Before trials, all subjects were given ample time to familiarize themselves with the purpose of the study and the equipment used. In each trial, the subject was allowed to walk on the treadmill until he felt secure, before the measurement started. Then subjects were asked to walk on a treadmill for 20 minutes at a speed of 1.1 m/s, a comfortable speed of walking for children (Malhotra and Sen Gupta, 1965) with school bags carried on the mid-back region (Figure 32.1).

The gait was filmed by a 3-CCD video camera (50 Hz) positioned lateral to the subject with the lens axis perpendicular to the movement plane. The distance of the camera to the movement plane was 7.5 m and the shutter speed was set at 1/250 s. The recorded video tapes were then digitized on a motion analysis system by using a two dimensional human body model consisting of 11 points, including the toe, knee, heel, ankle, hip, shoulder, elbow, wrist, finger, ear and the neck on the lateral side against the camera. A Butterworth low-pass filter was used to smooth the position-time data for each anatomical landmark.

The start of the gait cycle begins with heel strike of one foot and finishes when the same foot returns to heel strike. Stride distance is the liner distance between two consecutive ipsilateral heel strikes. Stance phase is between the time instant of heel strike and toe-off of the same foot. In double support phase both feet are in contact with the ground at the same time. Swing phase begins when foot leaves the ground and finishes when the heel of the foot strikes. In this study, stance duration, swing duration and double support duration were normalized as the percentage value within a walking cycle (Murray *et al.*, 1985). That is, the duration of the complete cycle was used as a base line (or 100 per cent), and values for the

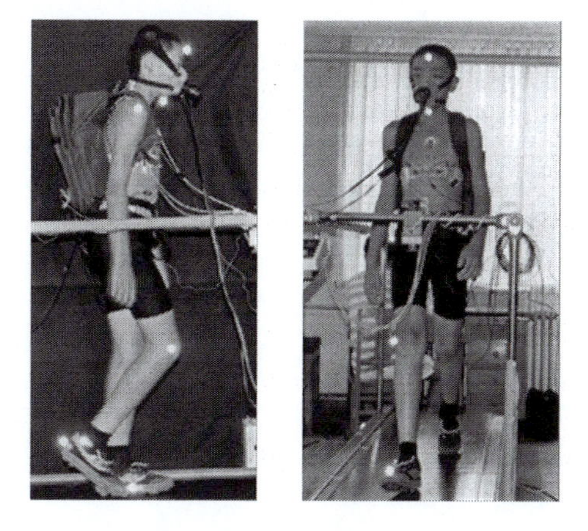

Figure 32.1 Subjects walked on the treadmill. The walk movement was video-filmed. The heart rate was recorded and expired gas was collected and analyzed using a cardiopulmonary function system.

other durations were expressed as a percentage of this base line. Stride frequency refers to the number of strides per second. Trunk inclination angle refers to the mean angle of the line connecting the shoulders and hips with the horizontal at the hips through all frames of one complete stride. Values less than 90° represent a forward lean of trunk, while values greater than 90° represent a backward leaning trunk position. Trunk motion range refers to the range of angular motion that was observed during the stride. In each trial, three complete gait cycles were taken every five minutes for analysis, with the first cycle being selected as the walk was visually observed to become normalized. Kinematic parameters of each three cycles were then averaged to represent the gait pattern at this time point.

The $\dot{V}O_{2max}$ in each subject was measured one week before the first trial. The $\dot{V}O_{2max}$ was used to calculate the relative per cent of maximum capacity an individual was working with the various loads. The $\dot{V}O_{2max}$ tests were conducted walking on the treadmill using a continuous incremental protocol until the children were exhausted. The modified Balke protocol for $\dot{V}O_{2max}$ max test of children (American College of Sports Medicine, 1995) was used in this investigation. Initially the children were habituated to the general environment and the technique of using the treadmill. Subjects warmed up for three minutes at a speed of 6 km h^{-1}, then walked at 6 km h^{-1} with a 2 per cent increase in gradient every three minutes beginning at 6 per cent. The $\dot{V}O_{2max}$ was the maximal index measured when the subjects could no longer continue and their heart rate was 200 beats per minute or above. The mean and standard deviation of the $\dot{V}O_{2max}$ of the subjects in this study was 44.03 ± 4.52 ml kg^{-1} min^{-1}.

Heart rate was continuously monitored and recorded automatically before walking, throughout and until 5 minutes after walking test using a cardiopulmonary function system (Oxycon, Champion, Jeager) (Figure 32.1).

Blood pressure was measured before, immediately afterward, and at three and five minutes after walking. The heart rate and blood pressure recorded before walking were used as baseline data to compare the cardiovascular responses during and after walking.

The expired air was collected by mask connected to the cardiopulmonary function system and also continuously analyzed automatically to provide oxygen consumption ($\dot{V}O_2$, ml kg^{-1} min^{-1} and ml min^{-1}) and respiratory quotient (RQ) (Figure 32.1). When beginning the measurements, subjects were asked to stand for a few minutes until the heart rate reached a steady state. The heart rate and $\dot{V}O_2$ recorded at this state were used as a baseline. Subjects then began the walking trial and, though all measurements were continuously monitored and averaged every thirty seconds, only the data at each five minutes interval of the 20 minutes walking and three and five minutes after walking were used for analysis. Energy expenditure during walking and at three and five minutes after walking was then calculated according to the absolute $\dot{V}O_2$ ($l\ min^{-1}$), the RQ and the formula by Weir (1949). The energy expenditure is expressed in kilocalorie per minute ($Cal\ min^{-1}$). The relative working intensity was calculated as the rate of $\dot{V}O_2$ to $\dot{V}O_{2max}$ (per cent $\dot{V}O_{2max}$). In this way, the energy expenditure, $\dot{V}O_2$ and relative working intensity over time and under each weight carried, was obtained.

Two-way analysis of variance (ANOVA) (load by time) with repeated measures was applied on each of the kinematics and physiological (dependent) variables to see significant effects by load and time. If significant interactive effect of load and time was found, stratified ANOVA was conducted to demonstrate the time effect at each load or vice versa with Scheffe's significant difference test. If no significant interactive, but significant main effect of load, was found, one-way ANOVA with Scheffe's post hoc test was used to locate the significant mean differences between loads on each of the dependent variables. A value of $\alpha = 0.05$ was used for all tests as the criterion to determine the presence or absence of significance. A value of $\alpha = 0.05$ was used for all tests as the criterion to determine the presence or absence of significance.

Results

Kinematics responses

The two-way ANOVA revealed significant main effects of load on trunk forward lean, trunk angular motion, double support duration and swing duration.

The results showed that trunk forward lean angle was significantly increased with the loads of 20 per cent and 15 per cent body weight when compared to 0 per cent ($p < 0.01$, Scheffé) and 10 per cent ($p < 0.05$, Scheffé) load conditions. No significant difference in trunk forward lean angle was observed between 20 per cent and 15 per cent load conditions (Figure 32.2).

Trunk angular motion was significantly decreased with carrying a load of 20 per cent body weight when compared to 0 per cent ($p < 0.05$, Scheffé) load condition. No significant differences in trunk motion range were found among other load conditions (Figure 32.3).

Carrying backpacks of 20 per cent body weight resulted in significant increase in the double support duration when compared to 0 per cent ($p < 0.01$, Scheffé) and 10 per cent ($p < 0.05$, Scheffé) load conditions (Figure 32.4), with a significant decrease in the swing duration when compared with 0 per cent ($p < 0.01$, Scheffé) and 10 per cent ($p < 0.05$, Scheffé) load conditions (Figure 32.5). No significant differences were found among other load conditions.

Physiological responses

Mean and standard deviation of heart rate, $\dot{V}O_2$, work intensity and energy expenditure during the 20 minutes of walking and until five minutes after walking is presented in Table 32.1. Among all load conditions, heart rate increased significantly in the first five minutes of walking ($p < 0.05$, Scheffé), and then gradually increased over time during walking.

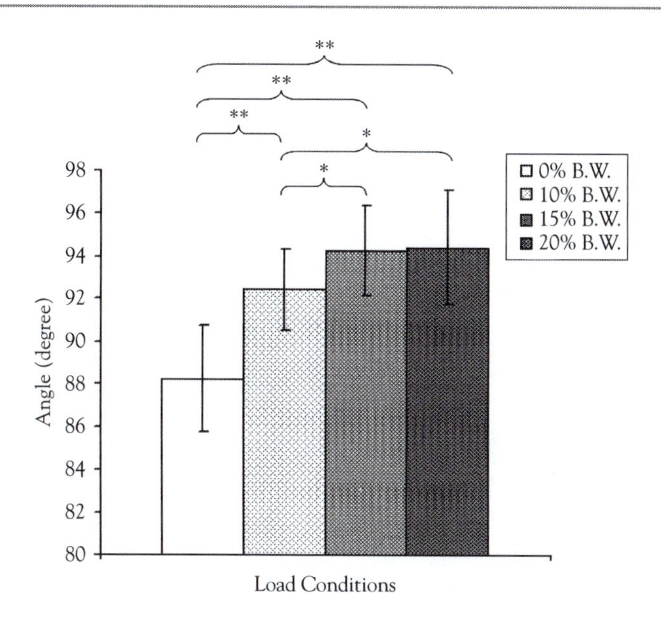

Figure 32.2 Trunk forward lean angle with different loads (Mean ± SD).
* and ** indicate significant difference with p<0.05 and p<0.01 respectively.

After three minutes of recovery, heart rate fell to a level approximate to the baseline. There was no significant difference in heart rate among different loads carried (F (3, 56) = 0.128, $p > 0.05$).

The effect of load carriage on systolic and diastolic blood pressure is presented in Table 32.2. Walking for 20 minutes significantly increased the systolic blood pressure at all work loads ($p < 0.05$, Scheffé). A significant increase in diastolic blood pressure measured after 20 minutes of walking was only found at the load conditions of 15 per cent and 20 per cent body weight

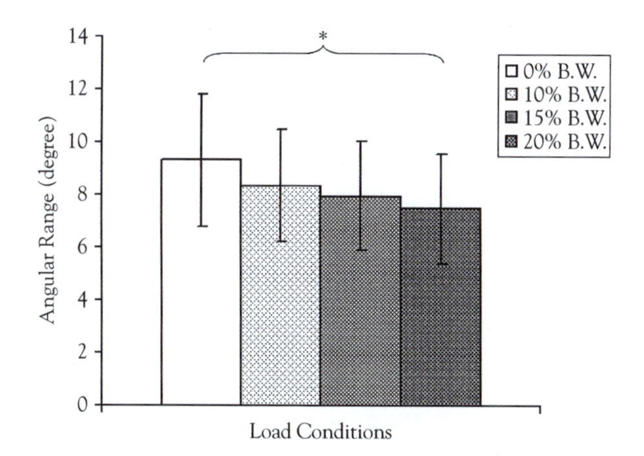

Figure 32.3 Trunk angular range with different loads (Mean ± SD).
* and ** indicate significant difference with p<0.05 and p<0.01 respectively.

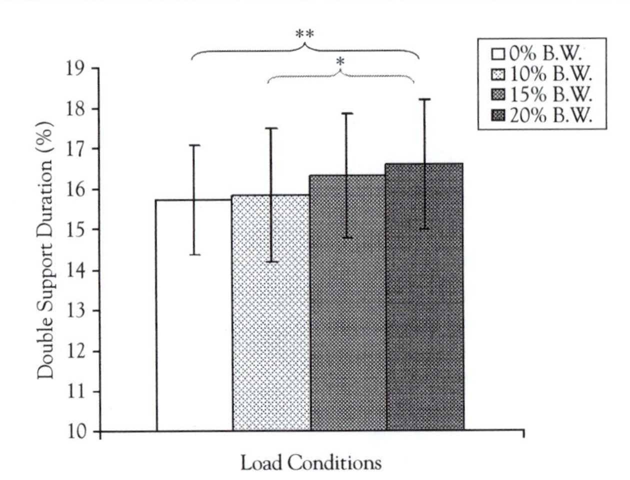

Figure 32.4 Normalized double support duration with different loads (Mean ± SD).
* and ** indicate significant difference with p<0.05 and p<0.01 respectively.

(p < 0.05, Scheffé). The recovery in systolic blood pressure showed significant differences among different loads. In carrying loads of 0 per cent and 10 per cent body weight, after three minutes of recovery the systolic blood pressure fell to the level recorded before walking. However, even after five minutes of recovery the systolic blood pressure in carrying loads of 15 per cent and 20 per cent body weight were still higher than that recorded before walking (p < 0.05, Scheffé). No significance was found in diastolic blood pressure between after three minutes of recovery and before walking in each load condition. Comparing the measurements among different loads, significant differences in systolic and diastolic blood pressure were found between the loads of 0 per cent and 20 per cent body weight (p < 0.05, Scheffé). While there was no significant difference in the alteration of blood pressure among the other load conditions (p > 0.05, Scheffé).

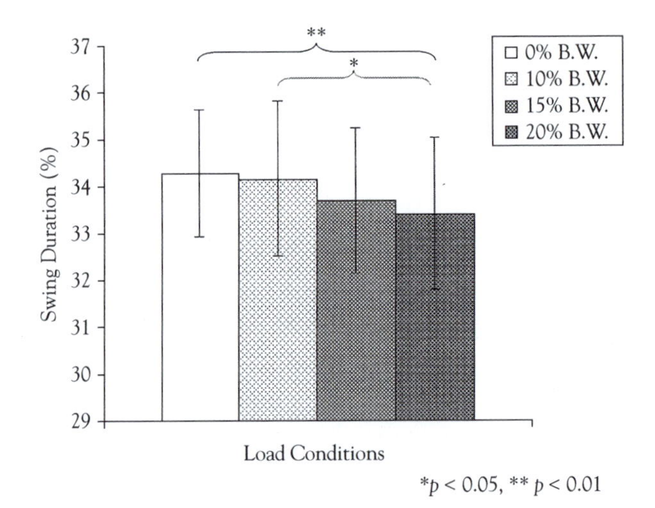

Figure 32.5 Normalized swing duration with different loads (Mean ± SD).

Table 32.1 Mean and standard deviation of heart rate, oxygen uptake, work intensity and energy expenditure under different loads with time

		Time (min)						
Variables	*Load %BW*	*0*	*5*	*10*	*15*	*20*	*23**	*25**
Heart rate (Beats min^{-1})	0	92.33	116.73†	119.93†	120.73†	120.80†	103.47	97.13
		(8.54)	(10.23)	(11.52)	(11.37)	(11.83)	(13.18)	(9.91)
	10	93.33	118.80†	121.27†	123.93†	124.07†	107.27	100.07
		(7.11)	(9.86)	(11.35)	(11.06)	(12.42)	(13.51)	(8.53)
	15	94.00	118.33†	122.67†	124.80†	125.13†	104.60	101.93
		(6.91)	(10.39)	(11.86)	(12.01)	(12.50)	(11.29)	(11.25)
	20	93.33	119.60†	123.07†	124.20†	125.00†	101.67	98.07
		(7.10)	(9.59)	(11.37)	(11.21)	(11.74)	(11.36)	(9.68)
$\dot{V}O_2$(ml kg^{-1} min^{-1})	0	5.02	17.05†	17.63†	17.35†	17.23†	5.42†	4.86
		(0.89)	(1.35)	(1.20)	(1.22)	(1.23)	(0.59)	(0.61)
	10	4.99	17.35†	18.08†	17.81†	17.67†	5.90†[a]	5.12
		(0.53)	(1.24)	(1.34)	(1.13)	(1.23)	(0.55)	(0.51)
	15	5.25	17.69†	18.51†	18.19†	18.15†	6.49†[a]	5.32
		(0.86)	(1.43)	(1.29)	(1.14)	(1.12)	(0.61)	(0.59)
	20	5.44	19.11†[ab]	19.61†[ab]	19.41†[ab]	19.30†[ab]	6.87†[ab]	5.55
		(0.87)	(1.47)	(1.39)	(1.29)	(1.23)	(0.80)	(0.83)
Work intensity (%VO_{2max})	0	11.43	38.86†	40.37†	39.65†	40.40†	12.38†	11.12
		(1.96)	(4.47)	(4.27)	(3.97)	(2.64)	(1.53)	(1.67)
	10	11.33	39.75†	41.38†	40.76†	40.41†	13.49†	11.67
		(1.33)	(4.46)	(4.30)	(4.08)	(3.89)	(1.57)	(1.13)
	15	11.39	40.56†	42.44†	41.73†	41.57†	14.84†[a]	12.29
		(3.32)	(5.10)	(5.15)	(4.67)	(4.45)	(1.69)	(1.88)
	20	12.43	43.76†[ab]	44.87†[ab]	44.42†[ab]	44.11†[ab]	15.69†[ab]	12.69
		(2.08)	(4.96)	(4.69)	(4.55)	(4.20)	(2.06)	(2.01)
Energy expenditure (Cal)	0	0.82	2.80†	2.92†	2.87†	2.86†	0.89†	0.81
		(0.12)	(0.29)	(0.32)	(0.25)	(0.28)	(0.14)	(0.14)
	10	0.82	2.88†	3.01†	2.97†	2.95†	0.98†	0.84
		(0.07)	(0.29)	(0.27)	(0.27)	(0.23)	(0.11)	(0.11)
	15	0.87	2.93†	3.08†	3.04†	3.00†	1.09†	0.91
		(0.10)	(0.29)	(0.27)	(0.23)	(0.27)	(0.12)	(0.11)
	20	0.83	3.13†	3.26†	3.23†[a]	3.23†[ab]	1.17†[ab]	0.92
		(0.23)	(0.35)	(0.32)	(0.32)	(0.29)	(0.13)	(0.13)

*denotes 3 and 5 min recovery after walking
†Significant difference from 0 min of the same load condition, $p < 0.05$
[a]Significant difference from 0% load condition of the same time point, $p < 0.05$
[b]Significant difference from 10% load condition of the same time point, $p < 0.05$
[c]Significant difference from 15% load condition of the same time point, $p < 0.05$

As shown in Table 32.1, the first five minutes of walking yielded a highly significant increase in $\dot{V}O_2$ for each load carried, F (3, 56) = 6.586, $p < 0.01$, after which the $\dot{V}O_2$ increased gradually with no significant differences during the walking period. After five minutes recovery, $\dot{V}O_2$ fell to the baseline for each load condition. When comparing the demand for oxygen between the different load conditions, walking with a load of 20 per cent body weight elicited significantly higher $\dot{V}O_2$ throughout and until three minutes after walking than the loads of 0 per cent and 10 per cent body weight ($p < 0.05$, Scheffé). No significant differences in $\dot{V}O_2$ were found among the loads of 0 per cent, 10 per cent and 15 per cent body weight during the walking time, but carrying the load of 15 per cent body

Table 32.2. Mean and standard deviation of blood pressure under different load conditions with time

Variables	*Load (%BW)*	*Time (min)* 0	20	23*	25*
Systolic pressure (mm Hg)	0	101.00 (6.51)	112.97† (9.40)	105.47 (8.42)	101.73 (7.07)
	10	99.27 (8.01)	114.33† (8.48)	104.53 (8.28)	102.27 (7.73)
	15	100.73 (10.07)	120.67† (8.77)	106.40† (8.06)	105.73† (7.78)
	20	103.53[a] (5.91)	122.00†[a] (7.38)	109.47†[a] (7.85)	108.10†[a] (6.24)
Diastolic pressure (mm Hg)	0	69.40 (6.24)	73.80 (4.57)	68.87 (5.57)	70.53 (6.10)
	10	69.80 (4.52)	73.13 (5.32)	71.93 (3.63)	71.67 (6.04)
	15	68.93 (6.05)	76.80† (6.54)	73.13 (6.22)	73.13 (6.40)
	20	70.73 (8.85)	77.53†[a] (7.34)	74.73[a] (4.61)	74.20 (5.31)

*denotes 3 and 5 min recovery after walking
†Significant difference from 0 min of the same load condition, p < 0.05
[a]Significant difference from 0% load condition of the same time point, p < 0.05
[b]Significant difference from 10% load condition of the same time point, p < 0.05
[c]Significant difference from 15% load condition of the same time point, p < 0.05

weight induced significant higher $\dot{V}O_2$ (6.49±0.61 ml kg^{-1} min^{-1}) at three minutes after walking than 0 per cent body weight (5.42±0.59 ml kg^{-1} min^{-1}) at this time ($p < 0.01$, Scheffé).

The changes of relative work intensity in each load condition showed the same trends as those found with $\dot{V}O_2$ (Table 32.1). Particularly, when standing, the work intensity in carrying the loads of 0 per cent, 10 per cent, 15 per cent and 20 per cent body weight were recorded as 11.43 per cent, 11.33 per cent, 11.39 per cent and 12.43 per cent $\dot{V}O_{2max}$ respectively. After 20 minutes walking on treadmill these figures were changed to 40.40 per cent, 40.41 per cent, 41.57 per cent and 44.11 per cent of their maximal aerobic power, respectively. The average work intensity during the period of steady state (5th to 20th minute) for each respective work load was 39.82 per cent, 40.57 per cent, 41.58 per cent and 44.29 per cent of $\dot{V}O_{2max}$. The average work intensity during the steady state conditions indicated a 4.6 per cent $\dot{V}O_{2max}$ increase in metabolism at the 20 per cent load condition above the baseline of 0 per cent load, compared to only a 0.8 per cent $\dot{V}O_{2max}$ increase for the 10 per cent load condition. The results implied that an increase in load caused a differential increase in metabolic cost.

As would be expected, the changes of energy expenditure showed similar trends as the $\dot{V}O_2$ (Table 32.1). For each load condition, the energy expenditure significantly increased ($p < 0.01$, Scheffé) after five minutes of walking and gradually increased thereafter with no significant differences during the walking period. The 20 per cent body weight work load yielded significantly higher ($p < 0.05$, Scheffé) energy expenditure than did the 0 per cent load during walking. Whereas, no significant difference was found among other load conditions, throughout walking. When walking for 20 minutes, a boundary significant difference ($p = 0.056$, Scheffé) was found between the loads of 10 per cent and 20 per cent body weight.

Discussion

The kinematics data obtained from this study demonstrated that the subjects made quick appropriate adjustments for the gait pattern to fit the loads at the beginning of each trial. Under the experimental conditions set in this study (max. load of 20 per cent body weight with walking duration of 20 minutes), the walking time was not the critical factor to significantly influence the gait pattern in the children.

Figure 32.2 shows that carrying load of 15 per cent or 20 per cent body weight induced a significant increase of trunk forward lean when compared with 0 per cent or 10 per cent load conditions. The heavy load carried on the back would force the subjects to alter their body position to counteract the deviation from the normal kinematics pattern when body posture and balance were disturbed by carrying the substantial additional load. These results were comparable to the previous findings of Malhotra and Sen Gupta (1965) who reported that the children who carried 2.6 kg backpack that represented 10 per cent to 12 per cent of their body weight did not show appreciable trunk forward bend. However, in Malhotra and Sen Gupta's study the relative load of their backpack to the subject's body weight may have been insufficient to cause a shift in the trunk posture. Pascoe *et al.* (1997) found that carrying a two-strap backpack of 17 per cent of the body weight of the youth subjects significantly promoted forward lean of head and trunk compared to walking without a bag. In Pascoe *et al.*'s study only a single weight was involved and, therefore, the influence of increased load on the gait pattern could not be examined. The present study examined four different loads and found that the load equal to or greater than 15 per cent of the subject's body weight could result in significant trunk forward lean.

Carrying a load of 20 per cent of body weight significantly reduced the range of trunk motion when compared with no load (Figure 32.3). The increased load on the back would generate extra pelvic moment and thus contraction of the abdomen muscles would increase. Furthermore, a significant increase in anterior-posterior swing of trunk at higher carrying loads would cause the abdomen, back and leg muscles to work harder to maintain the dynamic balance. This particularly involves the semispinalis, erector spinae, trapezius, tibialis anterior, vastus lateralis and hamstring muscles (Cook, Neumann, 1987). To avoid harmful muscle strain subjects need to make more effort in keeping the body more stable in the anterior-posterior direction. In this study the load of 15 per cent body weight did not induce significant difference in trunk angular motion. This result was comparable to that reported by Pascoe *et al.* (1997) who found no significant difference in the angular range of trunk motion between 0 per cent and 17 per cent load conditions.

Carrying a load equal to 20 per cent body weight while walking induced significant increase in double support duration (Figure 32.4) with a decrease in swing duration (Figure 32.5), as compared to the 0 per cent and 10 per cent body weight conditions. These observations also occurred with Martin and Nelson (1986) who stated that when the load became higher, subjects were forced to alter their locomotion biomechanics, and this resulted in higher actual power output to carry a given load. In this study the subjects walked on a treadmill with a fixed speed for controlling walking and any tendency to walk faster when carrying a load, common in over ground walking, was prevented

(Ghori, Luckwill, 1985). The higher load on the back would raise the subject's centre of gravity, making the walking subject even more unstable. Subjects were forced to adjust their gait to compensate for this change by shortening the swing phase, by delaying the takeoff, giving the gait cycle a greater proportion with the feet on the ground. Shortening the swing phase may be an attempt to minimize the duration of unsteady single-limb stance. Similar findings were reported by Ghori and Luckwill (1985) who compared the loads of 10 per cent, 20 per cent, 30 per cent, 40 per cent and 50 per cent of body weight and found that the load of 20 per cent body weight and above induced a significant decrease in swing duration and an increase in feet support duration in adult subjects.

While standing, there was no significant difference in heart rate with the different loads. However, the average heart rate increased significantly by 8.12 beats per minute from a resting value of 85.13 to 93.25 beats while standing with a load. This is comparable to the results of Holewjin (1990) with adults, who reported an increase of 9 beats per minute for young male subjects while standing, independent from type of support or mass of the backpack.

Walking for five minutes elicited a significant increase in heart rate from standing for carrying each of the four loads, as would be expected. However, during the next 15 minutes walking period heart rates did not show any difference and the children had reached a fairly steady exercise state at each workload. These findings coincided with the study by Malhotra and Sen Gupta (1965) on load carriage in children. Based on the measurements of heart rate, it was evident that the cardiovascular response to load carriage was substantially in homeostatic balance (steady state). This is consistent with Åstrand and Rodahl (1986), who found that when a fit subject is exercising at less than 65 per cent of $\dot{V}O_{2max}$ cardiovascular response is in a steady state after about five minutes of exercise. In evaluating the heart rate responses of children, it was found that walking for 20 minutes at 20 per cent body weight load elicited an average heart rate of 125 beats per minute. This figure is 30 beats per minute higher than the baseline and accounts for about 60 per cent of the maximum heart rate.

The trends of increasing heart rate during walking with load carriage were also found in the study on adults. Holewijn (1990) reported 7 per cent body weight load caused a significant increase of 8 beats per minute and 14 per cent body weight load caused a further significant increase of 6 beats per minute when subjects walked 20 minutes on the treadmill at a speed of 1.33 m/s, however, there was no significant difference in heart rate between the loads selected.

Although measurements of blood pressure immediately after walking did not show any significant difference among the four carrying conditions, a rise of systolic blood pressure by an average of 12 mm Hg for 0 per cent body weight load and an average of 19 mm Hg for 20 per cent body weight was found. Likewise the different effects of carrying different loads on the cardiovascular system were observed from the recovery of blood pressure. Carrying loads equal to 15 per cent and 20 per cent body weight required a longer time for blood pressure to return to the baseline. Systolic blood pressure increases in direct proportion to increases in exercise intensity (Rowell, 1986; Berger, 1982). Therefore, the changes of blood pressure in this study indicated that the loads of 15 per cent and 20 per cent body weight produce a relatively greater extra stress on the cardiovascular system than lighter loads.

In the present study, subjects showed a significant difference in the metabolic cost in terms of oxygen uptake and energy expenditure between the loads of 0 and 10 per cent and the load of 20 per cent body weight. This indicated that subjects performed differently and the metabolic requirement varied between low and high load conditions. It was clear that subjects had to work harder to carry the school bags of 20 per cent body weight. The relative work intensity (per cent $\dot{V}O_{2max}$) in 20 per cent load condition was significantly greater than that in 0 and 10 per cent load conditions. The increase in metabolic cost probably resulted from more muscles, not only large but also smaller muscle groups, being involved in walking (Berger, 1982). The kinematics analysis of the present study has shown that the increase of load forced the subjects to lean forward, which would bring the centre of gravity back over the base of support to maintain the balance. Likewise, the forward flexion would cause hamstring, semispinalis, tibialis anterior, vastus lateralis, erector spinae and trapezius muscles to work harder to support the movement (Cook and Neumann, 1987). As the load became higher, subjects recruited additional motor units and muscle groups and altered their gait to carry the load (Pascoe *et al.*, 1997). Martin and Nelson (1980) reported altered locomotion biomechanics, which resulted in higher actual power output to carry a given load. Moreover, the inclined body position and the altered locomotion biomechanics on a daily basis would increase the stresses of the back and leg muscles of the subjects. For subjects who were at the age of ten, these stresses might be harmful and influence their normal musculo-skeletal developmental growth.

Malhotra and Sen Gupta (1965) recommended that the weight usually carried by students is not likely to exceed 10–12 per cent of the body weight because in their study nobody was observed bending forward. This recommendation was then widely accepted as a criterion for students carrying school bags (Voll and Klimt, 1977; Sander, 1979). The biomechanical and physiological evidence of the present study has provided experimental evidences to support this recommendation because carrying load of 10 per cent body weight did not induce pronounced change in gait pattern, trunk posture, heart rate, metabolic cost, and recovery time of blood pressure after walking.

Carrying load of 20 per cent body weight induced pronounced increase of biomechanical and physiological strain detected when compared with 0 per cent and 10 per cent load conditions. The load of 15 per cent body weight, however, only caused significant increase in trunk forward lean and systolic blood pressure recovery time. Therefore, children carrying load of 15 per cent body weight still need to maintain a forward bending of their trunk and require more time for blood pressure to return to the baseline after 20 minutes walking. Holewjin (1990) suggested that the load of the pack on the shoulders provided pressure, strain on musculature, and skin irritation that are limiting factors to tolerate pack weight. Pascoe *et al.*, (1997) reported that school students transported all their materials in book bags and among these students there were complaints of physical muscle soreness (67.2 per cent), back pain (50.8 per cent), numbness (24.5 per cent), and shoulder pain (14.7 per cent) in their survey. Several case studies have reported medical complications from backpack use in children aged 12–14 years (Ford, 1966; Rothner *et al.*, 1975; White, 1968). For physically fit adult individuals, backpack load limits are considered to be 30 per cent of body weight (Hardin and Kelly, 1975). Children of ten-years-old are experiencing physical growth and motor development. The backpack limitations are sensitive to their age, weight, growth pattern and fitness level and therefore need to be carefully determined. From our observation, most subjects suffered from fatigue with heavy loads (15–20 per cent body weight) in prolonged walking (15–20 min) that was confirmed by the analysis of biomechanical and physiological data. However, systematic data collection on subjective feeling was not collected, which could be suggested in future study.

The treadmill was used as an experimental apparatus in this study. It allowed control of walk speed and facilitated use of biomechanical and physiological monitoring equipment that provided quantitative observation of carrying load in children. Research has found significant differences between treadmill and floor walking in the displacements of the head, hip, and ankle in the sagittal plane, periods of double-limb support (Myrray *et al.*, 1985), and knee motion (Strathy *et al.*, 1983). However, it was pointed out that, in general, treadmill walking did not differ markedly from floor walking in kinematic measurements and electromyography (EMG) at slow, free and fast speed, and heart rate at slow and free speed (Myrray *et al.*, 1985). Therefore, the results obtained from this study would reflect the general trend of biomechanical and physiological characteristics of children when carrying an increasing load.

Conclusion

For the participants in this study, carrying a load of up to 20 per cent of body weight while walking resulted in a steady state with about 60 per cent of the maximum heart rate of youth of this age. The loads of 20 per cent body weight induced higher biomechanical and physiological strain in terms of pronounced changes in trunk forward lean, trunk angular motion, leg support duration, swing duration, blood pressure recovery time, oxygen uptake, energy expenditure and work intensity when compared to 10 per cent and 0 per cent body weight loads in youth. Carrying load of 15 per cent body weight resulted in a significant increase in trunk forward lean and blood pressure recovery time. No significant difference was observed between 10 per cent and 0 per cent load condition with the measured biomechanical and physiological parameters.

References

1. *Bases of Exercise*, 3rd edn. (New York: McGraw-Hill), 308–311.
2. American College of Sports Medicine (1995), *Guidelines for Exercise Testing and Prescription*, 5th edn. (Baltimore: Williams & Wilkins).
3. Astrand, P. O., Rodahl, K. (1986), '*Textbook of Work Physiology: Physiological*'.
4. Berger, R. A. (1982) '*Applied Exercise Physiology*' (Philadelphia, PA: Lea & Febiger).
5. Chansirinukor, W., Wilson, D., Grimmer, K., Dansie, B. (2001) 'Effects of backpacks on students: measurement of cervical and shoulder posture'. *Aust J Physiother*, 47: 110–16.
6. Cook, T. M., Neumann, D. A. (1987) 'The effects of load placement on the activity of the low back muscles during load carrying by men and women'. *Ergonomics*, 30: 1413–1423.
7. Ford, F. R. (1966) 'Diseases of the Nervous System in Infancy', Childhood and Adolescence. Springfield, IL: Charles C Thomas: 1113.
8. Ghori, G. M. U., Luckwill, R. G. (1985) 'Responses of the lower limb to load carrying in walking man'. *European Journal of Applied Physiology*; 54: 145–150.
9. Hardin, D., Kelly, B. (1975) 'Your and your gear: physical fitness'. *Backpacker*; 3: 3–31.
10. Holewijn, M., (1990) 'Physiological strain due to load carrying'. *European Journal of Applied Physiology*, 61: 237–245.
11. Hong Kong Society for child health and development (1988) '*The weight of school bags and its relation to spinal deformity*'. (Hong Kong: The Department of Orthopaedic Surgery, University of Hong Kong, The Duchess of Kent Children's Hospital).
12. Malhotra, M. S., Sen Gupta, J. (1965) 'Carrying of school bags by children'. *Ergonomics*, 55–60.
13. Martin, P. E. and Nelson, R. C. (1986) 'The effect of carried loads on the walking patterns of men and women'. *Ergonomics*, 29: 1191–1202.
14. Myrray, M. P., Spurr, G. B., Sepic, S. B., Gardner, G. M. (1985) 'Treadmill vs. floor walking: kinematics, electromyogram, and heart rate'. *Journal of applied physiology*, 59: 87–91.
15. Negrini, S., Carabalona, R. (2002) 'Backpacks on! Schoolchildren's perceptions of load, associations with back pain and factors determining the load'. *Spine*, 27: 187–95.
16. Pascoe, D. D., Pascoe, D. E., Wang, Y. T., Shin, D. M., Kim, C. K. (1997) 'Kinematics analysis of book bag weight on gait cycle and posture of youth'. *Ergonomics*, 40: 631–641.

17. Rothner, D. A., Wilbourn, A., Mercer, R. D. (1975) 'Rucksack palsy'. *Pediatrics*, 56: 822–824.
18. Rowell, L. (1986) '*Human Circulation Regulation During Physical Stress*' (New York: Oxford University Press).
19. Sander, M. (1979) 'Weight of school bags in a Freiburg elementary school: Recommendations to parents and teachers'. *Offentliche Gesundheitswesen*, 41: 251–253.
20. Strathy, G. M., Chao, E. Y., Laughman, R. K. (1983) 'Changes in knee function associated with treadmill ambulation'. *Journal of Biomechanics*, 16: 517–522.
21. Voll, H. J. and Klimt, F., (1977) 'Strain in children caused by carrying school bags'. *Offentliche Gesundheitswesen*, 39: 369–378.
22. Weir, J. B. de V. (1949) 'New methods of calculating metabolic rate with special references to protein metabolism'. *Journal of Physiology*, 109: 1–9.
23. White, H. H. (1968) 'Pack palsy: a neurological complication of scouting'. *Pediatrics*, 41: 1001–1002.

33

Ankle proprioception in young ice hockey players, runners, and sedentary people

Jing Xian Li[1] *and Youlian Hong*[2]
[1]*University of Ottawa, Ottawa;* [2]*The Chinese University of Hong Kong, Hong Kong*

Abstract

The study examined if regular exercise with high demand in postural control and cyclic exercise could improve proprioception of the foot and ankle complex. A total of 38 young health people with different exercise habits for more than five years formed three groups: the ice hockey; running; and sedentary groups. Kinesthesia of the foot and ankle complex was measured in plantarflexion (PF), dorsiflexion (DF), inversion (IV) and eversion (EV) at 0.4°/s passive rotation velocity using a custom-made device. The results showed that the ice hockey group had significantly better kinesthesia in PF/DF, and IV/EV than did the running and sedentary groups. The running group did not show better kinesthesia compared with the sedentary group. It is concluded that propiorception of foot and ankle complex can be improved by long-term exercise that has high demand for postural stability, such as ice hockey.

Introduction

Proprioception is the sensory feedback that contributes to conscious sensation (muscle sense), total posture (postural equilibrium), and segmental posture (joint stability), and is mediated by proprioceptors that are located in the skin, muscles, tendons, ligaments, and joint capsules (Lephart, Pincivero, Giraldo, and Fu, 1997). Research has shown that postural control stability is significantly affected by proprioception in the lower limb (Lord, Clark, and Webster, 1991). Colledge *et al.* (1994) studied the relative contributions to balance of vision, proprioception, and the vestibular system with age by measuring body sway during standing. In four different age groups through 20- to 70-years-old, the relative contribution of each sensory input was the same, with proprioception being predominant throughout each age group. Moreover, the lack of proprioceptive feedback that results from injuries, such as ankle injury (Guskiewicz and Perrin, 1996), may allow the excessive or inappropriate loading of a joint (Co, Skinner, and Cannon, 1993), and is one of the factors that leads to progressive degeneration of the joint and continued deficits in joint dynamics, balance, and coordination (Riemann and Guskiewicz, 2000). Much clinical research has demonstrated that individuals with proprioception and neuromuscular response deficits as a result of injury, lesions, and joint degeneration are less capable of maintaining postural stability and equilibrium (Cornwall and Murrell, 1991; Forkin, Koczur, Battle, and Newton, 1996; Garn and Newton, 1988; Pintsaar, Brynhildsen, and Tropp, 1996).

Considering the importance of the proprioception in postural stability, movement control, and injury prevention, it would therefore be beneficial to understand the effects of exercise on proporioception function. The foot and ankle complex is a critical structure in postural stability and its injury rate is very high in all sports injuries of the body. To prevent foot and ankle injury, apart from prophylactic devices, exercise training in proprioception is needed to enable athletes to return to preinjury levels of activity following ligament and muscle injuries. However, it is still being debated upon whether exercise training can improve proprioception or not (Ashton-Miller, 2001). Some published research works have found that proprioception can be improved through exercise, especially proprioceptive exercise that requires three actions: the proprioception of the joints, balance capacity, and neuromuscular control (Irrgang and Neri 2000; Eils and Rosenbaum, 2001). In particular, long-term Tai Chi practice was found to be beneficial to the proprioception of the lower limb (Xu, *et al.* 2004). However, a recently published work which studied the influence of a five-month professional dance training without concurrent additional coordinative training found that such training did not lead to improvements in ankle joint position sense or improved measures of balance (Schmitt *et al.*, 2005). These studies suggest that exercise may benefit proprioception. However, the effect may be influenced by the form and modality of the exercise. Up to now, scientific evidence on the impact of exercise form and modality on proprioception of the lower extremity is limited. To bridge the research gap, we designed a cross-sectional study to examine the effects of exercise on ankle proprioception in young people through comparison of the passive motion sense, kinesthesia, among three groups, long-term regular ice hockey players, long-term regular runners, and sedentary people. Such efforts would add the understanding in the effects of exercise on proprioception and subsequently contribute to the development of the program in the enhancement of proprioception function in children, elderly people, and patients with deficits of proprioception.

Methods

Subjects: A total of 38 young healthy people were recruited based on their exercise habits. For the first group, 12 people (6 males, 6 females) with a regular running habit, running for more than three times each week for more than five years, formed the running (R) group. The next group, 13 males with an ice hockey playing habit, playing more than three times each week for more than five years, consisted of the hockey (H) group. Finally, 13 people with no regular exercise habit in the past five years served as the sedentary (S) group. The demographic data of the participants are presented in Table 33.1. All participants were predominantly healthy and they had no history of significant cardiovascular, pulmonary, metabolic, musculoskeletal, or neurological diseases and injuries. An informed consent form was given to each subject prior to participation. This study was approved by the Human Ethics office, University of Ottawa.

Data collection: The testing was performed in a well-lit and well-ventilated room. The room was sound-attenuated and isolated so as to reduce any auditory or visual interference that might distract the participants. After being measured for body weight and body height,

Table 33.1 The demographic data of the subjects in ice hockey group (H group), running group (R group), and sedentary group (S group) (Mean ± SD).

	Age (years)	*Body weight (kg)*	*Body height (cm)*	*Body mass index (kg/m²)*
H Group (n = 13, m)	20 ± 0.5	82.5 ± 7.6**	180.5 ± 7.0**	25.3 ± 2.0
R Group (m = 6, f = 6)	20 ± 1.9	70.1 ± 11.6§§	171.0 ± 8.4§§	24.0 ± 4.0
S Group (m = 5, f = 8)	21.5 ± 2.5	65.1 ± 8.4	169.3 ± 7.3	22.6 ± 1.7

m, males; f, females. **, different from S group, $P < 0.01$; §§, different from H group, $P < 0.01$.

each subject individually participated in one session of data collection. Data were collected using the instrumentation and procedures described by Lentell *et al.* (1995), but small changes were instituted (Xu *et al.*, 2004). The reliability tests were conducted in ten young healthy people through three times of measurements, on two consecutive days and a week after the first measurement. Interclass correlation coefficients (ICC) of these measurements had moderate to high test-retest reliability, ranging from 0.68–0.92 in plantar-flexion and dorsi-flexion movements, respectively. As illustrated in Figure 33.1, the custom-made device is a box with a movable platform that rotates about a single axis in two directions. With the foot resting on this platform, plantar-dorsiflexion of the foot and ankle complex movements can occur. This platform is moved by an electric motor that rotates the foot on an axis at a rate of 0.4°/sec. Movement can be stopped at any time with the use of a hand-held switch. The device is also equipped with a hanging scale and a fixed pulley supported by a trestle, which is outside the device. A thigh cuff attached to the lower end of the scale is wrapped around the lower thigh of the subjects. Through adjustments of the length of the cuff, the extremity is lifted by the scale and its weight is recorded when a subject fully relaxes his/her thigh. After this, the thigh cuff is attached to one end of the rope around the pulley and the other end is hung with weights. The extremity can then be adjusted to where the foot is in contact with the platform. Through the addition or reduction of the weights, the investigator can standardize and control the amount of the lower extremity weight resting on the platform during testing.

For data collection, each subject was seated on an adjustable chair and his/her dominant foot was so placed on the platform that the axis of the apparatus coincided with the plantar-dorsiflexion axis or inversion-eversion axis of the foot and ankle joint. The hip, knee, and ankle were positioned at 90°, respectively. Fifty per cent of each subject's lower extremity weight was rested on the platform by the use of the thigh cuff suspension system to control unwanted sensory cues from the contact between the instrument and the plantar surface of the foot. During testing, the subjects' eyes were closed to eliminate visual stimuli from the testing procedure and apparatus. Data collected in each test movement began with the foot placed in a starting position of 0°. The subjects were instructed to concentrate on their foot and to press the hand-switch when they could sense motion and identify the direction of the movement. After performing two practice trials, a formal data collection was conducted. The motor was engaged to rotate the foot into dorsiflexion or plantarflexion, at a random time interval between 2 and 10s after subject instruction. The researcher recorded the rotation angles of the platform and the direction of movements as passive motion sense. A total of six trials, three for plantar flexion and three for dorsiflexion,

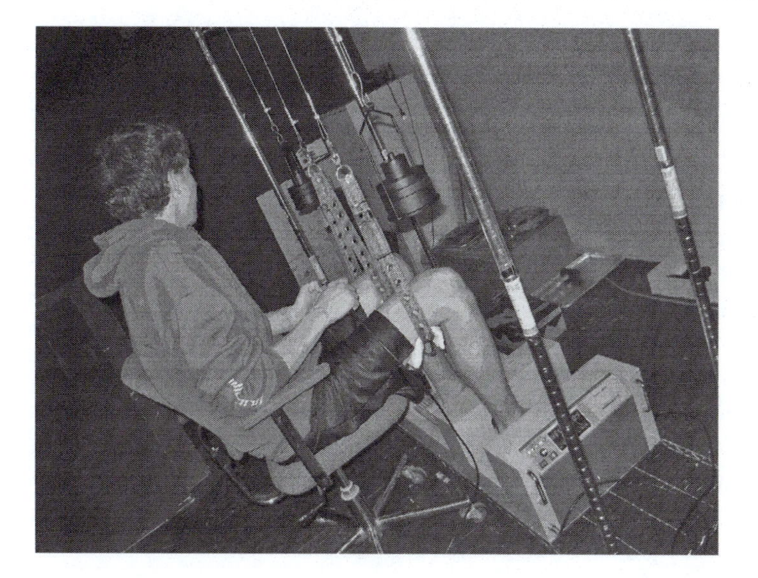

Figure 33.1 Testing apparatus and a subject positioned for evaluation of passive movement sense of ankle.

were randomly conducted. The same device was also used to measure the passive motion sense of foot and ankle complex for inversion and eversion by turning the device for 90 degree and following the same procedures described above. In this study, the kinesthesia of plantar flexion and dorsiflexion was measured first. The measurement of the kinesthesia of inversion and eversion followed.

Data analysis: All variables are presented as mean and standard deviation (SD). Paired t-test was used to determine if there was any significant difference between the data collected from the dominant and nondominant sides within group. Because there was no significant difference, the data from both sides was pooled together. Gender differences in the passive motion sense, in each movement direction, were examined by comparing the pooled data from all male and female participants using independent t-test. No significant gender differences in the measurements were found. Therefore, the data from males and females in each group were merged to reduce the mean of the measurements in the group.

In each group, values for passive motion sense of foot and ankle complex in paired motions, dorsiflexion vs plantarflexion, inversion vs eversion, were respectively compared using paired t-test. One-way analysis of variance was used to estimate significant differences in the values of each passive motion sense of the foot and ankle complex among groups. The post hoc Tukey tests were performed when necessary to isolate the differences and $P \leq 0.05$ was considered statistically significant.

Results

There are significant differences in body weight and body height between three groups. However, no significant difference in body mass index (BMI) was found among three groups. Ankle joint kinesthesia significantly differed among the three groups ($P < 0.01$ or 0.001).

The kinesthesia of 1.93±0.65° measured in dorsiflexion, 1.92±0.62° in plantarflexion, 2.79±0.64° in inversion, and 3.17±1.12° in eversion were perceived in the ice hockey group, and were significantly smaller than that in the running and sedentary groups. The most remarkably significant differences were found in the measurements in inversion and eversion among the three groups. The ice hockey group was able to perceive the 2.79±0.64° of inversion motion and 3.17±1.12° of eversion motion. The values of the perceived passive inversion and eversion motion were 4.33±2.04° and 4.76±2.59° in the running group, which was 31% to 35% larger than that in ice hockey group. The sedentary group perceived values of passive motion in inversion and eversion larger by 50% to 55% than the values of the ice hockey group. No significant difference in the passive motion sense was found between the running group and the sedentary control group. Table 33.2 lists the statistical analysis results of the values of kinesthesia measured in different direction movement in the three groups.

Discussion

The present study provides the evidence that long-term ice hockey practitioners not only showed significantly better ankle joint kinesthesia than sedentary controls but also that their ankle joint kinesthesia measured in dorsiflexion, inversion and eversion was significantly better than long-term running exercisers. Moreover, long-term regular running exercisers did not perform better in ankle joint kinesthesia compared with their sedentary counterparts.

The recognition of the effects of proprioception and neuromuscular control on joint and posture stability has led the orthopedic and sports medicine communities toward an emphasis in rehabilitation on the restoration of proprioception to enhance dynamic joint stability (Irrgang and Neri, 2000). However, a few publications examined the impacts of exercise on proprioception function, especially on the effects of different forms of exercise. According to Irrgang and Neri's description (2000), propriocepitive exercises require to take account of three parts: proprioception of joints, balance capacity, and neuromuscular control.

Table 33.2 Kinesthesia measured in dorsiflexion, plantarflexion, inversion, and eversion in ice hockey group (H group), running group (R group), and sedentary group (S group) (Mean ± SD)

	Kinesthesia (degrees)	*H group (degrees)*	*R group (degrees)*	*S group (degrees)*
Dorsiflexion	Mean ± SD	1.93 ± 0.65*	2.53 ± 0.88[§]	2.47 ± 0.83
	Maximum	3.35	4.64	5.99
	Minimum	1.07	1.30	1.27
Plantarflexion	Mean ± SD	1.92 ± 0.62	2.61 ± 1.12	2.51 ± 0.86
	Maximum	3.14	5.64	4.60
	Minimum	1.07	1.35	1.58
Inversion	Mean ± SD	2.79 ± 0.64**	4.33 ± 2.04**,[§§]	5.41 ± 2.71
	Maximum	4.90	7.98	13.23
	Minimum	1.30	1.03	2.59
Eversion	Mean ± SD	3.17 ± 1.12**	4.76 ± 2.59[§§]	5.18 ± 1.78
	Maximum	6.31	10.15	8.31
	Minimum	1.95	1.22	2.01

*, differs from S group, $P < 0.05$;**, differs from S group, $P < 0.01$. [§],differs from H group, $P < 0.05$. [§§],differs from H group, $P < 0.01$.

The key of exercise in the successful enhancement of proprioception is to choose the correct forms of exercise to progressively stimulate the demands required during functional activities in which the individual participates. A cross-sectional study of examining the kinesthesia of the knee and ankle joint among three groups of elderly people showed that long-term Tai Chi exercisers had significantly better kinesthesia of the knee and ankle joint than long-term runners/swimmers. Moreover, long-term runners/swimmers could not perform better in perceiving the passive motion of dorsiflexion and plantar flexion of ankle joint (Xu, *et al.*, 2004) than did a sedentary group. Tai Chi exercise is a series of individual graceful movements in a slow, continuous, circular pattern. The movements of Tai Chi are gracefully fluent and consummately precise because specificity of joint angles and body position is of critical importance in accurately and correctly performing each form (Jacobson, Chen, Cashel, and Guerrero, 1997). Conscious awareness of body position and movement is demanded by the nature of the activity. Such exercise form contains all components that are needed in training proprioception and benefits proprioception. Eils and Rosenbaum (2001) investigated the effects of a six-week multi-station proprioceptive exercise program on patients with chronic ankle instability. The proprioceptive exercise included exercise mats, a swinging platform, an ankle disk, exercise bands, an air squab, wooden inversion-eversion boards, a mini trampoline, an aerobic step, an uneven walkway, and a swinging and hanging platform. The results showed a significant improvement in ankle joint position sense tested in weight bearing conditions and postural sway as well as significant changes in muscle reaction times in the exercise group. The results demonstrated that proprioception can be trained by proproceptive exercise. However, a recently published work could not support the effect of exercise training on proprioception. Schmitt and co-workers (2005) studied the effects of five-month ballet training on ankle position sense. There were 42 dancers with more than 10 years dancing experience in professional training and 40 age-matched and gender-matched controls with no prior dance or specific sport training who participated in the study. Passive angle-replication tests (joint position sense tests) were conducted during the pre- and post-training program. The training consisted of 15 to 16 hours per week of classical ballet and 4.5 hours per week of modern dance, as well as traditional dances, jazz dance, and weight training for men. However, no significant differences in joint position sense were found either in the pre- or post-test of the training program. It is well known that ballet and other dances have a very high demand to proprioception, balance capacity, and neuromuscular control. Ballet training should be considered as a proprioceptive exercise. The possible cause of the undetectable effect of ballet training on proprioception in Schimitt's study may be related to the sensitivity and reliability of the testing method. Beynnon *et al.* (2000) published their research work in validation of the techniques used to measure knee proprioception. They compared the accuracy, repeatability, and precision of seven joint position sense techniques and one joint kinesthesia measurement technique in normal subjects with no history of knee injury. They found that joint kinesthesia was more repeatable and precise than each of the joint position sense techniques. Moreover, the measurement in weight bearing condition was also more repeatable. Therefore, they recommended that studies designed to evaluate proprioception should consider using kinesthesia, which should result in increased power and sensitivity to detect significant differences, if they truly exist. In Schmitt's study, proprioception was examined by measuring joint position sense under non-weight bearing conditions. The testing method may not have been sensitive enough to detect the proprioception function.

Another factor influencing the impact of exercise on proprioception is the duration of training. Xu, *et al.* (2003) examined the effects of a 15-week Tai Chi exercise program on the proprioception of the knee and foot and ankle complex in elderly people. The Tai Chi exercise program lasted 15 weeks, with three sessions a week, consisting of one hour for each session. The results demonstrated that the significant training effect of kinesthesia was found in the knee joint, but not in the foot and ankle complex. The study of Eils and Rosenbaum (2001) showed that six-week proprioceptive exercise gained effects on the joint position sense in young people. In the study, the subjects trained 20 minutes each day, and the intensity of the six-week training period was increased by small modifications every two weeks. These scientific evidences demonstrated that exercise form, training duration, and age of the participants, as well as the evaluation method used, should be considered as the examination of the effect of exercise on proprioception.

The present study shows that long-term ice hockey players had significantly better passive motion sense than long-term runners, suggesting that long-term ice hockey playing results in a training effect in proprioception. Ice hockey is a non-cyclic, but fast-paced sport, requiring good levels of skill, speed, agility, and endurance. It has very high demands with regard to postural control capacity, which would contain all components of the proprioceptive exercise. Movements in ice hockey, such as turning, sideward movement, and power stride, put a very big challenge to the ankle and foot complex. Moreover, the very narrow supporting base of the body has a very high demand for balance control capacity. Therefore, the awareness of joint position and movement is always emphasized during ice hockey exercise. Compared with ice hockey, running is a cyclic exercise. The awareness of joint position and movement is not specifically emphasized during this form of exercise. Additionally, most runners exercise only for the sake of enhancing their health and for recreation; they do not pay much attention to the techniques involved in the exercise, unless awareness of the joint position and angles is especially required in some exercises, such as ice hockey. This might be one of the reasons that the running exercise did not show benefits to ankle kinesthesia in this study. The study conducted by Xu *et al.* (2004) had denoted the fact that old long-term running exercisers did not show significantly better kinesthesia than the old sedentary people. The result of the present study from young health people showed a similar pattern as that in old people. It is suggested that choosing a proper exercise form in designing exercise program for improvement of proprioception is critical.

Since there is no published work reporting the study of kinesthesia measured in inversion and eversion, the present study for the first time profiled passive motion sense in inversion and eversion in the young people with different exercise habits. The ice hockey group had significantly better passive motion sense in inversion and eversion compared with the running and sedentary groups. The passive motion sense of inversion was significantly better than that in eversion in the hockey group. The cause leading to such difference is unclear. It might be associated to the movement's characteristics and the hockey boot design and lateral longitudinal arch of the foot. Ice hockey movements involve much more medial-lateral movements, which would train the proprioceptors in the foot and ankle. The lateral longitudinal foot arch might be more sensitive to passive motion because of a larger contact area with the contact surface. Additionally the current design of hockey boots provides more stability to the ankle joint and only a very small motion range of the ankle joint is allowed by the footwear. The range of supination of the bare foot inside the boot (between touchdown and maximum) was 12.3° +/–4.6° in a high-cut hockey boot (Avramakis, Stakoff, and Stussi, 2000). Such hockey boot design and the demand in movement and balance in ice hockey both need muscles, especially the small muscles, to do very fine contraction so as to accurately adjust tension and length. Working muscles in this way would result in training effects on muscle spindles and Golgi tendon organs as well as other proprioceptors.

The study findings provide evidence supporting the effects of regular exercise on proprioception. However, the exercises should emphasize balance control and should be characterized by multi-direction movements. Although no gender differences in the proprioception of the ankle joint complex were found in the participants within the running and the sedentary groups, the findings in the beneficial effects of regular ice hockey on the proprioception of the ankle joint complex in the present study have been restricted to young males. Therefore the next question is whether the findings would be replicated for females, elderly people, and children. Clearly, answers to these questions require further mixed-gender study and among different populations. Additionally, by the cross-sectional design of the study, we are unable to specify how much time is needed for regular exercise to result in beneficial effects on the proprioception in the ankle. Moreover, we are also unable to document whether the better proprioception acuity in ice hockey players is related to their larger foot contact area and the lever arm of ankle joint rotation as the body height of the ice hockey group was higher and subsequently their foot length was longer compared with those in the running and the sedentary groups. To evaluate the effects of different exercises and modalities of exercise on proprioception, further study with a longitudinal follow-up design and consideration to the foot contact area with the testing apparatus is needed.

Conclusion

Long-term ice hockey exercise results in training effects on kinesthesia of the foot and ankle complex in young males. The better passive motion sense was especially significant in inversion/eversion direction. Long-term running could not yield the training effects on kinesthesia of the ankle and foot complex in the young runners. However, long-term running can improve the passive motion sense in inversion/eversion direction, but did not reach a statistical level. The results suggest that proprioception could be improved by exercise. However, exercise form and training duration are critical to gaining training effects.

References

1. Ashton-Miller, J.A. Wojtys, E.M. Huston, L.J., and Fry-Welch, D. (2001) 'Can proprioception really be improved by exercises?' *Knee Surg Sport Tra*, 9(3): 128–36.
2. Avramakis, E. Stakoff, A., and Stussi, E. (2000) 'Effect of shoe shaft and shoe sole height on the upper ankle joint in lateral movements in floorball' (uni-hockey) (German). *Sportverletzung Sportschaden*, 14(3): 98–106.
3. Beynnon, B.D., *et al.* (2000) 'Validation of techniques to measure knee proprioception'. In S.M. Lephart and F.H. Fu. (eds) *Proprioception and neuromuscular control in joint stability* (pp. 127–38). Human Kinetics, Champaign.
4. Colledge, N.R., Cantley, P., Peaston, I., Brash, H., Lewis, S., and Wilson, J.A. (1994) 'Ageing and balance: the measurement of spontaneous sway by posturography'. *Gerontology*, 40(5): 273–8.
5. Co, F.H., Skinner, H.B., and Cannon, W.D. (1993) 'Effect of reconstruction of the anterior cruciate ligament on proprioception of the knee and the heel strike transient'. *J Orthop Res*, 11(5): 696–704.
6. Cornwall, M.W. and Murrell, P. (1991) 'Postural sway following inversion sprain of the ankle'. *J Am Podiatric Med Assoc*, 81(5): 243–7.
7. Eils, E. and Rosenbaum, D. (2001) 'A multi-station proprioceptive exercise program in patients with ankle instability'. *Med Sci Sports Exerc*, 33(12): 1991–8.

8. Forkin, D.M., Koczur, C., Battle, R., and Newton, R.A. (1996) 'Evaluation of kinesthetic deficits indicative of balance control in gymnasts with unilateral chronic ankle sprains'. *J Orthop Sports Phys Ther*, 23(4):245–50.
9. Garn, S.N. and Newton, R.A. (1988) 'Kinesthetic awareness in subjects with multiple ankle sprains'. *Phys Ther*, 68(11):1167–71.
10. Guskiewicz, K.M. and Perrin, D.H. (1996) 'Effect of orthotics on postural sway following inversion ankle sprain'. *J Orthop Sports Phys Ther*, 23(5):326–31.
11. Irrgang, J.J. and Neri, R. (2000) 'The rationale for open and closed kinetic chain activities for restoration of proprioception and neuromuscular control following injury'. In S.M. Lephart and F. H. Fu (eds.) *Proprioception and neuromuscular control in joint stability* (pp. 363–74) Human Kinetics.
12. Jacobson, B.H., Chen, H.C., Cashel, C., and Guerrero, L. (1997) 'The effect of Tai Chi Chuan training on balance, kinesthetic sense, and strength'. *Perceptual & Motor Skills*, 84:27–33.
13. Lephart, S.M. Pincivero, D.M., and Giraldo, J.L. *et al.* (1997) 'The role of proprioception in the management and rehabilitation of athletic injuries'. *Am J Sports Med*; 25:130–7.
14. Lentell, G., Baas, B., Lopez, D., McGuire, L., Sarrels, M., and Snyder, P., (1995) 'The contributions of proprioceptive deficits, muscle function, and anatomic laxity to functional instability of the ankle'. *J Orthop Sports Phys Ther*, 21(4):206–15.
15. Lord, S.R., Clark, R.D., and Webster, I.W. (1991b) 'Postural stability and associated physiological factors in a population of aged persons'. *J Geron*, 46(3):M69–M76.
16. Pintsaar, A., Brynhildsen, J., and Tropp, H. (1996) Postural corrections after standardized perturbations of single limb stance: effect of training and orthotic devices in patients with ankle instability'. *Brit J Sports Med*, 30(2):151–5.
17. Riemann, B.L. and Guskiewicz, K.M. (2000) 'Contribution of the peripheral somatosensory system to balance and postural equilibrium'. In S.M. Lephart and F.H. Fu (eds.) *Proprioception and neuromuscular control in joint stability*. Human Kinetics.
18. Schmitt, H., Kuni, B., and Sabo, D. (2005) 'Influence of professional dance training on peak torque and proprioception at the ankle'. *Clin J Sport Med*. 15(5):331–9.
19. Subotnick, S.I. (1989) 'Sports medicine of the lower extremity'. Churchill Livingstone, New York.
20. Xu, D.Q. (2003) 'The effects of Tai Chi exercise on proprioception and neuromuscular responses in the elderly people,' PhD thesis, The Chinese University of Hong Kong.
21. Xu, D.Q., Hong, Y., Li, J.X., and Chan, K.M. (2004) 'Effect of Tai Chi exercise on proprioception of ankle and knee joints in old people'. *Brit J Sport Med*, 38:50–4.

34

The plantar pressure characteristics during Tai Chi exercise

De Wei Mao[1], Youlian Hong[2] and Jing Xian Li[3]
[1]Shandong Institute of Physical Education and Sports, Jinan;
[2]The Chinese University of Hong Kong, Hong Kong; [3]University of Ottawa, Ottawa

Abstract

The aim of this study is to quantify the plantar pressure characteristics of five fundamental movements and all one-leg stances of the 42-form Tai Chi to explain why Tai Chi exercise benefits the balance control and muscle strength when compared with normal walking. Sixteen experienced Tai Chi practitioners participated in this study. The Novel Pedar-X insole system was used to record the plantar pressure. Results demonstrated that the loading of the first metatarsal head and the great toe were significantly greater than in other regions ($p < 0.05$). Compared with normal walking, the locations of the centre of pressure (COP) in the Tai Chi movements were significantly more medial and posterior at initial contact ($p < 0.05$), and were significantly more medial and anterior at the end of contact with the ground ($p < 0.05$). The displacements of the COP were significantly wider ($p < 0.05$) in the mediolateral direction in the forward, backward and sideways Tai Chi movements and in one-leg stances. The displacement was significantly larger ($p < 0.05$) in the anterposterior direction in the forward movement. The plantar pressure characteristics found in this study may be one of the important factors that Tai Chi exercise improves balance control and muscle strength.

Introduction

Tai Chi is an ancient Chinese martial art. The broad consensus is that Tai Chi exercise improves balance control and muscle strength in the lower extremities (Hong *et al.*, 2000; Schaller, 1996; Tsang *et al.*, 2003; Tse and Bailey, 1992). However, the mechanisms by which Tai Chi improves balance control and subsequently prevents falls in older people are still unclear. According to the theory of Tai Chi, foot posture and movement are the foundation of the whole body posture, and the concept of proper position and direction are always emphasized (Editor Group, 2000). Stepping forward, backward, sideways, up and down, and fixing are the five fundamental movements of Tai Chi (Editor Group, 2000). A whole set of Tai Chi contains these fundamental movements which are performed symmetrically between the left and right leg stance and frequently throughout the exercise. On the other hand, the one-leg stance is used in almost all individual movements and accounts for about 35.73 per cent of the whole set of 42-form Tai Chi in terms of performance time (Mao *et al.*, 2006). The ability to maintain a one-leg stance is an important predictor of injurious falls for elderly people (Richardson *et al.*, 1996), and has been shown to correlate strongly with falls (Richardson *et al.*, 1996; Schaller, 1996) and age (Bohannon *et al.*, 1984; Iverson *et al.*, 1990; Jonsson *et al.*, 2004).

There have been numerous studies on the plantar pressure during walking and showed it is a reliable and repeatable feature of gait (Cook *et al.*, 1997; Takahashi *et al.*, 2004). However, there have been few studies on the plantar pressure in Tai Chi movements. Wu and Hitt (Wu and Hitt, 2005) studied the foot ground contact characteristics using force platforms and pressure plates. They found that the impact force to the foot was significantly lower and the main pressure was located in the more medial and anterior of the foot when compared to walking. Unfortunately, only the stepping forward movement was selected and analyzed. Very few studies have investigated the one-leg stance duration during Tai Chi exercise or explained why Tai Chi exercise may improve balance in the one-leg stance.

Walking is the most common daily activity, for both feet, 80 per cent of a complete gait cycle is spent on one leg (Winter, 1991). Generally, walking is different from Tai Chi exercise, but several studies have demonstrated that Tai Chi and walking are both moderate forms of exercise that are suitable for elderly people (Shin, 1999; Li *et al.*, 2001), and both exercises have been found to have beneficial effects on balance control, muscle strength, and cardio-respiratory response in the elderly (Shin, 1999; Hong *et al.*, 2000; Li *et al.*, 2001; Melzer *et al.*, 2003).

Therefore, the objective of this study is to quantify the plantar pressure characteristics of five most representative fundamental movements and one-leg stances of the whole set of 42-form Tai Chi. It is hoped that the information obtained from this study would explain why Tai Chi exercise benefits balance control and muscle strength when compared with normal walking.

Methods

Subject

Sixteen gender-matched elite Tai Chi masters (8 women and 8 men, ages 23.1 ± 5.5 years, body height 166.0 ± 7.6 cm, body mass 62.2 ± 7.8 kg, experience of practicing 8.1 ± 5.7 years) with no previous diseases or injuries in the year before the study were recruited.

Procedure

Each subject visited the biomechanics laboratory to perform five selected fundamental movements of Tai Chi. The five most representative fundamental movements were: (1) Brush Knee and Twist Steps; (2) Step Back to Repulse Monkey; (3) Wave Hand in Cloud; (4) Kick Heel to Right; and (5) Grasping the Bird's Tail, each representing stepping forward, backward, sideways, up-down and fixing movements, respectively. The foot movements during these movements are described as follows: (1) forward movement: the left foot makes contact with the ground first, the right foot then steps forward, and the left foot leaves the ground at the end; (2) backward movement: the left foot makes contact with the ground, the right foot then steps backward, and the left foot then leaves the ground; (3) sideways movement: the left foot makes contact with the ground, the right foot then steps sideways, and the left foot then leaves the ground; (4) up-down movement: the right foot leaves the ground, moves upward, kicks in the air, and then moves back down to the ground; and (5) fixing movement: the left foot makes contact with the ground, both feet then make contact with the ground, and the movement is finished (Figure 34.1). Subjects performed three trials of each movement. In order to keep the continuous and smoothness of each measured movement, subject was asked to perform three consecutive movements which included the specific movement to be studied in the middle. As stated above, in a set of Tai Chi, most movements are performed left and right symmetrically. Due to the quasi-identity of both sides, only movements in which the left leg serves as the stance leg were selected and performed.

The subjects were then asked to perform the whole set of 42-form Tai Chi once at a self-selected training pace and to walk a 15 m pathway three trials at a normal self-selected speed. The 42-form Tai Chi was selected for this study because this style was designed on the bases of the Yang style and contains the most representative components of the other traditional schools (Editor Group, 2000). It is likewise the standard for national and international competitions and has become one of the most popular styles of Tai Chi exercise (Editor Group, 2000).

All subjects wore the identical socks and Chinese Tai Chi shoes. The sole of this kind of shoe is flat and has uniform rigidity. Before testing, the subjects completed consent forms and were given sufficient time to warm-up. An ample time was given to the subject between the movements to avoid possible fatigue.

Measurement

The Pedar-X insole system (Germany) was used to collect the plantar forces during each trial (fundamental movements, whole set of 42-form Tai Chi, 15 m pathway walk). Each insole has 99 sensors and the sampling rate was set at 50 Hz. The reliability of this system has been well documented in the previous studies (Boyd *et al.*, 1997; Kernozek *et al.*, 1996). With the aid of the trublu calibration device, all of the sensors of the system were individually calibrated before testing. For analysis, the insole was divided into nine distinct regions: the medial heel, lateral heel, medial midfoot, lateral midfoot, first metatarsal head, second

Figure 34.1 Foot movements during (a) forward, (b) backward, (c) sideways, (d) up-down, and (e) fixing Tai Chi movements.

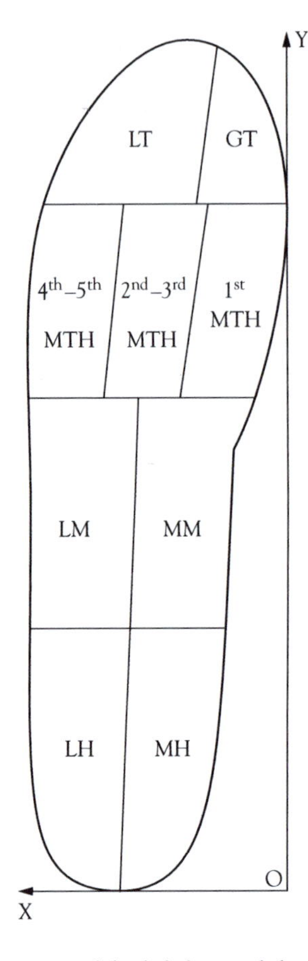

Figure 34.2 The divided plantar regions of the left foot and the definition of the coordinates system based on this insole system.
MH: medial heel; LH: lateral heel; MM: medial midfoot; LM: lateral midfoot; 1st MTH: first metatarsal head; 2nd–3rd MTH: second and third metatarsal heads; 4th–5th MTH: fourth and fifth metatarsal heads; GT: great toe; LT: lesser toes.

and third metatarsal heads, forth and fifth metatarsal heads, great toe and lesser toes, as illustrated in Figure 34.2.

In accordance with the definitions of the Pedar-X insole coordinates system, the most medial and posterior point of the left foot was defined as the origin (zero point). The parameters employed in this study included the pressure, pressure-time integral, and centre of pressure of the foot using Novel Database Pro software. The x and y coordinates of the COP were normalized to the maximum width and maximum length of the insole, respectively. The ground reaction force was normalized to body weight and drawn against the stance time.

The one-leg stance was defined as the interval spent in a one-footed stance when the ground reaction force of the other foot is below two Newtons, which is the initial threshold of the insole system.

Data analysis

The dependent variables included the pressure, pressure-time integral and the COP. For normal walking, the mean of each variable was obtained from the three walk trials; for each trial, the variable was the mean of the three complete gait cycles.

For the fundamental movements of Tai Chi, the mean of each variable was calculated from the three trials. All of the dependent variables of the left foot, which was the stance foot of Tai Chi movements selected, in Tai Chi movements and walking were statistically analyzed respectively using independent-T test to detect any differences between the male and female groups. A repeated-measures analysis of variance (ANOVA) was employed to compare each variable between the fundamental Tai Chi movements and normal walking. After finding the significant differences among Tai Chi movements and normal walking, the post hoc Bonferroni's adjustment was used to detect the differences between normal walking and each Tai Chi movement.

For the whole set of 42-form Tai Chi, the specific instances of one-leg stance of both feet were extracted from the whole set of Tai Chi. All of the dependent variables of both feet in Tai Chi set and walking were statistically analyzed respectively using independent-T test to detect any differences between the left and right foot and between the male and female groups. For each variable, a paired-t test was employed to detect the differences between the one-leg stance of Tai Chi set and normal walking.

For all movements, one hypothesis of this study is that the main plantar loadings are located on the first metatarsal head and the great toe region during Tai Chi exercise while they are located on the second and third and fourth and fifth metatarsal head regions during normal walking. To test the hypothesis, one-way ANOVA with planned comparisons was employed to detect: (1) whether the pressure-time integrals of the first metatarsal head and the great toe region were significantly greater than the remaining regions during each Tai Chi movement and one-leg stance phase; and (2) whether the pressure-time integrals of the second and third and fourth and fifth metatarsal head regions were significantly greater than the remaining regions during normal walking.

A significance level of 0.05 was chosen for all the statistical analysis.

Results

No significant difference was found in any of the parameters between the left and right foot ($p > 0.05$) for both the whole set of 42-form Tai Chi and normal walking. For the purpose of simplification, the data from both feet were averaged.

No significant difference was found in any of the parameters between the male and female groups ($p > 0.05$) for any movements under study and thus the data from the two groups were averaged for further analysis.

The fundamental movements of Tai Chi

Results show that the pressure-time integral of the first metatarsal head and the great toe were the two largest values (Table 34.1) and were significantly greater ($p < 0.05$, Cohen's $d \geq 0.7$) than those of the other regions across all five Tai Chi movements. However, during normal walking, the pressure-time integral of the second and third metatarsal heads and the fourth and fifth metatarsal heads were the two largest values (Table 34.1) and were significantly greater than those of the other regions ($p < 0.05$, Cohen's $d \geq 0.9$).

Table 34.1 The statistical results of pressure-time integral when the two largest regions are compared with the remaining regions for the five Tai Chi movements and during normal walking.

Movement	*Two largest regions*	*MH*		*LH*		*MM*		*LM*		*2nd–3rd MTH*		*4th–5th MTH*		*LT*	
		t	*d*	*t*	*d*	*t*	*d*	*t*	*d*	*t*	*d*	*t*	*d*	*t*	*d*
Forward	1st MTH	8.3	1.8	8.2	1.7	17.2	3.8	15.9	3.5	11.8	2.5	13.8	3.0	11.2	2.4
	GT	4.4	1.0	4.3	0.9	13.8	3.2	12.5	2.8	8.1	1.8	10.3	2.3	7.5	1.6
Backward	1st MTH	8.3	1.5	8.2	1.4	17.2	2.7	18.9	1.9	11.8	1.1	13.8	1.3	11.2	1.3
	GT	6.6	1.7	6.2	1.6	12.0	3.5	8.4	2.2	4.4	1.2	5.4	1.4	5.5	1.4
Sideways	1st MTH	4.0	1.0	4.1	1.0	8.9	2.4	6.9	1.8	2.4	0.9	3.5	0.9	3.5	0.9
	GT	6.3	1.4	6.4	1.4	10.7	2.6	8.9	2.1	4.8	1.1	5.8	1.3	5.8	1.3
Up-down	1st MTH	2.3	0.7	2.4	0.8	6.6	1.5	3.7	0.8	2.8	0.9	2.8	0.8	2.3	0.7
	GT	5.5	1.4	5.7	1.4	11.1	2.8	7.5	1.9	6.5	1.5	6.6	1.6	5.2	1.2
Fixing	1st MTH	9.9	2.4	9.6	2.2	16.1	3.9	15.1	3.7	9.7	2.1	12.8	3.1	10.8	2.4
	GT	4.0	1.4	3.2	1.1	12.2	4.3	11.0	3.9	2.8	1.0	7.7	2.7	4.1	1.5
		MH		LH		MM		LM		1st MTH		GT		LT	
Walking	2nd–3rd MTH	3.0	1.1	2.7	0.9	17.2	6.1	12.4	4.4	2.6	0.9	3.8	1.3	3.9	1.4
	4th–5th MTH	2.9	1.0	2.6	0.9	14.6	5.2	10.8	3.8	2.6	0.9	3.7	1.3	3.7	1.3

MH: medial heel; LH: lateral heel; MM: medial midfoot; LM: lateral midfoot; 1st MTH: 1st Metatarsal head; 2nd–3rd MTH: 2nd and 3rd metatarsal heads; 4th–5th MTH: 4th and 5th Metatarsal heads; GT: great toe; LT: lesser toes.

t: t value.

d: Cohen's *d* value.

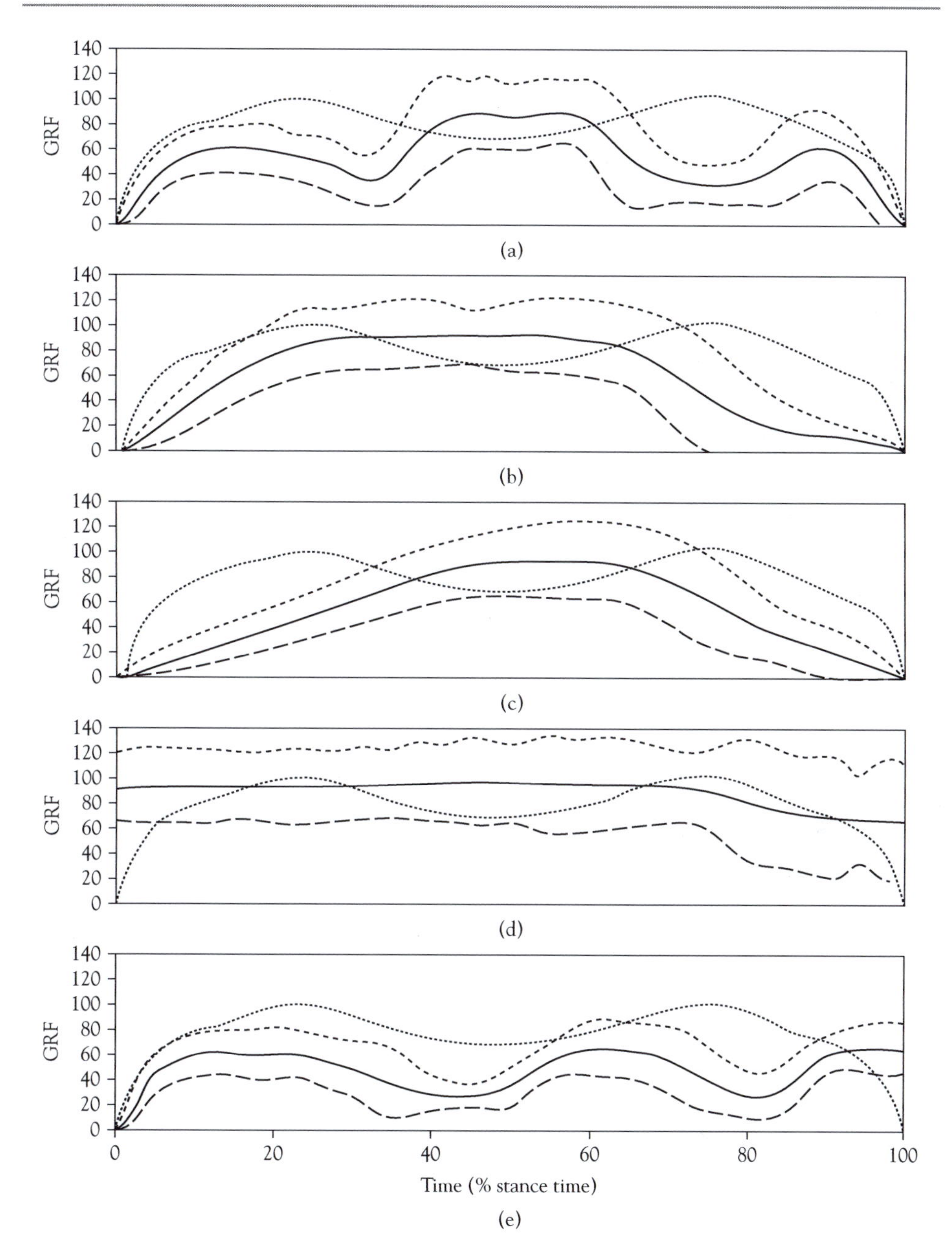

Figure 34.3 The normalized (% BW) ground reaction forces against the time (per cent stance time) during the stance phase in the Tai Chi movements and normal walking.
Solid lines indicate the (a) forward; (b) backward; (c) sideways; (d) up-down; (e) fixing Tai Chi movements. Dashed lines indicate one standard deviation from the mean of the Tai Chi movements. Dotted lines indicate normal walking.

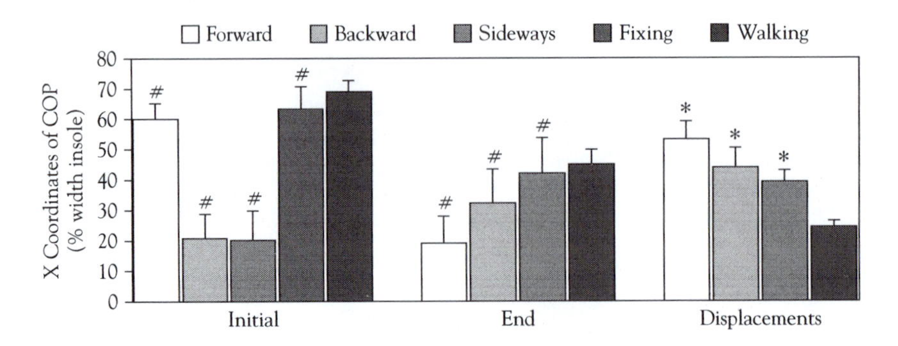

Figure 34.4 The comparisons of the x coordinates of the COP at initial and end contact with the ground and the displacements during contact in the Tai Chi movements and normal walking.
* indicates significantly greater than normal walking ($p < 0.05$), and # indicates significantly less than normal walking ($p < 0.05$) which were detected by one-ANOVA with repeated measures using Bonferroni's adjustment.

Figure 34.3 illustrates that the normalized reaction forces exerted on the foot varied between the five Tai Chi movements, which also shows lower magnitudes and different shapes of total pressure than in normal walking.

Figures 34.4 and 34.5 show the statistical results of x and y coordinates of the COP at initial and end contact with the ground and the displacement during contact when compared between normal walking and each Tai Chi movement. Compared with normal walking, the x coordinates (Figure 34.4) of the COP at initial and end contact were significantly less ($p < 0.05$) in the forward, backward, sideways and fixing movements, which indicates a more medial position of contact. The y coordinates (Figure 34.5) of the COP at initial contact were significantly greater ($p < 0.05$) in the backward and sideways movements, and the figures show that the contact positions were located in the forefoot region. The y coordinates

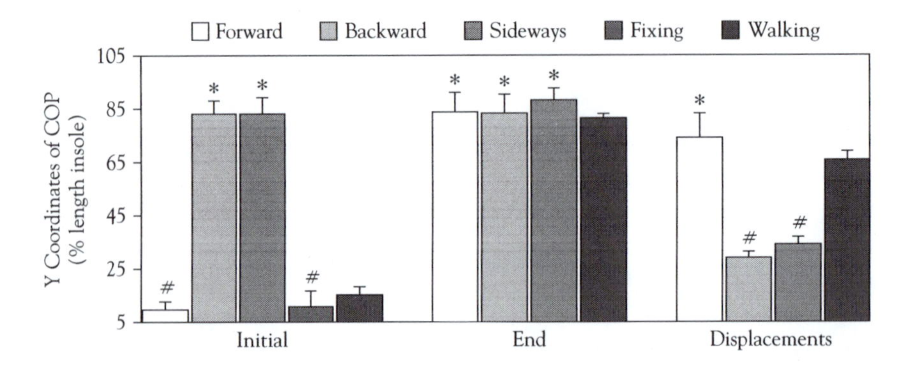

Figure 34.5 The comparisons of the y coordinates of the COP at initial and end contact with the ground and the displacements during contact in the Tai Chi movements and normal walking.
* indicates significantly greater than normal walking ($p < 0.05$), and # indicates significantly less than normal walking ($p < 0.05$) which were detected by one-ANOVA with repeated measures using Bonferroni's adjustment.

(Figure 34.5) of the COP at initial contact were significantly less ($p < 0.05$) in the forward and fixing movements, which indicates that the positions of contact were more posterior. The y coordinates (Figure 34.5) of the COP at end contact were significantly greater ($p < 0.05$) in the forward, backward, and sideways movements, and the figures show that the positions were more anterior. In addition, the forward, backward and sideways movements had significantly wider ($p < 0.05$) displacements in the mediolateral direction (Figure 34.4). The forward movement had a significantly larger ($p < 0.05$) displacement in the anterposterior direction (Figure 34.5).

The whole set of 42-form Tai Chi

When comparing the one-leg stance phases in whole set of 42-form Tai Chi with normal walking, pressure-time integral and peak pressure show a similar trend (Figure 34.6). Values were greater in great toe and first matatarsal heads than in other regions in Tai Chi, and greater in the second and third and fourth and fifth matatarsal heads than in other regions in walking. Other results include that the medial-lateral displacement of the COP in one-leg stance phases of the whole set of 42-form Tai Chi (25.2 ± 3.23 per cent) was significantly greater ($p < 0.05$) than in normal walking (21.8 ± 2.92 per cent) in terms of the maximum width of the insole.

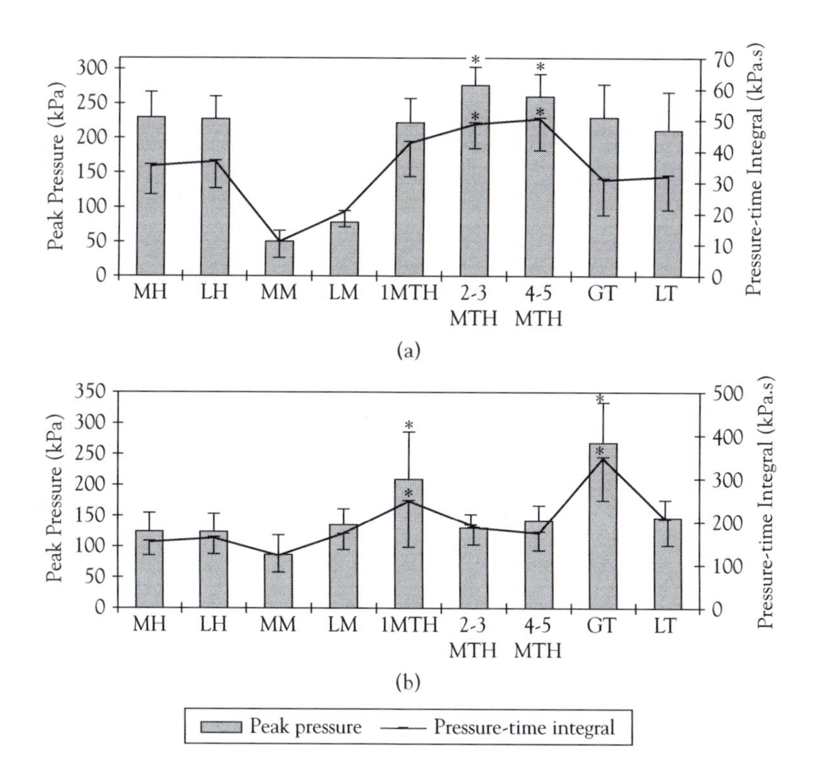

Figure 34.6 Comparison of the regions of peak pressure and the pressure-time integral during the one-leg stance in: (a) normal walking; and (b) Tai Chi exercise.
*indicates significantly greater than other regions.

Discussion

Pressure-time integral

In fundamental Tai Chi movements and one-leg stance phases in whole set of Tai Chi, the two greatest pressure-time integral were located in the first metatarsal head and great toe regions, but were located in the second and third, and fourth and fifth metatarsal heads regions in normal walking (Table 34.1 and Figure 34.6). The results further strengthen the findings of other researchers (Wu and Hitt, 2005), who reported that the main loading in Tai Chi movements occurs in the regions of the great toe and first metatarsal head, whereas the loading maintained in slow walking occurs in third metatarsal head region. The results are also consistent with Tanaka's report that the peak pressure under the great toe is significantly greater than the other four toes during the one-leg stance (Tanaka *et al.*, 1996a). These results reveal that the main plantar loading shifts from the second and third, and fourth and fifth metatarsal heads in normal walking to the first metatarsal head and the great toe in Tai Chi movements. Several studies have demonstrated that the great toe and the forefoot play a very important role in both cutaneous feedback and the muscle activity of the toe in maintaining balance during the one-leg stance in moveable or static platform and with eyes open or closed (Tanaka *et al.*, 1996b; Tanaka *et al.*, 1999; Meyer *et al.*, 2004). Nurse and Nigg (Nurse and Nigg, 1999, 2001) investigated the relationship between plantar pressure and the sensation of the great toe, and found that there is a negative relationship between plantar pressure and the sensation threshold under the great toe. The tactile sense of the great toe decreases with age, and elderly people are not able to sufficiently utilize the muscle of the great toe to maintain balance control when perturbations occur (Tanaka *et al.*, 1996b; Tanaka *et al.*, 1999). Tanaka *et al.* (Tanaka *et al.*, 1996b) suggested that not only the development of motor performance, but also the facilitation of sensory input should be considered in rehabilitation programmes to improve poor balance. Plantar loading is mainly located in the anterior and medial areas of the foot in Tai Chi movements, which presents a strong challenge to the exertion of the great toe, and subsequently has a training effect on the muscles controlling the great toe. Furthermore, the greater pressure in the anterior and medial regions may intensify the sensory input from the great toe (Tanaka *et al.*, 1996b; Tanaka *et al.*, 1999) and the first metatarsal head, because the first metatarsal head area is one of the most sensitive regions on the bottom of the foot (Nurse and Nigg, 1999). Thus, it is expected that long-term Tai Chi exercise not only enhances muscle strength, but also improves the somatosensory input and feedback of the great toe area to assist in balance control.

Ground reaction force

The shapes and amplitudes of the normalized ground reaction forces of the five Tai Chi movements depending on the movement (Figure 34.3). The amplitudes of the ground reaction forces during the Tai Chi movements were lower than in normal walking, which is consistent with the results of other studies (Wu and Hitt, 2005) that found that the peak vertical force is about the same as the body weight. The lower impact that is exerted on the foot may be explained by the fact that Tai Chi is characterized by slow and smooth motions, and light and steady steps (He, 1990). He (He, 1990) stated that the steps in Tai Chi are made 'as quietly as a cat walks', and the exertion is 'so mild that it looks like reeling raw silk from a cocoon'. This suggests that Tai Chi is a safe weight-bearing exercise (Wu and Hitt, 2005) and is suitable even for patients who suffer from rheumatoid arthritis (Kirsteins *et al.*, 1991).

The shapes of the ground reaction forces in the Tai Chi movements were varied from each other and different from those walking, which may reflect the complex gait of Tai Chi movements, which combine various foot support patterns and step directions (Editor Group, 2000). Various support patterns and step directions may be the reason for the differences in ground reaction forces between the different Tai Chi movements and normal walking.

Displacement of centre of pressure

The COP loci were significantly displaced posteriorly at initial contact in the forward and fixing movements, and were displaced anteriorly at end contact with the ground in the forward, backward and sideways movements compared with normal walking. McCaw and DeVita (1995) reported that, during stance phase of gait, shifting the COP posteriorly increases the flexor torque at the ankle, knee and hip joints of the present leg. Conversely, shifting the COP anteriorly increases the extensor torque at the ankle, knee, and hip joints of the present leg. Further, +/– 0.5 cm and +/– 1.0 cm shifts in the location of the COP cause about a 7 per cent and 14 per cent change in the maximum joint torque and angular impulse values, respectively. When the location of the COP is more posterior at initial contact, the ankle posture is more dorsiflexed, and when the location of the COP is more anterior at end contact, the ankle posture is more plantarflexed. The increase in the ROM of the ankle in the sagittal plane in Tai Chi movements may thus be expected. Some studies (Mecagni *et al.*, 2000; Whipple *et al.*, 1987) have reported that there is a positive relationship between the range of motion of the ankle joint and balance control and muscle strength in the lower extremities. This may partly explain why muscle strength was improved in the lower extremities after Tai Chi exercise (Hong *et al.*, 2000; Wu *et al.*, 2004). Further kinematic investigation on the range of motion of the ankle joint is required in the future.

The results show that the forward, backward and sideways Tai Chi movements have wider displacements in the mediolateral direction, and the forward movement has a greater displacement in the anterposterior direction than in normal walking. Szturm and Fallang (Szturm and Fallang, 1998) investigated the EMG and the displacement of the COP on the bottom of the foot when disturbance occurs. They found that there is a positive relationship between the displacement of the COP and the magnitude of the EMG in the lower extremities. Nakamura *et al.* (Nakamura *et al.*, 2001) found that an increase in the displacement of the COP not only increases the magnitude of the EMG, but also the number of muscles that are used in the lower extremities. Thus, the larger displacement of the COP in Tai Chi movements may induce the lower extremities to recruit more muscles that contract at a higher level (Wu *et al.*, 2004) than in normal walking.

Tropp and Odenrick (Tropp and Odenrick, 1988) found that ankle inversion-eversion is a critical mechanism in maintaining balance during the one-leg stance. Other investigators (Hoogvliet *et al.*, 1997; King and Zatsiorsky, 2002) further examined the relationship between foot inversion-eversion and centre of pressure displacement during the one-leg stance in healthy adults, and suggested that the tilting of the foot results in considerable changes in the pressure distribution, and is thus a major source of COP displacement. They concluded that medial-lateral displacements of the centre of pressure are good indicators of the tilting motions of the foot during the one-leg stance. King and Zatsiorsky (2002) suggested that balance is maintained during the one-leg stance through a relatively large displacement of the COP that is made possible by a rocking motion of the foot. This large displacement of the COP then in turn creates a large ankle torque and horizontal forces to restore the centre of gravity to a more balanced position. In this study, the medial-lateral displacement of the COP in the Tai Chi movements was 25.2 per cent, which is significantly greater than the displacement in normal walking of 21.8 per cent. These results reveal that the one-leg stance in Tai Chi exercise may force the ankle joint to produce more activities of motion and create more horizontal forces, while the other parts of the body are also moving to ensure that the COP is located within the base of support. It is likely that practicing Tai Chi may have a beneficial effect in training to maintain medial-lateral balance, which is considered to cause a higher risk of falls during the one-leg stance (Mak and Ng, 2003; Islam *et al.*, 2004).

Conclusion

The Tai Chi movements investigated in this study have greater anteromedial plantar loading, lower ground reaction forces, larger COP displacements both in the direction of anterposterior and mediolateral when compared with normal walking. It is speculated that the plantar pressure characteristics shown in Tai Chi exercise may benefit to intensify the plantar cutaneous tactile sensory input from the first metatarsal head and great toe areas, increase the muscle strength of the lower extremities, improve the ability to balance on one leg, and subsequently improve balance control.

Acknowledgement

The work described in this paper was fully supported by a grant from the Research Grants Council of the Hong Kong Special Administrative Region (Project no. CUHK4360/00H).

References

1. Bohannon, R. W., Larkin, P. A., Cook, A. C., Gear, J. and Singer, J. (1984) 'Decrease in timed balance test scores with aging.' *Phys Ther*, 64(7): 1067–1070.
2. Boyd, L. A., Bontrager, E. L., Mulroy, S. J. and Perry, J. (1997) 'The reliability and validity of the novel pedar system of in-shoe pressure measurement during free ambulation.' *Gait Posture*, 165.
3. Cook, T. M., Farrell, K. P., Carey, I. A., Gibbs, J. M. and Wiger, G. E. (1997) 'Effects of restricted knee flexion and walking speed on the vertical ground reaction force during gait.' *J Orthop Sports Phys Ther*, 25: 236–244.
4. Editor Group (2000) *Tai chi chuan exercise*: People's Sports Publishing House of China.
5. He, C. F. (1990) 'A new type of chinese taijiquan.' Hong Kong: Joint Publishing (H.K.) Co., Ltd.
6. Hong, Y., Li, J. X. and Robinson. (2000) 'Balance control, flexibility, and cardiorespiratory fitness among older tai chi practitioners.' *Br J Sports Med*, 34(1): 29–34.
7. Hoogvliet, P., Duyl, W. A., de Bakker, J. V., Mulder, P. G. and Stam, H. J. (1997) 'A model for the relation between the displacement of the ankle and the centre of pressure in the frontal plane, during one-leg stance.' *Gait Posture*, 6: 39–49.
8. Islam, M. M., Nasu, E., Rogers, M. E., Koizumi, D., Rogers, N. L. and Takeshima, N. (2004) 'Effects of combined sensory and muscular training on balance in Japanese older adults.' *Prev Med*, 39: 1148–1155.
9. Iverson, B. D., Gossman, M. R., Shaddeau, S. R. and Turner, M. E. (1990) 'Balance performance, force production, and activity levels in noninstitutionalized men 60 to 90 years of age.' *Phys Ther*, 70(6): 348–355.

10. Jonsson, E., Seiger, A. and Hirschfeld, H. (2004) 'One-leg stance in healthy young and elderly adults: A measure of postural steadiness.' *Clin Biomech*, 19: 688–694.
11. Kernozek, T. W., LaMott, E. E. and Dancisak, M. J. (1996) 'Reliability of an in-shoe pressure measurement system during treadmill walking.' *Foot & Ankle International*, 17(4): 204–209.
12. King, D. L. and Zatsiorsky, V. M. (2002) 'Periods of extreme ankle displacement during one-legged standing.' *Gait Posture*, 15: 172–179.
13. Kirsteins, A. E., Dietz, F. and Hwang, S. M. (1991) 'Evaluating the safety and potential use of a weight-bearing exercise, tai-chi chuan, for rheumatoid arthritis patients.' *Am J Phys Med Rehabil*, 70(3): 136–141.
14. Li, J. X., Hong, Y. and Chan, K. M. (2001) 'Tai chi: Physiological characteristics and beneficial effects on health.' *Br J Sports Med*, 35: 148–156.
15. Mak, M. K. and Ng, P. L. (2003) 'Mediolateral sway in single-leg stance is the best discriminator of balance performance for tai-chi practitioners.' *Arch Phys Med Rehabil*, 84: 683–686.
16. Mao, D.W., Hong, Y.L., Li, J.X. (2006) 'The characteristics of foot movement in Tai Chi exercise.' *Physical Therapy*, 86(2): 215–222.
17. McCaw, S. T. and DeVita, P. (1995) 'Errors in alignment of centre of pressure and foot coordinates after predicted lower extremity torques.' *Journal of Biomechanics*, 28(8): 985–988.
18. Mecagni, C., Smith, J. P., Roberts, K. E. and Susan, B. O. (2000) 'Balance and ankle range of motion in community-dwelling women aged 64 to 87 years: A correlation study.' *Phys Ther*, 80(10): 1004–1011.
19. Melzer, I., Benjuya, N. and Kaplanski, J. (2003) 'Effects of regular walking on postural stability in the elderly.' *Gerontology*, 49: 240–245.
20. Meyer, P. F., Oddsson, L. I. E., De Luca, C. J., (2004) 'The role of plantar cutaneous sensation in unperturbed stance.' *Exp Brain Res*, 156: 505–512.
21. Nakamura, H., Tsuchida, T. and Mano, Y. (2001) 'The assessment of posture control in the elderly using the displacement of the centre of pressure after forward platform translation.' *J Electromyogr Kinesiol*, 11: 395–403.
22. Nurse, M. A. and Nigg, B. M. (1999) 'Quantifying a relationship between tactile and vibration sensitivity of the human foot with plantar pressure distributions during gait.' *Clin Biomech*, 14: 667–672.
23. Nurse, M. A. and Nigg, B. M. (2001) 'The effect of changes in foot sensation on plantar pressure and muscle activity.' *Clin Biomech*, 16: 719–727.
24. Richardson, J. K., Ashton-Miller, J. A., Lee, S. G. and Jacobs, M. K. (1996) 'Moderate peripheral neuropathy impairs weight transfer and unipedal balance in the elderly.' *Arch Phys Med Rehabil*, 77: 1152–1156.
25. Schaller, K. J. (1996) 'Tai chi chih: An exercise option for older adults.' *J Gerontol Nurs*, 22(10): 12–17.
26. Shin, Y. H. (1999) 'The effects of a walking exercise program on physical function and emotional state of elderly korean women.' *Public Health Nurs*, 16(2): 146–154.
27. Szturm, T. and Fallang, B. (1998) 'Effects of varying acceleration of platform translation and toes-up rotations on the pattern and magnitude of balance reaction in humans.' *J Vestibular Res*, 8(5): 381–397.
28. Takahashi, T., Ishida, K., Hirose, D., Nagano, Y., Okumiya, K., Nishinaga, M., *et al.* (2004) 'Vertical ground reaction force shape is associated with gait parameters, timed up and go, and functional reach in elderly female.' *J Rehabil Med*, 36: 42–45.
29. Tanaka, T., Hashinoto, M., Nakata, M., Ito, T., Ino, S. and Ifulube, T. (1996a) 'Analysis of toe pressure under the foot while dynamic standing on one foot in healthy subjects.' *JOSPT*, 23(3): 188–193.
30. Tanaka, T., Seiji, N., Ino, S., Ifukube, T. and Nakata, M. (1996b) 'Objective method to determine the contribution of the great toe to standing balance and preliminary observations of age-related effects.' *IEEE Trans Rehabilitation Eng*, 4(2): 84–90.
31. Tanaka, T., Takeda, H., Izumi, T., Ino, S. and Ifukube, T. (1999) 'Effect on the location of the centre of gravity and the foot pressure contribution to standing balance associated with ageing.' *Ergonomics*, 42(7): 997–1010.

32. Tropp, H. and Odenrick, P. (1988) 'Postural control in single-limb stance.' *J Orthop Res*, 6(6): 833–839.
33. Tsang, W. W. N., and Hui-Chan, C. W. Y. (2003) 'Effects of Tai Chi on joint proprioception and stability limits in elderly subjects.' *Medicine and Science in Sports and* Exercise, 35:1962–1971.
34. Tse, A. K., and Bailey, D. M. (1992) 'Tai Chi and postural control in the well elderly.' *American Journal of Occupational Therapy*, 46: 295–300.
35. Whipple, R. H., Wolfson, L. I. and Amerman, P. M. (1987) 'The relationship of knee and ankle weakness to falls in nursing home residents: An isokinetic study.' *J Am Geriatr Soc*, 35: 13–20.
36. Winter, D. A. (1991) *The biomechanical and motor control of human gait*. Waterloo: University of Waterloo Press.
37. Wu, G. and Hitt, J. (2005) 'Ground contact characteristics of tai chi gait.' *Gait Posture*, 22(1): 32–39.
38. Wu, G., Liu, W. and Hitt, J. (2004) 'Spatial, temporal and muscle action patterns of tai chi gait.' *J Electromyogr Kinesiol*, 14: 343–354.

35

Biomechanical studies for understanding falls in older adults

Daina L. Sturnieks and Stephen R. Lord
University of New South Wales, Sydney

The issue of falls

Falls present a major threat to the well-being and quality of life of older people. Approximately one-third of older people living in the community fall at least once a year, with many suffering multiple falls. Fall rates are higher in older community-dwelling women (40 per cent) than in older men (28 per cent). Rates increase to over 50 per cent in people aged 85 years and over, in residents of intermediate care hostels and nursing homes, in those who have fallen in the past year and in those with particular medical conditions that affect muscle strength, balance and gait. Falls account for 4 per cent of hospital admissions, 40 per cent of injury-related deaths and 1 per cent of total deaths in persons aged 65 years and over. The major injuries that result from falls include fractures of the wrist, neck, trunk and hip. Falls may also result in disability, restriction of activity and fear of falling, which can reduce quality of life and independence and contribute to an older person being admitted to a nursing home. Furthermore, as many fall-related injuries require medical treatment including hospitalisation, falls constitute a condition requiring considerable health care expenditure.

Risk factors for falls

Identification of risk factors for falls is important to develop effective strategies for prevention. There are various psychosocial and demographic factors that are associated with a greater risk of falling, including advanced age, restrictions in activities of daily living, fear of falling, depression and a history of previous falls. Older people with multiple chronic illnesses have higher rates of falls than active older people without known pathology or impairments. However, attributing a degree of falls risk to a specific medical diagnosis is problematic because the relative severity of the pathological conditions may vary considerably among individuals. Furthermore, declines in sensorimotor function associated with age, inactivity, medication use, or minor pathology may be evident in older people with no documented medical illness. A more insightful strategy for understanding the intrinsic risk factors for falls is a 'physiological' approach that directly considers sensorimotor abilities and impairments irrespective of their cause.

The physiology of balance and falls

Poor balance, the inability to control the position of the body, is widely acknowledged as being a significant contributor to the increased incidence of falls in people aged 65 years and older. Balance requires the complex integration of sensory information regarding the position of the body relative to the surroundings, and the ability to generate appropriate motor responses to control body movement (Figure 35.1). The sensory component calls upon contributions from vision, peripheral sensation and vestibular sense, while the motor component requires muscle strength, neuromuscular control and reaction time. Linking these two components together are the higher level neurological processes enabling anticipatory mechanisms responsible for planning a movement, and adaptive mechanisms responsible for the ability to react to changing demands of the particular task.

With increased age, there is a progressive loss of functioning of sensory, motor and central processing systems and an increased likelihood of falls (Lord *et al.*, 1994). Instability and falls in older people can result from impairment in any of these systems. Furthermore, when one of the components of the postural control system is deficient, there is a greater reliance on the remaining components to maintain balance, increasing the demands on the sensorimotor systems and consequently increasing the likelihood of a fall. The extent to which one sensory input can compensate for the loss of another remains unclear, although there is some evidence that peripheral sensation is the most important sensory system in the regulation of standing balance (Fitzpatrick *et al.*, 1994). Components of the motor system found to be particularly important to balance are strength and reaction time.

Peripheral sensation

Sensory information from the limbs provides feedback regarding position, movement and touch. This information includes proprioception and tactile sensation. Proprioception is the awareness of body position, which comes from receptors in the muscles, tendons, and joints and is often assessed by measuring one's ability to determine joint position or joint movement. Two types of sensory receptors in skeletal muscles, muscle spindles and Golgi tendon organs, provide information regarding muscle tension and length (from which joint position can be determined), the velocity of movements and the force produced by to the muscle.

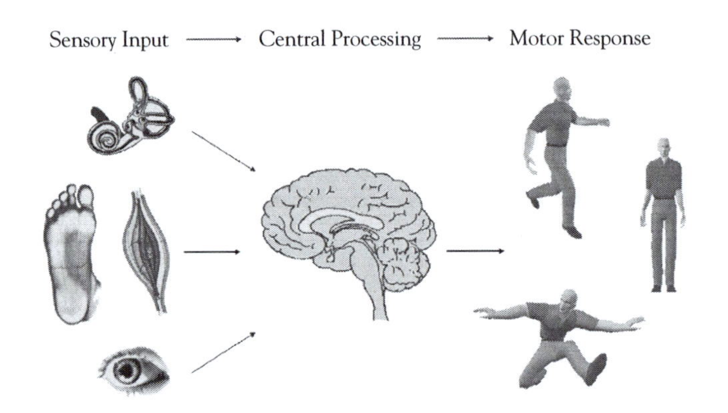

Figure 35.1 Key components in maintenance of the upright posture. Sensory information is received from vestibular, proprioceptive, tactile and visual organs, processed by the brain which accordingly develops an appropriate motor response.

Mechanoreceptors existing in joints and surrounding structures respond to distortion or pressure and also provide an indication of joint position, in addition to the degree of stretch, compression, tension, acceleration, and rotation. Tactile sensitivity is the awareness of touch and comes from receptors, generally in the skin that respond to variations in pressure (firm, brushing, sustained, etc). Plantar tactile sensitivity is reduced in older adults and correlated with measures of balance and functional test performance (Menz *et al.*, 2005). Furthermore, quantitatively assessed impairments in peripheral sensation, including tactile sensitivity at the ankle, vibration sense at the knee and knee joint position sense, are significant and independent risk factors for falls in populations of older people (Lord *et al.*, 1992,1994).

Vision

Vision is an important source of information for the control of balance. In addition to supplying continued information about the external environment, vision provides feedback about the position and movements of the body. While standing, visual information is used to monitor and moderate postural sway. Sway area increases approximately 30 per cent when closing the eyes while standing (Lord *et al.*, 1991). During gait, vision is important to identify environmental hazards and determine safe foot placement. In assessing visual competence for falls prevention, different aspects of vision, including acuity, contrast sensitivity and depth perception might be considered. Impaired depth perception is one of the strongest risk factors for multiple falls in community-dwelling older people (Lord *et al.*, 2001). Contrast sensitivity has also been found to be useful in identifying older people at risk of falling (Lord *et al.*, 2001). This suggests that the ability to accurately judge distances and perceive spatial relationships is important for maintaining balance.

Vestibular sensation

The vestibular system involves inner ear structures that detect position and motion of the head, relative to gravity, and is important for posture and coordination of head, eye and body movements. Recent studies have reported significant associations between vestibular impairment and fall-related fractures. Kristinsdottir and colleagues have used a head-shaking stimulus applied when subjects were in a supine position to induce nystagmus – a sign indicating asymmetry of the vestibular reflexes. In an initial study comprising 19 subjects (mean age 72 years) with hip fracture and 28 aged-matched controls, they found that 68 per cent of the hip fracture subjects demonstrated a nystagmus following the head-shake stimulus compared with 32 per cent of the controls (Kristinsdottir *et al.*, 2000). Similar findings were reported in a subsequent study of older wrist fracture patients (Kristinsdottir *et al.*, 2001). Vestibular function is less amenable to assessment with simple screening tests compared with vision and peripheral sensation. However, these recent studies provide preliminary evidence that when assessed with greater precision, impaired vestibular function may be an important risk factor for falls and fall-related fractures in older people.

Muscle strength

Starting in the mid-twenties, there is a progressive loss of muscle mass, diminishing by 35–40 per cent between the ages of 20 and 80 years (Evans *et al.*, 1995). Muscular strength is typically maintained at peak levels until the fifth or sixth decade, after which accelerated losses occur so that strength decreases approximately 50 per cent by the age of 80 years.

Not only is peak torque production affected with advancing age, but rate of torque development and muscular power also decline (Thelen *et al.*, 1996). The importance of muscular strength and power for balance is well recognised. A reduced capacity for rapid force generation would limit the ability to respond quickly to a loss of balance and increase the risk of falling. Appropriately timed and coordinated activation of motor units to finely adjust joint moments is required to control the centre of mass (COM) over the base of support. The ability to coordinate appropriate motor unit activation is difficult to assess, but this component of muscle function is likely to be associated with muscular strength.

Lower limb muscle weakness manifests in poor performance on tests of balance, abnormal gait patterns and reduced general mobility. Reduced knee extension strength (Lord *et al.*, 1994), ankle dorsiflexion strength (Whipple *et al.*, 1987), and hip strength (Robbins *et al.*, 1989) have been found to correlate with increased risk of falls. These consistent findings suggest that lower limb muscle weakness is a major risk factor for falling in older people (Lord *et al.*, 1991). Furthermore, improvements in hip, knee and ankle strength, following an exercise intervention in older women, were found to correspond with improvements in dynamic stability (Lord *et al.*, 1996).

Reaction time

The ability to react quickly and appropriately is an important component of balance and for avoiding a fall when subject to a postural challenge or threat. There is a 25 per cent increase in simple reaction time from the twenties to the sixties, with further significant slowing beyond this age (Fozard *et al.*, 1994). Increased simple reaction time is an independent risk factor for falls in populations of older people (Lord *et al.*, 1992, 1994). Fallers also have significantly slower reaction times than non-fallers in reaction time tests that involve more complicated motor responses, such as extending and flexing the knee, and stepping from either foot (Grabiner *et al.*, 1992; Lord *et al.*, 2001). Slower reaction times in older adults and fallers may be due to the slower muscle latencies, increased difficulty in producing and coordinating muscle force, and slower central processing times, particularly in the presence of distracting stimuli.

The biomechanics of balance and falls

In biomechanical terms, balance is a task of maintaining the body's centre of mass (COM, the point around which the body's mass is equally distributed) within the limits of the base of support (BOS, the area circumscribed by parts of the body that are in contact with a support surface). Mechanical factors affecting balance include: the mass of the body (according to Newton's Second Law of Motion [F = ma], which suggests the larger an object, the greater the force required to accelerate it); the size of the BOS or support area generally the larger the support area, the more stable an object is; the amount of friction between the body and the support surface, which must be sufficient to prevent slipping; the position of the COM relative to the BOS, in the horizontal plane and the vertical direction. Balance tasks may be grouped into four different categories: (1) maintaining a stable position, such as standing or sitting; (2) adjustments to voluntary movements, such as reaching, gait initiation or voluntary stepping; (3) reactions to expected external perturbations, such as catching a ball; and (4) reactions to unexpected perturbations, such as tripping. The numerous biomechanical studies of balance and falls in the elderly span each of these categories, however, the majority fit into the first and the last.

Posturography

Standing balance, or postural control, has been widely studied in relation to falls. Although quiet standing appears to be simple and somewhat static, it is in fact a mechanically challenging task, involving the constant monitoring and adjustment of body motion. Standing balance motion is often likened to an inverted pendulum, the body COM moving about a pivot point at the ankles. As an inverted pendulum is unstable, relatively minor perturbations (such as breathing) cause the swaying motion. In order to maintain the body COM within the BOS, an individual produces forces on the support surface/s (predominantly under the feet while standing), to control the COM position. The point at which the sum of these forces acts on the support surface is the centre of pressure (COP). Much like a constant process of catching and throwing, forces at the support surface act to accelerate the COM in a given direction and when this motion is detected as being adverse to stability, opposing forces act to catch the falling COM and throw it in another direction. Consequently, the displacement of the COP oscillates around that of the COM (Figure 35.2) and the COP can be considered a controlling variable for balance, as it governs the horizontal acceleration of the COM (Winter *et al.*, 1991).

In the biomechanics laboratory, postural sway may be measured in many different ways. For example, the displacements of an estimated body COM via motion analysis, excursions of the COP recorded by a force platform underfoot, or accelerations of the pelvis and head with accelerometers. Variables of interest from these data include displacement, velocity, acceleration, frequency or other statistical measures of sway magnitude or variance, such as the root mean square. Physiologists and clinicians have also measured postural sway using simple, portable equipment, such as a pen attached to a rod extending from the waist, which traces the movements of the body onto a sheet of paper.

The amount of postural sway increases after the age of ~30 years, as first studied by Hellebrandt and Braun in 1939 (Hellbrandt *et al.*, 1939) and supported by numerous studies since. With increasing age, the magnitude and velocity of postural sway increases, so that in a given period of time, the amount of postural sway (total sway path) in an older adult might easily double or triple that of a young healthy adult. In the lateral direction, this age-related difference is more pronounced, suggesting the control of balance in the lateral plane is particularly problematic for older adults. The decline in balance control with age is associated with sensory and motor functions that are necessary for the continuous detection and correction of postural position, including losses of lower extremity muscle strength, peripheral sensation, visual acuity and speed of reaction time (Lord *et al.*, 1991; Era *et al.*, 1996; Stelmach *et al.*, 1989). Indeed, impaired lower limb proprioception, quadriceps strength, and reaction time have been found to be important correlates of increased sway in a more challenging test of balance where BOS was reduced (Lord *et al.*, 1999).

Several cross-sectional studies have shown no differences in sway parameters between fallers and non-fallers while subjects are standing with normal sensory information available to them. A review of 9 prospective studies on falls found only five reported measures of postural sway to be significantly associated with falls outcomes (Piirtola *et al.*, 2006). However, significant differences in sway path have been identified in conditions where either visual or proprioceptive information was limited (eyes closed and standing on a compliant surface). Multiple fallers had 33–46 per cent greater body sway than those who did not experience a fall in the previous 12 months (Lord *et al.*, 1994). In a population of 100 adults aged 62–96 (predominately female), Maki *et al.* (1994) found the (root mean square) COP mediolateral displacement, while blindfolded, to predict those people who would fall in the next 12 months. This finding indicates an increased reliance on vision for maintaining stability in

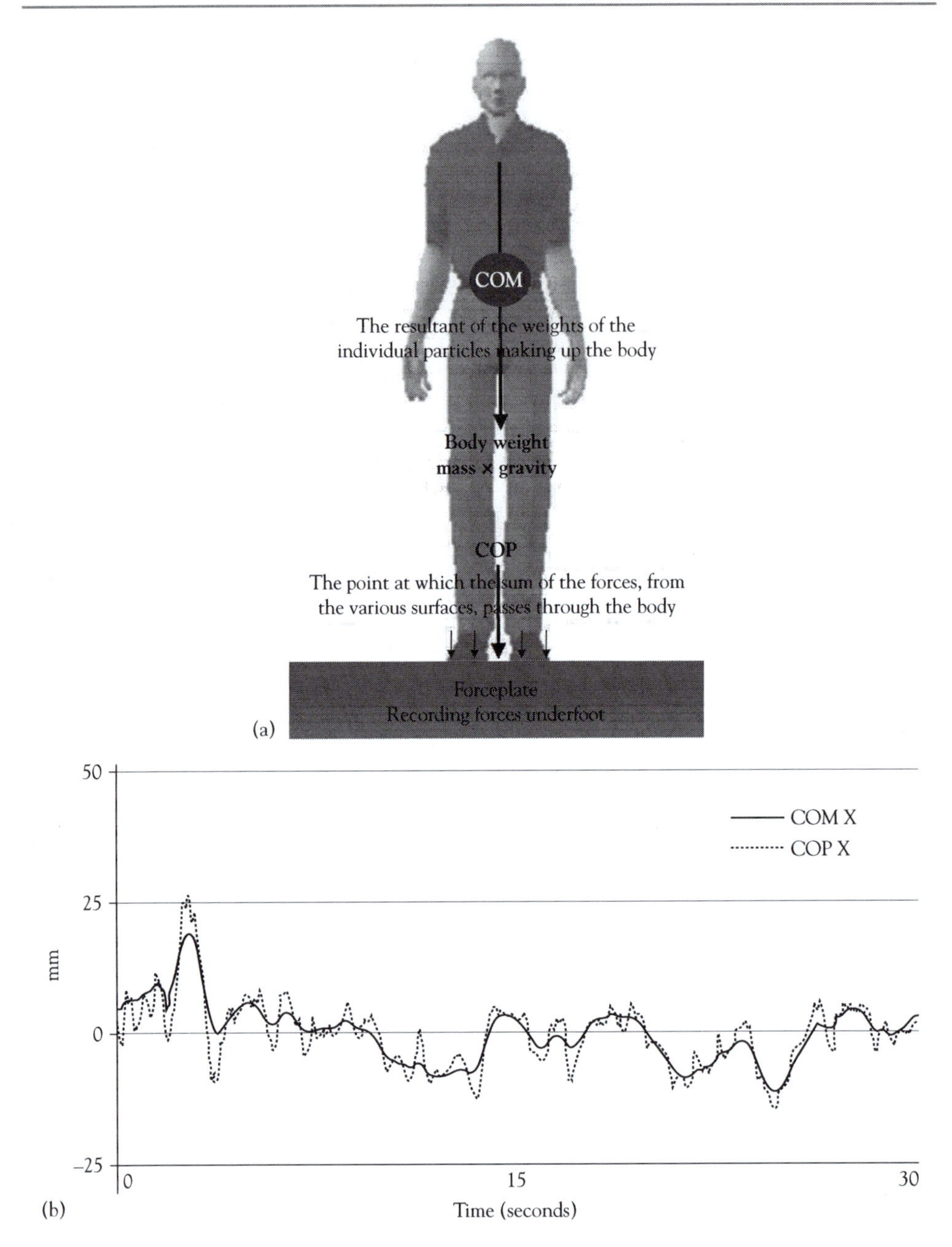

Figure 35.2 (a) person standing on a force plate, which measures forces underfoot, from which COP may be calculated; (b) typical sway path of the COP and the COM in the anteroposterior direction, over 30 seconds, showing the relationship between these variables.

older adults who are at risk of future falls. Similarly, a prospective study of 84-aged care hostel residents found increasing effect sizes between non fallers and multiple fallers in sway without vision and reduced proprioception (Lord *et al.*, 1991). While sway under normal standing conditions was not significantly different between fallers and non-fallers, the difference was over 60 per cent with eyes closed and while standing on a foam mat (see Figure 35.3).

Taken together these studies suggest that posturography variables are most sensitive to discriminating future fallers and non-fallers when sensory information is manipulated. It is probable that older adults who are likely to fall rely heavily on all sensory channels for maintaining normal stability, even while standing in a non-threatening environment. Those who are unlikely to fall seemingly maintain some redundancy in the sensory information required for normal stability, as is the case in young healthy adults. It is logical that older people who have less redundancy in their balance system are likely to fall in situations where balance is challenged.

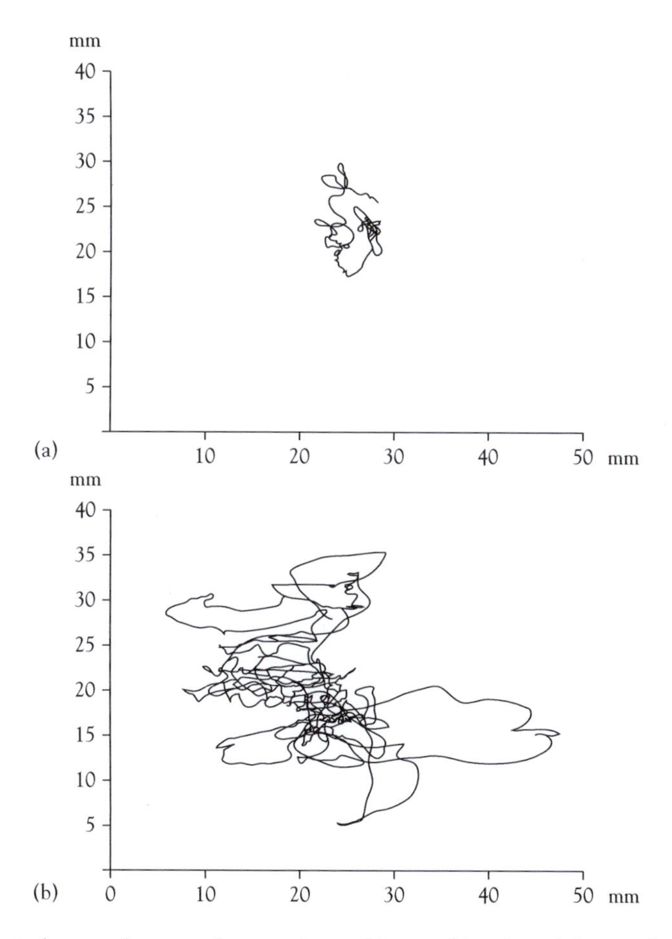

Figure 35.3 Typical postural sway path traces for an 80-year-old male in (a) normal; and (b) sensory deficient (eyes closed, foam mat) conditions, highlighting the importance of sensory information for minimising sway controlling standing balance.

Platform perturbations

In order to seek more functionally relevant quantitative measures to understand the mechanisms that cause falls in older adults, perturbation studies involving platforms that translate or rotate underfoot began in the late 1980s, to test an individual's ability to respond to an unexpected postural disturbance while standing. Measuring COP displacements, Stelmach and colleagues (1989) showed that elderly participants respond with more sway following rotation of the platform, and are therefore less able to control their balance following a perturbation, compared to young. These findings have been subsequently verified and appear to be due to longer latencies in reflexive and voluntary muscular responses in older, compared to younger adults. Older adults are less able to make use of higher level sensory integration for voluntary control of sway following perturbations. This may be due to poorer sensory acuity, increased central processing times, poorer generation of joint powers due to strength loss, as well as increased joint stiffness with age, which reduces the ability to respond to a perturbation using the same motor patterns as younger populations. These age-related changes are reflected in the perturbation studies more so than normal standing postural sway. However, little evidence exists regarding the value of platform perturbation studies for predicting fallers. One study monitored falls for 12 months following initial balance testing and showed the amount of postural sway in response to lateral platform perturbation, while blindfolded, to predict future falling risk with moderate accuracy (Maki *et al.*, 1994).

Stepping

Perturbations of larger magnitudes will require a step to recover balance. Successful balance recovery is dependent on step length, step direction, step execution time, and leg strength. Older adults, particularly those with balance impairment, step more frequently in response to a given perturbation compared to young and often require multiple steps before balance is recovered. In taking multiple steps, older adults are more likely to contact the contralateral limb, further increasing the risk of a fall (Maki *et al.*, 2000). Older adults step sooner, yet fail to properly arrest their momentum, partly due to the inadequate length of the step and a more laterally directed foot placement. In general, it seems that older adults and particularly those with balance impairment do not properly use the sensory information to determine when and how a step should be taken following an unexpected perturbation. As concluded by Mille *et al.* (2003), it may be that waiting to receive the necessary information for an appropriate step is a luxury afforded only by the young and that early stepping in older people is a necessary adaptation to diminished sensorimotor acuity.

The ability to control balance in the lateral plane is particularly problematic for older adults and most falls involve sideways motion. In responding to anteroposterior perturbations, older people, particularly those with recurrent falls, exhibit increased lateral motion and increased lateral foot placement, suggesting an initial lateral destabilisation, or that the older adult's response induces a lateral instability (Rogers *et al.*, 2001). It seems that fallers are particularly vulnerable to instability in the lateral direction, as also evidenced in lateral sway measures that are more sensitive in differentiating between fallers and non-fallers, compared to anterior-posterior sway.

Gait patterns

Falls in older people most often occur during tasks of locomotion or transfer. Locomotion relies upon the appropriate integration of afferent input and coordination of force generation for forward motion. Given that ageing is associated with declines in both peripheral sensation and neuromuscular function, it is not surprising that gait patterns change with age and are associated with postural instability and falling. Older adults tend to walk with slower velocity, shorter step length, wider step width and a relatively increased proportion of time spent in the double-support phase. Despite these differences being interpreted as a safer and more stable strategy, this pattern is more evident in fallers relative to non-fallers.

Spatiotemporal parameters

Stride length, step length, and cadence are reduced in people who have previously fallen, compared to non-fallers, while double support time is increased. This may be a case of fallers adapting their gait in response to a fall. However, in a prospective study, Kemoun *et al.* (2002) found that people who subsequently fell in the following 12 months walked with significantly reduced velocity and increased proportion of stride time spent in double-support, compared to those who did not fall. It is probable that prospective fallers adopt this kind of gait pattern with an awareness of their unsteadiness and likelihood of falling.

It is likely that the control of gait in older adulthood declines due to the deterioration of cognitive efficiency in integrating sensory information for the precise regulation of the motor pattern. Stride-to-stride variability in gait parameters has been examined in an attempt to identify abhorrent aspects of elderly gait that are likely to cause a trip and perhaps a fall. Step width has been found to be more variable in older adults than young adults (Owings *et al.*, 2004), while foot-lift asymmetries have been noted in older adults with high risk of falling, compared to those with low risk (Di Fabio *et al.*, 2004). It has also been found that older adults who fall show significantly greater variability in stride time and swing time, compared to non-fallers (Hausdorff *et al.*, 2001). It is suggested that a more variable gait pattern is reflective of poorer neuromuscular control and provides a greater possibility for inaccuracies leading to a fall.

Kinematics and kinetics

Kinematic and kinetic alterations apparent in older people, compared to young, include reduced hip motion (Murray *et al.*, 1969; Kerrigan *et al.*, 1998), reduced angular velocity of the lower trunk (Gill *et al.*, 2001), reduced ankle power generation (Kerrigan *et al.*, 1998; Winter *et al.*, 1990; Judge *et al.*, 1996) and range of motion (ROM) (Hageman *et al.*, 1986), increased anterior pelvic tilt (Kerrigan *et al.*, 1998; Winter *et al.*, 1990; Judge *et al.*, 1996), increased hip extension moment during swing phase (Mills *et al.*, 2001), increased mechanical energy demands of lower limb musculature (McGibbon *et al.*, 2001), reduced toe pressure (Kernozek *et al.*, 1995) and a larger toe-out angle (Murray *et al.*, 1969; Winter *et al.*, 1990; Murray *et al.*, 1964).

Kemoun *et al.* (2002) investigated kinematic and kinetic characteristics of gait in 54 healthy older adults who had not previously experienced a fall. Those who fell in the following 12 months walked with significantly reduced ankle ROM. Interestingly, they also walked with significantly delayed dorsiflexion, in preparation for foot contact, which might predispose to tripping. At the hip, future fallers had significantly reduced ROM, a reduced flexion moment and less power absorbed, for return during the swing phase, compared to non-fallers. Kerrigan *et al.* (2001) found that fallers exhibited reduced peak hip extension. These differences appear to be related to slower walking speeds.

Lee *et al.* (1999) reported kinetics in retrospective fallers and non-fallers. Despite the fact that the faller group walked only half the pace, significant increases were found in peak moments for hip flexion, hip adduction, knee extension, knee adduction, ankle dorsiflexion, and ankle eversion. These data suggest that the faller group co-contract antagonist muscles while walking, which might be an adaptive strategy adopted by those who have fallen in the past six months. Reduced power absorption at the knee and associated increased power absorption at the ankle, compared to the control group, indicates a strategy to account for poorer ability to efficiently control these joints.

Winter *et al.* (1990) selected healthy, active older adults and compared their gait patterns to healthy young adults, to determine the natural changes in gait due to ageing alone (before the impact of any obvious neural, muscular or skeletal disorders). They found gait differences that were related to a reduction in step length and consequently, gait velocity. If kinetics provide a window to the mechanisms of gait disturbances, reduced power during push off was responsible for these differences. However, it is unknown whether reduced push off, reduced step length, or slowed velocity was the end goal of the CNS. Interestingly, this study found little difference in the variation of gait parameters between young and older adults, suggesting that the contrary findings discussed above are the result of neural, muscular and/or skeletal degeneration, rather than healthy ageing.

Toe clearance is a variable of interest in investigating fall mechanisms for its potential role in the initiation of a trip. During swing phase the foot swings anteriorly with knee extension and minimum clearance with the ground occurs at about mid-swing. The height of the toe at midswing has been reported as being an average 13 mm for young adults and 11 mm for older adults (Winter *et al.*, 1991) so that even a small obstacle might cause a trip and fall.

The horizontal heel velocity at heel contact is another suspect variable in the mechanisms of falls, due to its association with slips. A faster horizontal heel velocity has been shown in older adults, compared to young adults (Winter *et al.*, 1990; Lockhart *et al.*, 2003) and is (paradoxically) associated with a slower gait speed (Mills *et al.*, 2001). Older adults have also been seen to walk with slower transitional acceleration of the body COM following HS, compared to young, further rendering them prone to slipping, due to a more posteriorly placed COM, relative to the BOS (Lockhart *et al.*, 2003).

The value of predicting falls using gait parameters in quiet and safe walking environments is limited, as falls generally occur when one experiences an unforeseen perturbation, such as a trip or a slip. It is supposed that investigating older adults' responses to unexpected perturbations while walking would provide a more sensitive indication of their ability to maintain balance and avoid falls.

Obstacle negotiation

The need to step over an obstacle while walking poses a greater threat to the postural control system due to the longer period of time spent on one leg, and the risk of the lead or trailing limb making contact with the obstacle. Indeed, a large proportion of falls in older people are attributed to tripping, and experiencing multiple 'stumbles' has been found to be a predictor of falls over a 12-month prospective period (Teno *et al.*, 1990). Older adults seem to adopt a more cautious method of negotiating obstacles. They cross with slower speed, shorter step lengths and shorter distance between the obstacle and subsequent heel strike than young adults, and have an increased risk for obstacle contact, particularly when the available response time is decreased (Chen *et al.*, 1994) and attention is divided (Chen *et al.*, 1996). Older adults are also less capable of incorporating the avoidance of an obstacle (turning, sidestepping, stopping) into their normal gait patterns. Compared to young, they slow down, take more steps and are less successful with shorter response times (Cao *et al.*, 1997). These results indicate that elderly people may be at greater risk of falls due to a difficulty in establishing proactive and reactive strategies to avoid obstacles.

Trips

A trip induces a forward rotation of the body over the base of support and is a common cause of falls in older adults. To avoid a fall following a trip, it is necessary to arrest the forward rotation by timely and appropriate muscular work. Several research groups have induced trips to study appropriate behaviours and limiting factors to recovery (see van Dieen *et al.*, 2005 for review).

Pavol and colleagues (1999, 2001) induced trips in a group of 79 healthy older adults using a mechanical obstacle that was revealed to impede the toe of the swinging leg. Two different strategies to compensate for an induced trip were identified: (1) a lowering strategy, in which the tripped foot is quickly lowered to the ground and the contralateral foot initiates a recovery step; and (2) an elevating strategy, in which the tripped limb is subsequently elevated over the obstacle in an attempt to continue the step. Irrespective of which strategy was adopted, subjects who fell following the trip were found to walk with faster velocity, increased cadence and a longer step length than subjects who successfully recovered. Faster walking velocity is also positively associated with an increased degree of trunk flexion displacement following a trip (Grabiner *et al.*, 1993). These findings corroborate the suggestion by Winter *et al.* (1990), that older adults adopt a safer gait pattern via slowed velocity and shortened step length to avoid a fall.

The support limb is of great importance for successful recovery of balance following a trip (Pijnappels *et al.*, 2004). A strong push-off reaction, prior to the recovery (stepping) limb contacting the ground enables time and clearance for correct positioning of the recovery limb. Furthermore, appropriate generation of joint moments in the support limb can help to arrest the angular momentum of the body. Older adults, particularly those with a history of falls are less successful in their recovery than young subjects, attributed to a slower generation of joint moments and a lower peak ankle moment in the support limb. When properly placed, the recovery limb can also generate a force and moment that counteract the body angular momentum (Grabiner *et al.*, 1993).

A slower development of mechanical responses seems to be a major factor limiting older adult's recovery from balance perturbations. The support limb moments necessary for adequate push-off reactions in young subjects have been found to be in the vicinity of 200 Nm for ankle plantar flexion and 50 Nm each for knee flexion and hip extension moments (Pijnappels *et al.*, 2005). It may be that older adults are unable to generate the muscle forces in the necessary time to achieve these joint moments. The maximum lean angle from which older adults are able to recover after a sudden release has been correlated with weight transfer time and step velocity (Thelen *et al.*, 1997), further suggesting that older adults are less able to recover due to a slower development of mechanical responses. Clearly, the recovery from a postural perturbation is dependent upon the rapid development of lower limb muscular force (power), in order to restore control of the flexing trunk. Larger perturbations necessitate stepping as the final protective option to prevent a fall. A reduced capacity for rapid force generation, particularly in frail older adults, might limit the ability to respond quickly to a loss of balance and increase the risk of falling.

Slips

Slips are another common gait-related mechanism of falls in older people and often result in injury due to the large impact forces. A slip occurs when the BOS moves relative to the COM. While walking, a slip occurs generally because the shear force at foot contact is greater than the frictional force at the surface, leading to a forward translation of the foot. This forward translation inhibits the deceleration force normally occurring at foot contact and the consequential forward COM motion over the BOS. Instead, the COM rotates backwards and without appropriate motor intervention, may fall behind the BOS, leading to a posterior fall. During normal gait on non-slippery surfaces, the heel may slide forwards between 1 and 3 cm without being perceived (Redfern *et al.*, 2001). On slippery surfaces, however, forward displacements of the heel are much larger and also occur at higher velocities, thereby requiring a rapidly executed response strategy to avoid a fall. The likelihood of falling following a slip depends on both the mechanics of the slip event and the efficacy of the response. Falling is more likely to occur with increased gait speed, increased forward heel displacement, increased posterior displacement of the body's COM relative to the BOS, and the a larger angle of the leg relative to the ground (representative of a longer step length prior to the slip) (Brady *et al.*, 2000).

Inducing slips with an unanticipated slippery floor surface has shown older adults slip longer and faster, and fall more often than younger participants (Lockhart *et al.*, 2003). It seems that gait changes associated with ageing, in particular the faster horizontal heel contact velocity and slower transition of the COM seen in older adults gait, affect the initiation of slips and lead to more falls. However, the incidence of slip initiation is similar between young and older adults, yet older adults have lower thresholds for recovery than young. Older adults' recovery process is much slower and less effective, seemingly due to changes in vision, reaction time and muscle strength with age (Lockhart *et al.*, 2005).

Summary

Falls are a major issue for older adults and the health system. Poor balance is an important risk factor for falls and is affected by the progressive loss of sensorimotor functioning with increasing age. Deficits in proprioception, vision, vestibular sense, muscle function and reaction time are particularly relevant to balance and falls. Poor balance has been traditionally measured by the swaying motion of the COM over the BOS while standing. However, more insightful information regarding balance and falls prevention is in responses to perturbations such as trips and slips. Understanding the sensory contributions to balance in different conditions, in addition to the kinetics and muscle activations of motor in response to perturbations, exposes deficits in the balance system that might predispose to falls. While kinetic and muscle activity data provide insight into how movement is achieved, the laboratory setting is a sterile environment, absent of the normal distractions and challenges that contribute to falls in everyday life. Investigating kinetics in real or simulated falling situations is possibly the best way to understand why older people are more likely to fall and how this risk can be moderated.

References

1. Brady, R.A., et al. (2000) 'Foot displacement but not velocity predicts the outcome of a slip induced in young subjects while walking'. *J Biomech*, 33(7): 803–8.

2. Cao, C., et al. (1997) 'Abilities to turn suddenly while walking: effects of age, gender, available response time'. J Gerontol, 52(2): M88–93.
3. Chen, H.C., et al. (1994) 'Effects of age and available response time on ability to step over an obstacle'. J Gerontology, 49(5): M227–33.
4. Chen, H.C., et al. (1996) 'Stepping over obstacles: dividing attention impairs performance of old more than young adults'. J Gerontol, 51(3): M116–22.
5. Di Fabio, R.P., et al., (2004) 'Footlift asymmetry during obstacle avoidance in high-risk elderly'. J Am Geriatr Soc, 52(12): 2088–93.
6. Era, P., et al. (1996) 'Postural balance and its sensory-motor correlates in 75-year-old men and women: a cross-national comparative study'. J Gerontol, 51(2): M53–63.
7. Evans, W.J. (1995) 'What is sarcopenia?' J Gerontol A Biol Sci Med Sci, 50 Spec No: 5–8.
8. Fitzpatrick, R., McCloskey, D.I. (1994) 'Proprioceptive, visual and vestibular thresholds for the perception of sway during standing in humans'. J Physiology, 478(Pt 1): 173–86.
9. Fozard, J.L., et al., (1994) 'Age differences and changes in reaction time: the Baltimore Longitudinal Study of Aging'. J Gerontology, 49(4): 179–89.
10. Gill, J., et al. (2001) 'Trunk sway measures of postural stability during clinical balance tests: effects of age'. J Gerontology, 56A(7): M438–47.
11. Grabiner, M.D., et al. (1993) 'Kinematics of recovery from a stumble'. J Gerontol, 48(3): M97–102.
12. Grabiner, M.D., Jahnigen, D.W. (1992) 'Modeling recovery from stumbles: preliminary data on variable selection and classification efficacy'. JAGS, 40(9): 910–13.
13. Hageman, P.A., Blanke, D.J. (1986) 'Comparison of gait of young women and elderly women'. Physical Therapy, 66(9): 1382–7.
14. Hausdorff, J., Rios, D. Edelberg, H. (2001) 'Gait variability and fall risk in community-living older adults: a 1-year prospective study'. Arch Phys Med Rehabil, 82: 1050–6.
15. Hellbrandt, F.A., Braun, G.L. (1939) 'The influence of sex and age on the postural sway of man'. American J Physical Anthropology, XXIV(3): 347–360.
16. Judge, J.O., Davis, R.B., 3rd, Ounpuu, S. (1996) 'Step length reductions in advanced age: the role of ankle and hip kinetics'. J Gerontol, 51(6): M303–12.
17. Kemoun, G., et al. (2002) 'Ankle dorsiflexion delay can predict falls in the elderly'. J Rehabilitation Medicine, 34(6): 278–83.
18. Kernozek, T.W., LaMott, E.E. (1995) 'Comparisons of plantar pressures between the elderly and young adults'. Gait Posture, 3: 143–8.
19. Kerrigan, D.C., et al. (1998) 'Biomechanical gait alterations independent of speed in the healthy elderly: evidence for specific limiting impairments'. Arch Phys Med Rehabil, 79: 317–22.
20. Kerrigan, D.C., et al. (2001) 'Reduced hip extension during walking: Healthy elderly and fallers versus young adults. Arch Phys Med Rehabil, 82: 26–30.
21. Kristinsdottir, E.K., et al., (2001) 'Observation of vestibular asymmetry in a majority of patients over 50 years with fall-related wrist fractures'. Acta Otolaryngol, 121(4): 481–5.
22. Kristinsdottir, E.K., Jarnlo, G.B., Magnusson, M. (2000) 'Asymmetric vestibular function in the elderly might be a significant contributor to hip fractures'. Scand J Rehabil Med, 32(2): 56–60.
23. Lee, L.W., Kerrigan, D.C. (1999) 'Identification of kinetic differences between fallers and nonfallers in the elderly'. American J Physical Medicine Rehabilitation'. 78(3): 242–6.
24. Lockhart, T.E., Smith, J.L., Woldstad, J.C. (2005) 'Effects of aging on the biomechanics of slips and falls'. Hum Factors, 47(4): 708–29.
25. Lockhart, T.E., Woldstad, J.C., Smith, J.L. (2003) 'Effects of age–related gait changes on the biomechanics of slips and falls'. Ergonomics, 46(12): 1136–60.
26. Lord, S.R., et al., (1994) 'Physiological factors associated with falls in older community-dwelling women'. JAGS, 42: 1110–17.
27. Lord, S.R., et al.(1994) 'Postural stability, falls and fractures in the elderly: results from the Dubbo Osteoporosis Epidemiology Study'. Medical J Australia, 160(11): 684–5, 688–91.
28. Lord, S.R., et al. (1999) 'Lateral stability, sensorimotor function and falls in older people'. JAGS, 47(9): 1077–81.

29. Lord, S., Dayhew, J. (2001) 'Visual risk factors for falls in older people'. JAGS, 49(5): 508.
30. Lord, S., Fitzpatrick, R. (2001) 'Choice stepping reaction time: A composite measure of falls risk in older people'. J Gerontol, 56(10): M627.
31. Lord, S.R., Clark, R.D., Webster, I.W. (1991) 'Physiological factors associated with falls in an elderly population'. JAGS, 39(12): 1194–200.
32. Lord, S.R., Clark, R.D., Webster, I.W. (1991) 'Visual acuity and contrast sensitivity in relation to falls in an elderly population'. Age Ageing, 20(3): 175–81.
33. Lord, S.R., McLean, D., Stathers, G. (1992) 'Physiological factors associated with injurious falls in older people living in the community'. Gerontology, 38(6): 338–46.
34. Lord, S.R., Ward, J.A, Williams, P. (1996) 'Exercise effect on dynamic stability in older women: a randomized controlled trial'. Arch Phys Med Rehabil, 77(3): 232–6.
35. Lord, S.R., Ward, J.A. (1994) 'Age-associated differences in sensori-motor function and balance in community dwelling women'. Age Ageing, 23(6): 452–60.
36. Maki, B.E., Edmondstone, M.A., McIlroy, W.E. (2000) 'Age-related differences in laterally directed compensatory stepping behavior'. J Gerontol, 55(5): M270-7.
37. Maki, B.E., Holliday, P.J., Topper, A.K. (1994) 'A prospective study of postural balance and risk of falling in an ambulatory and independent elderly population'. J Gerontology, 49(2): M72–84.
38. McGibbon, C., Puniello, M. Krebs, D. (2001) 'Mechanical energy transfer during gait in relation to strength impairment and pathology in elderly women'. Clinical Biomechanics, 16: 324–33.
39. Menz, H.B., Morris, M.E., Lord, S.R. (2005) 'Foot and ankle characteristics associated with impaired balance and functional ability in older people'. J Gerontol A Biol Sci Med Sci, 60(12): 1546–52.
40. Mille, M.L., et al. (2003) 'Thresholds for inducing protective stepping responses to external perturbations of human standing'. J Neurophysiol, 90(2): 666–74.
41. Mills, P., Barrett, R. (2001) 'Swing phase mechanics of healthy young and elderly men'. Human Movement Science, 20: 427–46.
42. Murray, M.P., Drought, A.B., Kory, R.C. (1964) 'Walking patterns of normal men'. J Bone Joint Surg, 46A(2): 335.
43. Murray, M.P., Kory, R.C., Clarkson, B.H. (1969) 'Walking patterns in healthy old men'. J Gerontology, 24(2): 169–78.
44. Owings, T.M., Grabiner, M.D. (2004) 'Variability of step kinematics in young and older adults'. Gait Posture, 20(1): 26–9.
45. Pavol, M.J., et al. (1999) 'Gait characteristics as risk factors for falling from trips induced in older adults'. J Gerontol, 54(11): M583–90.
46. Pavol, M.J., et al. (2001) Mechanisms leading to a fall from an induced trip in healthy older adults. J Gerontol A Biol Sci Med Sci, 56(7): M428–37.
47. Piirtola, M., Era, P. (2006) 'Force platform measurements as predictors of falls among older people – a review'. Gerontology, 52(1): 1–16.
48. Pijnappels, M., Bobbert, M.F., van Dieen, J.H. (2004) 'Contribution of the support limb in control of angular momentum after tripping'. J Biomech, 37(12): 1811–18.
49. Pijnappels, M., Bobbert, M.F., van Dieen, J.H. (2005) 'How early reactions in the support limb contribute to balance recovery after tripping'. J Biomech, 38(3): 627–34.
50. Redfern, M., et al. (2001) 'Biomechanics of slips'. Ergonomics, 44: 138–1166.
51. Robbins, A.S., et al., (1989) 'Predictors of falls among elderly people. Results of two population-based studies'. Arch Intern Med, 149(7): 1628–33.
52. Rogers, M., et al. (2001) 'Lateral stability during forward-induced stepping for dynamic balance recovery in young and older adults'. J Gerontology, 56A: M589–594.
53. Stelmach, G.E., et al. (1989) 'Age, functional postural reflexes, voluntary sway'. J Gerontology, 44(4): B100–6.
54. Teno, J., Kiel, D.P., Mor, V. (1990) 'Multiple stumbles: a risk factor for falls in community-dwelling elderly. A prospective study'. JAGS, 38(12): 1321–5.

55. Thelen, D.G., et al., (1996) 'Effects of age on rapid ankle torque development'. J Gerontol A Biol Sci Med Sci, 51(5): M226–32.
56. Thelen, D.G., et al. (1997) 'Age differences in using a rapid step to regain balance during a forward fall'. J Gerontol, 52(1): M8–13.
57. van Dieen, J.H., Pijnappels, M., Bobbert, M.F. (2005) 'Age-related intrinsic limitations in preventing a trip and regaining balance after a trip'. Safety Science, 43: 437–53.
58. Whipple, R.H., Wolfson, L.I., Amerman, P.M. (1987) 'The relationship of knee and ankle weakness to falls in nursing home residents: an isokinetic study'. JAGS, 35(1): 13–20.
59. Winter, D.A. (1991) 'The biomechanics and motor control of human gait: normal, elderly and pathological'. Waterloo, Ontario: University of Waterloo Press.
60. Winter, D.A., et al. (1990) 'Biomechanical walking pattern changes in the fit and healthy elderly'. Physical Therapy, 70(6): 340–7.

36

Postural control in Parkinson's disease

Stephan Turbanski
Johann Wolfgang Goethe-University of Frankfurt, Frankfurt

Introduction

At the beginning of the nineteenth century James Parkinson, an English neurologist, was the first to describe the disease that bears his name. Since he made first references to falls in his famous essay – Parkinson (1817) '*An essay on the shaking palsy*' – it is well known that postural instability is a hallmark of Parkinson's disease (PD). The loss of postural control leads to impairment of gait and a higher risk of falling. Therefore, postural instability is a major clinical problem.

Due to shortcomings of clinical evaluation and medical therapy, postural control in PD is of growing importance for biomechanics. On one hand it has been shown that clinical tests are inaccurate and do not identify possible factors influencing balance control in PD efficiently (Marchese *et al.* 2003) on the other hand it is proven that medication fails to affect balance problems in PD. Therefore, biomechanical tests for analyzing postural deficits in parkinsonian subjects need to be developed and assessed. Moreover, evaluation of physical exercises improving postural control is in order to optimize therapy.

In numerous studies it was shown that postural stability, balance correcting responses, and proprioception are impaired in parkinsonian subjects. Several alterations in postural control have been documented. However, there is still a controversy about the main causes of postural instability in PD. Nevertheless, increasing evidence suggests that exercise and physical therapy can improve postural control in PD significantly (Reuter 1999, Ellis *et al.* 2005, Pellecchia *et al.* 2004, Morris 2000). Further approaches to develop therapeutic strategies should be given a priority (Bloem *et al.* 2001a).

In this chapter these aspects will be presented and discussed in a brief overview.

Parkinson's disease

Parkinson's disease is a progressive neurodegenerative disorder of unknown cause. The majority of patients suffer from so-called 'idiopathic' PD. In the remaining patients the causes of the disease are known (drugs, metabolic disturbances, or infections). Lang and

Lozano (1998, 1044) pointed out that age is the single most consistent risk factor'. More than 90 per cent of the patients are over 60-years-old (Guttman *et al.* 2003). In fact, in elderly people the prevalence is about 1–3 per cent, and it is estimated that over one million people in the US are affected (Lang and Lozano 1998).

The hallmark of PD is the progressive loss of selected populations of neurons, especially the dopaminergic neurons of the *substantia nigra pars compacta* in basal ganglia. This pattern of neurodegeneration is specific to PD and differs from that of normal ageing and other diseases. Assumingly, at least 60 per cent of these neurons tend to be lost at the onset of clinical symptoms, which is problematic for potential therapy. Some reviews have emphasized the influences of genetic factors to cause the degeneration of dopaminergic neurons (Gwinn-Hardy 2002). However, in most cases an interaction of external and genetic factors is believed to result in PD (Guttman *et al.* 2003).

The diagnosis is based primarily on clinical criteria concerning present symptoms, but misdiagnosis is a very common problem because other neurological diseases show the same or similar symptoms. In retrospective studies misdiagnoses are reported in up to 35 per cent of cases (Jankovic *et al.* 2000) and Tanner *et al.* (1999) postulated that 10 per cent of parkinsonian subjects are not diagnosed as patients.

The classic triad of cardinal motor manifestations is made up of tremor, rigidity and akinesia. In modern neurology postural instability has been attributed to be the fourth cardinal symptom. The most known and obvious motor symptom in PD is tremor. Tremor is a rhythmic, involuntary movement that involves in most patients the upper and/or lower extremities. It predominantly occurs at rest and it begins unilateral in one extremity. Akinesia, also called bradykinesia, describes a group of difficulties in motor control like slowness of movements, disturbed skill in coordinated movements with fingers (e.g. micrographic of writing), reduced arm swing while walking, and freezing of gait. Rigidity, the third cardinal symptom, is an important cause of akinesia. Parkinsonian rigidity is present predominantly in the upper and lower extremities and in the neck. A characteristic feature in PD is a cogwheel type of rigidity on passive movements. Furthermore, a flexed posture and a shuffling gait are obvious in most patients with PD. Best diagnostic criteria are a combination of asymmetry of clinical symptoms, the presence of at least two cardinal symptoms, and a good response to Levodopa medication.

Levodopa still remains the most effective drug treatment for parkinsonian subjects but it may only be effective for a few years. Moreover, Levodopatherapy is associated with some motor complications in later stages of the disease. Drugs like Anticholinergics and Dopamine agonists are used in modern therapy as well. Although significant advances have been made in the last decades there are still no curative treatments and none have been proven to slow the progression of the disease.

Postural control in Parkinson's disease

Evaluation of postural control

To get information about the individual risk of falling, assessment of postural control is an important feature. But prediction of falls is difficult because of the multifactorial character of falls and postural disturbances (Bloem *et al.* 2001b, Bloem *et al.* 2006). There are two ways of evaluating postural stability, either by qualitative clinical testing or by quantitative biomechanical analyses.

Qualitative evaluation of postural control is widely used in routine clinical diagnosis by observing patients standing upright, rising from a chair, performing a reaching task, or responding to a push- or pulltest. Some validated balance tests can easily be performed in clinical examination like the *one-leg stance* test, the *Tandem Romberg stance* and the *functional reach test* (see in detail Smithson *et al.* 1998 and Morris *et al.* 1998). Newstead *et al.* (2005) demonstrated an excellent reliability for the *Berg balance scale* and the *Tinetti Balance Scale*. These two clinical examinations consist of 13 or 14 tasks respectively (Berg *et al.* 1992, Tinetti 1986). But in most cases a so-called *retropulsiontest* is used to identify postural deficits in elderly people, especially in PD. This test consists of a sudden pull at the shoulder performed by an examiner standing behind the subject. The examiner evaluates the ability to recover balance based on a rating score[1]. Although reliability of the *retropulsiontest* seems good (Martinez-Martin *et al.* 1994, Visser *et al.* 2003) the value of this test is limited by the lack of normative data, the lack of analysing postural control in medial-lateral direction and difficulties in standardisation across different subjects (Bloem *et al.* 1998, Marchese *et al.* 2003). Furthermore, Jacobs *et al.* (2006) reported that the *retropulsiontest* is often normal in patients with balance problems. And it fails to predict falls in PD patients (Bloem *et al.* 2001a).

Naturally, biomechanical evaluation of postural control leads to quantitative, more objective data. These tests not only assess balance during quiet standing but also the responses to perturbation stimuli in different conditions. In many studies posturography is used to evaluate postural control. In this test subjects are positioned on a force platform. The ability of controlling balance is commonly calculated by motions of COP. An increased sway of COP is considered to represent deficits in postural control: 'alterations of the postural control system are reflected in changes of COP characteristics' (Rocchi *et al.* 2006, 140). Yet, there are some shortcomings of posturography analyzing postural control in static conditions. In fact, several studies using posturography found no difference between parkinsonian patients and control subjects or even less sway in PD subjects (see Table 36.1).

As the postural threat is low in static conditions, we used an instable platform hanging on four springs (Coordex®) in our studies (Turbanski 2006, Turbanski *et al.* 2005, 2007). Subjects were tested in standardized conditions on their ability to maintain postural stability on this movable support surface. The platform displacements were measured by a 2D acceleration sensor which was attached to the platform. All trajectories in both directions (anterior-posterior and medial-lateral) were summed up to get an objective value of body sway reflecting postural stability.

The purpose of one study was to investigate the relationship between the *retropulsiontest* test and several biomechanical analyses of postural control in PD patients. In contrast to Adkin *et al.* (2003) and Horak *et al.* (2005) we found no correlations (range of results was

Table 36.1 Comparison of postural stability in posturography between PD patients and control subjects.

No difference in postural sway	*Less postural sway in PD patients*
Bronstein *et al.* 1990	Horak *et al.* 1992
Nardone 1991	Dietz *et al.* 1988
Waterson *et al.* 1993	Schieppati *et al.* 1994
Trenkwalder 1995	
Collins *et al.* 1995	
Marchese *et al.* 2003	

between r=–0.03 and r=0.02) between clinical examination and several quantitative data concerning postural control (Turbanski 2006, Turbanski *et al.* 2007). This result confirms a poor relationship between the clinical evaluation and the outcome of biomechanical analysis as it was reported by other studies, too (Marchese *et al.* 2003, Bloem *et al.* 1998). It is obvious that the clinical *retropulsiontest* and biomechanical analyses examine different aspects of postural control in PD.

Allum *et al.* (2002, 644) postulated that 'postural control is best probed using controlled perturbations of upright stance'. Standardized perturbations of postural stability consist of a sudden acceleration of support surface, whereas translations (pitch plane) and rotational stimuli (roll component) are used. Balance correcting responses are commonly examined by electromyography (EMG) recordings and support surface reaction forces (e.g. Dimitrova *et al.* 2004, Allum *et al.* 2002, Carpenter *et al.* 2004). It seems particularly important to evaluate multidirectional perturbations in analyzing postural abnormalities. Jacobs *et al.* (2006) demonstrated that lateral perturbations best differentiated PD patients from control subjects, for example. Furthermore, responses to lateral perturbations appears to correlate better with the risk of falling (Lord *et al.* 1999) because they require a more complex coordination (Carpenter *et al.* 1999, Allum *et al.* 2002).

Postural instability in Parkinson's disease

As mentioned above postural instability has been attributed to be a cardinal symptom in PD. Although it usually occurs in the advanced stage of the disease, falls have also been reported in the relatively early course of PD (Bloem *et al.* 2001a). Schrag *et al.* (2000) have demonstrated that postural instability has one of the greatest influences on patient's quality of life. Furthermore, mortality is increased after falls (Bennett *et al.* 1996). Therefore, Jankovic *et al.* (1990) speculated about a worse overall prognosis for patients with PD affected by a marked loss of postural control. Wood *et al.* (2002) reported that the risk of falling is approximately twice in parkinsonian subjects when compared to that of healthy older people. In other studies it is mentioned that PD patients fall at five times the rate of control subjects (Dimitrova *et al.* 2003, Bloem *et al.* 1998).

However, many factors have been identified to be associated with impaired postural control in PD. But 'the mechanisms of postural instability in PD are still uncertain and probably complex with involvement of different neural structures' (Marchese *et al.* 2003, 652). Moreover, the relationship between postural instability and falling is complex (Visser *et al.* 2003).

The following hypotheses are thought to contribute to postural instability in PD: 'Multiple studies have shown that postural reflexes in PD patients differ from those of normal subjects' (Carpenter *et al.* 2004, Horak *et al.* 1992 and 1996, Dietz *et al.* 1993). The balance corrective responses are often abnormally sized, inadequate and inflexible. But latencies of muscle activation seem to be unaffected (Horak *et al.* 1996). Particularly, the overproduction of corrective torque has been reported to be an underappreciated cause of postural instability (Peterka and Loughlin 2004). It is known that stretch reflexes are pathologically enhanced in PD patients. In particular, the later components of stretch reflexes are exaggerated (Cody *et al.* 1986). It was found that PD patients have increased postural reflexes in either distal lower leg muscles as well as in proximal muscles spanning the hip and the trunk (Carpenter *et al.* 2004).

The term postural inflexibility points to the fact that parkinsonian subjects have difficulties modifying postural responses for changing demands. They lack the ability to shape muscle activations adequately in order to maintain balance when conditions change

(Dimitrova *et al.* 2004). These stereotype reactions lead to directionally specific postural instability (Horak *et al.* 2005), specifically they do not modify their reflexes when changing the stand width (Dimitrova *et al.* 2004). Besides, impairment of protective arm movements during the loss of balance is described (Morris 2000, Grimbergen *et al.* 2004). These phenomena can be associated with the high incidence of wrist fractures in PD (Carpenter *et al.* 2004).

In addition, changes in muscle stiffness in PD have to be taken into account to explain postural instability (Dietz *et al.* 1988). In contrast to age-matched controls a larger co-activation was demonstrated in PD patients (Dimitrova *et al.* 2004). For example, PD patients co-activated antagonist muscle groups at the hip and at the trunk leading to trunk stiffness and axial rigidity (Jacobs *et al.* 2006, Horak *et al.* 2005, Van Emmerik *et al.* 1999). Lateral stability is primarily compromised in PD due to enhanced stiffness in trunk and hip muscles (Dimitrova *et al.* 2004, Horak *et al.* 2005). Interestingly, PD patients can show a reduced sway in static posturography because of rigidity (see above). This is to be considered when evaluating postural control with posturography.

A special feature in Parkinsonism is the fact that a simultaneous execution of a concurrent task induced a significant worsening of postural instability (Morris *et al.* 2000, Marchese *et al.* 2003). It has been demonstrated that a greater postural sway is present in parkinsonian subjects while completing a secondary cognitive task[2] (Ashburn *et al.* 2001) and it is reported that falls often occur in PD while patients attempt to perform multiple tasks at the same time (Grimbergen *et al.* 2004). A study of Marchese *et al.* (2003) proved that secondary tasks deteriorate postural control (considered the primary task) in PD not only during a cognitive task, but also during a motor task. It is hypothesized that the patients use a so-called '*posture second*' strategy (Bloem *et al.* 2006).

However, several studies have suggested that the basal ganglia are the main critical neural structures for postural impairments in PD. Firstly, the dopaminergic neurons of the *substantia nigra pars compacta* in basal ganglia are most affected by progressive cell death (see above), secondly, the basal ganglia modify muscle activation patterns quickly and control the agonist-antagonist relationship which is necessary to maintain balance. Moreover, these structures are important for adaptation of postural control by optimizing muscle activations during different perturbation stimuli and change set quickly (Chong *et al.* 2000, Horak *et al.* 2005, Dimitrova *et al.* 2004). Moreover, they are crucial for the integration of sensory feedback (Brown *et al.* 2006). In general, basal ganglia dysfunction causes impairment in motor control as well as postural inflexibility (Horak *et al.* 2005).

Furthermore, postural control is affected by the severity of PD symptoms (Gray and Hildebrand 2000) like rigor, freezing of gait (Grimbergen *et al.* 2004) and bradykinesia. Horak *et al.* (2005, 518) term the slow, weak postural responses to external perturbation '*postural bradykinesia*'.

In conclusion, several abnormalities in postural control have been documented in PD, but the principle reasons are uncertain and epidemiology of postural instability remains largely unknown (Bloem *et al.* 2001a).

Proprioception

Postural instability in PD may relate in part to impaired proprioception and kinaesthesia. An increased body sway is associated with poor tactile sensitivity and poor sense of joint position in PD (Lord *et al.* 1991). Several studies have proven an impaired sensitivity for proprioceptive information in parkinsonian subjects. One experiment revealed that parkinsonian subjects made overshooting errors with the more affected limb when attempting simultaneous and equivalent movements of both arms (Moore 1987). Demirci *et al.* (1997) demonstrated that PD patients underestimate the amplitude of passive angular displacements of the fingers more than control subjects. Moreover, it was shown that PD patients performed significantly worse in a test of kinaesthesia when they were denied visual guidance (Jobst *et al.* 1997). In another study it was demonstrated that PD patients undershoot targets in pointing movements when vision is occluded but not when complete visual feedback is provided (Klockgether *et al.* 1995). A recent study proved increased postural instability in PD when visual feedback is limited as well (Jacobs and Horak 2006).

These results are resembled by O'Suilleabhain *et al.* (2001) who observed a tendency to decreased accuracy in patients with PD when assessing proprioception in various tests.

It is proposed that a 'disturbance of proprioceptive guidance may be a generalized feature of Parkinson's disease' (Khudados *et al.* 1999, 508).

The main physiological causes of impaired proprioception and its contribution to postural control in PD are still debated. Demirci and colleagues (1997) proposed that the proprioceptive feedback may be present but patients are unable to use it properly for maintaining balance. Recently, Jacobs and Horak (2006) have proven inefficient proprioceptive-motor integration in PD. An abnormal sensory organization like a 'breakdown in the central hierarchy of postural control' is also discussed in advanced PD (Bronte-Stewart *et al.* 2002, 2107). Rickards and Cody (1997, 977) suggested the 'existence of an abnormality of higher-level proprioceptive integration in Parkinson's disease in which there is a mismatch of sensory (proprioceptive) and motor (corollary discharge) information'.

But it should be taken into consideration that most studies analyzing proprioception in PD measured slow voluntary movements only rather than ballistic or reflexive responses to perturbation of postural control.

Treatment

Levodopa is still the most effective drug treatment to reduce clinical symptoms in parkinsonian subjects. But, despite improvements in most symptoms, many PD patients have reported an impairment of postural control when on medication (see Bronte-Stewart *et al.* 2002). This observation is supported by growing evidence that levodopa, like all other antiparkinsonian drug treatment, fails to affect postural instability (Klawans 1986, Koller *et al.* 1990, Marsden and Obeso 1994, Bloem *et al.* 1996, Jankovic 2002, Guttman *et al.* 2003, Robert-Warrior *et al.* 2000, Frank *et al.* 2000). Other groups even associated levodopa with a worsening of postural stability (Bronte-Stewart *et al.* 2002, Rocchi *et al.* 2002). One study found that levodopa treatment particularly leads to increased difficulties in controlling lateral postural sway (Mitchell *et al.* 1995). In addition, some investigations suggested a worsening of proprioception and kinaesthesia in PD caused by the drug treatment (Moore 1987, Klockgether *et al.* 1995, Zia *et al.* 2000, O'Suilleabhain *et al.* 2001).

In conclusion, administration of levodopa and dopamine agonists seems to be associated with a decreased proprioceptive processing and increased postural instability, leading to a higher risk of falling – 'two-thirds of falls occurred when patients considered their symptoms to be well controlled' (Bloem *et al.* 1999, 955).

As disturbance of postural control is a severe problem in PD, particularly in advanced stages, that can not be sufficiently treated by medication, 'development of improved therapeutic strategies [...] is needed' (Bloem *et al.* 2001a, 956). It has been documented in numerous investigations that physical activities counteract postural instability in elderly people significantly (e.g. Perrin *et al.* 1999, Tinetti 2003, Tinetti *et al.* 2006, Gardner *et al.* 2000, Gardner *et al.* 2001). In addition, evidence is accumulating that PD patients can profit from physical exercises as well. Reuter *et al.* (1999) investigated the effects of an intensive exercise training. The main results clearly indicated significant changes in motor disability. Interestingly, dyskinesias also seemed to be under better control. More recently, it was proven that a combination of balance training and high-intensity resistance training improves postural control in PD to a greater extent than balance training only (Hirsch *et al.* 2003).

Other studies reported short-term and long-term benefits after a physical exercise programme (Ellis *et al.* 2005, Comella *et al.* 1994, Pellecchia *et al.* 2004). Morris has outlined a physical therapy model for PD and he proposed that 'falls prevention is a major goal in of physical therapy' (Morris 2000, 587). This assumption is confirmed by observations that physical therapy resulted in significant improvement of postural control in PD (Stankovic 2004). Furthermore, it was shown that a special training programme to improve compensatory steps is a therapeutic approach for postural instability. After the training session the length of compensatory steps increased and the step initiation shortened (Jöbges *et al.* 2004).

In two studies we investigated spontaneous effects of random whole-body vibration[3] on postural control in parkinsonian subjects. In the first study (Turbanski *et al.* 2005) we assessed the influence on the ability of maintaining postural stability. Balance was tested pre- and post-treatment in two standardized conditions (*narrow standing* and *tandem standing*). We found a significant reduction of postural sway in post-tests. But in condition *narrow standing* the result of group differences failed the level of significance.

In the second study (Turbanski *et al.* 2006) we used the identical design, e.g. regarding the treatment. But in pre- and post-tests we evaluated balance-correcting responses following postural perturbations consisting of standardized accelerations of the movable support

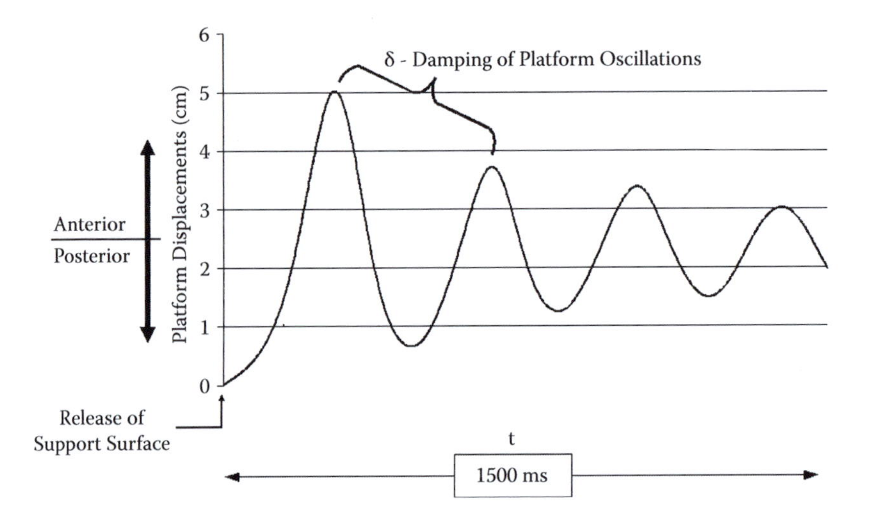

Figure 36.1 Evaluation of initial damping of platform oscillations using damping coefficient **δ**. This kinematic parameter represents the efficiency of balance correcting response following a standardized acceleration of support surface (Turbanski 2006).

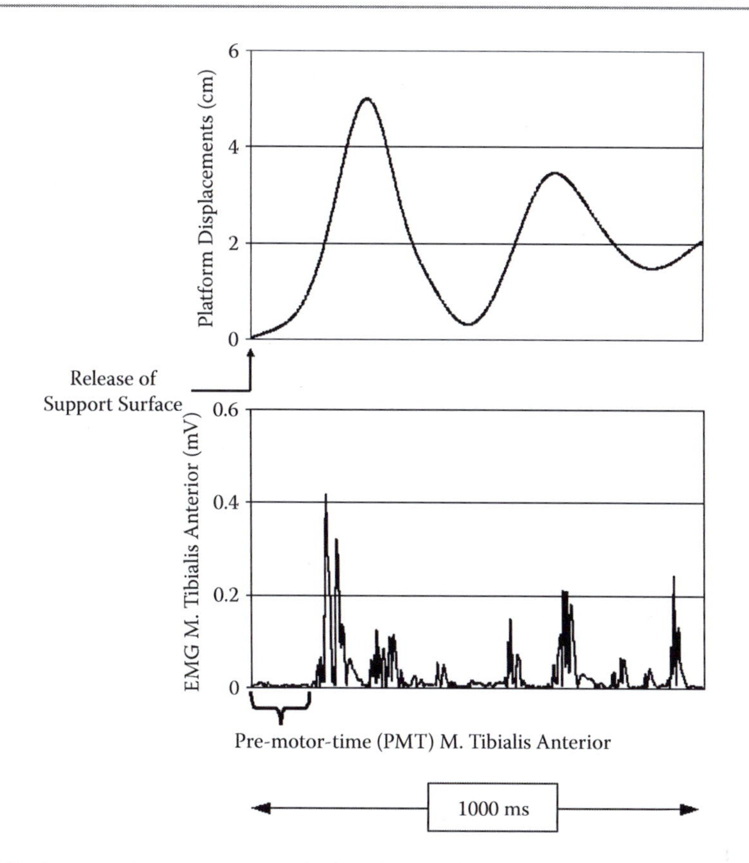

Figure 36.2 Evaluation of pre-motor-time which is the time between the perturbation stimulus and the onset of EMG-activity (Turbanski 2006).

surface in direction anterior. We analyzed the initial damping of platform oscillations representing the balance correcting response (see Figure 36.1).

In experimental group we found a significant improvement of kinematic balance correcting responses represented by a more efficient initial damping of platform oscillations. Besides analyses of platform displacements, EMG measurements were recorded from the M. Tibialis anterior. We assessed pre-motor time which is the time between the perturbation stimulus and the onset of EMG-activity (see Figure 36.2).

But in electromyographic pre-motor time we found no significant change. That indicates that postural reflexes were not generated earlier but more efficiently.

In conclusion, it is evident that patients suffering from PD profit from physical exercise and physical therapy, in particular with respect to their postural control.

Notes

1 Commonly the retropulsiontest is part of the Unified Parkinson's Disease Rating Score (UPDRS).
2 Cognitive tasks are, for example, reciting the days of the week backwards, counting backwards, etc.

3 We used a SRT® med System (amplitude: 3 mm, average frequency f: 6 Hz) and the subjects performed five series lasting 60 seconds each.

References

1. Adkin, A.L., Frank, J.S., Jog, M.S. (2003) 'Fear of falling and postural control in Parkinson's disease'. *Mov Disord*, 18: 496–502.
2. Allum, J.H., Carpenter, M.G., Honegger, F., Adkin, A.L., Bloem, B.R. (2002) 'Age-dependent variations in the directional sensitivity of balance corrections and compensatory arm movements in man'. *J Physiol*, 542: 643–663.
3. Ashburn, A., Stack, E., Pickering, R.M., Ward, C.D. (2001) 'A community-dwelling sample of people with Parkinson's disease: characteristics of fallers and non-fallers'. *Age and Aging*, 30: 47–52.
4. Bennett, D.A., Beckett, L.A., Murray, A.M., Shannon, M.D., Goetz, C.G., Pilgrim, D.M., Evans, D.A. (1996) 'Prevalence of Parkinsonian Signs and Associated Mortality in a Community Population of Older People'. *N Engl J Med*, 334: 71–76.
5. Berg, K.O., Wood-Dauphinee, S.L., Williams, J.L., Maki, B. (1992) 'Measuring balance in the elderly: validation of an instrument'. *Can J Public Health*, 83: Suppl 2, S7–S11.
6. Bloem, B.R., Beckley, D.J., van Dijk, J.G. (1999) 'Are automatic postural responses in patients with Parkinson`s disease normal to their stooped posture?' *Exp Brain Res*, 124: 481–488.
7. Bloem, B.R., Beckley, D.J., Van Dijk, J.G., Zwinderman, A.H., Remler, M.P., Roos, R.A. (1996) 'Influence of dopaminergic medication on automatic postural responses and balance impairment in Parkinson's disease'. *Mov Disord*, 11: 509–521.
8. Bloem, B.R., Beckley, D.J., Van Hilten, B.J., Roos, R.A. (1998) 'Clinimetrics of postural instability in Parkinson's disease'. *J Neurol*, 245: 669–673.
9. Bloem, B.R., Grimbergen, Y.A.M., Cramer, M., Willemsen, M., Zwinderman, A.H. (2001a) 'Prospective assessment of falls in Parkinson's disease'. *J Neurol*, 248: 950–958.
10. Bloem, B.R., Grimbergen, Y.A.M., Van Dijk, J.G., Muneke, M. (2006) 'The "posture second" strategy: A review of wrong priorities in Parkinson's disease'. *J Neurol Sci*, 248: 196–204.
11. Bloem, B.R., Valkenburg, V.V., Slabbekoorn, M., Willemsen, M.D. (2001b) 'The Multiple Task Test – Development and normal strategies'. *Gait and Posture*, 14: 191–202.
12. Bronstein, A.M., Hood, J.D., Gresty, M.A., Panagi, C. (1990) 'Visual control of balance in cerebellar and parkinsonian syndromes'. *Brain*, 113: 767–779.
13. Bronte-Stewart, H.M., Minn, A.Y., Rodrigues, K., Buckley, E.L., Nashner, L.M. (2002) 'Postural instability in idiopathic Parkinson's disease: the role of medication and unilateral pallidotomy'. *Brain*, 9: 2100–2112.
14. Brown, P., Chen, C.C., Wang, S., Kuhn, A.A., Doyle, L., Yarrow, K., Nuttin, B., Stein, J., Aziz, T., (2006) 'Involvement of human basal ganglia in offline feedback control of voluntary movement'. *Curr Biol*, 16: 2129–2134.
15. Carpenter, M.G., Allum, JHJ., Honegger, F. (1999) 'Directional sensitivity of stretch reflexes and balance corrections for normal subjects in the roll and pitch planes'. *Exp Brain Res*, 129: 93–113.
16. Carpenter, M.G., Allum, J.H.J., Honegger, F., Adkin, A.L., Bloem, B.R. (2004) 'Postural abnormalities to multidirectional stance perturbations in Parkinson's disease'. *J Neurol Neurosurg Psychiatry*, 75: 1245–1254.
17. Chong, R.K.Y., Jones, C.L., Horak, F.B. (2000) 'Parkinson's disease impairs the ability to change set quickly'. *J Neurol Sci*, 175: 57–70.
18. Cody, F.W., MacDermott, N., Matthews, P.B., Richardson, H.C. (1986) 'Observations on the genesis of stretch reflexes in Parkinson's disease'. *Brain*, 109: 229–249.
19. Cordo, P., Gurfinkel, V.S., Bevan, L., Kerr, G.K. (1995) 'Proprioceptive consequences of tendon vibration during movement'. *J Neurophysiol*, 74: 1675–1688.

20. Collins, J.J., De Luca, C.J. (1995) 'The effects of visual input on open-loop and closed-loop postural control mechnanisms'. *Exp Brain Res*, 103: 151–163.
21. Comella, C.L., Stebbins, G.T., Brown-Toms, N., Goetz, C.G. (1994) 'Physical therapy and Parkinson's disease: a controlled clinical trial'. *Neurology*, 44: 376–378.
22. Demirci, M., Grill, S., McShane, L., Hallett, M. (1997) 'A mismatch between kinesthetic and visual perception in Parkinson's disease'. *Ann Neurol.*, 41: 781–788.
23. Dietz, V., Berger, W., Hostmann, G.A. (1988) 'Posture in Parkinson's disease: impairment of reflexes and programming'. *Ann Neurol*, 24: 660–669.
24. Dietz, V., Zijlstra, W., Assaiante, C., Trippel, M., Berger, W. (1993) 'Balance control in Parkinson's disease'. *Gait and Posture*, 1: 77–84.
25. Dimitrova, D., Horak, F.B., Nutt, J.G. (2004) 'Postural Muscle Response to Multidirectional Translations in Patients With Parkinson's Diease'. *J Neurophysiol*, 91: 489–501.
26. Ellis, T., de Goede, C.J., Feldman, R.G., Wolters, E.C., Kwakkel, G., Wagenaar, R.C. (2005) 'Efficacy of a physical therapy program in patients with Parkinson's disease: a randomized controlled trial'. *Arch Phys Med Rehabil*, 86: 626–632.
27. Emry, C.A. (2003) 'Is there a clinical standing balance measurement appropriate for use in sports medicine? A review of the literature'. *J Sci Med Sport*, 6: 492–504.
28. Frank, J.S., Horak, F.B., Nutt, J. (2000) 'Centrally Initiated Postural Adjustments in Parkinsonian Patients On and Off Levodopa'. *J Neurophysiol*, 84: 2440–2448.
29. Gardner, M.M., Robertson, M.C., Campbell, A.J. (2000) 'Exercise in preventing falls and fall related injuries in older people: a review of randomised controlled trials'. *Br J Sports Med*, 34: 7–17.
30. Gray, P., Hildebrand, K. 'Fall risk factors in Parkinson's disease'. *J Neurosci Nurs*, 32: 222–228.
31. Grimbergen, Y., Munneke, M., Bloem, B.R. (2004) 'Falls in Parkinson's disease'. *Current Opinion in Neurology*, 17: 405–415.
32. Guttman, M., Kish, S.J., Furukawa, Y. (2003) 'Current concepts in the diagnosis and management of Parkinson's disease'. *CMAJ*, 168: 293–301.
33. Gwinn-Hardy, K. (2002) 'Genetics of Parkinsonism'. *Mov Disord*, 17: 645–656.
34. Hirsch, M.A., Toole, T., Maitland, C.G., Rider, R.A. 'The effects of balance training and high-intensity resistance training on persons with idiopathic Parkinson's disease'. *Arch Phys Med Rehabil*, 84: 1109–1117.
35. Horak, F.B., Dimitrova, D., Nutt, J.G. (2005) 'Direction-specific postural instability in subjects with Parkinson's disease'. *Exp Neurol.*, 193: 504–521.
36. Horak, F.B., Frank, J., Nutt, J. (1996) 'Effects of dopamine on postural control in parkinsonian subjects: scaling, set and tone'. *J Neurophysiol* 75: 2380–2396.
37. Horak, F.B., Nutt, J.G., Nashner, L.M. (1992) 'Postural inflexibility in parkinsonian subjects'. *J Neurol Sci*, 111: 46–58.
38. Jacobs, J.V., Horak, F.B. (2006) 'Abnormal proprioceptive-motor integration contributes to hypometric postural responses of subjects with Parkinson's disease'. *Neuroscience*, 141: 999–1009.
39. Jacobs, J.V., Horak, F.B., Tran, V.K., Nutt, J.G. (2006) 'An alternative clinical postural stability test for patients with Parkinson's disease'. *J Neurol Neurosurg Psychiatry*. 77: 322–326.
40. Jankovic, J. (2002) 'Levodopa – strengths and weaknesses'. *Neurology*, 58(Supplement 1): 19–32.
41. Jankovic, J., McDermott, M., Carter, J., Gauthier, S., Goetz, C., Golbe, L., Huber, S., Koller, W., Olanow, C., Shoulson, I. (1990) 'Variable expression of Parkinson's disease: a base-line analysis of the DATATOP cohort. The Parkinson Study Group'. *Neurology*, 40: 1529–1534.
42. Jöbges, M., Heuschkel, G., Pretzel, C., Illhardt, C., Renner, C., Hummelsheim, H. (2004) 'Repetitive training of compensatory Stepps: a therapeutic approach for postural instability in Parkinson's disease'. *J Neurol Neurosurg Psychiatry*, 75: 1682–1687.
43. Jöbges, M., Heuschkel, G., Pretzel, C., Illhardt, C., Renner, C., Hummelsheim, H. (2004) 'Repetitive training of compensatory steps: a therapeutic approach for postural instability in Parkinson's disease'. *J Neurol Neurosurg Psychiatry*, 75: 1682–1687.
44. Jobst, E.E., Melnick, M.E., Byl, N.N., Dowling, G.A., Aminoff, M.J. (1997) 'Sensory perception in Parkinson's disease'. *Arch Neurol*, 54: 450–454.

45. Khudados, E., Cody, FWJ., O'Boyle, D.J. (1999) 'Proprioceptive regulation of voluntray ankle movements, demonstrated using muscle vibration is impaired by Parkinson's disease'. *J Neurol Neurosurg Psychiatry*, 67: 504–510.
46. Klawans, H.L. (1986) 'Individual manifestations of Parkinson's disease after ten or more years of levodopa'. *Mov Disord* 1: 187–192
47. Klockgether, T., Borutta, M., Rapp, H., Spieker, S., Dichgans, J. (1995) 'A defect of kinesthesia in Parkinson's disease'. *Mov Disord* 10: 460–465.
48. Koller, W., Olanow, C., Shoulson, I. (1990) 'Variable expression of Parkinson`s disease. A baseline expression of the DATA-TOP study cohort'. *Neurol*, 40: 1529–1534.
49. Lang, A.E., Lozano, A.M. (1998) 'Parkinson's disease'. *N Engl J Med*, 339: 1044–1053 and 1130–1143.
50. Lord, S.R., Clark, R.D., Webster, I.W. (1991) 'Postural stability and associated physiological factors in a population of aged persons'. *J Gerontol*, 46: 69–76.
51. Lord, S.R., Rogers, M.W., Howland, A., Fitzpatrick, R. (1999) 'Lateral stability, sensorimotor function and falls in older people'. *J Am Geriatric Soc*, 47: 1077–1081.
52. Marchese, R., Bove, M., Abbruzzese, G. (2003) 'Effect of Cognitive and Motor Tasks on Postural Stability in Parkinson's disease: A Posturographic Study'. *Mov Disorder*, 18: 652–658.
53. Marsden, C.D., Obeso, J.A. (1994) 'The functions of the basal ganglia and the paradox of stereotaxic surgery in Parkinson's disease'. *Brain*, 117: 877–897
54. Mitchell, S.L., Collins, J.J., De Luca, C.J., Burrows, A., Lipsitz, L.A. (1995) 'Open-loop and closed-loop postural control mechanisms in Parkinson's disease: increased mediolateral activity during quiet standing'. *Neurosci Lett*, 197: 133–136.
55. Moore, A.P. (1987) 'Impaired sensorimotor integration in parkinsonism and dyskinesia: a role of corollary discharges'. *J Neurol Neurosurg Psychiatry*, 50: 544–552.
56. Morris, M.E., Iansek, R., Smithson, F., Huxham, F. (2000) 'Postural instabiltiy in Parkinson's disease: a comparison with and without a concurrent task'. *Gait and Posture*, 12: 205–216.
57. Morris, M.E. (2000) 'Movement Disorders in People with Parkinson Disease: A Model for Physical Therapy'. *Physical Therapy*, 80: 578–597.
58. Morris, S., Morris, M.E., Iansek R. (1998) 'Performance on clinical tests of balance in Parkinon`s disease'. *Phys Ther*, 81: 810–818.
59. Nardone, A., Siliotto, R., Grasso, M., Schieppati, M. (1995) 'Influence of aging on leg muscle reflex response to stance perturbations'. *Archives of Physical Medicine and Rehabilition*, 76: 158–165.
60. Newstead, A.H., Hinman, M.R., Tomberlini, J.A. (2005) 'Reliability of the Berg Balance Scale and Master Limits of Stability Test for individuals with brain injury'. *Journal of Neurologic Physical Therapy*, 29: 18–23.
61. O`Suilleablain, P., Bullard, J., Dewey, R.B. (2001) 'Proprioceptiption in Parkinson`s disease is actuely depressed by dopaminergic medications'. *J Neurol Neurosurg Psychiatry*, 71: 607–610.
62. Parkinson, J. (2002) 'An essay on the shaking palsy' 1817. *J Neuropsychiatry Clin Neurosci*, 14: 223–236.
63. Pellecchia, M.T., Grasso, A., Biancardi, L.G., Squillante, M., Bonavita, V., Barone, P. (2004) 'Physical therapy in Parkinson's disease: an open long-term rehabilitation trial'. *J Neurol*, 251: 595–598.
64. Perrin, P.P., Gauchard, G.C., Perrot, C., Jeandel, C. (1999) 'Effects of physical and sporting activities on balance control in elderly people'. *Br J Sports Med*, 33: 121–126.
65. Peterka, R.J., Loughlin, P.J. (2004) 'Dynamic regulation of sensorimotor integration in human postural control'. *J Neurophysiol*, 91: 410–423.
66. Reuter, I., Engelhardt, M., Stecker, K., Baas, H. (1999) 'Therapeutic value of exercise training in Parkinson's disease'. *Med Sci Sports Exerc*, 11: 1544–1549.
67. Rickards, C., Cody, FWJ. (1997) 'Proprioceptive control of wrist movements in Parkinson's disease'. *Brain*, 120: 977–990.
68. Robert-Warrior, D., Overby, A., Jankovic, J., Olsoon, S., Lai E.C., Krauss, J.K., Grossman, R. (2000) 'Postural control in Parkinson`s disease after unilateral posteroventral pallidotomy'. *Brain*, 123: 2141–2149.

69. Rocchi, L., Chiari, L., Cappello, A., Horak, F.B. (2006) 'Identification of distinct characteristics of postural sway in Parkinson's disease: a feature selection procedure based on principal component analysis'. *Neurosci Lett*, 394: 140–145.
70. Schieppati, M., Hugon, M., Grasso, M., Nardone, A., Galante, M. (1994) 'The limits of equilibrium in young and elderly normal subjects and in parkinsonians'. *Electroencephalogr Clin Neurophysiol.*, 93: 286–298.
71. Schrag, A., Jahanshahi, M., Quninn, N. (2000) 'What contributes to quality of life in patients with Parkinson's disease?' *J Neurol Neurosurg Psychiatry*, 69: 308–312.
72. Smithson, F., Morris, M.E., Jansek, R. (1998) 'Performance on clinical tests of balance in Parkinson's disease'. *Phy Ther*, 8: 577–592.
73. Stankovic, I. (2004) 'The effects of physical therapy on balance of patients with Parkinson's disease'. *International journal of rehabilitation research*, 27: 53–57.
74. Tanner, C.M., Ottman, R., Goldman, S.M., Ellenberg, J., Chan, P., Mayeux, R., Langston, J.W. (1999) 'Parkinson's disease in twins: an etiologic study'. *JAMA*, 281: 341–346.
75. Tinetti, M.E. (1986) 'Performance-oriented assessment of mobility problems in elderly patients'. *J Am Geriatr Soc*, 34: 119–126.
76. Tinetti, M.E. (2003) 'Preventing Falls in Elderly People'. *N Engl J Med*, 348: 42–49.
77. Tinetti, M.E., Gordon, C., Sogolow, E., Lapin, P., Bradley, E.H. 'Fall-risk evaluation and management: challenges in adopting geriatric care practices'. *Gerontologist* 46: 717–725.
78. Trenkwalder, C., Paulus, W., Krafczyk, S., Hawken, M., Oertel, W.H., Brandt, T. (1995) 'Postural stability differentiates "lower body" from idiopathic parkinsonism'. *Acta Neurol Scand*, 91: 444–452.
79. Turbanski, S. (2006) 'Zur posturalen Kontrolle bei Morbus Parkinson – Biomechanische Diagnose und Training'. Dissertation, 2006, Frankfurt/Main. (English: Postural control in Parkinson's disease – biomechanical analyses and training. PhD thesis, 2006, Frankfurt/Main, Germany).
80. Turbanski, S., Haas, C.T., Schmidtbleicher, D. (2005) 'Effects of random whole-body vibration on postural stability in Parkinson's disease'. *Research in Sports Medicine: An International Journal*, 3: 243–256.
81. Turbanski, S., Haas, C.T., Schmidtbleicher, D. (2006) 'Short term effects of sensorimotor training on postural reflexes in parkinsonian subjects'. *Abstractbook 11th Annual Congress ECSS Congress, Lausanne*, pp. 264–265.
82. Turbanski, S., Haas, C.T., Schmidtbleicher, D. (2007) 'Comparison of biomechanical and clinical assessment of postural stability in Parkinson's disease'. *Clinical Neurophysiology*, 1187(4): E104–E105.
83. Van Emmerik. R.E., Wagenaar, R.C., Winogrodzka, A., Wolters, E.C. 'Identification of axial rigidity during locomotion in Parkinson disease'. *Arch Phys Med Rehabil.* 80: 186–191.
84. Visser, M., Marinus, J., Bloem, B.R., Kisjes, H., Van Den Berg, B.M., Van Hilten, J.J. 'Clinical tests for the evaluation of postural instability in patients with parkinson's disease'. *Arch Phys Med Rehabil*, 84: 1669–1674.
85. Wade, D.T., Gage, H., Owen, C., Trend, P., Grossmith, C., Kaye, J. (2003) 'Multidisciplinary rehabilitation for people with Parkinson's disease: a randomised controlled study'. *J Neurol Neurosurg Psychiatry*, 74: 158–162.
86. Waterston, J.A., Hawken, M.B., Tanyeri, S., Jantti, P., Kennard, C. (1993) 'Influence of sensory manipulation on postural control in Parkinson's disease'. *J Neurol Neurosurg Psychiatry*, 56: 1276–1281.
87. Wood, B.H., Bilclough, J.A., Bowron, A., Walker, R.W. (2002 'Incidence and prediction of falls in Parkinson's disease: a prospective and multidisciplinary study'. *J Neurol Neurosurg Psychiatry*, 72: 721–725.
88. Zia, S., Cody, F., O'Boyle, D. (2000) 'Joint position sense is impaired by Parkinson's disease'. *Ann Neurol*, 47: 218–228.

Section VIII

Training, learning and coaching

37

Application of biomechanics in soccer training

W. S. Erdmann
Jedrzej Sniadecki University of Physical Education and Sport, Gdansk

Introduction

Soccer is based on several national games. But its modern version appeared in England in the XIX century. There in Cambridge in 1863 rules of the game were established (Lanfranchi *et al.* 2004: 17). The rules always dictated the skills of players and also technique and tactics of play.

In every sport discipline the most important are sport competition and sport training. In order to conduct proper training sessions one has to obtain a good knowledge on a sportsperson and on what he or she actually is doing while competing against the other sportsperson. This includes, among others, biomechanical, both kinematic and dynamic (static and kinetic) data. Biomechanics of play helps to understand such player's activity as using muscle strength, direction and velocity of player's and ball's movement, technical and tactical manner of playing with a ball, application of sport engineering to produce sporting goods and to maintain a pitch in proper condition.

During training sessions, while approaching a time of competition, sportspeople should obtain mechanical data of their movement similar or sometimes even better than those which are usually obtained during competition. For assessment of mechanical data a diagnosis of sportspeople should be performed. This has to be done when a change of skills is anticipated and one wants to know at what level skills are and when a coach needs to select proper sportspeople from a larger group for competition. The diagnosis may be performed by plain coach's observation, by administering fitness tests, or by using special equipment, especially that used by biomechanists, biochemists, physiologists, and psychologists.

From the biomechanical point of view diagnosis takes into account player's body build, his or her potential strength and movement possibilities (without interference of technique or tactics), technique of running, dribbling, tackling, heading, etc., tactics of play. It is worth checking the mental level of players. This can be associated with general knowledge of the human body and with knowledge of soccer rules. 'Cause–effect' chain of problems that can be diagnosed is as follows: match situation – information acquired by sensory organs – decision made – nervous stimulation – muscle contraction – movement of a body part – movement of the whole body – movement of a ball – match situation. How often to diagnose players? It depends at what level they play. A coach of Juventus (Turin, Italy) said: 'players are tested every ten days for strength, power, endurance and agility' (Reed 1998b).

For Lees (1993: 327) the most important in biomechanics of football were: equipment, technique of play, and sport injury investigations. A wider look at the discipline shows that biomechanics of soccer (association football) is an applied science which introduces mechanical principles to a sportsperson's body build, his or her potential function and control, technique of movement, individual, group and team tactics air characteristics, pitch and equipment, garments and protection accessories utilized by players, coaches, referees during both training and competition in soccer.

Body build of a soccer player

Body build has important role in a soccer player's activity. One may present the following problems of body build: (a) material; (b) structure; (c) construction; (d) geometry; (e) inertia (Erdmann 1988: 52).

Material of which a player is built may vary in density. The beginner has lower density of tissues such as bone, ligament, tendon. Also, a player after a prolonged break when he or she suffered contusion may have lower density of tissues. The density grows with greater load that acts on a body. Training influences development of tissue's density and its elasticity, hence its resistance to acting forces.

Structure of a body takes into account all links of which a kinematical chain representing a body is built. First of all a coach must decide whether a player without some parts of a body may be a good player (e.g. without a finger, or without a toe). In another situation he must decide whether a player with a contusion in a certain place of a body will manage to play at full intensity. It must be said that though for a spectator a movement involves sometimes just a few body parts, all body muscles are involved. Some muscles work to accelerate and decelerate moving body parts, while other muscles are responsible for keeping proper direction of movement, yet another set of muscles are involved in stabilizing the body. Lack of possibility of full function of some of them diminishes usefulness of a player.

Construction of a body gives possibility for moving in all directions with velocity up to about 10 m/s while moving forward. Unfortunately, bipedal support and small area of surface of feet often cause players to fall during active play. The human body has a construction which gives good security for many inner organs. So the most often contusions happen to joints, where one part of a body is stopped, usually by an opponent, while another part is still moving. This is a cause of contusion of soft tissues surrounding a joint. During vertical jumping a player can swing his arms in order to achieve greater momentum and then greater height of a jump. Construction of a body allows to make a swing by a player both in frontal and in sagittal plane. But doing this in the frontal plane with close presence of an opponent player is forbidden since this is dangerous to the opposing player. In this situation, when a contact with the opponent player exists, a referee may show a yellow card to the offending player.

Geometry of the body is an important factor in soccer. It is especially important to have a long body for a goalkeeper and for members of the defensive formation in order to reach a ball when it is high above the pitch. Long upper extremities of a goalkeeper allow him to reach a ball to greater extent. Taking into account attackers, their body is of a different length. Some of them are short but they possess the ability to obtain high acceleration of movement and high velocity by greater frequency during movement of extremities. They also possess

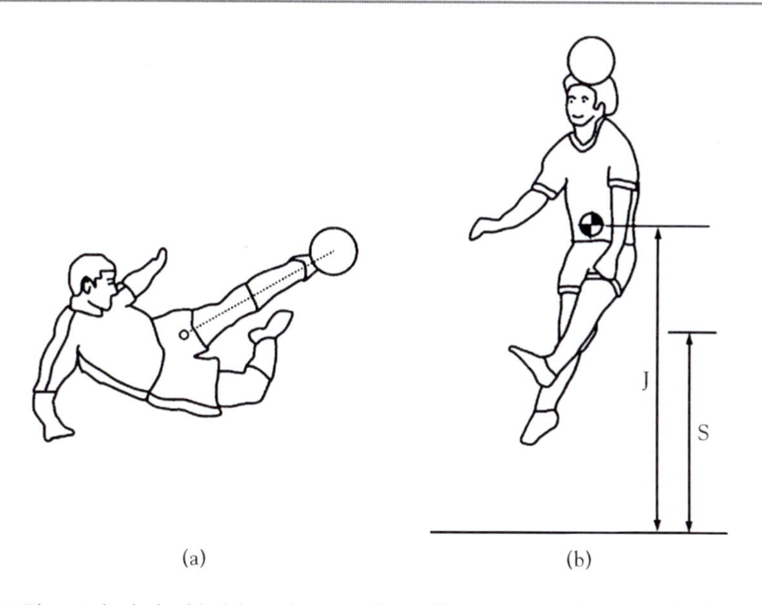

Figure 37.1 Player's body build: (a) – player with smaller moment of inertia of a lower extremity can achieve bigger angular velocity; (b) – knowledge of position of player's center of mass while standing (S) and jumping (J) gives a possibility of assessment of a height of a jump.

good agility of the body. On the other hand attackers of longer body are good in obtaining high velocity by having longer lower extremities and making longer strides. They are also good in fighting for a ball in a penalty area where they can reach a ball for heading.

Inertia of a body describes resistance that gives a body when it is accelerated by a force (during translatory movement) or by moment of force (during rotational movement). In translatory movement mass of a body is taken into account, e.g. when calculation of momentum is needed (mass × velocity) and in rotational movement MOI is taken into account, i.e. mass and its squared distance from the axis of rotation. Since soccer players need to obtain high acceleration of a body within a short distance their body should not be of a very high mass. Long and massive lower extremity has large MOI, i.e. it gives big resistance to the propelling muscles and players can achieve smaller angular velocity. Hence the kind of movement presented in the Figure 37.1a can be accomplished with high angular velocity only by players with smaller MOI.

Knowledge of position of COM of a player's body is valuable. Taking into account the position of COM in standing upright position of a body one can compare it with the height of elevation of COM in several situations during play when headings occur (see Figure 37.1b).

Fitness preparation

Obtaining of great potential possibilities of strength, velocity, acceleration, and power are the most important aims of fitness training. This kind of training is of greater intensity during while approaching the competition season.

Muscle strength is a basic feature of a person and of a sportsperson in particular. In soccer a player is characterized by different kinds of muscle strength. He or she needs explosive strength, i.e. big strength realized in a short time, like during kicks, especially those leading to scoring a goal or during jumps. He or she needs also endurance strength in order to play 90 minutes and sometimes to play 120 minutes and after that to execute penalty kicks. During very intensive play through 120 minutes sometimes players' muscles are subjected to cramps. Fortunately he or she does not need to perform movement with high a level of strength for full play-time. There are plenty of static pauses where a player can relatively relax. But even in these quasi-relax situations players have to maintain an upright position and have to stay or move by walking or jogging.

Diagnosis of muscle strength *F* is based on investigation of force executed against elastic element within a dynamometer. This device shows a resistance force *R* with which elastic element resists against its deformation. In order to obtain basic muscle strength with which a sportsperson acts onto a dynamometer, one has to make some measurements and calculations. If, for example, a strength of knee extensors is investigated, a subject takes a seat on a flat surface and holds with his hands to the sides of a seat. His lower leg is in vertical position at a right angle according to the thigh. Between a knee and an ankle, but usually closer to the ankle, a bar is attached to lower leg. A bar is connected to a dynamometer. Position of a connection link to dynamometer is at a right angle to lower leg. If a dynamometer acts on squeezing force then lower leg will push a bar, otherwise it will pull a bar and in this condition the lower leg has to be attached firmly to a bar. Next, a distance between the connection link and an axis of rotation of the knee has to be measured. This distance is called radius *r* of a force *R*. A product of multiplication of force *R* and a radius of force *r* is a moment of force *MR*. This moment of force of a dynamometer equals moment of force of acting muscle group *MF*. In order to calculate muscle force *F* its radius of force *f* should be obtained. According to Erdmann (1995: 52) radius of force *f* for knee extensors equals 12 per cent of a distance *tibiale* – *sphyrion* (*ti* – at a proximal edge of tibia bone, *sph* at a distal edge of tibia bone). Muscle force *F* equals the moment of force *MR* = *MF* divided by radius of muscle force *f*.

For comparison of several players of different body weight *W* one needs to obtain a moment of force index *IM*. It is obtained by dividing moment of force M by body weight *W*. This index says how a player can act against external moments of force taking into account his or her body weight. Another index *IF* is obtained by dividing muscle strength *F* by body weight *W*. This index says what is the player's muscle potential taking into account his or her body weight.

Explosive force and endurance force are investigated according to time. In the former a subject acts as strong as possible at a shortest interval of time, while in the latter a subject acts at a relatively short time, e.g. less than a second but repeatedly for a given interval of time, e.g. for five, ten, or more minutes at every five or ten seconds.

Velocity and acceleration of body parts depend on muscle explosive strength, on MOI of propelled body part and on resistance forces. Potential velocity and acceleration of body parts are investigated in isolated conditions, i.e. only a measured body part is moving while all other parts are stabilized. Sometimes additional load is added to the moving parts – as an additional MOI (weight is connected to the body part) or as additional force that must be overcome (changeable number of gums is connected with a body part through a wheel or a bar of a measuring device with preset resistance). Within this investigation few data are analyzed. There are absolute values of angular velocity and acceleration. Next, angular momentum is calculated by multiplying angular velocity by MOI. Then moment of force is obtained by multiplying MOI by angular acceleration. Above investigations are performed with a help of special stands and special measurement equipment, e.g. electrogoniometer, i.e. potentiometric device which measures changing of an electrical current which is proportional to the angle of joint in a function of time.

To obtain a value of mechanical work one needs data on a player's weight, horizontal distance covered and data on the vertical movement of a player's body. If a time data would be introduced one can calculate power of movement. Power of a short movement, e.g. jumps can be obtained during a diagnostic session using a force platform (dynamometric platform). This device possesses four force sensors (tensometric or piezoelectric) placed at four corners of a platform. A subject can perform different jumps in order to obtain the wide spectrum of player's possibilities.

Control of movement

To perform a movement a sportsman needs a program. The sources of obtaining a program are: (1) genetic transfer (e.g. for a heart work); (2) surroundings, e.g. as a coach's instruction, by reading a description, by watching a master performing a movement, by looking at a cinegram; and (3) information on own body build and function.

According to Bernstein (1965) performance of a movement is conducted on a basis of comparison of what was programmed and what was performed. In order to be informed on performed movement data of receptors are utilized. Among receptors that transmit mechanical data of a movement are extero-receptors and intero-receptors. The former are: eyes, ears, touch, the latter are receptors within muscles, tendons, skin. One needs to make an assumption that there is always a difference between programmed and performed mechanical data. The human being is not accurate. The errors still exist. They can be decreased by technical training. For example, Beckham perfected his efficient technique of free kicks during extra training hours. But the number of external situations (physical or psychical) influencing performance of a soccer player is so big that one can see, not rarely, a player while kicking a ball at a goal missed it during a play or during his or her execution of a free or penalty kicks.

Technique

In order to describe a technique one uses geometrical and mechanical quantities. For presenting body parts configuration in joints for static position, e.g. of a goalkeeper during penalty kick or for players in standing position angles between body parts, position of a COM, equilibrium angle, i.e. angle between lines connecting COM and edges of supporting area are used. For presenting a manner of movement one uses basic quantities – displacement and time, and derivative quantities, i.e. velocity, acceleration as kinematic quantities, and momentum, angular momentum, force, impulse of force, moment of force as kinetic quantities. These quantities are taken into account as instantaneous and as mean data. In coach's practice a word 'velocity' is sometimes substituted by a word 'tempo' (e.g. Reed 1998a).

For describing a movement of body parts during kicking, throw-in, tackling one uses angles in joints (see Figure 37.2a). Here, also, COMs of groups of body parts and MOIs are used. During locomotion movement like walking, jogging, running, sprinting the following quantities are investigated: length, time and frequency (or cadence) of a stride, velocity and acceleration. When a player wants to achieve greater velocity of movement at the

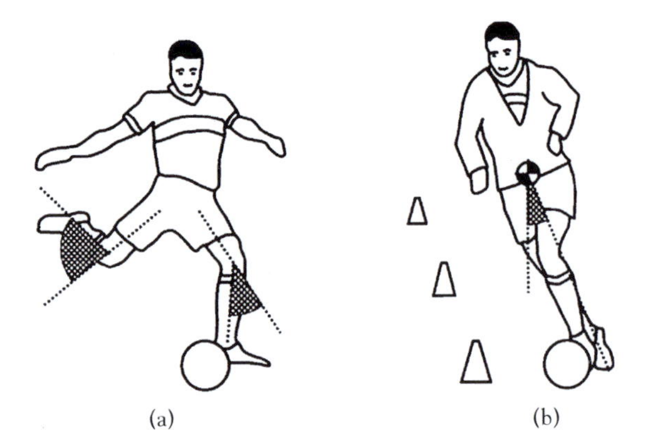

Figure 37.2 Assessment of technique of movement with a help of angles: (a) – of body parts during a kick; (b) – of the whole body during a dribbling.

beginning they lengthen their strides, thus higher frequency of strides is introduced. During analysis of player's movement a knowledge of a position of COM of the whole body is helpful (see Figure 37.2b).

During running with high velocity the feet of a player should be carried close to the buttocks, i.e. lower extremities should be flexed considerably in knees (more than 120°) in order to diminish MOI of lower extremities so they will have greater angular velocity during recovery. Also upper extremities must be moved vigorously in arm joints over a large range so this will help considerably in running, especially during a start, allowing acquiring bigger acceleration. In this situation a body should be leaned forward, initializing equilibrium receptors with attention information so that the brain would send impulses to the lower extremities' muscles to avoid falling down. As a result lower extremities would act faster and the body will achieve greater acceleration.

During passing a ball a player must consider the following variables: his or her own velocity, partner's velocity and direction of movement. If a low level passing is done a friction between a grass and a ball must be taken into account. A pitch can be dried, wet, covered with snow, etc. Also a ball can be wet and slippery and have bigger mass.

The idea of kicking a ball is transferring a momentum to the ball. If only a foot is considered in transferring a momentum it has a small mass; roughly 2 per cent of the body mass. In order to transfer high velocity to the ball a player accelerates his or her body through initial running, then performs rotation of the vertebral column in lumbar joints and rotation of the pelvis around the femur's head of a standing lower extremity, then he or she makes flexion of femur in hip joint (if forward movement in sagittal plane is taken into account), extension of a lower leg in knee joint and extension of a foot in ankle joint. Just before touching a ball a femur decelerates so that the lower leg can accelerate more. During instant of touching a ball muscles of a whole lower extremity are stretched so a hitting element has more mass than just the ending element. Lower extremity is straightened and a hitting place has high linear velocity that is transferred to the ball.

Every movement of a ball in the air is of ballistic manner. It means that a ball does not move straight-linear between a hitting source and a target. It must be aimed and propelled initially in a direction above a target. Then resistance of an air causes diminishing of ball's velocity in a forward direction and because of gravity a ball lowers its movement while approaching a target. A ball which is hit at the center moves transversally. But there are many occasions where a ball is given rotational movement. This is especially needed when a free kick is performed and a ball needs to be moved over opponent players, who stay as a fence in front of a kicker or a kick is performed from a corner. Within these situations a ball moves curve-linear.

When a ball moves forward the air at the front of a ball (sector A) becomes more dense while that at the rear (sector B) becomes thin. When a ball is hit in such a manner that it rotates around vertical axis then air layers which are closer to a ball rotate with it. Suppose that a ball was kicked from a right corner (looking from a goal) and a ball was rotating to the left. Those air layers which are at the side of a ball and further from a goal (sector C, at right hand side) are moving forward. Those air layers which are at the side of a ball and closer to a goal (sector D, at the left hand side) are moving backwards. Air layers from sector C moving forward encounter dense layers from sector A so they also become dense. Air layers from sector D moving backwards encounter thin layers from sector B so they can move there easily. As a result the air in sector D is of lower density than the air in sector C and a ball moves towards sector D, i.e. to the goal. The above situation with a curve-linear trajectory of a rotated object is called 'Magnus' effect'.

When a player needs to rotate his or her body around a longitudinal axis their body needs to have small MOI. In order to achieve this a body must be kept compact, i.e. upper extremities should be held close to the trunk. In this way a mass of the body would be closer to the axis of rotation and causing lower resistance.

During throw-in from behind a side line a movement around frontal axis is performed. In this case a player moves his body backwards causing straining of abdominal and pectoral muscles. Such a situation causes stronger stimulation of muscles for performing a stretch. It was revealed that in order to obtain a good result in such movement it should be performed in a way: 'counter – ipsi – counter – main movement'. More sweeping motions gave no better effect (Erdmann 1991: 307). One needs to say that a better result of a throw-in of a player can be achieved when a run-up is performed.

Taking into account goalkeepers, Suzuki *et al.* (1988: 468) analyzed their movements of diving and saving. They found that the more skilled keepers moved with bigger velocity (about 4 m/s) than the less skilled and more precise, i.e. more directly to the ball. More skilled keepers also introduced a counter-movement to obtain a better jump, i.e. they move at first downwards to strain their muscles to achieve, then a faster and stronger stretch of lower extremities' muscles.

Tactics

In soccer there are the tactics of: the individual player, groups of players (formation), and of the whole team. From a biomechanical point of view, a position in the pitch according to the coach's advice and according to a situation during a game is investigated. Also, the choice of specific locomotion, marking and passing of opponents, then direction, distance covered, velocity, and acceleration of movement of a player, formation and of the whole team are evaluated.

In order to obtain kinematic data of one player, Ohashi *et al.* since the 1980s (1993) used two video cameras positioned 20 m from the corners of a pitch. Each tripod which was attached to the camera was equipped with a potentiometer. A center of a view-finder

Figure 37.3 Assessment of a tactics' data. Position of every player at every square meter of a pitch, and at every second of a match is possible with a camera situated above and far from a pitch equipped with a wide-angle lens (Erdmann 1987, 1993: 174).

was always aimed at a player. All data from potentiometers were sent out to the computer where distance covered by a player in a function of time was calculated.

Erdmann (1987, 1993: 174) used a stationary video camera by placing it high above and in the farthest distance of the pitch. A camera was also equipped with a wide-angle lens having the whole pitch in the view-finder (see Figure 37.3). In this manner every player at every position in the pitch for the entire match was recorded.

At first during playback a transparent foil was put on a monitor's screen. All the pitch's lines were drawn on a foil. For easy recognition of a player's position squares on the picture of a pitch were also drawn. Such squares, formed by different formatting of a grass and put according to a pitch's lines, are seen now at many pitches. On separate foils a displacement line and a position of the player in the time intervals of 1 s were drawn. A foil was changed for every one minute. Displacement for all 22 players was obtained, then velocity and acceleration were calculated.

Since 1996 computerized versions of analysis are utilized with Kuzora's software (1996). Using side-lines as a reference system co-ordinates of every player are digitized every 0.1 s. Figure 37.4a presents a displacement curve representing the position of a player during the time of half of a match, while Figure 37.4b presents his instantaneous velocity.

For the first ten years finding of a position of a player was done by transferring video material to the computer and then during playback the position of a player was followed manually using the computer's mouse. Newer approaches use automatic tracking of a player by an image recognition system (e.g. Barros *et al.* 2001: 236). The problem appears when two or more players are very close to each other. Here again the manual part of work should be done.

For few players isochron, i.e. a line connecting all players of given formation, is used. Such isochron can be drawn for consecutive time instants. A mean position of these few players

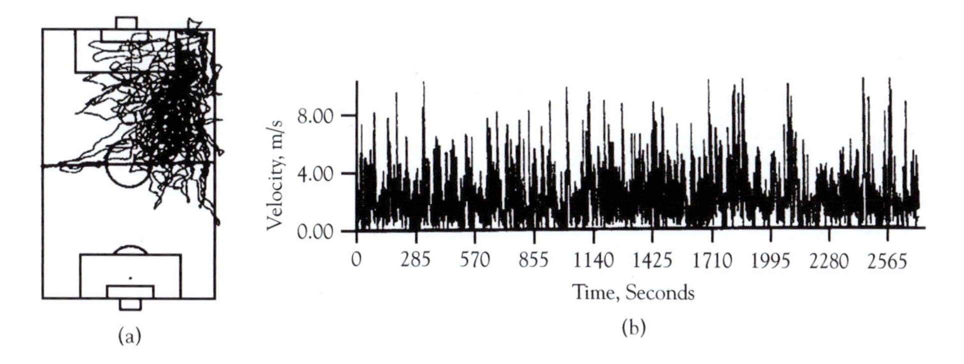

Figure 37.4 Individual point of position of left defensive midfielder through the Ist half of a match. (a) – displacement; (b) – instantaneous velocity (Dargiewicz *et al.* 2005: 36).

can be also obtained, e.g. by drawing a position of two players (center point between them) and then by drawing next point of position by including a third player.

For describing the position of a whole team (team point of position) on the pitch, a mean value and its standard deviation of 11 vertical and horizontal co-ordinates of every team member's position is used. Then displacement of this point is evaluated. With its help, direction of attack of a team – near side lines or at the center of a pitch is described. Also velocity and acceleration/deceleration of this point can be calculated. Figure 37.5 presents position of team point of position for every 15 minutes of the match's time.

Half a century ago players covered distances of about 5 km per match. In the 1970s they covered a distance of about 7–9 km, while in the 1980s and 1990s about 10–13 km and longer distances (Reilly and Thomas 1976, Ohashi *et al.* 1993: 124, Bangsbo 1994).

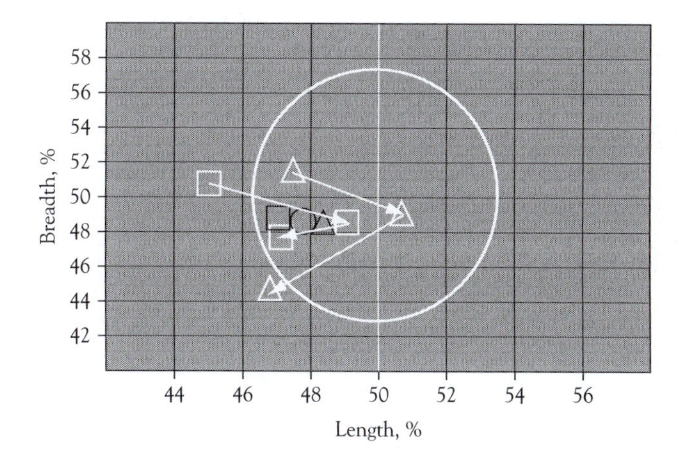

Figure 37.5 Team point of position of mostly defensive play. Position of a team was mostly less than 50 per cent of pitch's length. White triangles: 1st half quarters; white squares: 2nd half quarters; black triangle: entire 1st half; black square: entire 2nd half; black circle: entire match (Dargiewicz 2005, modified).

Table 37.1 Kinematic data of different groups according to age: U16, i.e. Junior teams – Germany and Poland; U21, i.e. Olympic teams – Norway and Poland; Senior teams – England, Italy and Poland.

Quantity	*Unit*	*Junior*		*Olympic*	*Senior*
Time	*min.*	80	90	90	90
Standing (v < 0.1 m/s)	s	489	550	403	452
Whole distance covered	m	8377	9425	9639	10866
Walking (v < 1.1 m/s)	m	1257	1414	1332	1374
Slow running (v < 4.0 m/s)	m	7090	7977	8203	9234
Fast running (v ≥ 4.0 m/s)	m	1287	1448	1436	1632
Sprinting (v ≥ 7.0 m/s)	m	213	240	206	267
Number of sprints		25	28	28	32
Mean velocity	m/s	1.74	1.74	1.76	2.01
Maximal velocity	m/s	8.1	8.1	8.6	8.7
Acceleration (a > 2.0 m/s^2)					
Time	s	105	118	138	153
Distance	m	549	618	695	762

Adapted from Dargiewicz (2005).

Notes: (1) Senior and Olympic teams play entire match 90 min. while Junior teams play 80 minutes Data of 90 minutes match for Junior teams were calculated.

(2) Velocity (v) of 4.0 m/s was established according to acquired anaerobic threshold (for Junior teams velocity of 3.7 m/s was established). Sprinting velocity for Junior teams was established over 6.0 m/s.

Dargiewicz (2005), using Erdmann's method and Kuzora's software, collected data of soccer teams during matches of three different groups based on years of age: Senior teams, U21 (Olympic teams), and U16 (Junior teams). Younger players stand longer during a match than older players. They also cover shorter distances, especially during slow movement and during sprinting. They achieve smaller maximal velocity and cover shorter distance with acceleration over 2 m/s^2. Mean velocity for the whole team was the biggest for older players and for the winning teams. For details see Table 37.1.

Comparing players' formations within Junior teams the shortest time of acquiring standing position during a match-play was recorded for midfielders and within Olympic teams for defenders. The longest distance during the whole match within all three age groups was covered by midfielders. Longer distance covered by walking was recorded mostly for attackers and sometimes for defenders (Polish seniors and German juniors). The longest distance covered with slower and faster velocity and also by sprinting was recorded for different formations but usually either for attackers or midfielders. The mean velocity was usually bigger for midfielder formations. The biggest difference between formations was recorded for Italian and German seniors combined. Their attackers achieved mean velocity 2.15 m/s, while their defenders achieved 1.86 m/s for the whole match.

Kinematic data of individual players, formations, and the whole teams obtained within a match-play can serve as reference data for providing training with the same or other players.

One of the newest methods of acquiring data on players' movement uses global positioning system's (GPS) instruments. They are very small (about 3 × 5 cm) and light. GPS can be attached to the player's body at the waist. The best procedure uses two instruments. One is attached to the player, while the other is immovable. The latter serves as a reference GPS. All errors caused by differences in signals coming from at least four satellites are provided by reference GPS (Henning and Briele 2000: 44). Currently Dargiewicz (2007) uses up to 25 GPS instruments to obtain data of training movement of all players of 2 teams including two substitutes and reference GPS. Another new method utilizes an electronic transmitter situated under the player's shin-guard. Detectors situated outside a pitch and connected with a computer can gather information on the player in situ.

Equipment and pitch

An engineering approach to soccer based on biomechanical research is present in players' garments, especially boots, protective elements, a ball, a pitch, and service equipment.

Biomechanical investigations concerning soccer boots are performed by several researchers – academic and manufacturers. Quantities that are investigated the most are forces that exist between a turf and a boot during different actions. There are pushing forces during accelerating and decelerating a movement in different directions, friction forces, moments of forces. For example, Lees and Kewley (1993: 335) gave data on forces acting on the boot during turns, runs, jumps, shots, etc. Forces during a turn achieved a value of about 16 kN through 37 ms. The severity index (impulse × time^{-1}) was 423 N. During modified pass force also achieved a value of 16 kN but was exerted through 115 ms, so the severity index was only 139 N. More accumulated forces (for several actions) were obtained for training than for match-play. Such great forces are not well distributed over the entire sole and players suffer pain at the areas where studs are placed.

A ball needs to be of proper mass – between 396 and 453 g and a circumference – between 68 and 71 cm. In order to have a proper knowledge on where a ball actually is in the pitch a radio transmitter is placed inside a ball. Every movement of a ball is acquired by detectors situated at the corners of a pitch and data are sent to a computer. Detectors could be situated also in the poles of a goal to detect whether a ball was inside a goal or not. A ball should not be of very smooth surface. An example of a golf ball says that a ball with uneven surface has smaller resistance while moving in the air. Layers of the air do not retain a smooth surface of a ball to the very end. They detach from a ball just after passing the ball's highest diameter. Uneven surface keeps air layers near a ball to the end, so behind a ball does not generate a thin air sector causing a braking force.

A pitch serves as a solid ground for a game. But this ground can be of different dimensions, even or uneven surface, different elasticity, friction. The main division of surfaces is onto natural and artificial surfaces. Artificial surface is more even, more durable, more homogeneous. Natural surface has better characteristics taking into account elasticity and friction. It is also better when in contact with player's skin.

Prophylaxis against contusions

It is very important to employ a proper moving up (warm up) before a main training or a match play. This should start from small joints of extremities, then bigger extremities' joints, and then centroid's joints, i.e. those belonging to the vertebral column. The movement should take into account all axes and at the beginning should apply smaller range, velocities and loads. They should be gradually increased along the course of training. Also strain exercises should be applied in order to prepare muscles for hard work. Unfortunately, recreation players especially like to start play immediately without warming-up.

Forces that act on a player's body are sometimes so strong that they may rupture a ligament, most often crucial ligaments of the knee joint, or may break a cartilage or a bone.

Overloading may take place in a short instance of time or may be prolonged during endured fatigue. Both may be a source of a contusion.

Especially dangerous are those situations when two opposing players strike each other in full running or when one player wants to stop another one by placing a leg in the direction of movement of the opponent. A leg of a running player is suddenly stopped while the rest of the body still moves forward. In such a situation soft tissues, mostly of a knee joint, cannot sustain overloading and tissue rupture can occur. During the sudden stop of a foot of an opposing player similar rupture can occur to the foot's tissues.

Prophylaxis may be based on mental explanation of the dangerous situations and then on avoiding sudden stops of opponent's body. A player who is in full run must oversee the possible situation when approaching an opponent and he or she must be prepared for dribbling a ball or to jump over the opponent's foot or entire lower extremity.

Closure

The game with a ball and two goals is very simple. But behind it are complicated problems. From the biomechanical point of view they can be solved with a sound understanding of the human body, human movement, technique and tactics. The future of the discipline will be based on continous knowledge of players' physiological and biomechanical preparation and possibilities; based on multi-sensory information, on coaching and refereeing that will use electronic and image equipment for diagnosis and decision making, and on development of the engineering of equipment and the pitch.

Bibliography

1. Bangsbo, J. (1994) *Fitness Training in Football – a Scientific Approach*, Copenhagen: University of Copenhagen.
2. Bernstein, N. (1967) *The Coordination and Regulation of Movements*, Oxford: Pergamon Press.
3. Barros, R.M.L., Figueroa, P.J., Anido, R., Cunha, S.A., Misuta, M.S., Leite, N., Lima Filho, E.C., Brenzikofer, R. (2001) 'Automatic Tracking of Soccer Players'. In R. Müller, H. Gerber, A. Stacoff (eds) *Book of Abstracts, XVIIIth Congress of the International Society of Biomechanics* (pp. 236–237), Zürich: Eidgenösische Technische Hochschule.
4. Dargiewicz, R. (2005) 'A football game at different levels of sport mastery in the light of kinematic analysis' (in Polish), Doctoral dissertation, Gdansk: Sniadecki University of Physical Education and Sport.
5. Dargiewicz, R., Erdmann, W.S., Jastrzebski, Z. (2005) 'Investigations on kinematics of play of competitors of soccer Polish First League teams' (in Polish). In W.S. Erdmann (ed.) *Locomotion 2003. Proceedings of the IInd All-Polish Conference 'Human Locomotion'* (pp. 36–41), Gdansk, May 2003. Gdansk: May Publisher.
6. Dargiewicz, R. (2007) Personal communication, Gdansk: Sniadecki University of Physical Education and Sport.
7. Erdmann, W.S. (1987) 'Assumption of investigations of players' movement in sport games by optical method' (in Polish), Unpublished Technical Report 1987–06, Dept. of Biomechanics, Gdansk: Sniadecki University of Physical Education and Sport.
8. Erdmann, W. S. (1988) 'Morphological biomechanics – complex view on mechanical body build' (in Polish). In L.B. Dworak (ed.) *VII School on Biomechanics, Abstracts* (pp. 52–53), Poznan – Dymaczewo, June 1988, Poznan: Academy of Physical Education.

9. Erdmann, W.S. (1991) 'Preliminary investigations of maximal angular velocity of lower extremity' (in Polish). In L.B. Dworak (ed.) *XII School on Biomechanics* (pp. 307–318), Poznan, September 1991, Poznan: Academy of Physical Education.
10. Erdmann, W.S. (1993) 'Quantification of games – preliminary kinematic investigations in soccer'. In T. Reilly, J. Clarys, A. Stibbe (eds) *Science and Football II. Proceedings of the Second World Congress of Science and Football* (pp. 174–179), Eindhoven – Veldhoven, May 1991, London: E & FN Spon.
11. Erdmann, W.S. (1995) 'Remarks on research of muscle strength' (in Polish). In S. Mazurkiewicz (ed.) *Biomechanics' 95, Proceedings of the All-Polish Conference of Biomechanics* (pp. 52–56), Cracow: Academy of Physical Education.
12. Henning, E.M., Briele, R. (2000) Game analysis by GPS satellite tracking of soccer players. In *CSB/SCB XI and SB XXV Canadian Society of Biomechanics Conference* (p. 44), Montréal: Université de Montréal.
13. Kuzora, P. (1996) 'Computer Aided Game Evaluation', unpublished paper, Gdansk.
14. Lanfranchi, P., Eisenberg, Ch., Mason, T., Wahl, A. (2004) *100 Years of Football. The FIFA Centennial Book*, London: Weidenfeld & Nicholson.
15. Lees, A. (1993) 'The biomechanics of football'. In T. Reilly, J. Clarys, and A. Stibbe (eds) *Science and Football II* (pp. 327–334), London: E & FN Spon.
16. Lees, A., Kewley, P. (1993) 'The demand on the soccer boot'. In T. Reilly, J. Clarys and A. Stibbe (eds) *Science and Football II* (pp. 335–340), London: E & FN Spon.
17. Ohashi, J., Isokawa, M., Nagahama, H., Ogushi, T. (1993) 'The ratio of physiological intensity of movements during soccer match-play'. In T. Reilly, J. Clarys and A. Stibbe (eds.) *Science and Football II* (pp. 124–128), London: E & FN Spon.
18. Reed, L. (1998a) 'Five Favourite Practices. Focusing on Passing and Control', *Insight. The F.A. Coaches Association Journal*, 2:1:4–7.
19. Reed, L. (1998b) 'International Symposium for Coaches – Paris, July 1998', *Insight. The F.A. Coaches Association Journal*, 2:1:8.
20. Reilly, T., Thomas, V. (1976) 'A motion analysis of work-rate in different positional role in professional football match-play', *Journal of Human Movement Studies*, 2: 87–97.
21. Suzuki, S., Togari, H., Isokawa, M., Ohashi, J., Ohgushi, T. (1988) 'Analysis of the goalkeeper's diving motion'. In T. Reilly, A. Lees, K. Davids and W.J. Murphy (eds) *Science and Football* (pp. 468–475), London: E. & F.N. Spon.

38

Exploring the perceptual-motor workspace: New approaches to skill acquisition and training

Chris Button[1], Jia-Yi Chow[2] and Robert Rein[3]
[1]University of Otago, Dunedin; [2]Nanyang Technological University, Singapore; [3]Ecole des Hautes Etudes en Sciences Sociales, Paris

Introduction

Humans spend much of their lives learning and executing motor skills. Consequently, the study of motor learning (skill acquisition) has potential implications across numerous fields, such as professional practice, medicine, education, sports, etc. In this chapter we analyze skill acquisition under the umbrella of the constraints-led approach and consider some implications of this framework for movement practitioners (e.g., coaches, physical education specialists, teachers, physiotherapists, and sports scientists). While different practitioners may have distinct objectives, each shares the important goal of enabling individuals to (re) acquire coordination and skilled control of movement. Researchers are becoming increasingly aware about the processes that may underpin learning as they observe performers experiment, explore, and exploit alternative movement solutions within the perceptual-motor workspace. Drawing upon the theoretical perspectives of ecological psychology and dynamical systems theory, we intend to show how awareness of the perceptual-motor workspace can help build a better understanding of key pedagogical issues such as practice organisation, the use of instructions, feedback, and demonstrations.

For much of the previous century, scientists have considered the question of how humans learn to control and coordinate their movements, with many different theories proposed (for an overview, see Davids, Button and Bennett, 2007). Although there is still considerable debate over which theory is most appropriate (Summers, 2004), the importance of developing a strong theoretical framework for studying skill acquisition and guiding practical activity remains clear. Many movement practitioners use models of human behaviour, either implicitly or explicitly, to plan their decision making for focused, effective practice activities (Lyle, 2002). Indeed, a conceptual model of coordination and control is not just important for designing optimal learning environments, it is also important for ensuring that learners gain positive experiences when acquiring motor skills. Given the alarm expressed at the lack of physical activity and poor movement competency shown in populations of affluent societies, this type of knowledge could be vital for the design of physical activity programmes to provide the fundamental skills necessary for sub-elite sport and exercise participation (Visser, Geuze and Kalverboer, 1998).

This chapter begins with a discussion of some important theoretical issues regarding skill acquisition, leading to a redefinition of this process from the constraints-led perspective. The foundations underpinning the perceptual-motor workspace analogy will then be explained with reference to realistic examples from sport and physical activities. We shall subsequently present some recent data collected in our laboratories which help to further illuminate the process of skill acquisition. Finally, we present a number of practical implications that arise from this approach which may be considered for use in pedagogy. Our assumption in compiling this chapter was that the reader may have a general interest in movement sciences and not necessarily have detailed knowledge of extant motor learning theories.

Skill acquisition redefined

Motor skill acquisition has traditionally been described as the internal processes that bring about 'relatively permanent' changes in the learner's movement capabilities (Ragert, Schmidt, Altenmüller and Dinse, 2004). For example, movement skills such as riding a bicycle, throwing a ball, driving a car, or touch-typing require a good deal of physical practice to allow us to improve and perform them consistently and effectively. Skill acquisition requires us to learn how to interact effectively with our environment, to detect important information, and time our responses appropriately. In other words, skill acquisition should result in coordination patterns that are adaptable to a range of varying performance characteristics. Adaptive behaviour is important because conditions like the environment, task requirements, and our motivations can change every time we perform a motor skill (Davids, Button, Renshaw, Araújo and Hristovski, 2006).

The process of skill acquisition is distinct from the execution of the skill (motor control) in that learning is a gradual process that occurs over many performance attempts, resulting in behaviour that is less vulnerable to transitory factors affecting performance such as fatigue, audience effects and anxiety. One way that researchers have tried to understand skill acquisition is by examining the performance changes that accompany practice. For example, in early research efforts, (Bryan and Harter, 1897) studied how learners' typing skills developed while practicing to send and translate Morse code. Over a period of 40 weeks, the telegraphers went through distinct phases of improvement and periods where performance levels plateaued. The inference taken from such performance curves was that learners initially construct simple elements of the skill, interspersed by periods of consolidation (i.e., development of automaticity) before linking individual parts of movements (in this case individual finger presses) into more integrated patterns of behaviour (linked sequences of finger presses typed as words)[1]. Performance curves have since been used in many studies of skill acquisition to examine the influence of different factors related to learning (e.g., modelling (Herbert and Landin, 1994); feedback (Vickers, 1996); practice organization (Ollis, Button and Fairweather, 2005).

Within the research literature there has been an overemphasis on the amount of change in performance outcomes associated with learning without sufficient analysis of the dynamic properties of change in movement coordination (Hong and Newell, 2006; Müller and Sternard, 2004). Undoubtedly this criticism may be partly attributed to the lack of sophisticated measuring devices for examining coordination in the past (e.g., 3D motion analysis systems). In fact the productive relationship that has been formed by the sub-disciplines of biomechanics and motor learning is proving extremely valuable in advancing current theoretical understanding. Another important issue is that, unlike Bryan and Harter's study, much skill acquisition research has typically been conducted over short, intense practice periods, e.g., a number of days (Hong and Newell, 2006). For obvious pragmatic reasons there are few examples in the research literature of motor learning experiments performed over longer periods of practice, as experienced in real life (e.g., the months and years of practice involved in many work and sports environments). Hence, conclusions from such snap-shot studies are likely to be based on the relatively transient effects of practice rather than the permanent consequences of learning.

Humans acquire skill by consistently coordinating relevant parts of the body together into functional synergies, for example when performing a triple salko jump in ice skating or, more mundanely, when stepping on to a surface. The search for simplistic laws of cause and effect has meant that movement scientists have often struggled to come to terms with the complexity humans must overcome to produce such skilled actions. This issue has become known as Bernstein's (Bernstein, 1967) 'degrees of freedom problem'. Degrees of freedom (DoF) refer to the independent components in the performer that can fit together in many different and redundant ways. For example, the large number of available muscles, joints, limb segments and bones of a human body exemplify parts of the DoF of the human motor system.

The huge number of motor system DoF actually represents a 'curse' for some traditional theories of skill acquisition (Kelso, Holt, Kugler and Turvey, 1980). This is because traditional theories have tended to invoke the coordination and control of movement to an executive controlling mechanism residing in the CNS. For example, the information processing approach views the performer as a sort of human communications channel in which the relationship between changes in input signals and system output are linearly related. As noise is an inherent feature of every biological system (Collins, 1999) it has been presumed that an important job for the performer is to eliminate or minimize noise (or movement variability) through practice and task experience. For this reason, the magnitude of the variability in performance has been viewed as an important feature for assessing the quality of system control (Schmidt, 1985). The role of repetitive practice is often conceived of as gradually reducing the amount of movement pattern variability viewed as noise. Further compounding this viewpoint, the selection of movement models to investigate motor system functioning has been biased away from dynamic, multi-joint actions prevalent in sports because of the view that experimental rigour could be better maintained in laboratory studies of simple movements (Davids, Button, Renshaw, Araújo and Hristovski, 2006).

However, is movement variability really always that bad? As we shall learn later in this chapter, researchers are increasingly recognizing the important structural role that variability might play in adapting successful goal-directed movement patterns to changing environments. Humans operate in dynamic environments, which are rich in information, requiring complex coordination patterns to interact with surfaces, objects and events. An important challenge for movement scientists is to identify a theoretical model that embraces these features. It has only recently become clear that a rigorous model of motor behaviour requires a multidisciplinary framework to capture the different interlocking scales of analysis (e.g., neural, behavioural, psychological) and the many different sub-systems (e.g., perceptual and movement) involved in producing human movement.

In recent years there has been increasing interest in the constraints that shape and influence the acquisition of movement skill in individuals (Al-Abood, Davids and Bennett, 2001; Bennett *et al.*, 1999; Davids, Glazier, Araujo and Bartlett, 2003; Chen, Liu, Mayer-Kress and Newell, 2005). Constraints are influential variables loosely categorized within the environment, the performer, and the task, that jointly determine how a movement pattern unfolds (Newell, 1986). With its roots in dynamical systems theory, the constraints-led perspective views the learner as a complex, biological system composed of many independent but interacting DoF or sub-systems. In many complex structures in nature, the dynamical interaction between components can actually enhance the intrinsic system organization. These interactions can produce ordered behaviour of the system through a process known as self-organization (Haken, 1996). In a reciprocal manner, the organization of the movement system is dependent upon the performer detecting and generating informational flows to regulate action in the desired manner. In this sense, self-organization should not be viewed as a mystical process in which the performer has no control over which movement patterns eventually emerge, instead intentionality (amongst other things) places a strong constraint over the movement system (Schöner and Kelso, 1988).

Because each individual is different and performance circumstances are constantly changing, individuals must explore a range of potential movement patterns in order to optimally adapt to the constraints placed on them. Variability permits flexible adaptation to the constraints of a dynamic environment, a useful characteristic in many sports performance contexts. Indeed, it will come as no surprise for practitioners to learn that an individual's movement preferences (intrinsic dynamics) play an influential role in the skill acquisition process. Consider a child who initially favours their right foot when kicking a ball. We often find that for this type of learner, kicking effectively is likely to take longer to learn with the left foot than with the right foot. In a series of research studies, Zanone and Kelso (Kelso and Zanone, 2002; Zanone and Kelso, 1992; Zanone and Kelso, 1997) have explored the influence of intrinsic dynamics on the learning of new skills. Amongst a number of interesting findings, it was shown that the degree of conflict or cooperation between the intrinsic dynamics and the to-be-learned task requirements plays a crucial role in the skill acquisition process. We will return to this important issue later in this chapter.

To summarize thus far, it is perhaps not surprising that traditional definitions of skill acquisition have been limited to vague descriptions of 'internal processes' associated with performance improvement. In the absence of sophisticated measurement tools to monitor the neuropsychological changes associated with learning, many theorists have based their assumptions on indirect sources of evidence. We advocate a broader definition which focuses on the journey the learner must take in becoming skilled. For example, using the constraints-led approach, Davids, Button and Bennett (2007) define skill acquisition as 'an ongoing dynamical process involving the search for, and stabilization of specific, functional movement patterns across the perceptual-motor workspace'. As we shall demonstrate in the remainder of the chapter, this alternative definition demands a thorough and detailed examination of coordination dynamics to better understand the process of skill acquisition.

The perceptual-motor workspace

How do long-term changes to the organization of human movement occur as a result of learning and practice? From a constraints-led perspective, skill acquisition can be characterized as a learner (a dynamical movement system) searching for stable and functional states of coordination or 'attractors' during goal-directed activity (Wenderoth and Bock, 2001; Zanone and Kelso, 1992; Zanone and Kelso, 1997). We can view different phases of learning as the creation of temporary states of coordination that resist constraints that could perturb the system's stability. As examples, one might think of early learners in activities such as swimming and ski-ing, and the cumbersome, inefficient coordination patterns that typically emerge. For skilled performance, individuals eventually need to develop a repertoire of movement attractors to satisfy the constraints of changing contexts. We can consider this repertoire of attractors as a kind of **perceptual-motor workspace** to denote that performers need to learn how to coordinate their actions with their environment in order to perform skills effectively.

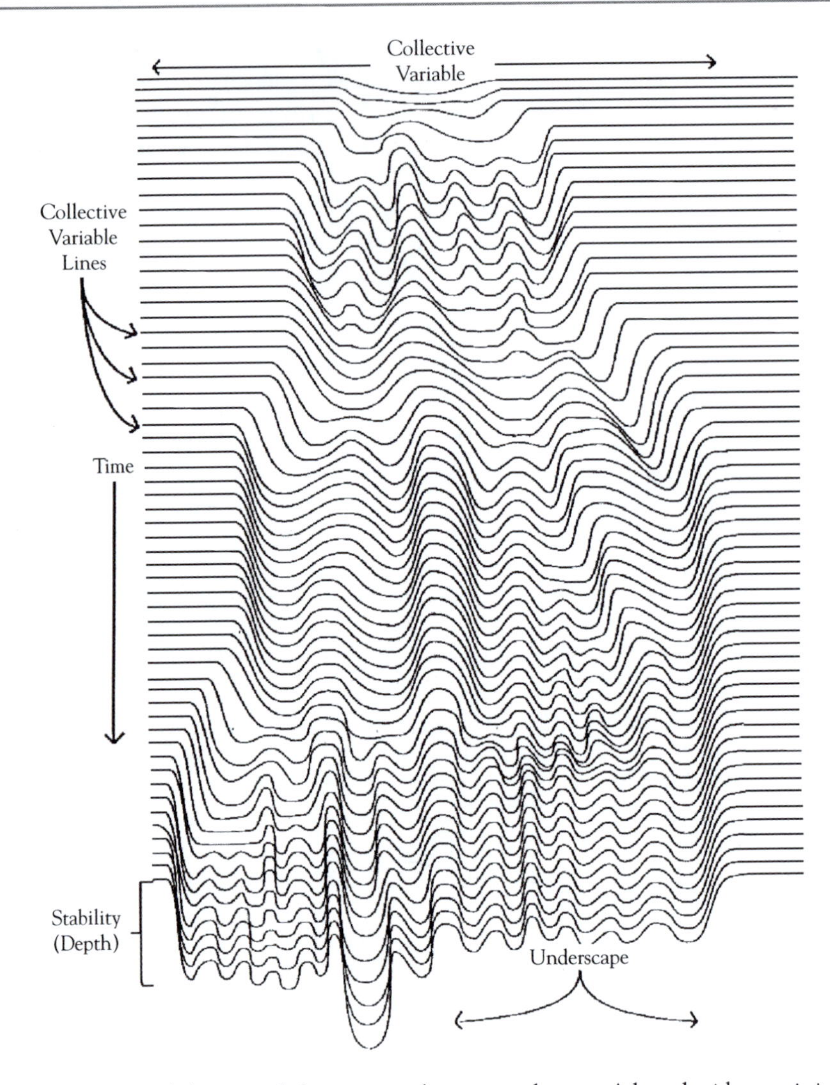

Figure 38.1 Conceptual diagram of the perceptual-motor workspace. Adapted with permission from Muchisky *et al.* (1996). N.B.: Alternative terms such as the 'epigenetic landscape' have been used to describe the workspace in relation to motor development over longer time periods (Waddington, 1966).

The perceptual-motor workspace is a useful metaphor for describing an individual learner's coordination dynamics. Its layout is constrained by genetic endowment, developmental status, past learning experiences and the task requirements. That is, each workspace is continually being shaped and altered by the interaction of an individual's genes, perceptions, and intentions, as well as physical constraints, surrounding information, and system dynamics (Muchisky, Stowe, Cole and Thelen, 1996; Thelen and Smith, 1994). Because these performance constraints are not static and fixed, we should view the workspace as undulating and ever-changing. As a learner's constraints change over time, the topology of the workspace alters to reflect the flow of information and of new experiences (see Figure 38.1).

For example, a perceptual-motor workspace for an experienced skateboarder might include a wide array of different jumps, tricks, landings and balances practiced in a skate-board park. For a child learning to ride a bicycle, specific areas of the workspace might include fledgling movement attractors for steering the handlebars, pedalling the wheels and controlling the force of braking.

For description purposes it is useful to try and imagine the workspace as if one were looking at it from above (i.e., a hypothetical bird's-eye view). The peaks and troughs of the workspace represent the relative stability of emerging coordination patterns where stable patterns occupy deep troughs (possibly well-established actions such as gait or prehension). In more rugged areas of the workspace, the coordination dynamics exhibit metastability (Kelso and Engstrom, 2006), and it is here where less stable and perhaps unfamiliar patterns are evolving. Different regions of the workspace are home to topographically similar classes of movement, e.g., throwing, kicking, etc. Therefore the movement system can draw upon multiple areas at any one time for the performance of more complex skills such as a gymnastics routine or a golf swing. Combining a number of movement elements together can be achieved through the specification of an appropriate task goal (that links the necessary attractor sites). Once a learner has specified a functional task goal that they wish to attain, then a process of continuous exploration eventually results in the emergence of a set of approximate solutions to the task.

The different solutions of a task can be represented in so-called task spaces which are formed by all the variables which determine the task result (Müller and Sternard, 2004). Relating task space and perceptual-motor workspace, the learner can begin to modify the workspace by experimenting with successful and unsuccessful coordination patterns through practice. As practice proceeds less successful patterns are gradually sacrificed by the learner and more successful actions are reinforced by strengthening connections between intentions and energy flows. Localized within a task space the performer may find several areas where successful solutions are closely situated, these regions are solution manifolds. Within these manifolds small fluctuations alter the task-solution only minimally providing some sort of task tolerance. Large solution manifolds have more tolerance for different movement solutions where smaller manifolds may only allow subtle modifications. For example, a penalty kick in soccer which can be achieved successfully in a number of different ways has a relatively large solution manifold, however, the basketball free throw shot which seems to have less tolerance for variation (Button, MacLeod, Sanders and Coleman, 2003), has a small solution manifold.

So, returning to a question raised earlier – how can movement variability play a useful role in learning? – by allowing the learner to search and find and subsequently refine appropriate solution manifolds for different performance contexts (Müller and Sternard, 2004). For example, in a recent study Chen, Liu, Mayer-Kress and Newell, (2005) investigated skill-acquisition of a pedalo locomotor task. On the first day of practice, the authors reported high fluctuations in the trial-to-trial kinematic variability which decreased over the course of learning. However, kinematic variability was still present late in practice even though related performance variables (e.g. movement time) had already levelled off. The high fluctuations in kinematic variables on the first day suggest that the participants were initially '[...] searching for a stable task-relevant coordination mode [...]' (p. 255). The presence of decreased kinematic variability further implied that some level of functional variability was maintained and required for further exploration of the workspace with extended practice.

Before we consider some empirical evidence, we must understand several key concepts regarding the workspace. First, it is important to understand that 'the extent to which specific behavioural information co-operates or competes with spontaneous self-organizing tendencies determines the resulting patterns and their relative stability'. (Kelso, 1995, p. 169). If the initial behavioural tendencies do not conform to what is required of the performer, learning will be more difficult than if there is a close match between the learner's intrinsic dynamics and the task's behavioural information. For this reason, it is clear that practitioners should expect to see substantial differences in the rates at which individual learners (re)acquire movement skills over time (Newell, Liu and Mayer-Kress, 2001).

A second important point is that the process of learning a new task involves not just one feature changing, but the whole workspace changing. As one attractor well is being carved out, nearby attractors can become less stable. For example, Zanone and Kelso's work (Kelso and Zanone, 2002; Zanone and Kelso, 1997) studying the acquisition of bimanual coordination patterns has confirmed that practice of a novel pattern (90° relative phasing between the fingers) can destabilize nearby learnt patterns (0 and 180°). Sports players may identify with this concept through their experiences of attempting to transfer existing techniques into different performance contexts (e.g., adapting from tennis to squash). In other words, when planning skill acquisition programmes, practitioners should be aware of the potential perturbing effects of constructing one attractor on the stability of existing attractors or skills. Reducing the competition between the existing pattern dynamics of the perceptual-motor workspace and the dynamics of a new task is desirable.

As highlighted above, when constraints change, learners may be required to adapt their coordination appropriately. Somewhat counter-intuitively, practitioners may find that manipulating constraints is a natural way to progress skill acquisition and to assess skill development. By learning to continuously adapt in dynamic performance environments, the performer is challenged to find the best solution at that moment in time (rather than to rely on a rehearsed action that was suited to different constraints). Importantly, the strategy of altering task constraints represents a more robust, active route to changing behaviour (than, for example, simply getting the learner to copy a demonstration or follow instructions). Early in learning, one might expect to see sudden changes in technique if a substantial alteration of the perceptual-motor workspace is required (i.e., such as the merger of a peak and a valley known as a saddle-node transition). Later in practice the nature of these changes is typically less abrupt and the pattern variability shown may reflect refinements and adaptations by learners (Hong and Newell, 2006).

It can often take longer to alter organismic or personal constraints and one may expect to see slower and more stable adaptations in coordination, for example, as a result of prolonged flexibility or endurance training. In terms of environmental constraints, cultural and social constraints also provide more persistent but subtle impacts on behaviour, although it has been argued that these constraints emerge and decay more rapidly than some organismic constraints such as effects of genetic constraints on population variation (Ehrlich, 2000). On a busy netball court, to use an example from sport, players need to adapt their actions quickly but their playing style may not be rapidly affected by societal views of them as players as dictated by changing cultural perspectives.

According to Newell (Newell, 1986), constraints may be relatively time dependent or time independent. That is, 'the rate with which constraints may change over time varies considerably with the level of analysis and parameter under consideration' (pp. 347). Newell, Liu and Mayer-Kress (2001) have pointed out that 'time scales in motor learning and development would include the influence of the time scales of phylogeny and ontogeny in motor learning and the related impact of culture and society on the development of human action. This broader context serves to highlight and emphasize the central role that some scales play in the study of motor learning and development' (pp. 64). An important consequence of different time scales in learning lies in the appreciation of inter-individual differences in learning behaviour. Investigating typical performance curves, Newell, Liu and Mayer-Kress (2001) showed how the ubiquitous power-law of learning can be a mere consequence of averaging group data, emphasizing the application of intra-individual analysis methods (see also Liu, Mayer-Kress and Newell, 2006).

To summarize thus far, the constraints-led approach to skill acquisition adopts a model of the learner as a dynamical movement system. The metaphor of the perceptual-motor workspace is used to describe the continual evolution of constraints that impact upon the learner's exploration process. Influential constraints such as task goals, intrinsic dynamics and feedback stabilize certain areas of the workspace, allowing the learner to experiment until effective movement solutions are found. In the next section we discuss some recent research that describes how constraints influence the acquisition of a soccer kicking skill.

Impact of constraints on skill acquisition in soccer

As discussed in the previous section, specific task constraints in a learning context can help shape the emergence of goal-directed behaviour (Chow, Davids, Button and Koh, 2006). The challenge for practitioners, then, is to effectively manipulate task constraints so that the individual learner's search process can be directed to a region of the perceptual-motor workspace where functional movement solutions can be achieved. In this section, we will exemplify the role of task constraints in shaping the emergence of coordination in a soccer-chipping action.

A recent programme of work in our human movement laboratory has examined coordination dynamics in a kicking task with extended practice (480 trials) alongside associated performance outcome changes. Four novice male participants were asked to kick a soccer ball with their dominant foot over a barrier to a skilled receiver. No explicit verbal or visual instructions were provided on how to kick the ball over the barrier. Participants were informed that the task goal was to kick the ball over the height barrier (from 1.5–1.7 m high) to land at the feet of a receiving player stood at different target zones (between 10–14 m) with appropriate force control. Performance was scored by experienced coaches with a 7-point Likert scale, emphasising appropriate weightage and accuracy of the kicks. Kinematic data of joint range of motion (ROM) for hip, knee and ankle of the kicking limb was captured with six infra red cameras (Proreflex, Model MCU 1000) using an online motion analysis system for further analysis (Qualysis, Gothenburg, Sweden).

After 12 sessions of practice (40 trials of kicks per session) over a four-weeks period, participants improved the accuracy and consistency of their performance scores (Figure 38.2). Moreover, performance scores to different target positions with varying height and distance constraints also improved. Kinematic analysis indicated that participants were able to manipulate higher order derivatives like foot velocity at ball contact to appropriately vary kicking height and distance. Interestingly, participants had not completely mastered the task at the end of practice as further improvement on the performance scores was possible (maximum score of seven points).

Further analysis using Normalized Root Mean Square Error (NoRMS) (see Mullineaux, Bartlett and Bennett, 2001) showed that variability in intra-limb coordination decreased

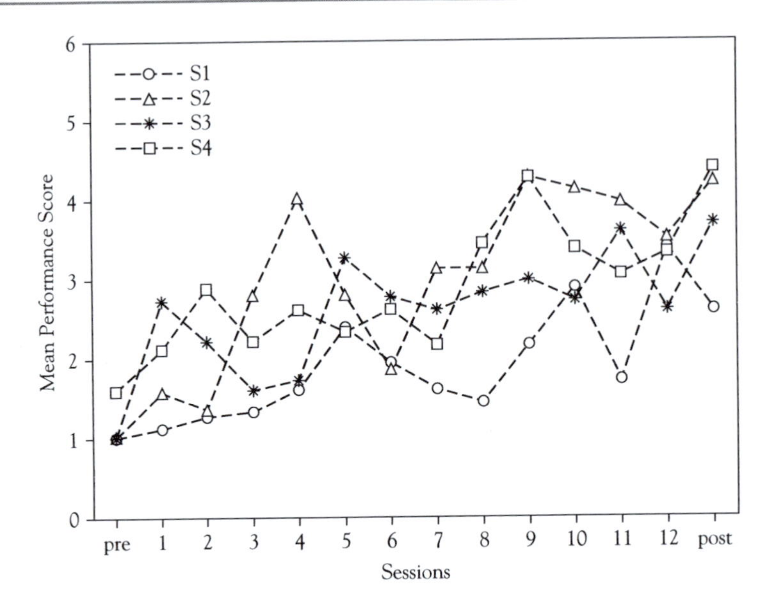

Figure 38.2 Individual mean performance scores across different target zones. *AD (pre: 1.0 ± 0; post: 2.6 ± 2.07); *CL (pre: 1.0 ± 0; post: 4.3 ± 1.81); *YH (pre: 1.0 ± 0; post: 3.7 ± 2.31); *KL (pre: 1.6 ± 1.26; post: 4.4 ± 2.01), *$p < 0.05$.

with practice. The increase in intra-limb consistency further highlighted the progression from early to later stages of learning for the novice participants (see Newell, 1985; Chow, Davids, Button and Koh, 2006). As can be seen in Figure 38.3, the reduction in variability across different trials blocks for each individual followed not a strict linear trend but showed a number of peaks and troughs. Since no explicit instructions on the specific technique for the kicking task were provided the improvement in performance and overall decrease in intra-limb movement variability occurred purely as a consequence of practice.

The multi-articular task provided a suitable research vehicle to examine how DoF can be controlled and coordinated which was not possible in many previous motor learning studies investigating single joint movements only (Chow, Davids, Button and Koh, 2006). Findings from recent studies on the acquisition of multi-articular actions have questioned a typically held belief that learners initially reduce ('freeze') DoF and gradually release ('free') joints as a function of practice (Bernstein, 1967; Vereijken, van Emmerik, Whiting and Newell, 1992). It seems that this process may not be uni-directional but is dependent on task constraints (e.g., Hong and Newell, 2006; Button, MacLeod, Sanders and Coleman, 2003; Newell, Broderick, Deutsch and Slifkin, 2003).

The four learners in the soccer chipping study demonstrated different trends in terms of joint motion changes while achieving a common task goal. There was evidence shown by some of the participants to alternate between movement patterns with reduced and increased active DoF. This observation is to be expected from learners exploring the perceptual-motor workspace for solution manifolds during skill acquisition. The lack of a common trend in joint involvement could be due to different inter-individual intrinsic dynamics. It was observed that one participant (S2) had configured motor system DoF to execute a mix of a drive (i.e., trying to maximize ball velocity) and a lifting technique early in learning and

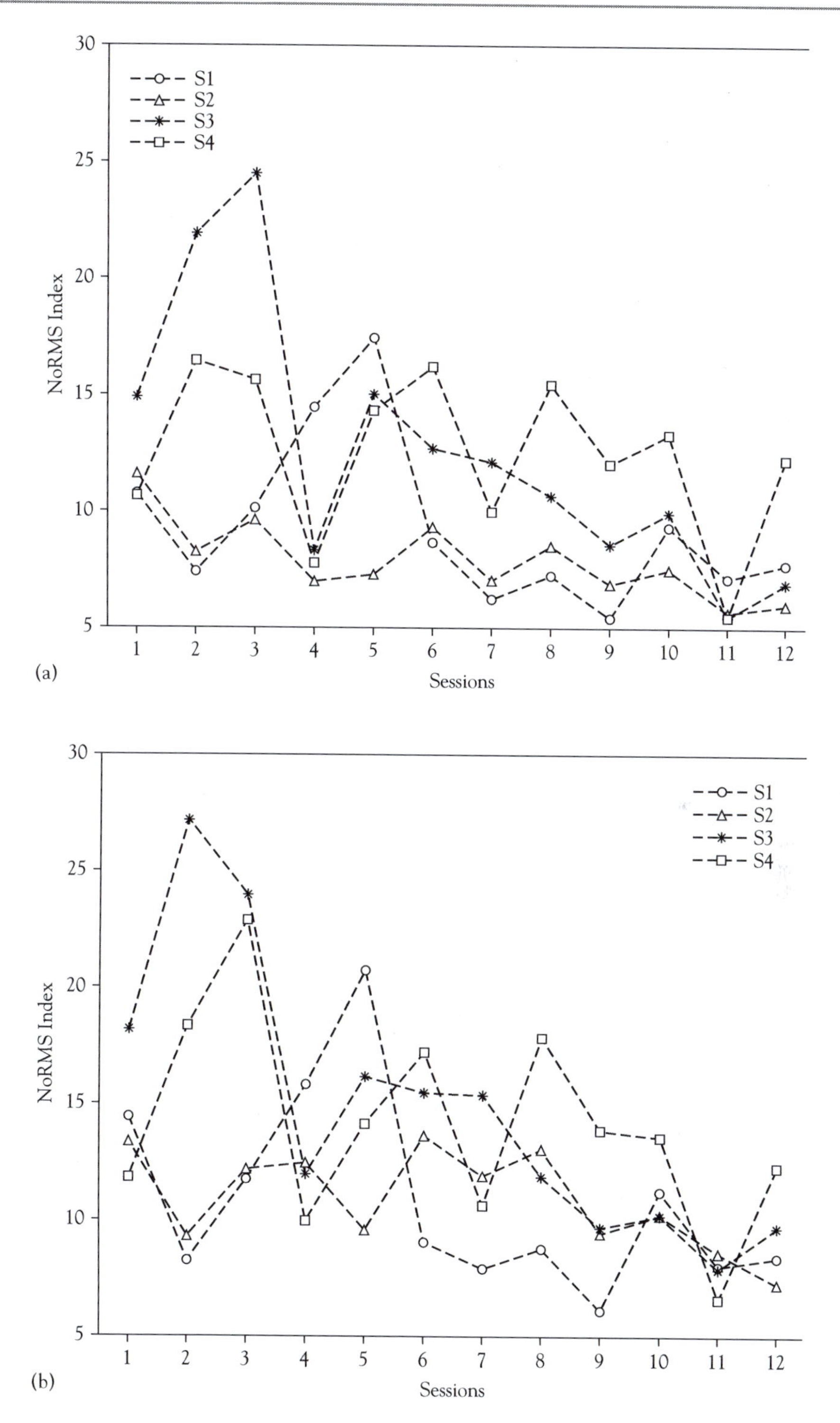

Figure 38.3 NoRMS indices: (a) hip-knee; and (b) knee-ankle intra-limb coordination for all participants from session 1 to 12.

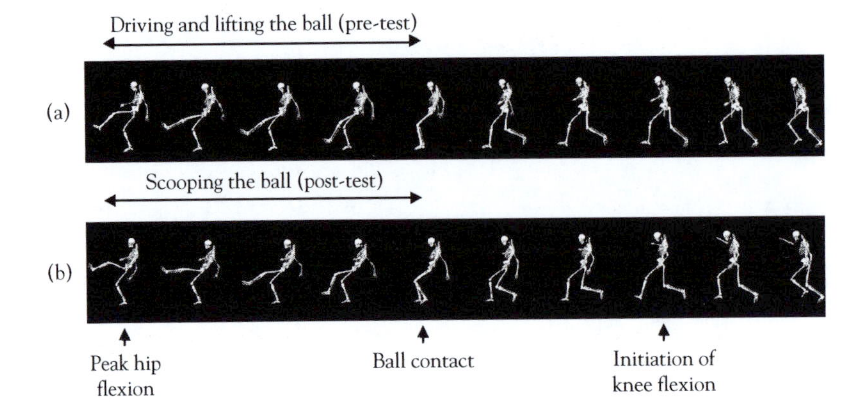

Figure 38.4 Skeletal model of a representative kicking trial at (a) pre-test; and (b) post-test for participant S2.

eventually acquired a scooping technique by the end of the study. In relation to perceptual-motor workspace changes, it can be postulated that the trough (or attractor) of the initial 'driving and lifting' action was deeply entrenched. With practice, the eventual emergence of a 'scooping' action (which is associated with a bigger range of joint motion) would seem to be a consequence of the alteration of the existing attractor rather than the creation of a completely new pattern (refer to Figure 38.4 for the general similarity in patterns in pre- and post-tests sessions). In this sense, the original intrinsic dynamics interacting with the specific task constraints have an over-riding influence on affecting the eventual organization of the scooping movement pattern seen later in learning.

The functional role of movement variability in effecting a transition between movement patterns was also investigated for the kicking task. A cluster analysis procedure was used to group similar movement patterns and a cluster movement switch ratio[2] (SR) (see Wimmers, Savelsbergh, Beek and Hopkins, 1998) provided an indication of the variability of the movement pattern clusters within a practice session. The cluster distribution over successive trials in Figure 38.5 highlighted the intra-individual different paths taken by the participant during the skill acquisition period. Participant S3 showed increased SR (SR = 0.53 for session 2 and SR = 0.41 for session 3) prior to a change from movement cluster 1 to 6 between sessions 1 and 4 (Figure 38.5). Similarly, increasing variability in movement clusters within session 5 (SR = 0.47) was observed before the use of cluster 2 appeared to be preferred at session 7 (SR = 0.12, 94% occurrence for cluster 2). However, for participant S2, a change in preferred movement pattern occurred even when SR was low prior to the change. For example, SR was zero in session 2 prior to a change in preferred pattern to cluster 6 in session 3. Clearly, an increase in movement pattern variability was not a prerequisite for a transition between preferred movement patterns across participants.

The findings suggest that informational and intentional constraints can play a role in effecting the search for pathways of change in movement patterns, especially in discrete, multi-articular actions such as kicking a ball. Unlike continuous cyclical movements, the trial-based nature of discrete movements potentially allow informational and intentional constraints to over-ride existing coordination dynamics of the learner. Certainly, the role of behavioural information in practice deserves further investigation, since such information

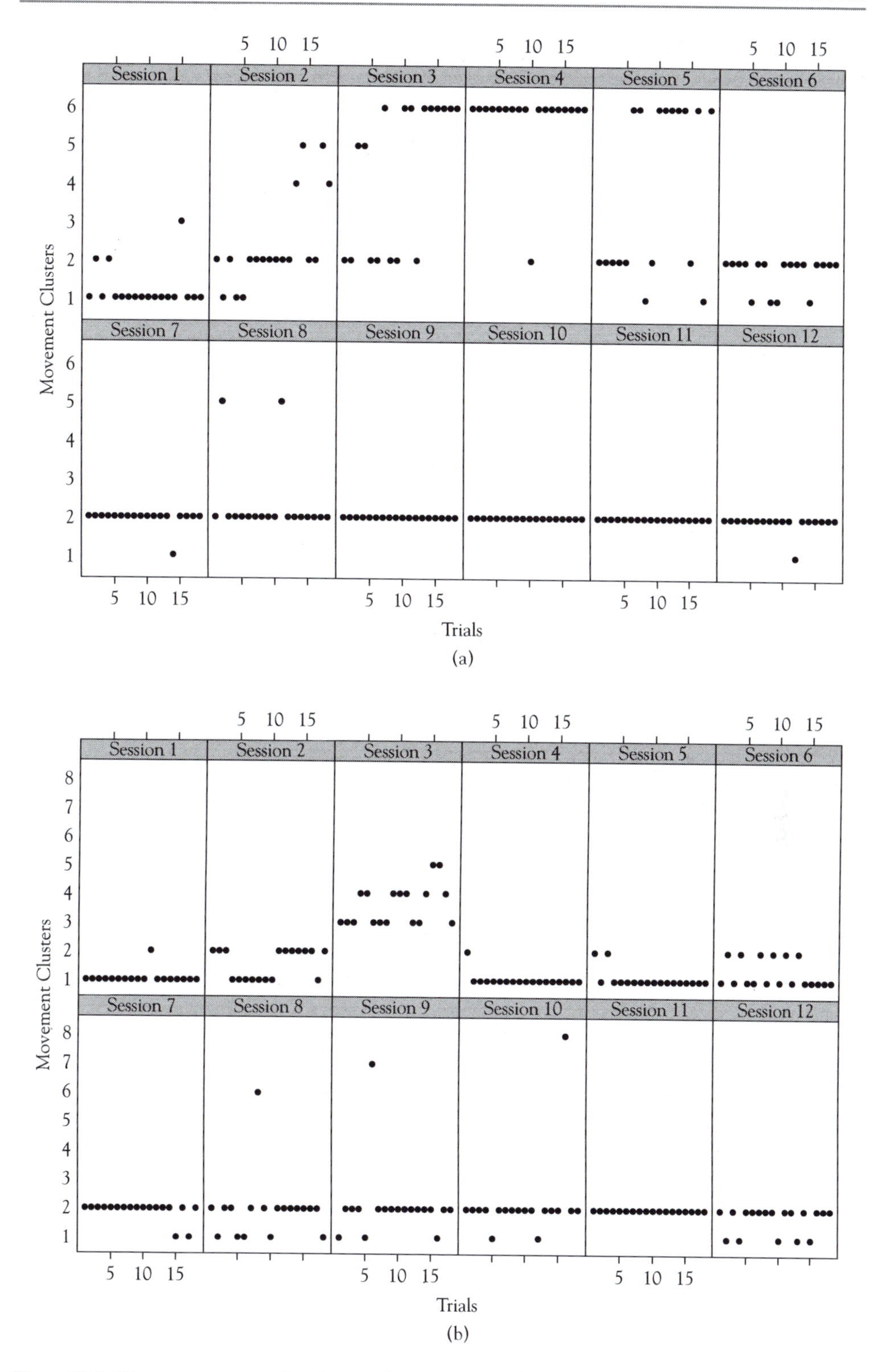

Figure 38.5 Cluster movement distribution for participants (a) S3; and (b) S2 over practice sessions.

from the task and from previous practice sessions/trials are likely to influence the search processes within the perceptual motor workspace of individual learners. The findings from the soccer kicking programme of work have augmented the theoretical discussions on how different pathways of change are dependent on the unique interactions among the constraints in the learning environment.

Conclusions and implications

This chapter has highlighted that the examination of skill acquisition from a constraints-led perspective requires a thorough and detailed analysis of coordination dynamics (as well as performance changes) as a function of practice. In this respect the synergy of the motor learning and biomechanics fields of study is a welcome and necessary progression to complement this multidisciplinary theoretical framework of the performer shifting the focus of investigations of learning processes from performance variables only to studying both changes in actual movement behaviour and performance.

Characterizing learner as complex, biological systems promotes awareness by practitioners that a learner's coordination solutions are the products of self-organization and that periods of movement variability (or instability) should be valued as part of the learning process (Chow, Davids, Button, Shuttleworth, Renshaw and Araújo, 2006). To encourage acquisition of functionally relevant coordination solutions performer-environment interactions should be manipulated through altering relevant task, environmental and performer constraints. Thereby, constraints operate on different timescales, which has important implications for the practitioner's judgment of the learner's rate of progress (see Table 38.1 for practical implications from a constraints-led perspective). When learning a new coordination

Table 38.1 Summary of practical implications from a constraints-led perspective.

Practical issue	*Implication from a constraints-led perspective*
Amount of practice	Practitioners should not overemphasize *quantity* of time over *quality*. The learner should not be engaged in activities where repetition of a movement pattern is the main goal.
Influential constraints	Sport scientists and practitioners need to work together to fully understand the nature of the personal, task and environmental constraints acting on each individual performer in different sports and physical activities.
Practice activities	Careful identification and manipulation of key constraints on learners is an effective way for practitioners to challenge them to search the perceptual-motor workspace.
Individual differences	Individuals learn at different rates due to the amount of competition~cooperation between intrinsic dynamics and behavioural information. Practitioners need to promote individuality and innovative problem solving amongst learners rather than to expect everyone to follow the same learning pathway.
Rate of learning	Learning can take the form of sudden jumps or gradual adaptation depending on the degree of change required in the perceptual-motor workspace.
Movement variability	Variability in movement performance needs to be carefully interpreted. For each performer, sometimes high levels of variability help in adapting to the environment, while on other occasions low levels help increase performance stability.

pattern, more permanent behavioural changes take longer to appear than immediate adaptations to task constraints during practice. Practitioners should understand that some behaviours might represent transient adaptations to immediate task constraints imposed during practice, which interact with organismic constraints related to developmental status. The concept of the perceptual-motor workspace provides an intuitive operationalization of this process, providing the necessary framework for experimentation and intervention programmes of work.

The challenge ahead for scientists and practitioners is to use this emerging information in the design of productive pedagogical activitities (Davids, Button and Bennett, 2007). Such activities should promote learners to actively explore the perceptual-motor workspace throughout their journey towards skilled performance.

Notes

1 One can readily identify in such work the traditional philosophical influence of reductionism that have since characterized many theoretical approaches to motor behaviour (Davids, Button and Bennett, 2007).

2 Cluster movement switch ratio (SR) provides an index on the strength of attraction of a movement pattern and is defined as the number of switches divided by the maximal possible number of switches within a practice session. A switch is possible between any two neighbouring trials. For example, in a series of three trials, the maximum number of switches would be two.

References

1. Al-Abood, S. A., Davids, K. and Bennett, S. J. (2001) 'Specificity of task constraints and the effect of visual demonstrations and verbal instructions on skill acquisition', *Journal of Motor Behaviour*, 33: 295–305.
2 Bennett, S., Davids, K., and Craig, T. (1999). The effect of temporal and informational constraints on one-handed catching performance. *Research Quarterly for Exercise and Sport*, 70(2), 206–211.
3. Bernstein, N. (1967). *The Co-ordination and Regulation of Movements*. Pergamon Press Ltd.
4. Bryan, W. L. and Harter, N. (1897) 'Studies in the physiology and psychology of the telegraphic language', *Psychological Review*, 4: 27–53.
5. Button, C., MacLeod, M., Sanders, R. and Coleman, S. (2003) 'Examining Movement Variability in the Basketball Free-Throw Action at Different Skill Levels', *Research Quarterly for Exercise and Sport*, 74(3): 257–269.
6. Chen, H. H., Liu, Y. T., Mayer-Kress, G. and Newell, K. M. (2005) 'Learning the pedalo locomotion task', *Journal of Motor Behaviour*, 37(3): 247–256.
7. Chow, J.-Y., Davids, K., Button, C. and Koh, M. (2006) 'Organisation of motor system degrees of freedom during the soccer chip: An analysis of skilled performance', *International Journal of Sport Psychology*, 2—3: 207–229.
8. Chow, J.-Y., Davids, K., Button, C. and Koh, M. (2007) 'Coordination changes in a discrete multi-articular action as a function of practice', *Acta Psychologica*, in press.
9. Chow, J.-Y., Davids, K., Button, C., Shuttleworth, R., Renshaw, I. and Araújo, D. (2006) 'Nonlinear Pedagogy: A Constraints-Led Framework to Understanding Emergence of Game Play and Skills', *Nonlinear Dynamics, Psychology and Life Sciences*, 10: 71–103.
10. Collins, J. J. (1999) 'Fishing for function in noise', *Nature*, 402: 241–242.
11. Davids, K., Button, C. and Bennett, S. J. (2007) *Dynamics of Skill Aquisition: A Constraints-Led Approach*. Champaign, Ill.: Human Kinetics.

12. Davids, K., Button, C., Renshaw, I., Araújo, D. and Hristovski, R. (2006) 'Movement models from sports provide representative task constraints for studying adaptive behaviour in human motor systems', *Adaptive Behaviour*, 14(1): 73–95.
13. Davids, K., Glazier, P., Araujo, D. and Bartlett, R. (2003) 'Movement Systems as Dynamical Systems', *Sports Medicine*, 33(4): 245–260.
14. Ehrlich, P. R. (2000) *Human Natures: Genes, Cultures, and the Human Prospect*. Washington, D.C.: Island Press.
15. Haken, H. (1996). *Principles of Brain Functioning*. Berlin: Springer.
16. Herbert, E. P. and Landin, D. (1994) 'The effects of learning model and augmented feedback on tennis skill acquisition', *Research Quarterly for Exercise and Sport*, 65(3): 250–257.
17. Hong, S. L. and Newell, K. M. (2006) 'Practice effects on local and global dynamics of the ski-simulator task', *Experimental Brain Research*, 169: 350–360.
18. Kelso, J. A. and Zanone, P. G. (2002) 'Coordination dynamics of learning and transfer across different effector systems', *J Exp Psychol Hum Percept Perform*, 28(4): 776–797.
19. Kelso, J. A. S. (1995) *Dynamic patterns: The self-organisation of brain and behaviour*. Cambridge, MA: MIT Press.
20. Kelso, J. A. S. and Engstrom, D. A. (2006) *The complementary nature*. Massachusetts: Bradford Books.
21. Kelso, J. A. S., Holt, K. G., Kugler, P. N. and Turvey, M. T. (1980) 'On The Conception Of Coordinative Structures As Dissipative Structures: II. Empirical Lines Of Convergence'. In G. E. Stelmach and J. Requin (eds.), *Tutorials in Motor Behaviour* (pp. 49–70).
22. Kelso, J. A. S. and Zanone, P. G. (2002) 'Coordination Dynamics of Learning and Transfer Across Different Effector Systems', *Journal of Experimental Psychology: Human Perception and Performance*, 28(4): 776–797.
23. Liu, Y.-T., Mayer-Kress, G. and Newell, K. M. (2006) 'Qualitative and quantitative change in the dynamics of motor learning', *Journal of Experimental Psychology: Human Perception and Performance*, 32(2): 380–393.
24. Lyle, J. (2002) *Sports Coaching Concepts: A Framework for Coaches' Behaviour*. Abingdon, Oxon: Routledge.
25. Muchisky, M., Stowe, L. G., Cole, E. and Thelen, E. (1996) 'The Epigenetic Landscape Revisited: A Dynamic Interpretation'. In C. Rovee-Collier and L. P. Lipsitt (eds.), *Advances in Infancy Research* (Vol. 10, pp. 121–159).
26. Müller, H. and Sternard, D. (2004) 'Decomposition of Variability in the Execution of Goal-Oriented Tasks: Three Components of Skill Improvement', *Journal of Experimental Psychology: Human Perception and Performance*, 30(1): 212–233.
27. Mullineaux, D. R., Bartlett, R. M. and Bennett, S. (2001) 'Research design and statistics in biomechanics and motor control', *Journal of Sports Sciences*, 19: 739–760.
28. Newell, K. M. (1985) 'Coordination, Control and Skill'. In D. Goodman, R. B. Wilberg and I. M. Franks (eds.), *Differing Perspectives in Motor Learning, Memory and Control* (pp. 295–317). Elsevier Science Publishers B.V.
29. Newell, K. M. (1986) 'Constraints on the Development of Coordination'. In M. G. Wade and H. T. A. Whiting (eds.), *Motor Development in Children: Aspects of Coordination and Control* (pp. 341–359). Nighoff: Dordrecht.
30. Newell, K. M., Broderick, M. P., Deutsch, K. M. and Slifkin, A. B. (2003) 'Task Goals and Change in Dynamical Degrees of Freedom With Motor Learning', *Journal of Experimental Psychology: Human Perception and Performance*, 29(2): 379–387.
31. Newell, K. M., Liu, Y.-T. and Mayer-Kress, G. (2001) 'Time scales in motor learning and development', *Psychological Review*, 108(1): 57–82.
32. Ollis, S., Button, C. and Fairweather, M. (2005) 'The influence of professional expertise and task complexity upon the potency of the CI effect', *Acta Psychologica*, 118(3): 229–244.
33. Ragert, P., Schmidt, A., Altenmüller, E. and Dinse, H. R. (2004) 'Superior tactile peformance and learning in professional pianists: evidence for meta-plasticity in musicians', *European Journal of Neuroscience*, 19: 473–478.

34. Schmidt, R. A. (1985) 'The search for invariance in skilled movement behaviour. The 1984 C.H. McCloy Research Lecture', *Research Quarterly for Exercise and Sport*, 56: 188–200.
35. Summers, J. J. (2004) 'A historical perspective on skill acquisition'. In A. M. Williams and N. J. Hodges (eds.), *Skill Acquisition in Sport: Research, Theory and Practice* (pp. 1–26). Abingdon, Oxon: Routledge.
36. Thelen, E. and Smith, L. B. (1994) *A Dynamic Systems Approach to the Development of Cognition and Action*. Cambridge, MA: MIT Press.
37. Vereijken, B., van Emmerik, R. E. A., Whiting, H. T. A. and Newell, K. M. (1992) 'Free(z)ing Degrees of Freedom in Skill Acquisition', *Journal of Motor Behaviour*, 24(1): 133–142.
38. Vickers, J. N. (1996) 'Control of Visual Attention During the Basketball Free Throw', *The American Journal of Sports Medicine*, 24(6): 93–97.
39. Visser, J., Geuze, R. H. and Kalverboer, A. F. (1998) 'The relationship between physical growth, the level of activity and the development of motor skills in adolescence: Differences between children with DCD and controls', *Human Movement Science*, 17(4–5): 573–608.
40. Waddington, C. H. (1966) *Principles of development and differentiation*. N.Y.: Macmillan.
41. Wenderoth, N. and Bock, O. (2001) 'Learning of a new bimanual coordination pattern is governed by three distinct processes', *Motor Control*, 1: 23–35.
42. Wimmers, R. H., Savelsbergh, G. J. P., Beek, P. J. and Hopkins, B. (1998) 'Evidence for a Phase Transition in the Early Development of Prehension', *Developmental Psychobiology*, 32(3): 235–248.
43. Zanone, P. G. and Kelso, J. A. S. (1992) 'Evolution of Behavioural Attractors with learning: Nonequilibrium Phase Transitions', *Journal of Experimental Psychology: Human Perception and Performance*, 18(2): 403–421.
44. Zanone, P. G. and Kelso, J. A. S. (1997) 'Coordination Dynamics of Learning and Transfer: Collective and Component Levels', *Journal of Experimental Psychology: Human Perception and Performance*, 23(5): 1454–1480.

39

Application of biomechanics in martial art training

Manfred M. Vieten
University of Konstanz, Konstanz

Abstract

An overview of the martial arts literature in English related to biomechanics is given. Unlike some popular sports like running, skiing, tennis, the percentage of biomechanics related papers among the literature in martial arts is very low (2 per cent). Most studies are devoted to a single martial art like Aikido, Judo, Karate, Wu Shu/Kung Fu, Tae Kwon Do or Tai Chi and are either concerned with injury probability and injury type or are of descriptive nature using cinematography, force plates, and EMG. The producers of descriptive papers try to find those parameters being significantly different between expert-practitioners and novices or they give values which most probably are important for the success of a technique as velocity, force, and muscle activity. We propose a future goal-oriented research with a combination of theoretical work, for example, in the form of computer models, experiments, and innovative analyzing methods.

Introduction

The *Encyclopaedia Britannica* online (2007) defines Martial art as: 'Any of several arts of combat and self-defense that are widely practiced as sport'. Examples of armed arts given are Kendo (sword fighting) and Kyudo (archery). The unarmed sports named are Aikido, Judo, Karate, Kung Fu and Tae Kwon Do. The *American Heritage Dictionary* (2004) offers a similar definition. In the broadest sense martial arts comprise the use of weaponry, unarmed combat and physical exercise to improve health, parts of traditional Chinese medicine, meditation and intellectual concepts influenced by religious and philosophical systems of the East. However, the biomechanics of martial art is concerned with the physical laws applied to the biological system, the human body. Martial arts in general comprise a huge variety of techniques and styles. However, there is a small group of procedures used in the martial arts' unarmed combat. These are kicking, punching, throwing, falling, joint locking, choking and the application of pressure-points. Martial arts can roughly be divided into two groups. The first group, the 'grappling arts' comprises mainly of throwing, falling, joint locking and choking. Prominent members are Aikido and Judo. The second group uses mainly kicking and punching, which might be called 'kick-punch arts'. Well known representatives are Tae Kwon Do and Karate. Karate does know throwing and falling but this is of minor interest in sport competitions. Wu Shu/Kung Fu is very diverse in terms of techniques but has a tendency toward kicking and punching. Grappling as well as kick-punch styles can involve the application of pressure-points. However, from the biomechanical standpoint it is of minor importance and the reflection of this topic in the scientific literature is almost zero. We are going to restrict this survey to the unarmed arts Aikido, Judo, Karate, Wu Shu/Kung Fu, Tae Kwon Do (TKD), Tai Chi/Taichichuan but do include papers containing the keyword martial art(s) regardless of the described art. Judo and Tae Kwon Do are Olympic sports and are together with Karate by far the best known and most practiced martial arts worldwide. According to the International Judo Federation there are 187 affiliated member nations, the World Taekwondo Federation lists 182 nations affiliated and Karate names 175 affiliated member nations. The named arts constitute the biggest contingent of the martial arts literature written in English devoted to a single art. In the sports data base SportDiscus we find on each of the above mentioned arts between 400 and 1400 papers and for the keyword Martial art(s) more than 3800 papers (see Table 39.1). If all listed arts and the term Martial art(s) are connected with an OR, SportDiscus delivers 8026 papers. Compared with the sum of all arts, which gives 8462, we find that just about 400 papers are listed multiply.

For a practitioner biomechanics can help to answer two important questions:

1. How to avoid injuries
2. How to improve technical skills and keep the ability at the top level

Successful execution of technically demanding martial arts motions on the highest level is based on two steps:

1. Knowledge about how to move appropriate (kinematics).
2. Ability to execute the movements (kinetics/dynamics).

To find studies on injury and technical skills (kinematics and dynamics) we searched for papers with the keywords biomechanics, injury, kinematics or kinetics/dynamics combined with the names of the above mentioned martial arts. The result is shown in Table 39.1.

Table 39.1 Number of papers related to Biomechanics/Injury/Kinematics/Kinetics or Dynamics

Art/Sport	*No of papers*	*Key word biomechanics*	*Key word injury*	*Key word kinematics*	*Key word kinetics / dynamics*
Aikido	407	1 (0.25%)	7 (1.7%)	0 (0.0%)	0 (0.0%)
Judo	951	36 (3.8%)	58 (6.1%)	9 (0.9%)	10 (1.1%)
Karate	1429	53 (2.7%)	101 (7.1)	13 (0.9%)	11 (0.8%)
Wu Shu/Kung Fu	590	11 (1.9%)	8 (1.4%)	2 (0.34%)	3 (0.5%)
Tae Kwon Do	667	27 (4.0%)	43 (6.4%)	5 (0.7%)	11 (1.6%)
Tai Chi/Taichichuan	576	28 (4.9%)	8 (1.4%)	6 (1.0%)	8 (1.4%)
Martial Arts	3842	28 (0.73%)	147 (1.8%)	5 (0.13%)	15 (0.4%)
Any of the keywords from above	8026	160 (2.0%)	311 (3.9%)	37 (0.46%)	47 (0.6%)

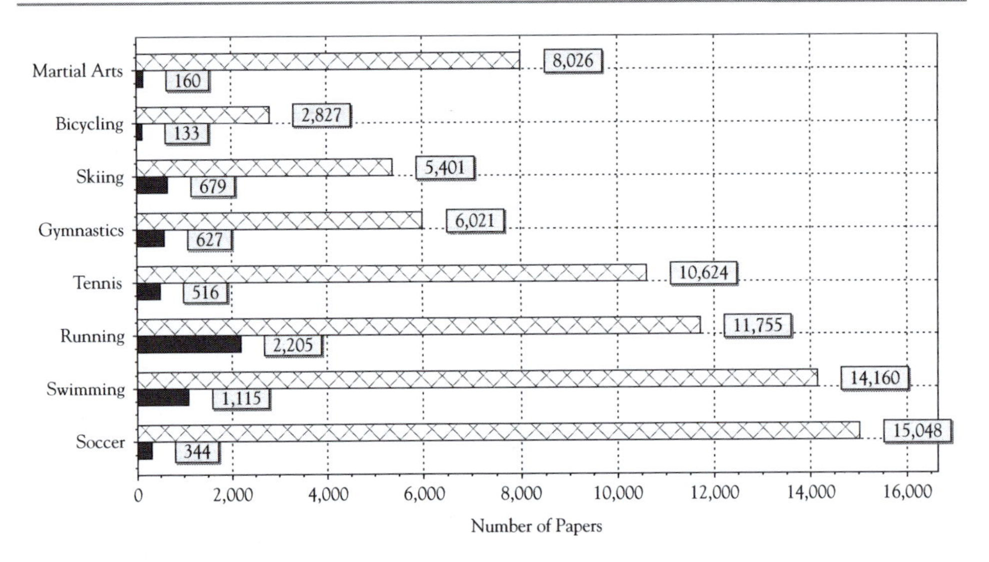

Figure 39.1 Number of papers (diagonal crossed) and number of papers concerned with biomechanics (solid).

Injury-related topics constitute between 1.4 per cent and 7.1 per cent of all papers of the particular art. In general biomechanics related are just 0.25 per cent to 4.9 per cent of a respective art. If we look for a combination with the keyword 'Kinematics' a mere 0.0 per cent to 1.0 per cent of all papers per art can be found. The combination with the words 'Kinetics/Dynamics' gets the meagre results between 0.0 per cent and 1.6 per cent. The comparison between the number of all martial art papers and the number of papers of other major sports shows (Figure 39.1) that the 8026 martial arts papers is about in the range of the average number of papers on the other listed sports. However, the 160 biomechanics related papers present the second lowest absolute number of papers concerned with a specific sport and biomechanics.

The fraction of biomechanics papers with 2 per cent (Figure 39.2) relative to the percentage of the other sports is by far the lowest. On the other hand, the total number of martial arts papers indicates a high interest in the field, the low percentage of biomechanics papers however makes clear how little is known about the science of martial arts.

What is already known – what research is done?

'Biomechanics of human movement can be defined as the interdiscipline which describes, analyzes, and assesses human movement' (Winter, 2005). This definition includes a benchmark to judge the maturity of the subject. Since the first papers on the biomechanics of martial arts at the end of the 60th and the beginning of the 70th centuries the collection grew up to the number of 160. Are these papers still of descriptive nature? Are they in the stage of analyzing or is even assessment on the agenda?

Injury

Zetaruk *et al.* (2005) presented a comparative study on martial artists from five different styles. They recruited 263 subjects, of which 114 practiced Shotokan Karate; 47 Aikido;

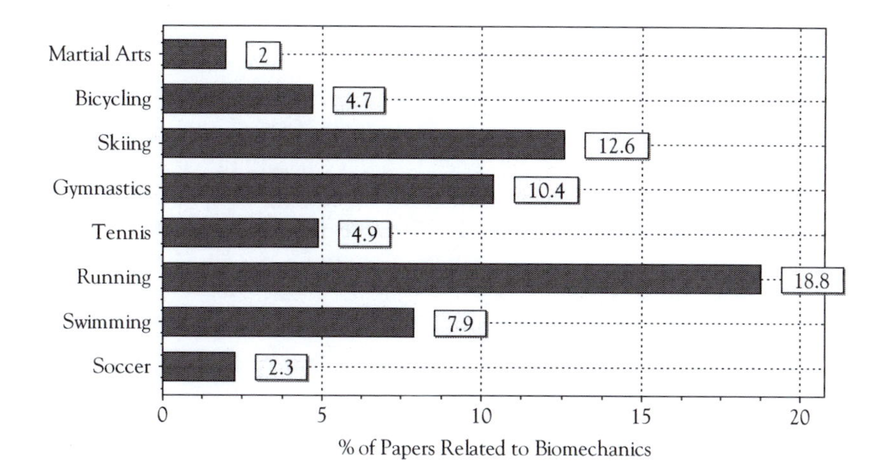

Figure 39.2 Biomechanics related papers in comparison to the total number of the specific sport.

49 Tae Kwon Do; 39 Kung Fu; and 14 were active in Tai Chi. They discriminated the groups according to age, sex, training frequency (two groups with training intensity either up to three hours a week or more than three hours a week), experience (below three years or above) and, of course, the five different martial art styles. An occasion was rated as an injury if a subject needed to require time off from training or competition. At least seven days off, immobilization or surgery defined a major injury and three or more injuries were termed multiple injury. The reported injury rates and locations by style are shown in Table 39.2.

The greatest risk of injuries were detected for the Tae Kwon Do practitioners (59.2 per cent) followed by those practicing Aikido (51.1 per cent). Aikido experiences a pretty high injury rate taking into account that Aikido does not participate in competitions. Judo, which was not included in the study, belongs to the same style family in which throwing, falling and

Table 39.2 Injury rates and locations by style

	No of participants	*Injuries*		*Major injury*	*Multiple injury*	*Upper extremity*	*Lower extremity*	*Groin*	*Trunk*	*Head/neck*
Style		*No*	*Per cent*	*Per cent*	*Per cent*	*Per cent*	*Per cent*	*Per cent*	*Per cent*	*Per cent*
Karate	114	34	29.8	16.7	18.4	16.7	22.8	0.9	14.9	9.6
TKD	49	29	59.2	26.5	44.9	40.8	57.1	18.4	24.5	30.6
Aikido	47	24	51.1	27.7	31.9	42.6	34.0	6.4	25.5	31.9
Kung Fu	39	15	38.5	17.9	23.1	20.5	35.9	5.1	12.8	10.3
Tai Chi	14	2	14.3	7.1	0.0	7.1	7.1	0.0	7.1	7.1
Total	263	104	39.5	20.2	25.5	25.9	32.3	5.7	17.9	17.5

Source: Zetaruk *et al.*, 2005.

joint locking are common and, in addition, competing is a major goal for many practitioners. Does this result in an even higher injury rate? In 1980 Kurland had already conducted 'a study on the comparison of Judo and Aikido injuries'. He mentioned the similarity of the techniques in Aikido and Judo but also named the different injury locations, with 54 per cent of all Aikido injuries at the lower extremities (Judo 19 per cent) and 59 per cent of all Judo injuries at the upper extremities (Aikido 27 per cent). Kurland presumed a higher injury rate in Judo compared to Aikido but the available data did not allow a confirmation. Most Aikido injuries he identified as mats-related while he cited findings that proposed improper falling techniques in Judo was a major cause of injuries.

The good news from the study of Zetaruk *et al.* (2005) was that younger participants under the age of 18 were at much lower risk of injury than adults. This is supported by a study of Finch *et al.* (1998), in which records of 51,203 children and 46,837 adults attending a National Injury Surveillance Unit (NISU) in Australia, collected during the years 1989 till 1993, were evaluated. They discriminated injuries into sports-related and not sports-related and classified according to the specific sports. In this study martial arts are not within the top ten injury-prone sports for children under the age of 15, whereas martial arts ranks number nine for adults and juveniles above 15. However, the data does not take into account the absolute number of practitioners in the specific sports nor, as they argue, is it possible to anticipate similar statistics for other countries or continents since Australia has a very specific spectrum of sports.

A relative new development is the growing number of mixed martial arts schools and studios. Here, techniques from different martial arts are combined in training and combat. Kochhar *et al.* (2005) analyzed four different takedowns, the o goshi (Judo), the suplex (Jujitsu), the souplesse (a variant of the suplex) and the guillotine drop (a choke hold). They concluded a high potential in cervical injury. From the literature of boxing it is known that the risk of traumatic brain injuries in professional boxing is high (Clausen *et al.*, 2005) but is decreasing because of the shorter boxing career and a significant lower number of bouts per athlete nowadays and also due to the 'increased clinical, neuropsychological, radiological, and genetic monitoring and screening'. Timm *et al.* (1993) presented a study conducted during a 15-years-period (1977–1992) of the injuries and illnesses at the US Olympic Training Center. They conclude that serious injuries present a small percentage (6.1 per cent) of all problems. Haglund and Eriksson (1993) in their retrospective study of 50 former Swedish amateur boxers did not find 'any sign of severe chronic brain damage'. Zetterberg *et al.* (2006), in a study searching for biomedical markers in the cerebrospinal fluid, identified acute neuronal injuries of amateur boxers after bouts. In mixed martial arts we have a combined problem with throwing, kicking, boxing and other techniques. Adequate risk estimation is not available. Regarding injuries in general we are still at the stage of describing.

Skill improvement/performance monitoring

Kinematics is the description of the motion of material objects. Video cinematography in the form of automatic digitizing systems as APAS, Vicon, Simi, Elite, Motion Analysis, etc., are the gold standard in acquiring (marker) coordinates and hence allow a quantitative description. Dynamics is the branch of mechanics that deals with forces and their application, while kinetics is a subset of dynamics dealing with the effects of forces upon the motion of material bodies. Both terms, very close-related, represent elements of analysis. A selection of literature with the keyword 'kinematics' is most probably a descriptive paper, while

Table 39.3 Collection of studies of the kinematics and/or dynamics of martial arts.

No	*Study*	*Year*	*Art(s)*	*Measurement system(s)*	*Parameter(s)*
1	(Andries *et al.*)	(1994)	Karate	3D, FP	c, v, F, CoG
2	(Sorensen *et al.*)	(1996)	TKD	2D, ID, EMG	c, v, a, L, T
3	(Kules and Mejovsek)	(1997)	Karate	3D, FP	c, v, a, CoG, F
4	(Kong *et al.*)	(2000)	TKD	3D	c, v
5	(Tsai and Huang)	(2000)	TWD	FP	F
6	(Lan *et al.*)	(2000)	TKD	3D	c, v
7	(Zhao *et al.*)	(2000)	Tai Chi	2D, FP	c, F
8	(Vieten and Riehle)	(2002)	Judo	3D, CS	c, v, a, CoG, p
9	(Chan *et al.*)	(2003)	Tai Chi	3D, EMG	c, CoG, EMG
10	(Degoutte *et al.*)	(2003)	Judo	BC (Blood chemistry)	E via chemical analysis
11	(Imamura and Johnson)	(2003)	Judo	2D	c, v, a, L
12	(Shahbazi *et al.*)	(2005)	Karate	3D, ID	c, v, a, F, E, P
13	(Emmermacher *et al.*)	(2005)	Karate	3D, ACC, EMG	c, v, a
14	(Witte *et al.*)	(2005)	Karate	3D, ACC, EMG	c, v, a, EMG
15	(Vieten and Riehle)	(2005)	TKD	3D, CS	c, v, a, E, P
16	(Huang *et al.*)	(2005)	Whusu	3D	c, v
17	(Kim *et al.*)	(2005)	Judo	3D	c, v, a, CoG
18	(Lee and Huang)	(2006)	TKD	3D	c, v, a
19	(Wong and Fok)	(2006)	Tai Chi	3D	c, v, a

'kinetics/dynamics' tends to be an indicator for a paper which includes at least two parts of Winter's definition of biomechanics 'describe' and 'analyze'. Table 39.3 shows a collection of papers resulted from searching with these three keywords. The abbreviations within the column 'measurement system(s)' are 2D – 2D motion capture system, 3D – 3D motion capture system, ACC – accelerometer, CS – computer simulation, EMG – electromyography, FP – force plate, ID – inverse dynamics and in the column 'parameter(s)' are *a* – acceleration, *c* – coordinate, CoG – Centre of Gravity, E – Energy, EMG – Readings, *F* – force, *L* – angular momentum, *p* – momentum, *P* – power, *T* – torque, *v* – velocity.

Almost all papers in Table 39.3 have cinematography as a common feature. Typically coordinates (c) as functions of time are the outcome. A numerical differentiation results in the velocity and a second derivative yields the acceleration. These are the variables of kinematics together with the CoG and deducted angular parameters as angle and its derivatives. With momentum *p*, force *F*, angular momentum *L*, torque *T*, energy *E*, and power *P* we have representatives of kinetics/dynamics.

Video cinematography can be used if one person is observed and no masking material objects are hiding the camera's view onto the markers. Tai Chi movements (Chan *et al.*, 2003), traditional forms, also called katas or hyungs and isolated kicks or punch techniques (Andries *et al.*, 1994; Lan *et al.*, 2000; Emmermacher *et al.*, 2005) can be analyzed very well. Using these digitizing systems in situations with two subjects involved is extremely cumbersome. Imamura and Johnson (2003) did a 2D study using one video camera on the osotogari judo throw. Vieten and Riehle (2002) published a 3D study on two judo throws hane-goshi and harai-goshi using three cameras. In both studies the movement directions were arranged before throwing to get the best views of the maneouvre and to get the maximum information recorded. Nevertheless, manual digitizing had to take place to approximate those markers blocked from view. Automated digitizing of combats is still out of range of today's technological possibilities.

Tai Chi

Chan *et al.* (2003) gave a kinematical description of the 'press and push movements'. They used 3D cinematography and EMG to investigate the track of the CoG and the muscle activity during the task. The one subject, an experienced Tai Chi master, showed a very good control of the CoG with very little sway. A sign of a very good movement control. It was concluded that the training effects of Tai Chi might prevent injuries resulting from poor balance or muscle strength. Wong and Fok (2006) also used one subject, a beginner, for their 3D study on the forward push and reported the power depending on the stand width. Zhao *et al.*(2000) analyzed the effect of Tai Chi on the gait pattern of 41 subjects aged 60 ± 4.5 years. All subjects did not show any lower extremity disease history. 21 subjects were Tai Chi practitioners for at least five years. The other 20 subjects of the control group 'took part in other kinds of physical activities such as jogging and aerobic dance'. Significant differences were found for the duration of the double support phase of a gait cycle. 29.7 per cent for Tai Chi practitioners and 31.8 per cent for the control group. There were also differences in the hip and leg movement and the average peak values of the vertical GRF were higher for the Tai Chi practitioners. Tai Chi-related research is mainly concerned with medicine and the implications on health. For example, a Medline search produced 247 hits exceeding those of the other martial arts. However, we are not further presenting those findings because those papers are out of the main topic of this article.

'Kick-punch arts' – Karate, Tae Kwon Do, Wu Shu/Kung Fu

All but one of the listed papers (Table 39.3) of the 'Kick-punch arts' use 2D/3D cinematography. Most analyzed is the roundhouse kick. In Karate this technique is known as mawashi gheri, while Tae Kwon Do calls it bal dung cha gi – variations in the technique might result in different naming (Andries *et al.*, 1994; Kong *et al.*, 2000; Lan *et al.*, 2000; Lee and Huang, 2006; Kules and Mejovsek, 1997). The maximal velocities of the kicking foot or the ankle are reported around 11 m/s with variations between 5–14 m/s. However, one paper reports a maximum of more than 26 m/s with a deviation of almost 9 m/s which, is a factor two higher than what is reported by the others. Maximum velocities of other kicks are hard to find. Sorensen *et al.* (1996) investigated the question if for high front kicks in Tae Kwon Do the 'proximal segment deceleration is performed actively by antagonist muscles or is a passive consequence of distal segment movement, and whether distal segment acceleration is enhanced by proximal segment deceleration'. They used a 2D motion capturing system, inverse dynamics, and EMG. Neither inverse dynamics calculations nor EMG recordings support muscle driven deceleration of the thigh. They associated one-third of the total shank and foot moment being produced by the knee extensor. Two-thirds were coming from the motion-dependent moment, which means being produced by muscles of the trunk and standing leg during the whole movement. Remarkable is the fact that a mathematical model was used (inverse dynamics), which produced a result that was supported by an independent second method (EMG). Vieten and Riehle (2005) presented a study assessing the quality of an outside kick. They processed data from a 3D motion capturing system within a computer simulation system. Such a system represents a mathematical model that has the advantage of build in laws of mechanics. There is no need for programming the differential equations explicitly. In addition the animated volume models allow a visual realization of the complete movement.

In Tae Kwon Do punches are of little interest since hand techniques almost surely do not score in the contemporary scoring system. As a consequence we did not find papers on

Tae Kwon Do punching. Two Karate-related papers from the same research group Emmermacher *et al.*, (2005) and Witte *et al.* (2005) are concerned with the straight fist punch. They report on the acceleration characteristic and the intramuscular coordination. Shahbazi *et al.* (2005) used two cameras to get the arm movements for tsuki techniques (straight fist punch). Depending on the different styles, beginning at the waist or midway to the target, they report velocities between 10 to 16 m/s. They used inverse dynamics to calculate force, energy, and power for the punching upper extremity.

'Grappling arts' – Aikido, Judo

In this section there are four papers on our list. Vieten and Riehle (2002) used video cinematography for a study with five participants. 14 throws were digitized manually to get the 3D curves of the major joints and some other landmarks. This data was put into a simulation programme which visualized the movement in the form of two Hanavan models performing the throws respectively to the falling. Momentums (as functions of time) and momentum transfer were calculated as indicators of the quality of the throws. Degoutte *et al.* (2003) used a method from physiology to calculate energy demands of judo. They recruited sixteen male competitors at interregional level. This non-biomechanical method can in future be compared with a biomechanical method e.g. computer simulation with input from cinematography. Imamura and Johnson's (2003) study represented a kinematical investigation on the osoto-gari, a throwing technique seen often in competition. They had two groups, ten black belt participants with at least five years of experience and ten novices with an experience between six- and twelve-months. The significant differences were found in the peak angular velocity of the trunk flexion and the foot plantar flexion. Kim *et al.* (2005) conducted a study on the uchimata (inner thigh reaping throw), the most successful Judo throw in competitions. They analyzed various traits variables including the movement of the CoG.

Conclusion

We do know several facts regarding injuries. Martial arts have the potential to inflict injuries. There is a hierarchy ranking the degree of danger of each art. Tae Kwon Do, a full contact sport, seems to generate a much higher injury rate than Karate, which uses similar techniques but is known as light or semi-contact sport. Wu Shu/Kung Fu might have rates somewhere between those of Tae Kwon Do and Karate. Judo, a highly competitive sport, seems to have a higher injury rate than Aikido. Tai Chi is an exercise derived from martial arts and nowadays mainly carried out to increase health and fitness. Naturally, there is a very low injury rate. On an absolute scale counting sports injuries establishes martial arts score low. Mixed martial arts on the other hand bear a much higher injury risk and might with increasing popularity elevate the number of injuries. We need more precise injury numbers per training hour or competition. What we also should know are the types of injuries, the reasons for injuries and the injury mechanisms. We need more research on protecting gear and grading material. For example, research by Ackland and Paterson (2006) allows specification of the properties on grading material to make sure braking occurs within an adequate force spectrum to avoid athletes suffering form traumas. New knowledge about the injury mechanism can lead to better training methods e.g., to instruct safe falling. It can also result in changes of competition rules. For example, after introducing the new karate rules in 2000 the injuries rate declined (Macan *et al.*, 2006).

We just begin to know basic facts about kicking, striking, and throwing movements. Facts we should find out are, for example: What is the duration of a specified kick? What is the force involved, peak velocity, acceleration, or what is the optimal segment coordination? What precisely and quantitatively are the factors of a successful throw and what are its principles?

The measurement methods that we will use in the near future are still those already in use today, e.g. 2D/3D cinematography, accelerometers, force plates, EMG, etc., and beside that there might be something new coming up. But the real change in (martial arts) biomechanics research will be how we choose our experimental design and how we analyze our data. Several of the above mentioned studies compare different groups of participants to find out the differences in the values of some parameters which might determine the efficiency of a movement. Meanwhile, we have enough information to start with mathematical models estimating the parameters and values predicting successful techniques. For example, Walker (1975) calculated impact forces of 'karate strikes' from the basic laws of mechanics as a demonstration for the physics classroom. Nonetheless, the underlying equations establish a model, which let us identify those parameters important for the execution and those that can be measured. The comparison of the calculation and the measurements tells us something about the validity of the model and its underlying mathematics. In any way it will lead to a more target-oriented research.

A promising method to analyze data is the 'system approach'. It was first described by Bernstein (1967) and since then used in various movement-related studies, as found in the collection of papers edited by Latash (Latash, 1998; Latash, 2002; Latash and Levin, 2004). Hamill *et al.* (1999) applied the method to sports in a study of the cyclic movement running. The general idea behind 'system approach' as applied to the human body is to treat the body as an interacting structure rather than a collection of individual acting segments.

Another analyzing method is 'computer simulation' (Zatsiorsky, 2002; Vieten, 2006). Computer simulation can serve as a way to model a movement. Especially when systems become increasingly complex, computer simulation with its build in rules of mechanics can help to keep the complication bearable.

Summary

From the more than 8000 articles on martial arts we find on SportDiscus, it is obvious the topic is of great interest worldwide. Of all these papers, 2 per cent are biomechanics-related, which is a very small number compared with the biomechanics related literature on gymnastics 10.4 per cent; skiing 12.6 per cent; or running 18.9 per cent. The biomechanics of the martial arts is still in its infancy. But, we do have some first data. Martial arts do not belong to the most injury-prone sports and especially young participants under the age of 15 are safe compared to other sports. However, there are differences between the arts – those with the attribute of competition and those belonging to the full contact combat sports have a higher injury rate. Martial arts movements are complex and there are many differences among them. Hence, there are very few quantitative descriptions explaining how techniques should be conducted. Naturally, many studies find out how experts' movements are and try to describe a kind of a role model movement. This is done in terms of kinematical data as well as through force readings and muscle activities. The complexity and the multifariousness of Martial arts calls for target-oriented researches. Work on theoretical models e.g., established with computer simulation and its verification by experiment, will give us a deeper insight, so in future we can ask the important research questions. Innovative analyzing methods as the 'system approach' will improve our understanding of martial arts further.

The biomechanics of martial arts is still in the state of describing and partly analyzing. Assessment of these diverse and complex movements in a quantitative and scientific way is not often done so far. However, the author is confident that, with the increasing number of studies coming up, we will be able to expand our knowledge beyond a quantitative description of the status quo. For example, by varying conditions within computer models we should be able to optimize movements, predict measurable values and check them experimentally.

How does the limited knowledge of biomechanics in martial arts help the practitioners today? There exist answers to specific problems regarding particular techniques, e.g., for the straight fist punch (Shahbazi *et al.*, 2005). However, the topics covered are sparse. A comprehensive scientifically motivated manual for martial arts practitioners still has to wait for much more research to be done.

References

1. Ackland, T. and Paterson, C. (2006) 'Destruction properties of grading materials used in the martial arts', *Proceedings of the International Society of Biomechanics in Sports*, 798–802.
2. American Heritage Dictionary (2004), Vol. 2006. Houghton Mifflin Company.
3. Andries, R., Van Leemputte, M., Nulens, I. and Desloovere, K. (1994) *Proceedings of the International Society of Biomechanics in Sports*, 260–265.
4. Bernstein, N. (1967) *The Co-ordination and Regulation of Movements*. Pergamon Press, Oxford.
5. Chan, S. P., Luk, T. C. and Hong, Y. (2003) 'Kinematics and electromyographic analysis of Push movement in Tai Chi: A case study', *British journal of sports medicine*, 37: 339–344.
6. Clausen, H., McCrory, P. and Anderson, V. (2005) 'The risk of chronic traumatic brain injury in professional boxing – change in exposure variables over the past century', *British journal of sports medicine*, 39: 661–664.
7. Degoutte, F., Jouanel, P. and Filaire, E. (2003) 'Energy demands during a judo match and recovery', *British journal of sports medicine*, 37: 245–249.
8. Emmermacher, P., Witte, K. and Hofmann, M. (2005) *Proceedings of the International Society of Biomechanics in Sports*, 844–847.
9. Encyclopedia_Britannica_Online (2007), Vol. 2007.
10. Finch, C., Valuri, G. and Ozanne-Smith, J. (1998) 'Sport and active recreation injuries in Australia: evidence from emergency department presentations', *British journal of sports medicine*, 32: 220–225.
11. Haglund, Y. and Eriksson, E. (1993) 'Does amateur boxing lead to chronic brain damage?', *American journal of sports medicine*, 21: 97–109.
12. Hamill, J., van-Emmerik, R. E., Heiderscheit, B. C. and Li, L. (1999) 'A dynamical systems approach to lower extremity running injuries', *Clinical Biomechanics*, 14: 297–308.
13. Imamura, R. and Johnson, B. (2003) 'A kinematic analysis of a judo leg sweep: major outer leg reap – *osoto-gari*', *Sports biomechanics*, 2: 191–201.
14. Kim, E., Yoon, H., Kim, S. and Chung, C. (2005) 'Biomechanical traits analysis when performing of JUDO UCHIMATA by posture and voluntary resistnance levels of UKE', *Proceedings of the International Society of Biomechanics in Sports*, 848–851.
15. Kochhar, T., Back, D. L., Mann, B. and Skinner, J. (2005) 'Risk of cervical injuries in mixed martial arts', *British journal of sports medicine*, 39: 444–447.

16. Kong, P., Luk, T. and Hong, Y. (2000) 'Difference between Taekwondo kick executed by the front and back leg – A biomechanical study', *Proceedings of the International Society of Biomechanics in Sports*, 268–272.
17. Kules, B. and Mejovsek, M. (1997) 'Kinematic and dynamic analysis of the Ushiro Mawashi Geri', *Kinesiology*, 29(2): 42–48.
18. Lan, Y. C., Wang, S. Y., Wang, L. L., Ko, Y. C. and Huang, C. (2000) 'The kinematic analysis of the three Taekwondo kicking movements', *Proceedings of the International Society of Biomechanics in Sports*, 277–280.
19. Latash, M. (1998) *Progress in Motor Control, Volume 1 – Bernstein's Traditions in Movement Studies*. Human Kinetics.
20. Latash, M. (2002) *Progress in Motor Control, Volume 2 – Structure-Function Relations in Voluntary Movements*. Human Kinetics.
21. Latash, M. and Levin, M. (2004) *Progress in Motor Control, Volume 3 – Effects of Age, Disorder, and Rehabilitation*. Human Kinetics.
22. Lee, C. and Huang, C. (2006) 'Biomechanical analysis of back kicks attack movement in Taekwondo', *Proceedings of the International Society of Biomechanics in Sports*, 803–806.
23. Macan, J., Bundalo-Vrbanac, D. and Romic, G. (2006) 'Effects of the new karate rules on the incidence and distribution of injuries Commentary', *British journal of sports medicine*, 40: 326–330.
24. Shahbazi, M., Sheikh, M. and Amini, A. (2005) 'Kinematic-kinetic comparison of Tsuki technique in performing from waist and from midway in classic and individual styles', *Proceedings of the International Society of Biomechanics in Sports*, 881–884.
25. Sorensen, H., Zacho, M., Simonsen, E. B., Dyhre-Poulsen, P. and Klausen, K. (1996) 'Dynamics of the martial arts high front kick', *Journal of sports sciences*, 14: 483–495.
26. Timm, K. E., Wallach, J. M., Stone, J. A. and Ryan, I. I. I. E. J. (1993) 'Fifteen years of amateur boxing injuries/illnessess at the United States Olympic Training Center', *Journal of athletic training*, 28, 330–334.
27. Vieten, M. (2006) In *http://www.ub.uni-konstanz.de/kops/volltexte/2006/1828/* University of Konstanz, Konstanz, pp. 1–276.
28. Vieten, M. and Riehle, H. (2002) In *International Research in Sports Biomechanics*. (Ed. Hong, Y.) Routledge–Taylor & Francis Group, London and New York, pp. 66–72.
29. Vieten, M. and Riehle, H. (2005) 'Movement quality of martial art outside kicks', *Proceedings of the International Society of Biomechanics in Sports*, 856–860.
30. Walker, J. D. (1975) 'Karate Strikes', *American journal of physics*, 43: 845–849
31. Winter, D. A. (2005) *Biomechanics and motor control of human movement*. Wiley: New York.
32. Witte, K., Emmermacher, P. and Hofmann, M. (2005) 'Electromyographic resistance of Gyaku-Zuki in karate kumite', *Proceedings of the International Society of Biomechanics in Sports*, 861–865.
33. Wong, T. and Fok, A. (2006) 'Body mechanics of Tai Chi Chuan', *Proceedings of the International Society of Biomechanics in Sports*, 506–509.
34. Zatsiorsky, V. M. (2002) *Kinetics of Human Motion*. Human Kinetics, Champaign.
35. Zetaruk, M. N., Violan, M. A., D., Z. and Micheli, L. J. (2005) 'Injuries in martial arts: a comparison of five styles', *British journal of sports medicine*, 39: 29–33.
36. Zetterberg, H., Hietala, M. A., Jonsson, M., Andreasen, N., Styrud, E., Karlsson, I., Edman, A., Popa, C., Rasulzada, A., Wahlund, L. O., Mehta, P. D., Rosengren, L., Blennow, K. and Wallin, A. (2006) 'Neurochemical Aftermath of Amateur Boxing', *Arch-Neurol.*, 63, 1277–80.
37. Zhao, F., Zhou, X., Wei, K. and Liu, W. (2000) 'Effects of Tai Chi execrise on gait pattern in the elderly', *Proceedings of the International Society of Biomechanics in Sports*, 925–929.

40

Developmental and biomechanical characteristics of motor skill learning

Jin H. Yan[1], *Bruce Abernethy*[2] *and Jerry R. Thomas*[3]
[1]*California State University, Hayward and Southwestern University, Chongqing;*
[2]*The University of Hong Kong, Hong Kong and The University of Queensland, Queensland;* [3]*Iowa State University, Ames*

Motor skill learning is one of the most fascinating human experiences. From early childhood to adolescence, from adolescence to early adulthood, and finally from adulthood to the later stages of life, humans continue to acquire motor skills through learning and practice. Cross-sectional experiments on sensory-motor skills suggest that the pattern of human motor and cognitive development over a lifetime resembles an inverted 'U' shape (e.g., Yan *et al.*, 2000). Humans improve motor and cognitive performance in early phases of life and most people reach a peak level of performance and maintain consistent performance during adulthood. In late adulthood, motor and cognitive performance generally deteriorates. A lifespan developmental approach provides a legitimate and natural framework for studying human motor behaviour because it reflects key changes and events that evolve over the course of human development (e.g., locomotion, growth, puberty, maturation, schooling, and aging). These changes influencing motor performance can be genetic or environmental in nature (or a combination of both). Knowledge of lifelong motor performance is required for the design and selection of developmentally appropriate activities to facilitate the teaching and learning of movement skills for people of all ages and ability levels.

During early stages of life, particularly throughout early childhood and adolescence, humans experience considerable changes in both body structure (e.g., gains in body weight and height, increases in length of arms and legs, shifts in the centre of mass (COM) of body segments) and function (e.g., improvements in balance control, motor coordination, movement flexibility, muscular strength). These structural and functional changes result in age-associated rather than age-dependent differences in the underlying mechanisms of movement production and control. A good understanding of developmental and biomechanical characteristics of children's motor performance is, therefore, essential to facilitate skill instruction and learning (Ivanchenko and Jacobs, 2003).

This chapter reviews relevant studies of motor control and learning from a developmental perspective and discusses some critical features of motor performance and the kinematic characteristics of children's motor skill acquisition. The chapter is organized into three major sections. The first section of the chapter summarizes evidence regarding children's motor performance and development as well as the developmental characteristics of motor control and skill learning. The second section uses the skill of overarm throwing as an example to illustrate developmental performance trends and underlying changes in the biomechanics and neural control of a fundamental motor skill. The final section of the chapter deals with some of the implications of the developmental motor control and learning literature for instruction or coaching and for future research directions.

Developmental motor performance

Age- and gender-related differences in children's motor performance

Developmental studies of children's motor performance date back to at least as early as the 1930s (e.g., Wild, 1938). Through research conducted over the past several decades in particular, a considerable volume of literature has now been accumulated concerning the quantitative and qualitative aspects of children's motor skills and development (e.g., Clark and Humphrey, 2002; Thomas and French, 1985). Significant age- and gender-related differences in motor performance have been reported in numerous studies (e.g., Butterfield and Loovis, 1998; Loovis and Butterfield, 1995; Morris *et al.*, 1982; Nelson, Thomas and Nelson, 1991; Sakurai and Miyasjita, 1983). Understanding gender differences in motor performance is one of the most important and interesting areas in the developmental study of motor skills, in that it provides a window into the impact of environmental and biological factors on children's physical activities and motor performance (e.g., Nelson *et al.*, 1986; Thomas and French, 1985).

In a study of throwing, balancing, running for speed, and standing long jump performance in children as young as three years of age, Morris *et al* (1982) found that, at very early ages, age effects on motor performance were greater than gender effects (except for balancing skill); however, as children get older, motor performance becomes better; young males outperform young females in throwing distance, speed of running, and long jump, while young females surpass young males in balance control. In a longitudinal study, Halverson, Roberton and Langendorfer (1982) examined changes in throwing velocity among age groups and between young females and males. They suggested that because of reduced opportunities for practicing throwing, young females were developmentally behind their age-matched male peers in ball velocity and throwing technique. Consistent with these observations, Sakurai and Miyasjita (1983) found no significant gender differences in overarm throwing speed and throwing patterns during the early years (ages 3 or 4 years) but observed that the differences became greater as children grew older (7+ years of age). Children's physical development and opportunities for practice therefore moderate the differences between males and females from, at least, three to nine years of age. Similar results have been observed for kindergarten to elementary school children in other fundamental motor tasks like kicking, catching, striking, and balancing (Butterfield and Loovis, 1998; Loovis and Butterfield, 1995).

In a large-scale meta-analysis of the fundamental motor skills literature, Thomas and French (1985) quantitatively reviewed gender differences in children's physical activities and motor performance from biological and environmental perspectives. Their analyses revealed significant gender differences for 12 of 20 motor tasks. After puberty, boys generally demonstrate superior motor performance to girls in terms of movement outcome on most tasks (e.g., the distance thrown, the throwing velocity, and running speed). However, girls outperform boys in motor performance in some tasks requiring balance control, flexibility, and timing accuracy. The gender differences in late childhood are primarily due to the interaction of biological (physical) and environmental (socio-cultural) factors. The data from the meta-analysis showed that before puberty the gender gap is small and basically influenced by environmental factors. The encouragement and expectations provided by parents, teachers, or peers, and the opportunities to engage in a variety of physical activities seem to be the major sources of gender differences in childhood (Yan and Thomas, 1995).

Developmental changes in motor control and learning

A number of different concepts and approaches that have been developed in an attempt to understand adult motor control and learning have also been applied to the problem of understanding developmental changes in motor skills. Expert-novice paradigms (e.g., Abernethy, 1988; McPherson, and Thomas, 1989; Ward and Williams, 2003), notions of information processing, motor programmes, schemas, and templates of motor memory (e.g., Kail, 1988; Schmidt, 1975; Thomas, 1980), models of the 'speed-accuracy trade-off' (e.g., Hay, 1981; Thomas, Yan and Stelmach, 2000; Yan *et al.*, 2000) and dynamical systems concepts (e.g., Clark and Phillips, 1993; Kugler, 1986; Thelen, Kelso and Fogel, 1987) have all been applied in an attempt to describe and, more importantly, explain and predict motor development. Addressing all these lines of research is beyond the scope of this chapter. From a developmental perspective, one fundamental question for the study of motor behaviour that each of these approaches must address, is 'how are children different from adults in motor control and acquisition?' (Stelmach and Thomas, 1997), or, conversely, 'how does age affect motor skill learning?' (Yan, Thomas and Thomas, 1998). Answers to these questions have practical implications for children's motor skill learning and instruction (Shapiro and Schmidt, 1982; Thomas *et al.*, 2000; Yan, Thomas and Payne, 2002).

Using a task requiring point-to-point fast aiming arm movements, Yan *et al.* (2000) examined the characteristics of motor control across the lifespan (from children as young six years to seniors close to 80 years of age). A remarkable difference was observed between the different age groups in the percentage of movement execution that was devoted to the ballistic (acceleration) phase compared to the corrective (deceleration or home-in) phase. Specifically, young children and the elderly controlled the movement primarily via 'online' correction, as reflected in a relatively short ballistic phase (20–25 per cent of the overall movement time (MT) and distance, and a relatively long corrective phase (75–80 per cent of the MT and or distance)). In addition, the movements of the young children and the seniors were less smooth and less consistent than those of the older children and young adults. This is most likely a consequence of the use of under-developed motor programmes by young children and compromised or deteriorated motor control by seniors. Older children (about 9 years of age) and young adults (about 24 years of age) demonstrated a longer ballistic phase (older children 30–60 per cent; young adults 40–80 per cent of the overall MT and distance, respectively), a shorter corrective phase, and a smoother acceleration component in their movements than did the young children and the elderly. These data can be interpreted as suggesting a greater use by older children and young adults of central mechanisms (e.g., motor programmes) and less dependence on feedback during motor execution by older children and young adults or, in other words, the adoption of a form of control that can result in a faster, smoother, and more coordinated motor performance.

Assuming variables such as the percentage of ballistic or acceleration phase, movement duration, and movement smoothness indeed accurately capture developmental variations in motor control, a subsequent question of interest is whether practice can enhance motor performance by changing the movement structures and reducing jerk (resulting in smoother movements). Using a similar motor task to Yan *et al.* (2000), Thomas *et al.* (2000) repeatedly measured the changes in the variables of percentage of ballistic phase and movement smoothness over five consecutive days of practice. Their study demonstrated that children of 6- and 9-years of age could benefit from practice more than young adults. Over the course of the five days of practice, the 9-year-olds increased by about 55 per cent, and the 6-year-olds by about 70 per cent, the proportion of time spent in the ballistic phase of movement (Figure 40.1),

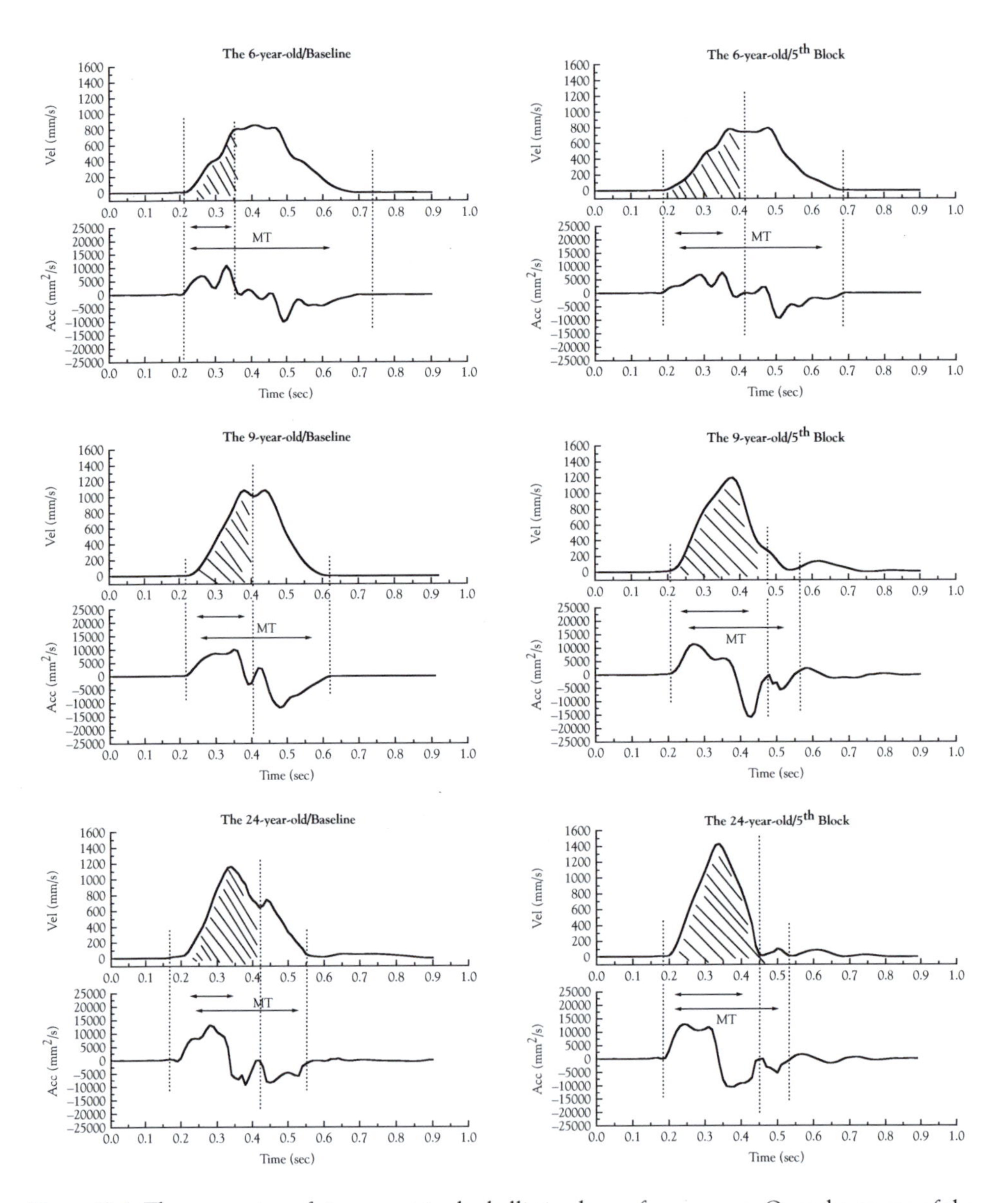

Figure 40.1 The proportion of time spent in the ballistic phase of movement. Over the course of the five days of practice, the 9-year-olds increased by about 55 per cent, and the 6-year-olds by about 70 per cent.

demonstrating that practice results in more programming control and less 'online' corrections. The 9- and 6-year-olds also reduced overall movement jerk by about 60 per cent (Figure 40.2), demonstrating that smoother movements also emerge from task-specific practice, regardless of age.

A further question of interest concerning motor practice relates to how young children differ from older children in learning various motor tasks (e.g., from laboratory to field settings, from slow to rapid movements, or from simple to complex motor tasks). A critical concern is that during childhood, development may moderate the effects of practice due to contemporaneous cognitive development and enriched movement experiences arising from participating in a variety of motor or physical activities (Shapiro and Schmidt, 1982). To examine this issue, Yan *et al.* (1998) quantitatively compared a number of studies examining the effects of variable practice on children from age 3 to 11 years. The results suggested that although the differences among age groups were moderate, age nevertheless exerts some influence on children's motor skill learning. Younger children were found to benefit more from variable practice, and/or have a higher potential to benefit more from variable practice, than older children. As younger children have less motor experience than older children, it is possible they may have more opportunities to develop new, effective motor programmes or motor schema than older children (Shapiro and Schmidt, 1982). Practice or experience, rather than age, also appears to have a greater impact on development of motor expertise (Abernethy, 1988) and motor memory (Thomas *et al.*, 1983). This parallels the observation with older adults that specific practice may allow skills to be either selectively optimized through compensatory adaptations (Baltes and Baltes, 1990) or at least selectively maintained (Ericsson, 2000).

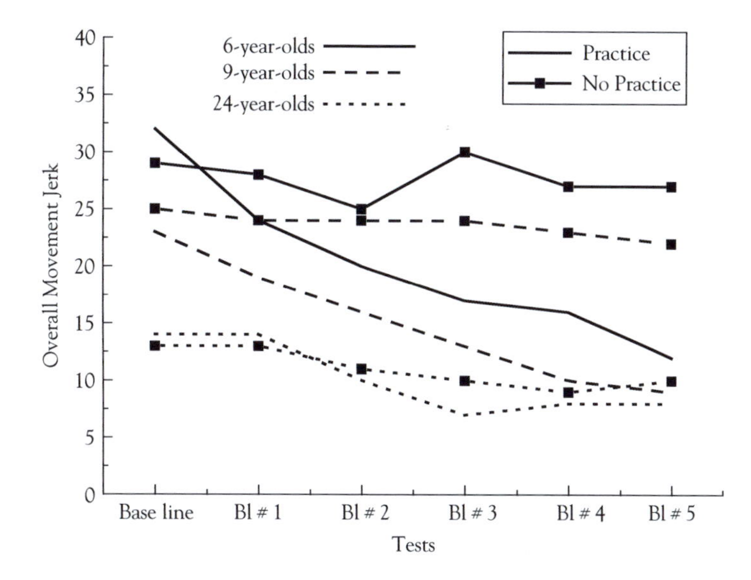

Figure 40.2 The 9- and 6-year-olds also reduced overall movement jerk by about 60 per cent.

New perspectives in developmental motor control and learning

Conceptually, the studies of motor control and learning are closely related. The former concerns understanding the underlying mechanisms or neurophysiological pathways responsible for the production of movement, while the latter primarily focuses on how these mechanisms are mediated by age and the processes or stages involved in motor skill acquisition and adaptation. In a practical sense, the ultimate goal of skill instruction and learning is to control the movements that are being practiced (Schmidt and Lee, 2005). At a behavioural level a number of aspects of movement are typically assessed to reflect changes in motor control as a function of skill learning. For example, measures of movement speed, smoothness, accuracy, and efficiency, as well as the observation of movement forms or techniques, are widely employed in research in human movement science. In this section, several contemporary views about motor control and skill learning will be discussed, with their significance and implications for children's motor control, skill learning and future research directions being addressed in the third section.

Movement variability has traditionally been considered as the manifestation of 'neural noise' (e.g., Newell and Corcos, 1993; Schmidt *et al.*, 1979). Movement development or practice effectively reduces the amount of 'noise' and consequently results in more stable or consistent motor performance (see Meyer *et al.*, 1988 for a review). The emerging contemporary view, however, is that movement variability may be functional, especially for perceptual-motor development (Savelsbergh, van der Kamp and Rosengren, 2006). In general, practice and experience seem to be associated with a developmental reduction in overall movement inconsistencies (e.g., Thomas, 1984; Thomas *et al.*, 2000; Thomas, Gallagher and Thomas, 2001).

In contrast, Deutsch and Newell (2001) suggested that a reduced level of 'noise' might not contribute to the age-related improvement in the consistent production of isometric force by children, suggesting that the adaptation of sensory-motor system to practice may be very task-specific. Recently, Ma *et al.* (2006) proposed that neural noise is fundamental in priming the brain for peak performance. According to Ma *et al*, 'noise' is an effective means of handling the 'uncertainties' of changing task demands or environments. Understanding noise and variability may well be central to further understanding of both motor control and developmental motor learning (Davids, Bennett and Newell, 2006).

In regard to motor memory – the retention of movement skills over time – there has recently been considerable research interest in the concept of motor memory consolidation and its significance for motor skill learning (e.g., Kuriyama, Stickgold and Walker, 2004; Stickgold and Walker, 2005; Walker and Stickgold, 2004). Empirical evidence suggests that newly learned motor skills require processes of memory consolidation to stabilize memorized information against interference and that sleep may be integral to successful consolidation (see Walker, 2005 for a review). Motor performance continues to improve following the completion of practice ('offline' learning) and this between-session skill improvement can be the consequence of memory consolidation and protein synthesis within the CNS (e.g., Cohen *et al.*, 2005; Robertson, Pascual-Leone and Miall, 2004; Rodriguez-Ortiz *et al.*, 2005; Robertson, Press and Pascual-Leone, 2005). However, the implications for sports skill learning across a variety of age groups or skill levels, especially in relation to the timing of practice within the daily sleep-wake cycle, remain to be determined.

New theories are constantly being proposed to account for skill learning. For example, Karni *et al.* (1998) suggested that skill acquisition takes place in three stages: fast 'online' learning during practice, memory consolidation while sleeping, followed by an 'offline' improvement after motor practice. Shadmehr and Brashers-Krug (1997) and Wolpert (1997) have taken a computational approach and proposed an internal model as fundamental to motor skill learning. In their views, motor practice or learning results in the formation of an internal model that contains the memory representation of the motor skill and determines the kinematics and dynamics of the movement, thus permitting a learned motor performance to be carried out automatically. Relatedly, there is growing interest in implicit learning (e.g., learning in the absence of concurrent accumulation of verbalizable knowledge) (e.g., see Abernethy *et al.*, in press). Again, whether the relative reliance on implicit or explicit approaches to learning may vary with age is currently unknown but important. Collectively, these new approaches, with more empirical data from participants of different developmental age, offer hope for shedding new light on the underlying processes or mechanisms of motor skill learning as well as on developmental brain-motor performance relationships.

The development of fundamental motor skills: the example of overarm throwing

Overview

A number of fundamental motor skills have been studied extensively in children, ranging from locomotion skills (e.g., walk, run, and jump) to object-control or projection movements (e.g., catching, throwing, kicking, and striking) (Clark and Humphrey, 2002; Keogh and Sugden, 1985; Thomas and French, 1985). Within the three common methods of object projection (kicking, striking, and throwing) and for the three typical throwing techniques (overarm, underarm, and sidearm), the overarm throw has received most attention and has been historically one of the most commonly used skills to study motor development (e.g., Halverson *et al.*, 1982; Nelson *et al.*, 1991; Yan, Payne and Thomas, 2000). The skill has been investigated for the explicit purpose of understanding the mechanisms of motor control (e.g., McNaughton *et al.*, 2004; Hore and Watts, 2005).

Overarm throwing is a complex and dynamic skill. A competent thrower actively controls fine movements (hand and fingers) and gross movements (arm, trunk, and legs) in producing accurate and forceful movement outcomes. To accomplish the multiple objectives of the task (velocity, distance, and accuracy), muscular strength, whole body coordination and precise timing control are essential. In this section, key biomechanical and developmental characteristics of overarm throwing will be reviewed in relation to the kinematics of motor control and age- and/or gender-associated differences.

Biomechanical characteristics of overarm throws

Overarm throwing requires the control of multi-joint movements (e.g., McDonald, van Emmerik and Newell, 1989). In most situations, throwing accuracy and ball speed are equally important to meet the task and environmental demands. For example, a baseball pitcher throws a ball toward a desirable spot or place as quickly as possible. In this case, both the thrower and the target are stationary. However, when a quarterback passes (throws) a ball to the receiver, both the thrower and the target are moving. Thus, in addition to the skill of aiming and anticipation for the target, how to produce forceful and accurate throws is a question of interest for both motor control and learning and sports biomechanics (e.g., Southard, 2006; Tillaar and Ettema, 2004). Anatomically and biomechanically, the generation of maximum torque and the precise sequencing and timing of each body segment directly contribute to optimal throwing performance (e.g., Cross, 2004; Hore, 1996).

By examining the movement sequence of joint torques in the upper arm, forearm, and hand (with the ball), Herring and Chapman (1992) suggested that skilled throwers typically initiate the joint torques in a timing sequence of 'proximal-to-distal' direction to reach a greatest distance of movement for the ball in the hand immediately before release. This pattern may help maximize ball release velocity. A study of overarm throwing in young females also suggested that a mature thrower executes the forward acceleration movement in the order of shoulder, elbow, and hand movements (Yan, Payne and Thomas, 2000). Furthermore, when angles of the middle finger, thumb, and arm positions at ball release were compared for various release speeds, Hore and Watts (2005) suggested that the timing precision of finger opening at ball release also directly affects throwing accuracy and release velocity. The spatial-temporal coupling of finger movements plays a critical role in reaching the goal of performance and, interestingly, these findings also hold true for cerebellar patients (McNaughton *et al.*, 2004).

Developmental studies of overarm throws

As noted earlier, developmental studies of overarm throwing were initiated by Wild (1938) and numerous studies have followed (e.g., Nelson *et al.*, 1991; Roberton *et al.*, 1982; Yan and Hinrichs *et al.*, 2000). Generally, as children age, grow, mature, and/or gain motor experience, they throw the ball faster and further (e.g., Halverson *et al.*, 1982; Nelson *et al.*, 1991; Yan, Payne and Thomas, 2000). A number of researchers have specifically investigated biomechanical characteristics and motor control in overarm throwing during early childhood (e.g., Yan, Payne and Thomas, 2000; Yan, Hinrichs *et al.*, 2000) with the goal of determining the primary biomechanics that are responsible for the age-associated (or gender-related) differences in children's overarm throwing. To capture the developmental features in children's throwing performance, a number of kinematic measures (e.g., resultant ball velocity and release velocity, joint angles, and angular velocities) are generally used. As indicated in the typical velocity profiles (Figure 40.3), young children (3-year-olds) release the ball at a slower speed than their older counterparts (4- and 6-year-olds) with angular velocities of elbow extension and shoulder horizontal adduction that are lower (Figure 40.4). In addition, older children execute overarm throws more smoothly (e.g., with lower jerk) than younger children (Yan, Hinrichs *et al.*, 2000) and this may, in part, explain why younger children's throwing distance was shorter, and ball velocity was slower, than was the case for the older children.

Another way of exploring age- or gender-related differences in overarm throwing is through the analysis of developmental patterns or skill levels. Studies comparing the 'processes' of overarm throwing among age groups and between genders, along with various developing skill levels, can be readily found in the literature of motor development (e.g., Roberton and Halverson, 1984; Roberton, 1978). Investigations of qualitative aspects in overarm throwing across various skill levels in upper arm, forearm, trunk, and foot movements have shown that developmental improvements in skill levels of individual body components (e.g., humerus or trunk action) or of the whole body movement significantly contribute to the increased horizontal ball velocities. In this regard, the movement 'processes' and 'products' of throwing are closely associated (e.g., Langendorfer and Roberton, 2002; Roberton and Konczak, 2001; Runion, Roberton and Langendorfer, 2003). Southard (2002) further suggested that for children, the consistency in relative timing between the wrist

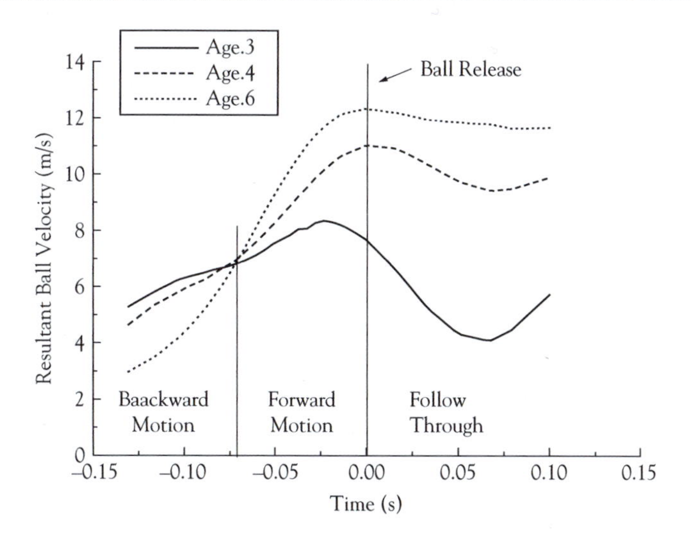

Figure 40.3 The typical velocity profiles.

and elbow movements differentiate throwers. Unskillful performers were more likely to display inconsistent joint lag of the wrist and elbow across trials than were skillful ones. Finally, differences in developmental throwing patterns were also noted between the dominant and non-dominant arm for throwers aged 4–10 years, although sustained practice may systematically reduce these laterality differences (Teixeira and Gasparetto, 2002).

Implications for developmental skill learning and future research

As discussed above, during childhood there are significant developmental changes in sensory-motor functioning, motor control, and skill learning, as well as age- and/or gender-related differences in motor performance. The biomechanical characteristics of motor performance that change with development are reasonably well documented for most fundamental motor skills. Children's development through growth, maturation, and motor experience collectively plays a significant role in improved motor performance. This section concentrates on several important implications for developmentally appropriate skill instruction and learning, drawing on notions variously derived from cognitive and dynamical systems viewpoints on motor control. (For detailed accounts of the contrasting cognitive and dynamical perspectives on motor control, see Abernethy and Sparrow, 1992; Abernethy, Burgess-Limerick and Parks, 1994; Newell, 1991.) This section also proposes some future research directions for the study of motor skill acquisition in children based on some of the new approaches and issues emerging in adult motor control and learning.

Motor memory consolidation

The challenge in ongoing learning of a skill is how to maintain an equivalent or a better performance after practice or in subsequent retention or transfer tests (Schmidt and Bjork, 1992). From a cognitive view, one of the key components of any skill learning over time

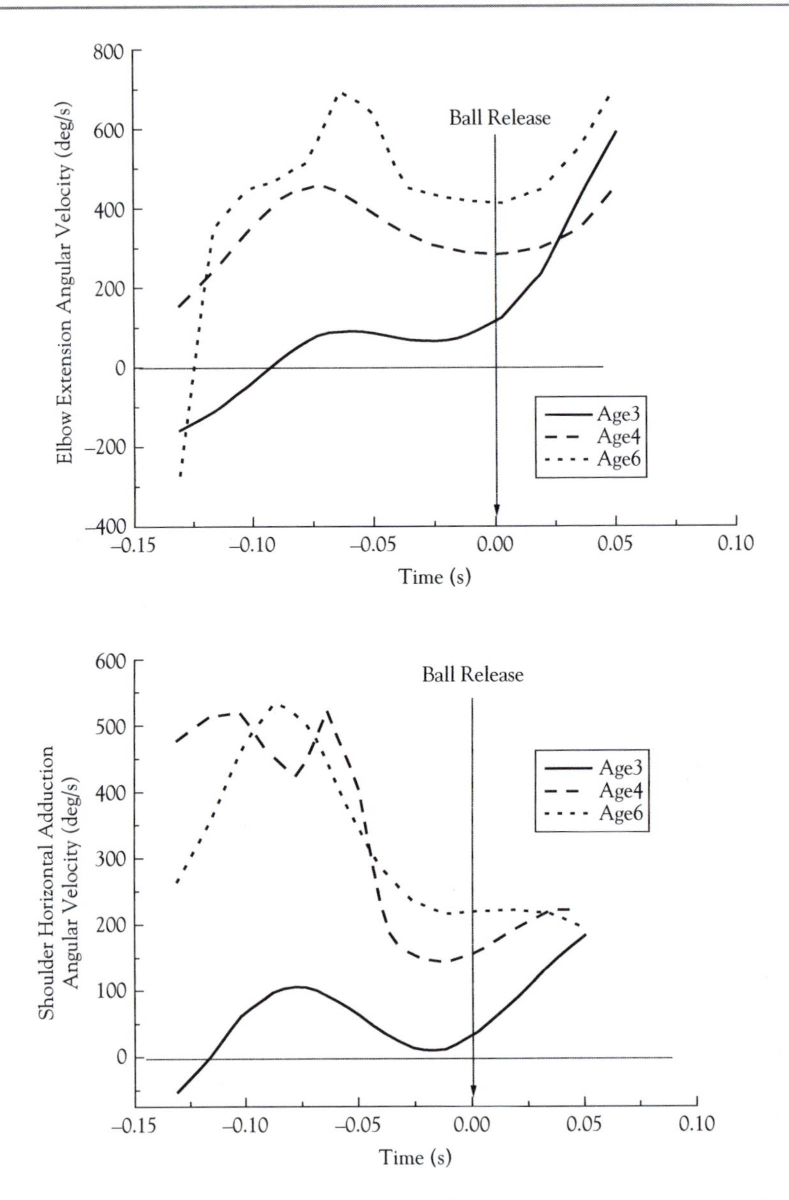

Figure 40.4 Young children (3-year-olds) release the ball at a slower speed than their older counterparts (4- and 6-year-olds) with angular velocities of elbow extension and shoulder horizontal adduction that are lower.

(be it via a motor programme, a cognitive representation, or an internal model) is the development of memory specifically related to production of a given action. The 'three-stage' framework proposed by Karni *et al.* (1998; namely viz., 'online' learning, memory consolidation, and 'off-line' improvement) suggests that each step has its unique functional characteristics and contributions to the learning processes. Learners conceptualize the skill that will be learned and experience a quick improvement during early phases of practice. A better performance would result from a consolidated memory after a period of rest (nap or sleep; Walker, 2005).

Children typically have a short attention span and increasing movement knowledge and experience (Thomas, 1980). One of the developmental implications for motor skill instruction is to take advantage of the unique neurophysiological or psychological changes that are associated with the processes of skill learning and memory consolidation (Walker and Stickgold, 2005). For instance, in addition to providing children with appropriate models, practice opportunities, and feedback during the initial learning phase, it may be equally imperative that children have sufficient time for daytime naps between practice sessions or have an overnight sleep before learning a new skill or practicing a previously learned skill. The skill instruction and rest periods theoretically could facilitate the formation of the internal model (Wolpert, 1997) and the 'offline' skill acquisition for children. The emphasis in 'online' learning is on the presentation of skill information, while the key for 'offline' learning is rest between practice sessions.

Motor coordination and variability

From a dynamical systems perspective, the control of 'degrees of freedom' (*cf.* Bernstein, 1967), the relations of movement noise and variability, and the coupling of perception with action, are all integral to understanding the changes or development in motor performance of multi-joint movements (Newell, 1991; Newell and Corcos, 1993). Because practice enhances movement speed and coordination (Kudo *et al.*, 2000; Schneider *et al.*, 1989), one important aspect for the design of the practice environment for a multi-joint motor task like overarm throwing is to help learners develop appropriate movement timing and sequencing (from proximal-to-distal joints; Herring and Chapman, 1992). For example, for children, a practice environment design that affords learning a sequence of joint movements in the order of trunk, shoulder, elbow, and hand for the overarm throw is important because of children's typically immature or under-developed abilities of motor coordination (Yan, Payne and Thomas, 2000). In addition, instruction and practice of a precise timing control of ball release is necessary. Young children tend to release the ball early before reaching a full body rotation and the completion of the forward acceleration movement (Yan, Payne and Thomas, 2000). The biomechanics literature would also suggest that special attention should be paid to the timing of finger release movements (Hore and Watts, 2005).

Developmental variations for individual body movements (trunk, upper arm, forearm, and hand) are common among children (Robertson and Halverson, 1984) and between- and within-subject variability also changes with practice in children (Thomas *et al.*, 2000). As motor variability seems to be of particular importance within development (Deutsch and Newell, 2001) it may be prudent that instead of attempting to reduce movement variability in practice, a variety of movement conditions or demands are provided or required to maximize the learning benefits of adaptability (Yan *et al.*, 1998). Finally, a recent review of resistance training programmes for overarm throwing suggests that improved muscular strength plays a part in better performance (Tillaar, 2004). Children are developmentally capable of receiving the benefits of weight training (Payne *et al.*, 1997), thus it may be useful to also include this within the training regimens for overam throwing.

Future directions

There are clearly a number of important theoretical and methodological issues associated with research in developmental motor control and learning that need to be addressed and a number of key areas in which empirical evidence is needed. For example, the theoretical concept of motor memory consolidation poses a number of questions that future research must attempt to address and answer. One important question relates to the interaction between development and motor memory consolidation or, in other words, how children may differ from adults in terms of the requirements of a time interval between practice and the duration of consolidation. Compared with adults, children typically have a shorter working memory and rely less on the use of strategies for rehearsal, yet it is not yet clear whether this affects children's memory consolidation and consequently undermines their motor skill learning. A related issue is the determination of factors, other than naps or sleep between practice sessions, which can optimize children's motor memory consolidation and/or aid the formation of an accurate and appropriate internal model. More information is also needed as to the relationship between practice conditions variability (or specificity) and type, and motor memory consolidation in general and in children in particular. Undoubtedly, answers to these questions have important practical implications for developmental motor skill learning.

A further important practical concern is how to determine, most appropriately, the representative, normative, and/or optimal levels of motor performance of children. This question is relevant to the quantitative or qualitative assessment of movement variability in children's developmental skill levels. Generally, children's motor performance is less consistent than adults, particularly during early childhood. Average values or the best performance trials are commonly used to represent an individual child's skill level, yet, given that the number of trials in biomechanical research is typically limited, the questions of what information should be used and how to use it are clearly issues requiring both an appropriate evidence base and conformity of standards between researchers. At the very least, it would appear prudent for researchers to routinely report the means, standard deviations, and the range of scores when describing the level of motor performance of an individual performer.

In summary, developmental and biomechanical research on children' motor control and learning is an active area of scholarship. With a variety of theoretical and practical approaches, the empirical information derived from the studies of the general population, and from children in particular, has the potential to shed light on the underlying processes or mechanisms of motor control and skill learning, as well as providing practical guidance for developmental skill instruction, practice, and learning.

Note

The chapter was written while the first author was supported by the 2006 Summer Research Fellowship from the Institute of Human Performance, University of Hong Kong.

References

1. Abernethy, B. (1988) 'The effects of age and expertise upon perceptual skill development in a racquet sport', *Research Quarterly for Exercise and Sport*, 59: 210–221.
2. Abernethy, B., Burgess-Limerick, R.J. and Parks, S. (1994) 'Contrasting approaches to the study of motor expertise', *Quest*, 46: 186–198.
3. Abernethy, B., Maxwell, J.P., Masters, R.S.W., van der Kamp, J. and Jackson, R.C. (in press) 'Attentional processes in skill learning and expert performance'. In G. Tenenbaum and R.C. Eklund (eds.), *Handbook of Sport Psychology* (3rd edn.). Wiley.

4. Abernethy, B. and Sparrow, W.A. (1992) 'The rise and fall of dominant paradigms in motor behaviour research'. In J. J Summers (ed.), *Approaches to the study of motor control and learning* (pp. 3–45). Amsterdam: North-Holland.
5. Baltes, P.B. and Baltes, M.M. (1990) 'Psychological perspectives on successful aging: The model of selective optimization with compensation'. In P.B. Baltes and M.M. Baltes (eds.), *Successful aging: Perspectives from the behavioural sciences* (pp. 1–34). Cambridge: Cambridge University Press.
6. Bernstein N. (1967) *The co-ordination and regulation of movements*. Oxford, U.K.: Pergamon Press.
7. Butterfield, S.A. and Loovis, E.M. (1998) 'Kicking, catching, throwing, and striking development by children in grades K-8: Preliminary findings', *Journal of Human Movement Studies*, 34: 67–81.
8. Clark, J.E. and Humphrey, J. (2002) *Motor development: Research and reviews* (vol. 2). Reston, VA: NASPE.
9. Clark, J.E. and Phillips, S.J. (1993) 'A longitudinal study of intralimb coordination in the first year of independent walking: A dynamical systems analysis', *Child Development*, 64: 1143–1157.
10. Cohen, D.A., Passual-Leone, A., Pree, D.Z. and Robertson, E.M. (2005) 'Off-line learning of motor skill memory: A double dissociation of goal and movement', *Proceedings of National Academy of Sciences*, 102: 18237–18241.
11. Cross, R. (2004) 'Physics of overarm throwing', *American Journal of Physics*, 72: 305–312.
12. Davids, K., Bennett, S.J. and Newell, K. (eds.) (2006) *Movement system variability*. Urbana-Champaign, IL: Human Kinetics.
13. Deutsch, K.M. and Newell, K.M. (2001) 'Age differences in noise and variability of isometric force production', *Journal of Experimental Child Psychology*, 80: 392–408.
14. Ericsson, K.A. (2000) 'How experts attain and maintain superior performance: Implications for the enhancement of skilled performance in older individuals', *Journal of Aging and Physical Activity*, 8: 366–372.
15. Halverson, L.E., Roberton, M.A. and Langendorfer, S. (1982) 'Development of the overarm throw: Movement and ball velocity changes by seventh grade', *Research Quarterly for Exercise and Sport*, 53: 198–205.
16. Hay, L. (1981) 'The effect of amplitude and accuracy requirements on movement time in children', *Journal of Motor Behaviour*, 13: 177–186.
17. Herring, R.M. and Chapman, A.E. (1992) 'Effects of changes in segmental values and timing of both torque and torque reversal in simulated throws', *Journal of Biomechanics*, 25: 1173–1184.
18. Hore, J. (1996) 'Motor control, excitement, and overarm throwing', *Canadian Journal of Physiology and Pharmacology*, 74: 385–389.
19. Hore, J. and Watts, S. (2005) 'Timing finger opening in overarm throwing based on a spatial representation of hand path', *Journal of Neurophysiology*, 93: 3189–3199.
20. Ivanchenko, V. and Jacobs, R.A. (2003) 'A developmental approach aids motor learning', *Neural Computation*, 15: 2051–2065.
21. Kail, R. (1988) 'Developmental functions for speeds of cognitive processes', *Journal of Experimental Child Psychology*, 45: 339–364.
22. Karni, A., Meyer, G., Rey-Hipolite, C., Jezzard, P., Adams, M.M., Turner, R. and Underleider, L.G. (1998) 'The acquisition of skilled motor performance: Fast and slow experience-driven changes in primary motor cortex', *Proceedings of National Academy of Sciences*, 95: 861–868.
23. Keogh, J. and Sugden, D. (1985) *Movement skill development*. New York: Macmillan.
24. Kudo, K., Ito, T., Tsutsul, S., Yamamoto, Y. and Ishikura, T. (2000) 'Compensatory coordination of release parameters in a throwing task', *Journal of Motor Behaviour*, 32: 337–345.
25. Kugler, P. (1986) 'A morphological perspective on the origin and evolution of movement patterns'. In M.G. Wade and H.T.A. Whiting (eds.), *Motor skill acquisition in children: Aspects of coordination and control* (pp. 459–525). The Hague: Martinus Nijhoff.
26. Kuriyama K., Stickgold R. and Walker M.P. (2004) 'Sleep-dependent learning and motor skill complexity', *Learning and Memory*, 11: 705–713.

27. Langendorfer, S. and Roberton, M.A. (2002) 'Individual pathways in the developmental of forceful throwing', *Research Quarterly for Exercise and Sport*, 73: 245–256.
28. Loovis, E.M. and Butterfield, S.A. (1995) 'Influence of age, sex, balance, and sport participation on development of side-arm striking by children grades K-8', *Perceptual and Motor Skills*, 81: 595–600.
29. Ma, W. J., Beck, J.M., Latham, P.E. and Pouget, A. (2006) 'Bayesian inference with probabilistic population codes', *Nature Neuroscience*, 9: 1432–1438.
30. McDonald, P. V., van Emmerik, R. E. A. and Newell, K. M. (1989) 'The effects of practice on limb kinematics in a throwing task', *Journal of Motor Behaviour*, 21: 245–264.
31. McNaughton, S., Timmann, D., Watts, S. and Hore, J. (2004) 'Overarm throwing speed in cerebellar subjects: effects of timing of ball release', *Experimental Brain Research*, 154: 470–478.
32. McPherson, S.L. and Thomas, J.R. (1989) 'Relation of knowledge and performance in boys' tennis: age and expertise', *Journal of Experimental Child Psychology*, 48: 190–211.
33. Meyer, D.E., Abrams, R.A., Kornblum, S., Wright, C.E. and Smith, J.E.K. (1988) 'Optimality in human performance: Ideal control of rapid aimed movements', *Psychological Review*, 95: 340–370.
34. Morris, A.M., Williams, J.M., Atwater, A.E. and Wilmore, J.H. (1982) 'Age and sex differences in motor performance of 3 through 6 year old children', *Research Quarterly for Exercise and Sport*, 53: 214–221.
35. Nelson, J.K., Thomas, J.R. and Nelson, J.K. (1991) 'Longitudinal change in throwing performance: Gender differences', *Research Quarterly for Exercise and Sport*, 62: 105–108.
36. Nelson, J.K., Thomas, J.R., Nelson, K.R. and Abraham, P.C. (1986) 'Gender differences in children's throwing performance: Biology and environment', *Research Quarterly for Exercise and Sport*, 57: 280–287.
37. Newell, K. M. (1991) 'Motor skill acquisition', *Annual Review of Psychology*, 42: 213–237.
38. Newell, K. M. and Corcos, D. M. (1993) 'Issues in variability and motor control'. In K. M. Newell and D. M. Corcos (eds.), *Variability and motor control* (pp. 1 –12). Campaign, IL: Human Kinetics.
39. Payne, V.G., Morrow, J.R., Johnson, L. and Dalton, S.N. (1997) 'Resistance training in children and youth: A meta-analysis', *Research Quarterly for Exercise and Sport*, 68: 80–88.
40. Roberton, M.A. (1978) 'Longitudinal evidence for developmental stages in the forceful overarm throw', *Journal of Human Movement Studies*, 4: 167–175.
41. Roberton, M.A. and Halverson, L.E. (1984) *Developing children–Their changing movement: A guide for teachers*. Philadelphia: Lea and Febiger.
42. Roberton, M.A. and Konczak, J. (2001) 'Predicting children's overarm throw ball velocities from their developmental levels in throwing', *Research Quarterly for Exercise and Sport*, 72: 91–103.
43. Robertson, E.M., Pascual-Leone, A. and Miall, R.C. (2004) 'Current concept in procedural consolidation', *Nature Reviews Neuroscience*, 5: 576–582.
44. Robertson, E.M., Press, D.Z. and Pascual-Leone, A. (2005) 'Off-line learning and the primary motor cortex', *The Journal of Neuroscience*, 25: 6372–6378.
45. Rodriguez-Ortiz, C.J., Cruz, V.D.L., Gutierrz, R. and Bermudez-Rattoni, F. (2005) 'Protein synthesis underlies post-retrieval memory consolidation to a restricted degree only when updated information is obtained', *Learning and Memory*, 12: 533–537.
46. Runion, B.P., Roberton, M.A. and Langendorfer, S. (2003) 'Forceful overarm throwing: A comparison of two cohorts measured 20 years apart', *Research Quarterly for Exercise and Sport*, 74: 324–330.
47. Sakurai, S. and Miyasjita, M. (1983) 'Developmental aspects of overarm throwing related to age and sex', *Human Movement Science*, 2: 67–76.
48. Savelsbergh, G.J.P., van der Kamp, J. and Rosengren, K.S. (2006) 'Functional variability in perceptual-movement development'. In K. Davids and S.J. Bennett and K. Newell (eds.), *Movement system variability* (pp. 185–198). Urbana-Champaign, IL: Human Kinetics.
49. Schmidt, R.A. (1975) 'A schema theory of discrete motor skill learning', *Psychological Review*, 82: 225–260.

50. Schmidt, R. A. and Bjork, R. A. (1992) 'New conceptualizations of practice: Common principles in three paradigms suggest new concepts for training', *Psychological Science*, 3: 207–217.
51. Schmidt, R.A. and Lee, T.D. (2005) *Motor control and learning: A behavioural emphasis* (4th ed.). Champaign, IL: Human Kinetics.
52. Schmidt, R. A., Zelaznik, H.N., Hawkins, B., Frank, J.S. and Quinn, J.T. (1979) 'Motor output variability: A theory for the accuracy of rapid motor acts', *Psychological Review*, 86: 415–451.
53. Schneider, K., Zernicke, R. F., Schmidt, R.A., Hart, T. J. (1989) 'Changes in limb dynamics during the practice of rapid arm movements', *Journal of Biomechanics*, 22: 805–817.
54. Shadmehr, R. and Brashers-Krug, T. (1997) 'Functional stages in the formation of human long-term motor memory', *The Journal of Neuroscience*, 17: 409–419.
55. Shapiro, D.C. and Schmidt, R.A. (1982) 'The schema theory: Recent evidence and developmental limitations'. In J.A.S. Kelso and J.E. Clark (eds.), *The development of movement control and coordination* (pp. 113–150). New York: Wiley.
56. Southard, D. (2002) 'Change in throwing pattern: Critical values for control parameter of velocity', *Research Quarterly for Exercise and Sport*, 73: 396–407.
57. Southard, D. (2006) 'Changing throwing pattern: Instruction and control parameter', *Research Quarterly for Exercise and Sport*, 77: 316–325.
58. Stelmach, E.G. and Thomas, J.R. (1997) 'What's different in the speed-accuracy trade-off in young and elderly subjects', *Behavioural and Brain Sciences*, 20: 321.
59. Stickgold, R. and Walker, M.P. (2005) 'Memory consolidation and reconsolidation: What is the role of sleep?', *Trends in Neuroscience*, 28: 408–415.
60. Teixeira, L.A. and Gasparetto, E.R. (2002) 'Lateral asymmetries in the development of the overarm throw', *Journal of Motor Behaviour*, 34: 151–160.
61. Thelen, E., Kelso, J.A.S. and Fogel, A. (1987) 'Self-organizing systems and infant motor development', *Developmental Review*, 7: 39–65.
62. Thomas, J.R. (1980) 'Acquisition of motor skills: Information processing differences between children and adults', *Research Quarterly for Exercise and Sport*, 51: 158–173.
63. Thomas, J.R. (1984) 'Planning "Kiddie" research: Little "Kids" but big problem'. In J.R. Thomas (ed.), *Motor development during childhood and adolescence* (pp. 260–273). Minneapolis, MN: Burgess.
64. Thomas, J.R. and French, K.E. (1985) 'Gender differences across age in motor performance: A meta-analysis', *Psychological Bulletin*, 98: 260–282.
65. Thomas, K.T., Gallagher, J. D. and Thomas, J. R. (2001) 'Motor development and skill acquisition during children and adolescence'. In R. N. Singer, H. A. Hausenbas and C. M. Janelle (eds.), *Handbook of sport psychology* (pp. 20–52). New York: John Wiley and Sons, Inc.
66. Thomas, J.R., Thomas, K.T., Lee, A.M., Testerman, E. and Ashy, M. (1983) 'Age differences in use of strategy for recall of movement in a large scale environment', *Research Quarterly for Exercise and Sport*, 54: 264–272.
67. Thomas, J.R., Yan, J.H. and Stelmach, G.E. (2000) 'Movement characteristics change as a function of practice in children and adults', *Journal of Experimental Child Psychology*, 75: 228–244.
68. Tillaar, R.V.D. (2004) 'Effect of different training program of the velocity of overarm throwing: A brief review', *Journal of Strength and Conditioning Research*, 18: 388–396.
69. Tillaar, R.V.D. and Ettema, G. (2004) 'A forceful-velocity relationship and coordination patterns in overarm throwing', *Journal of Sports Science & Medicine*, 3: 211–219.
70. Walker, M.P. (2005) 'A refined model of sleep and the time course of memory formation', *Behavioural and Brain Sciences*, 28: 51–104.
71. Walker, M.P. and Stickgold, R. (2004) 'Sleep-dependent learning and memory consolidation', *Neuron*, 44: 121–133
72. Walker, M.P. and Stickgold R. (2005) 'It's practice with sleep that makes perfect: Implications of sleep-dependent learning and plasticity for skill performance', *Clinics in Sports Medicine*, 24: 301–317.

73. Ward, P. and Williams, A.M. (2003) 'Perceptual and cognitive skill development in soccer: The multidimensional nature of expert performance', *Journal of Sport & Exercise Psychology*, 25: 93–111.
74. Wild, M.R. (1938) 'The behaviour pattern of throwing and some observations concerning its course of development in children', *Research Quarterly*, 9: 20–24.
75. Wolpert, D. M. (1997) 'Computational approaches to motor control', *Trends in Cognitive Sciences*, 1: 209–216.
76. Yan, J.H., Hinrichs, R.N., Payne, V.G. and Thomas, J.R. (2000) 'Normalized jerk: A measure to capture developmental characteristics of young girls' overarm throwing', *Journal of Applied Biomechanics*, 16: 196–202.
77. Yan, J.H., V.G. Payne. and Thomas, J.R. (2000) 'Developmental kinematics of young females' overarm throwing', *Research Quarterly for Exercise and Sport*, 71: 92–98.
78. Yan, J.H. and Thomas, J.R. (1995) 'Parents' assessment of physical activity in American and Chinese children', *Journal of Comparative Physical Education and Sports*, 17: 38–49.
79. Yan, J.H., Thomas, J.R. and Payne, G.V. (2002) 'How children and seniors differ from adults in controlling rapid aiming arm movements', In J.E. Clark and J. Humphrey (eds.), *Motor development: Research and reviews* (vol. 2, pp 191–217). Reston, VA: NASPE.
80. Yan, J.H., Thomas, J.R., Stelmach, G.E. and Thomas, K.T. (2000) 'Developmental features of rapid aiming arm movements across the lifespan', *Journal of Motor Behaviour*, 32: 121–140.
81. Yan, J.H., Thomas, J.R. and Thomas, K.T. (1998) 'Practice variability facilitates children's motor skill acquisition: A quantitative review', *Research Quarterly for Exercise and Sport*, 69: 210–215.

41

Using biomechanical feedback to enhance skill learning and performance

Bruce Abernethy, Richard S.W. Masters and Tiffany Zachry
The University of Hong Kong, Hong Kong and
The University of Queensland, Queensland

Introduction

The impressive technological advances of the past decades in biomechanics permit unprecedented opportunities for coaches and sport scientists to provide athletes with feedback information about their movements. Despite the obvious potential of biomechanical feedback to facilitate skill learning and performance, surprisingly little is yet known, either theoretically or empirically, about how best to utilize such information (Bartlett, 1997). The purpose of this chapter is to identify and briefly examine some key issues related to the question of how augmented feedback about movement might be used to enhance both the learning and performance of skilled actions of the type used in sports tasks. The perspective taken is one informed, not so much by applied biomechanics, but rather by the literature on how movement skills are controlled and learned.

In examining the issue of biomechanical feedback, a few simple, but important, distinctions need to be made from the outset. First, it needs to be recognized that the learning and performance of movement skills, while related, are not synonymous (Schmidt and Lee, 2005). Performance refers to an observable execution of a motor skill, quantifiable both in terms of its outcome and its form. Learning, in contrast, is an enduring, 'relatively permanent' improvement in performance that occurs as a consequence of the enhancement of underlying control processes. Consequently, learning is not directly observable or quantifiable and must be inferred indirectly from performance changes. Importantly, many variables, including those related to feedback provision, may influence learning and performance differentially. The demonstration that feedback about a particular biomechanical variable influences performance, for example, provides no guarantee of the same, enduring effect upon learning (Salmoni *et al.*, 1984).

Second, a distinction needs to be made between feedback derived or given about the technical or executional aspects of a movement (i.e., *movement pattern feedback* or, as it is more frequently labeled, *knowledge of performance*) and feedback about the success of the movement in achieving a specific pre-determined goal (i.e., *movement outcome feedback* or *knowledge of results*). In some sports (e.g., diving, gymnastics), where the performance score is a movement pattern measure, movement pattern and movement outcome feedback may be indistinguishable; however, in most activities, the two can be clearly differentiated. It does not necessarily follow that the same set of principles will underpin the effective provision of both types of feedback.

Third, a meaningful distinction can be made between augmented feedback and feedback intrinsic to the movement itself. All natural movements generate significant intrinsic feedback, both with respect to the overall movement pattern (largely obtained kinesthetically) and with respect to movement outcome (generally obtained visually). *Augmented feedback* (or *extrinsic feedback*; Schmidt and Wrisberg, 2000) refers to any additional or enhanced information provided to the performer, such as the verbal information provided by coaches or the biomechanical feedback information increasingly provided by sport scientists. The issue of how augmented feedback about movement patterns, in particular, might enhance skill learning or performance is given central focus in this chapter. In considering the likely effectiveness of any given type of augmented feedback it follows that the augmented information provided cannot be considered in isolation; rather it is the interaction between the augmented and the intrinsic information that may ultimately govern the effectiveness of augmented feedback in facilitating performance improvement.

This chapter is in two parts. The first part provides a very brief historical sketch of feedback research in the motor learning literature, sequentially considering the general findings related to information feedback about movement outcomes and those related to information feedback about movement patterns. The second part of the chapter identifies some key current and prospective issues related to the provision of augmented feedback and then seeks to establish potential research strategies to advance understanding about these key issues. The key issues are somewhat arbitrarily divided into those related to the design and selection of appropriate biomechanical feedback variables, to the implementation of biomechanical feedback schedules, and to the evaluation of the effectiveness of biomechanical feedback provision.

Feedback research in motor learning: a brief history

Research on movement outcome feedback

There is a long history of motor learning research examining the relationships between feedback, performance and learning (see Bilodeau, 1966; Newell, 1976; Newell *et al.*, 1985a; and Adams, 1987 for reviews). However, the bulk of this research has been rather homogenous, both in focus and method. The predominant focus has been upon the impact of information feedback on movement outcome (typically measured by absolute, constant or variable error scores) rather than upon the movement pattern used to produce the outcome. The experimental tasks that have been used have been typically simple; most frequently, blindfolded linear-positioning tasks. Such tasks are arguably trivial, are certainly atypical of those occurring in sport settings, and are far removed from those for which coaches and athletes seek practical advice and assistance (Whiting, 1982). The task variables most frequently under examination have been variables such as feedback precision and scheduling.

A persistent conclusion of the research on movement outcome has been that feedback of this type is critical for learning, given that little or no learning generally occurs in its absence (e.g., Trowbridge and Cason, 1932; Bennett and Simmons, 1984). It is obviously difficult, however, to extrapolate from these laboratory findings to the practical questions related to the provision of biomechanical feedback to athletes. The tasks used experimentally have been far removed from those in sport settings and the feedback provided is on movement outcome rather than movement pattern biomechanics. Situations in which sport skills are practiced are typically replete with sources of information for the athlete about movement outcome, but augmented movement pattern information is often difficult to obtain.

Research on movement pattern feedback

The advent of automated data acquisition systems has brought with it a massive increase in the potential for coaches and sport scientists to provide significant feedback to athletes about key biomechanical variables that need to be optimized for maximum performance. Accompanying this has been a marked increase in research interest in the effectiveness of such movement pattern feedback (e.g., Newell *et al.*, 1985b; Gregor *et al.*, 1992). While this research interest stands in contrast to the research on movement outcome feedback that dominated motor learning research in the 1960s, 1970s and early 1980s, it is important to recognize that research on movement pattern feedback is not an entirely recent phenomenon. For example, Lindahl (1945) examined the effectiveness of providing trainee operators of a mechanical disc-cutting machine with kinematic feedback (displacement-time plots) about their cutting movements. Howell (1956) attempted to improve sprint starting in athletics by providing kinetic feedback information about the force-time characteristics of the foot at push off (*cf.* Mendoza and Schöllhorn, 1993).

Several conclusions can be drawn from both the historical studies and many of the more recent studies on biomechanical feedback. First, determining which particular (set of) movement pattern variable(s) to provide feedback about is clearly extremely difficult, given the enormous array of potential variables that could be used. Nevertheless, it appears that, to date, the selection of the variable(s) upon which to provide feedback has been largely arbitrary and frequently driven by measurement convenience or convention, rather than a principled consideration of the role of the variable in movement control and learning. Second, the majority of the existing studies on the use of biomechanical feedback have been concerned with the impact of such information upon skill performance. There has been only very limited systematic use of retention and transfer designs to evaluate the longevity of any facilitatory effects. Hence, it is somewhat unclear whether movement pattern feedback actually enhances the learning of movement patterns or simply their performance in the short-term (Schmidt and Lee, 2005).

Current and prospective issues in providing augmented biomechanical feedback

Design issues

One fundamental notion in relation to the design of effective biomechanical feedback is that the content of the information provided should match what the performer *can* control (Schmidt, 1991, p. 238) and *needs* to control (e.g., Whiting and Vereijken, 1993). Feedback information should have the capability to help change the movement but such information will only be useful if the athlete can control the movement feature about which the information is provided. This fundamental notion, in turn, gives rise to some important practical and theoretical issues related to the optimal design of movement pattern information feedback. We address four of these issues here.

Issue 1: what are the essential control variables for the movement of interest?

A major impediment to the provision of feedback information about biomechanical variables that the learner can control is the identification of the essential control variables for the movement skill that we are seeking to improve. Such identification is far from a trivial research task. The predominant research strategy used in the motor control literature to attempt to identify essential control variables involves varying the parameters for the performance of the movement skill of interest through their widest possible range (e.g., by systematically and independently manipulating the speed, distance or force generation requirements) and then discriminating those movement features that change in direct response to the task conditions from those which remain immutable. Such a strategy permits the invariant features of the movement pattern to be distinguished from the parameters (Schmidt, 1985; Abernethy, 1993). Different theoretical interpretations exist regarding what the (task-specific) invariant features of movement represent – be they the essential cognitive framework for a motor program (Schmidt, 1980) or emergent properties of the unique dynamics of the human neuromusculoskeletal system (Kelso, 1981). Regardless of the theoretical significance, the practical significance with respect to skill learning of the differentiation of invariant from variant features of movement is that, for any augmented feedback to be effective, the feedback must help the performer either stabilize or consolidate the invariant features or gain more sensitive control over the variable parameters (*cf.* Hurley and Lee, 2006).

Issue 2: What is the relationship between the essential control variables and learning and performance?

If essential control variables for a skill can be identified then an equally important issue, from the perspective of feedback design, becomes that of knowing how these essential control variables alter, if at all, during skill acquisition. Such knowledge is important because one of the key roles of information feedback must be to assist the learner in progressing toward task expertise. The control structure and requirements for a skill may vary significantly between the beginner and the expert. An important research strategy in this regard is therefore systematic examination of how control of the skill of interest changes with learning. Such information may be gleaned from longitudinal learning-training studies or from cross-sectional comparisons of task experts and novices. Given that a minimum of ten years of daily, deliberate practice (or some 10,000 hours of practice) may be necessary to achieve expert performance (Chase and Simon, 1973; Ericsson *et al.*, 1993), comparative expert-novice studies offer the most pragmatic approach to gaining insight into control system changes with learning (Abernethy, 1993). Knowing the locus of expert-novice differences on a motor task can help identify the limiting factors to skill performance and provide a principled basis for the identification of features of the movement pattern that can be modified through appropriate practice and feedback provision. Having task-specific knowledge about the nature of expert-novice differences provides a starting point for making information feedback (in this case, augmented biomechanical feedback) skill-level appropriate.

A sport-specific example may help to illustrate this point. Abernethy and Neal (1991) examined the movement pattern kinematics of billiards and snooker players of different skill levels in an attempt to tease out the limiting factors to novice player performance. The billiard cue was instrumented with a triaxial accelerometer and the task presented to the players was

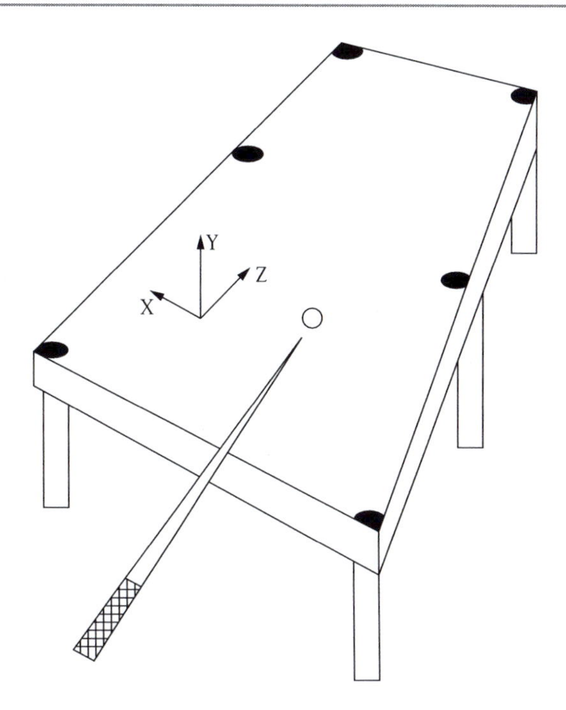

Figure 41.1 Schematic representation of the task setting in the Abernethy and Neal (1991) study. The cue was instrumented with a triaxial accelerometer from which kinematics could be resolved in three orthogonal directions.

to strike either a cue ball or an object ball (after collision with the cue ball) to a specified finishing position on the table (see Figure 41.1). Each task (cue ball or object ball) was repeated ten times to explore movement pattern consistency-variability. The kinematics of the cue were recorded horizontally (X direction), vertically (Y direction), and down the line of the table (Z direction) and trial-to-trial coefficients of variability were used to quantify kinematic consistency. In the dimensions down the line of the table and impact and horizontally across the table, variability diminished significantly with increasing skill expertise but there were no systematic expertise-related differences in variability in the vertical dimension of cue movement (Figure 41.2).

Armed with such knowledge it is possible to infer that feedback about kinematics in the X and Z dimensions is potentially valuable in augmenting learning of the control characteristics of this task, whereas there would appear to be little value in providing feedback about impact velocity variability in the Y dimension, as this is not a limiting or distinguishing factor for expert performance on this task. Such guidance for augmented feedback provision would not be possible in the absence of suitable task-specific expertise data. Similar links between understanding of the defining features for expert performance and the provision of feedback specifically on these key control and performance parameters have also been systematically developed for the sport of rowing (e.g., see Anderson *et al.*, 2005; Smith and Loschner, 2002; Smith and Spinks, 1995).

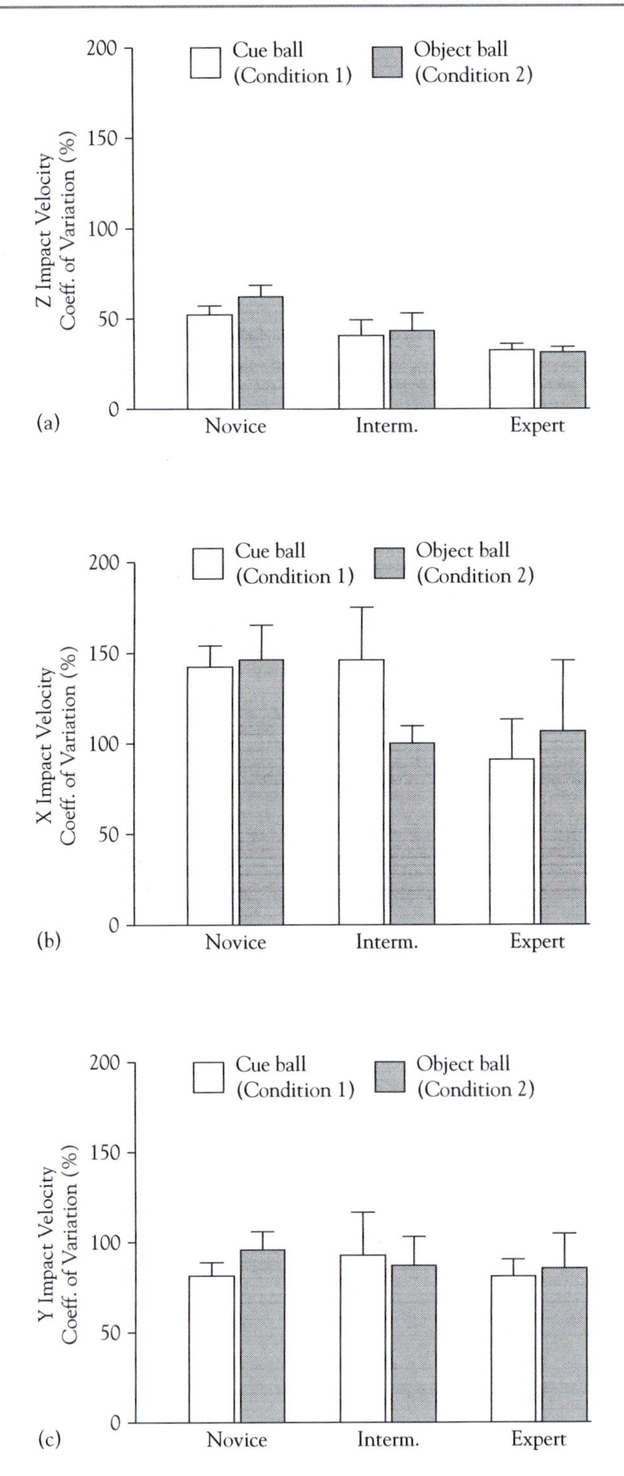

Figure 41.2 Trial-to-trial variability in the impact velocity of the cue: (a) along the long (*Z*) axis of the cue; (b) in the *X* (across the table) axis; and (c) in the *Y* (vertical) axis. Mean data are shown for six expert, seven intermediate, and fourteen novice players. (From Abernethy and Neal, 1991.)

Issue 3: What is the most appropriate level of biomechanical analysis about which to provide feedback?

One of the more vexing questions with respect to the provision of biomechanical feedback relates to the most appropriate level of control or analysis about which to supply information. The technology exists to provide augmented feedback from at least four different levels, viz:

- the neuromuscular level (e.g., feedback from electromyography (EMG) records);
- the kinetic level (e.g., feedback from force transducers);
- the kinematic level (e.g., feedback from high speed video or electromagnetic tracking devices); and
- the movement outcome level (feedback of knowledge of results).

These levels clearly differ in not only the type but also the precision and specificity of the information they may provide to the performer. For example, neuromuscular information can clearly be more local and more precise in control terms than movement outcome feedback, which is simply information about whether or not the global task objective (e.g., that of clearing the bar, getting the arrow in the target) was achieved. The key question is whether this greater precision indeed facilitates learning and performance?

At first approximation it might appear reasonable to suspect that the 'lower' and 'more local' neuromuscular and kinetic levels may be more suitable for the provision of augmented biomechanical feedback because of their greater proximity to the neuromotor control levels, their greater deterministic role in movement outcome, and their apparently greater potential to impact directly on movement production. (Movement kinematics and movement outcome are, after all, simply a consequence of the underlying forces, generated by carefully controlled motor unit recruitment, coupled with gravitational and other reactive forces.) However, while the 'lower' and 'more local' neuromuscular and kinetic levels may appear to have greater informative potential than more global (kinematic and movement outcome) levels, in practice the evidence suggests that the converse may be in fact true. At the neuromuscular level, in particular, one encounters much greater trial-to-trial variability as there may be multiple local solutions to the same global movement problem. Even in overlearned acts, such as signing one's own name, different motor units may be recruited from one execution of the skill to the next, even though the global movement pattern and the motor output are, for all intents and purposes, identical in each case (Rosenbaum, 1991). This local variability makes selection of the appropriate neuromuscular feedback pattern for any given skill problematic.

A second problem is that of conscious access to movement control mechanisms, and this becomes especially an issue at the neuromuscular feedback level. In many cases it may not be possible to consciously access the levels controlling a particular movement and, indeed, to attempt to do so, may fundamentally alter, or interfere with, usual control. Providing performers with highly precise feedback about local control variables may actually impede control and learning. By way of analogy, it would seem unlikely that providing a learner driver with precise feedback information regarding steering column strains or engine torques would facilitate performance in the same manner as more simple goal-related information, such as the proximity of the vehicle to the side of the road. Likewise, providing a billiards player with the precise geometric angle at which to strike the object ball with the cue ball, in order to make a particular shot, is not likely to be helpful.

The conclusions from analogies such as this have experimental support. Swinnen *et al* (1993) asked people to learn a novel task that required de-synchronization of the natural tendencies of the left and right hands to produce similar movement patterns. Groups of participants given detailed feedback about their limb kinematics (either limb displacements or velocities) learned the task no more rapidly than participants simply given global feedback about movement outcome.

More empirical evidence of the effectiveness of feedback about movement outcome (relative to feedback about one's own movement mechanics) has been observed by Wulf and her colleagues (Shea and Wulf, 1999; see also Wulf *et al.*, 2002). A group given feedback about the movement of a stabilometer balancing platform (i.e., feedback that induced an *external* focus of attention; see Wulf *et al.*, 1998) showed superior balance performance to a group given feedback about the movements of their feet on the platform (i.e., an *internal* focus of attention). The same effect carried over into a retention test of learning performed a day later. Wulf *et al.* (2001) proposed that instructions and feedback directed to a performer's movement mechanics cause the performer to actively interfere in the coordination of the system, whereas outcome-related feedback and instructions allow the motor system to naturally self-organize.

Issue 4: Is the key information feedback best provided explicitly or implicitly?

Consideration of the most appropriate way of presenting biomechanical feedback learning necessarily invokes consideration of two fundamentally different kinds of learning – *explicit* and *implicit* learning (Berry and Broadbent, 1988; Berry and Dienes, 1993; Reber, 1993). For those unfamiliar with the learning literature there is a common tendency to think that all learning occurs explicitly, consciously, and deliberately but this is not the case. Indeed, some of the more powerful and enduring aspects of learning occur implicitly, without the learner's direct knowledge or, often, intention.

In explicit learning the focus of the learner is upon *how* to best go about doing the task of interest, with any supporting instruction provided in a direct, usually verbal, form. Knowledge acquisition is intentional, relatively rapid, and conscious to the point that the learner is capable of easily describing the learning that has taken place. In contrast, in implicit learning, the focus is on problem solution without active aggregation of explicit knowledge about how the solution is reached. Implicit knowledge acquisition is incidental, slow, and difficult, if not impossible, for learners to describe (Magill, 1999), yet knowledge and skills learned implicitly may be more enduring and less prone to interference, forgetting, and stress (Hardy *et al.*, 1996; Masters, 1992; Masters and Maxwell, 2004). Importantly, the two types of learning may be influenced differentially by different variables and interference between implicit and explicit learning processes is possible (e.g., see Rossetti and Revonsuo, 2000; Sun *et al.*, 2005).

At least in some circumstances, it appears that relying on explicit processes (such as when verbal instructions are provided) can be detrimental to both performance and learning, whereas, relying on implicit processes may not be. Wulf and Weigelt (1997), for example, used a ski simulator task to demonstrate that, in contrast to the outcomes for a group given no instruction, people who were given explicit instructions performed poorly on the task, especially under stress conditions. Conversely, Masters (1992) showed that participants who learned a golf putting task implicitly (they were able to report almost no knowledge of how they performed each putt) exhibited stable putting performance under stress conditions.

In a review of the implicit learning literature, Magill (1999) concluded that it is probably best to provide explicit instructions and feedback to learners about the broad goals and movement characteristics needed for a particular task but that provision of explicit instructions and feedback about motor control and stimulus pattern rules may be contraindicated.

An extrapolation from this would be that rule-based, conscious processing of biomechanical feedback may not necessarily be advantageous, especially if the feedback provided is at the neuromuscular control level rather than at the more global kinematic or movement outcome level. However, even feedback at the more global level may be disadvantageous if it allows hypothesis testing behavior to occur during motor learning, because the tendency is for the learner to try different forms of a movement and, if the 'outcome' is positive, retain explicit knowledge of that particular movement. Over time, this process will cause the accumulation of a large amount of explicit knowledge and the learner may come to rely on explicit processing of the knowledge during performance (Masters and Maxwell, 2004).

Several implicit motor learning techniques have been devised to inhibit hypothesis testing (and the accumulation of explicit knowledge) during motor learning. Maxwell *et al.* (2001), for example, showed that learners who make few mistakes during learning (errorless learning) tend to learn implicitly (motivation to test hypotheses is low when a movement is to all intents and purpose already successful). A potential problem with this technique is that, despite the apparent success of a movement during errorless learning, the movement itself may not be biomechanically correct or efficient. Analogy learning is an alternative technique that allows some influence over the biomechanics of the movement. Liao and Masters (2001), for instance, showed that presenting information about movement biomechanics to learners in the form of an analogy or metaphor, rather than as verbal rules, allowed the performers to implicitly approximate the correct biomechanical form of the movement, without necessarily having detailed explicit knowledge of the movement.

Implementation issues

Even with the most appropriate source of information feedback, the actual benefits of such feedback will clearly depend upon critical implementation issues, such as the feedback precision, frequency, sequencing and scheduling. While relatively little is yet known of these issues for specific sport tasks, with appropriate experimentation, it should be possible to optimize implementation.

Some typical implementation issues that face the coach or sport scientist seeking to provide biomechanical feedback to athletes include:

1. What is the appropriate information feedback precision for the specific athlete and for the specific movement being practiced?
2. What is the appropriate bandwidth (tolerance) for athlete error on the specific parameter(s) being monitored?
3. Is continuous or intermittent feedback best?

Answering the first question requires knowledge of the sensitivity of learners to different types of feedback information. While suitable methodologies exist from psychophysics for the determination of perceptual sensitivities (*difference limens* or *jnd* – *just noticeable difference* measures) (see Shea and Northam, 1982 for an example), little recent research work of this type has been conducted specifically in the domain of the parameters upon which biomechanical feedback is likely to be provided. Nevertheless, it is known, for example, from old work on the *jnd* for joint angles that, for untrained people, a difference of some 2.2° (at a criterion of 90°) can only be reliably detected 50 per cent of the time. Therefore, it would seem illogical, if joint position feedback is to be provided on a task, to do so with levels of precision greater than 2–3° (Marteniuk *et al.*, 1972).

Like the issue of feedback precision, optimal bandwidth may also be determined experimentally, though again this would need to be done in a manner which was specific to the movement task, the feedback parameter, and the skill level of the performer receiving the feedback. Studies, of the type conducted by Smith *et al.* (1997) with golf pitching shots, indicate that learning appears to be greatest when players are only given feedback when their scores on key variables fall outside the accepted bandwidth (as compared to precise feedback on every trial) and, interestingly, the advantage is greatest when the bandwidth is broadest (i.e., least demanding in terms of precision). This suggests that feedback which is either too precise or too regular can impede learning.

The Smith *et al.* findings also clearly reflect on the issue of the relative merits of continuous versus intermittent feedback. Given that the ability to self-monitor is a ubiquitous characteristic of expert performance (Glaser and Chi, 1988), it is perhaps not surprising to find experimentally that it is advantageous to use feedback schedules that provide feedback intermittently and promote the performer's ability to derive their own feedback information without reliance on external sources. Development of the right kind of self-monitoring skills through intermittent (and progressively reducing) levels of augmented feedback is probably especially important in the domain of competitive sport. A related disadvantage of feedback that is too precise or regular is that it may also encourage an explicit mode of control given the opportunity for regular hypothesis-testing.

Evaluation issues

For biomechanical feedback to be valuable in sport it is necessary to demonstrate that such feedback brings about permanent rather than transitory changes in performance i.e., that the information feedback affects *learning* as well as *performance*. Demonstration that augmented biomechanical feedback actually enhances learning requires the systematic application of transfer designs, i.e., experimental designs in which all groups are compared under common conditions after a retention interval of appropriate length to permit any transitory effects to dissipate (see Schmidt and Lee, 2005 for further details). Convincing demonstration of the efficacy of biomechanical feedback also requires the systematic use of appropriate, rigorously-selected control groups. In practice, this equates to experimental designs that do not simply use just an experimental group (given the biomechanical feedback) and a control group (given no augmented feedback) but which also include a placebo group given irrelevant augmented feedback. Even in the majority of the well-cited research works on feedback efficacy it is not possible, in the absence of a suitable placebo group, to be completely certain that expectancy-placebo effects are not playing a significant role in the facilitatory effects seen for the augmented feedback group. Indeed, one of the more powerful means of demonstrating convincingly the strength of the augmented feedback-learning relationship would be by demonstrating that incorrect or misleading biomechanical feedback can actually impair learning. While there are obvious ethical and logistical difficulties with undertaking this kind of work with competitive athletes, such demonstrations with non-athletes would add to the credibility of existing contentions that appropriate biomechanical feedback can, in fact, enhance the performance and learning of athletes.

Conclusion

This chapter has attempted to present a framework upon which to build further understanding of the utility of augmented biomechanical feedback in facilitating motor learning and performance. How little is yet known, both theoretically and practically, about the use of biomechanical feedback to enhance motor learning and performance is striking. This is especially so when consideration is made of the potential importance and significance of such understanding. Applications of augmented biomechanical feedback to skill learning go far beyond those in elite sports performance into areas such as movement re-learning and rehabilitation (e.g., Cirstea *et al.*, 2006), the teaching of surgical and manual therapy skills (e.g., Berschback *et al.*, 2005; Descarreaux *et al.*, 2006), and the improvement of movement efficiency in the workplace (e.g., Kernozek *et al.*, 2006). The issues related to optimal use of biomechanical feedback stand right at the interface between the sub-disciplines of biomechanics and motor learning and control and are custom-made for the type of integrative, cross-disciplinary research endeavors that increasingly characterize modern science.

References

1. Abernethy, B. (1993) Searching for the minimal essential information for skilled perception and action. *Psychological Research*, 55, 131–138.
2. Abernethy, B. and Neal, R.J. (1991) *Perceptual-Motor Characteristics of Elite Performers in Aiming Sports*. Report to the Australian Sports Commission, Canberra.
3. Adams, J.A. (1987) Historical review and appraisal of research on the learning, retention, and transfer of human motor skills. *Psychological Bulletin*, 101, 41–74.
4. Anderson, R., Harrison, A. and Lyons, G.M. (2005) Accelerometry-based feedback: Can it improve movement consistency and performance in rowing? *Sports Biomechanics*, 4, 179–195.
5. Bartlett, R.M. (1997) Current issues in the mechanics of athletic activities: A position paper. *Journal of Biomechanics*, 30, 477–486.
6. Bennett, D.M. and Simmons, R.W. (1984) Effects of precision of knowledge of results on acquisition and retention of a simple motor skill. *Perceptual and Motor Skills*, 58, 785–786.
7. Berry, D.C. and Broadbent, D.E. (1988) Interactive tasks and the implicit-explicit distinction. *British Journal of Psychology*, 79, 251–272.
8. Berry, D.C. and Dienes, Z. (1993) *Implicit Learning: Theoretical and Empirical Issues*. Hove, UK: Erlbaum.
9. Berschback, J.C., Amadio, P.C., Zhao, C., Zobitz, M.E. and An, K.N. (2005) Providing quantitative feedback when teaching tendon repair: A new tool. *Journal of Hand Surgery*, 30, 626–632.
10. Bilodeau, I.M. (1966) Information feedback. In *Acquisition of Skill* (edited by E.A. Bilodeau), pp. 255–296. New York: Academic Press.
11. Chase, W.G. and Simon, H.A. (1973) Perception in chess. *Cognitive Psychology*, 4, 55–81.
12. Cirstea, C.M., Ptito, A. and Levin, M.F. (2006) Feedback and cognition in arm motor skill reacquisition after stroke. *Stroke*, 37, 1237–1242.
13. Descarreaux, M., Dugas, C., Lalanne, K., Vincelette, M. and Normand, M.C. (2006) Learning spinal manipulation: The importance of augmented feedback relating to various kinetic parameters. *Spine Journal*, 6, 138–145.
14. Ericsson, K.A., Krampe, R.T. and Tesch-Romer, C. (1993) 'The role of deliberate practice in the acquisition of expert performance. *Psychological Review*, 100, 363–406.
15. Glaser, R. and Chi, M.T.H. (1988) Overview. In *The Nature of Expertise* (edited by M.T.H. Chi, R. Glaser and M.J. Farr), pp. xv–xvii. Hillsdale, NJ: Erlbaum.
16. Gregor, R.J., Broker, J.P. and Ryan, M. (1992) Performance feedback and new advances in biomechanics. In *Enhancing Human Performance in Sport: New Concepts and Developments* (edited by H.M. Eckert), pp. 19–32. Champaign, IL: Human Kinetics.
17. Hardy, L., Mullen, R. and Jones, G. (1996) Knowledge and conscious control of motor actions under stress. *British Journal of Psychology*, 87, 621–636.

18. Howell, M.L. (1956) Use of force-time graphs for performance analysis in facilitating motor learning. *Research Quarterly*, 27, 12–22.
19. Hurley, S.R. and Lee, T.D. (2006) The influence of augmented feedback and prior learning on the acquisition of a new bimanual coordination pattern. *Human Movement Science*, 25, 339–348.
20. Kelso, J.A.S. (1981). Contrasting perspectives on order and regulation in movement. In *Attention and Performance IX* (edited by J. Long and A. Baddeley), pp. 437–457. Hillsdale, NJ: Erlbaum.
21. Kernozek, T., Iwasaki, M., Fater, D., Durall, C. and Langenhorst, B. (2006) Movement based feedback may reduce spinal moments in male workers during lift and lowering tasks. *Physiotherapy Research International*, 11, 140–147.
22. Liao, C.M. and Masters, R.S.W. (2001) Analogy learning: A means to implicit motor learning. *Journal of Sport Sciences*, 19, 307–319.
23. Lindahl, L.G. (1945) Movement analysis as an industrial training method. *Journal of Applied Psychology*, 29, 420–436.
24. Magill, R.A. (1999) What does implicit motor learning research tell us about practice conditions for motor skill learning? Paper presented at the *Congress of the German Association of Sport Science: Dimensions and Visions for Sport*, Heidelberg, Germany, September.
25. Marteniuk, R.G., Shields, K.W.D. and Campbell, S. (1972) Amplitude, position, timing and velocity cues in reproduction of movement. *Perceptual and Motor Skills*, 35, 51–58.
26. Masters, R.S.W. (1992) Knowledge, knerves and know-how: The role of explicit versus implicit knowledge in the breakdown of a complex motor skill under pressure. *British Journal of Psychology*, 83, 343–358.
27. Masters, R.S.W. and Maxwell, J.P. (2004) Implicit motor learning, reinvestment and movement disruption: What you don't know won't hurt you? In *Skill acquisition in sport: Research, theory and practice* (edited by A.M. Williams and N.J. Hodges), pp. 207–228. London: Routledge.
28. Maxwell, J. P., Masters, R. S. W., Kerr, E. and Weedon, E. (2001) The implicit benefit of learning without errors. *Quarterly Journal of Experimental Psychology* A, 54, 1049–1068.
29. Mendoza, L. and Schöllhorn, W. (1993) Training of the sprint start technique with biomechanical feedback. *Journal of Sports Sciences*, 11, 25–29.
30. Newell, K.M. (1976) Knowledge of results and motor learning. *Exercise and Sport Science Reviews*, 4, 195–228.
31. Newell, K.M. (1991) Motor skill acquisition. *Annual Review of Psychology*, 42, 213–237.
32. Newell, K.M., Morris, L.R. and Scully, D.M. (1985a) Augmented information and the acquisition of skill in physical activity. *Exercise and Sport Science Reviews*, 13, 235–261.
33. Newell, K.M., Sparrow, W.A. and Quinn, J.T. (1985b) Kinetic information feedback for learning isometric tasks. *Journal of Human Movement Studies*, 11, 113–123.
34. Reber, A.S. (1993) *Implicit Learning and Tacit Knowledge*. New York: Oxford University Press.
35. Rosenbaum, D.A. (1991) *Human Motor Control*. San Diego, CA: Academic Press.
36. Rossetti, Y. and Revonsuo, A. (Eds.) (2000) *Beyond Dissociation: Interaction Between Dissociated Implicit and Explicit Processing*. Amsterdam: John Benjamins.
37. Salmoni, A.W., Schmidt, R.A. and Walter, C.B. (1984) Knowledge of results and motor learning: A review and critical reappraisal. *Psychological Bulletin*, 95, 355–386.
38. Schmidt, R.A. (1980) Past and future issues in motor programming. *Research Quarterly for Exercise and Sport*, 51, 122–140.
39. Schmidt, R.A. (1985) The search for invariance in skilled movement behavior. *Research Quarterly for Exercise and Sport*, 56, 188–200.
40. Schmidt, R.A. (1991) *Motor Learning and Performance: From Principles to Practice*. Champaign, IL: Human Kinetics.
41. Schmidt, R.A. and Lee, T.D. (2005) *Motor Control and Learning: A Behavioral Emphasis* (4th Ed.). Champaign, IL: Human Kinetics.
42. Schmidt, R.A. and Wrisberg, C.A. (2000) *Motor Control and Performance: A Problem-Based Learning Approach* (2nd edn.). Champaign, IL: Human Kinetics.

43. Shea, C.M. and Northam, C. (1982) Discrimination of linear velocity. *Research Quarterly for Exercise and Sport*, 53, 222–225.
44. Shea, C.H. and Wulf, G. (1999) Enhancing learning through external-focus instructions and feedback. *Human Movement Science*, 18, 553–571.
45. Smith, P.J.K., Taylor, S.J. and Withers, K. (1997) Applying bandwidth feedback scheduling to golf putting. *Research Quarterly for Exercise and Sport*, 68, 215–221.
46. Smith, R.M. and Loschner, C. (2002) Biomechanics feedback for rowing. *Journal of Sports Sciences*, 20, 783–791.
47. Smith, R.M. and Spinks, W.L. (1995) Discriminant analysis of biomechanical differences between novice, good and elite rowers. *Journal of Sports Sciences*, 13, 377–385.
48. Sun, R., Slusarz, P. and Terry, C. (2005) The interaction of the explicit and the implicit in skill learning: A dual-process approach. *Psychological Review*, 112, 159–192.
49. Swinnen, S.P., Walter, C.B., Lee, T.D. and Serrien, D.J. (1993) Acquiring bimanual skills: Contrasting forms of information feedback for interlimb decoupling. *Journal of Experimental Psychology: Learning, Memory, and Cognition*, 19, 1328–1344.
50. Trowbridge, M.H. and Cason, H. (1932) An experimental study of Thorndike's theory of learning. *Journal of General Psychology*, 7, 245–260.
51. Wulf, G., Höß, M. and Prinz, W. (1998) Instructions for motor learning: Differential effects of internal versus external focus of attention. *Journal of Motor Behavior*, 30, 169–179.
52. Wulf, G., McConnel, N., Gärtner, M. and Schwarz, A. (2002) Feedback and attentional focus: Enhancing the learning of sport skills through external-focus feedback. *Journal of Motor Behavior*, 34, 171–182.
53. Wulf, G., McNevin, N.H. and Shea, C.H. (2001) The automaticity of complex motor skill learning as a function of attentional focus. *Quarterly Journal of Experimental Psychology*, 54A, 1143–1154.
54. Wulf, G. and Weigelt, C. (1997) Instructions about physical principles in learning a complex motor skill: To tell or not to tell ... *Research Quarterly for Exercise and Sport*, 68, 362–367.
55. Whiting, H.T.A. (1982) Skill in sport: A descriptive and prescriptive appraisal. In *New Paths of Sport Learning and Excellence* (edited by J.H. Salmela, J.T. Partington and T. Orlick), pp. 7–13. Ottawa, Ontario: Sport in Perspective Inc.
56. Whiting, H.T.A. and Vereijken, B. (1993) The acquisition of coordination in skill learning. *International Journal of Sport Psychology*, 24, 343–357.

Index

B

C

M

N

O

P